Universitext

Universitext

Universitext is a series of textbooks that presents material from a wide variety of mathematical disciplines at master's level and beyond. The books, often well class-tested by their author, may have an informal, personal even experimental approach to their subject matter. Some of the most successful and established books in the series have evolved through several editions, always following the evolution of teaching curricula, to very polished texts.

Thus as research topics trickle down into graduate-level teaching, first textbooks written for new, cutting-edge courses may make their way into *Universitext*.

More information about this series at http://www.springer.com/series/223

Gary R. Jensen • Emilio Musso • Lorenzo Nicolodi

Surfaces in Classical Geometries

A Treatment by Moving Frames

Springer

Gary R. Jensen
Department of Mathematics
Washington University
St. Louis, MO, USA

Emilio Musso
Dipartimento di Scienze Matematiche
Politecnico di Torino
Torino, Italy

Lorenzo Nicolodi
Dipartimento di Matematica e Informatica
Università di Parma
Parma, Italy

ISSN 0172-5939 ISSN 2191-6675 (electronic)
Universitext
ISBN 978-3-319-27074-6 ISBN 978-3-319-27076-0 (eBook)
DOI 10.1007/978-3-319-27076-0

Library of Congress Control Number: 2015957218

Mathematics Subject Classification (2010): 53-01, 53A05, 53A10, 53A20, 53A30, 53A35, 53A40, 53A55, 53C42

Printed on acid-free paper

This Springer imprint is published by Springer Nature
The registered company is Springer International Publishing AG Switzerland

We dedicate this book to Jen, Alice, Ettore, Giorgio, and Laura

Preface

Research in differential geometry requires a broad core of knowledge. For over half a century, this core has begun with the study of curves and surfaces in Euclidean space and has ended with the Gauss–Bonnet Theorem. Even this portion of core material is disappearing from many undergraduate programs. One goal of this book is to broaden core knowledge with an introduction to topics in classical differential geometry that are of current research interest. Another goal is to provide an elementary introduction to the method of moving frames. This method provides a unifying approach to each topic.

A space is homogeneous if a Lie group acts transitively on it. In this book a moving frame along a submanifold of a homogeneous space is a map from an open subset of the submanifold into the group of transformations of the ambient space. It is a very special case of the idea of sections of the principal bundle of linear frames of a manifold.

This exposition is restricted to curves and surfaces in order to emphasize the geometric interpretation of invariants and of other constructions. Working in these low dimensions helps a student develop a strong geometric intuition.

The book presents a careful selection of important results to serve two basic purposes. One is to show the reader how to prove the most important theorems in the subject, as this kind of knowledge is the foundation of future progress. Secondly, the method of moving frames is a natural means for discovering and proving important results. Its application in many areas helps to uncover many deep relationships, such as the Lawson correspondence. Finally, we think the topics chosen are very interesting. The more one studies them, the more fascinating they become.

A moving frame does not exist globally, in general. Its existence is obstructed by topology or by the existence of objects like umbilic points. Despite this, however, the method of moving frames often leads to global results. These results require arguments using covering space theory and cohomology theory, for which we give extra details and references; we do not assume the reader has a background in this material.

We have written this text for intermediate-level graduate students. The method of moving frames requires an elementary knowledge of Lie groups, which most students learn in their first or second year of graduate study. The exposition is detailed, especially in the earlier chapters. Moreover, the book pursues significant results beyond the standard topics of an introductory differential geometry course. A sample of these results includes the Willmore functional, the classification of cyclides of Dupin, the Bonnet problem, constant mean curvature immersions, isothermic immersions, and the duality between minimal surfaces in Euclidean space and constant mean curvature surfaces in hyperbolic space. The book concludes with Lie sphere geometry and its spectacular result that all cyclides of Dupin are Lie sphere congruent to each other.

We use *Mathematica*®, Matlab™, and Xfig to illustrate selected concepts and results.

There are nearly 300 problems and exercises in the text. They range from simple applications of what is being presented to open problems. The exercises are embedded in the text as essential parts of the exposition. The problems are gathered at the end of each chapter. Solutions to select problems are given at the end of the book.

It is the authors' pleasure to thank Joseph Hutchings for permission to use some illustrations and examples that he made in the summer of 2008 at Washington University while partially supported by an REU supplement of NSF Grant No. DMS-0312442.

This material is based upon work supported by the National Science Foundation under Grant No. DMS-0604236 and the Italian Ministry of University and Research (MIUR) via the PRIN project "Varietà reali e complesse: geometria, topologia e analisi armonica." The first author thanks GANG at the University of Massachusetts at Amherst for their support and hospitality in February and March of 2001; the Matematisk Institutt at the University of Bergen, Norway, for its generous hospitality during April and May 2001; the University of California at Berkeley for its generous hospitality from September 2009 through February 2010; the Politecnico di Torino for its support and hospitality and the support of the GNSAGA of the Istituto Nazionale di Alta Matematica "F. Severi" during May and June of 2010 and in July 2015; the Università di Parma for its support and hospitality in September 2013; and especially Washington University for valuing research and scholarship and providing sabbatical leaves during the spring semester of 2001 and the academic year 2009–2010. The second and third authors gratefully acknowledge the Mathematics Department of Washington University in St. Louis for its support and hospitality in February and March of 2008.

St. Louis, MO, USA — Gary R. Jensen
Torino, Italy — Emilio Musso
Parma, Italy — Lorenzo Nicolodi

Contents

Chapter 1
Introduction

This book evolved from the lecture notes of a graduate topics course given by the first author in the Fall of 1996. S.-S. Chern's University of Houston notes [46] formed the core of these lectures and influenced the philosophy of this book, which is to present the method of moving frames in the context of interesting problems in the highly intuitive geometric setting of three-dimensional space. Chern's article [47] and Bryant's papers [20] and [21] influenced the choice of topics. These latter two papers contain the first modern expositions of Möbius geometry and hyperbolic geometry treated by the method of moving frames.

By referring to the classical geometries we mean the three space forms plus Möbius geometry and Lie sphere geometry. We use the term *space form* for the simply connected spaces $\mathbf{R}^3$ of Euclidean geometry, $\mathbf{S}^3$ of spherical geometry, and $\mathbf{H}^3$ of hyperbolic geometry. Möbius geometry is the sphere $\mathbf{S}^3$ acted upon by its Lie group of all conformal diffeomorphisms. Lie sphere geometry is the unit tangent bundle of $\mathbf{S}^3$ acted upon by the Lie group of all contact transformations. The book does not cover some other classical geometries, such as projective geometry, Laguerre geometry, equiaffine geometry, and similarity geometry, all of which can be treated extensively by the method of moving frames.

We use the idea of congruence throughout the book. By this we mean the following. Immersions $\mathbf{x}, \mathbf{y} : M^m \to N^n$ into a manifold on which a Lie group G acts transitively are *congruent* if there exists a group element $A \in G$ such that $\mathbf{y}(p) = A\mathbf{x}(p)$ for every point $p \in M$. This is not the same as the notion of congruence in elementary Euclidean geometry, for which congruence of $\mathbf{x}(M)$ and $\mathbf{y}(M)$ means there exists $A \in G$ such that $A\mathbf{y}(M) = \mathbf{x}(M)$. For Euclid's notion we use the following. Immersions $\mathbf{x} : M \to N$ and $\mathbf{y} : \tilde{M} \to N$ are *equivalent* if there exists a diffeomorphism $F : M \to \tilde{M}$ such that $\mathbf{x}$ and $\mathbf{y} \circ F$ are congruent.

The book contains several fascinating threads that emerge as larger groups of transformations enter the picture. One of these is the story of Dupin immersions in Euclidean geometry. Such immersions are natural generalizations of isoparametric immersions (constant principal curvatures) in any of the space forms. Although the

G.R. Jensen et al., *Surfaces in Classical Geometries*, Universitext,
DOI 10.1007/978-3-319-27076-0_1

latter are easily classified, the Dupin immersions seem too abundant to classify. It turns out that the Dupin condition is invariant under conformal transformations. In Möbius geometry we discover that any Dupin immersion is Möbius congruent to some isoparametric immersion in a space form. The Dupin condition is also invariant under parallel transformations, so the notion of Dupin immersion makes sense in Lie sphere geometry, where it turns out that the classification is as simple as possible: All Dupin immersions are Lie sphere congruent to the Legendre lift of a great circle in the sphere.

Another thread running through the book is the story of isothermic immersions, which have enjoyed a recent renaissance in the field of integrable systems. Constant mean curvature immersions in the space forms are isothermic. Proper Bonnet immersions are isothermic off their necessarily discrete set of umbilic points. Any isothermic immersion generates a Bonnet pair by the Kamberov–Pedit–Pinkall (KPP) construction. An isothermic immersion remains isothermic after a conformal transformation. In Möbius geometry an isothermic immersion is special if it is Möbius congruent to a constant mean curvature immersion in a space form. A Willmore immersion in Möbius space is Möbius congruent to a minimal immersion in some space form precisely when it is isothermic.

The idea of associates of a given immersion weaves throughout the book. Constant mean curvature immersions in the space forms have a 1-parameter family of associates, which are noncongruent immersions of the same surface with the same induced metric and the same constant mean curvature. Willmore immersions in Möbius space have associates, which are noncongruent Willmore immersions of the same Riemann surface. Isothermic immersions in Möbius space have associates, historically called T-transforms.

The Lawson correspondence between minimal surfaces in $\mathbf{R}^3$ and CMC 1 surfaces in $\mathbf{H}^3$, as well as between CMC 1 surfaces in $\mathbf{R}^3$ and minimal surfaces in $\mathbf{S}^3$, is neatly described with moving frames adapted to a complex coordinate on a surface. Several chapters are devoted to understanding the first correspondence.

We include many illustrations to help the reader develop geometric intuition and to understand the concepts.

We proceed now to a chapter by chapter description of the text. It is our advice to the reader with the suggested background to begin with Chapter 4, Euclidean geometry, and refer back to Chapter 2, Lie groups, and Chapter 3, Theory of Moving Frames, as needed.

Chapter 2 presents a brief introduction to matrix Lie groups, their Lie algebras, and their actions on manifolds. We review left-invariant 1-forms and the Maurer–Cartan form of a Lie group, and the adjoint representation of the Lie group on its Lie algebra. The treatment of principal bundles is self-contained. We derive basic properties of transitive actions. We define the notion of a slice for nontransitive actions. In many instances this is just a submanifold cutting each orbit uniquely and transitively such that the isotropy subgroup at each point of the submanifold is the same. We use the general idea of a slice, however, for the ubiquitous action of conjugation by the orthogonal group on symmetric matrices. The chapter concludes with statements and proofs of the Cartan–Darboux uniqueness and existence theorems. Our proof of the global existence theorem is simpler than those proofs we have seen in the literature.

Chapter 3 presents an outline of the method of moving frames for any submanifold of an arbitrary homogeneous space. We explain how a Lie group acting transitively on a manifold N is related to the principal bundle of linear frames on N. We present a general outline of the frame reduction procedure after first describing the procedure for the elementary example of curves in the punctured plane acted upon by the special linear group $\mathbf{SL}(2,\mathbf{R})$. We re-emphasize that this book demonstrates the use of the method of moving frames to study submanifolds of homogeneous spaces. It is not a text on the theory of moving frames. This chapter concludes with basic theorems that characterize when a submanifold of a homogeneous space is itself homogeneous.

Chapter 4 begins with a standard elementary introduction to the theory of surfaces immersed in Euclidean space $\mathbf{R}^3$, whose Riemannian metric is the standard dot product. Section 4.2 is a review for readers who have studied basic differential geometry of curves and surfaces in Euclidean space. Geometric intuition is used to construct Euclidean frames on a surface. Section 4.3 repeats the exposition, but this time following the frame reduction procedure outlined in Chapter 3. The classical existence and congruence theorems of Bonnet are stated and proved as consequences of the Cartan–Darboux Theorems. A section on tangent and curvature spheres provides needed background for Lie sphere geometry. The Gauss map helps tie together the formalism of Gauss and that of moving frames. We discuss special examples, such as surfaces of revolution, tubes about a space curve, inversions in a sphere, and parallel transforms of a given immersion. These constructions provide many valuable examples throughout the book. The latter two constructions introduce for the first time Möbius, respectively Lie sphere, transformations that are not Euclidean motions. The section on elasticae contains material needed in our introduction of the Willmore problems.

Chapter 5 applies the method of moving frames to immersions of surfaces in spherical geometry, modeled by the unit three-sphere $\mathbf{S}^3 \subset \mathbf{R}^4$ with its group of isometries the orthogonal group, $\mathbf{O}(4)$. Stereographic projection from the sphere to Euclidean space appears in this chapter. It is our means to visualize geometric objects in $\mathbf{S}^3$. The existence of compact minimal immersions in $\mathbf{S}^3$, such as the Clifford torus, provide important examples of Willmore immersions. The chapter concludes with Hopf cylinders and Pinkall's Willmore tori in $\mathbf{S}^3$. Their construction uses the universal cover $\mathbf{SU}(2) \cong \mathbf{S}^3$ of $\mathbf{SO}(3)$.

Chapter 6 applies the method of moving frames to immersions of surfaces in hyperbolic geometry $\mathbf{H}^3$, for which we use the hyperboloid model with its full group of isometries $\mathbf{O}_+(3,1)$. Moving frames lead to natural expressions of the sphere at infinity and the hyperbolic Gauss map. The Poincaré ball model is introduced as a means to visualize surfaces immersed in hyperbolic space. As in the chapters on Euclidean and spherical geometry, the notions of tangent and curvature spheres of an immersed surface are described in detail as preparation for their fundamental role in Lie sphere geometry. The chapter concludes with many elementary examples.

Chapter 7 reviews complex structures on a manifold, then gives an elementary exposition of the complex structure induced on a surface by a Riemannian metric. In this way a complex structure is induced on any surface immersed into one of the space forms. Surfaces immersed into Möbius space inherit a complex structure. In all cases we use this structure to define a reduction of a moving frame to a unique frame associated to a given complex coordinate. Umbilic points do not hinder the existence of these frames, in contrast to the obstruction they can pose for the existence of second order frame fields. The Hopf invariant and the Hopf quadratic differential play a prominent role in the space forms as well as in Möbius geometry. Using the structure equations of the Hopf invariant h, the conformal factor e^u, and the mean curvature H of such frames, we give an elementary description of the Lawson correspondence between minimal surfaces in Euclidean geometry and constant mean curvature equal to one (CMC 1) surfaces in hyperbolic geometry; and between minimal surfaces in spherical geometry and CMC surfaces in Euclidean geometry.

Chapter 8 gives a brief history and exposition of minimal immersions in Euclidean space. We present the calculation of the first variation of the area functional and we derive the Enneper–Weierstrass representation. Scherk's surface is used to illustrate the problems that arise in integrating the Weierstrass forms. This integration problem is a simpler version of the monodromy problem encountered later in finding examples of CMC 1 immersions in hyperbolic geometry. We present results on complete minimal immersions with finite total curvature, which will be used in Chapter 14 to characterize minimal immersions in Euclidean space that smoothly extend to compact Willmore immersions into Möbius space. The final section on minimal curves applies the method of moving frames to the nonintuitive setting of holomorphic curves in $\mathbf{C}^3$ whose tangent vector is nonzero and isotropic at every point. An isotropic vector in $\mathbf{C}^3$ is one whose $\mathbf{C}$-bilinear dot product with itself is zero.

Chapter 9 gives a brief introduction to classical isothermic immersions in Euclidean space, a notion easily extended to immersions of surfaces into each of the space forms. The definition, which is the existence of coordinate charts that are isothermal and whose coordinate curves are lines of curvature, seems more analytic than geometric. We show that CMC immersions are isothermic away from their umbilics, which indicates that isothermic immersions are generalizations of CMC immersions. The Christoffel transform provides geometric content to the concept.

Chapter 10 presents the Bonnet Problem, which asks whether an immersion of a surface $\mathbf{x} : M \to \mathbf{R}^3$ admits a Bonnet mate, which is another noncongruent immersion $\tilde{\mathbf{x}} : M \to \mathbf{R}^3$ with the same induced metric and the same mean curvature at each point. Any immersion with constant mean curvature admits a 1-parameter family of Bonnet mates, all noncongruent to each other. These are its associates. The problem is thus to determine whether an immersion with nonconstant mean curvature has a Bonnet mate. The answer for an umbilic free immersion is whether it is isothermic or not. If it is nonisothermic, then it possesses a unique Bonnet mate. We believe that this is a new result. If it is isothermic, then only in special cases will it have a Bonnet mate, and if it does, it has a 1-parameter family of mates, similar to the CMC case. Such immersions are called proper Bonnet. A brief introduction to the notion of G-deformation is used to derive the KPP Bonnet pair construction of Kamberov, Pedit, and Pinkall. We state and prove a new result on proper Bonnet immersions that implies results of Cartan, Bonnet, Chern, and Lawson–Tribuzy. The chapter concludes with a summary of Cartan's classification of proper Bonnet immersions.

Chapter 11 is an introduction to immersions of surfaces in hyperbolic space with constant mean curvature equal to one (CMC 1 immersions in $\mathbf{H}^3$). The approach follows Bryant's paper [21], which replaces the hyperboloid model of $\mathbf{H}^3$ by the set of all 2×2 hermitian matrices with determinant one and positive trace. This model is acted upon isometrically by $\mathbf{SL}(2, \mathbf{C})$, the universal cover of the group of all isometries of hyperbolic space. The method of moving frames is applied to the study of immersed surfaces in this homogeneous space. Departing from Bryant's approach, we use frames adapted to a given complex coordinate to great advantage. A null immersion from a Riemann surface into $\mathbf{SL}(2, \mathbf{C})$ projects to a CMC 1 immersion into hyperbolic space. The null immersions are analogous to minimal curves of the Enneper–Weierstrass representation of minimal immersions into Euclidean space. A solution of these equations leads to a more complicated monodromy problem, which is described in detail. The chapter ends with some of Bryant's examples as well as more recent examples of Bohle–Peters and Bobenko–Pavlyukevich–Springborn.

Chapter 12 introduces conformal geometry and Liouville's characterization of conformal transformations of Euclidean space. Through stereographic projection these are all globally defined conformal transformations of the sphere $\mathbf{S}^3$. The Möbius group **Möb** is the group of all conformal transformations of $\mathbf{S}^3$. It is a ten-dimensional Lie group containing the group of isometries of each of the space forms as a subgroup. Möbius space $\mathscr{M}$ is the homogeneous space consisting of the sphere $\mathbf{S}^3$ acted upon by **Möb**. $\mathscr{M}$ has a conformal structure invariant under the action of **Möb**. The reduction procedure is applied to Möbius frames. The space forms are each equivariantly embedded into Möbius geometry. Conformally invariant properties, such as Willmore immersion, or isothermic immersion, or Dupin immersion, have characterizations in terms of the Möbius invariants. Oriented spheres in Möbius space provide the appropriate geometric interpretation of the vectors of a frame field.

Chapter 13 takes up the Möbius invariant conformal structure on Möbius space. It induces a conformal structure on any immersed surface, which in turn induces a complex structure on the surface. Möbius geometry is the study of properties of conformal immersions of Riemann surfaces into Möbius space $\mathscr{M}$ that remain invariant under the action of **Möb**. Each complex coordinate chart on an immersed surface has a unique Möbius frame field adapted to it, whose first-order invariant we call k and whose second-order invariant we call b. These are smooth, complex valued functions on the domain of the frame field. These frames are used to derive the structure equations for k and b, the conformal area, the conformal Gauss map, and the conformal area element. The equivariant embeddings of the space forms into Möbius space are conformal. Relative to a complex coordinate, the Hopf invariant, conformal factor, and mean curvature of an immersed surface in a space form determine the Möbius invariants k and b of the immersion into $\mathscr{M}$ obtained by applying the embedding of the space form into $\mathscr{M}$ to the given immersed surface. This gives a formula for the conformal area element showing that the Willmore energy is conformally invariant.

Chapter 14 introduces isothermic immersions of surfaces into Möbius space. An isothermic immersion of a surface into a space form, then composed with the equivariant embedding of the space form into Möbius space, becomes an isothermic immersion of the surface into Möbius space. Thus, an isothermic immersion in a space form remains isothermic under conformal transformations. An isothermic immersion into Möbius space is special if it comes from a CMC immersion into a space form. By a theorem of Voss, the Bryant quartic differential form of an umbilic free conformal immersion into Möbius space is holomorphic if and only if it is Willmore or special isothermic. A minimal immersion into a space form followed by the equivariant embedding of the space form into Möbius space becomes a Willmore immersion into Möbius space. Moreover, it is isothermic, with isolated umbilics, since this is true for minimal immersions in space forms and these properties are preserved by the embeddings into Möbius space. The theorem of Thomsen states that up to Möbius transformation, all isothermic Willmore immersions with isolated umbilics arise in this way.

Chapter 15 presents the method of moving frames in Lie sphere geometry. This involves a number of new ideas, beginning with the fact that some Lie sphere transformations are not diffeomorphisms of space $\mathbf{S}^3$, but rather of the unit tangent bundle of $\mathbf{S}^3$. This we identify with the set of pencils of oriented spheres in $\mathbf{S}^3$, which is identified with the set Λ of all lines in the quadric hypersurface $Q \subset \mathbf{P}(\mathbf{R}^{4,2})$. The set Λ is a five-dimensional subspace of the Grassmannian $G(2,6)$. The Lie sphere transformations are the projective transformations of $\mathbf{P}(\mathbf{R}^{4,2})$ that send Q to Q. This is a Lie group acting transitively on Λ. The Lie sphere transformations taking points of $\mathbf{S}^3$ to points of $\mathbf{S}^3$ are exactly the Möbius transformations, which form a proper subgroup of the Lie sphere group. In particular, the isometry groups of the space forms are natural subgroups of the Lie sphere group. There is a contact structure on Λ invariant under the Lie sphere group. A surface immersed in a space form with a unit normal vector field has an equivariant Legendre lift into Λ. A surface conformally immersed into Möbius space with an oriented tangent sphere map has an equivariant Legendre lift into Λ. This chapter studies Legendre immersions of surfaces into this homogeneous space Λ under the action of the Lie sphere group. A major application is a proof that all Dupin immersions of surfaces in a space form are Lie sphere congruent to each other. One of these Dupin immersions is the Legendre lift of a great circle of $\mathbf{S}^3$.

Chapter 2
Lie Groups

We present here a brief introduction to matrix Lie groups and their Lie algebras and their actions on manifolds. We review left-invariant 1-forms and the Maurer–Cartan form of a Lie group, and the adjoint representation of the Lie group on its Lie algebra. The treatment of principal bundles is self-contained. We derive basic properties of transitive actions. We define the notion of a slice for nontransitive actions. In many instances this is just a submanifold cutting each orbit uniquely and transitively such that the isotropy subgroup at each point of the submanifold is the same. The general idea of a slice is used, however, for the ubiquitous action by conjugation of the orthogonal group on symmetric matrices. We review the Frobenius theory of smooth distributions. The chapter concludes with statements and proofs of the Cartan–Darboux uniqueness and existence theorems. Our proof of the global existence theorem is simpler than those proofs we have seen in the literature. These theorems provide the principal analytic tools of the book.

2.1 Lie group actions

The *real general linear group* $\mathbf{GL}(n,\mathbf{R})$ of all $n \times n$ nonsingular matrices is a group under matrix multiplication and it is an open submanifold of the vector space $\mathbf{R}^{n\times n}$ of all $n \times n$ matrices. It is thus a Lie group of dimension n^2. In the same way, the complex general linear group $\mathbf{GL}(n,\mathbf{C})$ is a complex Lie group, which means it is a complex manifold and the group operations are holomorphic. (See Chapter 7 for the definition of complex manifold). A *matrix Lie group* is a closed subgroup of some $\mathbf{GL}(n,\mathbf{R})$. All Lie groups used in this book are matrix groups, so we restrict our exposition to this case.

G.R. Jensen et al., *Surfaces in Classical Geometries*, Universitext,
DOI 10.1007/978-3-319-27076-0_2

The Lie algebra of all left-invariant vector fields on $\mathbf{GL}(n,\mathbf{R})$ is naturally identified with $\mathfrak{gl}(n,\mathbf{R})$, the set $\mathbf{R}^{n\times n}$ with Lie bracket given by matrix commutations, $[X,Y] = XY - YX$. If $G \subset \mathbf{GL}(n,\mathbf{R})$ is a matrix subgroup, its Lie algebra $\mathfrak{g}$ of all left-invariant vector fields is a Lie subalgebra of $\mathfrak{gl}(n,\mathbf{R})$.

For a matrix group, the derivative of left or right multiplication is again left or right multiplication. That is, if $g \in G$ and $L_g : G \to G$ is left multiplication, $L_g(x) = gx$, for any $x \in G$, then the tangent space T_xG is a subspace of $\mathbf{R}^{n\times n}$ and

$$(dL_g)_x : T_xG \to T_{gx}G, \quad (dL_g)_x X = gX.$$

Since $\mathfrak{g} = T_1 G \subset \mathfrak{gl}(n,\mathbf{R})$, where $1 \in G$ is the identity element, it follows that $T_xG = x\mathfrak{g}$. For right multiplication $R_g(x) = xg$,

$$(dR_g)_x : T_xG \to T_{xg}G, \quad (dR_g)_x X = Xg.$$

In particular, $T_xG = \mathfrak{g}x$, for any $x \in G$. That $x\mathfrak{g} = \mathfrak{g}x$, for any $x \in G$, follows from the invariance of $\mathfrak{g}$ under the *adjoint representation* of G on $\mathfrak{g}$, which is

$$\mathrm{Ad}(g) : \mathfrak{g} \to \mathfrak{g}, \quad \mathrm{Ad}(g) = (dC_g)_1,$$

where for any $g \in G$, the *conjugation map* $C_g : G \to G$ is $C_g(x) = gxg^{-1}$. For a matrix group G, the adjoint representation is

$$\mathrm{Ad}(g)X = gXg^{-1},$$

for any $X \in \mathfrak{g} \subset \mathfrak{gl}(n,\mathbf{R})$, and so $g\mathfrak{g}g^{-1} = \mathfrak{g}$.

The Maurer–Cartan form ω of G is the $\mathfrak{g}$-valued left-invariant 1-form on G whose value at $g \in G$ is

$$\omega_g = g^{-1}dg,$$

where this is matrix multiplication. In more detail, if $X \in T_gG$, then $\omega_g(X) = g^{-1}X$. The Maurer–Cartan form of $\mathbf{GL}(2,\mathbf{R})$ is

$$\omega_A = A^{-1}dA = \begin{pmatrix} x^1_1 & x^1_2 \\ x^2_1 & x^2_2 \end{pmatrix}^{-1} \begin{pmatrix} dx^1_1 & dx^1_2 \\ dx^2_1 & dx^2_2 \end{pmatrix}$$

at the point $A \in \mathbf{GL}(2,\mathbf{R})$. Here dx^i_j is the differential of the coordinate function x^i_j on the open subset $\mathbf{GL}(2,\mathbf{R}) \subset \mathbf{R}^{2\times 2}$. On a closed subgroup G of $\mathbf{GL}(n,\mathbf{R})$, these forms are pulled back to G by the inclusion map, so on G they would satisfy the set of linear equations defining the subspace $\mathfrak{g} \subset \mathfrak{gl}(2,\mathbf{R})$. The Maurer–Cartan structure equations of G are

$$d\omega = d(g^{-1}dg) = -g^{-1}dg\, g^{-1} \wedge dg = -\omega \wedge \omega,$$

where in this matrix multiplication, the terms are multiplied by the wedge product of 1-forms.

Example 2.1. The *special orthogonal group* is

$$\mathbf{SO}(n) = \{A \in \mathbf{GL}(n,\mathbf{R}) : {}^tAA = I_n,\ \det A = 1\}.$$

Its Maurer–Cartan form is an $n \times n$ matrix of left-invariant 1-forms ω^i_j on $\mathbf{SO}(n)$ that satisfy $\omega^i_j = -\omega^j_i$, for all $i,j = 1,\dots,n$.

Example 2.2. The *special linear group* is

$$\mathbf{SL}(n,\mathbf{R}) = \{A \in \mathbf{GL}(n,\mathbf{R}) : \det A = 1\}.$$

Its Maurer–Cartan form satisfies trace $\omega = 0$.

Definition 2.3. A Lie group G *acts smoothly on the left* of a smooth manifold N if there is a smooth map $\mu : G \times N \to N$, which we denote by $\mu(g,x) = gx$, with the properties $1x = x$, for any $x \in N$, where $1 \in G$ is the identity element, and

$$(g_1 g_2)x = g_1(g_2 x),$$

for any $g_1, g_2 \in G$ and $x \in N$. The action is *from the right* if we write $\mu(g,x) = xg$ and this satisfies $x1 = x$ and $x(g_1 g_2) = (xg_1)g_2$. For any $g \in G$, the map $g : N \to N$ given by $x \mapsto gx$ (respectively, $x \mapsto xg$) is a diffeomorphism whose inverse is given by the action of g^{-1}. The action is *free* if, for any $g \in G$ unequal to the identity element 1 of G, this diffeomorphism has no fixed points. The action is *effective* if 1 is the only element of G that acts as the identity element. The action is *transitive* if, for any points $x, y \in N$, there exists $g \in G$ whose action sends x to y.

The adjoint representation of G is a smooth left action of G on its Lie algebra $\mathfrak{g}$.

An action of the additive group of real numbers $\mathbf{R}$ on a manifold N is a *global flow*. A global flow $\theta : \mathbf{R} \times N \to N$ generates a smooth vector field X on N, called the *infinitesimal generator* of the flow, by $X_{(m)} = \frac{d}{dt}\big|_{t=0} \theta(t,m)$. For each fixed $m \in N$, the curve $\theta_m(t) = \theta(t,m)$ is the integral curve of X starting at m. Conversely, any smooth vector field X on N generates a flow $\theta : W \to N$, where $W \subset \mathbf{R} \times N$ is an open set containing $\{0\} \times N$. See [16, pp. 127 ff] or [110, pp. 438 ff]. If $W = \mathbf{R} \times N$, then the vector field X is called complete.

A left-invariant vector field X on a Lie group G is complete and generates the global flow

$$\theta : \mathbf{R} \times G \to G, \quad \theta(t,g) = g\exp(tX),$$

where the *exponential map* $\exp(tX) = e^{tX}$ is the matrix exponential for any matrix group G. The curve $\exp(tX)$ is the integral curve of X through $1 \in G$. This is also the integral curve of the right-invariant vector field whose value at 1 is $X_{(1)}$. A right-invariant vector field generates the global flow $\theta(t,g) = \exp(tX)g$.

If G acts smoothly on the left (respectively, right) of N, then any $X \in \mathfrak{g}$ defines the global flow on N,

$$\theta : \mathbf{R} \times N \to N, \quad \theta(t,m) = \exp(tX)m,$$

(respectively, $\theta(t,m) = m\exp(tX)$). This flow generates a vector field $\hat{X}$ on N, called the *vector field induced by X on N*. If $\hat{X}_{(m)} = 0$ at some point $m \in N$, then its integral curve through m must be constant, by the uniqueness of the integral curve through a point. This means that m is a fixed point for $\exp(tX)$, for all $t \in \mathbf{R}$. In particular, if the action is free, then $\hat{X}$ has no zeros on N, for any $X \neq 0 \in \mathfrak{g}$.

Right actions play a fundamental role in the idea of principal bundles.

Definition 2.4. A *principal H-bundle over N* with smooth *total space P* and smooth *base space N* is a smooth, surjective submersion $\pi : P \to N$ and a free, right action of the Lie group H on P such that for each point $m \in N$, if p is a point in the *fiber over m*, $\pi^{-1}\{m\}$, then this fiber is the H-orbit of p,

$$\pi^{-1}\{m\} = pH.$$

The map π is called the *projection* of the bundle. A *local section* of the bundle on an open subset $U \subset N$ is a smooth map $\sigma : U \to P$ such that $\pi \circ \sigma : U \to U$ is the identity map. A *local trivialization* of the bundle over U is a smooth diffeomorphism

$$F : U \times H \to \pi^{-1}U \subset P,$$

satisfying $F(u,h) = F(u,1)h$ for any $(u,h) \in U \times H$. Any local section $\sigma : U \to P$ defines a local trivialization

$$F : U \times H \to \pi^{-1}U, \quad F(u,h) = \sigma(u)h.$$

If $X \in \mathfrak{h}$, the Lie algebra of H, then the vector field $\hat{X}$ it induces on P is the *fundamental vertical vector field on P induced by X*.

Remark 2.5. A smooth map $\pi : P \to N$ is a *submersion* if the rank of $d\pi_p$ equals the dimension of N at every point $p \in P$. If $\pi : P \to N$ is a submersion, then π is an open map and every point of P is in the image of a smooth local section of π. (See, for example, [110, Proposition 7.16, p. 169]). In particular, for any point $p \in P$ in the total space of a principal H-bundle $\pi : P \to N$, there is a local section whose image contains p. Moreover, given any point $n \in N$, there is a local section defined on a neighborhood of n.

2.2 Transitive group actions

Suppose G acts smoothly on the left of the manifold N. Choose a point $o \in N$, and call it the *origin* of N. Then the *projection map*

$$\pi : G \to N, \quad \pi(g) = go$$

is smooth. The action is *transitive* if and only if π is surjective for any choice of origin $o \in N$. If $X \in \mathfrak{g}$ induces the vector field $\hat{X}$ on N, then the integral curve of $\hat{X}$ through $o \in N$ is $\exp(tX)o = \pi \circ \exp(tX)$. In particular, $\hat{X}_{(o)} = d\pi_1 X$. The *isotropy subgroup* of G at $o \in N$ is

$$H = \{g \in G : go = o\}.$$

It is a closed subgroup of G. The *isotropy representation of H* is

$$H \to \mathbf{GL}(T_oN), \quad h \mapsto dh_o : T_oN \to T_oN. \tag{2.1}$$

Proposition 2.6. *If a Lie group G acts smoothly and transitively on a manifold N, then for any point $o \in N$, the projection map*

$$\pi : G \to N, \quad \pi(g) = go \tag{2.2}$$

is a surjective submersion. If G acts on the left of N and if H is the isotropy subgroup of G at o, then H acts freely on the right of G by right multiplication and G is a principal H-bundle over N with projection (2.2). *For any $X \in \mathfrak{h}$, the Lie algebra of H, the fundamental vertical vector field $\hat{X}$ induced on G by the right action of H coincides with the left-invariant vector field $X \in \mathfrak{h} \subset \mathfrak{g}$ on G.*

Proof. Suppose G acts on N on the left. For any $g \in G$, let $L_g : G \to G$ denote left multiplication by g. This is a diffeomorphism. Then

$$\pi \circ L_g(x) = \pi(gx) = (gx)o = g(xo) = g \circ \pi(x),$$

for any $x \in G$, shows that $\pi \circ L_g = g \circ \pi$. Since $L_g : G \to G$ and $g : N \to N$ are diffeomorphisms, it follows that the rank of π is constant on G. Then $\pi : G \to N$ is surjective and of constant rank, so it must be a submersion, that is, its rank must equal the dimension of N at every point of G. (See, for example, [110, Theorem 7.15, p. 168]). A similar argument proves the result when G acts on N on the right.

By Definition 2.4, it remains to prove that the right action of H on G is free and that for any $g \in G$, we have $\pi^{-1}\{go\} = gH$. This is elementary.

If $X \in \mathfrak{h}$, and if $g \in G$, then

$$\hat{X}_{(g)} = \left.\frac{d}{dt}\right|_0 g\exp(tX) = (dL_g)_1 \left.\frac{d}{dt}\right|_0 \exp(tX) = (dL_g)_1 X_{(1)} = X_{(g)}.$$

□

Corollary 2.7 (Lift Property). *If $f : M \to N$ is a smooth map from a manifold M and if $m_0 \in M$, then there exists a neighborhood U of m_0 in M on which there is a smooth map $g : U \to G$, such that $f = \pi \circ g$ on U.*

Proof. There exists a neighborhood V of $f(m_0)$ in N on which there exists a smooth section $\sigma : V \to G$ of the submersion (2.2). If $U = f^{-1}V$, then $g = \sigma \circ f : U \to G$ is the desired map. □

Corollary 2.8. *The kernel of $d\pi_1 : \mathfrak{g} \to T_oN$ is $\mathfrak{h}$, so $d\pi_1 : \mathfrak{g}/\mathfrak{h} \to T_oN$ is an isomorphism. For any $h \in H$, the diagram*

$$\begin{array}{ccc} \mathfrak{g}/\mathfrak{h} & \stackrel{\mathrm{Ad}(h)}{\to} & \mathfrak{g}/\mathfrak{h} \\ d\pi_1 \downarrow & & \downarrow d\pi_1 \\ T_oN & \stackrel{dh_o}{\to} & T_oN \end{array}$$

commutes.

Proof. It is clear that $\mathfrak{h} \subset \ker d\pi_1$. Conversely, if $X \in \mathfrak{g}$, then $0 = d\pi_1 X = \hat{X}_o$ implies that the integral curve $\exp(tX)o$ of $\hat{X}$ at o must be constant, so $\exp(tX) \in H$, for all $t \in \mathbf{R}$, and thus $X \in \mathfrak{h}$. This proves the first statement. For any $X \in \mathfrak{g}$ and $h \in H$, we have $h^{-1}o = o$ and $he^{tX}h^{-1} = e^{thXh^{-1}}$, so

$$\begin{aligned} dh_o d\pi_1 X &= \frac{d}{dt}\Big|_0 h\exp(tX)o = \frac{d}{dt}\Big|_0 h\exp(tX)h^{-1}o \\ &= \frac{d}{dt}\Big|_0 \exp(thXh^{-1})o = d\pi_1 \mathrm{Ad}(h)X. \end{aligned}$$

□

Corollary 2.9. *There exists a G-invariant Riemannian metric on N if and only if there exists an* $\mathrm{Ad}(H)$*-invariant inner product on* $\mathfrak{g}/\mathfrak{h}$.

Proof. By the preceding corollary, there exists an inner product on T_oN invariant under the isotropy representation of H if and only if there exists an $\mathrm{Ad}(H)$-invariant inner product on $\mathfrak{g}/\mathfrak{h}$. If N possesses a G-invariant Riemannian metric I, then a fortiori, I_o is invariant under the linear isotropy representation of H. Conversely, if I_o is an inner product on T_oN invariant under the linear isotropy representation of H, define an inner product on T_mN, for any $m \in N$, by $I_m = g^*I_o$, where $g \in G$ is any element for which $gm = o$. This is independent of the choice of such g because of the invariance of I_o. □

Example 2.10 ($\mathbf{O}(3)$ acting on $\mathbf{S}^2$). Label the standard basis of $\mathbf{R}^3$ by ϵ_i, for $i = 0,1,2$ and the entries of a matrix $A \in \mathbf{O}(3)$ by A^i_j, for $i,j = 0,1,2$. The standard action of $\mathbf{O}(3)$ on $\mathbf{R}^3$ induces a transitive action on the unit sphere $\mathbf{S}^2 \subset \mathbf{R}^3$ (by the Gram–Schmidt orthonormalization process). The isotropy subgroup of $\mathbf{O}(3)$ at $\epsilon_0 \in \mathbf{S}^2$ is

$$H = \{\begin{pmatrix} 1 & 0 \\ 0 & A \end{pmatrix} : A \in \mathbf{O}(2)\} \cong \mathbf{O}(2).$$

Its isotropy representation is just the standard representation of $\mathbf{O}(2)$ on $\mathbf{R}^2 = \epsilon_0^\perp \subset \mathbf{R}^3$. Consider the vector space direct sum $\mathfrak{o}(3) = \mathfrak{h} \oplus \mathfrak{m}$, where

$$\mathfrak{h} = \{\begin{pmatrix} 0 & 0 \\ 0 & X \end{pmatrix} : X \in \mathfrak{o}(2)\} \cong \mathfrak{o}(2)$$

is the Lie algebra of H in $\mathfrak{o}(3)$, and

$$\mathfrak{m} = \{\begin{pmatrix} 0 & -{}^t\mathbf{x} \\ \mathbf{x} & 0 \end{pmatrix} : \mathbf{x} \in \mathbf{R}^2\} \cong \mathbf{R}^2.$$

Then $\mathfrak{m} \cong \mathfrak{o}(3)/\mathfrak{h}$. The adjoint action of H on $\mathfrak{m}$ is

$$\begin{pmatrix} 1 & 0 \\ 0 & A \end{pmatrix} \begin{pmatrix} 0 & -{}^t\mathbf{x} \\ \mathbf{x} & 0 \end{pmatrix} \begin{pmatrix} 1 & 0 \\ 0 & A^{-1} \end{pmatrix} = \begin{pmatrix} 0 & -{}^t(A\mathbf{x}) \\ A\mathbf{x} & 0 \end{pmatrix},$$

which is just the standard action of $\mathbf{O}(2)$ on $\mathbf{R}^2$ under the above isomorphisms. In particular, $\mathfrak{m}$ is invariant under the adjoint action of H. The Maurer–Cartan form of $\mathbf{O}(3)$ is

$$\omega = A^{-1}dA = (\omega^i_j),$$

where $\omega^i_j + \omega^j_i = 0$, for all $i,j = 0,1,2$. The structure equations are

$$d\omega^i_j = -\sum_{k=0}^{2} \omega^i_k \wedge \omega^k_j.$$

The only $\mathrm{Ad}(\mathbf{O}(2))$-invariant inner products on $\mathfrak{m} \cong \mathfrak{o}(3)/\mathfrak{h} \cong \mathbf{R}^2$ are the constant positive multiples of

$$\langle X, Y \rangle = \mathrm{trace}\, {}^tXY = 2\mathbf{x} \cdot \mathbf{y},$$

if $X \leftrightarrow \mathbf{x}$ and $Y \leftrightarrow \mathbf{y}$. Up to constant positive multiple, these inner products induce the Riemannian metric on $\mathbf{S}^2 \subset \mathbf{R}^3$ induced from the standard inner product on $\mathbf{R}^3$.

Example 2.11 (Grassmannians). For $m < n$, let $\mathbf{R}^{n\times m*}$ be the set of all $n \times m$ matrices of rank m. Consider the equivalence relation on $\mathbf{R}^{n\times m*}$ given by $X \sim Y$ if and only if $Y = XA$, for some $A \in \mathbf{GL}(m,\mathbf{R})$, if and only if the columns of X span the same subspace of $\mathbf{R}^n$ as do the columns of Y. Let $[X]$ denote the equivalence class of $X \in \mathbf{R}^{n\times m*}$. The *Grassmannian of m-dimensional subspaces of* $\mathbf{R}^n$ is the set of all equivalence classes $G(m,n) = \mathbf{R}^{n\times m*}/\sim$. Left multiplication action of $\mathbf{GL}(n,\mathbf{R})$ on $\mathbf{R}^{n\times m*}$ preserves the equivalence relation, so induces an action on $G(m,n)$ given by $B[X] = [BX]$, for any $B \in \mathbf{GL}(n,\mathbf{R})$ and $[X] \in G(m,n)$. This action is transitive, since it takes the origin $P_0 = \begin{bmatrix} I_m \\ 0 \end{bmatrix}$ to any designated point of $G(m,n)$, because any basis of a given m-dimensional subspace can be extended to a basis of $\mathbf{R}^n$. The isotropy subgroup of $\mathbf{GL}(n,\mathbf{R})$ at P_0 is

$$G_0 = \{\begin{pmatrix} a & b \\ 0 & c \end{pmatrix} \in \mathbf{GL}(n,\mathbf{R}) : a \in \mathbf{GL}(m,\mathbf{R}),\ c \in \mathbf{GL}(n-m,\mathbf{R})\}.$$

Example 2.12 ($\mathbf{SL}(2,\mathbf{R}) \to \mathbf{C}^+$). Consider the smooth transitive action of

$$\mathbf{SL}(2,\mathbf{R}) = \{\begin{pmatrix} r & s \\ t & u \end{pmatrix} : ru - ts = 1\}$$

on the upper half-plane $\mathbf{C}^+ = \{z = x + iy \in \mathbf{C} : y > 0\}$ given by

$$\begin{pmatrix} r & s \\ t & u \end{pmatrix} z = \frac{rz+s}{tz+u}.$$

Choose $i \in \mathbf{C}^+$ for origin to define the submersion

$$\pi : \mathbf{SL}(2,\mathbf{R}) \to \mathbf{C}^+, \quad \pi \begin{pmatrix} r & s \\ t & u \end{pmatrix} = \frac{ri+s}{ti+u} = \frac{rt+su}{t^2+u^2} + \frac{i}{t^2+u^2}.$$

The isotropy subgroup H of $\mathbf{SL}(2,\mathbf{R})$ at i is

$$H = \{\begin{pmatrix} r & s \\ t & u \end{pmatrix} : \frac{ri+s}{ti+u} = i\} = \{\begin{pmatrix} r & -t \\ t & r \end{pmatrix} : r^2 + t^2 = 1\} \cong \mathbf{SO}(2).$$

Its Lie algebra $\mathfrak{h}$ has an $\mathrm{Ad}(H)$-invariant complementary subspace

$$\mathfrak{m} = \{X = \begin{pmatrix} x & y \\ y & -x \end{pmatrix} : x, y \in \mathbf{R}\} \cong \mathfrak{sl}(2,\mathbf{R})/\mathfrak{h}.$$

Then, under the identification $T_1\mathbf{SL}(2,\mathbf{R}) \cong \mathfrak{h} \oplus \mathfrak{m}$,

$$d\pi_1 : \mathfrak{m} \to T_i\mathbf{C}^+ = \mathbf{R}^2, \quad d\pi_1 X = 2y\epsilon_1 + 2x\epsilon_2.$$

The only $\mathrm{Ad}(H)$-invariant inner products on $\mathfrak{m}$ are

$$\langle X, \tilde{X} \rangle = c \operatorname{trace} {}^t X \tilde{X} = 2c(x\tilde{x} + y\tilde{y}) = \frac{c}{2}(d\pi_1 X) \cdot (d\pi_1 \tilde{X}),$$

the standard dot product on $T_i\mathbf{C}^+ = \mathbf{R}^2$, where c is any positive constant. Taking $c = 1/2$, we get for the metric on $T_i\mathbf{C}^+$,

$$I_i = dx^2 + dy^2 = dz d\bar{z},$$

where $z = x + iy$. Identify all tangent spaces of $\mathbf{C}^+$ with $\mathbf{R}^2$. If $A = \begin{pmatrix} r & s \\ t & u \end{pmatrix} \in \mathbf{SL}(2,\mathbf{R})$ is regarded as a diffeomorphism $A : \mathbf{C}^+ \to \mathbf{C}^+$, then the matrix of its differential at i is dA_i, which in the standard basis is

$$dA_i = \frac{1}{k}\begin{pmatrix} u^2 - t^2 & 2tu \\ -2tu & u^2 - t^2 \end{pmatrix}, \quad (dA_i)^{-1} = \begin{pmatrix} u^2 - t^2 & -2tu \\ 2tu & u^2 - t^2 \end{pmatrix},$$

where $k = (t^2 + u^2)^2$. For any $\mathbf{v}, \tilde{\mathbf{v}} \in T_{Ai}\mathbf{C}^+ = \mathbf{R}^2$, and $A \in \mathbf{SL}(2,\mathbf{R})$,

$$I_{Ai}(\mathbf{v}, \tilde{\mathbf{v}}) = I_i((dA_i)^{-1}\mathbf{v}, (dA_i)^{-1}\tilde{\mathbf{v}}) = k\, \mathbf{v} \cdot \tilde{\mathbf{v}},$$

and $y(Ai) = 1/\sqrt{k}$, so

$$I_z = \frac{dx^2 + dy^2}{y^2} \tag{2.3}$$

at any point $z = x + iy \in \mathbf{C}^+$. This Riemannian metric I on $\mathbf{C}^+$ is the upper half-plane model of the hyperbolic plane, which is discussed further in Example 16.

2.3 A slice theorem

In this section, group actions will be from the left, unless stated otherwise. Let G be a Lie group acting smoothly on a manifold N. The *orbit* of a point $x \in N$ is

$$Gx = \{gx : g \in G\},$$

which is an immersed submanifold of N. The *isotropy subgroup* of G at x is

$$G_x = \{g \in G : gx = x\},$$

which is a closed subgroup of G. If x and $y = gx$ are points in the same orbit, then their isotropy subgroups are conjugate in G, namely,

$$G_{gx} = gG_xg^{-1}.$$

An orbit is of *type* G/H, where H is a closed subgroup of G, if the isotropy subgroup at any point of the orbit is G-conjugate to H. We let N_H denote the set of all points in N lying on orbits of type G/H.

If H is a closed subgroup of G, then the set G/H of left cosets of H is a smooth manifold on which G acts smoothly and transitively (see [16, Theorem 9.2, pp 166 ff]). For any point $x \in N$, the orbit of G through x is an immersed submanifold of N diffeomorphic to G/G_x, whose dimension is $\dim G - \dim G_x$. Since conjugate subgroups have diffeomorphic quotients, it follows that any orbit of type G/H is diffeomorphic to G/H.

The smooth map

$$\pi : G \to G/H, \quad \pi(g) = gH, \tag{2.4}$$

is the projection map of a principal H-bundle, where H acts freely on the right on G by right multiplication. If H acts smoothly on the left on a manifold $\mathscr{Y}$, then H acts smoothly on $G \times \mathscr{Y}$ on the right by

$$(g,y)h = (gh, h^{-1}y),$$

for any $y \in \mathscr{Y}$, $g \in G$, and $h \in H$. Denote the orbit space of this action by $G \times_H \mathscr{Y}$, the *twisted product* of G with $\mathscr{Y}$ over H. Denote the projection map

$$\mu : G \times \mathscr{Y} \to G \times_H \mathscr{Y}, \quad \mu(g,y) = [g,y], \tag{2.5}$$

so $[gh, h^{-1}y] = [g,y]$ for any $h \in H$. Then $G \times_H \mathscr{Y}$ is a smooth manifold, μ is smooth, and the smooth map

$$\nu : G \times_H \mathscr{Y} \to G/H, \quad \nu[g,y] = gH$$

is the projection map of the *fiber bundle* over G/H with *standard fiber* $\mathscr{Y}$ associated to the principal H-bundle (2.4). A local section

$$\sigma : U \subset G/H \to G$$

of (2.4) defines a *local trivialization* of $G \times_H \mathscr{Y}$, which is a diffeomorphism

$$\varphi : U \times \mathscr{Y} \to \nu^{-1}U \subset G \times_H \mathscr{Y}, \quad \varphi(u,y) = [\sigma(u), y].$$

For details see [100, pp 54–55]. Now $\mu \circ (\sigma, \mathrm{id}_{\mathscr{Y}}) = \varphi$, for any such section σ, shows that μ of (2.5) is a submersion.

Definition 2.13. A *slice* of the smooth action of a Lie group G on a smooth manifold N is a pair $(\mathscr{Y}, H)$, where $\mathscr{Y}$ is a regular submanifold of N with $\dim \mathscr{Y} < \dim N$ and H is a closed subgroup of G such that

1. $H\mathscr{Y} = \mathscr{Y}$,
2. $G\mathscr{Y}$ is an open submanifold of N,
3. $\mathscr{Y}$ is closed in $G\mathscr{Y}$, and
4. $F : G \times_H \mathscr{Y} \to G\mathscr{Y}$, $F[g,y] = gy$ is a diffeomorphism.

Item (1) means that H is a subgroup of the *stabilizer of* $\mathscr{Y}$, which is the subgroup $\{g \in G : g\mathscr{Y} = \mathscr{Y}\}$ of G. If we call a submanifold $\mathscr{Y}$ a slice, without mention of the subgroup H, then it is to be understood that H is its stabilizer.

Remark 2.14. For most group actions in this book, a slice $\mathscr{Y}$ exists for which the isotropy subgroup $G_y = H$, for all $y \in \mathscr{Y}$. In this case the orbit space has a global trivialization $G \times_H \mathscr{Y} = (G/H) \times \mathscr{Y}$.

The following characterization of a slice involves derivatives only at points of $\mathscr{Y}$.

Theorem 2.15 (Slice Property). *If all assumptions of Definition 2.13 through items (1), (2), and (3) hold, and if*

(4a) $g \in G$ *and* $(g\mathscr{Y}) \cap \mathscr{Y} \neq \emptyset$ *implies* $g \in H$,
(4b) the Lie algebra of the isotropy subgroup G_y *is* $\mathfrak{h}$, *for every point* $y \in \mathscr{Y}$, *and*
(4c) $T_yN = T_y(Gy) \oplus T_y\mathscr{Y}$, *for every* $y \in \mathscr{Y}$,

then $(\mathscr{Y}, H)$ *is a slice of the action of* G *on* N.

Proof. We must prove that the map F of item (4) in Definition 2.13 is a diffeomorphism under the present hypotheses. F is certainly smooth and surjective. It remains to prove that it has a smooth inverse.

For any $y \in \mathscr{Y}$, the dimension of the orbit Gy is $\dim G/G_y = \dim G/H$, by items (4a) and (4b). The tangent space to the orbit Gy at a point y is

$$T_y(Gy) = \{\hat{X}_{(y)} : X \in \mathfrak{g}\},$$

where $\hat{X}$ is the vector field induced on N by the action of G, so $\hat{X}_{(y)} = \frac{d}{dt}\big|_0 \exp(tX)y$. The dimension of $G \times_H \mathscr{Y}$ is thus $\dim G/H + \dim \mathscr{Y} = \dim N$, by item (4c).

If $F[g,y] = F[\tilde{g},\tilde{y}]$, then $g^{-1}\tilde{g}\tilde{y} = y \in \mathscr{Y}$ implies $g^{-1}\tilde{g} = h \in H$, by item (4a). Then $\tilde{g} = gh$, so $gy = gh\tilde{y}$ implies $h\tilde{y} = y$. Hence, $[\tilde{g},\tilde{y}] = [gh, h^{-1}y] = [g,y]$, so F is injective. So the inverse of F exists, and will be smooth if dF is an isomorphism at every point of $G \times_H \mathscr{Y}$. Being a linear map between spaces of equal dimension, dF is an isomorphism if it is surjective.

Consider the composition $F \circ \mu : G \times \mathscr{Y} \to N$, where μ is defined in (2.5). For any $(g,y) \in G \times \mathscr{Y}$, we have $F \circ \mu(g,y) = gy = gF \circ \mu(1,y)$, so

$$d(F \circ \mu)_{(g,y)} = dg_y d(F \circ \mu)_{(1,y)},$$

and $dg_y : T_yN \to T_{gy}N$ is an isomorphism. It follows that $d(F \circ \mu)_{(g,y)}$ is surjective, for any $(g,y) \in G \times \mathscr{Y}$, provided that $d(F \circ \mu)_{(1,y)}$ is surjective for any $y \in \mathscr{Y}$. This latter map is surjective, since for any $(X,v) \in \mathfrak{g} \oplus T_y\mathscr{Y} = T_{(1,y)}(G \times \mathscr{Y})$,

$$d(F \circ \mu)_{(1,y)}(X,v) = \hat{X}(y) + v \in T_y(Gy) \oplus T_y\mathscr{Y}$$

is surjective, and the image is T_yN, by item (4c). It follows then that $dF_{[g,y]}$ is an isomorphism at every $[g,y] \in G \times_H \mathscr{Y}$. □

Remark 2.16. The reduction procedure in the method of moving frames requires an explicit slice of some action at each step. Our requirement that $\dim \mathscr{Y} < \dim N$ in the definition of slice ensures that each step of the reduction is nontrivial. It also implies that the action of a discrete group G has no slice.

The following action occurs in each of the space form geometries. It is the most complicated action in this book.

Example 2.17. Consider the action of $G = \mathbf{O}(2) \times \mathbf{O}(1)$ on the vector space $\mathscr{S}$ of all 2×2 symmetric real matrices,

$$G \times \mathscr{S} \to \mathscr{S}, \quad (A, \epsilon)X = \epsilon A X {}^t\!A,$$

where $\epsilon = \pm 1$. Notice that $\operatorname{trace}(A,\epsilon)X = \epsilon \operatorname{trace} X$. From elementary linear algebra we know that for each $S \in \mathscr{S}$, this action will diagonalize S. In the language of group actions, every G-orbit meets the hyperplane of all diagonal matrices $\mathscr{D}$ in $\mathscr{S}$, which is the hyperplane $z = 0$, if we make the identification of vector spaces

$$\mathscr{S} = \mathbf{R}^3, \quad \begin{pmatrix} x & z \\ z & y \end{pmatrix} = (x, y, z).$$

Consider the line $\mathscr{L} = \{tI_2 : t \in \mathbf{R}\}$ of scalar matrices in $\mathscr{S}$ and consider the submanifold

$$\mathscr{Y} = \mathscr{D} \setminus \mathscr{L} = \{\begin{pmatrix} x & 0 \\ 0 & y \end{pmatrix} : x \neq y\}$$

of all nonscalar, diagonal matrices in $\mathscr{S}$. Let

$$K = \{\pm I_2, \pm I_{1,1}, \pm \begin{pmatrix} 0 & 1 \\ 1 & 0 \end{pmatrix}, \pm \begin{pmatrix} 0 & -1 \\ 1 & 0 \end{pmatrix}\}, \tag{2.6}$$

a closed subgroup of $\mathbf{O}(2)$, so $H = K \times \mathbf{O}(1)$ is a closed subgroup of G. It is a useful exercise to prove:

(1) H is the stabilizer of $\mathscr{Y}$.

(2) For any $p = \begin{pmatrix} x & 0 \\ 0 & y \end{pmatrix} \in \mathscr{Y}$, the tangent spaces $T_p\mathscr{S} = \mathscr{S}$,

$$T_p(G\begin{pmatrix} x & 0 \\ 0 & y \end{pmatrix}) = \{z\begin{pmatrix} 0 & 1 \\ 1 & 0 \end{pmatrix} : z \in \mathbf{R}\},$$

since $\frac{d}{dt}\big|_0 e^{tX}\begin{pmatrix} x & 0 \\ 0 & y \end{pmatrix} e^{-tX} = z(x-y)\begin{pmatrix} 0 & 1 \\ 1 & 0 \end{pmatrix}$, for $X = \begin{pmatrix} 0 & -z \\ z & 0 \end{pmatrix} \in \mathfrak{o}(2)$, and

$$T_p\mathscr{Y} = \{\begin{pmatrix} u & 0 \\ 0 & v \end{pmatrix} : u, v \in \mathbf{R}\}.$$

(3) By Theorem 2.15 one proves that $\mathscr{Y}$ is a slice of this action.

Note that $G\mathscr{L} = \mathscr{L}$, and the action of G on $\mathscr{L}$ is the standard action of $\mathbf{O}(1)$ on $\mathbf{R}$.

Theorem 2.18 (Factor Property). *Let G be a Lie group acting smoothly on the manifold N. Suppose that $(\mathscr{Y},H)$ is a slice of this action. Given a point m_0 in a smooth manifold M, if $f : M \to N$ is a smooth map such that $f(m_0) \in G\mathscr{Y}$, then for any $g_0 \in G$ and $y_0 \in \mathscr{Y}$ for which $f(m_0) = g_0 y_0$, there exists a neighborhood U of m_0 in M and there exists a smooth map*

$$(g,y) : U \to G \times \mathscr{Y},$$

such that $f(m) = g(m)y(m)$, for every $m \in U$, and $g(m_0) = g_0$, $y(m_0) = y_0$.

Proof. Let $F : G \times_H \mathscr{Y} \to G\mathscr{Y} \subset N$ be the diffeomorphism of item (4) in the definition of slice. Suppose $g_0 \in G$ and $y_0 \in \mathscr{Y}$ satisfy $f(m_0) = g_0 y_0$. These exist, since $f(m_0) \in G\mathscr{Y}$. Since μ of (2.5) is a submersion, it has a local section $\tau : V \to G \times \mathscr{Y}$ on a neighborhood $V \subset G \times_H \mathscr{Y}$ of $[g_0, y_0]$ such that $\tau[g_0, y_0] = (g_0, y_0)$. Then $\mu \circ \tau = \mathrm{id}_V$, $U = f^{-1}F(V)$ is a neighborhood of m_0 in M, and

$$(g,y) = \tau \circ F^{-1} \circ f : U \to G \times \mathscr{Y}$$

is a smooth map satisfying

$$g(m)y(m) = F \circ \mu \circ (g,y)(m) = F \circ \mu \circ \tau \circ F^{-1} \circ f(m) = f(m),$$

for every $m \in U$, and $(g,y)(m_0) = \tau(F^{-1}(x_0)) = \tau[g_0, y_0] = (g_0, y_0)$. □

2.4 Distributions

Knowledge of smooth distributions on a manifold is a prerequisite of this book. In this section we will review the terminology and principal results of this theory in preparation for our use of it throughout the book. There are many excellent references, including Conlon [53, Chapter 4], Lee [110, Chapter 19], and Warner [166, pp 41–50].

Definition 2.19. Let M be an n-dimensional smooth manifold. Let k be an integer in the set $\{1,\dots,n\}$. A *k-dimensional distribution $\mathscr{D}$ on M* assigns to each point $p \in M$ a k-dimensional subspace $\mathscr{D}(p)$ of T_pM. The distribution is *smooth* if each point of M has a neighborhood U on which there exist smooth vector fields $X_1,\dots,X_k$, which span $\mathscr{D}(p)$ at every point $p \in U$. Such a set is called a *smooth local frame* of $\mathscr{D}$. A smooth vector field defined on an open subset of M *belongs to* $\mathscr{D}$ if $X_{(p)} \in \mathscr{D}(p)$ for every point in the domain of X. In modern terminology, a smooth, k-dimensional distribution is a smooth, rank k subbundle $\mathscr{D}$ of the tangent bundle TM.

The smooth distribution $\mathscr{D}$ satisfies the *Frobenius condition* if whenever a pair of smooth vector fields with a common domain belong to $\mathscr{D}$, their Lie bracket belongs to $\mathscr{D}$. An *integral manifold* of a smooth k-dimensional distribution $\mathscr{D}$ is a one-to-one immersion $f : N^k \to M$ such that

$$df_{(p)}T_pN = \mathscr{D}(f(p)),$$

for every $p \in N$. An integral manifold is *maximal* if it is connected and not a proper subset of any other connected integral manifold.

There is a dual way to define a distribution in terms of equations rather than spanning sets. If $\mathscr{D}$ is a k-dimensional distribution on M^n, then the subspace $\mathscr{D}(p) \subset T_pM$ has an annihilator $\mathscr{D}^\perp(p) \subset T_p^*M$, which is a subspace of dimension $n-k$ of the cotangent space of M at p. A smooth 1-form θ defined on an open subset U of M *belongs to* $\mathscr{D}^\perp$ if $\theta_{(p)} \in \mathscr{D}^\perp(p)$, for every $p \in U$. A *smooth local coframe* for $\mathscr{D}^\perp$ is a set $\theta^{k+1}, \dots, \theta^n$ of smooth 1-forms on an open set $U \subset M$ that spans $\mathscr{D}^\perp(p)$ at each point of U. $\mathscr{D}$ has smooth local frames if and only if $\mathscr{D}^\perp$ has smooth local coframes. In modern terms, $\mathscr{D}^\perp$ is a smooth, rank $n-k$ subbundle of T^*M. In this dual formulation, a one-to-one immersion $f : N^k \to M$ is an integral manifold of $\mathscr{D}$ if and only if $f^*\theta = 0$ for every smooth 1-form in $\mathscr{D}^\perp$.

Lemma 2.20. *A smooth k-dimensional distribution $\mathscr{D}$ on M^n satisfies the Frobenius condition if and only if any local coframe $\theta^{k+1}, \dots, \theta^n$ of $\mathscr{D}^\perp$ on $U \subset M$ satisfies*

$$d\theta^\alpha = \sum_{\beta=k+1}^{n} \theta^\beta \wedge \omega_\beta^\alpha, \tag{2.7}$$

for $\alpha = k+1, \dots, n$, for some smooth 1-forms ω_β^α on U. We shall express the conditions of (2.7) *by*

$$d\theta^\alpha \equiv 0 \mod \mathscr{D}^\perp.$$

We shall also call (2.7) *the Frobenius condition for $\mathscr{D}$.*

Proof. If θ is a smooth 1-form on an open set $U \subset M$, then for any smooth vector fields X, Y on U,

$$d\theta(X,Y) = X\theta(Y) - Y\theta(X) - \theta([X,Y]).$$

The proof follows from this formula. □

The Frobenius Theorem has a local and a global part. For the local part we use Warner's formulation [166, Theorem 1.60]. A coordinate chart $(U, x = (x^1, \dots, x^n))$ of a manifold M^n is *cubic* if $x(U)$ is the open unit cube $(0,1)^n \subset \mathbf{R}^n$.

Theorem 2.21 (Local Frobenius). *Let $\mathscr{D}$ be a smooth, k-dimensional distribution satisfying the Frobenius condition on the smooth manifold M^n. Let $p \in M$. There* *exists a cubic coordinate chart* $(U,(x^1,\dots,x^n))$, *centered at* p, *such that the integral manifolds of* $\mathscr{D}$ *contained in* U *are precisely the slices* $\mathbf{x}^\alpha = c^\alpha$, *for arbitrary constants* $0 < c^\alpha < 1$, *for* $\alpha = k+1,\dots,n$.

A k-dimensional distribution $\mathscr{D}$ is called *completely integrable* if for each point $p \in M$, there exists an integrable manifold of $\mathscr{D}$ containing this point. Theorem 2.21 shows that if $\mathscr{D}$ satisfies the Frobenius condition, then it is completely integrable. The converse is also true (see any of the references cited at the beginning of this section).

We use the following version of the global Frobenius theorem. It is stated as follows, with a complete proof, in Warner [166, pp. 42–49]. A one-to-one immersion $\iota : Y \to M$ is *quasi-regular* if, for every smooth map $f : Z \to M$ such that $f(Z) \subset \iota(Y)$, the induced map $F : Z \to Y$ is smooth, where $\iota \circ F = f$.

Theorem 2.22 (Frobenius). *Let* M^m *be a smooth manifold endowed with a* k*-dimensional distribution* $\mathscr{D} \subset T(M)$ *satisfying the Frobenius condition. Then for each point* $p \in M$ *there exists a unique maximal connected integral submanifold* $Y \subset M$, *such that* $p \in Y$. *Moreover (this is the nice part of the Warner approach)* Y *is quasi-regular.*

A fundamental application of the Frobenius Theorems is to the correspondence between Lie subgroups of a Lie group G and Lie subalgebras of the Lie algebra $\mathfrak{g}$ of G. See any of the three sources cited above for proofs of the statements in the following example.

Example 2.23. Let G be a Lie group of dimension n, with its Lie algebra $\mathfrak{g}$ of all left-invariant vector fields. A Lie subalgebra $\mathfrak{h}$ of $\mathfrak{g}$ of dimension k defines a k-dimensional smooth distribution $\mathscr{D}$ on G. Note that a vector field X can belong to $\mathscr{D}$ but not be left-invariant. An example of such would be a linear combination of vectors in $\mathfrak{h}$ with smooth, nonconstant function coefficients. $\mathscr{D}$ satisfies the Frobenius condition, since $\mathfrak{h}$ is a Lie subalgebra of $\mathfrak{g}$. The maximal integral submanifold of $\mathscr{D}$ through the identity element $1 \in G$ is the connected Lie subgroup H of G whose Lie algebra is $\mathfrak{h}$. The maximal integral submanifold through a point $g \in G$ is the right coset gH.

In the dual formulation, an $(n-k)$-dimensional subspace $\mathfrak{h}^*$ of $\mathfrak{g}^*$, the space of all left-invariant 1-forms on G, defines the k-dimensional smooth distribution $\mathscr{D}^\perp$ on G. It satisfies the Frobenius condition if and only if $d\theta \equiv 0 \mod \mathfrak{h}^*$, for every $\theta \in \mathfrak{h}^*$.

2.5 Cartan–Darboux

The Maurer–Cartan form ω of a matrix Lie group $G \subset \mathbf{GL}(n,\mathbf{R})$ is a left-invariant 1-form on G with values in the Lie algebra $\mathfrak{g} \subset \mathfrak{gl}(n,\mathbf{R})$ of G. More generally, let

α, β be any smooth $\mathfrak{g}$-valued 1-forms on a smooth manifold M^m. These are just $n \times n$ matrices of ordinary 1-forms on M, satisfying the linear relations defining $\mathfrak{g} \subset \mathfrak{gl}(n, \mathbf{R})$.

Their *wedge product*, $\alpha \wedge \beta$, is the $\mathfrak{gl}(n, \mathbf{R})$-valued 2-form defined on M by matrix multiplication of α and β, where elements multiply by the wedge product of 1-forms. In general, $\alpha \wedge \beta$ is not $\mathfrak{g}$-valued. The *bracket* defined by

$$[\alpha, \beta] = \frac{1}{2}(\alpha \wedge \beta + \beta \wedge \alpha)$$

is $\mathfrak{g}$-valued. To see this, let $\mathbf{e}_1, \ldots, \mathbf{e}_l$ be a basis of $\mathfrak{g}$. Then $\mathfrak{g}$-valued 1-forms on M have expansions

$$\alpha = \sum_1^l \alpha^i \mathbf{e}_i, \quad \beta = \sum_1^l \beta^j \mathbf{e}_j,$$

where α^i and β^j are ordinary 1-forms on M. Then

$$[\alpha, \beta] = \frac{1}{2} \sum_{i,j=1}^l \alpha^i \wedge \beta^j [\mathbf{e}_i, \mathbf{e}_j],$$

is $\mathfrak{g}$-valued, since $[\mathbf{e}_i, \mathbf{e}_j] \in \mathfrak{g}$, for $i,j = 1, \ldots, l$. Notice that $[\alpha, \beta] = [\beta, \alpha]$ and $[\alpha, \alpha] = \alpha \wedge \alpha$.

Let $f : M \to G$ be a smooth map from a smooth manifold M into a Lie group G with Lie algebra $\mathfrak{g}$ and Maurer–Cartan form ω. Then $f^*\omega$ is a $\mathfrak{g}$-valued 1-form on M. If $\mathbf{e}_1, \ldots, \mathbf{e}_l$ is a basis of $\mathfrak{g}$, with dual basis $\omega^1, \ldots, \omega^l$, then $\omega = \sum_1^l \omega^i \mathbf{e}_i$, $f^*\omega = \sum_1^l (f^*\omega^i)\mathbf{e}_i$, and we easily verify that

$$d f^*\omega = \sum_1^l f^* d\omega^i \otimes \mathbf{e}_i = f^* d\omega = -f^*\omega \wedge f^*\omega. \tag{2.8}$$

If $\eta = f^*\omega$, a $\mathfrak{g}$-valued 1-form on M, then

$$d\eta = -\eta \wedge \eta. \tag{2.9}$$

If we start with a $\mathfrak{g}$-valued 1-form η on M, then (2.9) is a necessary condition for the existence of a smooth map $f : M \to G$ such that $\eta = f^*\omega$.

Observe that if $f : M \to G$ is a smooth map such that $f^*\omega = \eta$, and if $a \in G$, then $L_a \circ f : M \to G$ is a smooth map such that $(L_a \circ f)^*\omega = f^* L_a^* \omega = \eta$, since $L_a^*\omega = \omega$.

Theorem 2.24 (Cartan–Darboux Congruence). *Let M be a smooth connected manifold, let G be a Lie group with Lie algebra $\mathfrak{g}$ and Maurer–Cartan form ω.*

If $f,h : M \to G$ are smooth maps such that $f^*\omega = h^*\omega$, then there exists an element $a \in G$ such that $h = L_a \circ f$.

Proof. Let $g : M \to G$ be the smooth map defined by $g(p) = h(p)f(p)^{-1}$, where $f(p)^{-1}$ denotes the map $M \to G$ given by $p \mapsto f(p)^{-1}$, the inverse of the matrix $f(p)$. From $d(f(p)^{-1}) = -f(p)^{-1}df_{(p)}f(p)^{-1}$ and $(f^*\omega)_{(p)} = f(p)^{-1}df_{(p)}$, we find $dg = 0$ at every point of M. Thus, g is constant, say $g(p) = a$ for every $p \in M$, and $h = L_a \circ f$. □

To prove the Cartan–Darboux Existence theorem we use the global version of the Frobenius Theorem 2.22.

Theorem 2.25 (Cartan–Darboux Existence). *Let G be a connected Lie group with Lie algebra $\mathfrak{g}$ and $\mathfrak{g}$-valued Maurer–Cartan form ω. If X is a smooth manifold endowed with a $\mathfrak{g}$-valued 1-form α satisfying*

$$d\alpha = -\alpha \wedge \alpha,$$

then for every $p_0 \in X$ and every $g_o \in G$ there exist a connected open neighborhood U of p_0 and a unique smooth map $A : U \to G$ such that $A^*(\omega) = \alpha$ and $A(p_0) = g_0$.

If, in addition, X is simply connected, then for every $p_0 \in X$ and every $g_0 \in G$ there exists a unique smooth map $A : X \to G$ such that $A^*(\omega) = \alpha$ and $A(p_0) = g_0$.

Proof. We divide the proof into three steps :

Step 1: Proof of the local statement.

The first part is just an application of the local Frobenius theorem. Consider on $M = X \times G$ the $r = \dim X$ dimensional distribution defined by the equation

$$\theta = \omega - \alpha = 0.$$

From

$$d\omega = -\omega \wedge \omega, \quad d\alpha = -\alpha \wedge \alpha,$$

we get

$$d\theta = -[\theta + \alpha, \omega] + [\omega - \theta, \alpha] = -[\theta, \omega + \alpha].$$

This implies that our equations define an r-dimensional distribution satisfying the Frobenius condition. For every $p_0 \in X$, there exists a unique connected maximal integral submanifold Y, such that $(p_0, 1) \in Y$. Restriction to Y of projection onto the first factor gives the smooth map

$$F : Y \subset X \times G \to X, \quad F(p,g) = p.$$

$F^*\alpha = \alpha = \omega$ on Y implies dF has maximal rank at every point of Y. Thus, F is a local diffeomorphism, so there exists an open neighborhood $W \subset Y$ of $(p_0, 1)$ such that $F : W \to X$ is a diffeomorphism onto the image. Set $U = F(W) \subset X$ and consider

$$(F|_W)^{-1} : U \subset X \to W \subset Y \subset X \times G.$$

This map is necessarily of the form

$$(F|_W)^{-1} : p \in U \to (p, A(p)) \in Y,$$

where $A : U \to G$ is a smooth map. Then θ pulled back to $Y \subset X \times G$ is zero, so

$$0 = ((F|_W)^{-1})^* \theta = A^* \omega - \alpha.$$

Moreover, $A(p_0) = 1$, since $(p_0, 1) \in Y$. The uniqueness of A follows at once from the uniqueness of the integral manifold passing through $(p_0, 1)$. For any $g \in G$, the map $A_g = L_g \circ A : U \to G$ is the unique map on U satisfying $A_g^* \omega = \alpha$ and $A_g(p_0) = g$, since left multiplication $L_g : G \to G$ preserves ω.

Step 2: The map F is surjective.

Let us now prove that $F : Y \subset X \times G \to X$ is surjective. Take any other point $p_1 \in X$ and let $\gamma : [0,1] \to X$ be a smooth path from $p_0 = \gamma(0)$ to $p_1 = \gamma(1)$. For each $t \in [0,1]$, Step 1 implies there exists a connected open neighborhood U_t of $\gamma(t)$ in X and a unique smooth map $A_t : U_t \to G$ such that $A_t^* \omega = \alpha$ and $A_t(\gamma(t)) = 1$. In particular, the graph of A_t,

$$\{(p, A_t(p)) : p \in U_t\},$$

is an integral manifold of our distribution. Use the Lebesgue number $\delta > 0$ of the open covering $\{\gamma^{-1} U_t\}_{t \in [0,1]}$ of the compact metric space $[0,1]$ to construct a partition $0 = t_0 < t_1 < \cdots < t_n = 1$ such that for each $k = 1, \ldots, n$ there exists $t(k) \in [0,1]$ with $\gamma[t_{k-1}, t_k] \subset U_{t(k)}$. For information about the Lebesgue number, see [122, Lemma 27.5 on page 175].

Then $\gamma[t_0, t_1] \subset U_{t(1)}$. Let $g_0 = A_{t(1)}(\gamma(t_0)) \in G$ and let

$$B_1 = L_{g_0^{-1}} \circ A_{t(1)} : U_{t(1)} \to G.$$

Then $B_1(p_0) = B_1(\gamma(t_0)) = 1$. The graph of $B_1 : U_{t(1)} \to G$ is a connected integral manifold passing through $(p_0, 1) \in M$, so must be contained in Y. In particular, $(\gamma(t_1), B_1(\gamma(t_1))) \in Y$. Next let $g_1 = A_{t(2)}(\gamma(t_1)) \in G$ and let

$$B_2 = L_{B_1(\gamma(t_1))} \circ L_{g_1^{-1}} \circ A_{t(2)} : U_{t(2)} \to G.$$

The graph of $B_2 : U_{t(2)} \to G$ is a connected integral manifold passing through the point $(\gamma(t_1), B_1(\gamma(t_1))) \in Y$, so it must lie entirely in Y. Proceeding inductively in this way define maps $B_k : U_{t(k)} \to G$, for $k = 1, \ldots, n$, for which one concludes that the point $(p_1, B_n(p_1)) \in Y$, which shows that p_1 is in the image of F. Hence, F is surjective.

Step 3: ***F* is a covering map.**

We need to prove that X is *evenly covered* by F. We do this by finding the group of *deck transformations* of this covering. Left multiplication of G on itself gives a left action of G on $X \times G$ by

$$a(p,g) = (p,ag),$$

for any $a \in G$ and $(p,g) \in X \times G$. The form θ is invariant under this action, $a^*\theta = a^*\omega - \alpha = \theta$, since ω is left-invariant. It follows that aY is a maximal integral manifold for any $a \in G$. If $(p,g) \in Y$, then aY is the maximal integral manifold through $a(p,g) = (p,ag)$. The stabilizer of Y,

$$G_Y = \{a \in G : aY = Y\},$$

is a closed Lie subgroup of G. For any $(p,g) \in Y$, the above comments imply that

$$G_Y = \{a \in G : a(p,g) \in Y\},$$

and thus

$$F^{-1}\{p\} = G_Y(p,g). \tag{2.10}$$

The action of G_Y on Y is smooth, since Y is quasi-regular. Since F is a local diffeomorphism, the fiber $F^{-1}(p)$ is discrete and hence G_Y is a discrete subgroup of G. Write $G_Y = \{g_j\}_{j\in J}$, where $J \subset \mathbf{N}$.

Given a point $x_0 \in X$, we want to find a neighborhood U of x_0 that is evenly covered by F. There exists $(x_0,g_0) \in Y$, since F is surjective. There exist open neighborhoods $W \subset Y \subset X \times G$ of (x_0,g_0) and $U \subset X$ of x_0 such that $F|_W$ is a diffeomorphism of W onto U. Then

$$F|_W^{-1} : U \subset X \to W \subset Y, \quad F|_W^{-1}(p) = (p,A(p)),$$

where $A : U \to G$ is a smooth map and $A^*\omega = \alpha$. For any $p \in U$,

$$F^{-1}\{p\} = G_Y(p,A(p)) = \cup_{j\in J} g_j(p,A(p)),$$

by (2.10). This, with $W = \{(p,A(p)) : p \in U\}$, implies

$$F^{-1}U = \cup_{j\in J} g_j W, \quad g_i W \cap g_j W = \emptyset, \quad i \neq j.$$

F invariant under the action of G_Y on Y implies F maps each open set[1] $g_j W \subset Y$ diffeomorphically onto U. Hence, F evenly covers U.

We have proved that $F : Y \to X$ is a covering whose group of deck transformations is G_Y.

[1]To ensure that $g_j W$ is actually an open subset of Y requires the quasi-regularity.

Conclusion

If X is simply connected, then the covering $F : Y \to X$ must be a diffeomorphism whose inverse is of the form

$$F^{-1} : X \to Y \subset X \times G, \quad F^{-1}(p) = (p, A(p)) \in Y,$$

where $A : X \to G$ is a smooth map satisfying $A^*(\omega) = \alpha$. The uniqueness of A with specified value g_0 at a specified point p_0 is a consequence of the uniqueness of the maximal integral submanifold passing through a given point of $X \times G$. □

Griffiths [79, pp 780–782], Malliavin [116, pp 167–172], and Sharpe [150, pp 116–125] contain other proofs. Spivak [154, Volume I, Chapter 10] proves the local result.

Problems

2.26. Prove that if $\pi : P \to N$ is a principal H-bundle, then for any $p \in P$,

$$\ker(d\pi_p) = \{\hat{X}_{(p)} : X \in \mathfrak{h}\},$$

where $\hat{X}$ is the fundamental vector field induced on N by $X \in \mathfrak{h}$. See Definition 2.4.

2.27. Use the notation of Example 2.11. Prove the following: The orthogonal group $\mathbf{O}(n)$ acts transitively on the Grassmannian $G(m,n)$. Its isotropy subgroup at P_0 is $G_0 = \mathbf{O}(m) \times \mathbf{O}(n-m)$. An $\mathrm{Ad}(G_0)$-invariant subspace of $\mathfrak{o}(n)$ complementary to $\mathfrak{g}_0 = \mathfrak{o}(m) \oplus \mathfrak{o}(n-m)$ is

$$\mathfrak{m} = \{\begin{pmatrix} 0 & -{}^tX \\ X & 0 \end{pmatrix} : X \in \mathbf{R}^{(n-m)\times m}\}.$$

2.28 (Poincaré disk model). Consider the Lie group

$$\mathbf{SU}(1,1) = \{A \in \mathbf{GL}(2,\mathbf{C}) : {}^t\bar{A} I_{1,1} A = I_{1,1}\} = \{\begin{pmatrix} z & \bar{w} \\ w & \bar{z} \end{pmatrix} : |z|^2 - |w|^2 = 1\},$$

where $I_{1,1} = \begin{pmatrix} 1 & 0 \\ 0 & -1 \end{pmatrix}$. Following Example 2.12, analyze the action of $\mathbf{SU}(1,1)$ on the unit disc $\mathbf{D} = \{\zeta \in \mathbf{C} : |\zeta| < 1\}$, given by

$$\begin{pmatrix} z & \bar{w} \\ w & \bar{z} \end{pmatrix} \zeta = \frac{z\zeta + \bar{w}}{w\zeta + \bar{z}}. \tag{2.11}$$

Prove the following: This action preserves the Riemannian metric

$$I_{\mathbf{D}} = 4\frac{d\zeta d\bar{\zeta}}{(1-|\zeta|^2)^2} = \frac{du^2+dv^2}{(1-u^2-v^2)^2}$$

on $\mathbf{D}$, where $\zeta = u+iv$ and $u, v \in \mathbf{R}$. The *Cayley transform*

$$f : \mathbf{C}^+ \to \mathbf{D}, \quad f(z) = \frac{z-i}{z+i}$$

pulls $I_{\mathbf{D}}$ back to the upper half-space metric I on $\mathbf{C}^+$ defined in Example 2.12. The pair $(\mathbf{D}, I_{\mathbf{D}})$ is the *Poincaré disk model* of the hyperbolic plane. It is discussed further in Example 4.54.

2.29. The *Iwasawa decomposition* of $\mathbf{SU}(1,1) = KAN$ (see [84, p. 234]) says that any element of $\mathbf{SU}(1,1)$ can be expressed as such a product of elements from the subgroups

$$K = \{\begin{pmatrix} e^{it} & 0 \\ 0 & e^{-it} \end{pmatrix} : t \in \mathbf{R}\},$$

$$A = \{\begin{pmatrix} \cosh t & \sinh t \\ \sinh t & \cosh t \end{pmatrix} : t \in \mathbf{R}\},$$

$$N = \{\begin{pmatrix} it+1 & -it \\ it & -it+1 \end{pmatrix} : t \in \mathbf{R}\}.$$

Describe the orbits in $\mathbf{D}$ of each of the subgroups K, A, and N, for the action (2.11).

2.30. The Iwasawa decomposition of $\mathbf{SL}(2,\mathbf{R})$ is KAN, for the subgroups $K = \mathbf{SO}(2)$,

$$A = \{\begin{pmatrix} e^{t} & 0 \\ 0 & e^{-t} \end{pmatrix} : t \in \mathbf{R}\}, \quad N = \{\begin{pmatrix} 1 & t \\ 0 & 1 \end{pmatrix} : t \in \mathbf{R}\}.$$

Describe the orbits in $\mathbf{C}^+$ of each of the subgroups K, A, and N, for the action of $\mathbf{SL}(2,\mathbf{R})$ on $\mathbf{C}^+$ discussed in Example 2.12.

2.31. Prove that if H is the stabilizer of a submanifold $\mathscr{Y}$ of N, and if $\mathscr{Y}$ is closed in $G\mathscr{Y}$, then H is a closed subgroup of G.

2.32. Consider the standard matrix multiplication action of $\mathbf{O}(2)$ on $\mathbf{R}^2$. Prove the following: The isotropy subgroup at any point of $\mathscr{Y} = \{r\epsilon_1 : r > 0\}$ is $H = \{I_2, I_{1,1}\}$, where I_2 is the 2×2 identity matrix and $I_{1,1}$ was defined in Problem 2.28. $\mathscr{Y}$ is a slice of this action for H. $\mathscr{X} = \{r\epsilon_1 : r \neq 0\}$ with the subgroup $K = \{\pm I_2, \pm I_{1,1}\}$ also is a slice.

Chapter 3
Theory of Moving Frames

We present here an outline of the method of moving frames for any submanifold of an arbitrary homogeneous space. We explain how a Lie group acting transitively on a manifold N is related to the principal bundle of linear frames on N. We present a general outline of the frame reduction procedure after first describing the procedure for the elementary example of curves in the punctured plane acted upon by the special linear group $\mathbf{SL}(2,\mathbf{R})$.

Elie Cartan's Method of Moving Frames determines when two immersions $\mathbf{x}, \tilde{\mathbf{x}} : M \to N$ are *G-congruent*, where G is a Lie group acting smoothly and transitively on N, and G-congruence means there exists an element $g \in G$ such that $\hat{\mathbf{x}} = g \circ \mathbf{x}$. The chapter concludes with basic theorems that characterize when a submanifold of a homogeneous space is itself homogeneous.

A general outline of the method is abstract and covers a multitude of cases. Its conceptual overview will guide an understanding of what is being done in the many applications given in the subsequent chapters. This book is about using the method of moving frames to study submanifolds of homogeneous space. It is not a text on the theory of moving frames.

Cartan [32] gave an elegant introduction to moving frames with emphasis on the notion of contact. Subsequent expositions, with additional examples, are in Griffiths [79] and Jensen [93].

3.1 Bundle of linear frames

Let a Lie group G act transitively on a manifold N^n. Choose an *origin* $o \in N$, and let $G_0 = \{g \in G : go = o\}$ be the *isotropy subgroup* of G at o. The smooth map

$$\pi : G \to N, \quad \pi(g) = go \tag{3.1}$$

G.R. Jensen et al., *Surfaces in Classical Geometries*, Universitext,
DOI 10.1007/978-3-319-27076-0_3

is the projection map of a principal G_0-bundle over N, by Proposition 2.6. Let $\mathfrak{g}$ be the Lie algebra of left-invariant vector fields on G. Use evaluation at the identity element $1 \in G$ to identify $\mathfrak{g}$ with T_1G, the tangent space of G at 1. In the present context we will usually denote left multiplication on G by an element $g \in G$ by $L_g : G \to G$, in order to distinguish it from the diffeomorphism $g : N \to N$ defined by the action of G on N.

Exercise 1. Prove that if $g \in G$, then $g \circ \pi = \pi \circ L_g : G \to N$.

Let $\mathfrak{g}_0$ denote the Lie subalgebra of $\mathfrak{g}$ that is the Lie algebra of G_0. If $\mathfrak{g}_0$ is identified with T_1G_0, then $\mathfrak{g}_0$ is the kernel of the linear map $d\pi_1 : T_1G \to T_oN$.

Recall that if $F : P \to Q$ is any smooth map between smooth manifolds, then the derivative map $dF : TP \to TQ$ is a smooth map between their tangent bundles. A *smooth vector field along* F is a smooth map $X : P \to TQ$ such that $\psi \circ X = F$, where we denote the bundle projections by $\varphi : TP \to P$ and $\psi : TQ \to Q$. An important class of smooth vector fields along F are those obtained by *pushing forward* by F a smooth vector field X on P, to get the vector field X^F along F defined by

$$X^F = dF \circ X : P \to TQ.$$

In detail, $X^F_{(p)} = dF_pX_{(p)} \in T_{F(p)}Q$, for any $p \in P$. Applying this to our smooth map $\pi : G \to N$, we get from any $X \in \mathfrak{g}$ a smooth vector field $X^\pi = d\pi \circ X$ along π.

Exercise 2. Use Exercise 1 to prove that, if $X \in \mathfrak{g}$, then the vector field X^π along π has value at any point $g \in G$ given by

$$X^\pi_{(g)} = dg_oX^\pi_{(1)}.$$

Let $\mathfrak{m}_0$ be any vector subspace of $\mathfrak{g}$ complementary to $\mathfrak{g}_0$, so $\mathfrak{g} = \mathfrak{m}_0 \oplus \mathfrak{g}_0$, as a vector space direct sum. The restriction, $d\pi_1 : \mathfrak{m}_0 \to T_oN$, is a linear isomorphism. Choose a basis

$$E_1, \dots, E_n \tag{3.2}$$

of $\mathfrak{m}_0$ and call it the *reference frame*.

Exercise 3. Prove that for any reference frame (3.2), the vector fields evaluated at any $g \in G$,

$$E^\pi_{1(g)}, \dots, E^\pi_{n(g)}$$

form a basis of $T_{\pi(g)}N$. In particular, their values at $1 \in G$ form a basis of T_oN. We also call this basis of T_oN the *reference frame*.

If $h \in G_0$, then $dh_o : T_oN \to T_oN$. Let $A(h) = (A^j_i) \in \mathbf{GL}(n, \mathbf{R})$ be the matrix of dh_o relative to this reference frame of T_oN. By Exercise 2,

$$dh_oE^\pi_{i(1)} = \sum_{j=1}^n A^j_iE^\pi_{j(1)}, \text{ for } i = 1, \dots, n.$$

Thus, $A(h)$ is the matrix of the isotropy representation (2.1) relative to this reference frame. It follows that

$$A : G_0 \to \mathbf{GL}(n,\mathbf{R}), \quad h \mapsto A(h), \tag{3.3}$$

is a homomorphism. It is the *linear isotropy representation* of G_0 relative to this reference frame.

Lemma 3.1. *The linear isotropy representation* (3.3) *equals the adjoint representation of* G_0 *on* $\mathfrak{g}/\mathfrak{g}_0 \cong \mathfrak{m}_0$ *relative to the basis* $E_1, \dots, E_n$ *of* $\mathfrak{g}/\mathfrak{g}_0$.

Proof. If $h \in G_0$, then for any $g \in G$, we have

$$h \circ \pi(g) = (hg)o = (hgh^{-1})o = \pi \circ C_h(g),$$

where $C_h : G \to G$, $C_h(g) = hgh^{-1}$, is conjugation by h. For each E_i,

$$\mathrm{Ad}(h)E_i = \sum_{j=1}^{n} B_i^j E_j + F_i, \text{for } i = 1, \dots, n,$$

for some constants B_i^j and some $F_i \in \mathfrak{g}_0$. Using (2.1), we get

$$\sum_{j=1}^{n} A(h)_i^j E_j^\pi{}_{(1)} = dh_o E_i^\pi{}_{(1)} = d(h \circ \pi)_1 E_i = d(\pi \circ C_h) E_i$$

$$= d\pi_1(\mathrm{Ad}(h)E_i) = \sum_{j=1}^{n} B_i^j E_j^\pi{}_{(1)},$$

so $\mathrm{Ad}(h)E_i = \sum_{j=1}^{n} A(h)_i^j E_j \mod \mathfrak{g}_0$, for every $h \in G_0$. □

Exercise 4. Recall the principal $\mathbf{GL}(n,\mathbf{R})$-bundle of all linear frames on N, denoted $\mu : L(N) \to N$ in Kobayashi-Nomizu [100, Example 5.2 pp 55–56]. The right action of $A \in \mathbf{GL}(n,\mathbf{R})$ on a frame $(v_1, \dots, v_n)$ of T_pN is

$$(v_1, \dots, v_n)A = (\sum_1^n v_i A_1^i, \dots, \sum_1^n v_i A_n^i).$$

The projection map μ sends a linear frame at $p \in N$ to the point $p \in N$. We also have the principal G_0-bundle $\pi : G \to N$ of (3.1). Prove that the map

$$F : G \to L(N), \quad F(g) = (E^\pi_{1(g)}, \dots, E^\pi_{n(g)})$$

is a principal bundle map, with the homomorphism between the structure groups being the linear isotopy map $A : G_0 \to \mathbf{GL}(n,\mathbf{R})$ of (3.3). This requires proving that $F(gh) = F(g)A(h)$, for any $g \in G$ and $h \in G_0$. Prove F is a bundle monomorphism if the linear isotropy representation of G_0 is faithful.

3.2 Moving frames

Consider an immersion $\mathbf{x} : M^m \to N^n$. This could include the case of the inclusion map of an open submanifold of N. Assume M connected.

Definition 3.2. An element $g \in G$ is a *frame at* $p \in M$ if $go = \mathbf{x}(p)$. A *moving frame* or *frame field* along $\mathbf{x}$ on an open subset $U \subset M$ is a smooth map

$$e : U \to G \tag{3.4}$$

such that $\pi \circ e = \mathbf{x}$ on U.

Exercise 5. Apply the Lift Property of Corollary 2.7 to the action (3.1) to prove that if $p \in M$, and if $g \in G$ is a frame at p, then there exists a neighborhood U of p in M on which there is a moving frame $e : U \to G$ such that $e(p) = g$.

Why is this called a moving frame? The answer requires a choice of reference frame (3.2). For any smooth vector field X on G, we have a smooth vector field $X^\pi = d\pi \circ X$ along π, and its composition $d\pi \circ X \circ e : U \to TN$ is a smooth vector field along $\pi \circ e = \mathbf{x} : U \to N$. It follows that

$$\mathbf{e}_1 = E_1^\pi \circ e, \dots, \mathbf{e}_n = E_n^\pi \circ e \tag{3.5}$$

is a collection of vector fields along $\mathbf{x}$ whose value at any point $p \in U$ is a basis of $T_{\mathbf{x}(p)}N$.

Exercise 6. Prove that if $e : U \subset M \to G$ is a smooth moving frame along $\mathbf{x}$, then any other smooth moving frame along $\mathbf{x}$ on U must be given by

$$\tilde{e} : U \to G, \quad \tilde{e}(p) = e(p)h(p),$$

where $h : U \to G_0$ is any smooth map. Prove that the frame (3.5) determined by the element $\tilde{e} = eh \in G$, for some $h \in G_0$, is

$$\tilde{\mathbf{e}}_1 = \sum_{j=1}^n A(h)^j_1 \mathbf{e}_j, \dots, \tilde{\mathbf{e}}_n = \sum_{j=1}^n A(h)^j_n \mathbf{e}_j, \tag{3.6}$$

where $A : G_0 \to \mathbf{GL}(n, \mathbf{R})$ is the adjoint representation of G_0 on $\mathfrak{g}/\mathfrak{g}_0$ relative to the reference frame (3.2).

The vector space direct sum $\mathfrak{g} = \mathfrak{m}_0 \oplus \mathfrak{g}_0$ decomposes the $\mathfrak{g}$-valued Maurer–Cartan form $\omega = g^{-1}dg$ of G as

$$\omega = \omega_{\mathfrak{m}_0} + \omega_{\mathfrak{g}_0},$$

where the subscript denotes projection into that subspace. Then

$$\omega_{\mathfrak{m}_0} = \sum_1^n \omega^i E_i, \tag{3.7}$$

for some left-invariant 1-forms $\omega^1, \ldots, \omega^n$ on G. They form a basis of the annihilator $\mathfrak{g}_0^\perp$ in the dual space of $\mathfrak{g}$.

Proposition 3.3. *If $e : U \subset M \to G$ is a moving frame along $\mathbf{x}$, if $p \in U$, and if $v \in T_pM$, then*

$$d\mathbf{x}_p v = \sum_{i=1}^n (e^*\omega^i)(v)\mathbf{e}_{i(p)},$$

where $\mathbf{e}_1, \ldots, \mathbf{e}_n$ are the vector fields along $\mathbf{x}$ defined in (3.5). *Evaluated at $p \in U$, the 1-forms $e^*\omega^1, \ldots, e^*\omega^n$ span the cotangent space T_p^*M.*

Proof. If $g, h \in G$, then $\pi \circ L_g = g \circ \pi$, by Exercise 1. If $X \in T_gG$, then by definition of the Maurer–Cartan form, $\omega(X) \in \mathfrak{g} = T_1G$ is the left-invariant vector field whose value at g is X, so $X = dL_g\omega(X)$. Now $\pi \circ e = \mathbf{x}$ on U and $\mathfrak{g}_0 = \ker(d\pi_1)$ imply that

$$\begin{aligned}
d\mathbf{x}_p v &= d(\pi \circ e)_p v = d\pi_{e(p)} \circ de_p v = d\pi_{e(p)} \circ dL_{e(p)}\omega(de_p v) \\
&= d(\pi \circ L_{e(p)})_1(e^*\omega)(v) = d(e(p) \circ \pi)_1(e^*\omega_{\mathfrak{m}_0} + e^*\omega_{\mathfrak{g}_0})(v) \\
&= d(e(p))_o d\pi_1(e^*\omega_{\mathfrak{m}_0})(v) = d(e(p))_o \circ d\pi_1 \sum_1^n (e^*\omega^i)(v)E_i \\
&= \sum_1^n (e^*\omega^i)(v) d(e(p))_o \circ d\pi_1 E_i = \sum_1^n (e^*\omega^i)(v)\mathbf{e}_{i(p)}.
\end{aligned}$$

Since $d\mathbf{x}_p$ is injective, $e^*\omega^1, \ldots, e^*\omega^n$ is a spanning set at each point of U. □

Given a moving frame $e : U \to G$ along $\mathbf{x}$, Exercise 6 asserts that any other moving frame $\tilde{e} : U \to G$ along $\mathbf{x}$ on U is given by $\tilde{e} = eh$, where $h : U \to G_0$ can be any smooth map. A frame field $e : U \to G$ pulls back the Maurer–Cartan form of G to the $\mathfrak{g}$-valued 1-form $e^*\omega = e^{-1}de$ on U.

In the special case when $G_0 = \{1\}$, which means G acts simply transitively on N, then effectively $G = N$ acting on itself by left multiplication. In this case, an immersion $\mathbf{x} : M \to G$ is itself the only frame field along it. The congruence problem is solved here by the Cartan–Darboux Congruence Theorem 2.24, which states that immersions $\mathbf{x}, \hat{\mathbf{x}} : U \to G$ are congruent if and only if $\mathbf{x}^{-1}d\mathbf{x} = \hat{\mathbf{x}}^{-1}d\hat{\mathbf{x}}$ on M. If M is simply connected, the Cartan–Darboux Existence Theorem 2.25 states that if η is any $\mathfrak{g}$-valued 1-form on M, then there exists an immersion $\mathbf{x} : M \to G$ satisfying $\mathbf{x}^{-1}d\mathbf{x} = \eta$ if and only if $d\eta = -\eta \wedge \eta$.

In the general case when G_0 is nontrivial, if $\mathbf{x}, \hat{\mathbf{x}} : M \to N$ are immersions for which there exists $g \in G$ such that $\hat{\mathbf{x}} = g \circ \mathbf{x}$, and if $e : M \to G$ is a frame field along $\mathbf{x}$, then $\hat{e} = ge$ is a frame field along $\hat{\mathbf{x}}$ for which

$$\hat{e}^{-1} d\hat{e} = (ge)^{-1} d(ge) = e^{-1} de$$

on M, since $dg = 0$ on M. Thus, congruence implies the existence of frame fields along each immersion that pull back the Maurer–Cartan form of G to the same $\mathfrak{g}$-valued 1-form on M.

Conversely, if $e : M \to G$ is a frame field along $\mathbf{x}$ and $\hat{e} : M \to G$ is a frame field along $\hat{\mathbf{x}}$ such that

$$\hat{e}^{-1} d\hat{e} = e^{-1} de \tag{3.8}$$

on M, then the Cartan–Darboux Uniqueness Theorem 2.24 implies there exists $g \in G$ such that $\hat{e} = ge$, and thus $\hat{\mathbf{x}} = g \circ \mathbf{x}$ on M. Frame fields satisfying (3.8) determine the element g, since the map $g : M \to G$ defined by $g(p) = \hat{e}(p)e(p)^{-1}$, has derivative

$$\begin{aligned} dg_p &= d\hat{e}_{(p)} e(p)^{-1} - \hat{e}(p)e(p)^{-1} de_{(p)} e(p)^{-1} \\ &= \hat{e}(p)(\hat{e}(p)^{-1} d\hat{e}_{(p)} - e(p)^{-1} de_{(p)})e(p)^{-1} = 0 \end{aligned}$$

on M, so g is constant. The ambiguity of the frame field along $\mathbf{x}$, however, prevents this from being a satisfactory solution to the congruence problem. If $e : M \to G$ is a frame field along $\mathbf{x}$, then any smooth map $h : M \to G_0$ gives another frame field $\tilde{e} = eh : M \to G$ along $\mathbf{x}$, and

$$\tilde{e}^{-1} d\tilde{e} = (eh)^{-1} d(eh) = \mathrm{Ad}(h^{-1}) \circ e^{-1} de + h^{-1} dh \neq e^{-1} de \tag{3.9}$$

in general. To use the Cartan–Darboux Uniqueness Theorem 2.24 to decide if an immersion $\hat{\mathbf{x}} : M \to N$ is congruent to $\mathbf{x}$, we need to find some frame field $\hat{e} : M \to G$ along $\hat{\mathbf{x}}$, and some frame field $e : M \to G$ along $\mathbf{x}$, such that

$$\hat{e}^{-1} d\hat{e} = e^{-1} de. \tag{3.10}$$

The method of moving frames gives a frame reduction procedure for removing the ambiguity in the choice of frame field along $\mathbf{x}$. This is a finite sequence of steps that produces the (nearly) unique *Frenet frame field* along $\mathbf{x}$. We think of the Frenet frame field as the best frame field along $\mathbf{x}$ in a sense that is related to order of contact. It determines a coframe field $\omega^1, \dots, \omega^m$ in M and a set of functions $\{k_1, \dots, k_l : M \to \mathbf{R}\}$ called the *invariants* of $\mathbf{x}$. Immersions $\mathbf{x} : M^m \to N$ and $\hat{\mathbf{x}} : \hat{M}^m \to N$ are congruent if and only if (3.10) holds for their Frenet frame fields.

3.3 Frame reduction procedure

Here is an outline of the frame reduction procedure. We begin with a simple example from centro-affine geometry.

Example 3.4 (Centro-affine curves). For recent research on this topic see Musso [125] and Pinkall [137]. Let $\dot{\mathbf{R}}^2 = \mathbf{R}^2 \setminus \{\mathbf{0}\}$. The special linear group $\mathbf{SL}(2,\mathbf{R})$ acts transitively on $\dot{\mathbf{R}}^2$ by its standard matrix multiplication on $\mathbf{R}^2$. Choose $\boldsymbol{\epsilon}_1$ to be the origin of $\dot{\mathbf{R}}^2$. The isotropy subgroup of $\mathbf{SL}(2,\mathbf{R})$ at $\boldsymbol{\epsilon}_1$ is

$$G_0 = \{\begin{pmatrix} 1 & u \\ 0 & 1 \end{pmatrix} : u \in \mathbf{R}\}.$$

We have the principal G_0 bundle projection

$$\pi : \mathbf{SL}(2,\mathbf{R}) \to \dot{\mathbf{R}}^2, \quad \pi(A) = A\boldsymbol{\epsilon}_1 = \mathbf{A}_1,$$

where $\mathbf{A}_1$ denotes the first column of A as a vector in the standard basis of $\mathbf{R}^2$. Choose

$$\mathfrak{m}_0 = \{\begin{pmatrix} u & 0 \\ v & -u \end{pmatrix} : u, v \in \mathbf{R}\} \subset \mathfrak{sl}(2,\mathbf{R})$$

as a vector space complement of $\mathfrak{g}_0$, so

$$\mathfrak{g} = \mathfrak{m}_0 + \mathfrak{g}_0. \tag{3.11}$$

Choose as a basis of $\mathfrak{m}_0$

$$E_1 = \begin{pmatrix} 1 & 0 \\ 0 & -1 \end{pmatrix}, \quad E_2 = \begin{pmatrix} 0 & 0 \\ 1 & 0 \end{pmatrix}.$$

The adjoint representation of G_0 on $\mathfrak{g}/\mathfrak{g}_0$ relative to the basis E_1, E_2 is the homomorphism $A = (A^i_j) : G_0 \to \mathbf{GL}(2,\mathbf{R})$ defined by

$$\mathrm{Ad}(K)E_i = \sum_{j=1}^{2} A^j_i E_j \mod \mathfrak{g}_0.$$

For $K = \begin{pmatrix} 1 & u \\ 0 & 1 \end{pmatrix} \in G_0$ we calculate $A(K) = \begin{pmatrix} 1 & u \\ 0 & 1 \end{pmatrix}$.

The vectors $E_i \in \mathfrak{m}_0$ generate vector fields E^π_i along π on $\dot{\mathbf{R}}^2$. If $A \in \mathbf{SL}(2,\mathbf{R})$, then

$$E^\pi_{i\,(A)} = dA_{\boldsymbol{\epsilon}_1} d\pi_1 E_i = \mathbf{A}_i, \tag{3.12}$$

for $i = 1,2$. Let $\omega = \begin{pmatrix} \omega_1^1 & \omega_2^1 \\ \omega_1^2 & -\omega_1^1 \end{pmatrix}$ be the Maurer–Cartan form of $\mathbf{SL}(2,\mathbf{R})$, so ω_1^1, ω_1^2, ω_2^1 are linearly independent, left-invariant 1-forms on $\mathbf{SL}(2,\mathbf{R})$. The decomposition (3.11) gives the decomposition $\omega = \omega_{\mathfrak{m}_0} + \omega_{\mathfrak{g}_0}$, where a subscript indicates projection onto that subspace. Then

$$\omega_{\mathfrak{m}_0} = \begin{pmatrix} \omega_1^1 & 0 \\ \omega_1^2 & -\omega_1^1 \end{pmatrix}, \quad \omega_{\mathfrak{g}_0} = \begin{pmatrix} 0 & \omega_2^1 \\ 0 & 0 \end{pmatrix},$$

and using our basis of $\mathfrak{m}_0$ we have

$$\omega_{\mathfrak{m}_0} = \omega_1^1 E_1 + \omega_1^2 E_2.$$

Let $\mathbf{x} : \mathbf{R} \to \dot{\mathbf{R}}^2$ be a smooth immersed curve. We shall carry out the frame reduction procedure for $\mathbf{x}$. A frame field along $\mathbf{x}$ is a smooth map $e : \mathbf{R} \to \mathbf{SL}(2,\mathbf{R})$ such that $\mathbf{x} = \pi \circ e = \mathbf{e}_1$. Thus, the frame field must have the form $e = (\mathbf{x}, \mathbf{y})$, where $\mathbf{y} : \mathbf{R} \to \dot{\mathbf{R}}^2$ is a smooth map such that $\det(e) = 1$. Let t be a coordinate function on $\mathbf{R}$ so that $e^*\omega_1^i = X_1^i dt$ for some functions $X_1^i : \mathbf{R} \to \mathbf{R}$, for $i = 1,2$. Then

$$d\mathbf{x} = \Big(\sum_{i=1}^{2} (e^*\omega_1^i) E_i\Big)^{\pi}_{(e)} = ((X_1^1 E_1 + X_1^2 E_2) dt)^{\pi}_{(e)}, \tag{3.13}$$

by (3.12). Since $\mathbf{x}$ is an immersion, the image of the linear map

$$e^*\omega_{\mathfrak{m}_0} = (X_1^1 E_1 + X_1^2 E_2) dt : T_t\mathbf{R} \to \mathfrak{m}_0 \tag{3.14}$$

is a 1-dimensional subspace of $\mathfrak{m}_0$.

Any other frame field along $\mathbf{x}$ must be given by $\tilde{e} = eK$, where

$$K = \begin{pmatrix} 1 & u \\ 0 & 1 \end{pmatrix} : \mathbf{R} \to G_0$$

is an arbitrary smooth map. Then $\tilde{e}^*\omega = \mathrm{Ad}(K^{-1}) e^*\omega + K^{-1} dK$ implies

$$\tilde{e}^*\omega_{\mathfrak{m}_0} = (\mathrm{Ad}(K^{-1}) e^*\omega_{\mathfrak{m}_0})_{\mathfrak{m}_0},$$

since $K^{-1} dK \in \mathfrak{g}_0$. Thus

$$\begin{aligned} \tilde{e}^*\omega_{\mathfrak{m}_0} &= \sum_1^2 e^*\omega_1^i \mathrm{Ad}(K^{-1}) E_i = \Big(\sum X_1^i A(K^{-1})^j_i E_j\Big) dt \\ &= ((X_1^1 - u X_1^2) E_1 + X_1^2 E_2) dt. \end{aligned}$$

The change of first order frame changes the coefficients of the 1-dimensional subspace that is the image of the linear map (3.14) relative to the fixed basis E_1, E_2 of $\mathfrak{m}_0$. In effect, we have an action of G_0 on the space $\mathbf{RP}^1$ of 1-dimensional subspaces of $\mathbf{R}^2$, given by

$$G_0 \times \mathbf{RP}^1 \to \mathbf{RP}^1, \quad (K, \begin{bmatrix} x \\ y \end{bmatrix}) \mapsto A(K) \begin{bmatrix} x \\ y \end{bmatrix} = \begin{bmatrix} x + uy \\ y \end{bmatrix}, \tag{3.15}$$

if $K = \begin{pmatrix} 1 & u \\ 0 & 1 \end{pmatrix}$. The goal of the first step of the frame reduction is to make the coefficients as simple as possible in the sense that we choose a slice for this action. This action has two orbit types. It has the fixed point $\begin{bmatrix} 1 \\ 0 \end{bmatrix}$ and it acts transitively on the complement of this point. A point $t \in \mathbf{R}$ is *radial for* $\mathbf{x}$ if $e^*\omega_1^2 = 0$ at t. Otherwise t is a *nonradial* point. We consider two types of curves. It is *nonradial* if every point of $\mathbf{R}$ is nonradial. It is *radial* if every point of $\mathbf{R}$ is radial for $\mathbf{x}$.

For the first type there exists a smooth map $K : \mathbf{R} \to G_0$ such that $\tilde{e}^*\omega_1^1 = 0$ on $\mathbf{R}$. In fact, let $u = X_1^1/X_1^2$. To see the general picture, we regard the point $\begin{bmatrix} 0 \\ 1 \end{bmatrix} \in \mathbf{RP}^1$ as a slice of the action (3.15) on the complement of $\begin{bmatrix} 1 \\ 0 \end{bmatrix}$. It is a single point here because the action is transitive on this set. In addition, the isotropy subgroup of G_0 at this point is the trivial group $G_1 = \{1\}$. We call a frame field $e : \mathbf{R} \to \mathbf{SL}(2, \mathbf{R})$ *first order* if it satisfies

$$e^*\omega_1^2 \neq 0, \quad e^*\omega_1^1 = 0.$$

First order frame fields exist and are unique, since $\mathfrak{g}_1 = \{0\}$. The frame reduction procedure ends here. The first order frame field is called the *Frenet frame* along the curve (Figure 3.1). Then $e^*\omega_1^2$ is a coframe field in $\mathbf{R}$. In terms of the standard coordinate on $\mathbf{R}$ we have $e^*\omega_1^2 = \lambda(t)dt$, where $\lambda = \det(\mathbf{x}, \dot{\mathbf{x}})$ is never zero. A solution s of $ds = \lambda dt$ is a *centro-affine arclength parameter* of $\mathbf{x}$. The remaining component of $e^*\omega$ is $e^*\omega_2^1$, which we express as $e^*\omega_2^1 = \kappa e^*\omega_1^2$, for some function $\kappa : \mathbf{R} \to \mathbf{R}$, which is called the *centro-affine curvature* of $\mathbf{x}$.

For the second type of curve, $e^*\omega_1^2 = 0$, for any frame field $e = (\mathbf{x}, \mathbf{y})$ along $\mathbf{x}$, so $\dot{\mathbf{x}} = X_1^1\mathbf{x}$, by (3.13), where dot indicates derivative with respect to the coordinate t. Such a curve is radial.

For the general case, let G be a Lie group acting transitively on a smooth manifold N of dimension n. Choose a point $o \in N$ as the origin and let G_0 be the isotropy subgroup of G at o. Let

$$\pi : G \to N, \quad \pi(g) = go \tag{3.16}$$

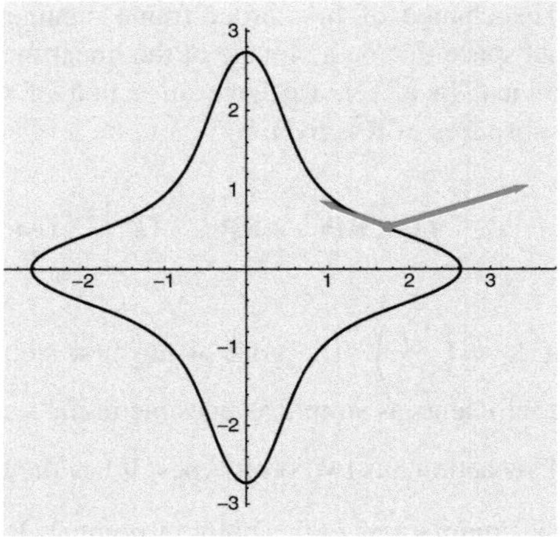

Fig. 3.1 First order centro-affine frame at a point of a non-radial curve **x**.

be the resulting principal G_0-bundle projection map. Let $\mathfrak{g}$ be the Lie algebra of G, and let $\mathfrak{g}_0 \subset \mathfrak{g}$ be the Lie algebra of G_0. Let $\mathfrak{m}_0 \subset \mathfrak{g}$ be a vector subspace complement to $\mathfrak{g}_0$, so $\mathfrak{g} = \mathfrak{m}_0 + \mathfrak{g}_0$ is a vector space direct sum. Let $E_1, \ldots, E_n$ be a basis of $\mathfrak{m}_0$. The adjoint representation of G_0 on $\mathfrak{g}/\mathfrak{g}_0$ relative to the basis $E_1, \ldots, E_n$ is the homomorphism $A = (A^i_j) : G_0 \to \mathbf{GL}(n, \mathbf{R})$ defined by

$$\mathrm{Ad}(K)E_i = A(K)E_i = \sum_{j=1}^{n} A^j_i E_j \mod \mathfrak{g}_0, \tag{3.17}$$

for $i = 1, \ldots, n$. The vectors $E_i \in \mathfrak{m}_0$ generate vector fields E^π_i along π. At a point $g \in G$,

$$E^\pi_i{}_{(g)} = dg_o d\pi_1 E_i,$$

where dg_o denotes the differential of the diffeomorphism $g : N \to N$ at the point o. Let ω be the Maurer–Cartan form on G. The direct sum $\mathfrak{g} = \mathfrak{m}_0 + \mathfrak{g}_0$ decomposes ω into the sum $\omega = \omega_{\mathfrak{m}_0} + \omega_{\mathfrak{g}_0}$, where subscripts indicate projection onto that subspace. Then $\omega_{\mathfrak{m}_0}$ is a left-invariant 1-form on G taking values in $\mathfrak{m}_0$, and $\omega_{\mathfrak{g}_0}$ is a left-invariant 1-form on G taking values in $\mathfrak{g}_0$. Restricted to the subgroup G_0 it is the Maurer–Cartan from of G_0. Using our basis of $\mathfrak{m}_0$, we have

$$\omega_{\mathfrak{m}_0} = \sum_{1}^{n} \omega^i E_i,$$

for some left-invariant 1-forms ω^i on G.

Let $\mathbf{x} : M^m \to N^n$ be an immersion with $m < n$. We shall carry out the frame reduction procedure for $\mathbf{x}$. A local frame field along $\mathbf{x}$ on an open set $U \subset M$ is a smooth map $e : U \to G$ such that $\pi \circ e = \mathbf{x}$; that is, e is a local section of the bundle (3.16). Such a section is called a *zeroth order frame field* along $\mathbf{x}$. Let $\varphi^1, \dots, \varphi^m$ be a coframe field on U and set

$$e^*\omega^i = \sum_{a=1}^{m} X^i_a \varphi^a,$$

for some smooth functions $X^i_a : U \to \mathbf{R}$, for $i = 1, \dots, n$ and $a = 1, \dots, m$. Then

$$e^*\omega_{\mathfrak{m}_0} = \sum_{i=1}^{n}\sum_{a=1}^{m} X^i_a E_i \varphi^a : T_pM \to \mathfrak{m}_0 \tag{3.18}$$

is a linear map. Since $\mathbf{x}$ is an immersion, this linear map sends the tangent space of M at a point of U to an m-dimensional subspace of $\mathfrak{m}_0$. The chosen basis of $\mathfrak{m}_0$ identifies it with $\mathbf{R}^n$. The image is the point $[X] \in G(m,n)$, where $X = (X^i_a) \in \mathbf{R}^{n\times m}$ has rank m. The matrix X is determined only up to multiplication on the right by an element of $\mathbf{GL}(m,\mathbf{R})$, because we allow an arbitrary choice of the coframe field $\varphi^1, \dots, \varphi^m$ on U.

We are using here the terminology and notation for the Grassmannian $G(m,n)$ of m-dimensional subspaces of $\mathbf{R}^n$ discussed in Example 2.11. There $G(m,n) = \mathbf{R}^{n\times m*}/\mathbf{GL}(m,\mathbf{R})$, where $\mathbf{R}^{n\times m*}$ is the space of all $n \times m$ real matrices of rank m, on which $\mathbf{GL}(m,\mathbf{R})$ acts by right multiplication. If $X \in \mathbf{R}^{n\times m*}$, then $[X]$ is its equivalence class in $G(m,n)$. $\mathbf{GL}(n,\mathbf{R})$ acts transitively on the left of $G(m,n)$ by left multiplication: $A[X] = [AX]$, for any $A \in \mathbf{GL}(n,\mathbf{R})$ and $X \in \mathbf{R}^{n\times m*}$.

Any other frame field along $\mathbf{x}$ on U must be given by $\tilde{e} = eK$, where $K : U \to G_0$ is an arbitrary smooth map. Then $\tilde{e}^*\omega = \mathrm{Ad}(K^{-1})e^*\omega + K^{-1}dK$ and $K^{-1}dK$ is $\mathfrak{g}_0$ valued implies that

$$\tilde{e}^*\omega_{\mathfrak{m}_0} = (\mathrm{Ad}(K^{-1})e^*\omega_{\mathfrak{m}_0})_{\mathfrak{m}_0} = \sum_{i,j=1}^{n}\sum_{a=1}^{m} A(K^{-1})^j_i X^i_a E_j \varphi^a : T_pM \to \mathfrak{m}_0,$$

where we have used the adjoint representation (3.17) of G_0 on $\mathfrak{m}_0$ relative to the basis $E_1, \dots, E_n$. This sends the tangent space of M at a point of U to the point $[A(K^{-1})X] \in G(m,n)$. The goal of the frame reduction at this step is to simplify the coefficients X in the sense that we seek a slice of this action of G_0 on $G(m,n)$. In general there can be more than one orbit type, in which case the reduction method requires one to assume that the points $[X(p)] \in G(m,n)$ are of the same orbit type for all $p \in U$.

To simplify the exposition at this stage, we assume that all points $[X(p)]$ lie in the orbit of the point $P_0 = \begin{bmatrix} I_m \\ 0 \end{bmatrix}$, where 0 is the $(n-m)\times m$ matrix of zeros.

Definition 3.5. A *first order frame field along* $\mathbf{x}$ is a frame field $e : U \to G$ along $\mathbf{x}$ for which

$$e^*\omega^{m+1} = \cdots = e^*\omega^n = 0, \quad e^*\omega^1 \wedge \cdots \wedge e^*\omega^m \neq 0 \tag{3.19}$$

at every point of U.

First order frame fields $e : U \to G$ exist by the Factor Property of Theorem 2.18. They are characterized by the fact that the map $X : U \to \mathbf{R}^{n\times m}$ that it defines in (3.18) has the property that $[X] = P_0$ at every point of U.

Let G_1 be the isotropy subgroup of the action of G_0 on $G(m,n)$ at the point P_0. This means that $A(K)P_0 = P_0$, for any $K \in G_1$, which implies that $A(K)$ has the block form $A(K) = \begin{pmatrix} A_1 & A_2 \\ 0 & A_3 \end{pmatrix}$, where $A_1 \in \mathbf{GL}(m,\mathbf{R})$, $A_3 \in \mathbf{GL}(n-m,\mathbf{R})$, and 0 is the $(n-m)\times m$ matrix of zeros. Let $\mathfrak{g}_1$ denote the Lie algebra of G_1. We assume that $\mathfrak{g}_1 \neq \{0\}$. If, to the contrary, $\mathfrak{g}_1$ were zero, then the frame reduction is complete. This is what happened in our centro-affine curve example.

Choose a decomposition $\mathfrak{g}_0 = \mathfrak{m}_1 + \mathfrak{g}_1$, where $\mathfrak{m}_1$ is a vector subspace of $\mathfrak{g}_0$. Then $\mathfrak{g} = \mathfrak{m}_0 + \mathfrak{m}_1 + \mathfrak{g}_1$ is a vector space direct sum which gives a decomposition $\omega = \omega_{\mathfrak{m}_0} + \omega_{\mathfrak{m}_1} + \omega_{\mathfrak{g}_1}$. The second assumption required for a continuation of the frame reduction at this step is that

$$e^*\omega_{\mathfrak{m}_1} \neq \{0\}, \tag{3.20}$$

at each point of U, for any first order frame field e along $\mathbf{x}$. If $e^*\omega_{\mathfrak{m}_1} = \{0\}$ identically on U, then the reduction stops here. Examples of this exceptional case are provided by totally umbilic immersions of a surface into a space form.

Assume (3.20) for any first order frame field e along $\mathbf{x}$. Let $E_{n+1},\dots,E_{n_1}$ be a basis of $\mathfrak{m}_1$. The adjoint representation of G_1 on $\mathfrak{g}/\mathfrak{g}_1 \cong \mathfrak{m}_0 + \mathfrak{m}_1$ relative to the basis $E_1,\dots,E_{n_1}$ is the homomorphism $A = (A^i_j) : G_1 \to \mathbf{GL}(n_1,\mathbf{R})$ defined by

$$\mathrm{Ad}(K)E_i = A(K)E_i = \sum_{j=1}^{n_1} A^j_i E_j \mod \mathfrak{g}_1, \tag{3.21}$$

for $i = 1,\dots,n_1$. Then $A(K)$ must have the block form

$$A(K) = \begin{pmatrix} A_1 & A_2 & 0 \\ 0 & A_3 & 0 \\ A_4 & A_5 & A_6 \end{pmatrix}, \tag{3.22}$$

where $A_1 \in \mathbf{GL}(m,\mathbf{R})$ and $A_6 \in \mathbf{GL}(n_1-n,\mathbf{R})$. Using this basis of $\mathfrak{m}_1$, we also have the expansion

$$\omega_{\mathfrak{m}_1} = \sum_{\mu=n+1}^{n_1} \omega^\mu E_\mu,$$

where ω^μ, for $\mu = n+1,\dots,n_1$, are left-invariant 1-forms on G.

Let $e : U \subset M \to G$ be a first order frame field along $\mathbf{x}$. Then $e^*\omega^1,\dots,e^*\omega^m$ is a coframe field on U and $e^*\omega^\alpha = 0$, for $\alpha = m+1,\dots,n$. Set

$$e^*\omega^\mu = \sum_{a=1}^{m} X^\mu_a e^*\omega^a, \tag{3.23}$$

for $\mu = n+1,\dots,n_1$. At a point $p \in U$, the image of the linear map

$$e^*(\omega_{\mathfrak{m}_0} + \omega_{\mathfrak{m}_1}) : T_pM \to \mathfrak{m}_0 + \mathfrak{m}_1 \tag{3.24}$$

is an m-dimensional subspace. With the given basis, it is the point

$$\begin{bmatrix} I_m \\ 0 \\ X \end{bmatrix} \in G(m,n_1),$$

where the smooth map

$$X = (X^\mu_a) \in \mathbf{R}^{(n_1-n)\times m} \tag{3.25}$$

is defined in (3.23). By our assumption (3.20) above, X is not the zero matrix.

Any other first order frame field along $\mathbf{x}$ on U is given by $\tilde{e} = eK$, where $K : U \to G_1$ is an arbitrary smooth map. Then the linear map $\tilde{e}^*(\omega_{\mathfrak{m}_0} + \omega_{\mathfrak{m}_1})$ is related to the map (3.23) of e by the adjoint action of K^{-1} on $\mathfrak{g}/\mathfrak{g}_1$. Using the block form (3.22) for $A(K^{-1})$, we see that the image becomes

$$\begin{bmatrix} I_m \\ 0 \\ (A_4 + A_6X)A_1^{-1} \end{bmatrix} \in G(m,n_1).$$

The goal of the frame reduction at this step is to simplify the coefficients X of the map (3.24) in the sense of finding a slice of the action of G_1 on $\mathbf{R}^{(n_1-n)\times m}$ given by

$$A(K)X = (A_4 + A_6X)A_1^{-1}. \tag{3.26}$$

The conditions defining the previous order of frames, in this case equations (3.19), restrict the possible values of X to an affine subspace of $\mathbf{R}^{(n_1-n)\times m}$, which is invariant under the action (3.26). We describe this affine subspace as follows. Let $E_{n_1+1},\dots,E_r$ be a basis of $\mathfrak{g}_1$, where $r = \dim G$. Then

$$E_a, E_\alpha, E_\mu, E_\sigma \tag{3.27}$$

is a basis of $\mathfrak{g} = \mathfrak{m}_0 \oplus \mathfrak{m}_1 \oplus \mathfrak{g}_1$, where we introduce the index ranges

$$1 \le a,b \le m,\ m+1 \le \alpha,\beta \le n,\ n+1 \le \mu,\nu \le n_1,\ n_1+1 \le \sigma,\tau \le r. \tag{3.28}$$

Consider the *structure constants* $C^i_{jk} = -C^i_{kj}$ of $\mathfrak{g}$ relative to this basis. These are defined by the equations

$$[E_j, E_k] = \sum_{i=1}^r C^i_{jk} E_i, \tag{3.29}$$

for $j,k \in \{1,\ldots,r\}$. The Maurer–Cartan structure equations of G are expressed in terms of the structure constants by

$$d\omega^k = -\frac{1}{2}\sum_{i,j=1}^{r} C^k_{ij}\omega^i \wedge \omega^j,$$

for all $k \in \{1,\ldots,r\}$.

Lemma 3.6. *Relative to a first order frame field $e : U \to G$ along $\mathbf{x}$, the map $X = (X^\mu_b) : U \to \mathbf{R}^{(n_1-n)\times m}$ in (3.25) takes values in the affine subspace*

$$\mathscr{X} = \{(X^\mu_b) : C^\alpha_{ab} + \sum_{\mu=n+1}^{n_1} (C^\alpha_{a\mu}X^\mu_b - C^\alpha_{b\mu}X^\mu_a) = 0\}, \tag{3.30}$$

where $a,b = 1,\ldots,m$ and $\alpha = m+1,\ldots,n$.

Proof. These restrictions on the entries of X come from the exterior derivative of the equations $e^*\omega^\alpha = 0$ defining a first order frame field. The forms $e^*\omega^a$, $a = 1,\ldots,m$ constitute a coframe field on U, and the pull-back of the remaining forms must be a linear combination of these. Then

$$0 = -2d\,e^*\omega^\alpha = \sum_{a,b=1}^{m} (C^\alpha_{ab} + C^\alpha_{a\mu}X^\mu_b - C^\alpha_{b\mu}X^\mu_a)e^*\omega^a \wedge e^*\omega^b,$$

since all other structure constants are zero. In fact, $C^\alpha_{\mu\nu} = C^\alpha_{\mu\sigma} = C^\alpha_{\sigma\tau} = 0$ because $E_\mu, E_\sigma \in \mathfrak{g}_0$ and $[\mathfrak{g}_0,\mathfrak{g}_0] \subset \mathfrak{g}_0$, so the brackets $[E_\mu,E_\nu]$, $[E_\mu,E_\sigma]$, and $[E_\sigma,E_\tau]$ have no E_α-components.

Finally, the coefficients $C^\alpha_{a\sigma} = 0$ because G_1 fixes P_0. For, if $K \in G_1$, then $\mathrm{Ad}(K)E_a$ has no E_α components, by (3.22). Then $E_\sigma \in \mathfrak{g}_1$ implies $\exp(tE_\sigma) \in G_1$, for all $t \in \mathbf{R}$, so

$$[E_\sigma, E_a] = \mathrm{ad}(E_\sigma)E_a = d\mathrm{Ad}(E_\sigma)E_a = \left.\frac{d}{dt}\right|_0 \mathrm{Ad}(\exp(tE_\sigma))E_a$$

has no E_α-component. See [110, Theorem 20.12, p 529] for information about the adjoint representations $\mathrm{Ad} : G \to \mathbf{GL}(\mathfrak{g})$ and $\mathrm{ad} : \mathfrak{g} \to \mathfrak{gl}(\mathfrak{g})$. □

As in the previous step, in order to proceed farther we must assume that the values of the map $X : U \to \mathscr{X} \subset \mathbf{R}^{(n_1-n)\times m}$ are all of the same orbit type under this action of G_1. Given that assumption, we then choose a slice of the action. To simplify this general exposition, we assume that there is a slice $\mathscr{Y} \subset \mathscr{X}$ for which the isotropy subgroup is the same subgroup G_2 of G_1 at every point. We then define a *second order frame field* $e : U \to G$ to be a first order frame field whose coefficient map

$$X : U \to \mathscr{Y} \subset \mathscr{X} \subset \mathbf{R}^{(n_1-n)\times m}$$

takes its values in the slice. We summarize this as follows.

Definition 3.7. A *second order frame field along* $\mathbf{x} : M \to N$ is any frame field $e : U \subset M \to G$ along $\mathbf{x}$ for which

1. $e^*\omega^\alpha = 0$, for $\alpha = m+1,\dots,n$; and $e^*\omega^1 \wedge \cdots \wedge e^*\omega^m \neq 0$, (first order);
2. $e^*\omega^\mu = \sum_{a=1}^m Y_a^\mu e^*\omega^a$, for $\mu = n+1,\dots,n_1$,

where $Y = (Y_a^\mu) \in \mathscr{Y}$ at every point of U.

The Factor Property of Theorem 2.18 applied to the action (3.26) ensures the existence of a second order frame field on some neighborhood of any point $p \in M$ at which the map (3.25) satisfies $X(p) \in G_1\mathscr{Y}$; i.e., $\mathbf{x}$ is of type $(G_2, \mathscr{Y})$ at p.

The next step of the frame reduction proceeds in the same way as the preceding step. Let $\mathfrak{g}_1 = \mathfrak{m}_2 + \mathfrak{g}_2$ be a vector space direct sum. The frame reduction is complete at this point if either $\mathfrak{g}_2 = \{0\}$ or if $e^*\omega_{\mathfrak{m}_2} = \{0\}$, for any second order frame field. In either case we then express the component forms of $\omega_{\mathfrak{g}_1}$ in terms of the coframe field $\epsilon^*\omega^1,\dots,\epsilon^*\omega^m$, like this

$$\omega^\sigma = \sum_1^m \kappa_a^\sigma e^*\omega^a,$$

where the smooth functions $\kappa_a^\sigma : U \to \mathbf{R}$ are the *invariants* of $\mathbf{x}$. The result of the final frame reduction is called a *Frenet frame field along* $\mathbf{x}$.

We conclude our general description of the frame reduction process here. Additional contingencies will be exhibited in the many examples that follow in the book. For more details about the theory of the method of moving frames see Cartan [32], Chern [45], do Carmo [30], Green [77], Jensen [93], and Fels and Olver [67].

Proposition 3.8. *Let* $\mathbf{x} : M^m \to N$ *and* $\hat{\mathbf{x}} : \hat{M}^m \to N$ *be immersions with Frenet frames of the same type* $e : M \to G$ *and* $\hat{e} : \hat{M} \to G$*, respectively. If* $F : M \to \hat{M}$ *is a diffeomorphism that pulls back the coframe field of* $\hat{e}$ *to the coframe field of* e*, and pulls back the invariants of* $\hat{e}$ *to the invariants of* e*, then there exists an element* $g \in G$ *such that* $\hat{\mathbf{x}} \circ F = g \circ \mathbf{x}$*.*

Proof. It is evident from the frame reduction procedure that under the given hypotheses, $F^*\hat{e}^*\omega = e^*\omega$. Application of the Cartan–Darboux Uniqueness Theorem to the maps $\hat{e} \circ F : M \to G$ and $e : M \to G$ yields the result. □

3.4 Homogeneous submanifolds

We continue with our smooth manifold N^n acted upon transitively by the Lie group G. We choose the point $o \in N$ as the origin.

Definition 3.9. An embedding $\mathbf{x} : M^m \to N^n$ is *homogeneous* if there exists a Lie subgroup H of G that stabilizes $\mathbf{x}(M)$ and acts transitively on this set.

If $\mathbf{x} : M \to N$ is a homogeneous embedding, we may assume, up to an action by an element of G, that there is a point $p_0 \in M$ such that $\mathbf{x}(p_0) = o$. Then

$$\mathbf{x}(M) = \{ho : h \in H\} = \pi(H),$$

where $\pi : G \to N$ is the bundle projection (3.16). Let $H_0 \subset G_0$ be the isotropy subgroup of H at o. Given a point $p \in M$, there exists a neighborhood $U \subset M$ of p such that there exists a smooth section $e : U \to H$ of the principal H_0-bundle projection $\pi : H \to \mathbf{x}(M)$. Because $\mathbf{x}$ is an embedding, we may identify M with $\mathbf{x}(M) \subset N$ as smooth manifolds. Let $\mathfrak{h} \subset \mathfrak{g}$ be the Lie algebra of H, and let $\mathfrak{h}_0 \subset \mathfrak{g}_0$ be the Lie algebra of H_0. If $\mathfrak{m}_0 \subset \mathfrak{h}$ is a vector space complement of $\mathfrak{h}_0$ in $\mathfrak{h}$, then

$$d\pi_1 : \mathfrak{m}_0 \to T_o\mathbf{x}(M) \tag{3.31}$$

is a linear isomorphism. Moreover, we have the direct sum of vector spaces

$$\mathfrak{g} = \mathfrak{m}_0 + \mathfrak{h}_0 + \mathfrak{h}', \tag{3.32}$$

where $\mathfrak{h}'$ is a vector space complement of $\mathfrak{h}$ in $\mathfrak{g}$. Let E_a, for $a = 1, \dots, m$ be a basis of $\mathfrak{m}_0$, let E_α, for $\alpha = m+1, \dots, r$ be a basis of $\mathfrak{h}'$, and let $\omega_{\mathfrak{h}_0}$ denote the $\mathfrak{h}_0$-component of the Maurer–Cartan form of G under the direct sum (3.32). Then we have an expansion

$$\omega = \sum_{a=1}^{m} \omega^a E_a + \sum_{\alpha=m+1}^{r} \omega^\alpha E_\alpha + \omega_{\mathfrak{h}_0},$$

where $\omega^1, \dots, \omega^r$ are left-invariant 1-forms linearly independent on G. With this preparation we can state the property of a frame field along $\mathbf{x}$ that takes values in H.

Proposition 3.10. *Given a point $p \in M$, there exists a local frame field $e : U \subset M \to H \subset G$ along* $\mathbf{x}$. *It has the properties*

1. $e^*(\omega^1 \wedge \cdots \wedge \omega^m) \neq 0$ *at every point of U,*
2. $e^*\omega^\alpha = 0$, *for $\alpha = m+1, \dots, r$, and*
3. *the k-dimensional distribution defined by*

$$\mathscr{D} = \operatorname{span}\{\omega^\alpha : \alpha = m+1, \dots, r\},$$

satisfies the Frobenius condition, where $k = \dim \mathfrak{h} = \dim(\mathfrak{g}) - (r - m)$.

Proof. The first item follows from the fact that the linear map (3.31) is an isomorphism. The second item follows from the fact that $e : U \to H$ and the ω^α annihilate $\mathfrak{h}$. For the third item we observe that the distribution $\mathscr{D}$ is spanned by left-invariant 1-forms on G and $\mathscr{D}^\perp = \mathfrak{h}$, which satisfies the Frobenius condition because it is a Lie subalgebra of $\mathfrak{g}$. □

The following proposition is an important converse to Proposition 3.10. It is applied in Proposition 4.39, Theorem 5.15, Problem 6.49, Theorem 12.51, the classification of totally umbilic submanifolds of the space form geometries, and in Theorem 15.51, which classifies all nonumbilic Dupin immersions in Euclidean space. It is the method Cartan [32, p. 155] developed to characterize homogeneous submanifolds of a homogeneous space. See also Jensen [93, pp. 41–44] and Sulanke [157, Theorem 4.1, p. 702].

Proposition 3.11. *Let G be a Lie group of dimension n, with Lie algebra $\mathfrak{g}$ and space of left-invariant 1-forms $\mathfrak{g}^*$. Let $\omega^1, \dots, \omega^r$ be a linearly independent subset of $\mathfrak{g}^*$. Let $e : M^m \to G$ be an immersion of a connected manifold M of dimension $m < r$. Let $\theta^i = e^*\omega^i$, for $i = 1, \dots, m$ and let $\eta^\alpha = e^*\omega^\alpha$, for $\alpha = m+1, \dots, r$. If*

1. *$\theta^1 \wedge \dots \wedge \theta^m \neq 0$ at each point of M, if*
2. *$\eta^\alpha = \sum_{i=1}^m A_i^\alpha \theta^i$ on M, for constants A_i^α, for each $\alpha = m+1, \dots, r$, and if*
3. *the $(n-r+m)$-dimensional smooth distribution $\mathscr{D}^\perp$ defined by the subspace*

$$\mathfrak{h}^\perp = \operatorname{span}(\omega^\alpha - \sum_{i=1}^m A_i^\alpha \omega^i : \alpha = m+1, \dots, r) \subset \mathfrak{g}^*$$

satisfies the Frobenius condition, then

$$\mathfrak{h} = \{X \in \mathfrak{g} : \omega(X) = 0, \forall \omega \in \mathfrak{h}^\perp\}$$

is a Lie subalgebra of $\mathfrak{g}$ and $e(M) \subset gH$, for some $g \in G$, where H is the connected Lie subgroup of G whose Lie algebra is $\mathfrak{h}$.

Proof. The integral manifolds of $\mathscr{D}$ are the right cosets gH, for every $g \in G$. If $r = n$, then it is clear that $e : M \to G$ is an integral manifold of $\mathscr{D}$, and thus is contained in some right coset gH of H. If $r < n$, then it is still true that $e^*\omega = 0$ for every smooth 1-form ω in $\mathscr{D}^\perp$. The local Frobenius Theorem then implies that $e(M)$ must be contained in some integral manifold of $\mathscr{D}$. □

Proposition 3.12. *Let $\mathbf{x} : M^m \to N^n = G/G_0$ be an immersion with Frenet frame field $e : M \to G$. If the invariants of $\mathbf{x}$ are all constant on M, then there exists a Lie subgroup H of G such that $e(M)$ is an open subset of the right coset gH, for some $g \in G$, so $\mathbf{x}(M)$ is congruent to an open subset of the homogeneous submanifold $\pi(H) \subset N$, where $\pi : G \to N$ is the principal bundle projection* (3.1).

Proof. Suppose the Frenet frame field is of order $k \geq 1$. Let

$$\mathfrak{g} = \mathfrak{m}_0 + \dots + \mathfrak{m}_{k-1} + \mathfrak{m}_k + \mathfrak{g}_k$$

be the decomposition obtained from the frame reduction procedure. Let

$$E_1, \dots, E_r$$

be the basis of the Lie algebra $\mathfrak{g}$ obtained from bases of each component chosen during the reduction, with the last $\dim \mathfrak{g}_k$ vectors being a basis of $\mathfrak{g}_k$. Then the hypotheses of Proposition 3.11 are satisfied and our result follows from that. See the comments preceding Proposition 3.11 for specific applications occurring later in the book. □

In the following chapters we will use the method of moving frames to study surfaces in the classical Euclidean, spherical, and hyperbolic geometries, with applications to a selection of important problems. The final four chapters use the method in classical Möbius geometry and Lie sphere geometry. This represents only a small number of important applications of the method.

It has been widely used in projective geometry, both real and complex. For a sample of this area see Liao [112], Jensen-Musso [95], and Yang [174].

For its use in Grassmannian geometries, real or complex, see Griffiths [79], Yang [173], Zheng [175], and Jensen-Rigoli [96].

Problems

3.13. Let $\mathbf{x} : \mathbf{R} \to \dot{\mathbf{R}}^2$ be a centro-affine curve of Example 3.4. Prove that a point t is radial if and only if $\lambda(t) = \det(\mathbf{x}(t), \dot{\mathbf{x}}(t)) = 0$, where dot is derivative with respect to the standard coordinate t in $\mathbf{R}$. In the nonradial case, prove that the first order frame field along $\mathbf{x}$ is $e = (\mathbf{x}, \frac{1}{\lambda}\dot{\mathbf{x}})$, that the centro-affine arclength parameter s satisfies $ds = e^*\omega_1^2 = \lambda dt$, and that the centro-affine curvature $\kappa = \frac{1}{\lambda^3}(\dot{\mathbf{x}}, \ddot{\mathbf{x}})$. Prove that if a curve $\mathbf{x}$ is radial, then $\mathbf{x}(t) = f(t)\mathbf{a}$, where $\mathbf{a} \in \dot{\mathbf{R}}^2$ is constant and f is a positive function.

3.14 (Parabolas). If p is a nonzero real constant, then the centro-affine curve $\mathbf{x}(t) = \begin{pmatrix} pt^2+1 \\ t \end{pmatrix}$ is a parabola. Find any radial points. Off the radial points, find its centro-affine curvature.

3.15 (Constant affine-centro curvature). Find the centro-affine curves in $\dot{\mathbf{R}}^2$, up to $\mathbf{SL}(2, \mathbf{R})$ congruence, that have constant centro-affine curvature κ.

3.16. Sketch some of the curves found in Problem 3.15. On the sketch, draw the centro-affine Frenet frame at several points. Give a geometric interpretation of $\lambda = \det(\mathbf{x}, \dot{\mathbf{x}})$.

Chapter 4
Euclidean Geometry

We begin with a standard elementary introduction to the theory of surfaces immersed in Euclidean space $\mathbf{R}^3$, whose Riemannian metric is the standard dot product. Section 4.2 will be review for readers who have studied basic differential geometry of curves and surfaces in Euclidean space. Geometric intuition is used to construct Euclidean frames on a surface. Section 4.3 repeats the exposition, but this time following the frame reduction procedure outlined in Chapter 3. The classical existence and congruence theorems of Bonnet are stated and proved as consequences of the Cartan–Darboux Theorems. A section on tangent and curvature spheres provides needed background for Lie sphere geometry. The Gauss map helps tie together the formalism of Gauss and that of moving frames. We discuss special examples, such as surfaces of revolution, tubes about a space curve, inversions in a sphere, and parallel transforms of a given immersion. These constructions provide many valuable examples throughout the book. The latter two constructions introduce for the first time Möbius, respectively Lie sphere, transformations that are not Euclidean motions. The section on elasticae contains material needed in our introduction of the Willmore problems.

Euclidean space is $\mathbf{R}^3$ with the Riemannian metric given by its standard dot product. The Euclidean group $\mathbf{E}(3)$ is the set of all isometries of this space. Euclidean geometry is the study of properties of subsets invariant under isometries. Immersions $\mathbf{x}, \hat{\mathbf{x}} : M \to \mathbf{R}^3$ are *congruent* if there is an isometry $T \in \mathbf{E}(3)$ such that $\hat{\mathbf{x}} = T \circ \mathbf{x}$. In Chapter 10, the related notion of equivalence plays a prominent role. Immersions $\mathbf{x} : M \to \mathbf{R}^3$ and $\hat{\mathbf{x}} : \hat{M} \to \mathbf{R}^3$ are *equivalent* if there is an isometry $T \in \mathbf{E}(3)$ and a diffeomorphism $F : M \to \hat{M}$ such that $\hat{\mathbf{x}} \circ F = T \circ \mathbf{x} : M \to \mathbf{R}^3$, that is, $\hat{\mathbf{x}} \circ F$ is congruent to $\mathbf{x}$. In the case when $\mathbf{x}$ and $\hat{\mathbf{x}}$ are embeddings, equivalence implies $T\mathbf{x}(M) = \hat{\mathbf{x}}(\hat{M})$, which is Euclid's notion of congruence.

G.R. Jensen et al., *Surfaces in Classical Geometries*, Universitext,
DOI 10.1007/978-3-319-27076-0_4

4.1 The Euclidean group

An *isometry* of $\mathbf{R}^3$ is a diffeomorphism $T:\mathbf{R}^3\to\mathbf{R}^3$ whose differential dT preserves the dot product at each point. An example is $T=(\mathbf{v},A)\in\mathbf{R}^3\times\mathbf{O}(3)$, where

$$(\mathbf{v},A)\mathbf{x}=\mathbf{v}+A\mathbf{x},$$

and the *orthogonal group*

$$\mathbf{O}(3)=\{A\in\mathbf{GL}(3,\mathbf{R}):{}^tAA=I\}.$$

Exercise 7. Prove that any isometry of $\mathbf{R}^3$ is of the form $(\mathbf{v},A)$, for some $\mathbf{v}\in\mathbf{R}^3$ and $A\in\mathbf{O}(3)$.

The set of all Euclidean isometries forms a Lie group, $\mathbf{E}(3)$, called the Euclidean group. As a manifold,

$$\mathbf{E}(3)=\mathbf{R}^3\times\mathbf{O}(3),$$

and the group structure defined by composition of maps is

$$(\mathbf{v},A)(\mathbf{w},B)=(\mathbf{v}+A\mathbf{w},AB),$$

which is a semi-direct product, $\mathbf{R}^3\rtimes\mathbf{O}(3)$, with $\mathbf{R}^3$ as the normal subgroup. The inverse transformation of $(\mathbf{v},A)$ is

$$(\mathbf{v},A)^{-1}=(-A^{-1}\mathbf{v},A^{-1}),$$

and conjugation by $\mathbf{E}(3)$ on its normal subgroup $\mathbf{R}^3$ is

$$(\mathbf{v},A)(\mathbf{w},I)(\mathbf{v},A)^{-1}=(A\mathbf{w},I),$$

which is the standard action of $\mathbf{O}(3)$ on $\mathbf{R}^3$. The connected component of $\mathbf{E}(3)$ containing the identity element is the subgroup $\mathbf{E}_+(3)=\mathbf{R}^3\rtimes\mathbf{SO}(3)$, which is called the *Euclidean group of rigid motions*. The Euclidean group $\mathbf{E}(3)$ has the faithful representation in $\mathbf{GL}(4)$,

$$\mathbf{E}(3)=\{\begin{pmatrix}1&0\\ \mathbf{x}&A\end{pmatrix}:A\in\mathbf{O}(3),\ \mathbf{x}\in\mathbf{R}^3\},$$

so it is a matrix Lie group. Its Lie algebra $\mathscr{E}(3)$ is faithfully represented in the Lie algebra $\mathfrak{gl}(4,\mathbf{R})$ by

$$\mathscr{E}(3)=\{\begin{pmatrix}0&0\\ \mathbf{x}&X\end{pmatrix}:X\in\mathfrak{o}(3),\ \mathbf{x}\in\mathbf{R}^3\}$$

where the Lie bracket is the usual matrix commutator. As vector space,

$$\mathscr{E}(3) = \mathbf{R}^3 + \mathfrak{o}(3), \quad \begin{pmatrix} 0 & 0 \\ \mathbf{x} & X \end{pmatrix} \leftrightarrow (\mathbf{x}, X). \tag{4.1}$$

The adjoint action of $\mathbf{O}(3)$ on $\mathbf{R}^3 \cong \mathscr{E}(3)/\mathfrak{o}(3)$ is

$$\mathrm{Ad}(\mathbf{0}, A)(\mathbf{x}, 0) = (A\mathbf{x}, 0),$$

which is the standard action of $\mathbf{O}(3)$ on $\mathbf{R}^3$. The Lie bracket is given by matrix commutation

$$\left[\begin{pmatrix} 0 & 0 \\ \mathbf{x} & X \end{pmatrix}, \begin{pmatrix} 0 & 0 \\ \mathbf{y} & Y \end{pmatrix}\right] = \begin{pmatrix} 0 & 0 \\ \mathbf{x} & X \end{pmatrix}\begin{pmatrix} 0 & 0 \\ \mathbf{y} & Y \end{pmatrix} - \begin{pmatrix} 0 & 0 \\ \mathbf{y} & Y \end{pmatrix}\begin{pmatrix} 0 & 0 \\ \mathbf{x} & X \end{pmatrix}$$
$$= \begin{pmatrix} 0 & 0 \\ X\mathbf{y} - Y\mathbf{x} & XY - YX \end{pmatrix},$$

which, in the identification (4.1), is

$$[(\mathbf{x}, X), (\mathbf{y}, Y)] = (X\mathbf{y} - Y\mathbf{x}, [X, Y]),$$

where $[X, Y] = XY - YX$ is the Lie bracket in $\mathfrak{o}(3) \subset \mathfrak{gl}(3, \mathbf{R})$. In particular,

$$\begin{aligned} [(\mathbf{0}, X), (\mathbf{0}, Y)] &= (\mathbf{0}, [X, Y]), \\ [(\mathbf{0}, X), (\mathbf{y}, 0)] &= (X\mathbf{y}, 0), \\ [(\mathbf{x}, 0), (\mathbf{y}, 0)] &= (\mathbf{0}, 0) = 0. \end{aligned} \tag{4.2}$$

The Maurer–Cartan form of $\mathbf{E}(3)$ is the $\mathscr{E}(3)$-valued left-invariant 1-form

$$\begin{pmatrix} 0 & 0 \\ \theta & \omega \end{pmatrix} = \begin{pmatrix} 1 & 0 \\ \mathbf{x} & A \end{pmatrix}^{-1} d\begin{pmatrix} 1 & 0 \\ \mathbf{x} & A \end{pmatrix} = \begin{pmatrix} 0 & 0 \\ A^{-1}d\mathbf{x} & A^{-1}dA \end{pmatrix},$$

so $\theta = (\omega^i)$ is $\mathbf{R}^3$-valued and $\omega = A^{-1}dA = (\omega^i_j)$ is $\mathfrak{o}(3)$-valued. Differentiation of these forms gives the Maurer–Cartan structure equations of $\mathbf{E}(3)$,

$$\begin{aligned} d\theta &= d(A^{-1}d\mathbf{x}) = -A^{-1}dAA^{-1} \wedge d\mathbf{x} = -\omega \wedge \theta, \\ d\omega &= d(A^{-1}dA) = -A^{-1}dAA^{-1} \wedge dA = -\omega \wedge \omega. \end{aligned}$$

In terms of the left-invariant component 1-forms on $\mathbf{E}(3)$, this is

$$d\omega^i = -\sum_{j=1}^{3} \omega^i_j \wedge \omega^j, \quad d\omega^i_j = -\sum_{k=1}^{3} \omega^i_k \wedge \omega^k_j.$$

The isometric action of $\mathbf{E}(3)$ on $\mathbf{R}^3$ is transitive, so the $\mathbf{E}(3)$-orbit of the origin $\mathbf{0} \in \mathbf{R}^3$ is all of $\mathbf{R}^3$. The isotropy subgroup G_0 of $\mathbf{E}(3)$ at $\mathbf{0}$ is

$$G_0 = \{(\mathbf{0}, A) \in \mathbf{E}(3) : A \in \mathbf{O}(3)\} \cong \mathbf{O}(3).$$

By Proposition 2.6, $\mathbf{E}(3)$ is a principal $\mathbf{O}(3)$-bundle over $\mathbf{R}^3$ with projection

$$\pi : \mathbf{E}(3) \to \mathbf{R}^3 \quad \pi(\mathbf{v}, A) = (\mathbf{v}, A)\mathbf{0} = \mathbf{v}. \tag{4.3}$$

For reference frame at $\mathbf{0}$ choose the standard basis $\epsilon_1, \epsilon_2, \epsilon_3$ of $\mathbf{R}^3$. Then any element $(\mathbf{v}, A) \in \mathbf{E}(3)$ defines a frame

$$d(\mathbf{v}, A)_{\mathbf{0}}(\epsilon_1, \epsilon_2, \epsilon_3) = (\mathbf{A}_1, \mathbf{A}_2, \mathbf{A}_3)$$

at $(\mathbf{v}, A)\mathbf{0} = \mathbf{v} \in \mathbf{R}^3$, where $\mathbf{A}_i$ denotes column i of A. Every orthonormal frame on $\mathbf{R}^3$ is obtained in this way. The basis of $\mathfrak{m}_0 = \mathbf{R}^3 \subset \mathscr{E}(3)$ that projects by $d\pi_1$ onto the standard basis of $\mathbf{R}^3$ is

$$E_i = (\epsilon_i, 0), \quad i = 1, 2, 3. \tag{4.4}$$

The adjoint representation of $G_0 = \mathbf{O}(3)$ on $\mathfrak{g}/\mathfrak{g}_0 \cong \mathbf{R}^3$ relative to E_1, E_2, E_3 is the standard representation $\mathbf{O}(3) \subset \mathbf{GL}(3, \mathbf{R})$. An orthonormal frame field on an open set $U \subset \mathbf{R}^3$ is a smooth section

$$(\mathbf{x}, e) : U \to \mathbf{E}(3)$$

of (4.3). It must be of the form (id_U, e), where $\mathrm{id}_U : U \to U$ is the identity map of U and $e : U \to \mathbf{O}(3)$ can be any smooth map. Smooth local sections exist on a neighborhood of any point of $\mathbf{R}^3$.

4.2 Surface theory of Gauss

Let M be a surface and let

$$\mathbf{x} : M \to \mathbf{R}^3$$

be an immersion in the three dimensional Euclidean space $\mathbf{R}^3$. If u, v are local coordinates on M, then $\mathbf{x}(u, v)$ is a smooth vector valued function. The condition that it is an immersion is that the tangent vectors $\mathbf{x}_u$ and $\mathbf{x}_v$ be linearly independent for every (u, v). The unit normal vector

$$\mathbf{e}_3(u, v) = \pm \frac{\mathbf{x}_u \times \mathbf{x}_v}{|\mathbf{x}_u \times \mathbf{x}_v|}$$

is then defined up to sign. Gauss [72] based his theory of surfaces on the first and second fundamental forms

$$I = d\mathbf{x}\cdot d\mathbf{x} = Edu^2 + 2Fdudv + Gdv^2,$$
$$II = -d\mathbf{x}\cdot d\mathbf{e}_3 = Ldu^2 + 2Mdudv + Ndv^2,$$

which are symmetric, bilinear form fields on M. The form I is positive definite and the form II is defined up to the choice of sign in $\mathbf{e}_3$. The ratio $II/I = k_N(u,v,\frac{dv}{du})$, which depends on the point (u,v) and a tangent line T through it, is called the *normal curvature*. Geometrically it is equal to the curvature at (u,v) of the curve of intersection of the plane spanned by the normal $\mathbf{e}_3$ and T with the surface. Explicitly, if the line T at (u,v) is tangent to the nonzero vector $X = a\mathbf{x}_u + b\mathbf{x}_v$, then the corresponding normal curvature is

$$k_N(u,v,T) = \frac{II(X,X)}{I(X,X)} = \frac{La^2 + 2Mab + Nb^2}{Ea^2 + 2Fab + Gb^2}.$$

If X is multiplied by any nonzero number t, then k_N remains unchanged, which shows that it depends only on the line T and not on the choice of vector tangent to the line.

Let the point (u,v) be fixed. The critical values of k_N as a function of the line T are called the *principal curvatures*, which we denote by a and c. Their elementary symmetric functions

$$H = \frac{1}{2}(a+c) \text{ and } K = ac$$

are called the *mean curvature* and *Gaussian curvature*, respectively. A *Weingarten surface*, or *W-surface*, is one satisfying a functional relationship

$$f(a,c) = 0 \tag{4.5}$$

between its principal curvatures. Special cases are

$H = 0$, *minimal surfaces*,

$H =$ nonzero constant, *constant mean curvature (CMC) surfaces*,

$K =$ constant, *constant curvature surfaces*.

If the surface is given as a graph

$$z = z(x,y),$$

then

$$H=\frac{1}{2}\frac{(1+z_y^2)z_{xx}-2z_xz_yz_{xy}+(1+z_x^2)z_{yy}}{(1+z_x^2+z_y^2)^{3/2}} \tag{4.6}$$

and

$$K=\frac{z_{xx}z_{yy}-z_{xy}^2}{(1+z_x^2+z_y^2)^2}.$$

The Weingarten equation (4.5) is essentially a geometric expression of a nonlinear partial differential equation in two independent variables. For example, the minimal surface equation $H=0$ is a quasi-linear elliptic PDE.

4.2.1 Surface theory of Darboux, Cartan, and Chern

A frame field $(\mathbf{x},\mathbf{e})$ in Euclidean space $\mathbf{R}^3$ defined on an open set $U\subset\mathbf{R}^3$ consists of the position vector $\mathbf{x}=\begin{pmatrix}x^1\\x^2\\x^3\end{pmatrix}$ and smooth vector fields $\mathbf{e}_1$, $\mathbf{e}_2$, $\mathbf{e}_3$ on U such that they form an orthonormal basis of $\mathbf{R}^3$ at each point. The exterior differential of each of $\mathbf{x},\mathbf{e}_1,\mathbf{e}_2,\mathbf{e}_3$ can be expressed as linear combinations of the orthonormal frame $e=(\mathbf{e}_1,\mathbf{e}_2,\mathbf{e}_3)$, where the coefficients are smooth 1-forms defined on U. Namely,

$$d\mathbf{x}=\sum_1^3\omega^i\mathbf{e}_i,\qquad d\mathbf{e}_i=\sum_1^3\omega_i^j\mathbf{e}_j,$$

where the coefficient 1-forms ω^i and ω_j^i are given by the dot products

$$\omega^i=d\mathbf{x}\cdot\mathbf{e}_i,\qquad \omega_i^j=d\mathbf{e}_i\cdot\mathbf{e}_j.$$

Differentiating $\mathbf{e}_i\cdot\mathbf{e}_j=\delta_{ij}$, we find that

$$\omega_i^j+\omega_j^i=0,$$

for $1\le i,j\le 3$. Since $dd\mathbf{x}=dd\mathbf{e}_i=0$, we arrive at the *structure equations*

$$d\omega^i=-\sum_{j=1}^3\omega_j^i\wedge\omega^j,\qquad d\omega_j^i=-\sum_{k=1}^3\omega_k^i\wedge\omega_j^k. \tag{4.7}$$

The latter equations show that the curvature forms of Euclidean space are zero,

$$\Omega_j^i=d\omega_j^i+\sum_1^3\omega_k^i\wedge\omega_j^k=0,\quad i,j=1,2,3.$$

Let M be a surface and let

$$\mathbf{x} = \begin{pmatrix} x^1 \\ x^2 \\ x^3 \end{pmatrix} : M \to \mathbf{R}^3$$

be a smooth immersion. The differential of $\mathbf{x}$ at $p \in M$ can be expressed as

$$d\mathbf{x}_p = \begin{pmatrix} dx^1_p \\ dx^2_p \\ dx^3_p \end{pmatrix} : T_pM \to \mathbf{R}^3,$$

where each dx^i is the ordinary differential of the function $x^i : M \to \mathbf{R}$. The condition that $\mathbf{x}$ be an immersion is that $d\mathbf{x}$ has rank two at each point of M; that is, the dimension of the span of $\{dx^1, dx^2, dx^3\}$ is two at every point of M.

A Euclidean frame field along $\mathbf{x}$ on an open set $U \subset M$ consists of smooth vector fields $\mathbf{e}_i : U \to \mathbf{R}^3$, $i = 1,2,3$, such that $e = (\mathbf{e}_1, \mathbf{e}_2, \mathbf{e}_3)$ is an orthonormal frame of $\mathbf{R}^3$ at each point of U. Denote such a frame field by $(\mathbf{x}, e)$, where $e = (\mathbf{e}_1, \mathbf{e}_2, \mathbf{e}_3)$. Then $d\mathbf{x}$ can be expressed in terms of e by

$$d\mathbf{x} = \sum_1^3 \omega^i \mathbf{e}_i,$$

where each ω^i is a smooth 1-form defined on U, given by

$$\omega^i = d\mathbf{x} \cdot \mathbf{e}_i.$$

It is a linear combination of the dx^j with coefficients being the smooth function entries of $\mathbf{e}_i$. Suppose that we can choose $\mathbf{e}_3$ to be a unit normal vector to the surface at each point. Such a smooth vector field exists on all of M if and only if M is orientable, in which case the normal is determined up to sign (assuming that M is connected). With such a choice for $\mathbf{e}_3$ it follows that the tangent plane, which is the image of $d\mathbf{x}$, must be the span of $\mathbf{e}_1$ and $\mathbf{e}_2$. Hence, choosing $\mathbf{e}_3$ normal at every point is equivalent to the condition that

$$\omega^3 = 0 \tag{4.8}$$

at every point of U. By the immersion condition, it follows then that ω^1, ω^2 must be linearly independent at every point of U, and hence they form a coframe field on U. Frame fields $(\mathbf{x}, \mathbf{e})$ satisfying this condition will be called *first order* frame fields along $\mathbf{x}$. By the structure equations (4.7), exterior differentiation of (4.8) gives

$$\omega^3_1 \wedge \omega^1 + \omega^3_2 \wedge \omega^2 = 0 \tag{4.9}$$

Lemma 4.1 (Cartan's Lemma [32], pp. 218–219). *Let $\alpha^1,\dots,\alpha^p$ be p linearly independent elements in a vector space V of dimension n. If p elements $\varphi_1,\dots,\varphi_p$ of V satisfy the equation in $\Lambda_2 V$,*

$$\sum_{j=1}^{p} \alpha^j \wedge \varphi_j = 0,$$

then $\varphi_i = \sum_{j=1}^{p} h_{ij}\alpha^j$, for some scalars h_{ij} satisfying $h_{ij} = h_{ji}$, for all $i,j = 1,\dots,p$.

Proof. Complete $\alpha^1,\dots,\alpha^p$ to a basis $\alpha^1,\dots,\alpha^p,\alpha^{p+1},\dots,\alpha^n$ of V. Then

$$\varphi_i = \sum_{j=1}^{n} h_{ij}\alpha^j,$$

for some scalars h_{ij}, and

$$0 = \sum_{i=1}^{p} \alpha^i \wedge \varphi_i = \sum_{1\le i<j\le p} (h_{ij} - h_{ji})\alpha^i \wedge \alpha^j + \sum_{i=1}^{p}\sum_{k=p+1}^{n} h_{ik}\alpha^i \wedge \alpha^k,$$

so $h_{ij} - h_{ji} = 0$ and $h_{ik} = 0$, for $i,j = 1,\dots,p$ and $k = p+1,\dots,n$, since the $\frac{p(p-1)}{2} + p(n-p)$ bivectors

$$\alpha^i \wedge \alpha^j, \quad i = 1,\dots,p, \quad j = 1,\dots,n, \quad i<j,$$

are linearly independent in $\Lambda_2 V$. □

The following exercise is also called Cartan's Lemma.

Exercise 8. Let $\omega^1,\dots,\omega^n$ be smooth 1-forms on a manifold N such that they are linearly independent at every point of N. Let $\theta_1,\dots,\theta_n$ be smooth 1-forms on N such that

$$\sum_{1}^{n} \omega^i \wedge \theta_i = 0,$$

at every point of N. Then $\theta_i = \sum_{j=1}^{n} h_{ij}\omega^j$, for each $i = 1,\dots,n$, for smooth functions h_{ij} on N satisfying $h_{ij} = h_{ji}$.

By Cartan's Lemma, a first order frame field satisfies

$$\omega_i^3 = \sum_{j=1}^{2} h_{ij}\omega^j, \text{ for } i = 1,2, \text{ and } h_{12} = h_{21}. \tag{4.10}$$

Comparing the situation with Gauss's formalism, we see that for our first order frame field $(\mathbf{x}, e)$,

$$I = d\mathbf{x}\cdot d\mathbf{x} = \omega^1\omega^1 + \omega^2\omega^2 \tag{4.11}$$

is the first fundamental form and

$$II = -d\mathbf{x}\cdot d\mathbf{e}_3 = \omega^1\omega_1^3 + \omega^2\omega_2^3 = h_{11}\omega^1\omega^1 + 2h_{12}\omega^1\omega^2 + h_{22}\omega^2\omega^2 \tag{4.12}$$

is the second fundamental form. The first fundamental form is a Riemannian metric I on M. It is called the Riemannian metric induced on M from the dot product by the immersion $\mathbf{x} : M \to \mathbf{R}^3$. If $(\mathbf{x}, e)$ is a first order frame field on an open set $U \subset M$, then (4.11) shows that ω^1, ω^2 is an orthonormal coframe field on U. From the structure equations (4.7) and (4.8)

$$d\omega^1 = -\omega_2^1 \wedge \omega^2, \qquad d\omega^2 = -\omega_1^2 \wedge \omega^1,$$

from which it follows that the Levi-Civita connection form relative to this orthonormal coframe field is

$$\omega_2^1 = -\omega_1^2.$$

Taking the exterior derivative of this, using the definition of the Gaussian curvature of I, and using the structure equations (4.7) we arrive at the *Gauss equation* for the Gaussian curvature K,

$$K\omega^1\wedge\omega^2 = d\omega_2^1 = \omega_1^3\wedge\omega_2^3 = (h_{11}h_{22} - h_{12}^2)\omega^1\wedge\omega^2, \tag{4.13}$$

where the last expression on the right is essentially Gauss's definition of curvature in terms of the second fundamental form. This proves Gauss's Theorema Egregium [72]: the Gaussian curvature depends only on the first fundamental form. The Gauss equation can be expressed as

$$K = \det S,$$

where the 2×2 symmetric matrix S is defined by

$$S = \begin{pmatrix} h_{11} & h_{12} \\ h_{21} & h_{22} \end{pmatrix}. \tag{4.14}$$

This is the matrix of the *shape operator* (also called the *Weingarten map*)

$$-d\mathbf{e}_3 : T_\mathbf{x}M \to T_\mathbf{x}M$$

relative to the orthonormal basis $\mathbf{e}_1, \mathbf{e}_2$, as can be seen from (4.12). Making the identification $d\mathbf{x}(T_mM) = T_{\mathbf{e}_3(m)}\mathbf{S}^2$, we see that the shape operator is the differential of the *Gauss map* $-\mathbf{e}_3 : T_\mathbf{x}M \to S^2$. Gauss defined curvature to be the Jacobian of this map. The *mean curvature* is half the trace of the shape operator

$$H = \frac{1}{2}\text{trace } S = \frac{1}{2}(h_{11} + h_{22}).$$

The *principal curvatures* are the principal values of S, thus the roots a and c of the quadratic equation in t,

$$\det(S - tI_2) = 0.$$

Definition 4.2. The *Hopf invariant* h relative to a first order frame field $(\mathbf{x}, e)$ on $U \subset M$ is the smooth function

$$h : U \to \mathbf{C}, \quad h = \frac{1}{2}(h_{11} - h_{22}) - ih_{12}.$$

For any first order frame field $(\mathbf{x}, e)$ we have

$$\omega_1^3 - i\omega_2^3 = h(\omega^1 + i\omega^2) + H(\omega^1 - i\omega^2),$$

where h is the Hopf invariant relative to $(\mathbf{x}, e)$ and H is the mean curvature.

Here are some of the basic properties of the Hopf invariant.

Exercise 9. Let $\mathscr{S}$ be the real vector space of all 2×2 symmetric matrices and let L be the real linear transformation

$$L : \mathscr{S} \to \mathbf{C}, \quad L(S) = \frac{1}{2}(S_{11} - S_{22}) - iS_{12}. \tag{4.15}$$

The Hopf invariant is then $h = L(S)$, where S is given by (4.14). Prove the following:

1. The kernel of L is the set of all scalar matrices, that is, scalar multiples of the identity matrix.
2. $L(S)$ is real if and only if S is a diagonal matrix.
3. If

$$A = \begin{pmatrix} \cos t & -\sin t \\ \sin t & \cos t \end{pmatrix} \tag{4.16}$$

is rotation through the angle t, then $L({}^tASA) = e^{i2t}L(S)$.

4. If

$$B = A\begin{pmatrix} 1 & 0 \\ 0 & -1 \end{pmatrix} = \begin{pmatrix} \cos t & \sin t \\ \sin t & -\cos t \end{pmatrix}, \tag{4.17}$$

is a rotation through an angle t composed with reflection through the horizontal axis, then $L({}^tBSB) = e^{-i2t}\overline{L(S)}$.

Since S of (4.14) is symmetric, its principal values are real and are called the *principal curvatures* of $\mathbf{x}$ at the point. The principal vectors of S are called the *principal directions* of $\mathbf{x}$. These are orthogonal whenever the principal curvatures are distinct. A point of M is called an *umbilic* of $\mathbf{x}$ if the principal curvatures are equal at this point. At an umbilic, every direction is principal.

Definition 4.3. A smooth curve $\gamma : J \to M$, where $J \subset \mathbf{R}$ is an interval, is a *line of curvature* of $\mathbf{x}$, if its tangent vector $\dot{\gamma}$ is a principal direction at each point of J.

Lemma 4.4 (Lines of Curvature). *If $(\mathbf{x}, e)$ is a first order frame field along $\mathbf{x}$ on an open set $U \subset M$, then a smooth curve $\gamma : J \to U$ is a line of curvature if and only if at every point of J*

$$\gamma^*(\omega^1\omega_2^3 - \omega^2\omega_1^3) = 0.$$

Proof. For the first order frame field $(\mathbf{x}, e)$ on U we have $\omega_i^3 = \sum_{j=1}^2 h_{ij}\omega^j$. Then $\dot{\gamma} = \omega^1(\dot{\gamma})\mathbf{e}_1 + \omega^2(\dot{\gamma})\mathbf{e}_2$ and $S\mathbf{e}_i = \sum_{j=1}^2 h_{ji}\mathbf{e}_j$, so $\dot{\gamma}$ is a principal vector if and only if $S\dot{\gamma} = \lambda\dot{\gamma}$, for some $\lambda \in \mathbf{R}$, that is,

$$\sum_{j=1}^{2}\omega_j^3(\dot{\gamma})\mathbf{e}_j = \sum_{i,j=1}^{2} h_{ji}\omega^i(\dot{\gamma})\mathbf{e}_j = \sum_{i=1}^{2}\omega^i(\dot{\gamma})S\mathbf{e}_i = S\dot{\gamma} = \lambda\dot{\gamma} = \lambda\sum_{j=1}^{2}\omega^j(\dot{\gamma})\mathbf{e}_j,$$

if and only if $\omega_j^3(\dot{\gamma}) = \lambda\omega^j(\dot{\gamma})$, for $j = 1, 2$, if and only if

$$\det\begin{pmatrix} \omega^1(\dot{\gamma}) & \omega^2(\dot{\gamma}) \\ \omega_1^3(\dot{\gamma}) & \omega_2^3(\dot{\gamma}) \end{pmatrix} = 0,$$

which is equivalent to the statement of the lemma. □

The two remaining structure equations are

$$d\omega_1^3 = -\omega_2^1 \wedge \omega_2^3, \qquad d\omega_2^3 = \omega_2^1 \wedge \omega_1^3. \tag{4.18}$$

If we take the exterior differential of the equations $\omega_i^3 = \sum_{j=1}^2 h_{ij}\omega^j$, and again apply Cartan's Lemma, we obtain the *Codazzi equations*

$$dh_{ij} - \sum_{k=1}^{2} h_{ik}\omega_j^k - \sum_{k=1}^{2} h_{kj}\omega_i^k = \sum_{k=1}^{2} h_{ijk}\omega^k, \tag{4.19}$$

where h_{ijk} are smooth functions on U, totally symmetric in all three indices, $1 \le i, j, k \le 2$.

Given a first order frame field $(\mathbf{x}, e)$ on a connected subset $U \subset M$, any other is given by

$$\tilde{\mathbf{e}}_3 = \epsilon\mathbf{e}_3, \quad \tilde{\mathbf{e}}_1 = \mathbf{e}_1 A_1^1 + \mathbf{e}_2 A_1^2, \quad \tilde{\mathbf{e}}_2 = \mathbf{e}_1 A_2^1 + \mathbf{e}_2 A_2^2, \tag{4.20}$$

where $\epsilon = \pm 1$ and $A = (A_j^i) : U \to \mathbf{O}(2)$ is a smooth map. In matrix notation we have

$$(\tilde{\mathbf{e}}_1, \tilde{\mathbf{e}}_2) = (\mathbf{e}_1, \mathbf{e}_2)A.$$

From $d\mathbf{x} = \tilde{\omega}^1\tilde{\mathbf{e}}_1 + \tilde{\omega}^2\tilde{\mathbf{e}}_2 = \omega^1\mathbf{e}_1 + \omega^2\mathbf{e}_2$ we see that

$$\omega^1 = A^1_1\tilde{\omega}^1 + A^1_2\tilde{\omega}^2, \quad \omega^2 = A^2_1\tilde{\omega}^1 + A^2_2\tilde{\omega}^2,$$

which in matrix notation is

$$\begin{pmatrix}\omega^1\\ \omega^2\end{pmatrix} = A\begin{pmatrix}\tilde{\omega}^1\\ \tilde{\omega}^2\end{pmatrix}.$$

Under this change of frame, the *area form* changes by

$$\omega^1\wedge\omega^2 = \det(A)\,\tilde{\omega}^1\wedge\tilde{\omega}^2.$$

Since $A \in \mathbf{O}(2)$, we know that $\det(A) = \pm 1$, so the orientation of M is preserved if and only if $A \in \mathbf{SO}(2)$. In the new frame, (4.10) becomes

$$\tilde{\omega}^3_i = \sum_{j=1}^{2}\tilde{h}_{ij}\tilde{\omega}^j, \tag{4.21}$$

for $i = 1,2$. From $\tilde{\omega}^3_i = d\tilde{\mathbf{e}}_i\cdot\tilde{\mathbf{e}}_3 = -\tilde{\mathbf{e}}_i\cdot d\tilde{\mathbf{e}}_3$, we have

$$\tilde{\omega}^3_i = \epsilon\sum_{j=1}^{2}A^j_i\omega^3_j. \tag{4.22}$$

By (4.10), (4.21) and (4.22) it follows that the 2×2 symmetric matrices $\tilde{S} = (\tilde{h}_{ij})$ and $S = (h_{ij})$ transform by

$$\tilde{S} = \epsilon\,{}^t\!ASA. \tag{4.23}$$

From linear algebra we know that at a point in U, we can diagonalize S by this action. The resulting diagonal entries are the principal curvatures of $\mathbf{x}$ at the point. If the principal curvatures are equal at the point, then S and $\tilde{S}$ are scalar matrices and the point is *umbilic*. The principal values of $\tilde{S}$ are ϵ times the principal values of S. Therefore, replacing the unit normal $\mathbf{e}_3$ by $-\mathbf{e}_3$ reverses the sign of the principal curvatures and the mean curvature, but leaves the Gaussian curvature invariant. By Exercise 9, the Hopf invariants relative to each frame are related by

$$\tilde{h} = \epsilon e^{2it}h, \tag{4.24}$$

if A is rotation by an angle $t \in \mathbf{R}$ given in (4.16), and by

$$\tilde{h} = \epsilon e^{-2it}\bar{h},$$

if A is the matrix B in (4.17).

Definition 4.5. A *second order frame* at a point $p \in U$ is a first order frame $(\mathbf{x}, e)$ at p for which $\mathbf{e}_1(p)$ and $\mathbf{e}_2(p)$ are principal directions of $\mathbf{x}$ at p.

That is, $(\mathbf{x}, e)$ is of second order at p if the matrix S of its shape operator, defined in (4.14), is diagonalized at p:

$$S = \begin{pmatrix} a & 0 \\ 0 & c \end{pmatrix},$$

where $a, c \in \mathbf{R}$ are the principal curvatures of $\mathbf{x}$ at p. This is equivalent to the conditions at p:

$$\omega^3 = 0, \quad \omega_1^3 = a\omega^1, \quad \omega_2^3 = c\omega^2.$$

If $a = c$, then the point is *umbilic* and any first order frame is automatically of second order there, because every vector in T_pM is then a principal vector of the shape operator.

A change of frame (4.20) at p preserves the second order property of $(\mathbf{x}, e)$ at p if and only if the matrix $\tilde{S}$ in (4.23) is also diagonalized. At a nonumbilic, there are just a finite number of changes of second order frame. See Problem 4.60.

A second order frame field along $\mathbf{x} : M \to \mathbf{R}^3$ is a first order frame field on an open set $U \subset M$ that is of second order at every point of U. Figure 4.1 shows a second order frame at a nonumbilic point of an ellipsoid with distinct principal axes.

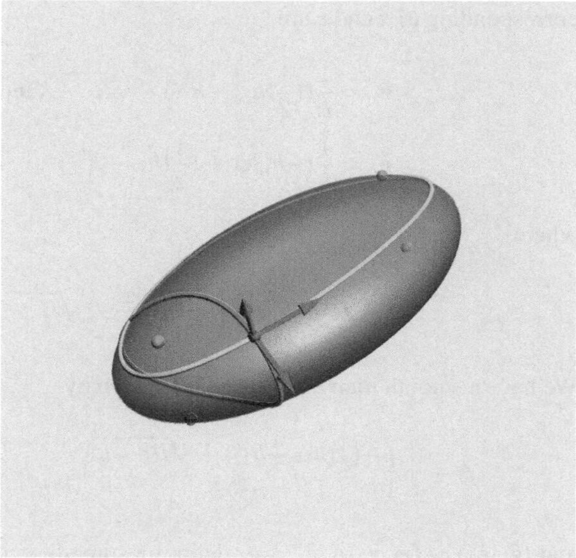

Fig. 4.1 A second order frame on a generic ellipsoid. The blue points are umbilics.

Lemma 4.6 (Existence of Second Order Frame Fields). *For the immersed surface* $\mathbf{x} : M \to \mathbf{R}^3$ *suppose that the point* $p \in M$ *is nonumbilic. Then there exists an open neighborhood* U *about* p *on which there is a smooth second order frame field.*

Proof. Let $(\mathbf{x}, e)$ be a smooth first order frame field on an open neighborhood V of p and let S be the coefficient matrix of the second fundamental form with respect to this frame field. The entries of S are smooth functions, and thus the mean curvature $H = \frac{1}{2}$ trace S and the Gaussian curvature $K = \det S$ are smooth functions on V. The principal curvatures are the solutions of the quadratic equation in t

$$0 = \det(S - tI) = t^2 - 2Ht + K$$

whose solutions a and c are

$$H + \sqrt{H^2 - K}, \qquad H - \sqrt{H^2 - K}. \tag{4.25}$$

The smooth function $H^2 - K$ on M is nonnegative and the umbilic points are characterized by the equation

$$H^2 - K = 0.$$

In particular, the set of umbilic points is closed in V and the set of nonumbilic points is open. The functions a and c are continuous on V, and smooth on the open set of all nonumbilic points. Let $U \subset V$ be an open neighborhood of p consisting only of nonumbilic points. It is standard linear algebra to verify that unit principal vectors corresponding to a and c are

$$\tilde{\mathbf{e}}_1 = \frac{1}{L}((\frac{1}{2}(h_{22} - h_{11}) - \sqrt{H^2 - K})\mathbf{e}_1 - h_{21}\mathbf{e}_2),$$
$$\tilde{\mathbf{e}}_2 = \frac{1}{L}(-h_{12}\mathbf{e}_1 + (\frac{1}{2}(h_{11} - h_{22}) + \sqrt{H^2 - K})\mathbf{e}_2),$$

where

$$L = (h_{12}^2 + \left(\frac{h_{11} - h_{22}}{2} + \sqrt{H^2 - K}\right)^2)^{\frac{1}{2}}.$$

We have a smooth map $A : U \to \mathbf{SO}(2)$ given by

$$A = \frac{1}{L}\begin{pmatrix} -(\frac{1}{2}(h_{11} - h_{22}) + \sqrt{H^2 - K}) & -h_{12} \\ -h_{12} & \frac{1}{2}(h_{11} - h_{22}) + \sqrt{H^2 - K} \end{pmatrix}$$

and $(\tilde{\mathbf{e}}_1, \tilde{\mathbf{e}}_2) = (\mathbf{e}_1, \mathbf{e}_2)A$, $\tilde{\mathbf{e}}_3 = \mathbf{e}_3$ defines a smooth second order frame field along $\mathbf{x}$ on U. □

4.3 Moving frame reductions

In this section we apply the method of moving frames, as outlined in Chapter 3, to surfaces immersed in Euclidean space. We believe it is instructive to see how this method arrives at the same invariants as we found in the preceding section.

Let $\mathbf{x} : M \to \mathbf{R}^3$ be an immersion of a surface M. A *frame field along* $\mathbf{x}$ is a smooth map $(\mathbf{x}, e) : U \to \mathbf{E}(3)$ from an open subset $U \subset M$ such that $\pi \circ (\mathbf{x}, e) = \mathbf{x}$, where π is the projection (4.3). In brief, the diagram

$$\begin{array}{ccc} & & \mathbf{E}(3) \\ & (\mathbf{x}, e) \nearrow & \downarrow \pi \\ & U \xrightarrow{\mathbf{x}} & \mathbf{R}^3 \end{array}$$

commutes. Given a point $m \in M$ and a point $(\mathbf{v}, A)$ in the fiber $\pi^{-1}\{\mathbf{x}(m)\}$, the Lift Property of Corollary 2.7 guarantees the existence of a neighborhood $U \subset M$ of m on which there is a frame field along $\mathbf{x}$ that passes through $(\mathbf{v}, A)$. Let

$$(\mathbf{x}, e) : U \subset M \to \mathbf{E}(3) \tag{4.26}$$

be a smooth frame field along $\mathbf{x}$. The pull-back of the Maurer–Cartan form of $\mathbf{E}(3)$ by this frame field is

$$(\mathbf{x}, e)^{-1} d(\mathbf{x}, e) = (e^{-1} d\mathbf{x}, e^{-1} de) = ((\omega^i), (\omega^i_j)), \quad i, j = 1, 2, 3,$$

now an $\mathscr{E}(3)$-valued 1-form on $U \subset M$. Here, and throughout the rest of this book, we omit $(\mathbf{x}, e)^*$ when writing the pull-back of forms to M. The same symbol is used for the form on $\mathbf{E}(3)$ and for its pull-back to M. The context will indicate the correct interpretation. We now carry out a reduction of the frames following the general procedure outlined in Section 3.3. We have $\mathfrak{g} = \mathscr{E}(3)$ and $\mathfrak{g}_0 = \mathfrak{o}(3)$. As a vector subspace complement of $\mathfrak{g}$ we choose $\mathfrak{m}_0 = \mathbf{R}^3$. The vector space direct sum $\mathfrak{g} = \mathfrak{m}_0 + \mathfrak{g}_0$ decomposes the Maurer–Cartan form of $\mathbf{E}(3)$ into $\theta + \omega$, where

$$\theta = \sum_1^3 \omega^i \boldsymbol{\epsilon}_i$$

denotes the $\mathfrak{m}_0$ component and

$$\omega = (\omega^i_j), \quad i, j = 1, 2, 3,$$

denotes the $\mathfrak{g}_0 = \mathfrak{o}(3)$ component. As we did in (4.4), we choose $E_i = \boldsymbol{\epsilon}_i$, for $i = 1, 2, 3$, for a basis of $\mathbf{m}_0$. The adjoint representation of $G_0 = \mathbf{O}(3)$ on $\mathfrak{g}/\mathfrak{g}_0 \cong \mathbf{R}^3$ relative to E_1, E_2, E_3 is the standard representation of $\mathbf{O}(3) \subset \mathbf{GL}(3, \mathbf{R})$, as we observed in Section 4.1 above. Then

$$d\mathbf{x} = \sum_{i=1}^{3} \omega^i \mathbf{e}_i(m)$$

on U, where $\mathbf{e}_1, \mathbf{e}_2, \mathbf{e}_3$ are the columns of $e \in \mathbf{O}(3)$. We saw in the preceding section that the frame can be chosen so that $\mathbf{e}_1$ and $\mathbf{e}_2$ span the tangent plane $d\mathbf{x}_m T_m M$ at each point $m \in U$. Let us see how that is accomplished through the general frame reduction procedure. If φ^1, φ^2 is a coframe field on U, then

$$\omega^i = \sum_{a=1}^{2} X^i_a \varphi^a, \quad i = 1,2,3,$$

where $X = (X^i_a) : U \to \mathbf{R}^{3\times 2*}$ is a 3×2 matrix whose rank is two at each point, since the linear map

$$\omega_{\mathfrak{m}_0} = \sum_{i=1}^{3}\sum_{a=1}^{2} X^i_a \epsilon_i \varphi^a : T_p M \to \mathfrak{m}_0$$

has rank two at every point $p \in U$, since $\mathbf{x}$ is an immersion. Its image at a point of U is a 2-dimensional subspace of $\mathfrak{m}_0 \cong \mathbf{R}^3$. It is the image of the map

$$[X] : U \to G(2,3), \tag{4.27}$$

where $G(2,3) = \mathbf{R}^{3\times 2*}/\mathbf{GL}(2,\mathbf{R})$ is the Grassmannian of 2-dimensional subspaces of $\mathbf{R}^3$. Any other frame field along $\mathbf{x}$ on U is given by

$$(\mathbf{x}, \tilde{e}) = (\mathbf{x}, e)(\mathbf{0}, A) = (\mathbf{x}, eA),$$

where $A : U \to \mathbf{O}(3)$ is any smooth map. The pull-back of the Maurer–Cartan form by this new frame field is

$$\begin{pmatrix} 0 & 0 \\ \tilde{\theta} & \tilde{\omega} \end{pmatrix} = \begin{pmatrix} 1 & 0 \\ \mathbf{x} & \tilde{e} \end{pmatrix}^{-1} d\begin{pmatrix} 1 & 0 \\ \mathbf{x} & \tilde{e} \end{pmatrix} = \begin{pmatrix} 1 & 0 \\ \mathbf{x} & eA \end{pmatrix}^{-1} d\begin{pmatrix} 1 & 0 \\ \mathbf{x} & eA \end{pmatrix},$$

so $\tilde{\theta} = A^{-1}\theta$ and $\tilde{\omega} = A^{-1}\omega A + A^{-1} dA$. Then

$$\sum_{a=1}^{2} X^i_a \varphi^a = \omega^i = \sum_{j=1}^{3} A^i_j \tilde{\omega}^j = \sum_{a=1}^{2}\sum_{j=1}^{3} A^i_j \tilde{X}^j_a \varphi^a \tag{4.28}$$

shows that $A\tilde{X} = X$. The action

$$\mathbf{O}(3) \times G(2,3) \to G(2,3), \quad (A, [Y]) \mapsto [AY]$$

of $\mathbf{O}(3)$ on the Grassmannian $G(2,3)$ is transitive. Choose $P_0 = \begin{bmatrix} I_2 \\ 0 \end{bmatrix}$ to be the origin of $G(2,3)$. The isotropy subgroup of $\mathbf{O}(3)$ at P_0 is

$$G_1 = \mathbf{O}(2) \times \mathbf{O}(1) = \{A = \begin{pmatrix} a & 0 \\ 0 & \epsilon \end{pmatrix} \in \mathbf{O}(3) : a \in \mathbf{O}(2),\ \epsilon = \pm 1\}, \tag{4.29}$$

whose Lie algebra is

$$\mathfrak{g}_1 = \mathfrak{o}(2) = \{\begin{pmatrix} Z & 0 \\ 0 & 0 \end{pmatrix} \in \mathfrak{o}(3) : Z \in \mathfrak{o}(2)\}. \tag{4.30}$$

Then $\mathbf{O}(3)$ is a principal G_1-bundle over $G(2,3)$ with projection map

$$\pi : \mathbf{O}(3) \to G(2,3), \quad \pi(A) = AP_0,$$

which has local sections.

Definition 4.7. A *first order frame field* along $\mathbf{x} : M \to \mathbf{R}^3$ is a frame field $(\mathbf{x}, e) : U \to \mathbf{E}(3)$ on an open subset U of M for which $[X] = P_0$. This is equivalent to the conditions

$$\omega^3 = 0, \quad \omega^1 \wedge \omega^2 \neq 0,$$

at every point of U.

Proposition 4.8. *Given any point $m \in M$, there exists a neighborhood U of m in M on which there is a first order frame field along* $\mathbf{x}$.

Proof. There exists a frame field (4.26) on a neighborhood U of m. Apply the Lift Property of Corollary 2.7 to the smooth map $[X] : U \to G(2,3)$ in (4.27), to get a neighborhood V of m in U on which there is a smooth map $A : V \to \mathbf{O}(3)$ such that

$$AP_0 = [X],$$

on V. If $(\mathbf{x}, \tilde{e}) = (\mathbf{x}, eA)$, then $[\tilde{X}] = [A^{-1}X] = P_0$ by (4.28), so $(\mathbf{x}, \tilde{e}) : V \to \mathbf{E}(3)$ is a first order frame field along $\mathbf{x}$. □

A first order frame field $(\mathbf{x}, e) : U \to \mathbf{E}(3)$ has $\omega^3 = 0$ and ω^1, ω^2 a coframe field on U. This coframe field is orthonormal for the Riemannian metric $I = d\mathbf{x} \cdot d\mathbf{x}$ induced on M by $\mathbf{x}$. Let

$$\mathfrak{m}_1 = \{\begin{pmatrix} 0 & 0 & x^1_3 \\ 0 & 0 & x^2_3 \\ x^3_1 & x^3_2 & 0 \end{pmatrix} \in \mathfrak{o}(3)\}.$$

be a vector space complement to $\mathfrak{g}_1$ (defined in (4.30)) in $\mathfrak{g}_0 = \mathfrak{o}(3)$. For a basis of $\mathfrak{m}_1$ we choose

$$E_4 = \begin{pmatrix} 0 & 0 & -1 \\ 0 & 0 & 0 \\ 1 & 0 & 0 \end{pmatrix}, \quad E_5 = \begin{pmatrix} 0 & 0 & 0 \\ 0 & 0 & -1 \\ 0 & 1 & 0 \end{pmatrix}.$$

The direct sum of vector spaces

$$\mathscr{E}(3) = \mathfrak{g} = \mathfrak{m}_0 + \mathfrak{m}_1 + \mathfrak{g}_1$$

decomposes the Maurer–Cartan form of $\mathbf{E}(3)$, to $\theta + \omega_{\mathfrak{m}_1} + \omega_{\mathfrak{g}_1}$, where

$$\omega_{\mathfrak{m}_1} = \omega_1^3 E_4 + \omega_2^3 E_5.$$

Relative to a first order frame field $(\mathbf{x}, e) : U \to \mathbf{E}(3)$, the Maurer–Cartan form pulled back to U satisfies $\omega^3 = 0$ and ω^1, ω^2 is a coframe on U, and

$$\omega_i^3 = \sum_{j=1}^{2} h_{ij}\omega^j, \quad i = 1, 2.$$

We now consider the rank two linear map

$$\theta + \omega_{\mathfrak{m}_1} = (\boldsymbol{\epsilon}_1 + \sum_1^2 h_{i1}E_{3+i})\theta^1 + (\boldsymbol{\epsilon}_2 + \sum_1^2 h_{i2}E_{3+i})\theta^2 : T_pM \to \mathfrak{m}_0 + \mathfrak{m}_1. \tag{4.31}$$

Its image is a 2-dimensional subspace of $\mathfrak{m}_0 + \mathfrak{m}_1 \cong \mathbf{R}^5$, represented as a point

$$\begin{bmatrix} I_2 \\ 0 \\ h \end{bmatrix} \in G(2, 5),$$

where $h = (h_{ij}) \in \mathbf{R}^{2\times 2}$. The exterior derivative of $\omega^3 = 0$ on U, combined with the structure equations of $\mathbf{E}(3)$, gives (4.9) $\sum_1^2 \omega_i^3 \wedge \omega^i = 0$, since $\omega^3 = 0$. By Cartan's Lemma 4.1, we saw (4.10) $\omega_i^3 = \sum_{j=1}^2 h_{ij}\omega^j$, for $i = 1, 2$, and the smooth functions $h_{ij} : U \to \mathbf{R}$ satisfy $h_{ij} = h_{ji}$. Let

$$S = (h_{ij}) : U \to \mathscr{S}, \tag{4.32}$$

a smooth map into the vector space $\mathscr{S}$ of all symmetric 2×2 matrices. It follows that the map (4.31) takes values only in the subspace

$$\begin{bmatrix} I_2 \\ 0 \\ \mathscr{S} \end{bmatrix} \subset G(2, 5).$$

Any other first order frame field on U is given by $(\mathbf{x},\tilde{e}) = (\mathbf{x},e)(\mathbf{0},A)$, where $A : U \to G_1 = \mathbf{O}(2) \times \mathbf{O}(1)$ is smooth of the form

$$A = \begin{pmatrix} B & 0 \\ 0 & \epsilon \end{pmatrix},$$

where $B : U \to \mathbf{O}(2)$ is smooth and $\epsilon = \pm 1$ is locally constant. They pull back the Maurer–Cartan form of $\mathbf{E}(3)$ to $(\tilde{\theta},\tilde{\omega}) = (A^{-1}\theta, A^{-1}\omega A + A^{-1}dA) =$

$$\left(\begin{pmatrix} B^{-1}\begin{pmatrix}\omega^1 \\ \omega^2\end{pmatrix} \\ 0 \end{pmatrix}, \begin{pmatrix} B^{-1}\begin{pmatrix} 0 & \omega^1_2 \\ \omega^2_1 & 0 \end{pmatrix} B + B^{-1}dB & \epsilon B^{-1}\begin{pmatrix}\omega^1_3 \\ \omega^2_3\end{pmatrix} \\ \epsilon(\omega^3_1,\omega^3_2)B & 0 \end{pmatrix} \right),$$

from which we conclude that

$$\begin{pmatrix}\tilde{\omega}^1 \\ \tilde{\omega}^2\end{pmatrix} = B^{-1}\begin{pmatrix}\omega^1 \\ \omega^2\end{pmatrix}, \quad (\tilde{\omega}^3_1,\tilde{\omega}^3_2) = \epsilon(\omega^3_1,\omega^3_2)B,$$

and

$$\tilde{\omega}^2_1 = (\det B)\omega^2_1 + \tau^2_1, \tag{4.33}$$

where $B^{-1}dB = \begin{pmatrix} 0 & -\tau^2_1 \\ \tau^2_1 & 0 \end{pmatrix}$ and τ^2_1 is a closed, smooth 1-form on U. This shows that the adjoint representation of G_1 on $\mathscr{E}(3)/\mathfrak{g}_1$ relative to the basis $E_1,\dots,E_5$ is

$$\mathrm{Ad}(A) = \begin{pmatrix} B & 0 & 0 \\ 0 & \epsilon & 0 \\ 0 & 0 & \epsilon B \end{pmatrix}.$$

The image of the map (4.31) thus transforms by

$$\begin{bmatrix} I_2 \\ 0 \\ \tilde{S} \end{bmatrix} = \mathrm{Ad}(A^{-1}) \begin{bmatrix} I_2 \\ 0 \\ S \end{bmatrix},$$

which implies

$$\tilde{S} = \epsilon\, {}^t\!BSB. \tag{4.34}$$

This is the action $(B,\epsilon)S = \epsilon BS\, {}^t\!B$ of $\mathbf{O}(2) \times \mathbf{O}(1)$ on $\mathscr{S}$ analyzed in Example 2.17. From there we know that a slice of this action is the set $\mathscr{Y}$ of all nonscalar diagonal matrices in $\mathscr{S}$ together with the closed subgroup

$$G_2 = K \times \mathbf{O}(1) \subset G_1, \tag{4.35}$$

where K is the finite subgroup of $\mathbf{O}(2)$ defined in (2.6). Following Definition 3.7 we would define second order frame fields along $\mathbf{x}$ as follows.

Definition 4.9. A *second order frame field* along $\mathbf{x}$ is a first order frame field $(\mathbf{x}, e) : U \to \mathbf{E}(3)$ for which the map $S : U \to \mathscr{S}$ of (4.32) takes all values in $\mathscr{Y}$. The *second order invariants of* $\mathbf{x}$ *on* U are the nonzero entries of this map.

In simpler terms, a *second order frame field* $(\mathbf{x}, e) : U \to \mathbf{E}(3)$ is characterized by

$$\omega^3 = 0, \quad \omega^1 \wedge \omega^2 \neq 0, \quad \text{(first order)},$$

$$\omega_1^3 = a\omega^1, \quad \omega_2^3 = c\omega^2, \quad \text{(second order)},$$

for smooth functions $a, c : U \to \mathbf{R}$ for which $a \neq c$ at every point of U. These functions are the second order invariants and are called the *principal curvatures* of $\mathbf{x}$ at each point of U. It is traditional to call the frame field second order even when $a = c$ at some points of U. These are the umbilic points of $\mathbf{x}$ in U. If all points of U are umbilic, then $\mathbf{x}$ is of a different type. This case is discussed below.

Lemma 4.10. *Let m_0 be a point in M. If m_0 is nonumbilic, or if there is an open set of umbilic points containing m_0, then there exists a smooth second order frame field on some neighborhood of m_0.*

Proof. There exists a smooth first order frame field $(\mathbf{x}, e) : U \to \mathbf{E}(3)$ on some neighborhood U of m_0. If there is an open neighborhood V of m_0 in U consisting entirely of umbilic points, then this frame field is of second order on V.

The smooth function $H^2 - K : M \to \mathbf{R}$ is zero precisely at the umbilic points of $\mathbf{x}$, so the set of umbilic points is closed in M. If m_0 is nonumbilic, then we may shrink U, if necessary, so that U contains only nonumbilic points. Consider the map (4.32) associated to the frame field $(\mathbf{x}, e)$. Let $a_0 \neq c_0$ be the principal curvatures at m_0. Using the notation of Example 2.17 for the slice $\mathscr{Y}$ of the action of G_1 on $\mathscr{S}$, we know that $S(U) \subset G_1\mathscr{Y}$. Apply the Factor Property of Theorem 2.18 to get an open neighborhood V of m_0 in U and smooth maps

$$A = (B, \epsilon) : V \to G_1 = \mathbf{O}(2) \times \mathbf{O}(1), \quad D = \begin{pmatrix} a & 0 \\ 0 & c \end{pmatrix} : V \to \mathscr{Y},$$

with $a(m_0) = a_0$ and $c(m_0) = c_0$, such that

$$S = (B, \epsilon)D = \epsilon BD{}^tB$$

on V. If $(\mathbf{x}, \tilde{e}) = (\mathbf{x}, eA) : V \to \mathbf{E}(3)$, then

$$\tilde{S} = \epsilon{}^tBSB = D,$$

on V, with $\tilde{S}(m_0) = D(m_0)$. Hence $(\mathbf{x}, \tilde{e})$ is a second order frame field. □

It is possible that a smooth, even continuous, second order frame field may not exist on any neighborhood of an umbilic point. Consider the following example, for which the origin in $M = \mathbf{R}^2$ is the only umbilic point.

Example 4.11 (Nonexistence). Consider the embedding of the circular paraboloid $z = x^2 + y^2$,

$$\mathbf{x} : \mathbf{R}^2 \to \mathbf{R}^3, \quad \mathbf{x}(x,y) = {}^t(x, y, x^2 + y^2).$$

In terms of polar coordinates $x = r\cos t$ and $y = r\sin t$ away from the origin, we can parametrize this surface as an immersion of revolution (see Example 4.40)

$$\mathbf{x}(r,t) = (r\cos t, r\sin t, r^2).$$

If $w = \sqrt{1+4r^2}$, then

$$\mathbf{e}_1 = \frac{1}{w}{}^t(\cos t, \sin t, 2r), \quad \mathbf{e}_2 = {}^t(-\sin t, \cos t, 0), \quad \mathbf{e}_3 = \mathbf{e}_1 \times \mathbf{e}_2$$

is a second order frame field for all $r > 0$ (see Example 4.40 below). Thus $\mathbf{e}_1$ is a principal vector field at every point of $\mathbf{R}^2 \setminus \{\mathbf{0}\}$ and for any fixed t,

$$\lim_{r\to 0} \mathbf{e}_1 = \lim_{r\to 0} \frac{1}{\sqrt{1+4r^2}}{}^t(\cos t, \sin t, 2r) = {}^t(\cos t, \sin t, 0),$$

which depends on t. Thus $\mathbf{e}_1$ cannot be extended continuously to the origin. Any second order frame field along $\mathbf{x}$ on a neighborhood of the origin must include $\pm\mathbf{e}_1$ away from the origin, so it cannot be extended continuously to the origin.

The obstruction to the existence of continuous second order frame fields is more subtle than just the existence of umbilic points. Example 4.41 in the next section has a second order frame field defined everywhere, even though there are whole curves of umbilic points.

Remark 4.12. If $(\mathbf{x}, e) : U \to \mathbf{E}(3)$ is a second order frame field on an umbilic free domain U, then any other second order frame field on U is given by $(\mathbf{x}, eA) : U \to \mathbf{E}(3)$, for any smooth map $A : U \to G_2$, where G_2 is the finite subgroup of G_1 defined in (4.35). Since $\mathfrak{g}_2 = 0$, the *Frenet frames* are the second order frame fields. The remaining Maurer–Cartan form of a Frenet frame is

$$\omega_1^2 = p\omega^1 + q\omega^2, \tag{4.36}$$

where the functions $p, q : U \to \mathbf{R}$ are the third order invariants.

4.3.1 Summary of frame reduction and structure equations

Let $\mathbf{x}: M \to \mathbf{R}^3$ be an immersed surface. At a nonumbilic point there exists a smooth second order frame field $e: U \to \mathbf{E}(3)$ on a neighborhood of the point. Its pull-back of the Maurer–Cartan form of $\mathbf{E}(3)$ satisfies:

$$\omega^3 = 0, \text{ (first order)}, \quad \omega^1 \wedge \omega^2 \neq 0$$
$$d\omega^1 = p\omega^1 \wedge \omega^2, \quad d\omega^2 = q\omega^1 \wedge \omega^2$$
$$\omega_1^3 = a\omega^1, \quad \omega_2^3 = c\omega^2, \quad \text{(second order)}$$
$$\omega_1^2 = p\omega^1 + q\omega^2, \quad \text{(third order).}$$

The structure equations of the immersion are the Codazzi and Gauss equations, which, because $\omega^1 \wedge \omega^2 \neq 0$ at each point, can be written as

$$\begin{aligned} a_2 &= (a-c)p, \quad c_1 = (a-c)q, \text{ (Codazzi equations)}, \\ p_2 - q_1 - p^2 - q^2 &= K = ac, \text{ (Gauss equation)}, \end{aligned} \tag{4.37}$$

where $da = \sum_1^2 a_i\omega^i$, $dc = \sum_1^2 c_i\omega^i$, $dp = \sum_1^2 p_i\omega^i$, and $dq = \sum_1^2 q_i\omega^i$, and K is the Gaussian curvature of the metric induced on M. The functions a and c are the *principal curvatures* of $\mathbf{x}$. They are continuous functions on M, smooth on an open neighborhood of any nonumbilic point.

If $\mathbf{x}$ is totally umbilic on U, then any first order frame is automatically second order and the above equations with $a = c$ give the structure equations for such a frame.

Exercise 10. Prove that for any first order frame field $(\mathbf{x}, e): U \to M$ the structure equations of $\mathbf{E}(3)$ imply

$$d\omega^1 = p\omega^1 \wedge \omega^2, \quad d\omega^2 = q\omega^1 \wedge \omega^2, \tag{4.38}$$

where $p, q: U \to \mathbf{R}$ are smooth functions satisfying

$$\omega_1^2 = p\omega^1 + q\omega^2. \tag{4.39}$$

If $(\mathbf{x}, e): U \to M$ is of second order, prove equations (4.37).

4.3.2 The criterion form

It is useful here to make use of the Hodge star operator, which we shall define only for a very specialized situation. For a fuller treatment of this operator, see [110, p 385].

Definition 4.13. The *Hodge star operator* $* : A^1(M) \to A^1(M)$ is a linear operator on the space $A^1(M)$ of smooth 1-forms on an oriented Riemannian surface (M, I) given relative to a positively oriented orthonormal coframe ω^1, ω^2 by

$$*\omega^1 = \omega^2, \quad *\omega^2 = -\omega^1.$$

Exercise 11. Prove that the Hodge star operator does not depend on the choice of positively oriented orthonormal coframe.

Definition 4.14. The *criterion form* of a first order frame field $(\mathbf{x}, e) : U \to \mathbf{E}(3)$ is the smooth 1-form

$$\alpha = - * \omega_1^2,$$

where U is oriented by the orthonormal coframe field ω^1, ω^2 of $(\mathbf{x}, e)$. Thus, if $\omega_1^2 = p\omega^1 + q\omega^2$ as in (4.39), then $\alpha = q\omega^1 - p\omega^2$.

Remark 4.15. From (4.38) and the structure equations, the criterion form of a first order frame field $(\mathbf{x}, e) : U \to \mathbf{E}(3)$ is characterized by the equations

$$d\omega^1 = \alpha \wedge \omega^1, \quad d\omega^2 = \alpha \wedge \omega^2.$$

Exercise 12. Prove that if α is the criterion form of a first order frame $(\mathbf{x}, e) : U \to \mathbf{E}(3)$, then the criterion form $\tilde{\alpha}$ of any other first order frame $(\mathbf{x}, \tilde{e}) = (\mathbf{x}, eA)$, where $A = \begin{pmatrix} B & 0 \\ 0 & \epsilon \end{pmatrix}$, $B : U \to \mathbf{O}(2)$, and $\epsilon = \pm 1$, satisfies

$$\tilde{\alpha} = \alpha + (\det B) * \tau_1^2,$$

where $*$ is the Hodge star operator defined by the orientation induced by the coframe field of $(\mathbf{x}, e)$, and $B^{-1}dB = \begin{pmatrix} 0 & -\tau_1^2 \\ \tau_1^2 & 0 \end{pmatrix}$ defines the 1-form τ_1^2, as in (4.33). In particular, $\tilde{\alpha} = \alpha$ if $B : U \to \mathbf{O}(2)$ is locally constant. Note: $\tilde{\alpha} = \tilde{*}\tilde{\omega}_1^2$, where $\tilde{*}$ is the Hodge star operator defined by the orientation of the orthonormal coframe field of $(\mathbf{x}, \tilde{e})$.

The criterion form of a second order frame field is independent of the choice of second order frame field. See Problem 4.62.

4.4 Bonnet's existence and congruence theorems

We reformulate Proposition 3.8 as follows for Euclidean geometry.

Proposition 4.16 (Congruence). *If* $(\mathbf{x}, e), (\hat{\mathbf{x}}, \hat{e}) : M \to \mathbf{E}(3)$ *are second order frame fields along immersions* $\mathbf{x}, \hat{\mathbf{x}} : M \to \mathbf{R}^3$, *respectively, on connected* M *such that* *at every point of M, $\hat{\omega}^1 = \omega^1$, $\hat{\omega}^2 = \omega^2$, $\hat{a} = a$, $\hat{c} = c$, $\hat{p} = p$, and $\hat{q} = q$, then there exists an isometry $(\mathbf{v}, A) \in \mathbf{E}(3)$ such that $(\hat{\mathbf{x}}, \hat{e}) = (\mathbf{v}, A)(\mathbf{x}, e)$, so $\hat{\mathbf{x}} = (\mathbf{v}, A) \circ \mathbf{x} = \mathbf{v} + A\mathbf{x}$ on M.*

Note that the isometry is determined explicitly by the Frenet frames evaluated at any point of M, since $(\mathbf{v}, A) = (\hat{\mathbf{x}}, \hat{e})(\mathbf{x}, e)^{-1}$ must be constant. In the light of (4.38), this proposition remains true without the hypotheses $\hat{p} = p$ and $\hat{q} = q$. Congruence follows from equal Frenet coframe fields and equal principal curvatures. From Problem 4.62 we see that the hypotheses can be relaxed to requiring the coframe fields be related as in (15.58) or (15.59). Ideally, one wants hypotheses that are global in that they would not require the existence of a Frenet frame field on all of M. For example, if M is assumed oriented, then we can specify the normal vector $\mathbf{e}_3$ in any frame and thus the principal curvatures are functions on M. Would the proposition remain true if we assumed equal principal curvatures and equal first fundamental forms, $I = \omega^1\omega^1 + \omega^2\omega^2$? This question, known as the Bonnet Problem, remains unresolved. It is the subject of Chapter 10. In 1867 Bonnet formulated congruence and existence theorems in terms of the first and second fundamental forms. These are Theorems 4.18 and 4.19 below.

Proposition 4.17 (Existence). *Given a coframe field ω^1, ω^2 and smooth functions a and c on a contractible domain $U \subset \mathbf{R}^2$, define smooth functions p and q on U by*

$$d\omega^1 = p\omega^1 \wedge \omega^2, \quad d\omega^2 = q\omega^1 \wedge \omega^2.$$

If $da = a_1\omega^1 + a_2\omega^2$, $dc = c_1\omega^1 + c_2\omega^2$, $dp = p_1\omega^1 + p_2\omega^2$, and $dq = q_1\omega^1 + q_2\omega^2$ satisfy

$$a_2 = (a-c)p, \quad c_1 = (a-c)q, \quad p_2 - q_1 = ac + p^2 + q^2 \tag{4.40}$$

on U, then there exists an immersion $\mathbf{x} : U \to \mathbf{R}^3$ with principal curvatures a and c and induced metric $I = \omega^1\omega^1 + \omega^2\omega^2$.

Proof. Let $\omega^2_1 = p\omega^1 + q\omega^2 = -\omega^1_2$, $\omega^3_1 = a\omega^1 = -\omega^1_3$, and $\omega^3_2 = c\omega^2 = -\omega^2_3$, to define the $\mathscr{E}(3)$-valued 1-form on U

$$\eta = \left(\begin{pmatrix} \omega^1 \\ \omega^2 \\ 0 \end{pmatrix}, \begin{pmatrix} 0 & \omega^1_2 & \omega^1_3 \\ \omega^2_1 & 0 & \omega^2_3 \\ \omega^3_1 & \omega^3_2 & 0 \end{pmatrix} \right).$$

Then $d\eta = -\eta \wedge \eta$, by (4.40), so Theorem 2.25 implies the existence of a smooth map $(\mathbf{x}, e) : U \to \mathbf{E}(3)$ such that $(\mathbf{x}, e)^{-1} d(\mathbf{x}, e) = \eta$ on U and $\mathbf{x} : U \to \mathbf{R}^3$ is the desired immersion. □

Theorem 4.18 (Bonnet's Congruence Theorem [15]). *Let $\mathbf{x}, \tilde{\mathbf{x}} : M \to \mathbf{R}^3$ be smooth immersions of a connected surface M. Let $I, \tilde{I}$ be the first fundamental forms of $\mathbf{x}$ and $\tilde{\mathbf{x}}$, respectively. Let $\mathbf{e}_3$ and $\tilde{\mathbf{e}}_3$ be unit normal vector fields along $\mathbf{x}$ and $\tilde{\mathbf{x}}$, respectively, and let*

$$II = -d\mathbf{e}_3 \cdot d\mathbf{x}, \quad \widetilde{II} = -d\tilde{\mathbf{e}}_3 \cdot d\tilde{\mathbf{x}}$$

be the second fundamental forms of $\mathbf{x}$ *and* $\tilde{\mathbf{x}}$ *relative to* $\mathbf{e}_3$ *and* $\tilde{\mathbf{e}}_3$*, respectively.*

If there exists an element $(\mathbf{v},A) \in \mathbf{E}(3)$ *such that* $\tilde{\mathbf{x}} = (\mathbf{v},A) \circ \mathbf{x}$ *on* M*, then* $\tilde{\mathbf{e}}_3 = \epsilon A\mathbf{e}_3$*, where* $\epsilon = \pm 1$*,* $I = \tilde{I}$*, and* $II = \epsilon \widetilde{II}$ *on* M*.*

Conversely, if $I = \tilde{I}$ *and* $II = \epsilon\widetilde{II}$*, where* $\epsilon = \pm 1$*, then there exists an element* $(\mathbf{v},A) \in \mathbf{E}(3)$ *such that* $\tilde{\mathbf{x}} = \mathbf{v} + A\mathbf{x}$ *and* $\tilde{\mathbf{e}}_3 = \epsilon A\mathbf{e}_3$*.*

Proof. If there exists an element $(\mathbf{v},A) \in \mathbf{E}(3)$ such that $\tilde{\mathbf{x}} = A\mathbf{x} + \mathbf{v}$, then $d\tilde{\mathbf{x}} = Ad\mathbf{x}$, which implies that $\tilde{I} = d\tilde{\mathbf{x}} \cdot d\tilde{\mathbf{x}} = Ad\mathbf{x} \cdot Ad\mathbf{x} = d\mathbf{x} \cdot d\mathbf{x} = I$ on M. In addition, both $\tilde{\mathbf{e}}_3$ and $A\mathbf{e}_3$ are smooth unit normal vector fields along $\tilde{\mathbf{x}}$ on the connected surface M, so $\tilde{\mathbf{e}}_3 = \epsilon A\mathbf{e}_3$ on M, where $\epsilon = \pm 1$. Thus,

$$\widetilde{II} = -d\tilde{\mathbf{e}}_3 \cdot d\tilde{\mathbf{x}} = -\epsilon A d\mathbf{e}_3 \cdot Ad\mathbf{x} = -\epsilon d\mathbf{e}_3 \cdot d\mathbf{x} = \epsilon II$$

on M.

Conversely, suppose $I = \tilde{I}$ and $II = \epsilon\widetilde{II}$ on M, where $\epsilon = \pm 1$. If $\epsilon = -1$, replace $\tilde{\mathbf{e}}_3$ by $-\tilde{\mathbf{e}}_3$, which will change $\widetilde{II}$ to $-\widetilde{II}$. Thus, for the converse it is sufficient to suppose that $I = \tilde{I}$ and $II = \widetilde{II}$ on M.

Let $p \in M$, and let U be a connected open neighborhood of p on which there exists a first order frame field $(\mathbf{x}, e) : U \to \mathbf{E}_+(3)$ whose third vector is $\mathbf{e}_3$. Then $d\mathbf{x} = \omega^1\mathbf{e}_1 + \omega^2\mathbf{e}_2$, where ω^1, ω^2 is an orthonormal coframe field for I on U. Since $\tilde{I} = I$, it follows that ω^1, ω^2 is also an orthonormal coframe field for $\tilde{\mathbf{x}}$ on U. There exists a first order frame field $(\tilde{\mathbf{x}}, \tilde{e}) : U \to \mathbf{E}(3)$, with third vector equal to $\tilde{\mathbf{e}}_3$, such that $d\tilde{\mathbf{x}} = \omega^1\tilde{\mathbf{e}}_1 + \omega^2\tilde{\mathbf{e}}_2$; that is, $\tilde{\omega}^1 = \omega^1$ and $\tilde{\omega}^2 = \omega^2$ on U. Then $\tilde{\omega}^1_2 = \omega^1_2$ on U, by (4.38) and (4.39), and

$$\tilde{\omega}^3_1\omega^1 + \tilde{\omega}^3_2\omega^2 = \widetilde{II} = II = \omega^3_1\omega^1 + \omega^3_2\omega^2$$

implies that $\tilde{\omega}^3_i = \omega^3_i$ on U, for $i = 1, 2$. Then $(\mathbf{x}, e)$ and $(\tilde{\mathbf{x}}, \tilde{e})$ satisfy

$$(\mathbf{x}, e)^{-1} d(\mathbf{x}, e) = \left(\begin{pmatrix} \omega^1 \\ \omega^2 \\ 0 \end{pmatrix}, (\omega^i_j) \right) = (\tilde{\mathbf{x}}, \tilde{e})^{-1} d(\tilde{\mathbf{x}}, \tilde{e})$$

on U. By the Cartan–Darboux Congruence Theorem 2.24, there exists an element $(\mathbf{v},A) \in \mathbf{E}(3)$ such that $(\mathbf{x}, e) = (\mathbf{v},A) \circ (\tilde{\mathbf{x}}, \tilde{e})$ on U. In particular, $\mathbf{x} = \mathbf{v} + A\tilde{\mathbf{x}}$ and $\mathbf{e}_3 = A\tilde{\mathbf{e}}_3$ on U.

There is no loss of generality in replacing $\tilde{\mathbf{x}}$ by the congruent immersion $\mathbf{v} + A\tilde{\mathbf{x}}$, in which case we then have the same hypotheses holding and now

$$(\mathbf{x}, e) = (\tilde{\mathbf{x}}, \tilde{e})$$

on U. This proves the theorem for the case when M possesses a global frame field. The existence of a global frame field on M implies M has a nowhere vanishing smooth vector field, and thus the Euler characteristic of M must be zero. In general, then, so far we have proved only a local result.

Let q be any point of M. We want to prove that $(\tilde{\mathbf{x}}(q), \tilde{\mathbf{e}}_3(q)) = (\mathbf{x}(q), \mathbf{e}_3(q))$. For this purpose, let $\gamma : [0,1] \to M$ be a continuous path from $p = \gamma(0)$ to $q = \gamma(1)$. For each $t \in [0,1]$, apply the argument above to the point $\gamma(t) \in M$ to conclude that there exists a connected neighborhood U_t of $\gamma(t)$ on which there are frame fields $(\mathbf{x}, e_t), (\tilde{\mathbf{x}}, \tilde{e}_t) : U_t \to \mathbf{E}(3)$, with third vector equal to $\mathbf{e}_3$ and $\tilde{\mathbf{e}}_3$, respectively, and an element $(\mathbf{v}_t, A_t) \in \mathbf{E}(3)$ such that

$$(\mathbf{x}, e_t) = (\mathbf{v}_t, A_t) \circ (\tilde{\mathbf{x}}, \tilde{e}_t)$$

on U_t. By a standard argument using the Lebesgue number of the open covering $\{\gamma^{-1}U_t\}_{t\in[0,1]}$ of $[0,1]$ (see [122, Lemma 27.5 on page 175]), there exists a partition $0 = t_0 < t_1 < \cdots < t_{k+1} = 1$, and connected open subsets $U_0 = U, U_1, \ldots, U_k$ of M such that

- $\gamma[t_i, t_{i+1}] \subset U_i$, for $i = 0, \ldots, k$;
- there exists an element $(\mathbf{v}_i, A_i) \in \mathbf{E}(3)$ such that

$$\mathbf{x} = \mathbf{v}_i + A_i\tilde{\mathbf{x}}, \quad \mathbf{e}_3 = A_i\tilde{\mathbf{e}}_3$$

on U_i, for $i = 0, \ldots, k$. By assumption, $\mathbf{v}_0 = \mathbf{0}$ and $A_0 = I_3$.

Let $p_i = \gamma(t_i)$, for $i = 0, \ldots, k+1$, so $p_0 = p$ and $p_{k+1} = q$.

On U_0 we have $\mathbf{x} = \tilde{\mathbf{x}}$ and $\mathbf{e}_3 = \tilde{\mathbf{e}}_3$ and on U_1 we have $\mathbf{x} = \mathbf{v}_1 + A_1\tilde{\mathbf{x}}$ and $\mathbf{e}_3 = A_1\tilde{\mathbf{e}}_3$. Thus, on the open neighborhood $U_0 \cap U_1$ of p_1 we have

$$\mathbf{x} = \mathbf{v}_1 + A_1\mathbf{x}, \quad \mathbf{e}_3 = A_1\mathbf{e}_3,$$

so

$$(I_3 - A_1)d\mathbf{x} = 0, \quad (I_3 - A_1)\mathbf{e}_3 = 0,$$

at every point of $U_0 \cap U_1$. Therefore, $A_1 = I_3$ and then $\mathbf{v}_1 = 0$, and

$$\mathbf{x} = \tilde{\mathbf{x}}, \quad \mathbf{e}_3 = \tilde{\mathbf{e}}_3$$

on $U_0 \cup U_1$. Repeating this argument for $U_2, \ldots, U_k$, we reach the conclusion that $\tilde{\mathbf{x}} = \mathbf{x}$ and $\tilde{\mathbf{e}}_3 = \mathbf{e}_3$ on U_k, so $\tilde{\mathbf{x}}(q) = \mathbf{x}(q)$ and $\tilde{\mathbf{e}}_3(q) = \mathbf{e}_3(q)$, as desired. We have proved that $\tilde{\mathbf{x}} = \mathbf{x}$ and $\tilde{\mathbf{e}}_3 = \mathbf{e}_3$ on all of M. □

Theorem 4.19 (Bonnet's Existence Theorem [14]). *Let (M, I) be a simply connected Riemannian surface with Gaussian curvature K. Let II be a symmetric bilinear form field on M. Suppose that II satisfies the Gauss and Codazzi equations in the sense that for any orthonormal coframe field θ^1, θ^2 in $U \subset M$, with Levi-Civita connection form $\omega^1_2 = -\omega^2_1$, the smooth function coefficients $h_{ij} = h_{ji}$, $i, j = 1, 2$ of II defined by*

$$II = h_{11}\theta^1\theta^1 + 2h_{12}\theta^1\theta^2 + h_{22}\theta^2\theta^2,$$

satisfy the Gauss equation

$$K = h_{11}h_{22} - h_{12}^2 \tag{4.41}$$

and the Codazzi equations

$$\sum_{k=1}^{2}(dh_{ik} - \sum_{j=1}^{2}(h_{ij}\omega_k^j + h_{jk}\omega_i^j)) \wedge \theta^k = 0, \tag{4.42}$$

for $i = 1,2$. Then there exists a smooth immersion $\mathbf{x} : M \to \mathbf{R}^3$ with unit normal vector field $\mathbf{e}_3$ such that $I = d\mathbf{x} \cdot d\mathbf{x}$ and $II = -d\mathbf{e}_3 \cdot d\mathbf{x}$.

Proof. It is known that a simply connected surface M is homeomorphic to the plane $\mathbf{R}^2$ or to the sphere $\mathbf{S}^2$.

If M is homeomorphic to $\mathbf{R}^2$, then it possesses a global orthonormal coframe field θ^1, θ^2 for I, with corresponding Levi-Civita connection form $\omega_2^1 = -\omega_1^2$. Let $h_{ij} = h_{ji}$ be the smooth coefficients of II relative to this coframe field on U, and define smooth 1-forms on U by

$$\omega_i^3 = \sum_{j=1}^{2} h_{ij}\theta^j = -\omega_3^i,$$

for $i = 1,2$. Consider the matrix valued 1-forms on M,

$$\theta = \begin{pmatrix} \theta^1 \\ \theta^2 \\ 0 \end{pmatrix}, \quad \omega = \begin{pmatrix} 0 & \omega_2^1 & \omega_3^1 \\ \omega_1^2 & 0 & \omega_3^2 \\ \omega_1^3 & \omega_2^3 & 0 \end{pmatrix}.$$

Then (θ, ω) is an $\mathscr{E}(3)$-valued 1-form on M. The Gauss and Codazzi equations (4.41) and (4.42) imply that

$$(d\theta, d\omega) = (-\omega \wedge \theta, -\omega \wedge \omega).$$

By the Cartan–Darboux Existence Theorem 2.25, there exists a smooth map

$$(\mathbf{x}, e) : M \to \mathbf{E}_+(3) = \mathbf{R}^3 \rtimes \mathbf{SO}(3),$$

such that $(e^{-1}d\mathbf{x}, e^{-1}de) = (\theta, \omega)$. In particular,

$$d\mathbf{x} = \theta^1\mathbf{e}_1 + \theta^2\mathbf{e}_2, \quad d\mathbf{e}_3 = \omega_3^1\mathbf{e}_1 + \omega_3^2\mathbf{e}_2,$$

shows that $\mathbf{x} : M \to \mathbf{R}^3$ is an immersion with smooth unit normal vector field $\mathbf{e}_3$, such that on M,

$$d\mathbf{x} \cdot d\mathbf{x} = \sum_{i=1}^{2} \theta^i\theta^i = I, \quad -d\mathbf{e}_3 \cdot d\mathbf{x} = \omega_i^3\theta^i = h_{ij}\theta^i\theta^j = II. \tag{4.43}$$

In the case when M is homeomorphic to $\mathbf{S}^2$, we know that for any point $m \in M$, the complement $U = M \setminus \{m\}$ is homeomorphic to $\mathbf{R}^2$, so the above proof gives a smooth map $(\mathbf{x}, e) : U \to \mathbf{E}_+(3)$ satisfying (4.43). Taking the complement $\tilde{U}$ of a different point $\tilde{m} \in M$, we obtain a smooth map $(\tilde{\mathbf{x}}, \tilde{e}) : \tilde{U} \to \mathbf{E}_+(3)$ satisfying

$$d\tilde{\mathbf{x}} \cdot d\tilde{\mathbf{x}} = I, \quad -d\tilde{\mathbf{e}}_3 \cdot d\tilde{\mathbf{x}} = II.$$

Apply the Bonnet Congruence Theorem 4.18 to $(\mathbf{x}, \mathbf{e}_3)$ and $(\tilde{\mathbf{x}}, \tilde{\mathbf{e}}_3)$ restricted to $U \cap \tilde{U}$, to get an element $(\mathbf{v}, A) \in \mathbf{E}(3)$ such that on $U \cap \tilde{U}$,

$$\tilde{\mathbf{x}} = \mathbf{v} + A\mathbf{x}, \quad \tilde{\mathbf{e}}_3 = A\mathbf{e}_3.$$

If we replace $(\mathbf{x}, e)$ by $(\mathbf{v}, A)(\mathbf{x}, e)$, then (4.43) continues to hold on U, and on $U \cap \tilde{U}$ we have $\tilde{\mathbf{x}} = \mathbf{x}$ and $\tilde{\mathbf{e}}_3 = \mathbf{e}_3$, thus showing that $\mathbf{x}$ and $\mathbf{e}_3$ extend smoothly to all of $M = U \cup \tilde{U}$ and satisfy (4.43) on M. □

Remark 4.20. By Cartan's Lemma, the Codazzi equations (4.42) are equivalent to the equations

$$h_{ijk} = h_{ikj}$$

for all i, j, k, where the functions $h_{ijk} = h_{jik}$ are defined by (4.19).

4.5 Tangent and curvature spheres

Example 4.21. The *oriented sphere with center* $\mathbf{p} \in \mathbf{R}^3$ and *signed radius* $0 \neq r \in \mathbf{R}$ is

$$S_r(\mathbf{p}) = \{\mathbf{x} \in \mathbf{R}^3 : |\mathbf{x} - \mathbf{p}|^2 = r^2\}$$

with unit normal vector field $\mathbf{n}(\mathbf{x}) = (\mathbf{p} - \mathbf{x})/r$. Thus, the orientation is by the inward pointing normal when $r > 0$, and by the outward normal when $r < 0$. The *unit sphere* is $S_1(\mathbf{0})$, which we denote by $\mathbf{S}^2$. Its default orientation is by the inward pointing normal $\mathbf{n}(\mathbf{x}) = -\mathbf{x}$. The spheres $S_r(\mathbf{p})$ are immersed surfaces. In a neighborhood of any point on $S_r(\mathbf{p})$ there is a first order frame field $(\mathbf{x}, e)$ with $\mathbf{e}_3 = \mathbf{n} = \frac{1}{r}(\mathbf{p} - \mathbf{x})$. Then $d\mathbf{e}_3 = d\mathbf{n} = -\frac{1}{r} d\mathbf{x}$, which implies that $\omega_3^i = -\frac{1}{r}\omega^i$, for $i, j = 1, 2$, and the principal curvatures are both $1/r$. The second fundamental form is

$$II = \omega_1^3 \omega^1 + \omega_2^3 \omega^2 = \frac{1}{r}(\omega^1 \omega^1 + \omega^2 \omega^2) = \frac{1}{r} I.$$

The Gaussian curvature is $K = 1/r^2$ and the mean curvature is $H = 1/r$.

Example 4.22. Fix $\mathbf{n} \in \mathbf{S}^2$ and $h \in \mathbf{R}$. The *oriented plane in* $\mathbf{R}^3$ *with unit normal* $\mathbf{n}$ and *signed height* h is

$$\Pi_h(\mathbf{n}) = \{\mathbf{x} \in \mathbf{R}^3 : \mathbf{x} \cdot \mathbf{n} = h\}.$$

These are immersed surfaces. There is a first order frame field $(\mathbf{x}, e)$ on all of $\Pi_h(\mathbf{n})$ with $\mathbf{e}_3 = \mathbf{n}$, which is constant, so $\omega^3_1 = 0 = \omega^3_2$, the principal curvatures are zero, and the second fundamental form is identically zero.

Theorem 4.23 (Totally umbilic case). *Suppose that every point of a connected immersed surface* $\mathbf{x} : M \to \mathbf{R}^3$ *is umbilic. Then either* $\mathbf{x}(M)$ *is an open subset of a sphere or it is an open subset of a plane.*

Proof. We may assume that the immersion $\mathbf{x}$ possesses a smooth unit normal vector field $\mathbf{n} : M \to \mathbf{S}^2$, for if it does not, then there is a double cover $\varphi : \tilde{M} \to M$ for which the immersion $\mathbf{x} \circ \varphi : \tilde{M} \to \mathbf{R}^3$ possesses a smooth unit normal vector field, and the images $\mathbf{x}(M) = \mathbf{x} \circ \varphi(M)$. We consider now only first order frame fields $(\mathbf{x}, e) : U \to \mathbf{E}_+(3)$ for which $\mathbf{e}_3 = \mathbf{n}$ on U. If $\mathbf{x}$ is totally umbilic, then for such a first order frame field we have $\omega^3_1 = a\omega^1$, and $\omega^3_2 = a\omega^2$, where $a : U \to \mathbf{R}$ is the principal curvature function. Taking the exterior derivative of these equations and using the structure equations of $\mathbf{E}_+(3)$, we find that a must be constant on M.

If $a \neq 0$, then $d(\mathbf{x} + \mathbf{n}/a) = 0$, so $\mathbf{x} + \mathbf{n}/a$ is constant on M and $\mathbf{x}(M)$ is a subset of the oriented sphere $S_{1/a}(\mathbf{x} + \mathbf{n}/a)$. This result has a more abstract proof, which we present now. It can be applied to submanifolds of homogeneous spaces whenever the invariants are constant. The equations

$$\omega^3 = 0, \quad \omega^1 = \frac{1}{a}\omega^3_1, \quad \omega^2 = \frac{1}{a}\omega^3_2,$$

define a 3-plane distribution on $\mathbf{E}_+(3)$ whose dual vector description

$$\mathfrak{h} = \{(\begin{pmatrix} s/a \\ t/a \\ 0 \end{pmatrix}, \begin{pmatrix} 0 & -r & -s \\ r & 0 & -t \\ s & t & 0 \end{pmatrix}) : r, s, t \in \mathbf{R}\} \subset \mathscr{E}(3)$$

is a Lie subalgebra, since the defining equations are all left-invariant 1-forms on $\mathbf{E}(3)$. We have a Lie algebra isomorphism

$$\mathfrak{o}(3) \cong \mathfrak{h}, \quad X \leftrightarrow (-X\frac{\epsilon_3}{a}, X).$$

Its corresponding Lie subgroup, obtained by exponentiation,

$$H = \{((I - A)\frac{\epsilon_3}{a}, A) : A \in \mathbf{SO}(3)\},$$

is the maximal integral submanifold of $\mathfrak{h}$ passing through the identity element $(\mathbf{0}, I) \in \mathbf{E}_+(3)$. Its projection by $\pi : \mathbf{E}_+(3) \to \mathbf{R}^3$ is

$$H\mathbf{0} = S_{1/a}(\frac{1}{a}\boldsymbol{\epsilon}_3).$$

Right cosets of H are the other maximal integral submanifolds of $\mathfrak{h}$. The set of all first order oriented frames over $\mathbf{x}$ is a connected 3-dimensional integral submanifold of $\mathfrak{h}$, so it must be contained in a right coset of H. For a point $m \in M$, this coset must be $(\mathbf{x}(m), e(m))H$, for a first order frame $(\mathbf{x}, e)(m) \in \mathbf{E}_+(3)$. Then

$$\mathbf{x}(M) \subset (\mathbf{x}(m), e(m))H\mathbf{0} = (\mathbf{x}(m), e(m))S_{1/a}(\frac{1}{a}\boldsymbol{\epsilon}_3) = S_{1/a}(\mathbf{x}(m) + \frac{1}{a}\mathbf{n}(m)).$$

If $a = 0$, then $\mathfrak{h}$ is defined by the equations $\omega^3 = 0$, $\omega^3_1 = 0$, $\omega^3_2 = 0$, so it is a Lie subalgebra of $\mathbf{E}_+(3)$ whose Lie subgroup is

$$H = \{\left(\begin{pmatrix} s \\ t \\ 0 \end{pmatrix}, \begin{pmatrix} A & \begin{matrix} 0 \\ 0 \end{matrix} \\ 0\ 0 & 1 \end{pmatrix}\right) : s, t \in \mathbf{R},\ A \in \mathbf{SO}(2)\}.$$

The set of all oriented first order frames along $\mathbf{x}$ must then be a coset $(\mathbf{x}(m), e(m))H$, for a first order frame at a point $m \in M$. Then

$$\mathbf{x}(M) \subset \{\mathbf{x}(m) + s\mathbf{e}_1(m) + t\mathbf{e}_2(m) : s, t \in \mathbf{R}\},$$

which is the plane through $\mathbf{x}(m)$ with unit normal $\mathbf{n}(m)$, that is, $\Pi_h(\mathbf{n}(m))$, where $h = (\mathbf{x}(m) + s\mathbf{e}_1(m) + t\mathbf{e}_2(m)) \cdot \mathbf{n}(m) = \mathbf{x}(m) \cdot \mathbf{n}(m)$. □

Definition 4.24. An *oriented tangent sphere* to an immersion $\mathbf{x} : M^2 \to \mathbf{R}^3$ at a point $m \in M$, with unit normal vector $\mathbf{n}$ at m, is any oriented sphere or plane through $\mathbf{x}(m)$ with unit normal $\mathbf{n}$ at $\mathbf{x}(m)$.

The set of all oriented tangent spheres to $\mathbf{x}$ at m with unit normal $\mathbf{n}$ is

$$\{S_r(\mathbf{x}(m) + r\mathbf{n}) : 0 \neq r \in \mathbf{R}\} \cup \{\Pi_{\mathbf{n}\cdot\mathbf{x}(m)}(\mathbf{n})\}.$$

Each of the oriented tangent spheres has its center on the oriented normal line $\{\mathbf{x}(m) + r\mathbf{n} : r \in \mathbf{R}\}$. It is convenient to refer to all the elements of the set of oriented tangent spheres as spheres, with the oriented tangent plane being thought of as an oriented sphere with infinite radius.

Definition 4.25. An *oriented curvature sphere* at $m \in M$ of an immersion $\mathbf{x} : M^2 \to \mathbf{R}^3$ with unit normal $\mathbf{n}$ at m is an oriented tangent sphere at m whose principal curvature is equal to a principal curvature of $\mathbf{x}$ at m for the normal $\mathbf{n}$.

If $m \in M$ is nonumbilic for $\mathbf{x}$, then there are two distinct oriented curvature spheres at m for a given unit normal vector $\mathbf{n}$. If m is umbilic, then there is only one, but we say it has *multiplicity two*. If a is a non-zero principal curvature of $\mathbf{x}$ at m relative to $\mathbf{n}$, then

$$S_{1/a}(\mathbf{x}(m) + \frac{1}{a}\mathbf{n})$$

is an oriented curvature sphere at m. If 0 is a principal curvature of $\mathbf{x}$ at m, then the oriented plane

$$\Pi_{\mathbf{x}(m)\cdot\mathbf{n}}(\mathbf{n})$$

is an oriented curvature sphere at m. If the unit normal vector $\mathbf{n}$ of $\mathbf{x}$ at m is replaced by $-\mathbf{n}$, then the curvature spheres at m remain unchanged, but with opposite orientation, as they will now have the orientation that equals $-\mathbf{n}$ at m.

For $r \neq 0$, and for first order frame field $(\mathbf{x}, e)$ along $\mathbf{x}$ on U, the smooth map

$$S = \mathbf{x} + r\mathbf{e}_3 : U \to \mathbf{R}^3 \tag{4.44}$$

determines the family $S_r(\mathbf{x} + r\mathbf{e}_3)$ of oriented tangent spheres at the points of U relative to the unit normals $\mathbf{e}_3$. The smooth map

$$S = \mathbf{e}_3 : U \to \mathbf{R}^3 \tag{4.45}$$

determines the oriented tangent planes $\Pi_{\mathbf{e}_3\cdot\mathbf{x}}(\mathbf{e}_3)$ with these normals.

Proposition 4.26. *If a family of oriented tangent spheres is determined by a smooth map* (4.44) *or* (4.45)*, then it is an oriented curvature sphere at a point $m \in U$ if and only if dS at m has rank less than two.*

Proof. If S is given by (4.44), then

$$dS = d\mathbf{x} + rd\mathbf{e}_3 = \sum_1^2 (\omega^i + r\omega^i_3)e_i,$$

which has rank less than two at a point $m \in U$ if and only if $\omega^i + r\omega^i_3 = 0$ at m, for $i = 1$ or $i = 2$, which holds if and only if $1/r$ is a principal curvature of $\mathbf{x}$ at m relative to $\mathbf{e}_3(m)$. The proof is similar for the case of the map (4.45) □

Example 4.27 (Curves on $\mathbf{S}^2$). Consider the transitive action of $\mathbf{SO}(3)$ on the unit sphere $\mathbf{S}^2 \subset \mathbf{R}^3$ obtained from the standard matrix multiplication action of $\mathbf{SO}(3)$ on $\mathbf{R}^3$. Let ϵ_1 be the origin of $\mathbf{S}^2$ and let $\pi : \mathbf{SO}(3) \to \mathbf{S}^2$ by the projection $\pi(A) = A\epsilon_1$. Let $\boldsymbol{\sigma} : J \to \mathbf{S}^2 \subset \mathbf{R}^3$ be a curve on the unit sphere, parametrized by arclength x, where J is an interval in $\mathbf{R}$. Its unit tangent vector is $\mathbf{T} = \dot{\boldsymbol{\sigma}}$, where dot indicates derivative with respect to x. The unit normal vector is $\mathbf{N} = \boldsymbol{\sigma} \times \mathbf{T}$. The frame field $e = (\boldsymbol{\sigma}, \mathbf{T}, \mathbf{N}) : J \to \mathbf{SO}(3)$ along $\boldsymbol{\sigma}$ satisfies

$$e^{-1}de = \begin{pmatrix} 0 & -1 & 0 \\ 1 & 0 & -\kappa \\ 0 & \kappa & 0 \end{pmatrix}$$

for some smooth function $\kappa : J \to \mathbf{R}$, called the *curvature* of $\boldsymbol{\sigma}$ in $\mathbf{S}^2$. This is called the *Frenet frame* along $\boldsymbol{\sigma}$. Its Serret-Frenet equations are

$$\dot{\boldsymbol{\sigma}} = \mathbf{T}, \quad \dot{\mathbf{T}} = -\boldsymbol{\sigma} + \kappa \mathbf{N}, \quad \dot{\mathbf{N}} = -\kappa \mathbf{T}. \tag{4.46}$$

Reversing the orientation of $\boldsymbol{\sigma}$, by reversing the sign of x, reverses the sign of κ.

Example 4.28 (Cones). A general cone in $\mathbf{R}^3$ is defined as follows. We may assume that the vertex is at the origin and the *profile curve*, which is the intersection of the cone with the unit sphere, has arclength parametrization $\boldsymbol{\sigma} : J \to \mathbf{S}^2$, where J is some open interval. We use the notation of Example 4.27. If $M = J \times \mathbf{R}$, then the cone is the immersed surface

$$\mathbf{x} : M \to \mathbf{R}^3, \quad \mathbf{x}(x,y) = e^{-y}\boldsymbol{\sigma}(x).$$

Then

$$d\mathbf{x} = e^{-y}\dot{\boldsymbol{\sigma}}(x)dx - e^{-y}\boldsymbol{\sigma}\,dy \tag{4.47}$$

from which we calculate the first fundamental form

$$I = d\mathbf{x} \cdot d\mathbf{x} = e^{-2y}(dx^2 + dy^2).$$

The dual coframe field is

$$\omega^1 = e^{-y}dx, \quad \omega^2 = e^{-y}dy.$$

We can also see from (4.47) that an oriented first order frame field along $\mathbf{x}$ is given by $(\mathbf{x}, e)$, where the columns of e are

$$\mathbf{e}_1 = \dot{\boldsymbol{\sigma}}, \quad \mathbf{e}_2 = -\boldsymbol{\sigma}, \quad \mathbf{e}_3 = \mathbf{e}_1 \times \mathbf{e}_2 = \boldsymbol{\sigma} \times \dot{\boldsymbol{\sigma}}.$$

Then $\ddot{\boldsymbol{\sigma}} = \mathbf{e}_2 + \kappa(x)\mathbf{e}_3$, by (4.46), so the second fundamental form is

$$II = -d\mathbf{x} \cdot d\mathbf{e}_3 = \kappa(x)e^{-y}dxdx = \kappa(x)e^{y}\omega^1\omega^1,$$

from which we see that the principal curvatures are $a = \kappa(x)e^y$ and $c = 0$, respectively. The Hopf invariant h relative to $(\mathbf{x}, e)$ and the mean curvature H are

$$h = \frac{\kappa(x)e^y}{2} = H.$$

The oriented curvature spheres at $(x, y) \in M$ relative to $\mathbf{e}_3$ are

$$S_{\frac{1}{a}}(\mathbf{x} + \frac{1}{a}\mathbf{e}_3) \quad \text{and} \quad \Pi_{\mathbf{x}\cdot\mathbf{e}_3}(\mathbf{e}_3). \tag{4.48}$$

The oriented plane passes through the origin, since $\mathbf{x} \cdot \mathbf{e}_3 = 0$.

4.6 The Gauss map

Let $\mathbf{n}$ be a smooth unit normal vector field along the immersed surface $\mathbf{x} : M \to \mathbf{R}^3$. The smooth map

$$\mathbf{n} : M \to \mathbf{S}^2 \subset \mathbf{R}^3$$

is called the *Gauss map* of $\mathbf{x}$. It is defined up to sign for a connected oriented surface. For an unoriented surface it is defined only locally, or must be regarded as a map into the real projective plane $\mathbf{RP}^2$. If $(\mathbf{x}, \mathbf{e}_1, \mathbf{e}_2, \mathbf{e}_3)$ is a first order frame field along $\mathbf{x}$ on $U \subset M$, with $\mathbf{e}_3 = \mathbf{n}$, then

$$d\mathbf{e}_3 = \omega_3^1 \mathbf{e}_1 + \omega_3^2 \mathbf{e}_2 \tag{4.49}$$

shows that $(\mathbf{n}, \mathbf{e}_1, \mathbf{e}_2, \mathbf{e}_3)$ is a first order frame field along $\mathbf{n}$. The Gauss map need not be an immersion. In fact, $d\mathbf{n}$ has rank two if and only if

$$0 \neq \omega_3^1 \wedge \omega_3^2 = K\omega^1 \wedge \omega^2;$$

that is, if and only if $K \neq 0$. The first fundamental form of $\mathbf{n}$ (restricting ourselves to the points of M where K is nonzero) is

$$III = d\mathbf{e}_3 \cdot d\mathbf{e}_3,$$

which is called the *third fundamental form* of $\mathbf{x}$.

Theorem 4.29. *Let $\mathbf{x} : M \to \mathbf{R}^3$ be an immersion of a connected surface M, and suppose $\mathbf{x}$ has a globally defined unit normal vector field $\mathbf{n} : M \to \mathbf{S}^2$. Its Gauss map is conformal if and only if the mean curvature of $\mathbf{x}$ is identically 0 on M or $\mathbf{x}$ is totally umbilic.*

Proof. That the Gauss map is conformal means that it pulls back the metric on the sphere to a multiple of the metric induced on M by $\mathbf{x}$; that is, III is a multiple of I on M. In general, $KI - 2H\,II + III = 0$ on M (see Problem 4.69), so III is a multiple of I if and only if $2H\,II$ is a multiple of I. This last condition is true if and only if at each point of M either $H = 0$ or $\mathbf{x}$ is umbilic. In particular, if H is identically 0 on M, or if $\mathbf{x}$ is totally umbilic, then the Gauss map is conformal. Conversely, if the Gauss map is conformal, suppose that H is not identically 0 on M. Then

$$W = \{m \in M : H(m) \neq 0\}$$

is a non-empty open subset of M. On a connected component W_0 of W, we must have $\mathbf{x}$ totally umbilic. Consequently, its principal curvatures and H must be constant on W_0. This constant H must be non-zero, and it must be the value of H on the closure of W_0 in M. Therefore, W_0 must equal its closure, so $W_0 = M$ and $\mathbf{x}$ is totally umbilic on M. □

Remark 4.30. Equation (4.49) and the structure equations (4.18) show that ω_2^1 is the Levi-Civita connection form of III with respect to this frame field along $\mathbf{n} = \mathbf{e}_3$. Thus, altering our view of the Gauss equation (4.13) slightly, we interpret $d\omega_2^1 = \omega_3^1 \wedge \omega_3^2$ to mean that the Gaussian curvature of III is 1. Looking again at the Gauss equation, we see that

$$\omega_3^1 \wedge \omega_3^2 = K\,\omega^1 \wedge \omega^2, \tag{4.50}$$

which shows that K is the ratio of the area element of III to the area element of I. This is a modern version of Gauss's definition of K in [72].

Definition 4.31. The *total curvature* of an immersion $\mathbf{x} : M \to \mathbf{R}^3$ of a connected, compact, oriented surface M is

$$\int_M K\,dA,$$

where dA is the area form of the induced metric on M.

If $\mathbf{n} : M \to \mathbf{S}^2$ is the Gauss map of $\mathbf{x}$, then a first order frame field $(\mathbf{x}, (\mathbf{e}_1, \mathbf{e}_2, \mathbf{e}_3))$ on an open subset U is *positively oriented* if $\mathbf{e}_3 = \mathbf{n}$ and $dA = \omega^1 \wedge \omega^2$ on U. Equation (4.50) implies that the total curvature of $\mathbf{x}$ is related to the area of the image of the Gauss map. In fact, recall a basic feature of integration on manifolds. If $g : M \to N$ is a diffeomorphism between connected oriented surfaces and if ν is a smooth 2-form on N with compact support, then

$$\int_M g^*\nu = \pm \int_N \nu,$$

where the sign is $+$ if g preserves orientation and is $-$ if g reverses orientation. To apply this to the Gauss map of an immersion $\mathbf{x} : M \to \mathbf{R}^3$, which is generally not a diffeomorphism, we need the concept of the degree of a map between surfaces.

Let M^2 and N^2 be compact, connected, oriented surfaces and let $g : M \to N$ be a smooth map. A point $y \in N$ is a *regular value* of g if $g^{-1}\{y\}$ contains no critical points, where $x \in M$ is a *critical point* of g if the rank of dg_x is less than two. Note that if y is not in the image of g, then it is a regular value, since the empty set contains no critical points of g. Regular values exist by Sard's Theorem (see Conlon [53, p. 80]). By the inverse function theorem, if $y \in N$ is a regular value of g, then for $x \in g^{-1}\{y\}$ there exists a neighborhood U of x that g maps diffeomorphically onto a neighborhood of y. In particular, g must be one-to-one on U, so $U \cap g^{-1}\{y\} = \{x\}$. Thus, $g^{-1}\{y\}$ is a set of isolated points in M, so is finite, since M is compact.

For a regular value $y \in N$ of g, suppose $g^{-1}\{y\} = \{x_1, \dots, x_k\}$, for some whole number $k \geq 1$. Let

$$\epsilon_j = \begin{cases} +1, & \text{if } dg_{x_j} \text{ preserves orientation,} \\ -1, & \text{if } dg_{x_j} \text{ reverses orientation,} \end{cases}$$

for $j = 1, \dots, k$.

Definition 4.32. The *local degree* of g at the regular value $y \in N$ is

$$\deg_y(g) = \sum_1^k \epsilon_j,$$

if $g^{-1}\{y\} \neq \emptyset$. Otherwise, the local degree of g at y is zero.

The following is a special case of [53, Proposition 8.7.2]. To prove it, we shall assume the result that for any compact, connected, oriented surface M, the linear functional defined on de Rham cohomology

$$\int_M : H^2(M) \to \mathbf{R}, \quad \int_M [\mu] = \int_M \mu$$

is an isomorphism. Here μ is a smooth 2-form on M representing the cohomology class $[\mu]$. See Corollary 8.6.5 in [53] for a proof. One consequence of this result is that if μ and ν are smooth 2-forms on M such that $\int_M \mu = \int_M \nu$, then there exists a smooth 1-form α on M such that $\mu - \nu = d\alpha$.

Proposition 4.33. *Let M and N be compact, connected, oriented surfaces and let $g : M \to N$ be a smooth map. If ν is a smooth 2-form on N, and if $y \in N$ is a regular value of g, then*

$$\int_M g^* \nu = \deg_y(g) \int_N \nu.$$

Proof. Given the regular value y of g, let $g^{-1}\{y\} = \{x_1, \dots, x_k\}$, for some $k \geq 1$. The case where y is not in the image of g will be left to Problem 4.70. There exists an open, connected, neighborhood $V \subset N$ of y such that

$$g^{-1}V = U_1 \cup \dots \cup U_k,$$

a union of disjoint open sets such that $x_j \in U_j$ and the restriction $g_j = g|_{U_j}$ maps U_j diffeomorphically onto V, for $j = 1, \dots, k$. There exists a smooth 2-form $\tilde{\nu}$ on N such that the support of $\tilde{\nu}$ is a subset of V and $\int_N \tilde{\nu} = \int_N \nu$. Thus $\nu - \tilde{\nu} = d\alpha$, for some smooth 1-form α on N. Now $g^*\tilde{\nu} = \sum_1^k \omega_j$, where the support of the smooth 2-form ω_j is a subset of U_j, and $\omega_j = g_j^*\tilde{\nu}$, for $j = 1, \dots, k$. Since $dg_{j x_j} = dg_{x_j}$, we have

$$\int_{U_j} \omega_j = \epsilon_j \int_V \tilde{\nu} = \epsilon_j \int_N \tilde{\nu},$$

for $j = 1, \dots, k$. Using Stokes's Theorem, we have

$$\begin{aligned}\int_M g^*\nu &= \int_M g^*(\tilde{\nu} + d\alpha) = \sum_1^k \int_M \omega_j + \int_M d(g^*\alpha) \\ &= \sum_1^k \int_{U_j} \omega_j = \sum_1^k \epsilon_j \int_V \tilde{\nu} = \deg_y(g) \int_N \nu.\end{aligned}$$

□

Remark 4.34. Since the two integrals in Proposition 4.33 are independent of the choice of regular value y of g, it follows that $\deg_y(g)$ is independent of y and we can write simply $\deg(g) = \deg_y(g)$, for any choice of regular value of g.

Corollary 4.35. *If $\mathbf{n} : M \to \mathbf{S}^2$ is the Gauss map of an immersion $\mathbf{x} : M \to \mathbf{R}^3$ of a compact, connected, oriented surface, then the total curvature of $\mathbf{x}$ is the degree of the Gauss map times the area of $\mathbf{S}^2$:*

$$\int_M K\, dA = 4\pi \deg(\mathbf{n}).$$

4.7 Isoparametric, Dupin, and canal immersions

Definition 4.36. An immersion is *isoparametric* if its principal curvatures are constant. A principal curvature satisfies the *Dupin condition* if it is constant along its lines of curvature. The immersion is *canal* if one of its principal curvatures satisfies the Dupin condition. It is *Dupin* if both principal curvatures satisfy the Dupin condition. A *cyclide of Dupin* is the image of a Dupin immersion.

A slight variation of Proposition 4.26 gives the following characterization of the Dupin condition in terms of oriented curvature spheres.

Proposition 4.37. *Let $\mathbf{x} : M \to \mathbf{R}^3$ be an umbilic free immersion with unit normal vector field $\mathbf{e}_3$. A principal curvature, a say, satisfies the Dupin condition if and only if the oriented curvature sphere $S = \mathbf{x} + \frac{1}{a}\mathbf{e}_3$ has rank one at every point of M.*

Proof. Let $(\mathbf{x},e): U \subset M$ be a second order frame field along $\mathbf{x}$ whose normal vector is $\mathbf{e}_3$. Then

$$dS = -\frac{a_1}{a^2}\mathbf{e}_3\omega^1 + (\frac{a-c}{a}\mathbf{e}_2 - \frac{a_2}{a^2}\mathbf{e}_3)\omega^2,$$

where $da = a_1\omega^1 + a_2\omega^2$ and ω^1, ω^2 is the coframe field dual to $(\mathbf{x},e)$. From this we see that dS has rank one if and only if $a_1 = 0$ on U if and only if a satisfies the Dupin condition on U. Since any point of M is in some such U, the proof is complete. □

Example 4.38 (Plane curves). A smooth curve $\boldsymbol{\gamma}: J \to \mathbf{R}^2$, $\boldsymbol{\gamma}(t) = f(t)\boldsymbol{\epsilon}_1 + g(t)\boldsymbol{\epsilon}_2$ in the oriented plane $\mathbf{R}^2$ is regular if

$$w = \sqrt{\frac{df}{dt}^2 + \frac{dg}{dt}^2} > 0$$

on the open interval $J \subset \mathbf{R}$. An arclength parameter $s = \int_{t_0}^t w(u)du$ satisfies $\frac{ds}{dt} = w$. The unit tangent vector field along $\boldsymbol{\gamma}$ is

$$\mathbf{T} = \dot{\boldsymbol{\gamma}} = \frac{d\boldsymbol{\gamma}}{ds} = \frac{1}{w}\frac{d\boldsymbol{\gamma}}{dt} = \dot{f}\boldsymbol{\epsilon}_1 + \dot{g}\boldsymbol{\epsilon}_2.$$

where dot denotes derivative with respect to s. The *principal normal* of $\boldsymbol{\gamma}$ is the unit vector field $\mathbf{N}$ along $\boldsymbol{\gamma}$ obtained by rotating $\mathbf{T}$ by $\pi/2$ in the positive direction,

$$\mathbf{N} = -\dot{g}\boldsymbol{\epsilon}_1 + \dot{f}\boldsymbol{\epsilon}_2.$$

The *curvature* of $\boldsymbol{\gamma}$ is

$$\kappa = \dot{\mathbf{T}} \cdot \mathbf{N} = \frac{f'g'' - g'f''}{w^3},$$

where prime denotes derivative with respect to t. The Euclidean group $\mathbf{E}(2) = \mathbf{R}^2 \rtimes \mathbf{O}(2)$ acts transitively on $\mathbf{R}^2$ by $(\mathbf{a},A)\mathbf{x} = \mathbf{a} + A\mathbf{x}$. A Frenet frame field along $\boldsymbol{\gamma}$ is $(\boldsymbol{\gamma},e): J \to \mathbf{E}_+(2)$, where the columns of $e \in \mathbf{SO}(2)$ are

$$\mathbf{e}_1 = \mathbf{T}, \quad \mathbf{e}_2 = \mathbf{N}.$$

It pulls back the Maurer–Cartan form to

$$(\mathbf{x},e)^{-1}d(\mathbf{x},e) = \left(\begin{pmatrix}1\\0\end{pmatrix}, \begin{pmatrix}0 & -\kappa\\ \kappa & 0\end{pmatrix}\right) ds,$$

where κ is the curvature of $\boldsymbol{\gamma}$.

Proposition 4.39 (Isoparametric surfaces, nonumbilic). *If* $\mathbf{x} : M \to \mathbf{R}^3$ *is a connected isoparametric surface with unit normal vector field* $\mathbf{n} : M \to \mathbf{S}^2$*, whose principal curvatures a and c are distinct with* $|a| > |c|$*, then* $c = 0$ *and* $\mathbf{x}(M)$ *is an open subset of a circular cylinder of radius* $1/|a|$.

Proof. The proof is a special case of Proposition 3.11. Replacing $\mathbf{n}$ by $-\mathbf{n}$, if necessary, we may assume $a > 0$. Having no umbilic points, $\mathbf{x}$ has a second order frame field about any point of M. Let$(\mathbf{x}, e) : U \to \mathbf{E}_+(3)$ be the second order frame field with $\mathbf{e}_3 = \mathbf{n}$ and

$$\omega_1^3 = a\omega^1 \quad \text{and} \quad \omega_2^3 = c\omega^2.$$

By the structure equations (4.37), $ac = 0$, so $c = 0$. The components of the pull back of the Maurer–Cartan form now look like

$$\omega^3 = 0, \quad \omega_2^1 = 0, \quad \omega_1^3 = a\omega^1, \quad \omega_2^3 = 0 \tag{4.51}$$

where a is a positive constant, and ω^1, ω^2 is an orthonormal coframe field on U. Regard (4.51) as the equations defining a 2-dimensional distribution $\mathfrak{h}$ on $\mathbf{E}_+(3)$. The structure equations imply that $\mathfrak{h}$ satisfies the Frobenius condition. Because the equations of $\mathfrak{h}$ are given in terms of left-invariant 1-forms with constant coefficients, it follows that it is a Lie subalgebra of $\mathscr{E}(3)$,

$$\mathfrak{h} = \left\{ \left(\begin{pmatrix} s \\ t \\ 0 \end{pmatrix}, \begin{pmatrix} 0 & 0 & -as \\ 0 & 0 & 0 \\ as & 0 & 0 \end{pmatrix} \right) : s, t \in \mathbf{R} \right\}.$$

If H is the maximal integral submanifold of $\mathfrak{h}$ through the identity element of $\mathbf{E}_+(3)$, then H is the Lie subgroup of $\mathbf{E}(3)$ given by exponentiation of $\mathfrak{h}$,

$$H = \{ \left(\begin{pmatrix} \frac{1}{a}\sin as \\ t \\ \frac{1}{a}(1 - \cos as) \end{pmatrix}, \begin{pmatrix} \cos as & 0 & -\sin as \\ 0 & 1 & 0 \\ \sin as & 0 & \cos as \end{pmatrix} \right) : s, t \in R \}.$$

The other integral surfaces of $\mathfrak{h}$ are the right cosets of H. There is a second order frame field $(\mathbf{x}, e) : M \to \mathbf{E}_+(3)$ for which $\mathbf{e}_1$ is the principal direction of the positive principal curvature and $\mathbf{e}_3 = \mathbf{n}$, because it is unique at each point. Since M is connected, we have $(\mathbf{x}, e)(M)$ contained in a right coset of H, which must be $(\mathbf{x}(m_0), e(m_0))H$ for a point $m_0 \in M$. Hence

$$\mathbf{x}(M) = (\mathbf{x}(m_0), e(m_0))H\mathbf{0} = (\mathbf{x}(m_0), e(m_0))C(a)$$

where $C(a)$ is the circular cylinder $x^2 + (z - \frac{1}{a})^2 = \frac{1}{a^2} \subset \mathbf{R}^3$, since

$$C(a) = H\mathbf{0} = \{ {}^t(\frac{1}{a}\sin as, t, \frac{1}{a}(1 - \cos as)) : s, t \in \mathbf{R} \}.$$

□

4.7.1 Surfaces of revolution

Example 4.40 (Surfaces of revolution). In the half-plane given by the x^1x^3-plane of $\mathbf{R}^3$ where $x^1 > 0$, oriented by $dx^1 \wedge dx^3 > 0$, consider a regular, smooth *profile curve* $\gamma(u) = (f(u), 0, g(u))$, for u in a connected open interval J, where $f > 0$ on J. The *surface of revolution* obtained by revolving this curve around the x^3-axis is the immersion

$$\mathbf{x} : J \times \mathbf{S}^1 \to \mathbf{R}^3, \quad \mathbf{x}(u,v) = {}^t\,(f(u)\cos v, f(u)\sin v, g(u)). \tag{4.52}$$

Notice that v is a local coordinate on the complement of any point of the circle $\mathbf{S}^1$ and that dv is a smooth 1-form defined on all of $\mathbf{S}^1$. Then

$$d\mathbf{x} = \mathbf{x}_u du + \mathbf{x}_v dv = {}^t(\dot{f}\cos v, \dot{f}\sin v, \dot{g})\,du + {}^t(-f\sin v, f\cos v, 0)\,dv.$$

where $\mathbf{x}_u = \frac{\partial \mathbf{x}}{\partial u}$ and $\mathbf{x}_v = \frac{\partial \mathbf{x}}{\partial v}$. A first order frame field along $\mathbf{x}$ is given by

$$\mathbf{e}_1 = \frac{1}{w}\mathbf{x}_u = \frac{1}{w}{}^t(\dot{f}\cos v, \dot{f}\sin v, \dot{g}), \quad \mathbf{e}_2 = \frac{1}{f}\mathbf{x}_v = {}^t(-\sin v, \cos v, 0),$$

where $w = \sqrt{\dot{f}^2 + \dot{g}^2}$, and unit normal vector

$$\mathbf{e}_3 = \mathbf{e}_1 \times \mathbf{e}_2 = \frac{1}{w}{}^t(-\dot{g}\cos v, -\dot{g}\sin v, \dot{f}).$$

The corresponding coframe field is

$$\omega^1 = d\mathbf{x}\cdot\mathbf{e}_1 = w\,du, \quad \omega^2 = d\mathbf{x}\cdot\mathbf{e}_2 = f\,dv,$$

and then

$$\omega^3_1 = d\mathbf{e}_1\cdot\mathbf{e}_3 = \frac{\dot{f}\ddot{g} - \dot{g}\ddot{f}}{w^3}\omega^1 = \kappa\omega^1, \quad \omega^3_2 = d\mathbf{e}_2\cdot\mathbf{e}_3 = \frac{\dot{g}}{wf}\omega^2,$$

where $\kappa(u)$ is the curvature of the profile curve (see Example 4.38). This frame field is second order, smooth on all of $J \times \mathbf{R}/2\pi$. The principal curvatures are

$$a = \kappa, \quad c = \frac{\dot{g}}{wf}.$$

The induced metric and the second fundamental form on $J \times \mathbf{S}^1$ are the symmetric bilinear form fields

$$\begin{aligned} I &= \omega^1\omega^1 + \omega^2\omega^2 = w^2 du^2 + f^2 dv^2 \\ II &= \omega^3_1\omega^1 + \omega^3_2\omega^2 = \kappa\omega^1\omega^1 + \frac{\dot{g}}{wf}\omega^2\omega^2 \end{aligned} \tag{4.53}$$

The level curves of u are called *circles of latitude* or *parallels of latitude*. The level curves of v are called *meridians* (all congruent to the profile curve). Tangents to these level curves are principal directions. These level curves are thus *lines of curvature* of $\mathbf{x}$. The Gaussian and mean curvatures are

$$K = \frac{\kappa \dot{g}}{wf}, \quad H = \frac{1}{2}(\kappa + \frac{\dot{g}}{wf}).$$

An immersion of revolution possesses a globally defined smooth second order frame field. This is possible because our definition of immersion of revolution has excluded the possibility that the immersion meets the axis of rotation. See Example 4.11 for what can happen when the surface meets the axis of rotation.

The following example has a second order frame field defined everywhere, even though there are whole curves of umbilic points.

Example 4.41 (Curves of umbilics). Fix $L > 0$ and rotate the curve $x^1 = L/(1 + (x^3)^2)$ about the x^3-axis. This is parametrized by the immersion of revolution

$$\mathbf{x} : \mathbf{R} \times \mathbf{S}^1 \to \mathbf{R}^3, \quad \mathbf{x}(u,v) = {}^t(\frac{L}{1+u^2}\cos v, \frac{L}{1+u^2}\sin v, u),$$

with profile curve $\gamma(u) = \frac{L}{1+u^2}\epsilon_1 + u\epsilon_3$, $u \in \mathbf{R}$. Using the formulas in Example 4.40, we find the principal curvatures to be

$$a = \frac{-2L(3u^2-1)(1+u^2)^3}{(4L^2u^2 + (1+u^2)^4)^{3/2}}, \quad c = \frac{(1+u^2)^3}{L(4L^2u^2 + (1+u^2)^4)^{1/2}}. \tag{4.54}$$

Therefore, when $u = 0$

$$a = 2L, \quad c = 1/L.$$

If $L = 1/\sqrt{2}$, then the circle of latitude $u = 0$ consists entirely of umbilic points. As with any surface of revolution, however, the second order frame field constructed in Example 4.40 is defined at every point of this surface.

Exercise 13. Prove that when u is arclength parameter of the profile curve of a surface of revolution, then

$$\omega_2^1 = -\dot{f} dv$$

and the Gaussian curvature

$$K = -\frac{\ddot{f}}{f}. \tag{4.55}$$

Example 4.42 (The pseudosphere). By (4.55), the surface of revolution, with profile curve parametrized by arclength, has constant Gaussian curvature equal to -1 if and only if $f(u)$ satisfies the differential equation

$$\ddot{f} - f = 0,$$

whose general solution is

$$f(u) = A\cosh u + B\sinh u,$$

where A and B are arbitrary constants, subject only to the requirements $f > 0$ and $\dot{f}^2 \leq 1$ on the interval J. Without loss of generality we may assume $0 \in J$ so that $A = f(0)$ and $B = \dot{f}(0)$ are the initial conditions on f. The *pseudosphere* is the solution obtained in the case $A = B = 1$, in which case $f(u) = e^u$ and

$$\dot{g}^2 = 1 - \dot{f}^2 = 1 - e^{2u}$$

requires that $J = -\infty < u \leq 0$, and thus $0 < f \leq 1$. There is no loss in generality in assuming the initial condition $g(0) = 0$. For convenience we assume that $g \geq 0$ on J, which amounts to taking the minus sign

$$\dot{g} = -\sqrt{1 - f^2} = -\sqrt{1 - e^{2u}}.$$

This profile curve $\gamma(u) = f(u)\epsilon_1 + g(u)\epsilon_3$ is called the *tractrix*. Since $df = f du$ and $dg = -\sqrt{1 - f^2}du$, the tractrix satisfies the differential equation

$$dg = -\frac{\sqrt{1 - f^2}}{f} df.$$

It is a simple exercise to show that 1 is the length of the segment of the tangent line at $\gamma(u)$ from $\gamma(u)$ to where it meets the ϵ_3-axis, for every $u \in J$. In order to solve for g, we abandon the arc-length parameter u and make the substitution $f = \sin t$, $0 < t \leq \pi/2$, in which case we get

$$g = -\int \frac{\cos^2 t}{\sin t} dt = \int (\sin t - \csc t)\, dt = -\cos t + \log|\csc t + \cot t| + C. \tag{4.56}$$

The constant of integration $C = 0$ in order to have $g = 0$ when $f = 1$, that is, when $t = \pi/2$. We arrive at the *pseudosphere* $\mathbf{x} : J \times S^1 \to \mathbf{R}^3$ given by (4.52) for $f(t) = \sin t$ and $g(t)$ given by (4.56).

Remark 4.43. An immersion $\mathbf{x} : M^2 \to \mathbf{R}^3$ whose induced metric has constant Gaussian curvature $K = -1$ is called a pseudospherical immersion. By an 1875 theorem of Bäcklund, these occur in pairs whose corresponding points are joined by tangent line segments of a fixed length and making a fixed angle with the normals.

This correspondence is related to the Bäcklund transformation of the sine-Gordon equation. For details and references see Chern [46] or Chern and Tenenblatt [49]. Mary Shepherd [151] used the method of moving frames to study the Bäcklund correspondence from the point of view of surfaces immersed in the four-dimensional space of all lines in $\mathbf{R}^3$, on which $\mathbf{E}(3)$ acts transitively.

Example 4.44. In 1841 Charles Delaunay [59] found all surfaces of revolution whose mean curvature is constant. The profile curve in Delaunay's examples is a *roulette of a conic*, which is the trace of a focus of a conic section as it rolls without slipping along one of its tangent lines. These consist of a catenary (from a parabola), undulary (from an ellipse), nodary (from an hyperbola), a straight line parallel to the axis (from a circle), or a semicircle centered on the axis of revolution (from a line segment).

Theorem 4.45 (Delaunay). *The complete immersed surfaces of revolution in* $\mathbf{R}^3$ *with constant mean curvature are those obtained by rotating about their axes the roulettes of the conics.*

See, for example, Eells [63] for a modern exposition, with proofs, of these examples. Given constants $a > b > 0$, the parametrized ellipse $\mathbf{x}(t) = a\cos t + b\sin t$ has *eccentricity* $e = \frac{\sqrt{a^2-b^2}}{a}$, so $0 < e < 1$. In the limit as $e \to 0$ the ellipse becomes a circle, and as $e \to 1$ it becomes a line segment. Figures 4.2 and 4.3 show unduloids coming from an ellipse with eccentricity close to 0 through eccentricity close to 1, respectively.

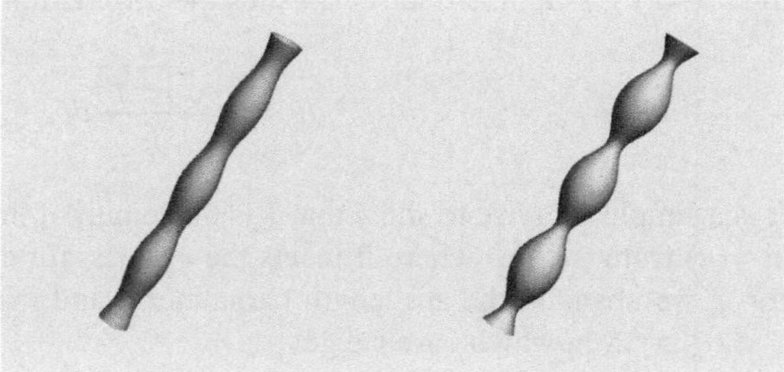

Fig. 4.2 Unduloids from an ellipse with $e = 0.222205$ and $e = 0.484123$, respectively.

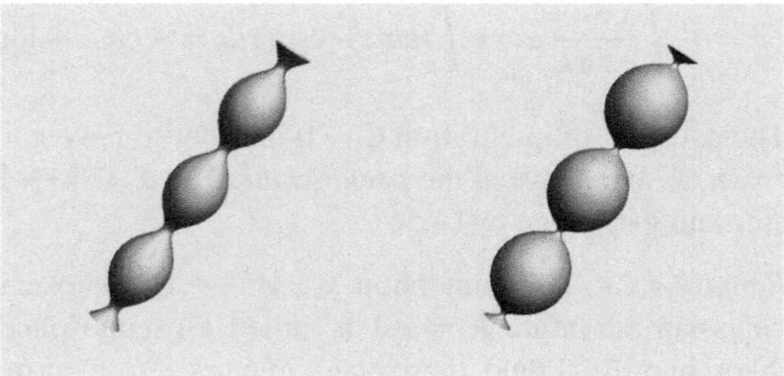

Fig. 4.3 Unduloids from an ellipse with $e = 0.661438$ and $e = 0.866025$, respectively.

4.8 New immersions from old

Given an immersion $\mathbf{x} : M^2 \to \mathbf{R}^3$ whose image is contained in an open subset $V \subset \mathbf{R}^3$, and given a diffeomorphism $F : V \to F(V) \subset \mathbf{R}^3$, one obtains a new immersion $\tilde{\mathbf{x}} = F \circ \mathbf{x} : M \to \mathbf{R}^3$. With a few exceptions, the geometry of $\tilde{\mathbf{x}}$ will be quite unrelated to the geometry of $\mathbf{x}$. One exception is when F is an isometry of $\mathbf{R}^3$, in which case the geometry of $\tilde{\mathbf{x}}$ is the same as that of $\mathbf{x}$, except for orientation dependent concepts, which generally change sign if F is orientation reversing. Another important exception is when F is inversion in a sphere.

Example 4.46 (Inversion). Inversion in the unit sphere $\mathbf{S}^2$ is

$$\mathscr{I} : \mathbf{R}^3 \setminus \{\mathbf{0}\} \to \mathbf{R}^3, \quad \mathscr{I}(\mathbf{x}) = \frac{\mathbf{x}}{|\mathbf{x}|^2}.$$

Then $\mathscr{I} \circ \mathscr{I} = \mathscr{I}$ shows that $\mathscr{I} = \mathscr{I}^{-1}$ and $\mathscr{I}$ is a diffeomorphism. Its differential at a point $\mathbf{x} \in \mathbf{R}^3 \setminus \{\mathbf{0}\}$ is

$$d\mathscr{I}_{\mathbf{x}} = \frac{1}{|\mathbf{x}|^2} d\mathbf{x} - \frac{2(\mathbf{x} \cdot d\mathbf{x})}{|\mathbf{x}|^4} \mathbf{x}.$$

Then

$$d\mathscr{I}_{\mathbf{x}} \cdot d\mathscr{I}_{\mathbf{x}} = \frac{1}{|\mathbf{x}|^4} d\mathbf{x} \cdot d\mathbf{x}$$

shows that $\mathscr{I}$ is a *conformal diffeomorphism*, as it satisfies Definition 12.1. Thus, $d\mathscr{I}_{\mathbf{x}}$ preserves angles and multiplies lengths by $1/|\mathbf{x}|^2$.

Example 4.47 (Inversion of a surface). Suppose an immersion $\mathbf{x} : M^2 \to \mathbf{R}^3$ never hits the origin of $\mathbf{R}^3$. Then we have the new immersion

$$\tilde{\mathbf{x}} = \mathscr{I} \circ \mathbf{x} = \frac{\mathbf{x}}{|\mathbf{x}|^2} : M \to \mathbf{R}^3.$$

If $(\mathbf{x}, (\mathbf{e}_1, \mathbf{e}_2, \mathbf{e}_3))$ is a first order frame field along $\mathbf{x}$ on an open subset $U \subset M$, let

$$\tilde{\mathbf{e}}_i = |\mathbf{x}|^2 d\mathscr{I}_{\mathbf{x}} \mathbf{e}_i = \mathbf{e}_i - \frac{2\mathbf{x} \cdot \mathbf{e}_i}{|\mathbf{x}|^2} \mathbf{x},$$

for $i = 1, 2, 3$. Then $(\tilde{\mathbf{x}}, (\tilde{\mathbf{e}}_1, \tilde{\mathbf{e}}_2, \tilde{\mathbf{e}}_3))$ is a first order frame field along $\tilde{\mathbf{x}}$ on U. From the calculation

$$d\tilde{\mathbf{e}}_3 = \sum_1^2 \omega_3^i \mathbf{e}_i - 2d\left(\frac{\mathbf{x} \cdot \mathbf{e}_3}{|\mathbf{x}|^2}\right) \mathbf{x} - \frac{2\mathbf{x} \cdot \mathbf{e}_3}{|\mathbf{x}|^2} \sum_1^2 \omega^i \mathbf{e}_i$$

we get

$$\tilde{\omega}_i^3 = \omega_i^3 + \frac{2\mathbf{x}\cdot\mathbf{e}_3}{|\mathbf{x}|^2}\omega^i,$$

for $i = 1,2$. If the functions a and c are the principal curvatures of $\mathbf{x}$, and if $(\mathbf{x}, e)$ is a second order frame field with $\omega_1^3 = a\omega^1$ and $\omega_2^3 = c\omega^2$, then

$$\tilde{\omega}_1^3 = (a|\mathbf{x}|^2 + 2\mathbf{x}\cdot\mathbf{e}_3)\tilde{\omega}^1, \quad \tilde{\omega}_2^3 = (c|\mathbf{x}|^2 + 2\mathbf{x}\cdot\mathbf{e}_3)\tilde{\omega}^2$$

shows that $(\tilde{\mathbf{x}}, \tilde{e})$ is a second order frame field along $\tilde{\mathbf{x}}$, the principal curvatures of $\tilde{\mathbf{x}}$ are

$$\tilde{a} = a|\mathbf{x}|^2 + 2\mathbf{x}\cdot\mathbf{e}_3, \quad \tilde{c} = c|\mathbf{x}|^2 + 2\mathbf{x}\cdot\mathbf{e}_3,$$

and $\tilde{\mathbf{x}}$ has the same lines of curvature as $\mathbf{x}$. It also follows from these formulae that $m \in M$ is an umbilic point of $\mathbf{x}$ if and only if it is an umbilic point for $\tilde{\mathbf{x}}$, provided that m is in the closure of the set of non-umbilic points of $\mathbf{x}$. Any point in the complement of this closure must be contained in an open set of umbilic points, on which there is thus a second order frame field and so again the statement is true.

If $\mathbf{x}$ is a canal immersion, say with a constant along its lines of curvature, which are the $\omega^2 = 0$ curves, then $da = a_1\omega^1 + a_2\omega^2$ implies that $a_1 = 0$ on the domain of the second order frame field $(\mathbf{x}, e)$, and thus

$$d\tilde{a} = (|\mathbf{x}|^2 a_2 + 2(a-c)\mathbf{x}\cdot\mathbf{e}_2)\omega^2$$

shows that $\tilde{a}_1 = 0$, so $\tilde{\mathbf{x}}$ is a canal immersion also. By the same reasoning, if $\mathbf{x}$ is a Dupin immersion, then $\tilde{\mathbf{x}}$ is a Dupin immersion. Inversion of a nonumbilic isoparametric immersion is not isoparametric, for if $a \neq c$ are constant on M, then $\tilde{a}$ and $\tilde{c}$ are constant on M only if $|\mathbf{x}|^2$ is constant, which is not possible if $\mathbf{x}$ is nonumbilic. An isoparametric immersion is, a fortiori, Dupin, so its inversion is Dupin.

Example 4.48. The circular cylinder of radius $R > 0$,

$$\mathbf{x} : \mathbf{S}^1 \times \mathbf{R} \to \mathbf{R}^3, \quad \mathbf{x}(s,t) = {}^t(R\cos s, R\sin s, t),$$

is isoparametric with principal curvatures $a = -1/R$ and $c = 0$ relative to the normal in the direction of $\mathbf{x}_s \times \mathbf{x}_t$. The axis of this cylinder passes through the origin. Its inversion is the immersion of revolution whose profile curve is the circle of radius $1/R$ with one point omitted, $\gamma(s) = {}^t(\frac{1}{R}(1+\cos s), 0, \sin s)$, where $-\pi < s < \pi$, as shown in Figure 4.4.

If the cylinder is translated so that the origin lies outside of it, say it becomes the circular cylinder $\mathbf{x}(s,t) = {}^t(2+\cos s, \sin s, t)$ then its inversion is illustrated in Figure 4.5.

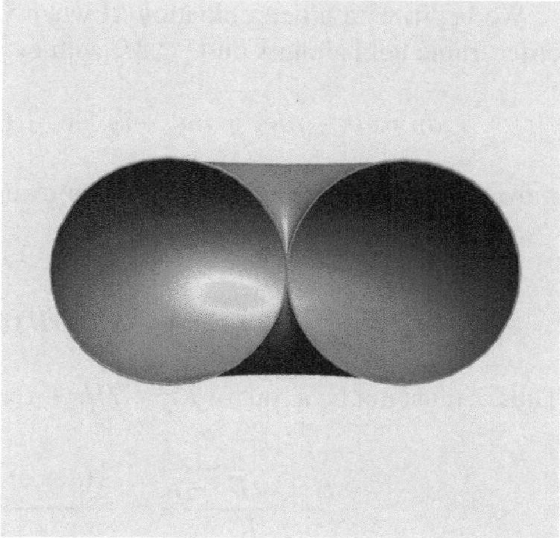

Fig. 4.4 Inversion of $\mathbf{x}(s,t) = {}^t(R\cos s, R\sin s, t)$, opened to show detail.

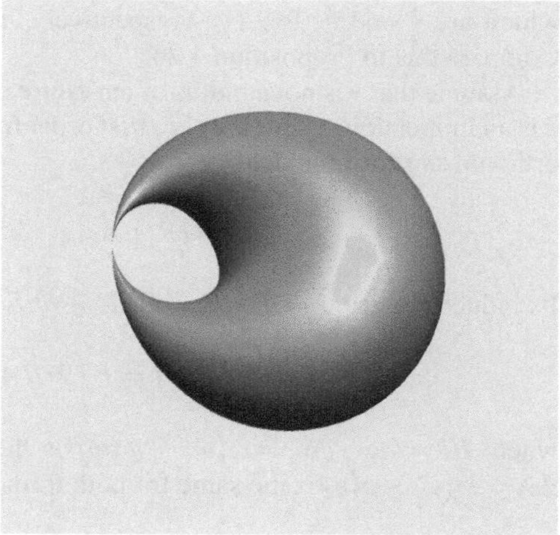

Fig. 4.5 Inversion of the cylinder $\mathbf{x}(s,t) = {}^t(2+\cos s, \sin s, t)$

4.8.1 Parallel transformations

Let $\mathbf{x} : M \to \mathbf{R}^3$ be an immersion with a smooth unit normal vector field $\mathbf{n}$. For any constant $r \in \mathbf{R}$, the *parallel transformation* of this oriented immersion by r is the map

$$\tilde{\mathbf{x}} = \mathbf{x} + r\mathbf{n} : M \to \mathbf{R}^3. \tag{4.57}$$

In general, a parallel transformation of $\mathbf{x}$ does not come from composing a diffeomorphism of $\mathbf{R}^3$ with $\mathbf{x}$. It is a special case of a Lie sphere transformation. These are discussed in detail in Section 15.3.

We begin with a determination of when $\tilde{\mathbf{x}}$ is an immersion. If $(\mathbf{x}, e)$ is any first order frame field along $\mathbf{x}$ on $U \subset M$, with $\mathbf{e}_3 = \mathbf{n}$ on U, then

$$d\tilde{\mathbf{x}} = d\mathbf{x} + rd\mathbf{e}_3 = (\omega^1 + r\omega_3^1)\mathbf{e}_1 + (\omega^2 + r\omega_3^2)\mathbf{e}_2 = \tilde{\omega}^1\mathbf{e}_1 + \tilde{\omega}^2\mathbf{e}_2$$

shows that $\tilde{\mathbf{x}}$ is an immersion at any point where

$$\begin{aligned} 0 \neq \tilde{\omega}^1 \wedge \tilde{\omega}^2 &= \omega^1 \wedge \omega^2 + r(\omega^1 \wedge \omega_3^2 + \omega_3^1 \wedge \omega^2) + r^2 \omega_3^1 \wedge \omega_3^2 \\ &= (1 + r^2 K - 2rH)\, \omega^1 \wedge \omega^2. \end{aligned}$$

Thus, r must not be a root of $Kr^2 - 2Hr + 1 = 0$. These roots are

$$\frac{H \pm \sqrt{H^2 - K}}{K} = \frac{\frac{1}{2}(a+c) \pm \sqrt{\frac{1}{4}(a+c)^2 - ac}}{ac},$$

which are $\frac{1}{a}$ and $\frac{1}{c}$, for $+/-$ respectively. These are the *radii of curvature* of $\mathbf{x}$. Compare this to Proposition 4.26.

Assume that r is not a radius of curvature of $\mathbf{x}$, so that the parallel transformation $\tilde{\mathbf{x}}$ is an immersion. Then $(\tilde{\mathbf{x}}, e)$ is a first order frame field along $\tilde{\mathbf{x}}$ on U with associated orthonormal coframe field

$$\tilde{\omega}^1 = \omega^1 + r\omega_3^1, \quad \tilde{\omega}^2 = \omega^2 + r\omega_3^2. \tag{4.58}$$

Its induced metric $\tilde{I} = d\tilde{\mathbf{x}} \cdot d\tilde{\mathbf{x}} = \tilde{\omega}^1\tilde{\omega}^1 + \tilde{\omega}^2\tilde{\omega}^2$ is

$$\tilde{I} = I - 2rII + r^2 III$$

where $III = d\mathbf{e}_3 \cdot d\mathbf{e}_3 = \omega_3^1\omega_3^1 + \omega_3^2\omega_3^2$ is the *third fundamental form* of $\mathbf{x}$. Since $d\mathbf{e}_3 = \omega_3^1\mathbf{e}_1 + \omega_3^2\mathbf{e}_2$ is the same for both frames, we have

$$\tilde{\omega}_1^3 = \omega_1^3, \quad \tilde{\omega}_2^3 = \omega_2^3. \tag{4.59}$$

By (4.10) and (4.58), we then have

$$h_{ik}\omega^k = \omega_i^3 = \tilde{\omega}_i^3 = \tilde{h}_{ij}\tilde{\omega}^j = \tilde{h}_{ij}(\omega^j + r\omega_3^j) = \tilde{h}_{ij}(\delta_{jk} - rh_{jk})\omega^k,$$

which implies that the symmetric matrices $S = (h_{ij})$ and $\tilde{S} = (\tilde{h}_{ij})$ satisfy

$$S = \tilde{S}(I - rS).$$

We can solve for $\tilde{S}$ provided that $\det(I - rS) \neq 0$, which is equivalent to the condition that r^{-1} not be a principal curvature of $\mathbf{x}$. In that case, $\tilde{S} = S(I - rS)^{-1}$, so the principal curvatures of $\tilde{\mathbf{x}}$ are the solutions $\tilde{a}$ and $\tilde{c}$ of the quadratic equation in t,

$$0 = \det(\tilde{S} - tI) = \det((I - rS)^{-1}(1 + rt)(S - \frac{t}{1+rt}I)).$$

If a and c are the principal curvatures of $\mathbf{x}$, then

$$\tilde{a} = \frac{a}{1 - ra}, \quad \tilde{c} = \frac{c}{1 - rc}. \tag{4.60}$$

In particular, a point $m \in M$ is umbilic for the parallel surface $\tilde{\mathbf{x}} = \mathbf{x} + r\mathbf{e}_3$ if and only if it is umbilic for $\mathbf{x}$. A frame field $(\mathbf{x}, \mathbf{e})$ is second order for $\mathbf{x}$ if and only if $(\tilde{\mathbf{x}}, \mathbf{e})$ is second order for $\tilde{\mathbf{x}}$, as can be seen from (4.59) and (4.58). The mean and Gauss curvatures of $\mathbf{x}$ and $\tilde{\mathbf{x}}$ are related by

$$\tilde{H} = \frac{H - rK}{1 - 2rH + r^2K}, \quad \tilde{K} = \frac{K}{1 - 2rH + r^2K}. \tag{4.61}$$

Formulas (4.60) show that the immersions parallel to an isoparametric (respectively, Dupin or canal) immersion are also isoparametric (respectively, Dupin or canal).

Definition 4.49. A point $\mathbf{y} \in \mathbf{R}^3$ is a *focal point* of $\mathbf{x}$ if $\mathbf{y} = \tilde{\mathbf{x}}(m) = \mathbf{x}(m) + r\mathbf{e}_3(m)$ for some point $m \in M$ and some $r \in \mathbf{R}$ for which $d\tilde{\mathbf{x}}_{(m)}$ is singular. The *multiplicity* of the focal point is the dimension of the kernel of $d\tilde{\mathbf{x}}_{(m)}$. The set of all focal points of $\mathbf{x}$ is called the *focal locus* of $\mathbf{x}$.

We have the following remarkable result of Bonnet's.

Theorem 4.50 (Bonnet [15]). *Let $\mathbf{x} : M \to \mathbf{R}^3$ be an immersion with constant mean curvature $H = (a + c)/2 \neq 0$ relative to the normal field $\mathbf{e}_3$. Consider the parallel surface $\tilde{\mathbf{x}} = \mathbf{x} + r\mathbf{e}_3$.*

If $r = \frac{1}{2H}$, then $\tilde{\mathbf{x}}$ has constant Gaussian curvature $\tilde{K} = 4H^2$.

If $r = 1/H$, then $\tilde{\mathbf{x}}$ has constant mean curvature $\tilde{H} = -H$.

If $\tilde{\mathbf{x}}$ is a parallel surface of $\mathbf{x}$ then $\mathbf{x}$ is a parallel surface of $\tilde{\mathbf{x}}$. Consequently, we can restate the result as follows. If $\mathbf{x}$ has constant positive Gaussian curvature K, then $\tilde{\mathbf{x}} = \mathbf{x} \pm \frac{1}{\sqrt{K}}\mathbf{e}_3$ has constant mean curvature $\tilde{H} = -\pm\frac{\sqrt{K}}{2}$.

Proof. The results follow from (4.61). □

Remark 4.51. In the case of a constant mean curvature immersion $\mathbf{x}$, the mean curvature H is a principal curvature at a point if and only if the point is umbilic, while $2H$ is a principal curvature if and only if the point is parabolic (meaning $0 = a < c$). In the case of constant positive Gaussian curvature K, we have $\sqrt{K}$ a principal curvature at a point if and only if the point is umbilic, while $-\sqrt{K}$ is never a principal curvature.

Fig. 4.6 A tube about a space curve.

4.8.2 Tubes

Let $\mathbf{f}(s)$ be a smooth immersed curve in $\mathbf{R}^3$ parametrized by arclength $s \in J$, for some connected interval J. Let $\mathbf{T} = \dot{\mathbf{f}}$, $\mathbf{N}$, $\mathbf{B}$ be its Frenet frame along this curve. We assume that the curvature $\kappa(s)$ is positive at every point to insure that the Frenet frame is defined and smooth on all of J. The Frenet-Serret equations for $\mathbf{f}$ are

$$\dot{\mathbf{f}} = \mathbf{T}, \quad \dot{\mathbf{T}} = \kappa\mathbf{N}, \quad \dot{\mathbf{N}} = -\kappa\mathbf{T} + \tau\mathbf{B}, \quad \dot{\mathbf{B}} = -\tau\mathbf{N} \tag{4.62}$$

where $\tau : J \to \mathbf{R}$ is the torsion. Let r be a positive constant and define the *tube about* $\mathbf{f}$ of radius r to be the map

$$\mathbf{x} : J \times \mathbf{R} \to \mathbf{R}^3, \quad \mathbf{x}(s,t) = \mathbf{f}(s) + r(\cos t\,\mathbf{N}(s) + \sin t\,\mathbf{B}(s)). \tag{4.63}$$

See figure 4.6.

Then

$$\begin{aligned}
\mathbf{x}_s &= (1 - r\kappa\cos t)\mathbf{T} + r\tau(-\sin t\,\mathbf{N} + \cos t\,\mathbf{B}),\\
\mathbf{x}_t &= r(-\sin t\,\mathbf{N} + \cos t\,\mathbf{B}),\\
\mathbf{x}_s \times \mathbf{x}_t &= -r(1 - r\kappa\cos t)(\cos t\,\mathbf{N} + \sin t\,\mathbf{B}).
\end{aligned}$$

Assume that $r < 1/\kappa$, so that $1 - r\kappa\cos t > 0$ and $\mathbf{x}$ will be an immersion. A first order frame field $(\mathbf{x}, \mathbf{e})$ is defined along $\mathbf{x}$ by

$$\begin{aligned}
\mathbf{e}_3 &= -\cos t\,\mathbf{N} - \sin t\,\mathbf{B},\\
\mathbf{e}_1 &= \mathbf{T} = \frac{1}{1 - r\kappa\cos t}(\mathbf{x}_s - \tau\mathbf{x}_t),\\
\mathbf{e}_2 &= \mathbf{e}_3 \times \mathbf{e}_1 = -\sin t\,\mathbf{N} + \cos t\,\mathbf{B} = \frac{1}{r}\mathbf{x}_t.
\end{aligned} \tag{4.64}$$

Then

$$\mathbf{x}_s = (1 - r\kappa\cos t)\mathbf{e}_1 + r\tau\mathbf{e}_2, \quad \mathbf{x}_t = r\mathbf{e}_2$$

shows that the coordinate curves are orthogonal if and only if τ is identically zero. The associated orthonormal coframe field is

$$\omega^1 = d\mathbf{x}\cdot\mathbf{e}_1 = (1 - r\kappa\cos t)ds, \quad \omega^2 = d\mathbf{x}\cdot\mathbf{e}_2 = r(\tau ds + dt), \tag{4.65}$$

so that the area form is

$$\omega^1 \wedge \omega^2 = r(1 - r\kappa\cos t)ds \wedge dt$$

and the orientation induced on $M = J \times \mathbf{R}$ is that of $ds \wedge dt$. If $\mathbf{f}$ has finite length L, then the area of $\mathbf{x}(M)$ is

$$\int_{J\times[0,2\pi]} \omega^1 \wedge \omega^2 = \int_0^{2\pi}\int_0^L r(1 - r\kappa\cos t)\,ds dt = 2\pi rL,$$

a version of Pappus's Theorem. Moreover,

$$\begin{aligned} \omega_1^3 &= d\mathbf{e}_1\cdot\mathbf{e}_3 = -\kappa\cos t ds = \frac{-\kappa\cos t}{1 - r\kappa\cos t}\omega^1, \\ \omega_2^3 &= d\mathbf{e}_2\cdot\mathbf{e}_3 = dt + \tau ds = \frac{1}{r}\omega^2 \end{aligned} \tag{4.66}$$

imply that the frame field $(\mathbf{x},\mathbf{e})$ is second order along $\mathbf{x}$ and the principal curvatures are

$$a = \frac{-\kappa\cos t}{1 - r\kappa\cos t}, \quad c = \frac{1}{r}. \tag{4.67}$$

There are no umbilic points. The principal curvature c is constant on M, so $\mathbf{x}$ is a canal immersion. If $\mathbf{f}$ is a simple closed analytic curve, then the tube is a compact, analytic, canal surface. By (4.65), the lines of curvature of c are the coordinate curves $s = s_0$, for $s_0 \in J$ any constant. The lines of curvature of the principal curvature a are the integral curves of $dt + \tau ds = 0$, by (4.66).

In general, the focal locus associated to the principal curvature c, of a tube (4.63) about a curve $\mathbf{f}$, is $\mathbf{x} + c^{-1}\mathbf{e}_3$, which is just $\mathbf{f}$, the curve we began with. It is special when a focal locus of an immersion degenerates into a curve. In fact, this characterizes Dupin immersions. See Problem 4.81.

4.8.3 Curvature spheres along canal immersions

Let $\mathbf{x} : M \to \mathbf{R}^3$ be a canal immersion for which the principal curvature a is constant along its connected lines of curvature. In this subsection we want to prove that the curvature spheres relative to a are constant along the connected lines of curvature of a, and these lines of curvature are just the intersection of the curvature sphere with $\mathbf{x}(M)$. Moreover, $\mathbf{x}$ of these lines of curvature are line segments or arcs of circles. In particular, they are plane curves. These results are easily pictured for the case of circular tori of revolution, circular cylinders, and circular cones.

Exercise 14. Use the Rank Theorem to prove the following technical result needed in the proof of the next proposition. If a smooth map $f : M^2 \to N^n$ has rank equal to one at every point of M, then any connected level set of f is an embedded curve $\gamma : J \subset \mathbf{R} \to M$ for which there exists a non-zero vector $v \in T_qN$, where $q = f(s)$ for every $s \in J$, such that

$$df_{\gamma(s)} T_{\gamma(s)} M = \text{span } v \subset T_qN,$$

for every $s \in J$. For a statement and proof of the Rank Theorem, see Lee [110, Theorem 7.13, p. 167]. It is not true, in general, that $f(M)$ is a smooth curve in N. See Cecil-Ryan [40, Remark 4.7, pp. 143–144] for a counterexample.

Proposition 4.52. *Let $\mathbf{x} : M \to \mathbf{R}^3$ be an immersion with unit normal vector field $\mathbf{n} : M \to \mathbf{S}^2 \subset \mathbf{R}^3$ and with distinct principal curvatures at each point of M. Let $\gamma : J \to M$ be a connected line of curvature for the principal curvature a, where $J \subset \mathbf{R}$ is connected and contains 0. Then a is constant on $\gamma(J)$ if and only if its curvature sphere is constant on $\mathbf{x} \circ \gamma(J)$.*

If the principal curvature a is constant on each of its connected lines of curvature, then $\mathbf{x}$ sends its lines of curvature to circles or lines in $\mathbf{R}^3$.

See Cecil-Ryan [40, Chapter 2, Section 4] for a statement and proof of this proposition in arbitrary dimensions.

Proof. Let $(\mathbf{x}, e) : U \to \mathbf{E}(3)$ be a second order frame field on a neighborhood U containing $\gamma(J)$ such that $\mathbf{e}_1(\gamma(s)) = (\mathbf{x} \circ \gamma)'(s)$ for every $s \in J$ and $\mathbf{e}_3 = \mathbf{n}$ on U. Using the notation of Remark 4.12 and Exercise 10, we have

$$\begin{aligned}(\mathbf{x} \circ \gamma)'(s) &= d\mathbf{x}(\gamma'(s)) = \mathbf{e}_1(\gamma(s)),\\ (\mathbf{x} \circ \gamma)''(s) &= d\mathbf{e}_1(\gamma'(s)) = (p\mathbf{e}_2 + a\mathbf{e}_3)(\gamma(s))\end{aligned} \tag{4.68}$$

Suppose that a is never zero on U and consider the *focal map*

$$f : U \to \mathbf{R}^3, \quad f = \mathbf{x} + \frac{1}{a}\mathbf{n} = \mathbf{x} + \frac{1}{a}\mathbf{e}_3,$$

whose derivative is, by the structure equations

$$df = -\frac{a_1}{a^2}\omega^1 \mathbf{e}_3 + ((1 - \frac{c}{a})\mathbf{e}_2 - \frac{a_2}{a^2}\mathbf{e}_3)\omega^2, \tag{4.69}$$

where $da = a_1\omega^1 + a_2\omega^2$ on U. Then $f(\gamma(s)) = \mathbf{x}(\gamma(s)) + \frac{1}{a(\gamma(s))}\mathbf{e}_3(\gamma(s))$ is the center of the curvature sphere through $\mathbf{x}(\gamma(s))$. By (4.69) and the fact that $\omega^2(\gamma') = 0$, we have

$$(f \circ \gamma)' = -\frac{a_1 \circ \gamma}{(a \circ \gamma)^2} \mathbf{e}_3 \circ \gamma = \left(\frac{1}{a \circ \gamma}\right)' \mathbf{e}_3 \circ \gamma$$

and $(a \circ \gamma)' = a_1 \circ \gamma$ on J. Thus, the centers, $f(\gamma(s))$, of the curvature spheres along γ are constant if and only if their radii, $1/a(\gamma(s))$, are constant. But the curvature spheres along γ are constant if and only if their centers and radii are constant.

At a point $m \in M$ where $a(m) = 0$, the curvature sphere at $\mathbf{x}(m)$ is the tangent plane $\{\mathbf{y} \in \mathbf{R}^3 : \mathbf{y} \cdot \mathbf{n}(m) = \mathbf{x}(m) \cdot \mathbf{n}(m)\}$. If $a(\gamma(s)) = 0$ for all $s \in J$, then

$$(\mathbf{e}_3 \circ \gamma)'(s) = -(a \circ \gamma)(s)(\mathbf{e}_1 \circ \gamma)(s) = 0,$$

for all $s \in J$, which shows that $\mathbf{e}_3 \circ \gamma$ is constant on J and thus the tangent planes along $\mathbf{x} \circ \gamma$ are all parallel. It follows that they must coincide along the connected curve γ.

We have now proved that if $\gamma : J \to M$ is a connected line of curvature of a and if $a \circ \gamma$ is constant on J, then $\mathbf{x} \circ \gamma(J) \subset \mathbf{x}(M) \cap S$, where S is the necessarily constant curvature sphere (or plane) along $\mathbf{x} \circ \gamma$. If $a \circ \gamma$ is a non-zero constant, then the curve $\mathbf{x} \circ \gamma(J)$ lies in a sphere. A spherical curve is a circle if and only if it is a planar curve. By (4.69),

$$\begin{aligned} df_{\gamma(s)} T_{\gamma(s)} M &\subset \operatorname{span}\{\left((1 - \frac{c}{a})\mathbf{e}_2 - \frac{a_2}{a^2}\mathbf{e}_3\right)(\gamma(s))\} \\ &= \operatorname{span}\{(a\mathbf{e}_2 - p\mathbf{e}_3)(\gamma(s))\}, \end{aligned} \tag{4.70}$$

since $a_1 \circ \gamma = (a \circ \gamma)' = 0$ on J, and where we use the structure equation $a_2 = (a - c)p$ in (4.37). From (4.68) we see that $(\mathbf{x} \circ \gamma)'(s)$ and $(\mathbf{x} \circ \gamma)''(s)$ are both orthogonal to $(a\mathbf{e}_2 - p\mathbf{e}_3)(\gamma(s))$, so $\mathbf{x} \circ \gamma$ is a planar curve if and only if the vectors

$$(a\mathbf{e}_2 - p\mathbf{e}_3)(\gamma(s))$$

are all parallel, for all $s \in J$. Without further assumptions about the principal curvature a, these vectors are not all parallel, in general. See Problems 4.78 and 4.82.

Assume now that a is never zero and that it is constant on each of its connected lines of curvature. Then $a_1 = 0$ on all of U, and (4.69) shows that the focal map f has constant rank equal to 1 on M. Use the notation above for a connected line of curvature $\gamma : J \to M$ for the principal curvature a and for a second order frame field $(\mathbf{x}, e) : U \to \mathbf{E}(3)$, where $\gamma(J) \subset U$. Then $f(\gamma(J)) = \mathbf{y}_0 \in \mathbf{R}^3$ is a single point and by Exercise 14, there is a unit vector $\mathbf{v} \in \mathbf{R}^3$ such that

$$df_{\gamma(s)} T_{\gamma(s)} M = \mathbf{R}\mathbf{v},$$

for every $s \in J$. Then the vectors $(a\mathbf{e}_2 - p\mathbf{e}_3)(\gamma(s))$ are all non-zero multiples of $\mathbf{v}$, by (4.70), and therefore the curve $\mathbf{x} \circ \gamma(J)$ is in a plane orthogonal to $\mathbf{v}$. Being also a curve in the curvature sphere along $\gamma(J)$, this curve must be an arc of a circle.

We have yet to consider the case when a is constant on each of its connected lines of curvature and it is zero at some points of M. This case requires a different proof, which uses the same theorem for surfaces immersed in $\mathbf{S}^3$ (Proposition 5.16) together with stereographic projection (Proposition 5.23). See Problem 5.51.

For now, we will complete the proof under the added assumption that a is non-zero on a dense open subset M' of M. Any connected line of curvature of a in M' must be sent by $\mathbf{x}$ to an arc of a circle, by the argument above. From (4.68), the curvature κ of such an arc must satisfy

$$\kappa^2 = p^2 + a^2$$

and be constant on J, so $p \circ \gamma$ must be constant on J. Hence, $p_1 = 0$ at every point of M', and thus at every point of M, by continuity. This implies that $p \circ \gamma$ is constant for any connected line of curvature γ. If $a = 0$ at every point of $\gamma(J)$, then $\mathbf{x} \circ \gamma$ is in the plane orthogonal to the constant normal vector $\mathbf{e}_3 \circ \gamma$ and has constant curvature $p \circ \gamma$, by (4.68). It is thus an arc of a circle, if $p \neq 0$, or a segment of a line, if $p = 0$.

□

Exercise 15. Assume the preceding Proposition proved in general. Prove that if the principal curvature a is constant on each of its connected lines of curvature, and if $p^2 + a^2 \neq 0$ on each such line of curvature, then for each connected line of curvature $\gamma : J \to M$ of a, $\mathbf{x} \circ \gamma(J)$ is an arc of a circle, whose center is the point

$$\mathbf{x} \circ \gamma(s) + \frac{1}{p^2 + a^2}(p\mathbf{e}_2 + a\mathbf{e}_3) \circ \gamma(s),$$

and whose radius is $1/\sqrt{p^2 + a^2}$, independent of $s \in J$.

4.9 Elasticae

We briefly introduce elasticae here in preparation for their role in the Willmore problem. They are critical curves of a functional that is a one-dimensional version of the Willmore functional. Special cases of these curves first arose as solutions to a problem proposed by James Bernoulli in 1691. The modern definition of an elastica, given in 1744 by Euler, is

> Among all curves γ of the same length passing through points A and B tangent to given lines at A and B, the elasticae minimize $\int_\gamma \kappa^2 ds$.

For additional history see the dissertation [111] by R. Levien and the expository article [161] by C. Truesdell.

For our purposes here, the constraint will be a given free homotopy class.

Definition 4.53. A *free elastic curve* in a Riemannian surface (M^2, I) is a smooth immersion that minimizes the functional

$$\mathscr{F}(\gamma) = \int_\gamma \kappa^2 ds,$$

over all smooth immersions $\gamma : \mathbf{S}^1 \to M$ in a free homotopy class. Here κ is the curvature and s is arclength parameter of γ.

Free elastic curves in space form geometries have been studied by Bryant and Griffiths [22], Griffiths [80], and Langer and Singer [105]. In Euclidean geometry, a circle σ of radius $R > 0$ has curvature $\kappa = \pm 1/R$, so

$$\mathscr{F}(\sigma) = \int_\sigma \kappa^2 ds = \frac{2\pi}{R}$$

has no minimum on circles. This contrasts rather surprisingly with the situation in the hyperbolic plane.

Example 4.54 (Circles in the Poincaré disc). The unit disk $\mathbf{D}^2 = \{(x,y) \in \mathbf{R}^2 : x^2 + y^2 < 1\}$ with the Riemannian metric

$$I = \frac{4}{(1 - x^2 - y^2)^2}(dx^2 + dy^2)$$

is the *Poincaré disk model* of the hyperbolic plane. Orient it by $dx \wedge dy > 0$. For any angle $\theta \in \mathbf{R}$, the radial curve

$$\gamma^\theta : \mathbf{R} \to \mathbf{D}, \quad \gamma^\theta(s) = (\tanh\frac{s}{2})(\cos\theta, \sin\theta),$$

is the geodesic starting at $\gamma(0) = \mathbf{0}$ with initial velocity $\dot{\gamma}(0) = (\cos\theta, \sin\theta)$. If we fix $r > 0$, then the hyperbolic circle of hyperbolic radius r and center $\mathbf{0}$ is

$$C = \{\gamma^\theta(r) : \theta \in \mathbf{R}\} \subset \mathbf{D}^2,$$

which is the Euclidean circle centered at $\mathbf{0}$ with radius $\tanh\frac{r}{2}$. C is parametrized by the embedding

$$\sigma : \mathbf{S}^1 = \mathbf{R}/2\pi \to \mathbf{D}, \quad \sigma(t) = (\tanh\frac{r}{2})(\cos t, \sin t), \tag{4.71}$$

whose arclength parameter is $s = (\sinh r)t$ and whose geodesic curvature is the constant $\kappa = \frac{\cosh r}{\sinh r} = \coth r$. The integral $\int_\sigma \kappa^2 ds$ is minimized when r satisfies $\sinh r = 1$. See Problem 4.83.

4.10 Willmore problems

In his 1965 paper [171], T. Willmore introduced the non-negative functional $\widetilde{\mathscr{W}}$ on the set of all immersions $\mathbf{x} : M \to \mathbf{R}^3$ of a given compact oriented surface M,

$$\widetilde{\mathscr{W}}(\mathbf{x}) = \int_M H^2 dA,$$

where H is the mean curvature and dA is the area element of the immersion $\mathbf{x}$. He asked for the infimum of $\widetilde{\mathscr{W}}(\mathbf{x})$ over all immersions of a given surface M. If a and c are the principal curvatures of $\mathbf{x}$, then $H^2 = K + \frac{1}{4}(a-c)^2$ and the Gauss-Bonnet Theorem imply

$$\widetilde{\mathscr{W}}(\mathbf{x}) = 2\pi\chi(M) + \frac{1}{4}\int_M (a-c)^2 dA, \tag{4.72}$$

where $\chi(M)$ is the Euler characteristic of M. A compact oriented surface M is determined up to homeomorphism by a nonnegative integer g, called its *genus*, which is related to its Euler characteristic by $\chi(M) = 2-2g$. The Euler characteristic of M is non-negative only in the cases $g = 0$, when M is homeomorphic to the sphere, or $g = 1$, when M is homeomorphic to the torus $T^2 = S^1 \times S^1$.

In the $g = 0$ case, we conclude from (4.72) that

$$\widetilde{\mathscr{W}}(\mathbf{x}) \geq 4\pi,$$

with equality if and only if $\mathbf{x}$ is totally umbilic, in which case $\mathbf{x}(M)$ must be a Euclidean round sphere by Theorem 4.23. That this is independent of the radius of the sphere suggests that the Willmore functional is invariant under *homotheties*, that is, under transformations of $\mathbf{R}^3$ given by multiplication by a positive constant.

In the $g = 1$ case, Willmore calculated his functional on a circular torus of revolution and arrived at the following result.

Example 4.55 ([171]). For constants $R > r > 0$, consider the circular torus of revolution $\mathbf{x} : \mathbf{S}^1 \times \mathbf{S}^1 \to \mathbf{R}^3$,

$$\mathbf{x}(u,v) = {}^t((R + r\cos\frac{u}{r})\cos v, (R + r\cos\frac{u}{r})\sin v, r\sin\frac{u}{r}), \tag{4.73}$$

obtained by rotating the profile circle ${}^t(R + r\cos\frac{u}{r}, 0, r\sin\frac{u}{r})$ about the ϵ_3-axis (see Figure 4.7). The calculations of Section 4.40 show that the principal curvatures of $\mathbf{x}$ are $a = \frac{1}{r}$ and $c = \frac{\cos\frac{u}{r}}{R+r\cos\frac{u}{r}}$, its area form is $dA = (R + r\cos\frac{u}{r})du \wedge dv$, and thus

$$\widetilde{\mathscr{W}}(\mathbf{x}) = \frac{R^2}{4r^2}\int_0^{2\pi}\int_0^{2\pi r} \frac{1}{R + r\cos\frac{u}{r}} du dv = \frac{\pi^2(\frac{R}{r})^2}{\sqrt{(\frac{R}{r})^2 - 1}},$$

whose minimum $2\pi^2$ occurs if and only if $\frac{R}{r} = \sqrt{2}$; that is, the right triangle in Figure 4.7 is isosceles.

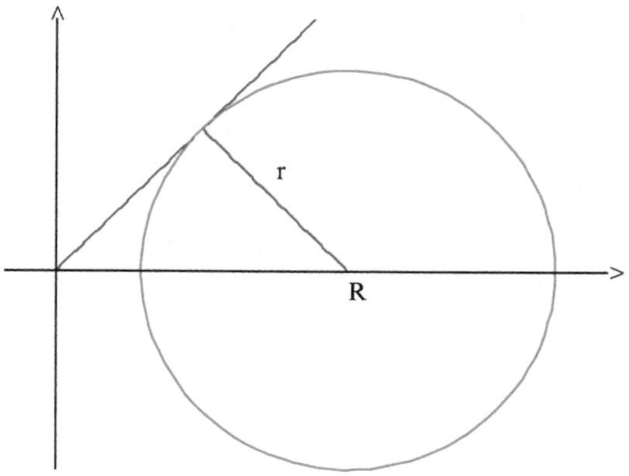

Fig. 4.7 Profile circle of a torus of revolution with $R/r = \sqrt{2}$.

Willmore then asked if $2\pi^2$ is the absolute minimum of his functional over all immersions of a torus. His question is known as the

Willmore Conjecture 1. If T^2 is a compact oriented surface of genus 1, then for any smooth immersion $\mathbf{x} : T^2 \to \mathbf{R}^3$,

$$\widetilde{\mathscr{W}}(\mathbf{x}) \geq 2\pi^2,$$

with equality if and only if $\mathbf{x} : T^2 \to \mathbf{R}^3$ is a circular torus of revolution (4.73) with $\frac{R}{r} = \sqrt{2}$.

Of course, Willmore also asked what is the minimum value of $\widetilde{\mathscr{W}}$ on immersions of surfaces of genus $g \geq 2$, but he did not offer a conjecture as to the value of these minima. In the light of (4.72), it seems natural to replace $\widetilde{\mathscr{W}}$ with the functional

$$\mathscr{W}(\mathbf{x}) = \int_M (H^2 - K)dA = \frac{1}{4}\int_M (a-c)^2 dA, \tag{4.74}$$

which is positive for all $\mathbf{x}$ except for totally umbilic immersions of the sphere. In [170], J. White proved that the integrand itself, $(H^2 - K)dA$, is invariant under the transformation

$$\mathscr{I} : \mathbf{R}^3 \setminus \{\mathbf{0}\} \to \mathbf{R}^3, \quad \mathscr{I}(\mathbf{x}) = \frac{\mathbf{x}}{|\mathbf{x}|^2},$$

which is inversion in the unit sphere with center at the origin (see Examples 4.46 and 12.5). By the Liouville Theorem 12.7 and Problem 4.84, it follows that $(H^2 - K)dA$ is invariant under any local conformal diffeomorphism of $\mathbf{R}^3$. The study of the functional $\mathscr{W}(\mathbf{x})$ belongs naturally in Möbius geometry. It is taken up in Section 13.6.

In his Math Review of [170], Willmore reported that in the 1923 paper [160], G. Thomsen had proved that the integrand of $\mathscr{W}$ is invariant under conformal transformations. Actually the concept of the Willmore energy appears already in the 1821 work of S. Germain [73]. Thomsen proved important results about this functional, which he called the *conformal area*. He worked in the context of Möbius geometry, which we shall do in Chapters 12 through 14. Section 14.5 contains an exposition of a global version of Thomsen's results.

In the calculus of variations, the first step in finding the minima of $\mathcal{W}(\mathbf{x})$ on immersions $\mathbf{x} : M \to \mathbf{R}^3$, for a given compact oriented M, is to find its critical points. An immersion $\mathbf{x} : M \to \mathbf{R}^3$ is a *critical point* of $\mathcal{W}(\mathbf{x})$ if for any 1-parameter family of immersions $\mathbf{x}_t : M \to \mathbf{R}^3$, for $|t| < \epsilon$, for some $\epsilon > 0$,

$$\left.\frac{d}{dt}\right|_{t=0} \mathcal{W}(\mathbf{x}_t) = 0.$$

Since $\widetilde{\mathcal{W}}(\mathbf{x}) - \mathcal{W}(\mathbf{x})$ is a constant depending only on M, it follows that the critical points of $\widetilde{\mathcal{W}}$ are the same as the critical points of $\mathcal{W}$. Thomsen states that $\mathbf{x} : M \to \mathbf{R}^3$ is a critical point of $\mathcal{W}$ if and only if the mean curvature H, Gauss curvature K, and Laplace-Beltrami operator Δ of $\mathbf{x}$ satisfy

$$\Delta H + 2H(H^2 - K) = 0, \tag{4.75}$$

on M. This is the *Euler-Lagrange equation* of the Willmore functional $\mathcal{W}$. In a footnote containing no publication citation, Thomsen attributes the derivation of this Euler-Lagrange equation to work done in 1922 by W. Schadow.

Definition 4.56. A *Willmore immersion* $\mathbf{x} : M \to \mathbf{R}^3$ of a compact oriented surface M is a critical point $\mathbf{x} : M \to \mathbf{R}^3$ of $\mathcal{W}$.

By Schadow's result, an immersion $\mathbf{x} : M \to \mathbf{R}^3$ of a compact oriented surface M is Willmore if and only if (4.75) holds for $\mathbf{x}$. The condition (4.75) does not require that M be compact or oriented. The following is Thomsen's terminology for this more general case.

Definition 4.57. A *conformally minimal immersion* $\mathbf{x} : M \to \mathbf{R}^3$ of a surface M (not necessarily compact or oriented) is an immersion that satisfies (4.75) on M. Following present usage, we shall use the term *Willmore immersion* instead of *conformally minimal immersion.*

An immersion $\mathbf{x} : M \to \mathbf{R}^3$ for which the mean curvature $H = 0$ at every point of M is called a *minimal immersion.* We show in Theorem 8.5 that $H = 0$ on M if and only if $\mathbf{x}$ is a critical point of the *area functional* $A(\mathbf{x}) = \int_M dA$. A minimal immersion clearly satisfies (4.75), so it is also a Willmore immersion. Part of Thomsen's results discussed in Section 14.5 characterize when a Willmore immersion is just a conformal transformation of a minimal immersion.

The Willmore problem has evolved into two separate problems: *Prove the Willmore conjecture* and *Find all Willmore immersions.*

4.10.1 Willmore conjecture

F. Marques and A. Neves [117] have recently confirmed the conjecture using methods that lie outside the scope of this book. Hertrich-Jeromin and Pinkall [87] proved the conjecture for all canal immersions of the torus (see Definition 4.36). Their proof uses the classification up to conformal transformations of isothermic canal immersions. These are cylinders, cones, and immersions of revolution. Of these, the only compact ones are immersions of revolution for which the profile curve is closed. The key to understanding the Willmore functional on immersions of revolution is to view the profile curve as a curve in the upper half-plane model of hyperbolic geometry.

Exercise 16 (Geodesic curvature in upper half-plane $\mathbf{H}^2$). The upper half-plane $\mathbf{H}^2 = \{(x,y) \in \mathbf{R}^2 : y > 0\}$ with the Riemannian metric

$$I = \frac{dx^2 + dy^2}{y^2}$$

is the *upper half-plane model* of hyperbolic geometry. Use the orientation $dx \wedge dy > 0$. If

$$\gamma : J \to \mathbf{H}^2, \quad \gamma(s) = (x(s), y(s))$$

is a regular curve on connected $J \subset \mathbf{R}$ parametrized by arclength, then $1 = |\dot{\gamma}|^2 = \frac{\dot{x}^2+\dot{y}^2}{y^2}$, where dot denotes derivative with respect to s. Prove that the oriented normal vector of γ is $\mathbf{N} = -\dot{y}\epsilon_1 + \dot{x}\epsilon_2$, its acceleration vector is the covariant derivative of $\dot{\gamma}$ with respect to $\dot{\gamma}$,

$$D_{\dot{\gamma}}\dot{\gamma} = (\ddot{x} - \frac{2\dot{x}\dot{y}}{y})\epsilon_1 + (\ddot{y} + \frac{\dot{x}^2 - \dot{y}^2}{y})\epsilon_2,$$

and its geodesic curvature $\kappa = I(\mathbf{N}, D_{\dot{\gamma}}\dot{\gamma})$ is

$$\kappa = \frac{\dot{x}}{y} + \frac{\dot{x}\ddot{y} - \ddot{x}\dot{y}}{y^2}.$$

Example 4.58 (Tori of revolution). Recall the surfaces of revolution discussed above in Example 4.40. Now, instead of regarding the profile curve as living in the Euclidean x^1x^3-plane, we shall regard it as living in the upper half-plane model of hyperbolic space $\mathbf{H}^2$ given by this same plane with $x^1 > 0$, the Riemannian metric $I = \frac{dx^3dx^3+dx^1dx^1}{x^1x^1}$, and orientation $dx^3 \wedge dx^1 > 0$. The profile curve $\gamma(s) = {}^t(f(s), 0, g(s))$ lies in $\mathbf{H}^2$. It is parametrized by hyperbolic arclength if its hyperbolic norm satisfies

$$1 = |\dot{\gamma}|^2 = \frac{\dot{f}^2 + \dot{g}^2}{f^2}.$$

In $\mathbf{H}^2$, its unit tangent vector is $\mathbf{T} = \dot{\gamma} = \dot{f}\epsilon_1 + \dot{g}\epsilon_3$, the principal normal is $\mathbf{N} = \dot{g}\epsilon_1 - \dot{f}\epsilon_3$, and the hyperbolic geodesic curvature is

$$\kappa = \frac{\ddot{f}\dot{g} - \dot{f}\ddot{g}}{f^2} + \frac{\dot{g}}{f},$$

as derived in Exercise 16. Assume now that this curve is periodic of period $L > 0$, its hyperbolic length, so the immersion of revolution is

$$\mathbf{x} : \mathbf{R}/L \times \mathbf{R}/2\pi \to \mathbf{R}^3, \quad \mathbf{x}(s,t) = {}^t(f(s)\cos t, f(s)\sin t, g(s)).$$

The unit tangent vector fields (in $\mathbf{R}^3$)

$$\mathbf{e}_1 = \frac{\mathbf{x}_s}{f} = \frac{\dot{f}}{f}\cos t\epsilon_1 + \frac{\dot{f}}{f}\sin t\epsilon_2 + \frac{\dot{g}}{f}\epsilon_3, \quad \mathbf{e}_2 = \frac{\mathbf{x}_t}{f} = -\sin t\epsilon_1 + \cos t\epsilon_2,$$

and unit normal

$$\mathbf{e}_3 = \mathbf{e}_1 \times \mathbf{e}_2 = -\frac{\dot{g}}{f}\cos t\epsilon_1 - \frac{\dot{g}}{f}\sin t\epsilon_2 + \frac{\dot{f}}{f}\epsilon_3$$

define a first order frame field along $\mathbf{x}$ with dual coframe field

$$\omega^1 = f ds, \quad \omega^2 = f dt.$$

Calculating $d\mathbf{e}_3$, we get

$$\omega_3^1 = d\mathbf{e}_3 \cdot \mathbf{e}_1 = \frac{1}{f}(\kappa - \frac{\dot{g}}{f})\omega^1, \quad \omega_3^2 = d\mathbf{e}_3 \cdot \mathbf{e}_2 = -\frac{\dot{g}}{f^2}\omega^2,$$

so the frame is second order with principal curvatures

$$a = \frac{1}{f}(\frac{\dot{g}}{f} - \kappa), \quad c = \frac{\dot{g}}{f^2}.$$

Here κ is the **hyperbolic curvature** of the profile curve. Thus, the Willmore functional on $\mathbf{x}$ is

$$\mathscr{W}(\mathbf{x}) = \int_M \frac{1}{4}(a-c)^2\omega^1 \wedge \omega^2 = \frac{1}{4}\int_0^L \int_0^{2\pi} \frac{\kappa^2}{f^2} f^2 ds dt = \frac{\pi}{2}\int_0^L \kappa^2 ds.$$

By Definition 4.53, a *free elastic curve* in the hyperbolic plane is a closed curve that minimizes $\int \kappa^2 ds$, where κ is its hyperbolic curvature. Langer and Singer in [105] proved that for periodic immersions, this integral is $\geq 4\pi$, with equality precisely for the hyperbolic circle of hyperbolic radius

$$s_0 = \sinh^{-1} 1 = \log(1 + \sqrt{2}),$$

discussed in Problem 4.83.

Example 4.59 (Tubes). Recall the discussion in Subsection 4.8.2 of the tube $\mathbf{x}(s,t)$ defined in (4.63) about a given space curve $\mathbf{f}(s)$ parametrized by arclength parameter s in $\mathbf{R}^3$. If this curve is closed, meaning that $\mathbf{f}$ is periodic of period L, its length, then the surface $M = \mathbf{R}/L \times \mathbf{R}/2\pi$ is diffeomorphic to a torus. The principal curvatures of $\mathbf{x}$ are given in (4.67) to be

$$a = \frac{-\kappa \cos t}{1 - r\kappa \cos t}, \qquad c = \frac{1}{r}.$$

From this we calculate

$$\mathscr{W}(\mathbf{x}) = \frac{1}{4r} \int_0^{2\pi} \int_0^L \frac{1}{1 - r\kappa \cos t} ds dt = \frac{1}{2r} \int_0^L \int_0^\pi \frac{1}{1 - r\kappa \cos t} dt ds.$$

For given s, the inner integral is

$$\int_0^\pi \frac{1}{1 - r\kappa \cos t} dt = \frac{\pi}{\sqrt{1 - (r\kappa)^2}},$$

which leads us to

$$\mathscr{W}(\mathbf{x}) = \frac{\pi}{2r} \int_0^L \frac{ds}{\sqrt{1 - (r\kappa)^2}}.$$

In their 1970 paper [152], Shiohama and Takagi prove that for given $r > 0$, this integral is minimized over all periodic curves $\mathbf{f}$ of length L by a circle of radius $R = \frac{L}{2\pi}$, for which $\mathscr{W}(\mathbf{x})$ becomes

$$\mathscr{W}(\mathbf{x}) = \frac{\pi^2 \frac{R}{r}}{\sqrt{1 - (\frac{r}{R})^2}}.$$

Willmore showed the minimum value of this is $2\pi^2$, achieved when $\frac{R}{r} = \sqrt{2}$.

4.10.2 Willmore immersions

The search for Willmore immersions is an active field of current research. The goal is to find Willmore immersions $\mathbf{x} : M \to \mathbf{R}^3$ of compact surfaces M. Since a conformal transformation composed with a Willmore immersion remains Willmore, we want to know whether two Willmore immersions are conformally distinct. We will do this in the chapters on Möbius geometry.

One collection of Willmore immersions is the stereographic projection of those Hopf tori in $\mathbf{S}^3$ found by Pinkall [136] to be Willmore. These are derived in Sections 5.7 and 5.8.

Minimal immersions in the space form geometries are Willmore immersions. Thomsen's Theorem in Section 14.5 uses the concept of isothermic immersion to identify when a Willmore immersion is Möbius congruent to a minimal immersion in a classical geometry. One application of his result is Corollary 14.36, which verifies that, except for the Clifford torus, none of Pinkall's Willmore tori is Möbius congruent to a minimal surface in a classical geometry.

Problems

4.60. If $S = \begin{pmatrix} a & 0 \\ 0 & c \end{pmatrix}$ as in (4.2.1), with $a \neq c$, and if $A \in \mathbf{O}(2)$ and $\epsilon = \pm 1$, prove that $\tilde{S} = \epsilon {}^t\!ASA = \begin{pmatrix} \tilde{a} & 0 \\ 0 & \tilde{c} \end{pmatrix}$ if and only if

$$A \in \left\{ \pm I_2, \pm \begin{pmatrix} 1 & 0 \\ 0 & -1 \end{pmatrix}, \pm \begin{pmatrix} 0 & 1 \\ 1 & 0 \end{pmatrix}, \pm \begin{pmatrix} 0 & -1 \\ 1 & 0 \end{pmatrix} \right\}.$$

Combining this with (4.20), prove that if $(\mathbf{x}, e) : U \to \mathbf{E}(3)$ is a second order frame field on an open connected set U of nonumbilic points of $\mathbf{x}$, then any other on U is given by $(\mathbf{x}, \tilde{e})$, where $\tilde{e} = (\tilde{\mathbf{e}}_1, \tilde{\mathbf{e}}_2, \tilde{\mathbf{e}}_3)$ is one of the sixteen cases

$$(\delta \mathbf{e}_1, \delta \mathbf{e}_2, \epsilon \mathbf{e}_3), \quad (\delta \mathbf{e}_1, -\delta \mathbf{e}_2, \epsilon \mathbf{e}_3), \quad (\delta \mathbf{e}_2, \delta \mathbf{e}_1, \epsilon \mathbf{e}_3), \quad (\delta \mathbf{e}_2, -\delta \mathbf{e}_1, \epsilon \mathbf{e}_3),$$

where $\delta = \pm$ and $\epsilon = \pm$.

4.61. Let $\mathbf{x} : M \to \mathbf{R}^3$ be an oriented immersion with induced metric I. Prove that if (U, z) is a complex chart of (M, I) and if $(\mathbf{x}, e) : U \to \mathbf{E}(3)$ is the frame field adapted to $z = x + iy$, then

$$d * dt = 4t_{z\bar{z}} dx \wedge dy = (t_{xx} + t_{yy}) dx \wedge dy,$$

for any smooth function $t : U \to \mathbf{R}$.

4.62. If $(\mathbf{x}, e) : U \to \mathbf{E}(3)$ is a second order frame field on a connected, open, umbilic free subset U, then any other on U is given by $\tilde{e} = eD$, where the constant matrix $D \in G_2 = K \times \mathbf{O}(1) \subset \mathbf{O}(3)$ is one of the sixteen possibilities

$$D \in \left\{ \begin{pmatrix} \delta & 0 & 0 \\ 0 & \rho\delta & 0 \\ 0 & 0 & \epsilon \end{pmatrix}, \begin{pmatrix} 0 & \rho\delta & 0 \\ \delta & 0 & 0 \\ 0 & 0 & \epsilon \end{pmatrix} : \delta, \epsilon, \rho \in \{\pm 1\} \right\},$$

The subgroup $K \subset \mathbf{O}(2)$ was defined in Example 2.17. Prove that the criterion form $\alpha = q\omega^1 - p\omega^2$ is independent of the choice of second order frame field on U.

4.63. Let $\mathbf{x} : M \to \mathbf{R}^3$ be an immersion of a connected surface M. Let $M' \subset M$ be the set of all nonumbilic points of $\mathbf{x}$. Prove: If M' is dense in M and if every point of M possesses a neighborhood on which there exists a smooth second order frame field, then the criterion form of $\mathbf{x}$ on M' extends uniquely to a smooth 1-form on all of M.

4.64 (Space curves). Carry out the moving frame reduction on an immersion $\mathbf{x} : J \to \mathbf{R}^3$, where $J \subset \mathbf{R}$ is any connected open subset.

4.65 (Cylinders). Let $\boldsymbol{\gamma} : J \to \mathbf{R}^2$, $\boldsymbol{\gamma}(s) = f(s)\boldsymbol{\epsilon}_1 + g(s)\boldsymbol{\epsilon}_2$ be a smooth, regular curve in the plane, parametrized by arclength s. Let $\mathbf{T} = \dot{\boldsymbol{\gamma}}$ be its unit velocity vector, $\mathbf{N}$ its principal normal, and $\kappa(s)$ its curvature. The *cylinder on* $\boldsymbol{\gamma}$ is the immersion

$$\mathbf{x} : J \times \mathbf{R} \to \mathbf{R}^3, \quad \mathbf{x}(s,t) = \boldsymbol{\gamma}(s) - t\boldsymbol{\epsilon}_3. \tag{4.76}$$

Prove that $\mathbf{x}$ is always canal. Prove that $\mathbf{x}$ is Dupin if and only if κ is constant on J if and only if γ is (an open subset of) a circle or a line. Thus, $\mathbf{x}$ is isoparametric if and only if it is an open subset of a plane or of a circular cylinder.

4.66 (Constant curvature). Prove that a smooth connected curve in $\mathbf{S}^2$ with constant curvature κ can be transformed by an element of $\mathbf{SO}(3)$ to an open submanifold of the circle

$$\boldsymbol{\gamma} : \mathbf{R} \to \mathbf{S}^2, \quad \boldsymbol{\gamma}(s) = (\cos(\frac{s}{\sin\alpha})\boldsymbol{\epsilon}_1 + \sin(\frac{s}{\sin\alpha})\boldsymbol{\epsilon}_2)\sin\alpha + \boldsymbol{\epsilon}_3\cos\alpha,$$

where $0 < \alpha < \pi$ satisfies $\cot\alpha = \kappa$.

4.67. Carry out the calculations of Example 4.28 for the *circular cone*, which is the case when the profile curve is a circle, say

$$\Pi_{\sin\alpha}(\boldsymbol{\epsilon}_3) \cap \mathbf{S}^2,$$

for some angle $0 < \alpha < \pi/2$. Show that the center of any nonplanar curvature sphere of this cone lies on the x^3-axis.

4.68. Consider the hyperboloid of one sheet $x^2 - z^2 = 1$ with the parametrization

$$\mathbf{x} : \mathbf{R}^2 \to \mathbf{R}^3, \quad \mathbf{x}(s,t) = {}^t(\cosh s\cos t, \cosh s\sin t, \sinh s).$$

Find a second order frame field along $\mathbf{x}$ and the principal curvatures. Show that the oriented curvature spheres at a point $\mathbf{x}(s,t)$ are on opposite sides of the tangent plane there. Show that one of the curvature spheres has its center on the z-axis and its intersection with the surface is the curvature line through the point $\mathbf{x}(s,t)$. See Figure 4.8.

Fig. 4.8 Oriented curvature spheres and tangent plane at a point of the hyperboloid.

4.69. Let $\mathbf{x} : M \to \mathbf{R}^3$ be an immersion with induced metric I, second fundamental form II, third fundamental form III, mean curvature H, and Gauss curvature K. Prove that $III = 2H\,II - KI$.

4.70 (Degree theorem). Prove Proposition 4.33 for the case when the regular value y is not in the image of $g : M \to N$.

4.71 (Cones). Prove that any cone

$$\mathbf{x} : J \times \mathbf{R} \to \mathbf{R}^3, \quad \mathbf{x}(x,y) = e^{-y}\boldsymbol{\sigma}(x),$$

of Example 4.28 is a canal immersion. Prove that it is Dupin if and only if the curvature κ of $\boldsymbol{\sigma}$ is constant if and only if $\boldsymbol{\sigma}$ is a circle. Prove that $\mathbf{x}$ is isoparametric if and only if $\kappa = 0$ if and only if $\boldsymbol{\sigma}$ is a great circle. See Problem 4.66.

4.72. Prove that a surface of revolution is always canal. Prove that it is Dupin if and only if the curvature of the profile curve is constant. Describe the possible surfaces when the curvature of the profile curve is constant. Thus, a circular torus of revolution is a cyclide of Dupin.

4.73 (Curves of umbilics). Prove that in (4.54) the equation $a = c$ has a solution for some value of u if and only if $L \geq 1/\sqrt{2}$. Prove that if $L > 1/\sqrt{2}$, then there are two solutions, which means that there are two circles of latitude that consist of umbilic points.

4.74. Prove that the mean curvature is identically zero for the immersion of revolution with profile curve given by the catenary $f(u) = \cosh u$, $g(u) = u$ on $\mathbf{R}$.

4.75 (Multiplicity). Prove that $\tilde{\mathbf{x}}(m) = \mathbf{x}(m) + r\mathbf{e}_3(m)$ is a focal point of $\mathbf{x}$ of multiplicity $k \geq 1$ if and only if $1/r$ is a principal curvature of $\mathbf{x}$ of multiplicity k.

4.76 (Parallel surface of a torus). Find the parallel surfaces of a circular torus of revolution. Find the focal loci.

4.77 (Focal locus). Let $\mathbf{x} : M \to \mathbf{R}^3$ be an immersion with principal curvatures a and c. Consider a focal locus of $\mathbf{x}$, for example, the image of $f = \mathbf{x} + \frac{1}{a}\mathbf{e}_3$. Show that for a second order frame field $(\mathbf{x}, e)$ on $U \subset M$,

$$df = \left((1 - \frac{c}{a})\mathbf{e}_2 - \frac{a_2}{a^2}\mathbf{e}_3\right)\omega^2 - \frac{a_1}{a^2}\mathbf{e}_3\omega^1,$$

where $da = a_1\omega^1 + a_2\omega^2$ on U. Explain how to conclude that f is an immersion at a point $m \in U$ if and only if $\mathbf{x}$ is nonumbilic at m and $a_1(m) \neq 0$. Show that if f is an immersion at a point $m \in U$, then $\mathbf{e}_3(m)$ is tangent to f at this point. See Figure 4.9.

4.78 (Tubes). For the tube (4.63), use (4.65) and (4.67) to verify that $da = a_1\omega^1 + a_2\omega^2$, where

$$a_1 = \frac{\dot{\kappa}\cos t + \tau\kappa\sin t}{(1 - r\kappa\cos t)^3}, \qquad a_2 = \frac{\kappa\sin t}{r(1 - r\kappa\cos t)^2}.$$

Prove that a is constant along all of its lines of curvature if and only if a_1 is identically zero on $J \times \mathbf{R}$ if and only if the curve $\mathbf{f}$ is a line ($\kappa = 0$) or a circle of radius $1/\kappa$ ($\tau = 0$ and κ nonzero constant). Prove further that if $\tau = 0$ on J, then the coordinate curves $t = \pm\pi/2$ are lines of curvature of a. If also $\dot{\kappa} \neq 0$, then $a_1 \neq 0$ if $t \neq \pm\pi/2$, so a is constant along only the two lines of curvature $t = \pm\pi/2$, which are planar curves, but not lines or circles. Compare with Proposition 4.52.

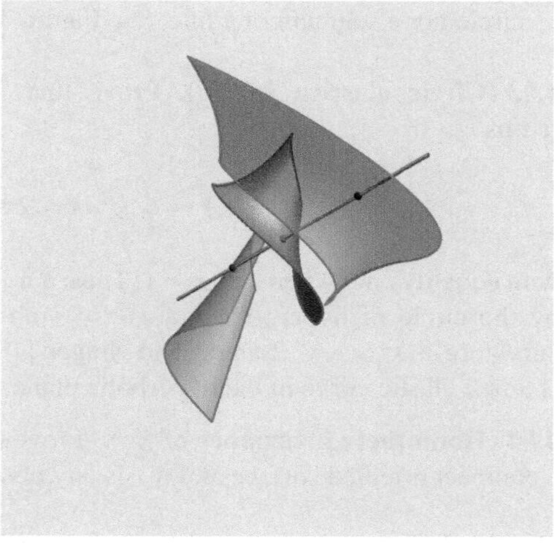

Fig. 4.9 Green and blue surfaces are the focal loci of the orange surface.

4.79 (Dupin tubes). Use Problem 4.78 and (4.67) to prove that a tube about a space curve is Dupin if and only if the curve is a line or a circle.

4.80 (Constant torsion). Does there exist a closed curve $\mathbf{f} : \mathbf{R} \to \mathbf{R}^3$ (closed means periodic) with constant positive curvature, which is not a circle? Does there exist a closed curve with constant torsion? See [106] or [29].

4.81 (Focal locus of Dupin). Prove that for a Dupin immersion, both focal loci are curves.

4.82 (Counterexample). Here is a tube on which the principal curvature a is constant along one of its lines of curvature, but the image of that line of curvature is neither a line segment nor an arc of a circle in $\mathbf{R}^3$. This does not contradict the last statement of Proposition 4.52, because a is nonconstant along any of its other nearby lines of curvature. Consider the smooth curve $\mathbf{f} : J \to \mathbf{R}^3$ with Frenet frame field $(\mathbf{T}, \mathbf{N}, \mathbf{B}) : J \to \mathbf{SO}(3)$ satisfying the Frenet-Serret equations (4.62), with

$$\kappa(s) = \sec s, \quad \tau(s) = -1, \quad J = \{-\frac{\pi}{2} < s < \frac{\pi}{2}\} \subset \mathbf{R}.$$

It exists by the Cartan–Darboux Existence Theorem 2.25. For any constant $0 < r < 1$, let $\mathbf{x} : M = J \times \mathbf{R} \to \mathbf{R}^3$ be the tube (4.63) about $\mathbf{f}$.

1. Find the open subset $U \subset J \times \mathbf{R}$ consisting of all (s,t) such that the rank of $d\mathbf{x}_{(s,t)}$ is two.
2. Show that if U' is the connected component of U containing $(0,0)$, then $\{(s,s) : s \in J\} \subset U'$. Moreover, prove that if $b \in J$, then the curve $\gamma_b(s) = (s, b+s)$ is contained in U' on some open interval about 0.
3. For each $b \in J$, prove that γ_b is a line of curvature of $\mathbf{x}$ for the principal curvature a. Prove that a is constant along γ_b if and only if $\sin b = 0$.
4. Let $\gamma = \gamma_0 : J \to U'$. Prove that $\mathbf{x} \circ \gamma = (1-r)\mathbf{f} + \mathbf{y}_0$, for some constant point $\mathbf{y}_0 \in \mathbf{R}^3$. Conclude that $\mathbf{x} \circ \gamma$ is not a plane curve, in particular, neither an arc of a circle nor a segment of a line. See Figure 4.10.

4.83 (Circle elastica in $\mathbf{H}^2$). Prove that for the circle (4.71), of hyperbolic radius $r > 0$,

$$\mathscr{F}(\sigma) = \int_\sigma \kappa^2 ds = 2\pi \frac{\cosh^2 r}{\sinh r} \geq 4\pi,$$

with equality only when $\sinh r = 1$. Thus, among hyperbolic circles, $\mathscr{F}$ is minimized by the circle of hyperbolic radius $r = \sinh^{-1} 1 = \log(1 + \sqrt{2})$, whose geodesic curvature is $\kappa = \sqrt{2}$. Langer and Singer [105, Theorem 1] prove that this circle is a free elastic curve in the hyperbolic plane.

4.84 (Homothety invariance of $\mathscr{W}$). Prove that if $\mathbf{x} : M \to \mathbf{R}^3$ is an immersion of a compact oriented surface and if r is any positive constant, then

$$\widetilde{\mathscr{W}}(\tilde{\mathbf{x}}) = \widetilde{\mathscr{W}}(\mathbf{x}),$$

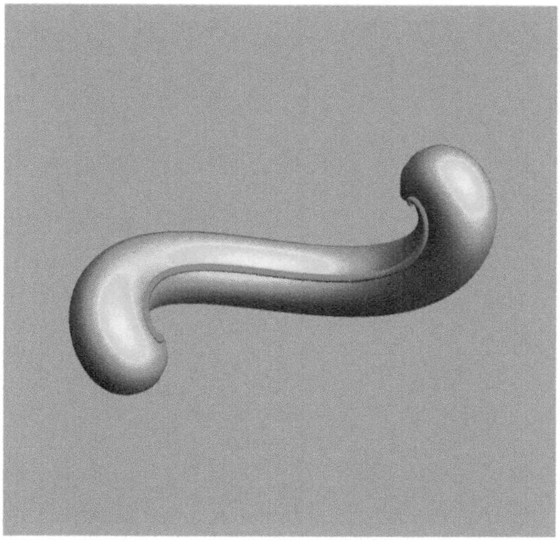

Fig. 4.10 The curve $\mathbf{x} \circ \gamma$ satisfies the Dupin condition.

where $\tilde{\mathbf{x}} = r\mathbf{x} : M \to \mathbf{R}^3$. Actually, prove the stronger result that

$$(\tilde{H}^2 - \tilde{K})d\tilde{A} = (H^2 - K)dA.$$

4.85 (Inversion invariance of $\mathscr{W}$). Prove that if $\mathbf{x} : M \to \mathbf{R}^3 \setminus \{\mathbf{0}\}$ is a smooth immersion of a compact oriented surface, then $\tilde{\mathbf{x}} = \mathscr{I} \circ \mathbf{x} : M \to \mathbf{R}^3$ is a smooth immersion and

$$(\tilde{H}^2 - \tilde{K})d\tilde{A} = (H^2 - K)dA$$

on M. Here $\tilde{H}$, $\tilde{K}$, and $d\tilde{A}$ are the mean curvature, Gauss curvature, and area element, respectively, of $\tilde{\mathbf{x}}$.

Chapter 5
Spherical Geometry

This chapter applies the method of moving frames to immersions of surfaces in spherical geometry, modeled by the unit three-sphere $\mathbf{S}^3 \subset \mathbf{R}^4$ with its group of isometries the orthogonal group, $\mathbf{O}(4)$. Stereographic projection from the sphere to Euclidean space appears in this chapter. It is our means to visualize geometric objects in $\mathbf{S}^3$. The existence of compact minimal immersions in $\mathbf{S}^3$, such as the Clifford torus, provide important examples of Willmore immersions. The chapter concludes with Hopf cylinders and Pinkall's Willmore tori in $\mathbf{S}^3$. Their construction uses the universal cover $\mathbf{SU}(2) \cong \mathbf{S}^3$ of $\mathbf{SO}(3)$.

5.1 Constant positive curvature geometry of the sphere

Consider the unit sphere in $\mathbf{R}^4$,

$$\mathbf{S}^3 = \{\mathbf{x} \in \mathbf{R}^4 : |\mathbf{x}| = 1\}$$

with the Riemannian metric induced from the standard inner product on $\mathbf{R}^4$. The standard action of the orthogonal group $\mathbf{O}(4)$ on $\mathbf{R}^4$ sends $\mathbf{S}^3$ to itself and acts as isometries on $\mathbf{S}^3$. Using the Gram-Schmidt orthonormalization process, one proves that $\mathbf{O}(4)$ acts transitively on $\mathbf{S}^3$. By essentially the same argument used to prove that $\mathbf{E}(3)$ is the full group of isometries of $\mathbf{R}^3$, one proves that $\mathbf{O}(4)$ is the full group of isometries of $\mathbf{S}^3$.

As an origin of $\mathbf{S}^3$ we choose the point $\epsilon_0 = {}^t(1,0,0,0)$. The isotropy subgroup of $\mathbf{O}(4)$ at this point is

$$G_0 = \{\begin{pmatrix} 1 & 0 \\ 0 & A \end{pmatrix} : A \in \mathbf{O}(3)\} \cong \mathbf{O}(3)$$

G.R. Jensen et al., *Surfaces in Classical Geometries*, Universitext,
DOI 10.1007/978-3-319-27076-0_5

Hence, $\mathbf{S}^3 \cong \mathbf{O}(4)/G_0 \cong \mathbf{O}(4)/\mathbf{O}(3)$ and the natural projection map defines a principal $\mathbf{O}(3)$-bundle

$$\pi : \mathbf{O}(4) \to \mathbf{S}^3, \quad \pi(e) = e\epsilon_0 = \mathbf{e}_0, \tag{5.1}$$

where $\mathbf{e}_i$ denotes column i of the orthogonal matrix e, for $i = 0, \dots, 3$. The Lie algebra $\mathfrak{o}(4)$ of $\mathbf{O}(4)$ has a decomposition

$$\mathfrak{o}(4) = \mathfrak{g}_0 + \mathfrak{m} \tag{5.2}$$

where the Lie subalgebra $\mathfrak{g}_0 \cong \mathfrak{o}(3)$ is the Lie algebra of G_0 and

$$\mathfrak{m} = \{\begin{pmatrix} 0 & -{}^t\mathbf{x} \\ \mathbf{x} & 0 \end{pmatrix} : \mathbf{x} \in \mathbf{R}^3\} \cong \mathbf{R}^3$$

is a complementary vector subspace, which is $\mathrm{Ad}(\mathbf{O}(3))$-invariant in the sense that

$$\mathrm{Ad}(A)(\mathbf{x}, 0) = (A^{-1}\mathbf{x}, 0) \in \mathfrak{m}.$$

Using this decomposition, we can write an element of $\mathfrak{o}(4)$ as

$$(x, X) = \begin{pmatrix} 0 & -{}^t\mathbf{x} \\ \mathbf{x} & X \end{pmatrix}, X \in \mathfrak{o}(3), \mathbf{x} \in \mathbf{R}^3 \tag{5.3}$$

in which case the bracket structure can be described by the formulas

$$\begin{aligned} [(0, X), (0, Y)] &= (0, [X, Y]), \\ [(0, X), (\mathbf{y}, 0)] &= (X\mathbf{y}, 0), \\ [(\mathbf{x}, 0), (\mathbf{y}, 0)] &= (0, -\mathbf{x}\,{}^t\mathbf{y} + \mathbf{y}\,{}^t\mathbf{x}), \end{aligned} \tag{5.4}$$

which put altogether is

$$[(\mathbf{x}, X), (\mathbf{y}, Y)] = (X\mathbf{y} - Y\mathbf{x}, -\mathbf{x}\,{}^t\mathbf{y} + \mathbf{y}\,{}^t\mathbf{x} + [X, Y]).$$

Compare this Lie algebra structure with that of $\mathscr{E}(3)$ described in equation (4.2). The Maurer–Cartan form of $\mathbf{O}(4)$ is the $\mathfrak{o}(4)$-valued 1-form

$$e^{-1}de = \begin{pmatrix} 0 & \omega_1^0 & \omega_2^0 & \omega_3^0 \\ \omega_0^1 & 0 & \omega_2^1 & \omega_3^1 \\ \omega_0^2 & \omega_1^2 & 0 & \omega_3^2 \\ \omega_0^3 & \omega_1^3 & \omega_2^3 & 0 \end{pmatrix} = \begin{pmatrix} 0 & -{}^t\theta \\ \theta & \omega \end{pmatrix} = (\theta, \omega) \tag{5.5}$$

in the notation of (5.3), where

$$\theta = \begin{pmatrix} \omega^1 \\ \omega^2 \\ \omega^3 \end{pmatrix}, \qquad \omega = (\omega^i_j),$$

where $\omega^j_i = -\omega^i_j$, for $i,j = 1,\dots,3$ and we have introduced the convenient notation

$$\omega^i = \omega^i_0 = -\omega^0_i, \tag{5.6}$$

for $i = 1,2,3$. The structure equations are

$$d\omega^i_j = -\sum_{k=0}^{3} \omega^i_k \wedge \omega^k_j \tag{5.7}$$

for $i,j,k = 0,1,2,3$, which in the matrix notation is

$$d\theta = -\omega \wedge \theta, \quad d\omega = -\omega \wedge \omega + \theta \wedge {}^t\theta.$$

We define a local orthonormal frame field on an open subset U of $\mathbf{S}^3$ to be a smooth local section of (5.1). Explicitly, it is a smooth map $e : U \to \mathbf{O}(4)$ such that $\pi \circ e = \mathrm{id}_U$. Geometrically, at a point $\mathbf{x} \in U$, we have $e(\mathbf{x}) = (\mathbf{e}_0, \mathbf{e}_1, \mathbf{e}_2, \mathbf{e}_3)$ (designating the columns) with $\mathbf{e}_0 = \mathbf{x}$. Since $T_{\mathbf{x}}\mathbf{S}^3$ is naturally identified with the subspace orthogonal to $\mathbf{x}$, it follows that $\mathbf{e}_1, \mathbf{e}_2, \mathbf{e}_3$ is an orthonormal basis of this tangent space. If we abuse notation slightly and retain the same letters for the pull-back of the Maurer–Cartan form by our frame field e, then we have from (5.5)

$$d\mathbf{x} = d\mathbf{e}_0 = \omega^i \mathbf{e}_i \tag{5.8}$$

Therefore, the Riemannian metric on $\mathbf{S}^3$ has the local expression

$$I = d\mathbf{x} \cdot d\mathbf{x} = \omega^i \omega^i = {}^t\theta\theta \tag{5.9}$$

and $\omega^1, \omega^2, \omega^3$ is an orthonormal coframe field in U. Any other frame field on U must be given by $\tilde{e} = eK$, where

$$K = \begin{pmatrix} 1 & 0 \\ 0 & A \end{pmatrix} : U \to G_0$$

is a smooth map, so $A : U \to \mathbf{O}(3)$ is a smooth map, and

$$(\tilde{\mathbf{e}}_1, \tilde{\mathbf{e}}_2, \tilde{\mathbf{e}}_3) = (\mathbf{e}_1, \mathbf{e}_2, \mathbf{e}_3)A$$

while $\tilde{\mathbf{e}}_0 = \mathbf{e}_0 = \mathbf{x}$. In order to compare the pull-back of the Maurer–Cartan form by each of these frame fields, let us use the decomposition (5.2) to write

$$e^{-1}de = \begin{pmatrix} 0 & -{}^t\theta \\ \theta & 0 \end{pmatrix} + \begin{pmatrix} 0 & 0 \\ 0 & \omega \end{pmatrix}.$$

For the pull-back of the Maurer–Cartan forms by $\tilde{e}$ use the same letters with tildes. Then

$$\begin{aligned} \tilde{e}^{-1}d\tilde{e} &= A^{-1}e^{-1}(deA + e\,dA) \\ &= A^{-1}\begin{pmatrix} 0 & -{}^t\theta \\ \theta & \omega \end{pmatrix}A + \begin{pmatrix} 0 & 0 \\ 0 & A^{-1}dA \end{pmatrix} \\ &= \begin{pmatrix} 0 & -{}^t\theta A \\ A^{-1}\theta & A^{-1}\omega A + A^{-1}dA \end{pmatrix}, \end{aligned} \tag{5.10}$$

from which it follows that

$$\tilde{\theta} = A^{-1}\theta. \tag{5.11}$$

One consequence of this is that the Riemannian metric of $\mathbf{S}^3$ comes from the group $\mathbf{O}(4)$ in the sense that, by (5.9) and (5.11) we have

$$d\mathbf{x}\cdot d\mathbf{x} = {}^t\theta\theta = {}^t\tilde{\theta}\tilde{\theta}.$$

Consider the frame field $e : U \to \mathbf{O}(4)$ and recall the notational convention (5.6). From (5.7) again we have

$$d\omega^i = -\omega^i_j \wedge \omega^j, \qquad \omega^i_j = -\omega^j_i$$

It follows that ω^i_j, for $i,j = 1,2,3$, are the Levi-Civita connection forms of I with respect to the orthonormal coframe field $\omega^1, \omega^2, \omega^3$. The curvature form Ω^i_j is then obtained from (5.7) to be

$$\Omega^i_j = d\omega^i_j + \omega^i_k \wedge \omega^k_j = -\omega^i_0 \wedge \omega^0_j = \omega^i \wedge \omega^j$$

This shows that the metric $d\mathbf{x}\cdot d\mathbf{x}$ has constant sectional curvature equal to one. Recall that $\Omega^i_j = R^i_{jkl}\omega^k \wedge \omega^l$, where the functions R^i_{jkl} are the components of the Riemann curvature tensor. The space has constant sectional curvature c if and only if $R^i_{jkl} = c(\delta_{ik}\delta_{jl} - \delta_{il}\delta_{jk})$, which is equivalent to the condition that $\Omega^i_j = c\,\omega^i \wedge \omega^j$. For background reference, see [154, Vol. I].

5.2 Moving frame reductions

Consider a smooth immersion

$$\mathbf{x} : M^2 \to \mathbf{S}^3$$

of a smooth surface M. A smooth frame field along $\mathbf{x}$ on an open subset $U \subset M$ is a smooth map $e : U \to \mathbf{O}(4)$ such that $\pi \circ e = \mathbf{x}$; that is, $\mathbf{e}_0 = \mathbf{x}$. We use the evident geometric interpretation of these frame fields to define a *first order frame field along* $\mathbf{x}$ to be a frame field for which $\mathbf{e}_3$ is normal to $\mathbf{x}$ (but tangent to $\mathbf{S}^3$). This is equivalent to the condition that $d\mathbf{x} = d\mathbf{e}_0 = \omega^1\mathbf{e}_1 + \omega^2\mathbf{e}_2$, namely, at each point of U

$$\omega^3 = \omega_0^3 = 0, \quad \omega^1 \wedge \omega^2 \neq 0. \tag{5.12}$$

Given a point $m \in M$, there exists a neighborhood $U \subset M$ of m on which there is a first order frame field along $\mathbf{x}$. See Problem 5.33 below.

Let e be a first order frame field along $\mathbf{x}$ on U. The induced metric on M, also called the *first fundamental form* of $\mathbf{x}$, is

$$I = d\mathbf{x} \cdot d\mathbf{x} = \omega^1\omega^1 + \omega^2\omega^2$$

and thus ω^1, ω^2 is an orthonormal coframe field for it. By (5.7) we have

$$d\omega^1 = -\omega_2^1 \wedge \omega^2, \quad d\omega^2 = -\omega_1^2 \wedge \omega^1 \tag{5.13}$$

since $\omega^3 = 0$ for a first order frame field. From (5.13) we conclude that ω_2^1 is the Levi-Civita connection form for I with respect to the orthonormal coframe field ω^1, ω^2. Its Gaussian curvature K is given by an application of (5.7),

$$K\omega^1 \wedge \omega^2 = d\omega_2^1 = -\omega_3^1 \wedge \omega_2^3 - \omega_0^1 \wedge \omega_2^0 = \omega_1^3 \wedge \omega_2^3 + \omega^1 \wedge \omega^2 \tag{5.14}$$

Taking the exterior derivative of (5.12) and using (5.7) together with Cartan's Lemma, we find that

$$\omega_1^3 = h_{11}\omega^1 + h_{12}\omega^2, \quad \omega_2^3 = h_{21}\omega^1 + h_{22}\omega^2$$

for smooth functions h_{ij} on U such that $h_{12} = h_{21}$. These functions give a smooth matrix valued map

$$S = (h_{ij}) : U \to \mathscr{S}$$

where $\mathscr{S}$ is the vector space of all 2×2 symmetric matrices. Thus,

$$\begin{pmatrix} \omega_1^3 \\ \omega_2^3 \end{pmatrix} = S \begin{pmatrix} \omega^1 \\ \omega^2 \end{pmatrix}.$$

The *second fundamental form* of $\mathbf{x}$ is the symmetric bilinear form defined on the tangent space of M by

$$II = -d\mathbf{x}\cdot d\mathbf{e}_3 = \omega_1^3\omega^1 + \omega_2^3\omega^2 = (\omega 1, \omega^2) S \begin{pmatrix} \omega^1 \\ \omega^2 \end{pmatrix}$$

It is clear from the first equality that II reverses sign when $\mathbf{e}_3$ is replaced by $-\mathbf{e}_3$ and otherwise does not depend on the choice of first order frame field e. From (5.14) we obtain the Gauss equation

$$K = 1 + \det(S)$$

where the 1 is coming from the sectional curvature of $\mathbf{S}^3$. The mean curvature of $\mathbf{x}$ is

$$H = \frac{1}{2}\text{trace}(S) = \frac{1}{2}(h_{11} + h_{22}) \tag{5.15}$$

From equation (5.16) below we see that H depends only on $\mathbf{e}_3$, changing sign when $\mathbf{e}_3$ changes sign. It otherwise does not depend on the choice of first order frame field.

If $e : U \to \mathbf{O}(4)$ is a first order frame field along $\mathbf{x}$, then any other is given by $\tilde{e} = eC$, where $C : U \to G_0$ is such that $\tilde{\mathbf{e}}_3$ is normal to $d\mathbf{x}$, so $\tilde{\mathbf{e}}_3 = \pm\mathbf{e}_3$. Hence $C : U \to G_1$, where

$$G_1 = \left\{ \begin{pmatrix} 1 & 0 \\ 0 & \begin{pmatrix} B & 0 \\ 0 & \epsilon \end{pmatrix} \end{pmatrix} \in G_0 : B \in \mathbf{O}(2), \epsilon = \pm 1 \right\}$$

From (5.10) with now $A = \begin{pmatrix} B & 0 \\ 0 & \epsilon \end{pmatrix}$ we can calculate that

$$\tilde{\theta} = \begin{pmatrix} \tilde{\omega}^1 \\ \tilde{\omega}^2 \end{pmatrix} = B^{-1} \begin{pmatrix} \omega^1 \\ \omega^2 \end{pmatrix}, \qquad \tilde{S} \begin{pmatrix} \tilde{\omega}^1 \\ \tilde{\omega}^2 \end{pmatrix} = \epsilon B^{-1} \begin{pmatrix} \omega_1^3 \\ \omega_2^3 \end{pmatrix} = \epsilon B^{-1} S \begin{pmatrix} \omega^1 \\ \omega^2 \end{pmatrix}.$$

Therefore,

$$\tilde{S} = \epsilon B^{-1} S B. \tag{5.16}$$

This is the same action as that of (4.34). The principal values a and c of S are the *principal curvatures* of $\mathbf{x}$ at the point. As in the Euclidean case, each orbit of the action (5.16) contains a unique element in the set

$$D = \{ \begin{pmatrix} a & 0 \\ 0 & c \end{pmatrix} : a \geq |c| \} \subset \mathscr{S}$$

A point of M is *umbilic* if $a = c$ and otherwise it is *nonumbilic*. The set of umbilic points ($a = c$) is closed in M.

Definition 5.1. A *second order frame field* $e : U \to \mathbf{O}(4)$ along $\mathbf{x}$ is a first order frame field for which

$$\omega_1^3 = a\omega^1, \quad \omega_2^3 = c\omega^2 \tag{5.17}$$

for some functions $a, c : U \to \mathbf{R}$.

Lemma 5.2. *Let $m \in M$. If m is nonumbilic, or if m belongs to an open set of umbilic points, then there exists a smooth second order frame field on some open neighborhood of m.*

Proof. Same as for the Euclidean case, either Lemma 4.6 or Lemma 4.10. □

Taking the exterior derivative of (5.17), we arrive at the Codazzi equations

$$\begin{aligned} da\wedge\omega^1 + (a-c)\omega_1^2\wedge\omega^2 &= 0 \\ dc\wedge\omega^2 + (a-c)\omega_1^2\wedge\omega^1 &= 0 \end{aligned} \tag{5.18}$$

In the nonumbilic cases, the isotropy subgroup is finite, same as in the Euclidean case, so $\mathfrak{g}_2 = 0$ and second order frame fields are Frenet. In the umbilic case the isotropy subgroup G_2 is all of G_1, so $\mathfrak{m}_2 = 0$ and first order frame fields are also second order frame fields and they are Frenet. The Maurer–Cartan form ω_1^2 remains to be expressed as a linear combination of the coframe ω^1, ω^2. We let

$$\omega_1^2 = p\omega^1 + q\omega^2 \tag{5.19}$$

for some smooth functions p and q on U. By equations (5.13), these functions are determined by the equations

$$d\omega^1 = p\omega^1\wedge\omega^2, \quad d\omega^2 = q\omega^1\wedge\omega^2$$

Taking the exterior derivative of (5.19) completes the structure equations.

Definition 5.3. For any function f on M, we set $df = f_1\omega^1 + f_2\omega^2$, where the smooth functions f_1 and f_2 are called the covariant derivatives of f with respect to the given coframe field.

5.2.1 Summary of frame reduction and structure equations

Let $\mathbf{x} : M \to \mathbf{S}^3$ be an immersed surface. At a nonumbilic point there exists a smooth second order frame field $e : U \to \mathbf{O}(4)$ on a neighborhood of the point. Its pull-back of the Maurer–Cartan form of $\mathbf{O}(4)$ satisfies:

$$\begin{aligned}
&\omega^3 = 0, \text{ (first order)}, \quad \omega^1\wedge\omega^2 \neq 0\\
&d\omega^1 = p\omega^1\wedge\omega^2, \quad d\omega^2 = q\omega^1\wedge\omega^2\\
&\omega_1^3 = a\omega^1, \quad \omega_2^3 = c\omega^2, \quad \text{(second order)}\\
&\omega_1^2 = p\omega^1 + q\omega^2, \quad \text{(third order)}.
\end{aligned} \tag{5.20}$$

The structure equations of the immersion include (5.18), the Codazzi equations, which, because $\omega^1\wedge\omega^2 \neq 0$ at each point, can be written as

$$a_2 = (a-c)p, \quad c_1 = (a-c)q, \text{ (Codazzi equations)}, \tag{5.21}$$

and the Gauss equations obtained by differentiating (5.19)

$$p_2 - q_1 - p^2 - q^2 = K = ac + 1, \text{ (Gauss equation)}, \tag{5.22}$$

where K is the Gaussian curvature of the metric induced on M. The functions a and c are the *principal curvatures* of $\mathbf{x}$. They are continuous functions on M, smooth on an open neighborhood of any non-umbilic point.

If $\mathbf{x}$ is totally umbilic on U, then any first order frame is automatically second order and the above equations with $a = c$ give the structure equations for such a frame.

5.3 Tangent and curvature spheres

For points $\mathbf{x}, \mathbf{m} \in \mathbf{S}^3$, the *distance* from $\mathbf{m}$ to $\mathbf{x}$ in $\mathbf{S}^3$ is

$$r = \cos^{-1}(\mathbf{x}\cdot\mathbf{m}) \in [0, \pi].$$

Definition 5.4. The sphere $S_r(\mathbf{m})$ in $\mathbf{S}^3$ with *center* $\mathbf{m} \in \mathbf{S}^3$ and radius $r \in [0, \pi]$ is

$$S_r(\mathbf{m}) = \{\mathbf{x} \in \mathbf{S}^3 : \mathbf{x}\cdot\mathbf{m} = \cos r\}. \tag{5.23}$$

Spheres of radius $r = \pi/2$ are called *great spheres*. Spheres of radius $r = 0$ or $r = \pi$ are called *point spheres*, since $S_0(\mathbf{m}) = \{\mathbf{m}\}$ and $S_\pi(\mathbf{m}) = \{-\mathbf{m}\}$.

Note that $S_r(\mathbf{m})$ is the intersection of $\mathbf{S}^3$ with the hyperplane in $\mathbf{R}^4$ whose equation is $\mathbf{y}\cdot\mathbf{m} = \cos r$. This hyperplane intersects $\mathbf{S}^3$ in a great sphere if and only if it passes through the origin, which happens if and only if $r = \pi/2$.

Proposition 5.5. *If Σ denotes the set of all nonpoint spheres in $\mathbf{S}^3$, then*

$$\Sigma = (\mathbf{S}^3 \times (0, \pi))/\mathbf{Z}_2,$$

Fig. 5.1 A point **x** on $S_r(\mathbf{m}) = S_{\pi-r}(-\mathbf{m})$ and on $S(\mathbf{y})$.

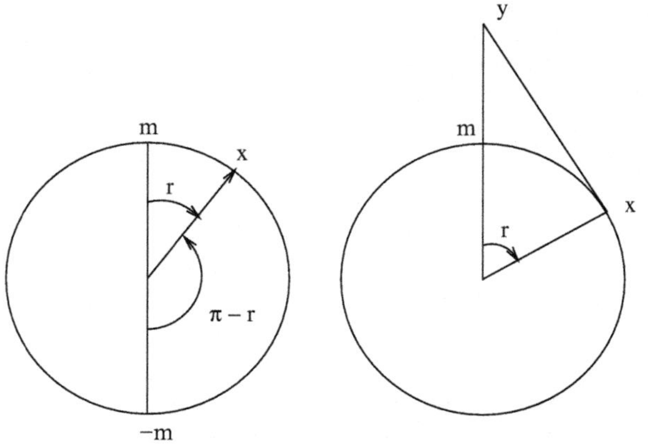

where $\mathbf{Z}_2$ *acts on* $\mathbf{S}^3 \times (0, \pi)$ *by*

$$-1(\mathbf{m}, r) = (-\mathbf{m}, \pi - r).$$

Proof. Any sphere in Σ is given by $S_r(\mathbf{m})$, for some center $\mathbf{m} \in \mathbf{S}^3$ and radius satisfying $0 < r < \pi$. There is duplication, however, as it is evident from the first diagram in Figure 5.1 that

$$S_{\pi-r}(-\mathbf{m}) = S_r(\mathbf{m}),$$

and this is the only duplication. □

Definition 5.6. An *oriented sphere* in $\mathbf{S}^3$ is a sphere $S_r(\mathbf{m})$ of radius $0 < r < \pi$ together with a choice of continuous unit normal vector field on it. The *canonical orientation* of $S_r(\mathbf{m})$ is by the unit normal vector field $\mathbf{n}$ whose value at $\mathbf{x} \in S_r(\mathbf{m})$ is

$$\mathbf{n}(\mathbf{x}) = \frac{\mathbf{m} - \cos r\, \mathbf{x}}{\sin r}. \tag{5.24}$$

Geometrically, the canonical orientation of $S_r(\mathbf{m})$, for $0 < r < \pi$, is by the unit normal pointing *towards* the center $\mathbf{m}$. See Figure 5.2 and Figure 5.3. From now on, when $0 < r < \pi$, we let $S_r(\mathbf{m})$ denote this sphere with its canonical orientation.

Exercise 17. Prove that $S_r(\mathbf{m})$ with unit normal (5.24) is totally umbilic with constant principal curvature $a = \cot r$.

Exercise 18 (Totally Umbilic Case). Prove that if every point of a connected surface M^2 is umbilic for the immersion $\mathbf{x} : M \to \mathbf{S}^3$, then the principal curvature $a = c$ relative to a unit normal vector field $\mathbf{e}_3$ is constant on M, and $\mathbf{x}(M) \subset S_r(\mathbf{m})$, where

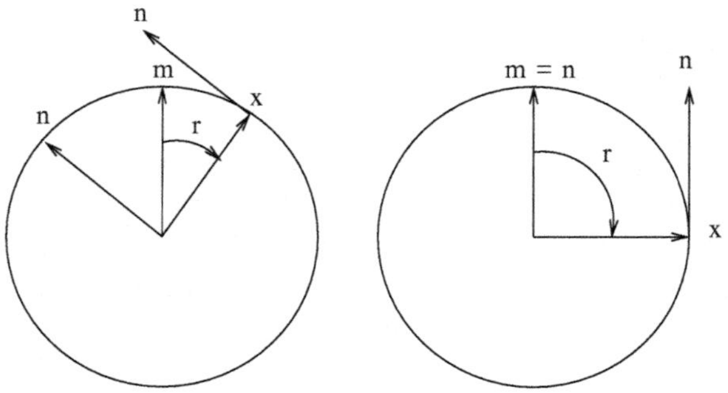

Fig. 5.2 Canonical orientations when $0 < r < \pi/2$ and $r = \pi/2$, respectively.

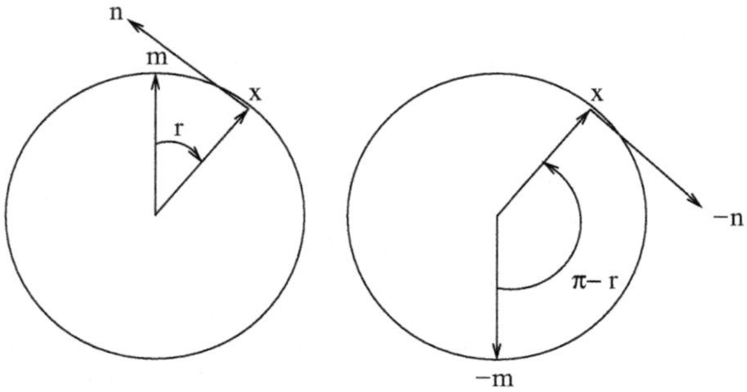

Fig. 5.3 Canonical orientations of $S_r(\mathbf{m})$ and $S_{\pi-r}(-\mathbf{m})$, respectively.

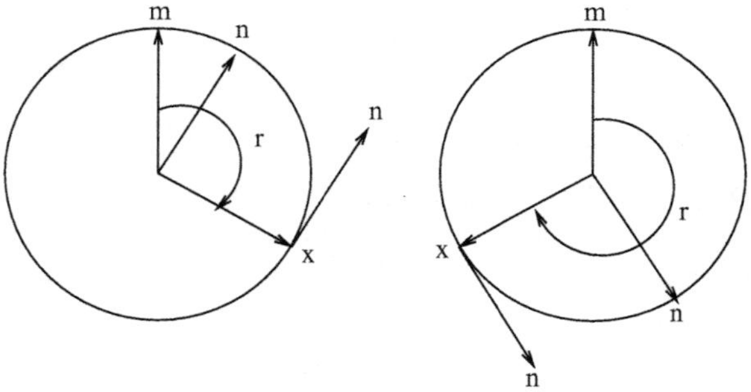

Fig. 5.4 $S_r(\mathbf{m})$ with normal **n** by (5.24) and angle $r = \theta$, for $0 < \theta < \pi$, on left; and $r = 2\pi - \theta$, on right.

$$\cos r = \frac{a}{\sqrt{1+a^2}}, \quad \mathbf{m} = \frac{1}{\sqrt{1+a^2}}(\mathbf{e}_3 + a\mathbf{x}).$$

Exercise 19. Prove that the set of all oriented spheres in $\mathbf{S}^3$ is the smooth manifold

$$\tilde{\Sigma} = \{S_r(\mathbf{m}) : \mathbf{m} \in \mathbf{S}^3,\ 0 < r < \pi\} = \mathbf{S}^3 \times (0, \pi).$$

Prove that the map $\tilde{\Sigma} \to \Sigma$ which sends an oriented sphere to that sphere without orientation is two-to-one and onto.

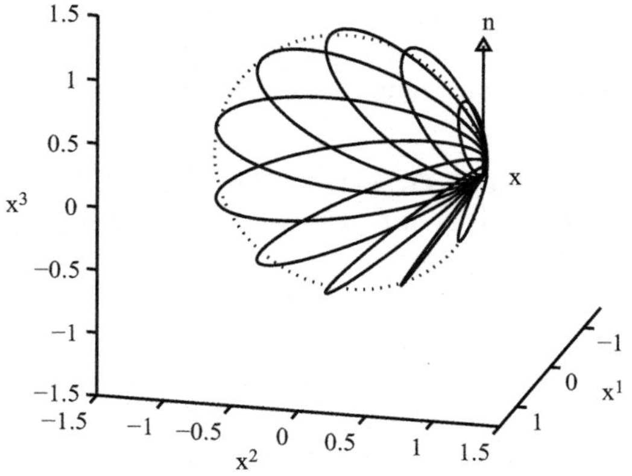

Fig. 5.5 The pencil determined by $\mathbf{x} = \epsilon_2$ and $n = \epsilon_3$.

Definition 5.7. For any point $\mathbf{x} \in \mathbf{S}^3$ and unit tangent vector $\mathbf{n} \in T_{\mathbf{x}}\mathbf{S}^3$, the *pencil of oriented spheres* in $\mathbf{S}^3$ determined by the pair $(\mathbf{x}, \mathbf{n})$ is the set of all oriented spheres through $\mathbf{x}$ with unit normal $\mathbf{n}$ at $\mathbf{x}$.

An oriented sphere through $\mathbf{x}$ with normal $\mathbf{n}$ at $\mathbf{x}$ must have its center on the open great semi-circle that runs from $\mathbf{x}$ to $-\mathbf{x}$ tangent to $\mathbf{n}$ at $\mathbf{x}$. The pencil determined by $(\mathbf{x}, \mathbf{n})$ is the set of spheres

$$\{S_r(\cos r\, \mathbf{x} + \sin r\, \mathbf{n}) : 0 < r < \pi\}, \tag{5.25}$$

each with its canonical orientation. See Figure 5.5.

Definition 5.8. An *oriented tangent sphere* to an immersion $\mathbf{x} : M^2 \to \mathbf{S}^3$ at a point $m \in M$, with unit normal vector $\mathbf{n}$ at m, is an oriented sphere through $\mathbf{x}(m)$ with unit normal $\mathbf{n}$ at $\mathbf{x}(m)$.

Each oriented tangent sphere has its center on the oriented normal line

$$\{\cos r\, \mathbf{x}(m) + \sin r\, \mathbf{n} : r \in \mathbf{R}\},$$

so the set of oriented tangent spheres at $\mathbf{x}(m)$ with normal $\mathbf{n}$ must be given by (5.25).

Definition 5.9. An *oriented curvature sphere* of an immersion $\mathbf{x} : M^2 \to \mathbf{S}^3$ at $m \in M$ relative to a unit normal vector $\mathbf{n}$ of $\mathbf{x}$ at m is an oriented tangent sphere whose principal curvature is equal to a principal curvature of $\mathbf{x} : M \to \mathbf{S}^3$ at m relative to the normal $\mathbf{n}$.

Remark 5.10. If $m \in M$ is nonumbilic for $\mathbf{x}$, then there are two distinct oriented curvature spheres at m relative to $\mathbf{n}$. If m is umbilic, then there is only one, but we say it has *multiplicity two*. If a is a principal curvature of $\mathbf{x}$ at m relative to $\mathbf{n}$, and if $r = \cot^{-1} a \in (0, \pi)$, then the oriented curvature sphere relative to $\mathbf{n}$ is

$$S_r(\cos r\, \mathbf{x}(m) + \sin r\, \mathbf{n}),$$

with its canonical orientation, since for $0 < r < \pi$, the principal curvatures of $S_r(\mathbf{m})$ relative to its canonical orientation are both $\cot r$.

Curvature spheres have another characterization, which generalizes to Möbius and Lie sphere geometries, where there is no concept of principal curvatures. This characterization uses the idea of an oriented tangent sphere map along an immersion.

Proposition 5.11. *Let $e : U \to \mathbf{O}(4)$ be a first order frame field along an immersion $\mathbf{x} : M^2 \to \mathbf{S}^3$ on a neighborhood $U \subset M$. Fix $r \in (0,\pi)$, and consider the oriented tangent sphere map*

$$\begin{aligned} &S : U \to \tilde{\Sigma} = \mathbf{S}^3 \times (0,\pi), \\ &S(m) = S_r(\cos r\, \mathbf{x}(m) + \sin r\, \mathbf{e}_3(m)) = \cos r\, \mathbf{x}(m) + \sin r\, \mathbf{e}_3(m) + r\epsilon_4. \end{aligned}$$

Then $S(m)$ is an oriented curvature sphere of $\mathbf{x}$ at $m \in U$ relative to the unit normal $\mathbf{e}_3(m)$ if and only if the rank of dS at m is less than two. The kernel of dS_m is the space of principal vectors at m for the principal curvature $\cot r$. The dimension of the kernel is the multiplicity of this principal curvature.

Proof. By the structure equations,

$$dS = \cos r\, d\mathbf{x} + \sin r\, d\mathbf{e}_3 = (\cos r\, \omega_0^1 + \sin r\, \omega_3^1)\mathbf{e}_1 + (\cos r\, \omega_0^2 + \sin r\, \omega_3^2)\mathbf{e}_2$$

has rank less than two at a point $m \in U$ if and only if

$$\begin{aligned} 0 &= (\cos r\, \omega_0^1 + \sin r\, \omega_3^1) \wedge (\cos r\, \omega_0^2 + \sin r\, \omega_0^2) \\ &= \cos^2 r\, \omega_0^1 \wedge \omega_0^2 + \cos r \sin r(\omega_0^1 \wedge \omega_3^2 + \omega_3^1 \wedge \omega_0^2) + \sin^2 r\, \omega_3^1 \wedge \omega_3^2 \\ &= (\cos^2 r - 2H \cos r \sin r + K \sin^2 r)\omega_0^1 \wedge \omega_0^2 \end{aligned}$$

at m, where H is the mean curvature of $\mathbf{x}$ relative to $\mathbf{e}_3$ and K is the Gaussian curvature. This last expression is zero at m if and only if $\cot r$ is a principal curvature of $\mathbf{x}$ relative to $\mathbf{e}_3$ at m, which is equivalent to $S(m)$ being a curvature sphere at m relative to $\mathbf{e}_3(m)$. See Remark 5.10. From the expression above for dS, we see that $v \in T_mM$ is in its kernel if and only if

$$-d\mathbf{e}_3 v = \cot r\, d\mathbf{x}_m v,$$

which means that v is a principal vector for the principal curvature $\cot r$. □

Example 5.12 (Circular Tori). Fix α to satisfy $0 < \alpha < \pi/2$ and let $r = \cos\alpha$ and $s = \sin\alpha$, so that $r^2 + s^2 = 1$. Consider the immersion

$$\mathbf{x}^{(\alpha)} : \mathbf{R}^2 \to \mathbf{S}^3, \quad \mathbf{x}^{(\alpha)}(x,y) = {}^t(r\cos\frac{x}{r}, r\sin\frac{x}{r}, s\cos\frac{y}{s}, s\sin\frac{y}{s}). \tag{5.26}$$

Then $\mathbf{x}^{(\alpha)}(\tilde{x},\tilde{y}) = \mathbf{x}^{(\alpha)}(x,y)$ whenever

$$(\tilde{x},\tilde{y}) - (x,y) \in 2\pi r\mathbf{Z} \times 2\pi s\mathbf{Z} = \Gamma_\alpha$$

a lattice in $\mathbf{R}^2$ (see Boothby [16, Example 8.7, page 99]), so $\mathbf{x}^{(\alpha)}$ descends to an immersion of the torus $T_\alpha = \mathbf{R}^2/\Gamma_\alpha$ into $\mathbf{S}^3$. For simplicity write $\mathbf{x} = \mathbf{x}^{(\alpha)}$. Let $e^{(\alpha)} = (\mathbf{e}_0, \mathbf{e}_1, \mathbf{e}_2, \mathbf{e}_3) : T_\alpha \to \mathbf{SO}(4)$ be the smooth frame field along $\mathbf{x}$, where

$$\mathbf{e}_0 = \mathbf{x}, \quad \mathbf{e}_1 = \mathbf{x}_x, \quad \mathbf{e}_2 = \mathbf{x}_y, \quad \mathbf{e}_3 = {}^t(s\cos\frac{x}{r}, s\sin\frac{x}{r}, -r\cos\frac{y}{s}, -r\sin\frac{y}{s}).$$

Then $d\mathbf{e}_0 = \mathbf{e}_1 dx + \mathbf{e}_2 dy$ implies that e is first order with $\omega^1 = dx$ and $\omega^2 = dy$, so $\omega^1_2 = 0$.

$$\begin{aligned} \omega^3_1 &= d\mathbf{e}_1 \cdot \mathbf{e}_3 = -\frac{s}{r}\omega^1 = -\tan\alpha\,\omega^1 \\ \omega^3_2 &= d\mathbf{e}_2 \cdot \mathbf{e}_3 = \frac{r}{s}\omega^2 = \cot\alpha\,\omega^2, \end{aligned} \tag{5.27}$$

so $e^{(\alpha)}$ is second order and the principal curvatures of $\mathbf{x}$ are the constants

$$a = -\tan\alpha = \cot(\frac{\pi}{2} + \alpha) \quad \text{and} \quad c = \cot\alpha.$$

The oriented curvature spheres at $\mathbf{x}$ relative to $\mathbf{e}_3$ are

$$S_{\alpha+\frac{\pi}{2}}(-s\mathbf{x} + r\mathbf{e}_3) \quad \text{and} \quad S_\alpha(r\mathbf{x} + s\mathbf{e}_3), \tag{5.28}$$

respectively. By (5.22) the Gaussian curvature of $\mathbf{x}$ is identically zero. This contrasts with the geometry of a compact surface in Euclidean space, which must have a point of positive Gaussian curvature. By (5.15), the mean curvature of $\mathbf{x}$, relative to the above unit normal vector field $\mathbf{e}_3$, is

$$H = \frac{1}{2}(\cot\alpha - \tan\alpha) = \cot 2\alpha = \frac{r^2 - s^2}{2rs}.$$

In particular, $H = 0$ if and only if $\alpha = \pi/4$. The *Clifford torus* is the circular torus $\mathbf{x}^{(\pi/4)}$.

5.4 Isoparametric, Dupin, and canal immersions

Definition 5.13. The definitions of isoparametric, Dupin, and canal immersions into $\mathbf{S}^3$ are identical to those given in Definition 4.36 for immersions into $\mathbf{R}^3$.

We shall look briefly at the simple examples of cylinders, cones, and surfaces of revolution in $\mathbf{S}^3$.

Example 5.14 (Cylinders in $\mathbf{S}^3$). We want our definition of cylinders in $\mathbf{S}^3$ to be analogous to their definition in $\mathbf{R}^3$ given in Problem 4.65. The analog of a plane in $\mathbf{R}^3$ is a great sphere $\mathbf{S}^2 \subset \mathbf{S}^3$, which we take to be $\boldsymbol{\epsilon}_0^\perp \cap \mathbf{S}^3 = \{\sum_1^3 x^i \boldsymbol{\epsilon}_i \in \mathbf{R}^4 : \sum_1^3 (x^i)^2 = 1\}$. Let $\boldsymbol{\sigma} : J \to \mathbf{S}^2$ be a curve parametrized by arclength s, with unit tangent $\dot{\boldsymbol{\sigma}}$ and unit normal $\mathbf{N}$, which satisfies $\ddot{\boldsymbol{\sigma}} = -\boldsymbol{\sigma} + \kappa \mathbf{N}$, where $\kappa : J \to \mathbf{R}$ is the curvature of $\boldsymbol{\sigma}$. See Example 4.27 for more details on curves in $\mathbf{S}^2$. The cylinder in $\mathbf{S}^3$ defined over $\boldsymbol{\sigma}$ is generated by the geodesics (great circles) through each point of $\boldsymbol{\sigma}$ and perpendicular to the great sphere $\mathbf{S}^2$. These lines all pass through the center of $\mathbf{S}^2$, which is $\pm\boldsymbol{\epsilon}_0$. Thus, a cylinder in $\mathbf{S}^3$ is always a cone as well. For reasons that are apparent from Problem 5.54, we parametrize the cylinder by

$$\mathbf{y} : J \times \mathbf{R} \to \mathbf{S}^3, \quad \mathbf{y}(s,t) = \operatorname{sech} t\, \boldsymbol{\sigma}(s) + \tanh t\, \boldsymbol{\epsilon}_0. \tag{5.29}$$

Then $d\mathbf{y} = \operatorname{sech} t\, \dot{\boldsymbol{\sigma}}\, ds + \operatorname{sech} t(-\tanh t\, \boldsymbol{\sigma} + \operatorname{sech} t\, \boldsymbol{\epsilon}_0) dt$, and $\mathbf{N}$ is a unit normal vector field along $\mathbf{y}$. The second fundamental form is

$$II = -d\mathbf{y} \cdot d\mathbf{N} = \kappa(s) \cosh t\, \omega^1 \omega^1,$$

where $\omega^1 = \operatorname{sech} t\, ds$, $\omega^2 = \operatorname{sech} t\, dt$ is an orthonormal coframe field on $J \times \mathbf{R}$.

Theorem 5.15 (Isoparametric surfaces, nonumbilic). *Let M be a connected surface and let $\mathbf{x} : M \to \mathbf{S}^3$ be an immersion with unit normal vector field $\mathbf{n} : M \to \mathbf{S}^3$ and constant distinct principal curvatures a and c. Then $ac = -1$ and, replacing $\mathbf{n}$ by $-\mathbf{n}$ if necessary to make $c > 0$, there exists $A \in \mathbf{SO}(4)$ such that*

$$\mathbf{x}(M) \subset A\mathbf{x}^{(\alpha)}(\mathbf{R}^2),$$

where $\mathbf{x}^{(\alpha)} : \mathbf{R}^2 \to \mathbf{S}^3$ is a circular torus of Example 5.12 with

$$\cot\alpha = c, \quad 0 < \alpha < \pi/2.$$

Proof. The proof is a special case of Proposition 3.11. From the assumption that the principal curvatures are distinct, it follows that every point of M is nonumbilic and of the same type. There exists a smooth second order frame field $e : M \to \mathbf{SO}(3)$ along $\mathbf{x}$ with $\mathbf{e}_3 = \mathbf{n}$. Then a and c constant implies

$$\omega_1^2 = 0$$

by the Codazzi equations (5.21). By the Gauss equation (5.22), we have $0 = 1 + ac$ and (5.17) becomes

$$\omega_1^3 = -\frac{1}{c}\omega^1, \quad \omega_2^3 = c\omega^2,$$

on M. Replacing $e = (\mathbf{x}, \mathbf{e}_1, \mathbf{e}_2, \mathbf{e}_3)$ by $e = (\mathbf{x}, \mathbf{e}_2, \mathbf{e}_1, -\mathbf{e}_3)$, if necessary, we may assume $c > 0$. Let $\mathfrak{h}$ be the 2-dimensional distribution on $\mathbf{SO}(4)$ defined by the left-invariant 1-form equations

$$\omega^3 = 0,\ \omega_1^2 = 0,\ \omega_1^3 = \frac{1}{c}\omega^1,\ \omega_2^3 = c\omega^2.$$

The structure equations imply $\mathfrak{h}$ satisfies the Frobenius condition. Being also left invariant, it is a Lie subalgebra of $\mathfrak{o}(4)$. The Lie subgroup H of $\mathbf{SO}(4)$ whose Lie algebra is $\mathfrak{h}$ is the maximal integral surface of $\mathfrak{h}$ through the identity element I_4 of $\mathbf{SO}(4)$. All other integral surfaces of $\mathfrak{h}$ are the left cosets of H. By Example 5.12, the second order frame field $e^{(\alpha)} : \mathbf{R}^2 \to \mathbf{SO}(4)$ along $\mathbf{x}^{(\alpha)}$ is an integral surface of $\mathfrak{h}$, if $\cot\alpha = c$ and $0 < \alpha < \pi/2$. Hence,

$$e^{(\alpha)}(\mathbf{R}^2) = \text{ left coset of } H \ = e^{(\alpha)}(0,0)H,$$

for an arbitrarily chosen point $e^{(\alpha)}(0,0) \in e^{(\alpha)}(\mathbf{R}^2)$. Then $H = e^{(\alpha)}(0,0)^{-1}e^{(\alpha)}(\mathbf{R}^2)$ is a way to find H, and

$$\mathbf{x}^{(\alpha)}(\mathbf{R}^2) = e^{(\alpha)}(0,0)H\boldsymbol{\epsilon}_0 \subset \mathbf{S}^3$$

is the projection of this coset. To complete the proof, we observe that the second order frame field $e : M \to \mathbf{SO}(4)$ along $\mathbf{x}$ is an integral surface of $\mathfrak{h}$, so $e(M) \subset e(m_0)H$, for an arbitrarily chosen point $m_0 \in M$. Therefore,

$$\mathbf{x}(M) = e(M)\boldsymbol{\epsilon}_0 \subset e(m_0)H\boldsymbol{\epsilon}_0 = e(m_0)e^{(\alpha)}(0,0)^{-1}\mathbf{x}^{(\alpha)}(\mathbf{R}^2),$$

which is $A\mathbf{x}^{(\alpha)}(\mathbf{R}^2)$, where $A = e(m_0)e^{(\alpha)}(0,0)^{-1} \in \mathbf{SO}(4)$. □

The following proposition is identical to Proposition 4.52, but with no separate cases as arise in Euclidean space due to the distinction between spheres and planes.

Proposition 5.16. *Let $\mathbf{x} : M \to \mathbf{S}^3 \subset \mathbf{R}^4$ be an immersion with unit normal vector field $\mathbf{e}_3 : M \to \mathbf{S}^3$ and with distinct principal curvatures a and c at each point of M. Let $\gamma : J \to M$ be a connected line of curvature parametrized by arclength for the principal curvature a, where $J \subset \mathbf{R}$ is connected and contains 0. Then a is constant on $\gamma(J)$ if and only if its curvature sphere is constant on $\mathbf{x} \circ \gamma(J)$.*

If the principal curvature a is constant on each of its connected lines of curvature, then $\mathbf{x}$ sends its connected lines of curvature to arcs of circles in $\mathbf{S}^3$.

See Cecil-Ryan [40, Chapter 2, Section 4] for a statement and proof of this proposition in arbitrary dimensions.

Proof. Let $(\mathbf{x}, e) : U \to \mathbf{SO}(4)$ be a second order frame field on a neighborhood U containing $\gamma(J)$ such that $\mathbf{e}_1(\gamma(s)) = (\mathbf{x} \circ \gamma)'(s)$ for every $s \in J$ and $\mathbf{e}_3$ is the given unit normal on U. By the structure equations (5.20) and (5.21), we have

$$\begin{aligned}(\mathbf{x} \circ \gamma)'(s) &= d\mathbf{x}(\gamma'(s)) = \mathbf{e}_1(\gamma(s)),\\(\mathbf{x} \circ \gamma)''(s) &= d\mathbf{e}_1(\gamma'(s)) = (-\mathbf{x} + p\mathbf{e}_2 + a\mathbf{e}_3)(\gamma(s)).\end{aligned}$$

For the smooth function $t = \cot^{-1} a : M \to \mathbf{R}$, define the *focal map*

$$f : M \to \mathbf{S}^3, \quad f = \cos t\, \mathbf{x} + \sin t\, \mathbf{e}_3 = \sin t\, (a\mathbf{x} + \mathbf{e}_3).$$

Then $f(\gamma(s))$ is the center of the curvature sphere through $\mathbf{x}(\gamma(s))$ for a. By the structure equations,

$$\begin{aligned} df &= a_1 \sin^2 t\ (\sin t\ \mathbf{x} - \cos t\ \mathbf{e}_3)\omega^1 \\ &\quad + (a_2 \sin^2 t\ (\sin t\ \mathbf{x} - \cos t\ \mathbf{e}_3) + (\cos t - c \sin t)\mathbf{e}_2)\omega^2, \end{aligned} \tag{5.30}$$

where $da = a_1\omega^1 + a_2\omega^2$ on U. Then $a_1 \sin^2 t = -t_1$, where $dt = t_1\omega^1 + t_2\omega^2$ on U. By (5.30) and the fact that $\omega^2(\gamma') = 0$, we have

$$(f \circ \gamma)' = df(\gamma') = -(t_1(\sin t\ \mathbf{x} - \cos t\ \mathbf{e}_3)) \circ \gamma.$$

Hence, $(f \circ \gamma)' = 0$ on J if and only if $(t \circ \gamma)' = 0$ on J if and only the curvature spheres along $\gamma(J)$ are constant.

We have now proved that if $\gamma : J \to M$ is a connected line of curvature of a and if $a \circ \gamma$ is constant on J, then $\mathbf{x} \circ \gamma(J) \subset \mathbf{x}(M) \cap S$, where S is the necessarily constant curvature sphere along $\mathbf{x} \circ \gamma$, whose center is $f(\gamma(s)) = \mathbf{y}_0 \in \mathbf{S}^3$, independent of $s \in J$.

Assume now that a is constant on each of its connected lines of curvature. Then $a_1 = 0$ on all of U, and (5.30) shows that the focal map f has constant rank equal to 1 on M. Use the notation above for a connected line of curvature $\gamma : J \to M$ for the principal curvature a and for a second order frame field $(\mathbf{x}, e) : U \to \mathbf{SO}(4)$, where $\gamma(J) \subset U$. Then $f(\gamma(J)) = \mathbf{y}_0 \in \mathbf{S}^3$ is a single point and by Exercise 14, there is a unit vector $\mathbf{v} \in T_{\mathbf{y}_0}\mathbf{S}^3 = \mathbf{y}_0^\perp \subset \mathbf{R}^4$ such that

$$df_{\gamma(s)} T_{\gamma(s)} M = \mathbf{R}\mathbf{v},$$

for all $s \in J$. It follows from this and (5.30) that $\mathbf{v}$ must be a multiple of

$$(a_2 \sin^2 t\ (\sin t\ \mathbf{x} - \cos t\ \mathbf{e}_3) + (\cos t - a \sin t)\mathbf{e}_2)(\gamma(s)),$$

which is orthogonal to $(\mathbf{x} \circ \gamma)'(s) = \mathbf{e}_1(s)$, for all $s \in J$. Therefore, $\mathbf{x}(\gamma(s)) \cdot \mathbf{v}$ is constant on J, which means that $\mathbf{x} \circ \gamma(J)$ lies in a sphere centered at $\mathbf{v}$ (or, equivalently, in a hyperplane of $\mathbf{R}^4$ orthogonal to $\mathbf{v}$). Since $\mathbf{v}$ and $\mathbf{y}_0$ are linearly independent, $\mathbf{x} \circ \gamma(J)$ lies in the intersection of two distinct spheres of $\mathbf{S}^3$, and hence must be an arc of a circle. □

5.4.1 Surfaces of revolution

Example 5.17 (Surfaces of revolution). Begin with a *profile curve* in the great sphere $\boldsymbol{\epsilon}_1^\perp \cap \mathbf{S}^3 = \mathbf{S}^2$,

$$\sigma : J \to \mathbf{S}^2, \quad \sigma(y) = f(y)\boldsymbol{\epsilon}_0 + g(y)\boldsymbol{\epsilon}_2 + l(y)\boldsymbol{\epsilon}_3,$$

parametrized by arclength y. Assume that $f(y) > 0$ for all $y \in J$. Then $\mathbf{T} = \dot{\sigma}$ is the unit tangent vector, $\ddot{\sigma} = -\sigma + \kappa \mathbf{N}$, where $\kappa : J \to \mathbf{R}$ is the curvature and

$$\mathbf{N}(y) = p(y)\epsilon_0 + q(y)\epsilon_2 + m(y)\epsilon_3$$

is the unit normal, which satisfies $\dot{\mathbf{N}} = -\kappa \mathbf{T}$. The surface of revolution generated by this curve is

$$\mathbf{x} : J \times \mathbf{R} \to \mathbf{S}^3, \quad \mathbf{x}(x,y) = A(x)\sigma(y) = {}^t(f(y)\cos x, f(y)\sin x, g(y), l(y)),$$

where

$$A(x) = \begin{pmatrix} \cos x & -\sin x & 0 \\ \sin x & \cos x & \\ 0 & & I_2 \end{pmatrix} \in \mathbf{SO}(4),$$

for each $x \in \mathbf{R}$. A second order frame field along $\mathbf{x}$ is given by $e = (\mathbf{e}_0, \dots, \mathbf{e}_3)$, where

$$\mathbf{e}_0 = \mathbf{x},\ \mathbf{e}_1 = \frac{1}{f}\mathbf{x}_x,\ \mathbf{e}_2 = \mathbf{x}_y,\ \mathbf{e}_3 = A(x)\mathbf{N}(y).$$

The orthonormal coframe is $\omega^1 = f(y)dx$, $\omega^2 = dy$. The coordinate curves are the lines of curvature. The principal curvatures are $a = -p(y)/f(y)$, $c = \kappa(y)$. So, a is always constant along its lines of curvature, but c is constant along its lines of curvature if and only if κ is constant.

5.5 Tubes, parallel transformations, focal loci

Exercise 20 (Curves). Let $\mathbf{f} : J \to \mathbf{S}^3$ be a smooth curve, with arc-length parameter s, where J is an interval in $\mathbf{R}$. Let dot denote derivative with respect to s. The smooth function $\kappa = |\ddot{\mathbf{f}} \text{ and } \mathbf{f}| : J \to \mathbf{R}$ is called the *curvature* of $\mathbf{f}$. Use the method of moving frames to show that, if $\kappa > 0$ on J, then there exists a unique frame field

$$E = (\mathbf{f}, \mathbf{T}, \mathbf{N}, \mathbf{B}) : J \to \mathbf{SO}(4)$$

along $\mathbf{f}$, called the *Frenet frame*, such that

$$\dot{\mathbf{f}} = \mathbf{T}, \quad \dot{\mathbf{T}} = \kappa \mathbf{N} - \mathbf{f}, \quad \dot{\mathbf{N}} = -\kappa \mathbf{T} + \tau \mathbf{B}, \quad \dot{\mathbf{B}} = -\tau \mathbf{N},$$

where the smooth function $\tau : J \to \mathbf{R}$ is called the *torsion* of $\mathbf{f}$.

Definition 5.18 (Tubes). Fix a real number r satisfying $0 < r < \pi/2$. The *tube of radius r about the curve* $\mathbf{f}$ is the map

$$\mathbf{x} : J \times \mathbf{R} \to \mathbf{S}^3, \quad \mathbf{x}(s,t) = \cos r \mathbf{f}(s) + \sin r (\cos t \mathbf{N}(s) + \sin t \mathbf{B}(s)). \tag{5.31}$$

Conditions on r and κ are needed in order for $\mathbf{x}$ to be an immersion. See Problem 5.46.

Definition 5.19 (Parallel transformation). Let $\mathbf{x} : M \to \mathbf{S}^3$ be a smoothly immersed surface with unit normal vector field $\mathbf{e}_3$. For any constant $r \in \mathbf{R}$, the *parallel transformation of* $\mathbf{x}$ *by* r *in direction* $\mathbf{e}_3$ is the map

$$\tilde{\mathbf{x}} : M \to \mathbf{S}^3, \quad \tilde{\mathbf{x}} = \cos r \mathbf{x} + \sin r \mathbf{e}_3.$$

If r is an even multiple of π, then $\tilde{\mathbf{x}} = \mathbf{x}$, while if it is an odd multiple of π, then $\tilde{\mathbf{x}} = -\mathbf{x}$. We call these cases *trivial parallel transformations*. Otherwise, we call them *non-trivial*. If $\tilde{\mathbf{x}}$ is *singular* at a point $m \in M$, meaning that the rank of $d\tilde{\mathbf{x}}$ at m is less than two, then $\tilde{\mathbf{x}}(m)$ is called a *focal point* of $\mathbf{x}$.

Exercise 21 (Focal loci). Prove that the parallel transformation $\tilde{\mathbf{x}}$ of $\mathbf{x} : M \to \mathbf{S}^3$ by r in direction $\mathbf{e}_3$ is singular at $m \in M$ if and only if $\cot r$ is a principal curvature of $\mathbf{x}$ at m.

Definition 5.20. Let $\mathbf{x} : M \to \mathbf{S}^3$ be an immersion with unit normal $\mathbf{e}_3 : M \to \mathbf{S}^3$. If $a : M \to \mathbf{R}$ is a principal curvature of $\mathbf{x} : M \to \mathbf{S}^3$ relative to $\mathbf{e}_3$, then a is continuous on M, and smooth off the umbilic set. Let $r = \cot^{-1}(a) : M \to \mathbf{R}$, with $0 < r < \pi$. The image of

$$f : M \to \mathbf{S}^3, \quad f(m) = \cos r \, \mathbf{x} + \sin r \, \mathbf{e}_3$$

is called a *focal locus* of $\mathbf{x}$.

5.6 Stereographic projection

Given a point $\mathbf{p} \in \mathbf{S}^3$, express the orthogonal direct sum of the span of $\mathbf{p}$, denoted $\{\mathbf{p}\}$, and its orthogonal complement by

$$\mathbf{R}^4 = \{\mathbf{p}\} \oplus \mathbf{p}^\perp \cong \mathbf{R} \oplus \mathbf{R}^3, \quad \mathbf{x} = x^0\mathbf{p} + \mathbf{y},$$

where $x^0 \in \mathbf{R}$ and $\mathbf{y} \in \mathbf{p}^\perp$ is the orthogonal projection of $\mathbf{x}$ onto $\mathbf{p}^\perp$.

Definition 5.21. Stereographic projection from the point $\mathbf{p}$ is the map

$$\mathscr{S}_\mathbf{p} : \mathbf{S}^3 \setminus \{\mathbf{p}\} \to \mathbf{p}^\perp, \quad \mathscr{S}_\mathbf{p}(\mathbf{x}) = \frac{1}{1 - x^0}\mathbf{y}. \tag{5.32}$$

Stereographic projection has an inverse,

$$\mathscr{S}_{\mathbf{p}}^{-1}(\mathbf{y}) = \mathbf{p} + \frac{2}{|\mathbf{y}-\mathbf{p}|^2}(\mathbf{y}-\mathbf{p}) = \frac{1}{|\mathbf{y}|^2+1}\left((|\mathbf{y}|^2-1)\mathbf{p}+2\mathbf{y}\right),$$

for any $\mathbf{y} \in \mathbf{p}^{\perp} \cong \mathbf{R}^3$. Its differential at $\mathbf{x} \in \mathbf{S}^3$ is

$$d\mathscr{S}_p = \frac{1}{(1-x^0)^2}\mathbf{y}dx^0 + \frac{1}{1-x^0}d\mathbf{y},$$

so $d\mathscr{S}_p \cdot d\mathscr{S}_p = \frac{d\mathbf{x}\cdot d\mathbf{x}}{(1-x^0)^2}$ shows that $\mathscr{S}_p$ is a conformal diffeomorphism with conformal factor $1/(1-x^0)$ (see Definition 12.1).

If $f : U \subset \mathbf{R}^3 \to \mathbf{R}^3$ is a local conformal diffeomorphism of $\mathbf{R}^3$, then

$$\mathscr{S}_{\mathbf{p}}^{-1} \circ f \circ \mathscr{S}_{\mathbf{p}}$$

is a local conformal diffeomorphism of $\mathbf{S}^3$.

Definition 5.22. The *default stereographic projection* is from $-\boldsymbol{\epsilon}_0$,

$$\mathscr{S} = \mathscr{S}_{-\boldsymbol{\epsilon}_0} : \mathbf{S}^3 \setminus \{-\boldsymbol{\epsilon}_0\} \to \mathbf{R}^3,$$

where we identify $\boldsymbol{\epsilon}_0^{\perp}$ with $\mathbf{R}^3$ by

$$\boldsymbol{\epsilon}_0^{\perp} = \{\mathbf{y} \in \mathbf{R}^4 : y^0 = 0\} \cong \mathbf{R}^3.$$

If $\mathbf{x} = x^0\boldsymbol{\epsilon}_0 + \sum_1^3 x^i\boldsymbol{\epsilon}_i \in \mathbf{S}^3 \setminus \{-\boldsymbol{\epsilon}_0\}$, and if $\mathbf{y} = \sum_1^3 x^i\boldsymbol{\epsilon}_i$, then

$$\mathscr{S}(\mathbf{x}) = \frac{1}{1+x^0}\mathbf{y}, \quad \mathscr{S}^{-1}(\mathbf{y}) = \frac{1}{1+|\mathbf{y}|^2}\left((1-|\mathbf{y}|^2)\boldsymbol{\epsilon}_0 + 2\mathbf{y}\right).$$

In the following we use the notation introduced in Examples 4.21 and 4.22 for oriented spheres and planes in $\mathbf{R}^3$. Stereographic projection being a conformal diffeomorphism implies that it and its inverse take a vector normal to an immersed surface to a vector normal to the image surface. The following proposition shows that stereographic projection and its inverse send oriented spheres to oriented spheres (thinking of oriented planes as oriented spheres with infinite radius). They do not take the center of the oriented sphere to the center of the image sphere, however. This feature makes the proof somewhat more difficult than one might expect.

Proposition 5.23. *Stereographic projection $\mathscr{S}_{\mathbf{p}}$ from a point $\mathbf{p} \in \mathbf{S}^3$ sends an oriented sphere in $\mathbf{S}^3$ to an oriented sphere or plane in $\mathbf{R}^3 \cong \mathbf{p}^{\perp}$ oriented by the image normal vector. The image is a plane if and only if the sphere in $\mathbf{S}^3$ passes through $\mathbf{p}$. If $\mathbf{m} = m^0\boldsymbol{\epsilon}_0 + \sum_1^3 m^i\boldsymbol{\epsilon}_i \in \mathbf{S}^3$, if $\mathbf{m}' = \sum_1^3 m^i\boldsymbol{\epsilon}_i$, and if $0 < r < \pi$, then*

$$\mathscr{S}(S_r(\mathbf{m})) = \begin{cases} S_{\frac{\sin r}{m^0+\cos r}}\left(\frac{\mathbf{m}'}{m^0+\cos r}\right), & \text{if } m^0+\cos r \neq 0. \\ \Pi_{\cot r}\left(\frac{\mathbf{m}'}{\sin r}\right), & \text{if } m^0+\cos r = 0. \end{cases}$$

Conversely, inverse stereographic projection $\mathscr{S}_{\mathbf{p}}^{-1}$ *sends any oriented sphere in* $\mathbf{R}^3$ *to an oriented sphere in* $\mathbf{S}^3$ *not passing through* $\mathbf{p}$*, with image normal, and it sends any oriented plane in* $\mathbf{R}^3$ *to an oriented sphere passing through* $\mathbf{p}$*, with image normal. If* $0 \neq R \in \mathbf{R}$ *and if* $\mathbf{c} \in \mathbf{R}^3$*, then*

$$\mathscr{S}^{-1}(S_R(\mathbf{c})) = S_r(\mathbf{m}),$$

where

$$\mathbf{m} = \frac{1+R^2-|\mathbf{c}|^2}{D}\boldsymbol{\epsilon}_0 + \frac{2}{D}\mathbf{c}, \quad r = \cos^{-1}\frac{1+|\mathbf{c}|^2-R^2}{D} \in (0,\pi), \tag{5.33}$$

and

$$D = \frac{R}{|R|}\sqrt{(1+R^2-|\mathbf{c}|^2)^2+4|\mathbf{c}|^2}. \tag{5.34}$$

If $h \in \mathbf{R}$ *and if* $\mathbf{n}$ *is a unit vector in* $\mathbf{R}^3$*, then*

$$\mathscr{S}^{-1}(\Pi_h(\mathbf{n})) = S_r(\mathbf{m}),$$

where

$$\mathbf{m} = \frac{-h\boldsymbol{\epsilon}_0+\mathbf{n}}{\sqrt{h^2+1}}, \quad r = \cos^{-1}\frac{h}{\sqrt{h^2+1}} \in (0,\pi). \tag{5.35}$$

Proof. Recall that $S_r(\mathbf{m}) = \{\mathbf{x} \in \mathbf{S}^3 : \mathbf{x}\cdot\mathbf{m} = \cos r\}$ is oriented by

$$\mathbf{n}(\mathbf{x}) = \frac{\mathbf{m} - \cos r\,\mathbf{x}}{\sin r},$$

so $-\boldsymbol{\epsilon}_0 \in S_r(\mathbf{m})$ if and only if $m^0 + \cos r = 0$. It suffices to prove the theorem for the default stereographic projection $\mathscr{S}$ (see Problem 5.50). Now

$$\begin{aligned}\mathscr{S}(S_r(\mathbf{m})) &= \{\mathbf{y} \in \mathbf{R}^3 : \mathscr{S}^{-1}(\mathbf{y}) \in S_r(\mathbf{m})\}\\ &= \{\mathbf{y} \in \mathbf{R}^3 : (\frac{1-|\mathbf{y}|^2}{1+|\mathbf{y}|^2}\boldsymbol{\epsilon}_0 + \frac{2}{1+|\mathbf{y}|^2}\mathbf{y})\cdot\mathbf{m} = \cos r\}\\ &= \{\mathbf{y} \in \mathbf{R}^3 : (m^0+\cos r)|\mathbf{y}|^2 - 2\mathbf{m}'\cdot\mathbf{y} + \cos r - m^0 = 0\}.\end{aligned}$$

If $m^0 + \cos r \neq 0$, then we can divide through by it, complete the square, and keep in mind that $(m^0)^2 + |\mathbf{m}'|^2 = |\mathbf{m}|^2 = 1$, to conclude that

$$\mathscr{S}(S_r(\mathbf{m})) = \{\mathbf{y} \in \mathbf{R}^3 : |\mathbf{y} - \frac{\mathbf{m}'}{m^0+\cos r}|^2 = \frac{\sin^2 r}{(m^0+\cos r)^2}\} = S_R(\mathbf{c}),$$

where

$$\mathbf{c} = \frac{\mathbf{m}'}{m^0 + \cos r}, \quad R = \frac{\delta \sin r}{m^0 + \cos r},$$

where $\delta = \pm 1$. To determine the value of δ, let $\mathbf{x} = x^0 \boldsymbol{\epsilon}_0 + \mathbf{y} \in S_r(\mathbf{m})$, where $\mathbf{y} \in \boldsymbol{\epsilon}_0^\perp$, and compare the image normal $d\mathscr{S}_{\mathbf{x}}\mathbf{n}(\mathbf{x})$ with

$$\frac{1}{R}(\mathbf{c} - \mathscr{S}(\mathbf{x})) = \frac{1}{R(m^0 + \cos r)}(\mathbf{m}' - \frac{m^0 + \cos r}{1 + x^0}\mathbf{y}).$$

For this, apply

$$d\mathscr{S}_{\mathbf{x}} = \frac{-1}{(1+x^0)^2}\mathbf{y}dx^0 + \frac{1}{1+x^0}d\mathbf{y}$$

to $\mathbf{n}(\mathbf{x})$ given above to get

$$d\mathscr{S}_{\mathbf{x}}\mathbf{n}(\mathbf{x}) = \frac{1}{(1+x^0)\sin r}(\mathbf{m}' - \frac{m^0 + \cos r}{1 + x^0}\mathbf{y}).$$

Hence, the two normals point in the same direction if and only if $R(m^0 + \cos r)$ is positive, so $\delta = +1$. If $m^0 + \cos r = 0$, then

$$\begin{aligned}
\mathscr{S}(S_r(\mathbf{m})) &= \{\mathbf{y} \in \mathbf{R}^3 : -\mathbf{m}' \cdot \mathbf{y} + \cos r = 0\} \\
&= \{\mathbf{y} \in \mathbf{R}^3 : \frac{\mathbf{m}'}{\sin r} \cdot \mathbf{y} = \cot r\} = \Pi_{\cot r}(\frac{\mathbf{m}'}{\sin r}),
\end{aligned}$$

since now $|\frac{\mathbf{m}'}{\sin r}| = 1$, and the calculation above shows that now

$$d\mathscr{S}_{\mathbf{x}}\mathbf{n}(\mathbf{x}) = \frac{1}{1+x^0}\frac{\mathbf{m}'}{\sin r},$$

which is a positive multiple of the unit normal $\frac{\mathbf{m}'}{\sin r}$.

For the converse, if $R \neq 0$ and if $\mathbf{c} \in \mathbf{R}^3$, then

$$S_R(\mathbf{c}) = \{\mathbf{y} \in \mathbf{R}^3 : |\mathbf{y} - \mathbf{c}|^2 = R^2\}$$

is oriented by the normal vector $(\mathbf{c} - \mathbf{y})/R$ at $\mathbf{y} \in S_R(\mathbf{c})$. If $\mathbf{x} \in \mathbf{S}^3$, with $\mathbf{x} = x^0\boldsymbol{\epsilon}_0 + \mathbf{y}$, where $\mathbf{y} \in \boldsymbol{\epsilon}_0^\perp$, then $|\mathbf{y}|^2 = (1 + x^0)(1 - x^0)$, so

$$\begin{aligned}
\mathscr{S}^{-1}(S_R(\mathbf{c})) &= \{\mathbf{x} \in \mathbf{S}^3 \setminus \{-\boldsymbol{\epsilon}_0\} : \frac{\mathbf{y}}{1+x^0} = \mathscr{S}(\mathbf{x}) \in S_R(\mathbf{c})\} \\
&= \{\mathbf{x} \in \mathbf{S}^3 : |\mathbf{y}|^2 - 2(1+x^0)\mathbf{y} \cdot \mathbf{c} + (1+x^0)^2(|\mathbf{c}|^2 - R^2) = 0\} \\
&= \{\mathbf{x} \in \mathbf{S}^3 : 2\mathbf{y} \cdot \mathbf{c} + (1 + R^2 - |\mathbf{c}|^2)x^0 = 1 + |\mathbf{c}|^2 - R^2\} \\
&= \{\mathbf{x} \in \mathbf{S}^3 : \mathbf{x} \cdot ((1 + R^2 - |\mathbf{c}|^2)\boldsymbol{\epsilon}_0 + 2\mathbf{c}) = 1 + |\mathbf{c}|^2 - R^2\} = S_r(\mathbf{m}),
\end{aligned}$$

where

$$\mathbf{m} = \frac{(1+R^2-|\mathbf{c}|^2)\boldsymbol{\epsilon}_0 + 2\mathbf{c}}{D}, \qquad \cos r = \frac{1+|\mathbf{c}|^2-R^2}{D},$$

and $D = \pm\sqrt{(1+R^2-|\mathbf{c}|^2)^2+4|\mathbf{c}|^2}$. In order to determine the sign of D, we compare the normal $\mathbf{n}(\mathscr{S}^{-1}(\mathbf{y}))$ to $S_r(\mathbf{m})$ at $\mathscr{S}^{-1}(\mathbf{y})$ with the normal $d\mathscr{S}_{\mathbf{y}}^{-1}(\frac{\mathbf{c}-\mathbf{y}}{R})$. It suffices to do this at just one point $\mathbf{y} \in S_R(\mathbf{c})$, which for simplicity we choose to be $\mathbf{y} = \mathbf{c} + R\boldsymbol{\epsilon}_1$, where the unit normal is $\frac{\mathbf{c}-\mathbf{y}}{R} = -\boldsymbol{\epsilon}_1$. Then

$$d\mathscr{S}_{\mathbf{y}}^{-1}(-\boldsymbol{\epsilon}_1) = \frac{2}{(1+|\mathbf{y}|^2)^2}(2y^1\boldsymbol{\epsilon}_0 + 2y^1\mathbf{y} - (1+|\mathbf{y}|^2)\boldsymbol{\epsilon}_1),$$

where $y^1 = \mathbf{y}\cdot\boldsymbol{\epsilon}_1$, and

$$\mathbf{n}(\mathscr{S}^{-1}(\mathbf{y})) = \frac{\mathbf{m} - \cos r\, \mathscr{S}^{-1}(\mathbf{y})}{\sin r} = \frac{2R(2y^1\boldsymbol{\epsilon}_0 + 2y^1\mathbf{y} - (1+|\mathbf{y}|^2)\boldsymbol{\epsilon}_1)}{D(1+|\mathbf{y}|^2)\sin r},$$

since $\mathbf{y} = \mathbf{c} + R\boldsymbol{\epsilon}_1$ implies $|\mathbf{c}|^2 = |\mathbf{y}|^2 - 2Ry^1 + R^2$. These two normals point in the same direction if and only if R/D is positive; that is, D has the same sign as R.

Finally, if $h \in \mathbf{R}$ and $\mathbf{n}$ is a unit vector in $\mathbf{R}^3$, then

$$\begin{aligned}
\mathscr{S}^{-1}\Pi_h(\mathbf{n}) &= \mathscr{S}^{-1}\{\mathbf{y}\in\mathbf{R}^3 : \mathbf{y}\cdot\mathbf{n} = h\} = \{\mathbf{x}\in\mathbf{S}^3\setminus\{-\boldsymbol{\epsilon}_0\} : \mathscr{S}(\mathbf{x})\cdot\mathbf{n} = h\} \\
&= \{\mathbf{x}\in\mathbf{S}^3 : \mathbf{x}\cdot\mathbf{n} = (1+x^0)h\} \\
&= \{\mathbf{x}\in\mathbf{S}^3 : \mathbf{x}\cdot\frac{-h\boldsymbol{\epsilon}_0+\mathbf{n}}{\delta\sqrt{h^2+1}} = \frac{h}{\delta\sqrt{h^2+1}}\} = S_r(\mathbf{m}),
\end{aligned}$$

where

$$\mathbf{m} = \frac{-h\boldsymbol{\epsilon}_0+\mathbf{n}}{\delta\sqrt{h^2+1}}, \qquad r = \cos^{-1}\frac{h}{\delta\sqrt{h^2+1}} \in (0,\pi),$$

and $\delta = \pm 1$. The sign must be chosen so that, if $\mathbf{y}\in\Pi_h(\mathbf{n})$, then

$$d\mathscr{S}_{\mathbf{y}}^{-1}\mathbf{n} = \frac{2}{(1+|\mathbf{y}|^2)^2}(-2h\boldsymbol{\epsilon}_0 + (1+|\mathbf{y}|^2)\mathbf{n} - 2h\mathbf{y})$$

points in the same direction as

$$\mathbf{n}(\mathscr{S}^{-1}(\mathbf{y})) = \frac{1}{\delta\sqrt{h^2+1}(1+|\mathbf{y}|^2)}(-2h\boldsymbol{\epsilon}_0 + (1+|\mathbf{y}|^2)\mathbf{n} - 2h\mathbf{y}),$$

which implies that $\delta = +1$. □

Proposition 5.24. *Stereographic projection from a point in* $\mathbf{S}^3$ *and its inverse take oriented curvature spheres of an immersed surface to oriented curvature spheres of the image surface. Any curvature vector of the oriented curvature sphere is also a curvature vector for the image surface.*

Remark 5.25. In the light of Proposition 5.23, this proposition is obvious, since diffeomorphisms always preserve tangency and rank, and stereographic projection and its inverse also send oriented spheres to oriented spheres. Nevertheless, we think a detailed proof is instructive.

Proof. It suffices to prove this result for the default stereographic projection $\mathscr{S}$ and its inverse. Let $\mathbf{x} : M^2 \to \mathbf{S}^3$ be an immersed surface with unit normal vector $\mathbf{n}$ at a point $p \in M$ such that $\mathbf{x}(p) \neq -\boldsymbol{\epsilon}_0$. Let $e : U \subset M \to \mathbf{SO}(4)$ be a first order frame field along $\mathbf{x}$ with $\mathbf{e}_3(p) = \mathbf{n}$. If S is an oriented curvature sphere of $\mathbf{x}$ at p relative to $\mathbf{n}$, then

$$S = S_r(\mathbf{x}\cos r + \mathbf{e}_3 \sin r)(p),$$

where $r \in (0, \pi)$ is a constant for which $\cot r$ is a principal curvature at p relative to $\mathbf{n}$ and the map

$$\mathbf{m} = \mathbf{x}\cos r + \mathbf{e}_3 \sin r : U \to \mathbf{S}^3$$

is singular at p; that is, the dimension of the kernel of

$$d\mathbf{m}_p = dm^0_p \boldsymbol{\epsilon}_0 + d\mathbf{m}'_p$$

is at least one, where $\mathbf{m} = m^0 \boldsymbol{\epsilon}_0 + \sum_1^3 m^i \boldsymbol{\epsilon}_i = m^0 \boldsymbol{\epsilon}_0 + \mathbf{m}'$. Thus, S is an oriented curvature sphere if and only if there exists a nonzero vector $v \in T_pM$ such that

$$dm^0_p v = 0 \text{ and } d\mathbf{m}'_p v = 0. \tag{5.36}$$

Now $\mathscr{S}(S_r(\mathbf{m}))$ is an oriented tangent plane or sphere to $\mathscr{S} \circ \mathbf{x}$ at a point of U, depending on whether $m^0 + \cos r$ is zero or not, respectively.

If $m^0(p) + \cos r \neq 0$, then we may choose U small enough so that this is true at every point of U. Then $\mathscr{S}(S_r(\mathbf{m})) = S_R(\mathbf{c})$ is an oriented sphere tangent to $\mathscr{S} \circ \mathbf{x} : M \to \mathbf{R}^3$ at each point of U, where

$$\mathbf{c} = \frac{\mathbf{m}'}{m^0 + \cos r}, \qquad R = \frac{\sin r}{m^0 + \cos r} \neq 0.$$

Let $(\mathscr{S} \circ \mathbf{x}, \tilde{\mathbf{e}}_1, \tilde{\mathbf{e}}_2, \tilde{\mathbf{e}}_3) : \tilde{U} \subset U \to \mathbf{E}(3)$ be a first order frame field along $\mathscr{S} \circ \mathbf{x}$ such that

$$\tilde{\mathbf{e}}_3 = \frac{\mathbf{c} - \mathscr{S} \circ \mathbf{x}}{R} = \frac{1}{\sin r}(\mathbf{m}' - (m^0 + \cos r)\mathscr{S} \circ \mathbf{x}) \tag{5.37}$$

at every point of the connected set $\tilde{U}$. Consider the parallel transformation of $\mathscr{S} \circ \mathbf{x}$ on $\tilde{U}$ given by

$$\mathbf{y} = \mathscr{S} \circ \mathbf{x} + R_0 \tilde{\mathbf{e}}_3 : \tilde{U} \to \mathbf{R}^3,$$

where $R_0 = R(p) = \frac{\sin r}{m^0(p)+\cos r}$. Then

$$\begin{aligned} d\mathbf{y}_p &= d(\mathscr{S} \circ \mathbf{x})_p + \frac{R_0}{\sin r}(d\mathbf{m}'_p - (m^0(p) + \cos r) d(\mathscr{S} \circ \mathbf{x})_p - \mathscr{S} \circ \mathbf{x}(p) dm^0_p) \\ &= \frac{R_0}{\sin r}(d\mathbf{m}'_p - \mathscr{S} \circ \mathbf{x}(p) dm^0_p) \end{aligned}$$

shows that $d\mathbf{y}_p v = 0$, if (5.36) holds, so $S_{R_0}(\mathbf{c}(p))$ is an oriented curvature sphere of $\mathscr{S} \circ \mathbf{x}$ at p relative to $\tilde{\mathbf{e}}_3(p)$ and that v is one of its principal curvature vectors.

If $m^0(p) + \cos r = 0$, then

$$\mathscr{S}(S_r(\mathbf{m}(p))) = \Pi_{\cot r}(\frac{\mathbf{m}'(p)}{\sin r})$$

is an oriented tangent plane to $\mathscr{S} \circ \mathbf{x}$ at p with unit normal $\frac{\mathbf{m}'}{\sin r} = \tilde{\mathbf{e}}_3(p)$, by (5.37), and

$$d\tilde{\mathbf{e}}_{3p} = \frac{1}{\sin r}(d\mathbf{m}'_p - \mathscr{S} \circ \mathbf{x}(p) dm^0_p)$$

has rank less than two, by (5.36). Thus zero is a principal curvature of $\mathscr{S} \circ \mathbf{x}$ at p and this oriented tangent plane is an oriented curvature sphere at p. This formula also shows that any principal curvature vector $v \in T_pM$ of $\mathbf{x}$ is also a principal curvature vector for $\mathscr{S} \circ \mathbf{x}$.

Now consider the inverse stereographic projection $\mathscr{S}^{-1} : \mathbf{R}^3 \to \mathbf{S}^3$ applied to the immersed surface $\mathbf{y} : M^2 \to \mathbf{R}^3$ to give an immersion $\mathbf{x} = \mathscr{S}^{-1} \circ \mathbf{y} : M \to \mathbf{S}^3$. Suppose $S_R(\mathbf{c}(p))$ is an oriented curvature sphere of $\mathbf{y}$ at $p \in M$, relative to the unit normal $\mathbf{n}$ at p. Let

$$(\mathbf{y}, \tilde{\mathbf{e}}_1, \tilde{\mathbf{e}}_2, \tilde{\mathbf{e}}_3) : U \to \mathbf{E}(3) \tag{5.38}$$

be a first order frame field along $\mathbf{y}$ on a neighborhood U containing p, such that $\tilde{\mathbf{e}}_3(p) = \mathbf{n}$. Then

$$\mathbf{c} = \mathbf{y} + R\tilde{\mathbf{e}}_3 : U \to \mathbf{R}^3$$

defines an oriented tangent sphere map $S_R(\mathbf{c})$ along $\mathbf{y}$ on U whose value at p is the given curvature sphere. Thus, there exists a nonzero vector $v \in T_pM$ such that $d\mathbf{c}_p v = 0$.

$\mathscr{S}^{-1}(S_R(\mathbf{c})) = S_r(\mathbf{m})$ is an oriented tangent sphere map along $\mathbf{x}$, where $\mathbf{m} : U \to \mathbf{S}^3$, $r : U \to (0, \pi)$, and $D : U \to \mathbf{R} \setminus \{0\}$ are the smooth maps defined in equations (5.33) and (5.34). The unit normal to $S_r(\mathbf{m})$ at $\mathbf{x}$ is

$$\mathbf{e}_3 = \frac{\mathbf{m} - \cos r\, \mathbf{x}}{\sin r}. \tag{5.39}$$

To obtain an oriented tangent sphere map along $\mathbf{x}$ on U with constant radius, we let $r_0 = r(p)$ and let

$$S = \cos r_0\, \mathbf{x} + \sin r_0\, \mathbf{e}_3 = (\cos r_0 - \sin r_0 \cot r)\mathbf{x} + \frac{\sin r_0}{\sin r}\mathbf{m}. \tag{5.40}$$

By the calculation requested in Exercise 22 below, it follows that for any curvature vector $v \in T_pM$ for $S_R(\mathbf{c})$, we have $dS_p v = 0$, which means that $S_{r(p)}(\mathbf{m}(p))$ is an oriented curvature sphere of $\mathbf{x}$ at p relative to $\mathbf{e}_3(p)$, and that v is a curvature vector of it.

Finally, suppose that $\Pi_{h_0}(\mathbf{n})$ is an oriented curvature sphere of $\mathbf{y}$ at $p \in M$ relative to the unit normal $\mathbf{n}$ at p, where $h_0 = \mathbf{y}(p) \cdot \mathbf{n}$. In terms of the first order frame field (5.38), there must exist a non-zero vector $v \in T_pM$ such that $d\tilde{\mathbf{e}}_{3p} v = 0$, so v is a principal curvature vector for the principal curvature zero.

Now $\mathscr{S}^{-1}(\Pi_h(\mathbf{e}_3)) = S_r(\mathbf{m})$ is an oriented tangent sphere map along $\mathbf{x}$ on U, where $\mathbf{m} : U \to \mathbf{S}^3$ and $r : U \to (0, \pi)$ are the smooth maps defined in (5.35). Using the unit normal $\mathbf{e}_3$ to $S_r(\mathbf{m})$ at $\mathbf{x}$ defined in (5.39), we obtain the oriented tangent sphere map (5.40) along $\mathbf{x}$ on U with constant radius $r_0 = r(p)$. By the calculation requested in Exercise 23 below, it follows that for a principal curvature vector $v \in T_pM$ for which $d\tilde{\mathbf{e}}_{3p} v = 0$, we have $dS_p v = 0$ as well. Thus, $\mathscr{S}^{-1}(\Pi_{h_0}(\mathbf{n})) = S_{r_0}(\mathbf{m}_0)$ is an oriented curvature sphere of $\mathbf{x}$ at p relative to the unit normal $\mathbf{e}_3(p)$, and v is a principal curvature vector of it. □

Exercise 22. In the notation of the preceding proof, use equations (5.33) and (5.34) to calculate $d\mathbf{m}_p$ and dD_p in order to verify that $dS_p v = 0$ for any $v \in T_pM$ for which $d\mathbf{c}_p v = 0$.

Exercise 23. In the notation of the preceding proof, use equation (5.35) to calculate $d\mathbf{m}_p$ and dr_p in order to verify that $dS_p v = 0$ for any $v \in T_pM$ for which $d\tilde{\mathbf{e}}_{3p} v = 0$.

5.7 Hopf cylinders

The Hermitian inner product on $\mathbf{C}^2$ is given by $\langle \mathbf{z}, \mathbf{w} \rangle = {}^t\mathbf{z}\bar{\mathbf{w}}$. Its real part is a Euclidean inner product on $\mathbf{C}^2$ as a vector space over $\mathbf{R}$, which makes the map

$$\psi : \mathbf{C}^2 \to \mathbf{R}^4, \quad \psi \begin{pmatrix} x^0 + ix^1 \\ x^2 + ix^3 \end{pmatrix} = {}^t(x^0, x^1, x^2, x^3) \tag{5.41}$$

an isometric isomorphism onto $\mathbf{R}^4$ with its standard dot product. Under this identification, the sphere

$$\mathbf{S}^3 = \{\mathbf{z} \in \mathbf{C}^2 : \langle \mathbf{z}, \mathbf{z} \rangle = 1\}$$

is the standard unit sphere in $\mathbf{R}^4$. The complex projective line $\mathbf{CP}^1$ is the set of orbits of $\mathbf{C}^2 \setminus \{\mathbf{0}\}$ under the right multiplication action of $\mathbf{C}^*$, the multiplicative group of nonzero complex numbers. Denote the orbit of the vector $\mathbf{z} \in \mathbf{C}^2$ by $[\mathbf{z}] \in \mathbf{CP}^1$. For more detail see Example 7.14. Restriction to $\mathbf{S}^3$ of the standard projection

$$\mathbf{C}^2 \setminus \{\mathbf{0}\} \to \mathbf{CP}^1, \quad \mathbf{z} \mapsto [\mathbf{z}]$$

is the *Hopf fibration* when we identify $\mathbf{CP}^1$ with $\mathbf{S}^2$ as explained below. This fibration is a principal $\mathbf{S}^1$ bundle whose fiber over a point $[\mathbf{z}] \in \mathbf{CP}^1$ is the circle

$$\{\mathbf{z}a : a \in \mathbf{S}^1 \subset \mathbf{C}\} \subset \mathbf{S}^3.$$

The complex projective line $\mathbf{CP}^1$ is biholomorphically equivalent to the Riemann sphere $\widehat{\mathbf{C}} = \mathbf{C} \cup \{\infty\}$ by

$$\zeta : \mathbf{CP}^1 \to \widehat{\mathbf{C}}, \quad \zeta \begin{bmatrix} z \\ w \end{bmatrix} = \frac{z}{w}, \quad \zeta \begin{bmatrix} 1 \\ 0 \end{bmatrix} = \infty.$$

The inverse of stereographic projection from $\boldsymbol{\epsilon}_1 \in \mathbf{S}^2$ onto the Riemann sphere $\widehat{\mathbf{C}}$ is a biholomorphic equivalence

$$\mathscr{S}^{-1} : \widehat{\mathbf{C}} \to \mathbf{S}^2, \quad \mathscr{S}^{-1}(x+iy) = {}^t(x^2+y^2-1, 2x, 2y)/(x^2+y^2+1).$$

Definition 5.26. The *Hopf fibration* $\pi_h : \mathbf{S}^3 \to \mathbf{S}^2$ is the map

$$\pi_h(\mathbf{z}) = \mathscr{S}^{-1} \circ \zeta \circ [\mathbf{z}] = \mathscr{S}^{-1}(\frac{z}{w}) = \begin{pmatrix} |z|^2 - |w|^2 \\ 2\Re(z\bar{w}) \\ 2\Im(z\bar{w}) \end{pmatrix},$$

where $\mathbf{z} = {}^t(z, w) \in \mathbf{S}^3 \subset \mathbf{C}^2$, so $|z|^2 + |w|^2 = 1$.

We continue with another description of this Hopf fibration that better accommodates the action of the group of isometries, $\mathbf{SO}(3)$, of $\mathbf{S}^2$. The *special unitary group*

$$\mathbf{SU}(2) = \{A \in \mathbf{GL}(2, \mathbf{C}) : A\,{}^t\!\bar{A} = I, \ \det A = 1\}.$$

The map

$$\mathbf{S}^3 \subset \mathbf{C}^2 \to \mathbf{SU}(2), \quad \mathbf{z} = \begin{pmatrix} z \\ w \end{pmatrix} \mapsto (\mathbf{z}, \mathbf{z}^*) = \begin{pmatrix} z & -\bar{w} \\ w & \bar{z} \end{pmatrix} \tag{5.42}$$

is a diffeomorphism. In addition, for any $a \in \mathbf{S}^1$,

$$\mathbf{z}a \mapsto \begin{pmatrix} z & -\bar{w} \\ w & \bar{z} \end{pmatrix} \begin{pmatrix} a & 0 \\ 0 & \bar{a} \end{pmatrix}.$$

The Lie algebra of $\mathbf{SU}(2)$ is

$$\mathfrak{su}(2) = \{X \in \mathfrak{gl}(2,\mathbf{C}) : X + {}^t\bar{X} = 0,\ \text{trace}\, X = 0\} = \{\begin{pmatrix} it & -\bar{z} \\ z & -it \end{pmatrix} : t \in \mathbf{R},\ z \in \mathbf{C}\}.$$

The *Killing form* of $\mathfrak{su}(2)$ is the positive definite inner product

$$\langle X, Y\rangle = -\frac{1}{2}\text{trace}\, XY. \tag{5.43}$$

The set of vectors

$$\mathscr{I} = \begin{pmatrix} i & 0 \\ 0 & -i \end{pmatrix}, \quad \mathscr{J} = \begin{pmatrix} 0 & i \\ i & 0 \end{pmatrix}, \quad \mathscr{K} = \begin{pmatrix} 0 & -1 \\ 1 & 0 \end{pmatrix},$$

is an orthonormal basis of $\mathfrak{su}(2)$ for the Killing form. The map

$$\mathbf{R}^3 \to \mathfrak{su}(2), \quad {}^t(t,x,y) \mapsto t\mathscr{I} + x\mathscr{J} + y\mathscr{K} \tag{5.44}$$

is a linear isometry, where $\mathbf{R}^3$ has the standard dot product and $\mathfrak{su}(2)$ has the Killing form inner product (5.43). In particular, this map sends the standard sphere $\mathbf{S}^2 \subset \mathbf{R}^3$ isometrically onto

$$\{\begin{pmatrix} it & ix-y \\ ix+y & -it \end{pmatrix} : t,x,y \in \mathbf{R},\ t^2+x^2+y^2 = 1\} \subset \mathfrak{su}(2). \tag{5.45}$$

The adjoint action of $\mathbf{SU}(2)$ on $\mathfrak{su}(2)$ is

$$\text{Ad}(A)X = AXA^{-1}.$$

It preserves the Killing form (5.43). The orbit of $\mathscr{I}$ under this action is the standard 2-sphere (5.45). Under the identifications $\mathbf{S}^3 = \mathbf{SU}(2)$ in (5.42) and $\mathbf{R}^3 = \mathfrak{su}(2)$ in (5.44), the Hopf fibration $\pi_h(\mathbf{z})$ of Definition 5.26 satisfies

$$Ad(\mathbf{z},\mathbf{z}^*)\mathscr{I} = (|z|^2 - |w|^2)\mathscr{I} + \Re(2z\bar{w})\mathscr{J} + \Im(2z\bar{w})\mathscr{K} = \pi_h(\mathbf{z}).$$

Since the adjoint action of $\mathbf{SU}(2)$ on $\mathfrak{su}(2)$ is isometric and $\mathbf{SU}(2)$ is connected, the matrix of $\text{Ad}(A)$ relative to the orthonormal frame $\mathscr{I}, \mathscr{J}, \mathscr{K}$ must be an element of $\mathbf{SO}(3)$. Define the map

$$\pi_s : \mathbf{SU}(2) \to \mathbf{SO}(3), \tag{5.46}$$

by $\pi_s(\mathbf{z},\mathbf{z}^*)$ is the matrix of $\mathrm{Ad}(\mathbf{z},\mathbf{z}^*)$ relative to the orthonormal frame $\mathscr{I}, \mathscr{J}, \mathscr{K}$,

$$\pi_s(\mathbf{z},\mathbf{z}^*) = \begin{pmatrix} |z|^2 - |w|^2 & -\Re(2zw) & \Im(2zw) \\ \Re(2z\bar{w}) & \Re(z^2 - \bar{w}^2) & \Im(\bar{z}^2 + w^2) \\ \Im(2z\bar{w}) & \Im(z^2 - \bar{w}^2) & \Re(\bar{z}^2 + w^2) \end{pmatrix},$$

where $\mathbf{z} = {}^t(z,w) \in \mathbf{S}^3$. The map (5.46) is a 2:1 Lie group homomorphism whose induced Lie algebra isomorphism

$$d\pi_s : \mathfrak{su}(2) \to \mathfrak{o}(3)$$

sends a skew-hermitian matrix $X \in \mathfrak{su}(2)$ to the matrix of the skew-symmetric operator on $\mathfrak{su}(2) \cong \mathbf{R}^3$ given by

$$d\pi_s(X)\mathscr{L} = X\mathscr{L} + \mathscr{L}{}^t\bar{X} = X\mathscr{L} - \mathscr{L}X,$$

for any $\mathscr{L} \in \mathfrak{su}(2)$. The matrix of the operator $d\pi_s X$ relative to the orthonormal frame $\mathscr{I}, \mathscr{J}, \mathscr{K}$ is

$$d\pi_s \begin{pmatrix} it & -\bar{z} \\ z & -it \end{pmatrix} = 2 \begin{pmatrix} 0 & -x & y \\ x & 0 & -t \\ -y & t & 0 \end{pmatrix} \in \mathfrak{o}(3), \tag{5.47}$$

where $t, x, y \in \mathbf{R}$ and $z = x + iy$.

Consider an immersed curve $\gamma : N \to \mathbf{S}^2$ with parameter s satisfying $|\dot{\gamma}(s)| = 1$. Here $N \subset \mathbf{R}$ is an interval. A frame field along γ is a smooth map $\mathscr{G} : N \to \mathbf{SO}(3)$ such that $\pi \circ \mathscr{G} = \gamma$, where the projection

$$\pi : \mathbf{SO}(3) \to \mathbf{S}^2, \quad \pi(A) = A\epsilon_1 = A_1,$$

is the first column of A, where $\epsilon_1, \epsilon_2, \epsilon_3$ is the standard basis of $\mathbf{R}^3$. The *Frenet frame* along γ is

$$\mathscr{G} = (\gamma, \dot{\gamma}, \gamma \times \dot{\gamma}),$$

which is the unique frame field along γ satisfying

$$\mathscr{G}^{-1} d\mathscr{G} = \begin{pmatrix} 0 & -1 & 0 \\ 1 & 0 & -\kappa \\ 0 & \kappa & 0 \end{pmatrix} ds,$$

for some function $\kappa : N \to \mathbf{R}$, called the *curvature* of γ.

Definition 5.27. A *spinor lift of* a frame field $\mathscr{G} : N \to \mathbf{SO}(3)$ along γ, is a smooth map $\Gamma : N \to \mathbf{S}^3 \cong \mathbf{SU}(2)$ such that $\mathscr{G} = \pi_s \circ \Gamma : N \to \mathbf{SO}(3)$.

Since $\pi_s : \mathbf{S}^3 \cong \mathbf{SU}(2) \to \mathbf{SO}(3)$ is a 2:1 covering projection, and since N is simply connected, any frame field $\mathcal{G} : N \to \mathbf{SO}(3)$ along γ has a spinor lift Γ, and the only other one is $-\Gamma$. Notice that $\pi_h \circ \Gamma = \gamma$, so that $\Gamma : N \to \mathbf{S}^3$ is at the same time a section of the Hopf fibration over γ. We have the commuting diagram

$$\begin{array}{ccccc} & & \mathbf{SU}(2) & & \\ & \overset{\Gamma}{\nearrow} & \downarrow \pi_s & \overset{\cong}{\searrow} & \\ N \subset \mathbf{R} & \overset{\mathcal{G}}{\to} & \mathbf{SO}(3) & \overset{\pi_s}{\leftarrow} & \mathbf{S}^3 \\ & \underset{\gamma}{\searrow} & \downarrow \pi & \underset{\pi_h}{\swarrow} & \\ & & \mathbf{S}^2 & & \end{array}$$

Proposition 5.28. *If $\Gamma : N \to \mathbf{SU}(2)$ is a spinor lift of the Frenet frame $\mathcal{G} : N \to \mathbf{SO}(2)$ of a unit speed curve $\gamma : N \to \mathbf{S}^2$, and if $d\Gamma = \dot{\Gamma} ds$, then*

$$\Gamma^{-1}\dot{\Gamma} = \begin{pmatrix} i\kappa/2 & -1/2 \\ 1/2 & -i\kappa/2 \end{pmatrix} : N \to \mathfrak{su}(2), \tag{5.48}$$

where $\kappa : N \to \mathbf{R}$ is the curvature of γ. Writing $\Gamma = (\Gamma_1, \Gamma_2)$, the columns of Γ, so $\Gamma_2 = \Gamma_1^*$, (5.48) says

$$\dot{\Gamma}_1 = \frac{i\kappa}{2}\Gamma_1 + \frac{1}{2}\Gamma_2, \quad \dot{\Gamma}_2 = -\frac{1}{2}\Gamma_1 - \frac{i\kappa}{2}\Gamma_2. \tag{5.49}$$

Proof. We have $\pi_s \circ \Gamma = \mathcal{G}$, the Frenet frame along γ, so

$$d\pi_s(\Gamma^{-1}\dot{\Gamma}) = \mathcal{G}^{-1}\dot{\mathcal{G}} = \begin{pmatrix} 0 & -1 & 0 \\ 1 & 0 & -\kappa \\ 0 & \kappa & 0 \end{pmatrix} = d\pi_s \begin{pmatrix} i\kappa/2 & -1/2 \\ 1/2 & -i\kappa/2 \end{pmatrix},$$

by (5.47). The result now follows since $d\pi_s$ is an isomorphism. □

Relative to the hermitian inner product on $\mathbf{C}^2$ given by $\langle \mathbf{z}, \mathbf{w} \rangle = {}^t\mathbf{z}\bar{\mathbf{w}}$, the columns Γ_1, Γ_2 of a map $\Gamma : N \to \mathbf{SU}(2)$ form a unitary frame of $\mathbf{C}^2$ at each $s \in N$. If Γ is the spinor lift of the Frenet frame of a unit speed curve $\gamma : N \to \mathbf{S}^2$, then (5.49) implies $\Gamma_1 : N \to \mathbf{S}^3$ is a curve satisfying

$$|\dot{\Gamma}_1|^2 = \frac{1}{4}(\kappa^2 + 1).$$

Definition 5.29 (Pinkall [136]). If $\gamma : N \to \mathbf{S}^2$ is a unit speed curve and if $\Gamma = (\Gamma_1, \Gamma_2) : N \to \mathbf{SU}(2)$ is a spinor lift of the Frenet frame field along γ, then the *Hopf cylinder over γ* is

$$f_\gamma : N \times \mathbf{R} \to \mathbf{S}^3, \quad f_\gamma(s, \theta) = e^{i\theta}\Gamma_1(s). \tag{5.50}$$

The Hopf cylinder (5.50) satisfies

$$df_\gamma = \dot{\Gamma}_1 e^{i\theta} ds + i\Gamma_1 e^{i\theta} d\theta. \tag{5.51}$$

Using the fact that the standard metric on $\mathbf{S}^3$ is the real part of the hermitian inner product on $\mathbf{C}^2$, we find its first fundamental form is

$$I = \langle df_\gamma, df_\gamma \rangle = \frac{1}{4}(\kappa ds + 2d\theta)^2 + \frac{1}{4} ds^2.$$

We conclude that an orthonormal coframe field for I on $N \times \mathbf{R}$ is given by the closed 1-forms

$$\alpha^1 = \frac{1}{2} ds, \quad \alpha^2 = d\theta + \frac{\kappa}{2} ds = \frac{1}{2} d\varphi, \tag{5.52}$$

where

$$\varphi = 2\theta + \int_0^s \kappa(u) du.$$

The metric I is flat (its Gaussian curvature is zero) and

$$df_\gamma = e^{i\theta}\Gamma_2 \alpha^1 + ie^{i\theta}\Gamma_1 \alpha^2,$$

by (5.49), (5.51), and (5.52). Then

$$e = (\psi \circ f_\gamma, \psi \circ \mathbf{e}_1, \psi \circ \mathbf{e}_2, \psi \circ \mathbf{e}_3) : N \times \mathbf{R} \to \mathbf{SO}(4),$$

is a first order frame field along f_γ, where $\psi : \mathbf{C}^2 \to \mathbf{R}^4$ is the isomorphism (5.41), and

$$\mathbf{e}_1 = e^{i\theta}\Gamma_2, \quad \mathbf{e}_2 = ie^{i\theta}\Gamma_1, \quad \mathbf{e}_3 = -ie^{i\theta}\Gamma_2 : N \times \mathbf{R} \to \mathbf{C}^2.$$

Then

$$d\mathbf{e}_3 = -ie^{i\theta}\dot{\Gamma}_2 ds + e^{i\theta}\Gamma_2 d\theta = (\alpha^2 - 2\kappa\alpha^1)\mathbf{e}_1 + \alpha^1 \mathbf{e}_2 = \alpha^1_3 \mathbf{e}_1 + \alpha^2_3 \mathbf{e}_2,$$

and the mean curvature H and Hopf invariant h relative to e of f_γ are

$$H = \frac{1}{2}(h_{11} + h_{22}) = \frac{1}{2}(2\kappa + 0) = \kappa, \quad h = \frac{h_{11} - h_{22}}{2} - ih_{12} = \kappa + i. \tag{5.53}$$

Suppose now that γ is a closed curve of length $l > 0$, by which we mean that $N = \mathbf{R}$ and $\gamma(s + l) = \gamma(s)$, for all $s \in \mathbf{R}$. Then $f_\gamma : \mathbf{R}^2 \to \mathbf{S}^3$ is doubly periodic, $f_\gamma(s + l, \theta + 2\pi) = f_\gamma(s, \theta)$, for all $(s, \theta) \in \mathbf{R}^2$, so it is an immersion of the torus $T^2 = \mathbf{R}^2/(l\mathbf{Z} \times 2\pi\mathbf{Z})$. The Frenet frame $\mathscr{G} : \mathbf{R} \to \mathbf{SO}(3)$ is also periodic of period l, but its spinor lift $\Gamma : \mathbf{R} \to \mathbf{SU}(2)$ is periodic of period l or $2l$.

5.8 Willmore tori

Definition 5.30 (Willmore [171], Bryant [20]). The *Willmore functional* for an immersion $f : M^2 \to \mathbf{S}^3$ of a compact surface M is

$$\mathscr{W}(f) = \int_M (H^2 + 1) dA,$$

where H is the mean curvature and dA is the induced area element of f. A *Willmore immersion* of M is a critical point f of W.

In [171], Willmore proved that a torus of revolution in $\mathbf{R}^3$ is a Willmore immersion if and only if the torus is obtained by revolving a circle of radius b about a line a distance a from its center with $a/b = \sqrt{2}$.

The Willmore functional of a Hopf torus $f_\gamma : T^2 \to \mathbf{S}^3$ of an l-periodic curve $\gamma : \mathbf{R} \to \mathbf{S}^2$ is

$$\mathscr{W}(f_\gamma) = \frac{1}{2} \int_0^{2\pi} \int_0^l (1 + \kappa(s)^2) ds d\theta = \pi \int_0^l (1 + \kappa(s)^2) ds,$$

by (5.53) and since $dA = \alpha^1 \wedge \alpha^2 = \frac{1}{2} ds \wedge d\theta$. By Palais's *Principle of symmetric criticality* [134], the immersion f_γ is an extremal of the Willmore functional if and only if the unit speed curve $\gamma : \mathbf{R}/l\mathbf{Z} \to \mathbf{S}^2$ is an extremal curve of the functional

$$\mathscr{E}(\gamma) = \int_0^l (1 + \kappa(s)^2) ds.$$

As derived in Griffiths [80, (I.d.35), p. 73]), the Euler-Lagrange equation for this functional is

$$\ddot{\kappa} + \frac{1}{2}(\kappa^3 + \kappa) = 0. \tag{5.54}$$

Curves in $\mathbf{S}^2$ whose curvature satisfies this equation are called *elastic curves*. As our interest is in periodic solutions, we choose a maximum point of κ as the initial point, so the initial conditions are

$$\kappa(0) = \kappa_0 \geq 0, \quad \dot{\kappa}(0) = 0.$$

The solution to this initial value problem is expressed in terms of a Jacobian elliptic function as

$$\kappa(t) = \sqrt{\frac{2m}{1 - 2m}} \mathrm{cn}\left(\frac{t}{\sqrt{2 - 4m}}\right), \tag{5.55}$$

where

$$m = \frac{\kappa_0^2}{2(1+\kappa_0^2)} \in [0, \frac{1}{2})$$

and $\sqrt{m}$ is the modulus of the Jacobian elliptic function cn. This solution is periodic of period

$$\omega_m = 4\sqrt{2-4m}K(m), \tag{5.56}$$

where

$$K(m) = \int_0^1 \frac{dt}{\sqrt{1-t^2}\sqrt{1-mt^2}} = \int_0^{\pi/2} \frac{d\theta}{\sqrt{1-m\sin^2\theta}}$$

is an elliptic integral of the first kind.

Exercise 24. The Jacobian elliptic functions sn$u$, cn$u$, and dn$u$ of modulus $\sqrt{m}$ satisfy the elementary identities

$$\begin{aligned}
&\mathrm{sn}u = x, \quad \Leftrightarrow \quad u = \int_0^x \frac{dt}{\sqrt{1-t^2}\sqrt{1-mt^2}}, \\
&\mathrm{cn}u = \sqrt{1-\mathrm{sn}^2 u}, \quad \mathrm{dn}u = \sqrt{1-m\mathrm{sn}^2 u}, \\
&\mathrm{sn}' u = \mathrm{cn}u\, \mathrm{dn}u, \quad \mathrm{cn}' u = -\mathrm{sn}u\, \mathrm{dn}u, \quad \mathrm{dn}' u = -m\mathrm{sn}u\, \mathrm{cn}u,
\end{aligned}$$

for $-K(m) < u < K(m)$. They extend as periodic functions to all $u \in \mathbf{R}$, with sn$u$ and cn$u$ of period $4K(m)$ and dn$u$ of period $2K(m)$. See, for example, Bowman [17, Chapter I] for an elementary exposition of these and many other properties of Jacobian elliptic functions.

Using these identities, prove that $\kappa(t)$ defined in (5.55) is the solution of (5.54) satisfying the given initial conditions and that it is periodic of period ω_m defined in (5.56).

For a general value of $m \in (0, 1/2)$, the curve $\gamma : \mathbf{R} \to \mathbf{S}^2$ of curvature κ defined in (5.55) is not periodic. We search for values of m for which γ is periodic. To do this, we consider the Maurer–Cartan equation (5.48) for the spinor lift $\Gamma : \mathbf{R} \to \mathbf{SU}(2)$ of the Frenet frame of γ. It is the solution $\Gamma : \mathbf{R} \to \mathbf{SU}(2)$ of the initial value problem (IVP)

$$\Gamma'(t) = \Gamma(t)A(t), \quad \Gamma(0) = I_2, \tag{5.57}$$

where

$$A(t) = \begin{pmatrix} i\kappa(t)/2 & -1/2 \\ 1/2 & -i\kappa(t)/2 \end{pmatrix} : \mathbf{R} \to \mathfrak{su}(2)$$

is periodic of period $\omega_m = 4\sqrt{2-4m}K(m)$. The curve $\gamma = \pi_h \circ \Gamma_1$ is periodic if and only if there exists a positive number ν such that $\Gamma(t+\nu) = \pm\Gamma(t)$, for all $t \in \mathbf{R}$, since $\pi_h(-\mathbf{z}) = \pi_h(\mathbf{z})$, for every $\mathbf{z} \in \mathbf{S}^3$. The existence of ν is governed by the proper values of the *monodromy operator* of the IVP,

$$M_m = \Gamma(\omega_m) \in \mathbf{SU}(2).$$

Lemma 5.31. *Let G be a Lie group, and let $A : \mathbf{R} \to G$ be a smooth, ω-periodic curve in G; i.e., $\omega > 0$ and $A(t+\omega) = A(t)$, for every $t \in \mathbf{R}$. If $\Gamma : \mathbf{R} \to G$ is the solution to the IVP*

$$\Gamma'(t) = \Gamma(t)A(t), \quad \Gamma(0) = I,$$

and if $M = \Gamma(\omega)$ is its monodromy operator, then

$$\Gamma(n\omega) = M^n, \tag{5.58}$$

and, thus

$$\Gamma[n\omega, (n+1)\omega] = M^n\Gamma[0,\omega], \tag{5.59}$$

for any integer $n \in \mathbf{Z}$.

Proof. Recall from the Cartan–Darboux Theorem, that the solution to this IVP is unique, and for any $B \in G$, the solution to this IVP with initial condition $\tilde{\Gamma}(0) = B$ is $\tilde{\Gamma}(t) = B\Gamma(t)$. Consider the curve

$$\tilde{\Gamma}(t) = \Gamma(t+\omega).$$

Then $\tilde{\Gamma}(0) = M$ and

$$\tilde{\Gamma}'(t) = \Gamma'(t+\omega) = \Gamma(t+\omega)A(t+\omega) = \tilde{\Gamma}(t)A(t),$$

by the ω-periodicity of $A(t)$. Hence, $\tilde{\Gamma}(t) = M\Gamma(t)$, by our initial remark, and so

$$\Gamma(2\omega) = \tilde{\Gamma}(\omega) = M^2.$$

The result for any whole number follows by induction on n. For the case of negative integers, consider the curve $\tilde{\Gamma}(t) = \Gamma(t-\omega)$, which is a solution to the IVP with initial condition $\tilde{\Gamma}(0) = \Gamma(-\omega)$, so

$$\tilde{\Gamma}(t) = \Gamma(-\omega)\Gamma(t),$$

for all $t \in \mathbf{R}$. In particular,

$$I = \Gamma(0) = \tilde{\Gamma}(\omega) = \Gamma(-\omega)\Gamma(\omega)$$

implies that $\Gamma(-\omega) = M^{-1}$. The result for any negative integer $-n$ now follows by induction on n. This completes the proof of (5.58). For the second result, for any integer $n \in \mathbf{Z}$, the curve

$$\tilde{\Gamma}(t) = \Gamma(t + n\omega)$$

is the solution to the IVP with initial condition $\tilde{\Gamma}(0) = M^n$, and therefore

$$\tilde{\Gamma}(t) = M^n \Gamma(t),$$

for all $t \in \mathbf{R}$, which implies (5.59). □

Exercise 25. Prove that the proper values of any matrix $A \in \mathbf{SU}(2)$ are $e^{\pm i\varphi}$, for some $\varphi \in \mathbf{R}$. Thus, A is diagonalizable, that is, there exists $P \in \mathbf{SU}(2)$ such that

$$A = P^{-1} \begin{pmatrix} e^{i\varphi} & 0 \\ 0 & e^{-i\varphi} \end{pmatrix} P.$$

In particular, $A^q = I_2$, for some whole number q, if and only if $\varphi = 2\pi\frac{p}{q}$ for some integer p.

The proper values of M_m are $e^{\pm i\varphi_m}$. The dependence of φ_m on $m \in [0, 1/2)$ can be determined numerically to be continuous, strictly decreasing from $\varphi_0 = (2 - \sqrt{2})\pi$ to 0. Its graph as a function of m is shown in Figure 5.6.

Exercise 26. Prove that $\varphi_0 = (1 - \frac{\sqrt{2}}{2})2\pi$.

We conclude from this brief digression that the solution $\Gamma(t)$ of the IVP (5.57) is periodic if and only if

$$\varphi_m = 2\pi\frac{p}{q}, \tag{5.60}$$

for some integers p and q. For any value $2\pi\frac{p}{q} \in (0, (1 - \frac{\sqrt{2}}{2})2\pi)$, the graph of φ_m in Figure 5.6 shows that there exists a value of $m \in (0, 1/2)$ such that (5.60) holds. In effect, one needs to invert the function $m \mapsto \varphi_m$. This can be done numerically. The evidence from these calculations is that the resulting periodic curve $\gamma = \pi_h \circ \Gamma_1$ is simple only if $p = 1$ and $q = 2\tilde{q}$ for any whole number $\tilde{q} \geq 2$. If $\varphi_m = \pi/\tilde{q}$, then

$$\gamma(\mathbf{R}) = \cup_{n \in \mathbf{Z}_{\tilde{q}}} \pi_s(M^n)\gamma[0, \omega_m],$$

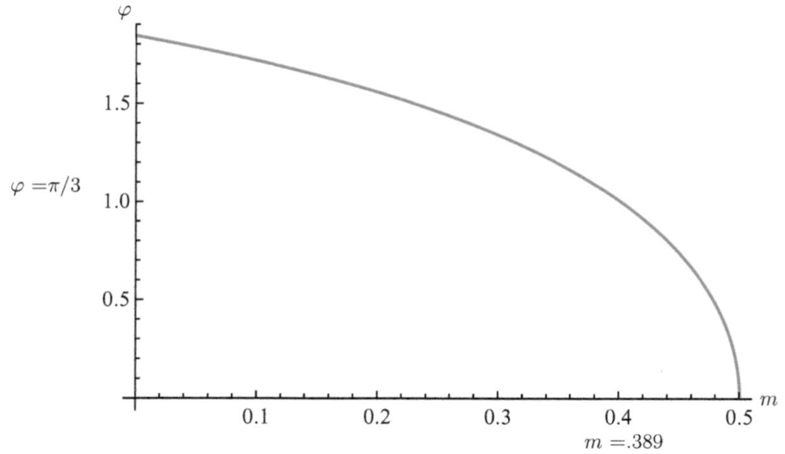

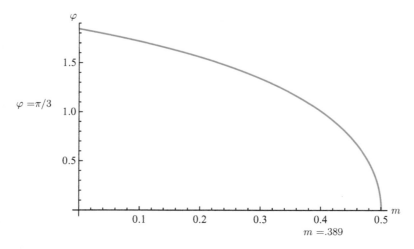

Fig. 5.6 Graph of φ_m as function of $m \in [0, 1/2)$.

by Lemma 5.31, Exercise 25, and the property of the covering projection π_s : $\mathbf{SU}(2) \to \mathbf{SO}(3)$ that $\pi_s(-B) = \pi_s(B)$. Thus, the curve γ is $\tilde{q}\omega_m$-periodic with a $\mathbf{Z}_{\tilde{q}}$ rotational symmetry coming from the invariance of $\gamma(\mathbf{R})$ under the action of the group $\{\pi_s(M^n) : n \in \mathbf{Z}_{\tilde{q}}\}$.

Figure 5.7 shows the embedded elastic curve $\gamma = \pi_h \circ \Gamma_1$ coming from the solution of the IVP (5.57) for the case $\varphi_m = 2\pi\frac{1}{6}$. The numerical estimate for m is 0.38922046026524204. The picture shows the three-fold rotational symmetry and antipodal symmetry of the curve.

The Clifford cylinder over this $3\omega_m$-periodic curve $\gamma = \pi_h \circ \Gamma_1$ is thus a Willmore immersion of a torus,

$$f_\gamma : \mathbf{R}^2/(2\pi\mathbf{Z} \times 3\omega_m\mathbf{Z}) \to \mathbf{S}^3, \quad f_\gamma(s,\theta) = e^{i\theta}\Gamma_1(s),$$

found by Pinkall [136]. Figure 5.8 shows $\mathscr{S}_{\epsilon_1} \circ f_\gamma(\mathbf{R}^2)$, its stereographic projection from $\boldsymbol{\epsilon}_1 \in \mathbf{S}^3$ into $\mathbf{R}^3$.

Eventually we shall see that if an immersion of a surface into $\mathbf{S}^3$ is Willmore, then this property is preserved under any conformal transformation of $\mathbf{S}^3$ (see Chapter 12). Any minimal isometric immersion of a surface into $\mathbf{S}^3$ is Willmore, so we want to check whether Pinkall's Willmore tori are conformally equivalent to minimal immersions. With the exception of the Clifford torus, they are not, but to prove this we need the concept of isothermic immersions into Möbius space, a topic covered in Chapter 14. In Corollary 14.36 of Section 14.6 we prove that, except for the case when $\gamma(\mathbf{R})$ is a great circle as in Problem 5.57, the Hopf torus f_γ is not isothermic and therefore it is not conformally equivalent to a minimal immersion in $\mathbf{S}^3$.

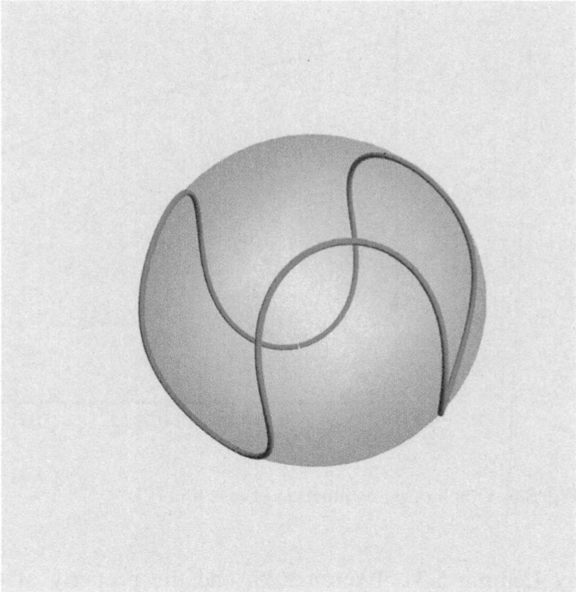

Fig. 5.7 Graph of $\gamma_3 = \pi_h \circ \Gamma_1 : \mathbf{R} \to \mathbf{S}^2$ when $\varphi_m = 2\pi/6$.

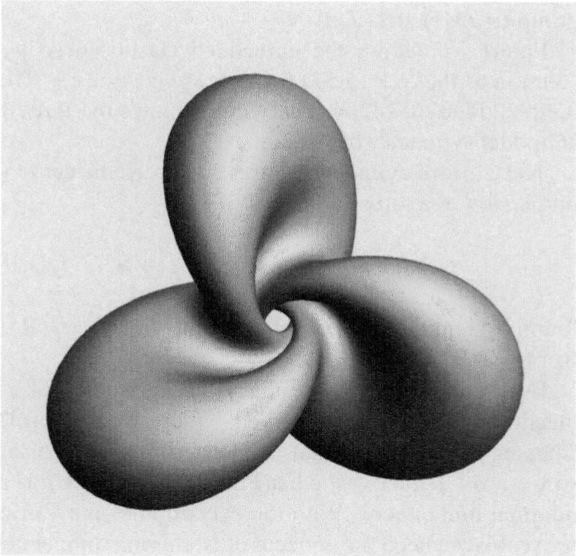

Fig. 5.8 The Pinkall torus over γ from $\varphi_m = \pi/3$.

Problems

5.32. Prove that the *volume form* $dV = \omega^1 \wedge \omega^2 \wedge \omega^3$ defined by the orthonormal coframe field ω^1, ω^2, ω^3 of (5.8) of any local frame field $e : U \subset \mathbf{S}^3 \to \mathbf{SO}(3)$ is independent of the choice of such a local frame field. Prove that

$$\int_{\mathbf{S}^3} \omega^1 \wedge \omega^2 \wedge \omega^3 = 2\pi^2.$$

5.33. Let $(U, (x,y))$ be a coordinate chart in M. Explain why $\mathbf{x}$ is an immersion if and only if $\mathbf{x}, \mathbf{x}_x, \mathbf{x}_y$ are linearly independent at each point of U. Prove that given a point $m \in M$ there exists a first order frame field along $\mathbf{x}$ on some neighborhood of m.

5.34. The set $S_r(\mathbf{m})$ in (5.23) can be defined for every $r \in \mathbf{R}$. Given $r \in \mathbf{R}$, prove there exists a unique integer n such that $|r - 2\pi n|$ is in the closed interval $[0,\pi]$, so $S_r(\mathbf{m})$ is the sphere with center $\mathbf{m}$ and radius $|r - 2\pi n|$.

5.35. To a point $\mathbf{y} \in \mathbf{R}^4$ *outside of* $\mathbf{S}^3$ (meaning $|\mathbf{y}| > 1$), associate the sphere (see the second diagram in Figure 5.1)

$$S(\mathbf{y}) = \{\mathbf{x} \in \mathbf{S}^3 : \mathbf{x} \cdot \mathbf{y} = 1\}.$$

Prove that $S(\mathbf{y}) = S_r(\mathbf{y}/|\mathbf{y}|)$, where $r = \operatorname{arcsec}|\mathbf{y}| \in (0, \pi/2)$.

For any point $\mathbf{p} \in \mathbf{RP}^3$, let

$$S(\mathbf{p}) = \{\mathbf{x} \in \mathbf{S}^3 : \mathbf{x} \cdot \mathbf{y} = 0 \text{ for any } 0 \neq \mathbf{y} \in \mathbf{p}\}.$$

Prove that $S(\mathbf{p})$ is the great sphere centered at $\mathbf{y}/|\mathbf{y}|$, for any $\mathbf{y} \in \mathbf{p}$, and that we have the decomposition and one-to-one correspondence

$$\begin{aligned}\Sigma &= \{S(\mathbf{y}) : \mathbf{y} \in \mathbf{R}^4,\ |\mathbf{y}| > 1\} \cup \{S(\mathbf{p}) : \mathbf{p} \in \mathbf{RP}^3\} \\ &\leftrightarrow \{\mathbf{y} \in \mathbf{R}^4 : |\mathbf{y}| > 1\} \cup \mathbf{RP}^3.\end{aligned}$$

5.36. Prove that for given r satisfying $0 < r < \pi$, the vector field $\mathbf{n}$ defined at $\mathbf{x} \in S_r(\mathbf{m})$ by (5.24) is a smooth, normal vector field on $S_r(\mathbf{m})$. Hint: If $\mathbf{x}$ is restricted to $S_r(\mathbf{m})$, then $\mathbf{x} \cdot \mathbf{m} = \cos r$. A vector $\mathbf{n}$ tangent to $\mathbf{S}^3$ at $\mathbf{x}$ is normal to $T_{\mathbf{x}}(S_r(\mathbf{m}))$ if and only if $\mathbf{n} \cdot d\mathbf{x} = 0$.

5.37. For $\mathbf{m} \in \mathbf{S}^3$ and $0 < r < \pi$, we know that $S_r(\mathbf{m}) = S_{\pi - r}(-\mathbf{m})$. Prove that the canonical orientation of $S_r(\mathbf{m})$ is opposite to the canonical orientation of $S_{\pi - r}(-\mathbf{m})$.

5.38. Prove that the tangent space to $S_r(\mathbf{m})$ at a point $\mathbf{x}_0 \in S_r(\mathbf{m})$ is

$$T_{\mathbf{x}_0} S_r(\mathbf{m}) = \{\mathbf{z} \in \mathbf{R}^4 : \mathbf{z} \cdot \mathbf{x}_0 = 0 = \mathbf{z} \cdot \mathbf{m}\}.$$

5.39. If $\mathbf{m} \in \mathbf{S}^3$ and if r is a real number not an integral multiple of π, then $\mathbf{n}(\mathbf{x})$ defined in (5.24) defines an orientation on the sphere $S_r(\mathbf{m})$, which by Problem 5.34 has center $\mathbf{m}$ and radius $|r-2\pi j|$, for some unique integer j.

(1) Prove that if $r-2\pi j > 0$, then $S_r(\mathbf{m})$ has the canonical orientation of $S_{r-2\pi j}(\mathbf{m})$, whereas if $r-2\pi j < 0$, it has the orientation opposite to the canonical orientation.
(2) For the special case illustrated in Figure 5.4, prove that if $\pi < r < 2\pi$, then $S_r(\mathbf{m}) = S_{2\pi-r}(\mathbf{m})$ as spheres, but the orientations defined by (5.24) are opposite.

5.40. Prove that the oriented curvature spheres of $\mathbf{x} : M \to \mathbf{S}^3$ at $m \in M$ relative to the unit normal $-\mathbf{n}$ at m are the same spheres as those relative to $\mathbf{n}$, but with opposite orientations.

5.41. Prove that the area $\int_{T_\alpha} dx \wedge dy$ of the circular torus $\mathbf{x}^{(\alpha)}$ is $2\pi^2 \sin 2\alpha$. Among the circular tori, the Clifford torus has the maximum area, but is the only minimal immersion in the sense that its mean curvature is identically zero.

5.42. Prove that the cylinder (5.29) has principal curvatures $a = \kappa(s)\cosh t$ and $c = 0$. Conclude that $\mathbf{y}$ is always canal and that it is Dupin if and only if κ is constant. It is never isoparametric.

5.43. Observe that the coordinate curves of the circular torus $\mathbf{x}^{(\alpha)}$ of Example 5.12 are the lines of curvature. Prove that each curvature sphere (5.28) is constant along its lines of curvature.

5.44. Prove that the surface of revolution $\mathbf{x}$ is Dupin if and only if it is isoparametric if and only if it is a circular torus.

5.45. Find all curves in Exercise 20 for which the curvature κ and the torsion τ are constant.

5.46. Prove that if $\cot r > \kappa(s)$, for all $s \in J$, then the tube $\mathbf{x} : J \times \mathbf{R} \to \mathbf{S}^3$ defined in (5.31) about the curve $\mathbf{f} : J \to \mathbf{S}^3$ with curvature $\kappa > 0$ is an immersion. Find the principal curvatures of $\mathbf{x}$ when it is an immersion.

5.47. Find the focal loci of the circular torus $\mathbf{x}^{(\alpha)} : \mathbf{R}^2 \to \mathbf{S}^3$ defined in Example 5.12. Prove that a generic parallel transformation of a circular torus $\mathbf{x}^{(\alpha)}$ is another circular torus. What constitutes *generic*?

5.48 (Principal curvatures). If $e = (\mathbf{x}, \mathbf{e}_1, \mathbf{e}_2, \mathbf{e}_3)$ is a first order frame field along an immersion $\mathbf{x} : M \to \mathbf{S}^3$, prove that $\tilde{e} = (\tilde{\mathbf{x}}, \mathbf{e}_1, \mathbf{e}_2, \tilde{\mathbf{e}}_3)$, where

$$\tilde{\mathbf{e}}_3 = -\sin r\mathbf{x} + \cos r\mathbf{e}_3,$$

is a first order frame field along the parallel transformation $\tilde{\mathbf{x}}$ of $\mathbf{x}$ by r in direction $\mathbf{e}_3$ (see Definition 5.19). Find the principal curvatures of $\tilde{\mathbf{x}}$ in terms of the principal curvatures of $\mathbf{x}$.

5.49. Prove that parallel transformation of an immersion $\mathbf{x} : M^2 \to \mathbf{S}^3$ with unit normal $\mathbf{e}_3$ by $r \in \mathbf{R}$ is conformal if and only if the mean curvature H of $\mathbf{x}$ is constant. Conformality of the parallel transformation means

$$d\tilde{\mathbf{x}} \cdot d\tilde{\mathbf{x}} = e^{2u} d\mathbf{x} \cdot d\mathbf{x}$$

for some smooth function $u : M \to \mathbf{R}$.

5.50. Prove that if $\mathbf{p} \in \mathbf{S}^3$ and if $A \in \mathbf{O}(4)$ is any isometry of $\mathbf{S}^3$, then

$$\mathscr{S}_{\mathbf{p}} \circ A = A \circ \mathscr{S}_{A\mathbf{p}}.$$

In particular, given any points $\mathbf{p}, \mathbf{q} \in \mathbf{S}^3$, if A is chosen so that $A\mathbf{q} = \mathbf{p}$, then

$$\mathscr{S}_{\mathbf{p}} = A \circ \mathscr{S}_{\mathbf{q}} \circ A^{-1}.$$

5.51. Use Propositions 5.16 and 5.23, to give an alternate, complete proof of Proposition 4.52.

5.52 (Cyclides of Dupin). Prove that $\mathscr{S} \circ \mathbf{x}^{(\alpha)} : T_\alpha \to \mathbf{R}^3$ is a Dupin immersion (see Definition 4.36), where $\mathbf{x}^{(\alpha)} : \mathbf{R}^2 \to \mathbf{S}^3$ is the circular torus defined in Example 5.12, and $\mathscr{S}$ is any stereographic projection.

5.53. Prove equivariance of the Hopf fibration π_h in the sense that

$$\pi_h(A\mathbf{z}) = \pi_s(A)\pi_h(\mathbf{z}),$$

for any $A \in \mathbf{SU}(2)$ and $\mathbf{z} \in \mathbf{S}^3$.

5.54. Find the default stereographic projection of the cylinder in $\mathbf{S}^3$ defined in Example 5.14.

5.55. Prove that with the right-action of the circle group $\mathbf{S}^1 \subset \mathbf{C}$ on $\mathbf{S}^3$ given by

$$\mathbf{S}^3 \times \mathbf{S}^1 \to \mathbf{S}^3, \quad \begin{pmatrix} z \\ w \end{pmatrix} e^{it} = \begin{pmatrix} ze^{it} \\ we^{it} \end{pmatrix},$$

the Hopf fibration is a principal $\mathbf{S}^1$-bundle.

5.56. For the map (5.46), find $\pi_s(\mathbf{z})$ when $\mathbf{z} = {}^t(e^{it}, 0)$, for any real value of t.

5.57. The unit speed curve $\gamma : \mathbf{R} \to \mathbf{S}^3$, $\gamma(s) = {}^t(0, \cos s, \sin s)$, is periodic of period 2π and has curvature $\kappa = 0$. Show its Frenet frame is $\mathscr{G}(s) = \begin{pmatrix} 0 & 0 & 1 \\ \cos s & -\sin s & 0 \\ \sin s & \cos s & 0 \end{pmatrix}$, whose spinor lift $\Gamma(s) = \frac{1}{\sqrt{2}} \begin{pmatrix} e^{i\frac{s}{2}} & -e^{i\frac{s}{2}} \\ e^{-i\frac{s}{2}} & e^{-i\frac{s}{2}} \end{pmatrix}$ is periodic of period 4π. Show that

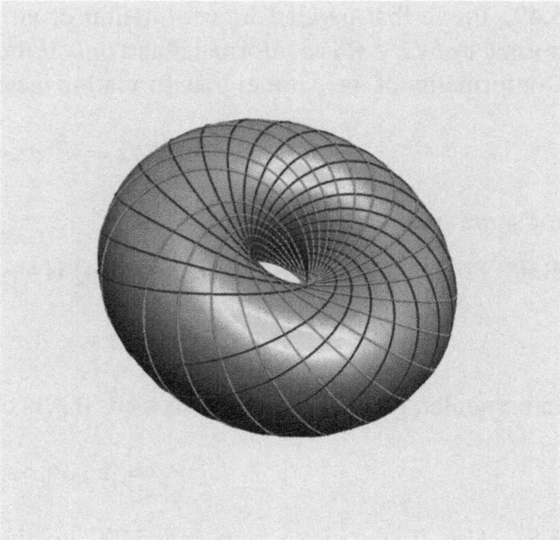

Fig. 5.9 Stereographic projection of Clifford torus from $-\boldsymbol{\epsilon}_0 \in \mathbf{S}^3$ showing the $s\theta$-parameter curves.

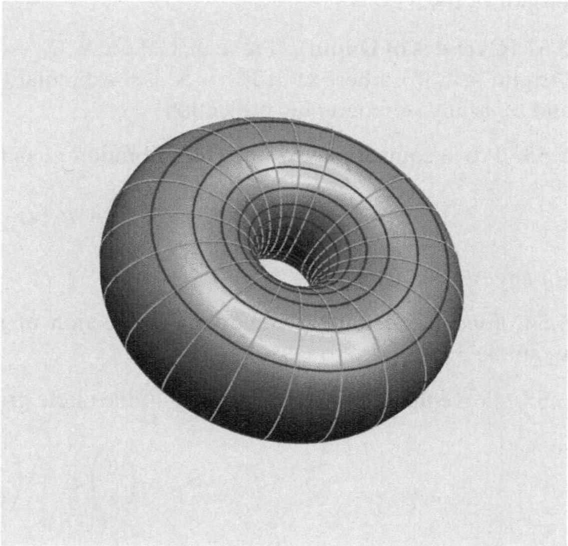

Fig. 5.10 Stereographic projection of Clifford torus from $-\boldsymbol{\epsilon}_0 \in \mathbf{S}^3$ showing the xy-parameter curves.

$$f_\gamma(s,\theta) = \Gamma_1(s)e^{i\theta} = \frac{1}{\sqrt{2}}\begin{pmatrix} e^{i(\theta+s/2)} \\ e^{i(\theta-s/2)} \end{pmatrix} = \frac{1}{\sqrt{2}}\begin{pmatrix} e^{ix} \\ e^{iy} \end{pmatrix},$$

where $x = \theta + s/2$, $y = \theta - s/2$. With the parameters x, y, this is the Clifford torus of Example 5.12. Figure 5.9 shows $\mathscr{S} \circ f_\gamma(s,\theta)$, its stereographic projection from $-\epsilon_0$, with the $s\theta$-parameter curves. The s-constant curves are the fibers, and the θ-constant curves are the sections. Use its stereographic projection from the point $-\boldsymbol{\epsilon}_0 \in \mathbf{S}^3$ in the xy-parameters,

$$\mathbf{x}(x,y) = \frac{1}{\sqrt{2}+\cos x}\,{}^t(\sin x, \cos y, \sin y),$$

to show that the image is a torus obtained by rotating a circle of radius $b = 1$ about a line in its plane a distance $a = \sqrt{2}$ from its center, so $a/b = \sqrt{2}$. See Figure 5.10.

The example in this problem illustrates that a rotation of 2π in Euclidean space is distinguishable from a rotation of 4π. From 0 to 2π, the corresponding closed path $\mathscr{G}([0,2\pi])$ in $\mathbf{SO}(3)$ is not null-homotopic, while from 0 to 4π it is. In other words, the spinor lift of rotation through 2π is not closed, while the spinor lift of the rotation through 4π is closed.

Chapter 6
Hyperbolic Geometry

This chapter applies the method of moving frames to immersions of surfaces in hyperbolic geometry $\mathbf{H}^3$, for which we use the hyperboloid model with its full group of isometries $\mathbf{O}_+(3,1)$. Moving frames lead to natural expressions of the sphere at infinity and the hyperbolic Gauss map. The Poincaré ball model is introduced as a means to visualize surfaces immersed in hyperbolic space. As in the chapters on Euclidean and spherical geometry, the notions of tangent and curvature spheres of an immersed surface are described in detail as preparation for their fundamental role in Lie sphere geometry. The chapter concludes with many elementary examples.

6.1 The Minkowski space model

Let $\mathbf{R}^{3,1}$ denote Minkowski space, which is $\mathbf{R}^4$ with the inner product of signature $(3,1)$ defined by

$$\langle \mathbf{x}, \mathbf{y} \rangle = \sum_1^3 x^i y^i - x^4 y^4 \tag{6.1}$$

where vectors in $\mathbf{R}^{3,1}$ are expressed in terms of the standard orthonormal basis $\boldsymbol{\epsilon}_1, \boldsymbol{\epsilon}_2, \boldsymbol{\epsilon}_3, \boldsymbol{\epsilon}_4$ of $\mathbf{R}^{3,1}$, with $\boldsymbol{\epsilon}_4$ the *time-like* vector, $\langle \boldsymbol{\epsilon}_4, \boldsymbol{\epsilon}_4 \rangle = -1$. A basis $\mathbf{x}_1, \mathbf{x}_2, \mathbf{x}_3, \mathbf{x}_4$ of $\mathbf{R}^{3,1}$ is *orthonormal* if

$$\langle \mathbf{x}_i, \mathbf{x}_j \rangle = g_{ij}, \quad i,j = 1,2,3,4,$$

G.R. Jensen et al., *Surfaces in Classical Geometries*, Universitext,
DOI 10.1007/978-3-319-27076-0_6

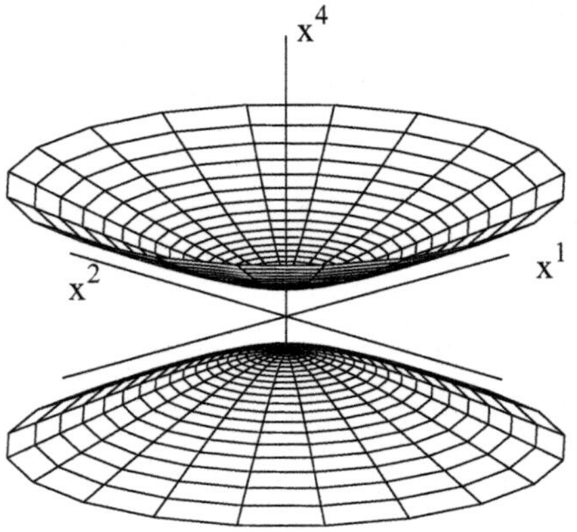

Fig. 6.1 $\mathbf{H}^3$ in the 3-d slice $\mathbf{x}^3 = 0$

where the g_{ij} are the entries of

$$I_{3,1} = \begin{pmatrix} 1 & 0 & 0 & 0 \\ 0 & 1 & 0 & 0 \\ 0 & 0 & 1 & 0 \\ 0 & 0 & 0 & -1 \end{pmatrix}.$$

Hyperbolic space $\mathbf{H}^3$ is the upper component of the hyperboloid,

$$\mathbf{H}^3 = \{\mathbf{x} \in \mathbf{R}^{3,1} : \langle \mathbf{x}, \mathbf{x} \rangle = -1 \text{ and } x^4 \geq 1\}.$$

The lower component is $-\mathbf{H}^3$. See Figure 6.1.

For any $\mathbf{x} \in \mathbf{H}^3$, the Minkowski inner product restricted to $\mathbf{x}^\perp$ is positive definite. To see this, let $\mathbf{R}^3$ denote the subspace of $\mathbf{R}^{3,1}$ spanned by $\epsilon_1, \epsilon_2, \epsilon_3$, so that we have a direct sum decomposition

$$\mathbf{R}^{3,1} = \mathbf{R}^3 \oplus \mathbf{R}\epsilon_4, \quad \mathbf{x} = \mathbf{x}' + x^4 \epsilon_4.$$

where $\mathbf{x}'$ is the orthogonal projection of $\mathbf{x}$ onto $\mathbf{R}^3$. If $\mathbf{x}' \in \mathbf{H}^3$, then

$$(x^4)^2 - |\mathbf{x}'|^2 = 1, \quad x^4 \geq 1.$$

If $\mathbf{y} \in \mathbf{x}^\perp$, then $\mathbf{y} = \mathbf{y}' + y^4 \epsilon_4$ and

$$0 = \langle \mathbf{x}, \mathbf{y} \rangle = \mathbf{x}' \cdot \mathbf{y}' - x^4 y^4.$$

The Cauchy–Schwarz inequality implies

$$|x^4 y^4| = |\mathbf{x}' \cdot \mathbf{y}'| \leq |\mathbf{x}'||\mathbf{y}'|,$$

so

$$(y^4)^2 \leq \frac{|\mathbf{x}'|^2|\mathbf{y}'|^2}{(x^4)^2}.$$

Therefore,

$$\langle \mathbf{y}, \mathbf{y} \rangle = |\mathbf{y}'|^2 - (y^4)^2 \geq |\mathbf{y}'|^2 - \frac{|\mathbf{x}'|^2|\mathbf{y}'|^2}{(x^4)^2} = \frac{|\mathbf{y}'|^2}{(x^4)^2} \geq 0$$

with equality if and only if $\mathbf{y}' = 0$ if and only if $\mathbf{y} = 0$ (because $x^4 y^4 = \mathbf{x}' \cdot \mathbf{y}'$ and $x^4 > 0$).

Exercise 27. Prove that the tangent space of $\mathbf{H}^3$ at any point $\mathbf{x} \in \mathbf{H}^3$ is

$$T_{\mathbf{x}}\mathbf{H}^3 = \mathbf{x}^\perp = \{\mathbf{y} \in \mathbf{R}^{3,1} : \langle \mathbf{y}, \mathbf{x} \rangle = 0\} \tag{6.2}$$

so the Minkowski inner product restricted to $T_{\mathbf{x}}\mathbf{H}^3$ is positive definite, for any $\mathbf{x} \in \mathbf{H}^3$, which means that the induced metric on $\mathbf{H}^3$ is a Riemannian metric.

The Lie group of linear transformations of $\mathbf{R}^{3,1}$ that preserve the Minkowski inner product is the *Lorentz group*, which is represented in the standard basis by the matrix group

$$\mathbf{O}(3,1) = \{A \in \mathbf{GL}(4, \mathbf{R}) : {}^t\!A I_{3,1} A = I_{3,1}\}. \tag{6.3}$$

Remark 6.1. The group of isometries of $\mathbf{R}^{3,1}$ is the analog of the Euclidean group given by the semi-direct product $\mathbf{E}(3,1) = \mathbf{R}^4 \rtimes \mathbf{O}(3,1)$. For an exposition of the method of moving frames applied to the study of certain surfaces immersed in $\mathbf{R}^{3,1}$ acted upon by $\mathbf{E}(3,1)$ see Elghanmi's paper [64].

An element $A \in \mathbf{O}(3,1)$ sends a sheet of the hyperboloid into a sheet of the hyperboloid. It will thus send $\mathbf{H}^3$ to itself if and only if A maps $\boldsymbol{\epsilon}_4$ into $\mathbf{H}^3$. Since $A\boldsymbol{\epsilon}_4 = A_4$, the last column of A, it follows that the subgroup of $\mathbf{O}(3,1)$ that sends $\mathbf{H}^3$ to itself is

$$\mathbf{O}_+(3,1) = \{A \in \mathbf{O}(3,1) : A_4^4 \geq 1\}. \tag{6.4}$$

The group $\mathbf{O}_+(3,1)$ acts as isometries on $\mathbf{H}^3$ with its induced Riemannian metric. This action is transitive if for any point $\mathbf{x} \in \mathbf{H}^3$, there exists a matrix $A \in \mathbf{O}_+(3,1)$ such that

$$A\boldsymbol{\epsilon}_4 = \mathbf{x};$$

that is, $\mathbf{x}$ must be the last column of A. The induced metric on $\mathbf{x}^\perp$ is positive definite, so there exists an orthonormal basis $A_1, \ldots, A_3$ of $\mathbf{x}^\perp$. Then $A = (A_1, A_2, A_3, \mathbf{x}) \in \mathbf{O}_+(3,1)$ sends $\boldsymbol{\epsilon}_4$ to $\mathbf{x}$. If $\det A = -1$, then interchanging the first two columns gives a matrix A in

$$\mathbf{SO}_+(3,1) = \{A \in \mathbf{O}_+(3,1) : \det A = 1\}, \tag{6.5}$$

satisfying $A\epsilon_4 = \mathbf{x}$, thus showing that $\mathbf{SO}_+(3,1)$ acts transitively on $\mathbf{H}^3$. Therefore, $\mathbf{H}^3$ with the induced Riemannian metric is homogeneous, so it is a complete Riemannian manifold.

For the *origin* in $\mathbf{H}^3$ we take ϵ_4. We denote the isotropy subgroup of $\mathbf{O}_+(3,1)$ at the origin by a slight abuse of notation as

$$\mathbf{O}(3) = \{\begin{pmatrix} B & 0 \\ 0 & 1 \end{pmatrix} : B \in \mathbf{O}(3)\}. \tag{6.6}$$

We have a principal $\mathbf{O}(3)$-bundle

$$\pi : \mathbf{O}_+(3,1) \to \mathbf{H}^3, \quad e \mapsto e\epsilon_4 = \mathbf{e}_4, \tag{6.7}$$

where $\mathbf{e}_i$ denotes column i of the matrix e, for $i = 1,\dots,4$. The Lie algebra of $\mathbf{O}_+(3,1)$ is

$$\begin{aligned} \mathfrak{o}(3,1) &= \{X \in \mathfrak{gl}(4,\mathbf{R}) : {}^tXI_{3,1} + I_{3,1}X = 0\} \\ &= \{\begin{pmatrix} X & \mathbf{x} \\ {}^t\mathbf{x} & 0 \end{pmatrix} : X \in \mathfrak{o}(3),\ \mathbf{x} \in \mathbf{R}^3\}. \end{aligned} \tag{6.8}$$

It has a vector space direct sum

$$\mathfrak{o}(3,1) = \mathfrak{o}(3) + \mathfrak{m}$$

where $\mathfrak{o}(3)$ is the Lie algebra of $\mathbf{O}(3)$ contained in $\mathfrak{o}(3,1)$ as shown in (6.8), and

$$\mathfrak{m} = \{\begin{pmatrix} 0 & \mathbf{x} \\ {}^t\mathbf{x} & 0 \end{pmatrix} : \mathbf{x} \in \mathbf{R}^3\} \cong \mathbf{R}^3$$

is a complementary vector subspace. Using this decomposition, we can write an element of $\mathfrak{o}(3,1)$ as

$$(X,\mathbf{x}) = \begin{pmatrix} X & \mathbf{x} \\ {}^t\mathbf{x} & 0 \end{pmatrix}, \quad X \in \mathfrak{o}(3), \quad \mathbf{x} \in \mathbf{R}^3, \tag{6.9}$$

in which case the bracket structure can be described by the formulas

$$\begin{aligned} [(X,0),(Y,0)] &= ([X,Y],0) \\ [(X,0),(0,\mathbf{x})] &= (0,X\mathbf{x}) \\ [(0,\mathbf{x}),(0,\mathbf{y})] &= (\mathbf{x}{}^t\mathbf{y} - \mathbf{y}{}^t\mathbf{x},0) \end{aligned}$$

Compare this Lie algebra structure with that of $\mathscr{E}(3)$ described in (4.2) and of $\mathfrak{o}(4)$ in (5.4). The Maurer–Cartan form of $\mathbf{O}_+(3,1)$ is the $\mathfrak{o}(3,1)$-valued 1-form

$$\eta = e^{-1}de = \begin{pmatrix} \omega & \theta \\ {}^t\theta & 0 \end{pmatrix} = (\omega,\theta) \tag{6.10}$$

in the notation of (6.9), where

$$\theta = {}^t(\omega^1, \omega^2, \omega^3), \quad \omega = (\omega^i_j) = -{}^t\omega,$$

$\omega^i = \omega^i_4 = \omega^4_i$ and $\omega^i_j = \eta^i_j$, for $i,j = 1,2,3$. The structure equations are

$$d\eta^i_j = -\sum_{k=1}^{4} \eta^i_k \wedge \eta^k_j, \quad i,j = 1,\dots,4, \tag{6.11}$$

which in the matrix notation is

$$d\theta = -\omega \wedge \theta, \quad d\omega = -\omega \wedge \omega - \theta \wedge {}^t\theta.$$

We define a local orthonormal frame field on an open subset U of $\mathbf{H}^3$ to be a smooth local section of (6.7). Explicitly, it is a smooth map

$$e : U \to \mathbf{O}_+(3,1), \tag{6.12}$$

such that $\mathbf{e}_4(\mathbf{x}) = \mathbf{x}$. Since $T_\mathbf{x}\mathbf{H}^3$ is naturally identified with the subspace orthogonal to $\mathbf{x}$, it follows that $\mathbf{e}_1, \mathbf{e}_2, \mathbf{e}_3$ is an orthonormal basis of this tangent space. If we abuse notation slightly and use the same letters for the pull-back of the Maurer–Cartan form by our frame field e, then we have from (6.10)

$$d\mathbf{x} = d\mathbf{e}_4 = \sum_{i=1}^{3} \omega^i \mathbf{e}_i.$$

The Riemannian metric on $\mathbf{H}^3$ has the local expression

$$I = \langle d\mathbf{x}, d\mathbf{x} \rangle = \sum_{i=1}^{3} \omega^i \omega^i, \tag{6.13}$$

and $\omega^1, \omega^2, \omega^3$ is an orthonormal coframe field in U. Incidentally, this also shows that the induced metric I is positive definite. Any other frame field on U must be given by $\tilde{e} = eA$, where

$$A = \begin{pmatrix} B & 0 \\ 0 & 1 \end{pmatrix}, \quad B : U \to \mathbf{O}(3) \tag{6.14}$$

is a smooth map, so

$$(\tilde{\mathbf{e}}_1, \tilde{\mathbf{e}}_2, \tilde{\mathbf{e}}_3) = (\mathbf{e}_1, \mathbf{e}_2, \mathbf{e}_3)B,$$

while $\tilde{\mathbf{e}}_4 = \mathbf{e}_4 = \mathbf{x}$. In order to compare the pull-back of the Maurer–Cartan form by each of these frame fields, we write

$$\tilde{e}^{-1}d\tilde{e} = (\tilde{\omega}, \tilde{\theta}).$$

Then, with A given by (6.14),

$$\tilde{e}^{-1}d\tilde{e} = A^{-1}e^{-1}(deA + e\,dA)$$

implies that

$$\tilde{\theta} = B^{-1}\theta, \;\; \tilde{\omega} = B^{-1}\omega B + B^{-1}dB. \tag{6.15}$$

One consequence of this is that the metric I on $\mathbf{H}^3$ actually comes from the group $\mathbf{O}_+(3,1)$, in the sense that, from (6.13) and (6.15) we have

$$\langle d\mathbf{x}, d\mathbf{x}\rangle = {}^t\theta\theta = {}^t\tilde{\theta}\tilde{\theta}$$

For the frame field $e : U \to \mathbf{O}_+(3,1)$, we get from (6.11), for $i,j = 1,2,3$,

$$d\omega^i = -\sum_{j=1}^{3} \omega^i_j \wedge \omega^j, \qquad \omega^i_j = -\omega^j_i,$$

for $j = 1,2,3$. It follows that these ω^i_j are the Levi-Civita connection forms of I relative to the orthonormal coframe field $\omega^1, \omega^2, \omega^3$. They determine the covariant derivative of a local vector field $X = \sum_1^3 \xi^i \mathbf{e}_i$ on U, where $\mathbf{e}_i$ is column $i = 1,2,3$ of e, with respect to a tangent vector $\mathbf{v} \in T_{\mathbf{x}}\mathbf{H}^3$ by

$$D_{\mathbf{v}}X = \sum_{j=1}^{3}(\mathbf{v}(\xi^j) + \sum_{i=1}^{3}\xi^i\omega^j_i(\mathbf{v}))\mathbf{e}_j.$$

In spherical geometry we used without comment the fact that the intrinsic distance between points $\mathbf{p}, \mathbf{q} \in \mathbf{S}^3$ is $r = \cos^{-1}(\mathbf{p}\cdot\mathbf{q}) \in [0, \pi]$. An analogous formula is true in hyperbolic geometry (see Problem 6.38). The intrinsic distance between points $\mathbf{p}, \mathbf{q} \in \mathbf{H}^3$ is

$$d(\mathbf{p}, \mathbf{q}) = \cosh^{-1}(-\langle \mathbf{p}, \mathbf{q}\rangle) \in (0, \infty).$$

Definition 6.2. For any non-zero real number r and any point $\mathbf{m} \in \mathbf{H}^3$, the *oriented sphere of signed radius r and center* $\mathbf{m}$ is the set $S_r(\mathbf{m})$ of all points in $\mathbf{H}^3$ a distance $|r|$ from $\mathbf{m}$,

$$S_r(\mathbf{m}) = \{\mathbf{x} \in \mathbf{H}^3 : \langle \mathbf{x}, \mathbf{m}\rangle = -\cosh r\},$$

oriented by the unit normal vector field

$$\mathbf{e}_3(\mathbf{x}) = \frac{\mathbf{m} - \cosh r\,\mathbf{x}}{\sinh r}.$$

The curvature forms Ω^i_j of the Levi-Civita connection forms ω^i_j on $\mathbf{H}^3$ are, by (6.11),

$$\Omega^i_j = d\omega^i_j + \sum_{k=1}^{3} \omega^i_k \wedge \omega^k_j = -\omega^i_4 \wedge \omega^4_j = -\omega^i \wedge \omega^j.$$

This shows that I has constant sectional curvature equal to minus one. For any local frame field (6.12) in $\mathbf{H}^3$, the 3-form $\omega^1 \wedge \omega^2 \wedge \omega^3$ is non-zero at every point of U. If $\tilde{e} = eA$, where A is given by (6.14), then

$$\tilde{\omega}^1 \wedge \tilde{\omega}^2 \wedge \tilde{\omega}^3 = (\det B)\omega^1 \wedge \omega^2 \wedge \omega^3, \tag{6.16}$$

where $\det B = \pm 1$, since $B \in \mathbf{O}(3)$. Thus, an orientation is defined on $\mathbf{H}^3$ by the *volume form* defined in (6.16), if we insist that $B \in \mathbf{SO}(3)$. This is accomplished by taking only frames (6.12) that take values in $\mathbf{SO}_+(3,1)$, defined in (6.5), which is the connected component of the identity of $\mathbf{O}_+(3,1)$. The isotropy subgroup of $\mathbf{SO}_+(3,1)$ at $\boldsymbol{\epsilon}_4$ is

$$\mathbf{SO}(3) \cong \left\{ \begin{pmatrix} B & 0 \\ 0 & 1 \end{pmatrix} \in \mathbf{SO}_+(3,1) : B \in \mathbf{SO}(3) \right\},$$

by the same abuse of notation used in (6.6).

6.2 The sphere at infinity

In Euclidean space, we can use parallel translation to identify the unit sphere $\mathbf{S}^2$ with center at the origin with the unit sphere $S^2_{\mathbf{x}}$ with center at an arbitrary point $\mathbf{x}$. With these identifications, we can then say that two lines, parametrized by arclength as $\mathbf{x} + t\mathbf{u}$ and $\mathbf{y} + t\mathbf{v}$, where $\mathbf{u} \in S^2_{\mathbf{x}}$ and $\mathbf{v} \in S^2_{\mathbf{y}}$, are parallel if and only if $\mathbf{v} = \pm\mathbf{u}$. It follows that parallelism is an equivalence relation on the set of all lines.

Similar identifications of the unit tangent spheres can be made in Hyperbolic space, but their role in identifying parallel lines is very different. For one thing, parallelism is not an equivalence relation on the set of all lines in Hyperbolic space. As in $\mathbf{R}^3$, a line in $\mathbf{H}^3$ is the trace of a geodesic. Given a line l and a point $\mathbf{x}$ not on the line, then in Euclidean geometry there exists a unique line l' through $\mathbf{x}$ and parallel to l. Indeed, this line is obtained by dropping the unique line m through $\mathbf{x}$ perpendicular to l at the point $\mathbf{b}$ on l, and then constructing the unique line l' through

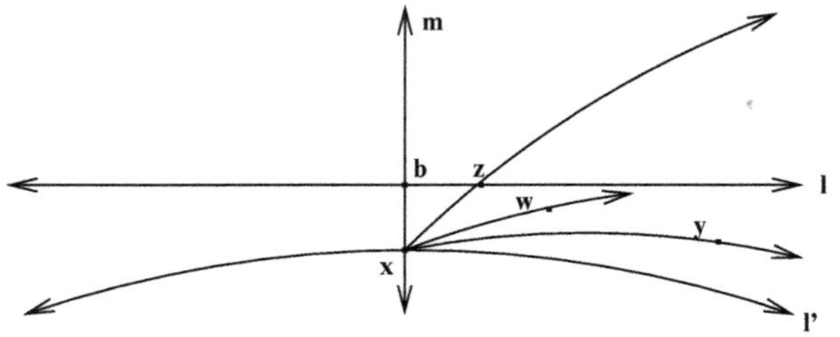

Fig. 6.2 Ray $\overrightarrow{xw}$ is a limiting parallel ray.

$\mathbf{x}$ perpendicular to m. In Hyperbolic geometry, these constructions remain valid and the line l' through $\mathbf{x}$ is parallel to l, but the hyperbolic parallel postulate requires that there must be at least one other line through $\mathbf{x}$ parallel to l. This line must contain a ray $\overrightarrow{\mathbf{xy}}$ parallel to l and making an acute angle with m. If the angle is small enough, then the ray $\overrightarrow{\mathbf{xz}}$ will meet l at some point $\mathbf{z} \in l$. There is a *limiting ray* $\overrightarrow{\mathbf{xw}}$ parallel to l such that any ray between $\overrightarrow{\mathbf{xb}}$ and $\overrightarrow{\mathbf{xw}}$ meets l, while any ray $\overrightarrow{\mathbf{xy}}$ such that $\overrightarrow{\mathbf{xw}}$ is between $\overrightarrow{\mathbf{xy}}$ and $\overrightarrow{\mathbf{xb}}$ does not intersect l. See Figure 6.2.

There is also a limiting ray $\overrightarrow{\mathbf{xw}'}$ on the opposite side of m, symmetric to $\overrightarrow{\mathbf{xw}}$ in the sense that the angle $\mathbf{wxb}$ is congruent to angle $\mathbf{w'xb}$. See [78, Theorem 6.6, p. 196] for details.

A line l' is *asymptotically parallel* to a line l, if l' contains a limiting ray to l. Otherwise, they are called *divergently parallel.* Parallel lines of this second type are characterized by having a common perpendicular. Asymptotical parallelism is an equivalence relation on the set of lines of Hyperbolic space. See [78, Theorem 6.7, p. 199].

The following identification of the unit tangent sphere $S^2_{\mathbf{x}}$, at any point $\mathbf{x} \in \mathbf{H}^3$, with a single sphere S^2_∞, produces a criterion for when two lines in $\mathbf{H}^3$ are asymptotically parallel.

The *light cone* in $\mathbf{R}^{3,1}$ is

$$N^3 = \{\mathbf{n} \in \mathbf{R}^{3,1} \setminus \{\mathbf{0}\} : \langle \mathbf{n}, \mathbf{n} \rangle = 0\}. \tag{6.17}$$

The multiplicative group of nonzero real numbers $\mathbf{R}^\times$ acts smoothly on N by

$$\mathbf{R}^\times \times N \to N, \quad (t, \mathbf{n}) \mapsto t\mathbf{n}.$$

Definition 6.3. The *sphere at infinity* of H^3 is the quotient space

$$S^2_\infty = N/\mathbf{R}^\times. \tag{6.18}$$

S^2_∞ is diffeomorphic to a 2-sphere because each equivalence class in it has a unique representative in the intersection of N with the hyperplane $x^4 = 1$. Dropping this intersection down to the coordinate hyperplane $x^4 = 0$, we get the diffeomorphism

$$S^2_\infty \to S^2_{\epsilon_4}, \quad [\mathbf{n}] \mapsto \mathbf{y} = \frac{1}{n^4}\mathbf{n} - \epsilon_4 = \frac{1}{n^4}\sum_1^3 n^i \epsilon_i,$$

whose inverse mapping is $\mathbf{y} \mapsto [\mathbf{y} + \epsilon_4]$. Here the unit sphere in the hyperplane $x^4 = 0$ is identified with the unit tangent sphere at ϵ_4 to H^3,

$$S^2_{\epsilon_4} = \{\mathbf{y} \in \epsilon_4^\perp \subset \mathbf{R}^{3,1} : \langle \mathbf{y}, \mathbf{y} \rangle = 1\}.$$

For any point $\mathbf{x} \in \mathbf{H}^3$, the unit tangent sphere in $T_\mathbf{x}\mathbf{H}^3 = \mathbf{x}^\perp$ is

$$S^2_\mathbf{x} = \{\mathbf{v} \in \mathbf{x}^\perp \subset \mathbf{R}^{3,1} : \langle \mathbf{v}, \mathbf{v} \rangle = 1\}.$$

We have a smooth diffeomorphism

$$\mathfrak{b} : S^2_\mathbf{x} \to S^2_\infty, \quad \mathfrak{b}(\mathbf{v}) = [\mathbf{n}] = [\mathbf{x} + \mathbf{v}], \tag{6.19}$$

whose inverse is $\mathfrak{b}^{-1}[\mathbf{n}] = \mathbf{v} = -\frac{1}{\langle \mathbf{n}, \mathbf{x} \rangle}\mathbf{n} - \mathbf{x}$. In Proposition 6.17 below we will define a topology on the disjoint union $\mathbf{H}^3 \cup S^2_\infty$ for which

$$\lim_{t \to \infty} \gamma(t) = [\mathbf{x} + \mathbf{v}] \in S^2_\infty,$$

where $\gamma(t)$ is the geodesic in $\mathbf{H}^3$ starting at $\gamma(0) = \mathbf{x}$ with unit velocity $\dot\gamma(0) = \mathbf{v} \in S^2_\mathbf{x}$.

Remark 6.4. 1). If $\mathbf{x} \in \mathbf{H}^3$ and if $\mathbf{v} \in S^2_\mathbf{x}$, then $\mathbf{x} + \mathbf{v} \in N$.

2). If $\mathbf{x} \in \mathbf{H}^3$ and if $\mathbf{n} \in N$, then $\langle \mathbf{n}, \mathbf{x} \rangle \neq 0$, since we know that the inner product is positive definite on $\mathbf{x}^\perp$.

6.2.1 *Conformal structure on S^2_∞*

The linear action of $\mathbf{O}_+(3,1)$ on $\mathbf{R}^{3,1}$ sends N to N, and it commutes with the action of $\mathbf{R}^\times$ on N. Thus, we have an action

$$\mathbf{O}_+(3,1) \times S^2_\infty \to S^2_\infty, \quad (e, [\mathbf{n}]) \mapsto [e\mathbf{n}]. \tag{6.20}$$

Since $\mathbf{O}_+(3,1)$ acts by isometries on $\mathbf{H}^3$, it sends the unit tangent sphere $S^2_\mathbf{x}$ at $\mathbf{x} \in \mathbf{H}^3$ to the unit tangent sphere $S^2_{e\mathbf{x}}$ at $e\mathbf{x} \in \mathbf{H}^3$, for any $e \in \mathbf{O}_+(3,1)$. This action and the action (6.20) commute with the diffeomorphisms (6.19) in the sense that if $\mathbf{x} \in \mathbf{H}^3$ and $\mathbf{y} \in S^2_\mathbf{x}$, then $[\mathbf{x} + \mathbf{y}] \in S^2_\infty$, and if $e \in \mathbf{O}_+(3,1)$, then $e\mathbf{y} \in S^2_{e\mathbf{x}}$ and

$$[e\mathbf{x} + e\mathbf{y}] = e[\mathbf{x} + \mathbf{y}].$$

Lemma 6.5. *The action* (6.20) *of* $\mathbf{SO}_+(3,1)$ *on* S^2_∞ *is transitive and its isotropy subgroup at* $[\boldsymbol{\epsilon}_3 + \boldsymbol{\epsilon}_4]$ *is*

$$G_0 = \left\{ K(t,\mathbf{u},A) = \begin{pmatrix} A & -A\mathbf{u}/t & A\mathbf{u}/t \\ {}^t\mathbf{u} & \frac{t^2+1-|\mathbf{u}|^2}{2t} & \frac{|\mathbf{u}|^2+t^2-1}{2t} \\ {}^t\mathbf{u} & \frac{t^2-|\mathbf{u}|^2-1}{2t} & \frac{|\mathbf{u}|^2+t^2+1}{2t} \end{pmatrix} : \begin{array}{c} t>0, \quad \mathbf{u}\in\mathbf{R}^2 \\ A\in\mathbf{SO}(2) \end{array} \right\}.$$

Proof. Exercise. □

Thus, S^2_∞ is the homogeneous space $\mathbf{SO}_+(3,1)/G_0$ and we have the principal G_0-bundle

$$\pi : \mathbf{SO}_+(3,1) \to S^2_\infty, \quad \pi(e) = [\mathbf{e}_3 + \mathbf{e}_4] = e[\boldsymbol{\epsilon}_3 + \boldsymbol{\epsilon}_4]. \tag{6.21}$$

The Lie algebra of G_0 is

$$\mathfrak{g}_0 = \left\{ \begin{pmatrix} a & -\mathbf{v} & \mathbf{v} \\ {}^t\mathbf{v} & 0 & s \\ {}^t\mathbf{v} & s & 0 \end{pmatrix} : s\in\mathbf{R},\ \mathbf{v}\in\mathbf{R}^2,\ a\in\mathfrak{o}(2) \right\}.$$

A complementary subspace to $\mathfrak{g}_0$ in $\mathfrak{o}(3,1)$ is

$$\mathfrak{n}_0 = \left\{ X(\mathbf{v}) = \begin{pmatrix} 0 & \mathbf{v} & \mathbf{v} \\ -{}^t\mathbf{v} & 0 & 0 \\ {}^t\mathbf{v} & 0 & 0 \end{pmatrix} : \mathbf{v}\in\mathbf{R}^2 \right\} \cong \mathbf{R}^2.$$

The adjoint action of G_0 does not leave $\mathfrak{n}_0$ invariant. This action by $K = K(t,\mathbf{u},A) \in G_0$ followed by projection onto $\mathfrak{n}_0$ takes $X = X(\mathbf{v})$ to

$$(K^{-1}XK)_{\mathfrak{n}_0} = X(t{}^tA\mathbf{v}).$$

The $\mathfrak{n}_0$-component of the Maurer–Cartan form of $\mathbf{SO}_+(3,1)$ is determined by the left-invariant $\mathbf{R}^2$-valued 1-form

$$\varphi = \begin{pmatrix} \varphi^1 \\ \varphi^2 \end{pmatrix}, \quad \varphi^i = \frac{1}{2}(\omega^i + \omega^i_3),\ i=1,2.$$

That is,

$$(e^{-1}de)_{\mathfrak{n}_0} = \begin{pmatrix} 0 & \varphi & \varphi \\ -{}^t\varphi & 0 & 0 \\ {}^t\varphi & 0 & 0 \end{pmatrix}.$$

Exercise 28 (Conformal structure on S^2_∞). Given a point of S^2_∞, there exists an open neighborhood U containing this point on which there exists a frame field, that is, a section of (6.21),

$$e : U \to \mathbf{SO}_+(3,1). \tag{6.22}$$

Then $e^*\varphi^1, e^*\varphi^2$ is a coframe field on U and

$${}^t\varphi\varphi = \varphi^1\varphi^1 + \varphi^2\varphi^2 \tag{6.23}$$

is a Riemannian metric on U. Conclude that the sections of (6.21) define a conformal structure on S^2_∞, which is a collection of Riemannian metrics any two of which are conformally related on their common domain of definition. Prove that the action of $\mathbf{SO}_+(3,1)$ on S^2_∞ is conformal. Hint: any other frame field on U is given by $\tilde{e} = eK(t,\mathbf{u},A)$, where $t : U \to \mathbf{R}^+$, $\mathbf{u} : U \to \mathbf{R}^2$, $A : U \to \mathbf{SO}(2)$ are any smooth maps. If $\tilde{\varphi} = \tilde{e}^*\varphi$, prove that $\tilde{\varphi} = t{}^tA\varphi$, so

$${}^t\tilde{\varphi}\tilde{\varphi} = t^2\,{}^t\varphi\varphi. \tag{6.24}$$

Remark 6.6. If the frame field (6.22) is $e = (\mathbf{e}_1, \mathbf{e}_2, \mathbf{e}_3, \mathbf{e}_4)$, then

$$\mathbf{n} = \mathbf{e}_3 + \mathbf{e}_4 : U \to N$$

is a section of the projection

$$N \to N/\mathbf{R}^\times = S^2_\infty, \tag{6.25}$$

and

$$d\mathbf{n} = 2\varphi^1\mathbf{e}_1 + 2\varphi^2\mathbf{e}_2,$$

so

$$\langle d\mathbf{n}, d\mathbf{n}\rangle = 4\,{}^t\varphi\varphi.$$

This shows that the sections of (6.25) are conformal if N is given the degenerate pseudometric induced from its embedding $N \subset \mathbf{R}^{3,1}$.

6.3 Surfaces in $\mathbf{H}^3$

Consider a smooth immersion

$$\mathbf{x} : M \to \mathbf{H}^3$$

of a smooth surface M. A smooth frame field along $\mathbf{x}$ on an open subset $U \subset M$ is a smooth map $e : U \to \mathbf{O}_+(3,1)$ such that $\pi \circ e = \mathbf{x}$; that is, $\mathbf{e}_4 = \mathbf{x}$. We define a *first order frame field along* $\mathbf{x}$ to be a frame field for which $\mathbf{e}_3$ is normal to $\mathbf{x}$ (but tangent to $\mathbf{H}^3$). This is equivalent to the condition that $d\mathbf{x} = d\mathbf{e}_4 = \omega^1\mathbf{e}_1 + \omega^2\mathbf{e}_2$, namely, that

$$\omega^3 = \omega_4^3 = 0. \tag{6.26}$$

Let e be a first order frame field along $\mathbf{x}$ on U. The metric induced on M by $\mathbf{x}$ and the Riemannian metric on $\mathbf{H}^3$, also called the first fundamental form of $\mathbf{x}$, is

$$I = \langle d\mathbf{x}, d\mathbf{x}\rangle = \omega^1\omega^1 + \omega^2\omega^2,$$

so $\omega^1 = \omega_4^1$, $\omega^2 = \omega_4^2$ is an orthonormal coframe field for it. By (6.11)

$$d\omega^1 = -\omega_2^1 \wedge \omega^2, \quad d\omega^2 = -\omega_1^2 \wedge \omega^1, \tag{6.27}$$

since $\omega^3 = 0$ for a first order frame field. From (6.27) we conclude that $\omega_2^1 = -\omega_1^2$ is the Levi-Civita connection form for I with respect to the orthonormal coframe field ω^1, ω^2. The Gaussian curvature is then K, given as an application of (6.11), by

$$K\omega^1 \wedge \omega^2 = d\omega_2^1 = -\omega_3^1 \wedge \omega_2^3 - \omega_4^1 \wedge \omega_2^4 = \omega_1^3 \wedge \omega_2^3 - \omega^1 \wedge \omega^2. \tag{6.28}$$

Taking the exterior derivative of (6.26) and using (6.11) together with Cartan's Lemma, we find that

$$\omega_1^3 = h_{11}\omega^1 + h_{12}\omega^2, \quad \omega_2^3 = h_{21}\omega^1 + h_{22}\omega^2, \tag{6.29}$$

for smooth functions h_{ij} on U such that $h_{12} = h_{21}$. Consider the smooth map

$$S = (h_{ij}) : U \to \mathscr{S},$$

where $\mathscr{S}$ is the vector space of all 2×2 symmetric matrices. We define the second fundamental form of $\mathbf{x}$ to be the symmetric bilinear form

$$II = \omega_1^3\omega^1 + \omega_2^3\omega^2 = h_{ij}\omega^i\omega^j. \tag{6.30}$$

The dependence of II on the choice of first order frame field e will be determined from equations (6.31) below. We see that (6.28) becomes the Gauss equation

$$K = -1 + \det S,$$

where the -1 is coming from the sectional curvature of $\mathbf{H}^3$.

6.4 Moving frame reductions

If $e : U \to \mathbf{O}_+(3,1)$ is a first order frame field along $\mathbf{x}$, then any other is given by $\tilde{e} = eA$, where $A : U \to \mathbf{O}(3)$ is such that $\tilde{\mathbf{e}}_3$ is normal to $d\mathbf{x}$, so $\tilde{\mathbf{e}}_3 = \pm\mathbf{e}_3$. Hence $A : U \to G_1 \subset \mathbf{O}(3)$, where

$$G_1 = \left\{ \begin{pmatrix} C & 0 & 0 \\ 0 & \epsilon & 0 \\ 0 & 0 & 1 \end{pmatrix} : C \in \mathbf{O}(2),\ \epsilon = \pm 1 \right\}.$$

From (6.15) with $B = \begin{pmatrix} C & 0 \\ 0 & \epsilon \end{pmatrix}$, we have

$$\begin{pmatrix} \tilde{\omega}^1 \\ \tilde{\omega}^2 \end{pmatrix} = C^{-1} \begin{pmatrix} \omega^1 \\ \omega^2 \end{pmatrix}, \qquad \begin{pmatrix} \tilde{\omega}^3_1 \\ \tilde{\omega}^3_2 \end{pmatrix} = \epsilon C^{-1} \begin{pmatrix} \omega^3_1 \\ \omega^3_2 \end{pmatrix}.$$

Hence, by (6.30),

$$\widetilde{II} = (\tilde{\omega}^3_1, \tilde{\omega}^3_2) \begin{pmatrix} \tilde{\omega}^1 \\ \tilde{\omega}^2 \end{pmatrix} = \epsilon (\omega^3_1, \omega^3_2)^t C^{-1} C^{-1} \begin{pmatrix} \omega^1 \\ \omega^2 \end{pmatrix} = \epsilon II, \tag{6.31}$$

and

$$\tilde{S} = \epsilon C^{-1} S C. \tag{6.32}$$

This is the same action as (4.34) and (5.16). Let the functions a and c on M denote the principal values of S. These are the *principal curvatures* of $\mathbf{x}$ at the point. As in the Euclidean case, each orbit of the action (6.32) contains a unique element in the set

$$D = \{ \begin{pmatrix} a & 0 \\ 0 & c \end{pmatrix} : a \geq |c| \} \subset \mathscr{S}. \tag{6.33}$$

As in that case a point of M is *umbilic* if $a = c$ and otherwise it is *nonumbilic*. The set of umbilic points ($a = c$) is closed in M.

We define a *second order frame field* $e : U \to \mathbf{O}_+(3,1)$ along $\mathbf{x}$ to be a first order frame field for which

$$\omega^3_1 = a\omega^1, \quad \omega^3_2 = c\omega^2, \tag{6.34}$$

for some functions $a, c : U \to \mathbf{R}$.

Lemma 6.7. *Let $m \in M$. If m is nonumbilic, or if m belongs to an open set of umbilic points, then there exists a smooth second order frame field on some open neighborhood of m.*

Proof. Same as the proofs of Lemma 4.6 or Lemma 4.10 in the Euclidean case. □

Taking the exterior derivative of (6.34), we arrive at the Codazzi equations

$$da\wedge\omega^1 + (a-c)\omega_1^2\wedge\omega^2 = 0, \quad dc\wedge\omega^2 + (a-c)\omega_1^2\wedge\omega^1 = 0.$$

The isotropy subgroup at this stage is finite, except in the totally umbilic case, so the frame reduction is completed. The totally umbilic case is considered below. The remaining Maurer–Cartan form is $\omega_2^1 = -\omega_1^2$, which we set equal to a linear combination of the coframe field

$$\omega_1^2 = p\omega^1 + q\omega^2,$$

for some smooth functions $p, q : U \to \mathbf{R}$, determined from

$$d\omega^1 = p\omega^1\wedge\omega^2, \quad d\omega^2 = q\omega^1\wedge\omega^2.$$

Taking the exterior derivative of this equation completes the structure equations. For any function f on M, we set $df = f_1\omega^1 + f_2\omega^2$, where the smooth functions f_1 and f_2 are called the derivatives of f with respect to the given coframe field.

6.4.1 Summary of frame reduction and structure equations

Proposition 6.8. *Let $\mathbf{x} : M \to \mathbf{H}^3$ be an immersed surface. At a nonumbilic point there exists a smooth second order frame field $e : U \to G$ on a neighborhood of the point. Its pull-back of the Maurer–Cartan form of G satisfies:*

$$\begin{aligned}
&\omega_4^3 = 0, \textit{ (first order)}, \quad \omega_4^1\wedge\omega_4^2 \neq 0,\\
&\omega_1^3 = a\omega_4^1, \quad \omega_2^3 = c\omega_4^2, \textit{ (second order)},\\
&\omega_1^2 = p\omega_4^1 + q\omega_4^2,\\
&a_2 = (a-c)p, \quad c_1 = (a-c)q, \textit{ (Codazzi equations)},\\
p_2 - q_1 - p^2 - q^2 &= K = ac - 1, \textit{ (Gauss equation)}
\end{aligned}$$

The functions a and c are the principal curvatures of $\mathbf{x}$. They are continuous functions on M, smooth on an open neighborhood of any nonumbilic point ($=$ point where $a \neq c$). At a nonumbilic point we may assume that $a > c$ on U. The Gaussian curvature of the metric induced on M is $K = -1 + ac$.

If $\mathbf{x}$ *is totally umbilic on* U*, then any first order frame is automatically second order and the above equations with* $a = c$ *give the structure equations for such a frame.*

6.5 Totally umbilic immersions

The set of totally umbilic surfaces in $\mathbf{H}^3$ includes spheres and analogs of planes, but also horospheres and ultraspheres, which have no analogs in Euclidean or spherical geometry.

Lemma 6.9. *If* $a \in \mathbf{R}$, $\mathbf{y} \in \mathbf{H}^3$, *and* $\mathbf{v} \in T_\mathbf{y}\mathbf{H}^3 = \mathbf{y}^\perp$, *then the set*

$$S(a\mathbf{y}+\mathbf{v}) = \{\mathbf{x} \in \mathbf{H}^3 : \langle \mathbf{x}, a\mathbf{y}+\mathbf{v}\rangle = -a\} \tag{6.35}$$

is a smoothly embedded, totally umbilic, surface in $\mathbf{H}^3$ *passing through the point* $\mathbf{y}$*, with unit normal vector* $\mathbf{v}$ *at* $\mathbf{y}$*. A unit normal vector field at* $\mathbf{x} \in S(a\mathbf{y}+\mathbf{v})$ *is*

$$\mathbf{e}_3 : S(a\mathbf{y}+\mathbf{v}) \to \mathbf{R}^{3,1}, \quad \mathbf{e}_3(\mathbf{x}) = a\mathbf{y}+\mathbf{v}-a\mathbf{x}. \tag{6.36}$$

Proof. Let $\mathbf{p} = a\mathbf{y}+\mathbf{v}$, so $\langle \mathbf{p}, \mathbf{p}\rangle = 1-a^2$. Consider the smooth map

$$F : \mathbf{R}^{3,1} \to \mathbf{R}^2, \quad F(\mathbf{x}) = (\langle \mathbf{x}, \mathbf{p}\rangle + a, \langle \mathbf{x}, \mathbf{x}\rangle + 1).$$

Then $S(\mathbf{p}) = F^{-1}(0,0)$ and

$$dF_\mathbf{x} = (\langle d\mathbf{x}, \mathbf{p}\rangle, 2\langle \mathbf{x}, d\mathbf{x}\rangle) = (-p^0dx^0 + \sum_1^3 p^i dx^i, -2x^0dx^0 + 2\sum_1^3 x^i dx^i)$$

has rank two at every point $\mathbf{x} \in S(\mathbf{p})$. In fact, if it has rank less than two at $\mathbf{x} \in S(\mathbf{p})$, then $\mathbf{x} = t\mathbf{p}$ for some $t \in \mathbf{R}$, so $-a = \langle \mathbf{x}, \mathbf{p}\rangle = t(1-a^2)$, which implies that $a^2 \neq 1$, $t = \frac{a}{a^2-1}$, and $-1 = \langle \mathbf{x}, \mathbf{x}\rangle = \frac{a^2}{1-a^2}$, a contradiction. The map $\mathbf{e}_3 : S(\mathbf{p}) \to \mathbf{R}^{3,1}$ defined in (6.36) is smooth. At $\mathbf{x} \in S(\mathbf{p})$, the vector $\mathbf{e}_3(\mathbf{x})$ has unit length, and $\langle \mathbf{x}, \mathbf{e}_3\rangle = 0$ shows that $\mathbf{e}_3(\mathbf{x}) \in T_\mathbf{x}\mathbf{H}^3$. It is normal to $S(\mathbf{p})$ at $\mathbf{x}$, since

$$\langle d\mathbf{x}, \mathbf{e}_3(\mathbf{x})\rangle = \langle d\mathbf{x}, \mathbf{p}\rangle - a\langle d\mathbf{x}, \mathbf{x}\rangle = 0,$$

where both of the last two terms are zero because $\langle \mathbf{x}, \mathbf{p}\rangle$ and $\langle \mathbf{x}, \mathbf{x}\rangle$ are constant on $S(\mathbf{p})$. Then $d\mathbf{e}_{3\mathbf{x}} = -ad\mathbf{x}$ shows that $\mathbf{x}$ is totally umbilic with principal curvature equal to a at every point. Note $\mathbf{e}_3(\mathbf{y}) = \mathbf{v}$. □

Remark 6.10. For the given $\mathbf{y} \in \mathbf{H}^3$, if we replace a with $-a$ and $\mathbf{v}$ with $-\mathbf{v}$, then $S(-a\mathbf{y}-\mathbf{v}) = S(a\mathbf{y}+\mathbf{v})$, but with the opposite orientation, whose value at $\mathbf{y}$ is now $-\mathbf{v}$.

Lemma 6.11. *If* $|a| > 1$, $\mathbf{y} \in \mathbf{H}^3$, *unit vector* $\mathbf{v} \in \mathbf{y}^\perp$, *then with its orientation* (6.36), $S(a\mathbf{y}+\mathbf{v})$ *is the oriented sphere of signed radius* r *and center* $\mathbf{m} \in \mathbf{H}^3$, *where*

$$\cosh r = \frac{|a|}{\sqrt{a^2-1}}, \quad \sinh r = \frac{\cosh r}{a}, \quad \mathbf{m} = \cosh r\,\mathbf{y} + \sinh r\,\mathbf{v}. \tag{6.37}$$

Proof. Follows from Definition 6.2 and Lemma 6.9. The center $\mathbf{m}$ lies on the geodesic starting at $\mathbf{y}$ with initial unit tangent vector $\mathbf{v}$. □

Definition 6.12. Given $a \in \mathbf{R}$, $\mathbf{y} \in \mathbf{H}^3$, unit vector $\mathbf{v} \in \mathbf{y}^\perp$:

If $|a| > 1$, then $S(a\mathbf{y}+\mathbf{v})$ is the oriented sphere of radius r and center $\mathbf{m}$ defined in (6.37), whose normal vector at $\mathbf{y}$ is $\mathbf{v}$.

If $|a| = 1$, then $S(a\mathbf{y}+\mathbf{v})$ is a *horosphere* whose orientation at $\mathbf{y}$ is $\mathbf{v}$.

If $0 < |a| < 1$, then $S(a\mathbf{y}+\mathbf{v})$ is an *ultrasphere* whose orientation agrees with $\mathbf{v}$ at $\mathbf{y}$.

If $a = 0$, then $S(a\mathbf{y}+\mathbf{v})$ is a *plane* with constant unit normal vector field $\mathbf{e}_3(\mathbf{x}) = \mathbf{v}$.

These surfaces are illustrated in the Poincaré Ball in Figure 6.4 below. The surfaces $S(a\mathbf{y}+\mathbf{v})$ constitute all the totally umbilic surfaces in $\mathbf{H}^3$.

Theorem 6.13 (Totally umbilic case). *If* M *is a connected surface and if* $\mathbf{x} : M \to \mathbf{H}^3$ *is a totally umbilic immersion, oriented by the normal vector field* $\mathbf{n}$ *along* $\mathbf{x}$, *then* $\mathbf{x}(M) \subset S(a\mathbf{y}+\mathbf{v})$, *where* $a \in \mathbf{R}$ *is the principal curvature relative to* $\mathbf{n}$, $\mathbf{y} = \mathbf{x}(m) \in \mathbf{H}^3$ *for an arbitrarily chosen point* $m \in M$, *and* $\mathbf{v} = \mathbf{n}(m) \in \mathbf{y}^\perp$.

Proof. Let e be a first order frame field along the totally umbilic $\mathbf{x}$ on a connected open set $U \subset M$ for which $\mathbf{e}_3$ is the given orientation on U. From the Codazzi equations, the derivatives $a_1 = a_2 = 0$, which means that the principal curvature a relative to $\mathbf{e}_3$ is constant on U. Then $d\mathbf{e}_3 = -ad\mathbf{x}$, so $\mathbf{e}_3 + a\mathbf{x} = \mathbf{p} \in \mathbf{R}^{3,1}$ is constant on U. Fix $m \in U$, and let $\mathbf{y} = \mathbf{x}(m)$ and $\mathbf{v} = \mathbf{e}_3(m)$. Then $\mathbf{e}_3 + a\mathbf{x} = a\mathbf{y}+\mathbf{v}$ on U. Since $\langle \mathbf{x}, \mathbf{e}_3 \rangle = 0$ and $\langle \mathbf{x}, \mathbf{x} \rangle = -1$, we have $\langle \mathbf{x}, a\mathbf{y}+\mathbf{v} \rangle = -a$ on U, which shows that $\mathbf{x}(U) \subset S(a\mathbf{y}) + \mathbf{v}$. The result now follows from the connectedness of M. □

6.6 Poincaré Ball model

We shall use the Poincaré Ball model of hyperbolic space as a means to illustrate the geometry of hyperbolic space. The strength of the hyperboloid model lies in the simplicity of the action of its group of isometries as the standard linear action of $\mathbf{O}_+(3,1)$ on $\mathbf{R}^{3,1}$. As a hypersurface in a four dimensional space, however, it is not convenient for illustrations. The group of isometries of the ball model, on the other hand, is by linear fractional transformations of $\mathbf{B}^3 \subset \mathbf{R}^3$ by $\mathbf{O}_+(3,1)$.

Definition 6.14. The *Poincaré Ball model* of hyperbolic space is the open submanifold

$$\mathbf{B}^3 = \{\mathbf{y} \in \mathbf{R}^3 : |\mathbf{y}| < 1\},$$

with the Riemannian metric

$$I_{\mathbf{B}} = \frac{4d\mathbf{y}\cdot d\mathbf{y}}{(1-|\mathbf{y}|^2)^2}.$$

To see that $(\mathbf{B}^3, I_{\mathbf{B}})$ is isometric to $\mathbf{H}^3$, we use a map analogous to stereographic projection from $-\boldsymbol{\epsilon}_4$. We decompose a point $\mathbf{x} \in \mathbf{R}^{3,1}$ as $\mathbf{x} = \sum_1^4 x^a \boldsymbol{\epsilon}_a = \mathbf{x}' + x^4 \boldsymbol{\epsilon}_4$, where $\mathbf{x}' = \sum_1^3 x^i \boldsymbol{\epsilon}_i \in \boldsymbol{\epsilon}_4^\perp \cong \mathbf{R}^3$.

Definition 6.15. *Hyperbolic stereographic projection* is the map

$$\mathfrak{s} : \mathbf{H}^3 \to \mathbf{B}^3 \subset \mathbf{R}^3, \quad \mathfrak{s}(\mathbf{x}) = \mathbf{y} = \frac{1}{1+x^4}\mathbf{x}', \tag{6.38}$$

whose inverse map is

$$\mathbf{x} = \mathfrak{s}^{-1}(\mathbf{y}) = \frac{1}{1-|\mathbf{y}|^2}\left(2\mathbf{y} + (1+|\mathbf{y}|^2)\boldsymbol{\epsilon}_4\right).$$

Here is a geometric description of $\mathfrak{s}$. Given a point $\mathbf{x} \in \mathbf{H}^3$, the affine line joining it to $-\boldsymbol{\epsilon}_4$ in $\mathbf{R}^{3,1}$ meets the subspace $\boldsymbol{\epsilon}_4^\perp \subset \mathbf{R}^{3,1}$ at the point $\mathfrak{s}(\mathbf{x}) \in \mathbf{B}^3 \subset \mathbf{R}^3 = \boldsymbol{\epsilon}_4^\perp$. See Figure 6.3.

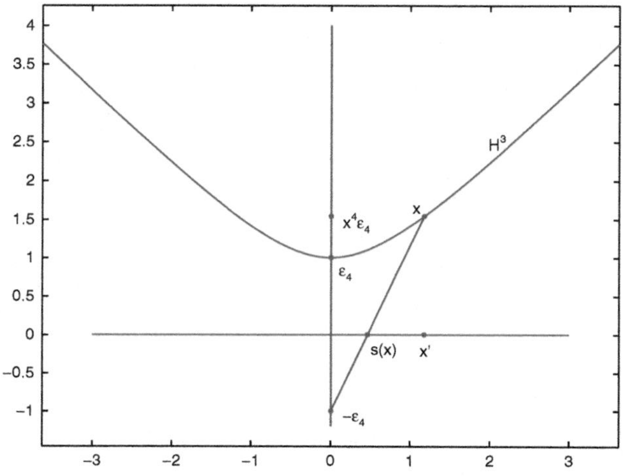

Fig. 6.3 $\mathbf{H}^3 \ni \mathbf{x} \mapsto \mathfrak{s}(\mathbf{x}) \in \mathbf{B}^3$ in the plane spanned by $\mathbf{x}$ and $\boldsymbol{\epsilon}_4$.

The formulas for $\mathfrak{s}$ and $\mathfrak{s}^{-1}$ show it is a diffeomorphism. Differentiating (6.38), we get

$$d\mathfrak{s}_{\mathbf{x}} = d\mathbf{y} = \frac{d\mathbf{x}'}{1+x^4} - \frac{\mathbf{x}'}{(1+x^4)^2}dx^4,$$

from which we get

$$\mathfrak{s}^* I_B = \frac{4d\mathbf{y}\cdot d\mathbf{y}}{(1-|\mathbf{y}|^2)^2} = \langle d\mathbf{x}, d\mathbf{x}\rangle.$$

This shows that the map (6.38) is an isometry onto $(\mathbf{B}^3, I_B)$ and conformal as a map $\mathfrak{s}: \mathbf{H}^3 \to \mathbf{R}^3$.

Let us see what the totally umbilic immersed surfaces look like in $(\mathbf{B}^3, I_B)$. In $\mathbf{H}^3$ these were given, up to $\mathbf{SO}_+(3,1)$ congruence, by $S(\mathbf{p})$, where $\mathbf{p} = a\boldsymbol{\epsilon}_4 + \boldsymbol{\epsilon}_3$, for any constant $a \geq 0$. All of these $S(\mathbf{p})$ pass through $\boldsymbol{\epsilon}_4 \in \mathbf{H}^3$ with unit normal vector $\boldsymbol{\epsilon}_3$ there, so their images by $\mathfrak{s}$ will pass through $\mathfrak{s}(\boldsymbol{\epsilon}_4)$, which is the origin 0 of $\mathbf{R}^3$, with unit normal vector $d\mathfrak{s}_{\boldsymbol{\epsilon}_4}\boldsymbol{\epsilon}_3 = \boldsymbol{\epsilon}_3$ there. Then

$$\begin{aligned}\mathfrak{s}(S(\mathbf{p})) &= \{\mathbf{y} \in \mathbf{B}^3 : \langle \mathfrak{s}^{-1}(\mathbf{y}), \mathbf{p}\rangle = -a\} \\ &= \{\mathbf{y} \in \mathbf{B}^3 : y^3 = a|\mathbf{y}|^2\}.\end{aligned} \tag{6.39}$$

If $a > 0$, then this is

$$\{\mathbf{y} \in \mathbf{B}^3 : (y^1)^2 + (y^2)^2 + (y^3 - \frac{1}{2a})^2 = \frac{1}{4a^2}\},$$

which is the intersection of $\mathbf{B}^3$ with $S^{\mathrm{Euc}}_{\frac{1}{2a}}(\frac{1}{2a}\boldsymbol{\epsilon}_3)$, the Euclidean sphere in $\mathbf{R}^3$ of radius $\frac{1}{2a}$ and center at the point $\frac{1}{2a}\boldsymbol{\epsilon}_3$.

When $a > 1$, $S^{\mathrm{Euc}}_{\frac{1}{2a}}(\frac{1}{2a}\boldsymbol{\epsilon}_3)$ lies entirely within B^3 and is the image of the sphere $S_r(\mathbf{m}) \subset \mathbf{H}^3$ of radius r satisfying $\cosh r = \frac{a}{\sqrt{1+a^2}}$ and center $\mathbf{m} = \cosh r\ \boldsymbol{\epsilon}_4 + \sinh r\ \boldsymbol{\epsilon}_3 \in \mathbf{H}^3$ given by (6.37). As with the case of stereographic projection, $\mathfrak{s}(S_r(\mathbf{m})) = S^{\mathrm{Euc}}_{\frac{1}{2a}}(\frac{1}{2a}\boldsymbol{\epsilon}_3)$, but it does not send the center $\mathbf{m}$ to the center $\frac{1}{2a}\boldsymbol{\epsilon}_3$, since

$$\mathfrak{s}(\mathbf{m}) = \frac{\sinh r}{1+\cosh r}\boldsymbol{\epsilon}_3 = \frac{1}{a+\sqrt{a^2-1}}\boldsymbol{\epsilon}_3 \neq \frac{1}{2a}\boldsymbol{\epsilon}_3.$$

When $a = 1$, $S^{\mathrm{Euc}}_{\frac{1}{2a}}(\frac{1}{2a}\boldsymbol{\epsilon}_3)$ passes through $\boldsymbol{\epsilon}_3$, but all of its other points lie within $\mathbf{B}^3$. Its intersection with $\mathbf{B}^3$ is the image of the horosphere $S(\boldsymbol{\epsilon}_4 + \boldsymbol{\epsilon}_3) \subset \mathbf{H}^3$.

As a decreases, with $1 > a > 0$, progressively more of $S^{\mathrm{Euc}}_{\frac{1}{2a}}(\frac{1}{2a}\boldsymbol{\epsilon}_3)$ lies outside of $\mathbf{B}^3$. Its intersection with $\mathbf{B}^3$ is the image of the ultrasphere $S(a\boldsymbol{\epsilon}_4 + \boldsymbol{\epsilon}_3) \subset \mathbf{H}^3$.

When $a = 0$, then the image is the totally geodesic plane

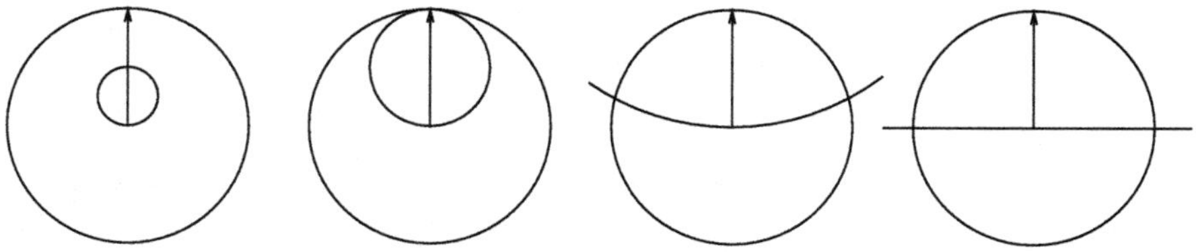

Fig. 6.4 $\mathfrak{s}$ of $S(2\epsilon_4+\epsilon_3)$, $S(\epsilon_4+\epsilon_3)$, $S(\frac{1}{2}\epsilon_4+\epsilon_3)$, and $S(\epsilon_3)$, with ϵ_3.

$$\{\mathbf{y} \in \mathbf{B}^3 : y^3 = 0\}.$$

In this case $\mathbf{p} = \epsilon_3$ is a unit normal vector field along $S(\mathbf{p})$, and its image ϵ_3 is a unit normal vector field along $\mathfrak{s}(S(\mathbf{p}))$.

These various cases are illustrated in Figure 6.4. To be more precise, these figures show the intersection of $\mathfrak{s}(S(a\mathbf{y}+\mathbf{v}))$ with the plane through the origin tangent to ϵ_3 there.

Remark 6.16. It is clear that $\mathfrak{s}(S(\mathbf{p})$ is always connected, and thus $S(\mathbf{p})$ is always connected.

For more details about totally umbilic surfaces in $\mathbf{H}^n$, for any $n \geq 3$, see Spivak [154, Vol. IV, pp. 10–26 and Theorem 29, pp. 114–117].

The boundary of $\mathbf{B}^3$ in $\mathbf{R}^3$ is the standard unit sphere,

$$\partial\mathbf{B}^3 = \mathbf{S}^2 \subset \mathbf{R}^3.$$

We shall define an identification of each unit tangent sphere of $\mathbf{B}^3$ with $\partial\mathbf{B}^3$, and then define a conformal diffeomorphism between S^2_∞ and $\partial\mathbf{B}^3$ that will show that $\partial\mathbf{B}^3$ is the sphere at infinity of the Poincaré ball model.

If $\mathbf{y} \in \mathbf{B}^3$ and $S^2_{\mathbf{y}} \subset \mathbf{R}^3$ is the unit sphere in the tangent space at $\mathbf{y}$, let

$$\mathfrak{b}_{\mathbf{B}} : S^2_{\mathbf{y}} \to \partial\mathbf{B}^3, \quad \mathfrak{b}_{\mathbf{B}}(\mathbf{u}) = \lim_{t\to\infty} \gamma_{\mathbf{u}}(t), \tag{6.40}$$

where $\gamma_{\mathbf{u}}$ is the geodesic in $\mathbf{B}^3$ starting at $\mathbf{y}$ in the direction $\mathbf{u}$.

Proposition 6.17. *Decompose* $\mathbf{n} = \sum_1^4 n^a\epsilon_a \in N^+$ *into* $\mathbf{n} = \mathbf{n}' + n^4\epsilon_4$, *where* $\mathbf{n}' = \sum_1^3 n^i\epsilon_i$. *The map*

$$\mathfrak{s}_\infty : S^2_\infty \to \partial\mathbf{B}^3 = \mathbf{S}^2, \quad \mathbf{v} = \mathfrak{s}_\infty[\mathbf{n}] = \frac{1}{n^4}\mathbf{n}', \tag{6.41}$$

is a conformal diffeomorphism, with inverse $\mathfrak{s}_\infty^{-1}(\mathbf{v}) = [\mathbf{v}+\epsilon_4]$. *The diagram*

$$\begin{array}{ccc} S^2_{\mathbf{x}} & \stackrel{\mathfrak{b}}{\to} & S^2_\infty \\ d\mathfrak{s}_{\mathbf{x}} \downarrow & & \downarrow \mathfrak{s}_\infty \\ S^2_{\mathfrak{s}(\mathbf{x})} & \stackrel{\mathfrak{b}_{\mathbf{B}}}{\to} & \partial \mathbf{B}^3 \end{array}$$

commutes. In particular, $\mathfrak{b}_B$ is a smooth diffeomorphism, since the other three maps are. The map

$$\hat{\mathfrak{s}} : \mathbf{H}^3 \cup S^2_\infty \to \mathbf{B}^3 \cup \partial \mathbf{B}^3 = \overline{\mathbf{B}^3} = \text{closed unit ball in } \mathbf{R}^3,$$

defined by $\hat{\mathfrak{s}}|_{\mathbf{H}^3} = \mathfrak{s}$ and $\hat{\mathfrak{s}}|_{S^2_\infty} = \mathfrak{s}_\infty$ is a bijection. Let $\mathbf{H}^3 \cup S^2_\infty$ have the unique topology that makes $\hat{\mathfrak{s}}$ a homeomorphism. If $\mathbf{x} \in \mathbf{H}^3$ and $\mathbf{v} \in S^2_{\mathbf{x}}$, then

$$\lim_{t\to\infty} \cosh t\, \mathbf{x} + \sinh t\, \mathbf{v} = [\mathbf{x} + \mathbf{v}] \in S^2_\infty. \tag{6.42}$$

Proof. It is evident from their formulas that $\mathfrak{s}_\infty$ and its inverse are smooth. By Lemma 6.5, the conformal structure on S^2_∞ comes from sections of the principal bundle (6.21). A point in S^2_∞ has a neighborhood U on which there is a section of the form

$$e([\mathbf{n}' + n^4 \boldsymbol{\epsilon}_4]) = (\mathbf{e}_1, \mathbf{e}_2, \frac{\mathbf{n}'}{n^4}, \boldsymbol{\epsilon}_4).$$

For this frame field, $\omega^i = 0$, since $\mathbf{e}_4$ is constant, so

$$\varphi^i = \frac{1}{2}(\omega^i + \omega^i_3) = \frac{1}{2}\omega^i_3,$$

for $i = 1,2$, and $\sum_1^2 \varphi^i \varphi^i$ is a metric in the conformal structure of S^2_∞. Now

$$\mathbf{e}_1, \mathbf{e}_2, \mathbf{e}_3 = \frac{\mathbf{n}'}{n^4} \in \boldsymbol{\epsilon}_4^\perp = \mathbf{R}^3,$$

so

$$A : U \to \mathbf{SO}(3), \quad A([\mathbf{n}' + n^4 \boldsymbol{\epsilon}_4]) = (\mathbf{e}_1, \mathbf{e}_2, \frac{\mathbf{n}'}{n^4})$$

is a frame field along $\mathfrak{s}_\infty$ on U. Thus, $\mathfrak{s}_\infty$ pulls back the standard metric on $\mathbf{S}^2$ to

$$d\mathbf{e}_3 \cdot d\mathbf{e}_3 = \sum_1^2 \omega^i_3 \omega^i_3 = 4 \sum_1^2 \varphi^i \varphi^i,$$

which shows that $\mathfrak{s}_\infty$ is conformal. The geodesic $\boldsymbol{\gamma}$ in $\mathbf{H}^3$ with initial value $\boldsymbol{\gamma}(0) = \mathbf{x} = \mathbf{x}' + x^4 \boldsymbol{\epsilon}_4$ and initial unit velocity vector $\dot{\boldsymbol{\gamma}}(0) = \mathbf{v} = \mathbf{v}' + v^4 \boldsymbol{\epsilon}_4 \in S^2_{\mathbf{x}} \subset T_{\mathbf{x}}\mathbf{H}^3 = \mathbf{x}^\perp$ is the curve

$$\boldsymbol{\gamma}(t) = \mathbf{x} \cosh t + \mathbf{v} \sinh t.$$

It is a geodesic because $\ddot{\boldsymbol{\gamma}}(t) = \boldsymbol{\gamma}(t) \in (T_{\gamma(t)}\mathbf{H}^3)^\perp$, for all t. The isometry (6.38) sends this geodesic to the geodesic

$$\mathfrak{s} \circ \boldsymbol{\gamma}(t) = \frac{1}{1 + x^4 \cosh t + v^4 \sinh t}(\mathbf{x} \cosh t + \mathbf{v}' \sinh t), \tag{6.43}$$

in $\mathbf{B}^3$ that starts at $\mathfrak{s}(\mathbf{x}) = \frac{1}{1+x^4}\mathbf{x}'$, with initial velocity

$$d\mathfrak{s}_\mathbf{x}\mathbf{v} = (\mathfrak{s} \circ \boldsymbol{\gamma})\dot{}(0) = \frac{1}{(1+x^4)^2}(-v^4\mathbf{x}' + (1+x^4)\mathbf{v}').$$

Hence, by (6.40),

$$\mathfrak{b}_\mathbf{B} \circ d\mathfrak{s}_\mathbf{x}(\mathbf{v}) = \lim_{t\to\infty} \mathfrak{s}(\boldsymbol{\gamma}(t)) = \frac{\mathbf{x}' + \mathbf{v}'}{x^4 + v^4} = \mathfrak{s}_\infty([\mathbf{x} + \mathbf{v}]) = \mathfrak{s}_\infty \circ \mathfrak{b}(\mathbf{v})$$

shows the diagram commutes. The diagram proves (6.42) by

$$\lim_{t\to\infty} \hat{\mathfrak{s}}(\gamma(t)) = \lim_{t\to\infty} \mathfrak{s}(\gamma(t)) = \mathfrak{s}_\infty([\mathbf{x} + \mathbf{v}]) = \hat{\mathfrak{s}}([\mathbf{x} + \mathbf{v}]).$$

□

Recall that the arclength parameter of a geodesic $\gamma(t)$ in any complete Riemannian manifold is determined up to transformations $s = \pm t + c$, for any constant $c \in \mathbf{R}$. An *orientation* of the geodesic is a choice of the sign.

Proposition 6.18. *Geodesics γ and σ in $\mathbf{B}^3, I_\mathbf{B}$ are asymptotically parallel if and only if for some choice of orientation of each*

$$\lim_{t\to\infty} \gamma(t) = \lim_{t\to\infty} \sigma(t) \in \partial\mathbf{B}^3.$$

Thus, geodesics γ and σ in $\mathbf{H}^3$ are asymptotically parallel if and only if for some choice of orientation of each

$$[\gamma(0) + \dot{\gamma}(0)] = [\sigma(0) + \dot{\sigma}(0)] \in S^2_\infty.$$

Proof. See Problem 6.47. □

6.7 Tangent and curvature spheres

In this section we use the term oriented sphere in $\mathbf{H}^3$ to mean an oriented sphere, horosphere, or ultrasphere.

Definition 6.19. An *oriented tangent sphere* to an immersion $\mathbf{x}: M^2 \to \mathbf{H}^3$ at a point $m \in M$, with unit normal vector $\mathbf{v} \in \mathbf{x}(m)^\perp$, is an oriented sphere through $\mathbf{x}(m)$ with unit normal $\mathbf{v}$ at $\mathbf{x}(m)$.

The set of all oriented tangent spheres to $\mathbf{x}: M^2 \to \mathbf{H}^3$ at $m \in M$ with unit normal $\mathbf{v}$ at $\mathbf{x}(m)$ is

$$\{S(a\mathbf{x}(m) + \mathbf{v}) : a \in \mathbf{R}\},$$

as defined in (6.35) of Lemma 6.9. When $a^2 > 1$, so the oriented tangent sphere is a genuine hyperbolic sphere, its center must lie on the normal line

$$\{\cosh t\, \mathbf{x}(m) + \sinh t\, \mathbf{v} : t \in \mathbf{R}\},$$

as observed at the end of the proof of Lemma 6.11.

Definition 6.20. An *oriented curvature sphere* of an immersion $\mathbf{x}: M^2 \to \mathbf{H}^3$ at $m \in M$ relative to a unit normal vector $\mathbf{v} \in \mathbf{x}(m)^\perp$ is an oriented tangent sphere whose principal curvature is equal to a principal curvature of $\mathbf{x}: M \to \mathbf{S}^3$ at m relative to the unit normal $\mathbf{v}$ at $\mathbf{x}(m)$.

Remark 6.21. If $m \in M$ is nonumbilic for $\mathbf{x}$, then there are two distinct oriented curvature spheres at m relative to $\mathbf{v}$. If m is umbilic, then there is only one, but we say it has *multiplicity two*. If a is a principal curvature of $\mathbf{x}$ at m relative to $\mathbf{v}$, then the corresponding oriented curvature sphere at m is $S(a\mathbf{x}(m) + \mathbf{v})$, by Lemma 6.9.

As a conformal map, hyperbolic stereographic projection $\mathfrak{s}: \mathbf{H}^3 \to \mathbf{B}^3 \subset \mathbf{R}^3$ of Definition 6.15, sends curvature spheres to curvature spheres.

Proposition 6.22. *If S_0 is an oriented curvature sphere of the immersion $\mathbf{x}: M^2 \to \mathbf{H}^3$ at $m \in M$, then $\mathfrak{s}(S_0)$ is an oriented curvature sphere of the immersion into Euclidean space $\mathfrak{s} \circ \mathbf{x}: M \to \mathbf{R}^3$ at m. The converse is true for any immersion $\mathbf{y}: M^2 \to \mathbf{R}^3$ for which $\mathbf{y}(M) \subset \mathbf{B}^3$.*

Proof. If $S: U \subset M \to \mathbf{H}^3$ is a tangent sphere map along $\mathbf{x}$ such that $S(m) = S_0$, then $\mathfrak{s} \circ S: U \to \mathbf{R}^3$ is a tangent sphere map along $\mathfrak{s} \circ \mathbf{x}$ in the sense that each image is an open subset of a Euclidean sphere, which is tangent to $\mathfrak{s} \circ \mathbf{x}$. In addition, the rank of $d(\mathfrak{s} \circ S)_m$ is the same as the rank of dS_m, which is less than two. The proof of the converse is similar. □

Corollary 6.23. *The conformal map*

$$\mathscr{S}^{-1} \circ \mathfrak{s}: H^3 \to \mathbf{S}^3,$$

where $\mathscr{S}: \mathbf{S}^3 \setminus \{-\epsilon_0\} \to \mathbf{R}^3$ is stereographic projection from $-\epsilon_0 \in \mathbf{R}^4$, sends spheres, horospheres, ultraspheres, and planes to spheres. It sends an oriented curvature sphere of an immersion $\mathbf{x}: M \to \mathbf{H}^3$ at $m \in M$ to an oriented curvature sphere of $\mathscr{S}^{-1} \circ \mathfrak{s} \circ \mathbf{x}: M \to \mathbf{S}^3$ at m.

6.8 Dupin and isoparametric immersions

Definition 6.24. A principal curvature a of an immersion $\mathbf{x} : M^2 \to \mathbf{H}^3$ satisfies the *Dupin condition* if it is constant on each of its connected lines of curvature. If both principal curvatures satisfy the Dupin condition, then $\mathbf{x}$ is called a *Dupin immersion*. It is *proper Dupin* if, in addition, the number of distinct principal curvatures is constant on M. If both principal curvatures are constant on M, then $\mathbf{x}$ is called an *isoparametric immersion*.

Any isoparametric immersion is a fortiori a proper Dupin immersion.

If M is a connected surface and if $\mathbf{x} : M \to \mathbf{H}^3$ is an immersion with constant principal curvatures a and c, then either $a = c$ and $\mathbf{x}$ is totally umbilic, or $ac = 1$ and $K = 0$. If $0 < a < 1$ and if $b = \sqrt{1 - a^2}$, then up to congruence by $\mathbf{SO}_+(3,1)$ the isoparametric immersion with principal curvatures a and $1/a$ is $\mathbf{x} : \mathbf{R}^2 \to \mathbf{H}^3$, where

$$\mathbf{x}(s,t) = \frac{a}{b}(\epsilon_1 \cos\frac{b}{a}t + \epsilon_2 \sin\frac{b}{a}t) + \epsilon_3 \frac{\sinh bs}{b} + \epsilon_4 \frac{\cosh bs}{b}. \tag{6.44}$$

See Problem 6.49 for details.

Figures 6.5 and 6.6 show $\mathfrak{s} \circ \mathbf{x}$ in $\mathbf{B}^3$, for the cases $a = 1/2$ and $a = 9/10$. Figure 6.7 shows the congruent copy of the $a = 1/2$ case obtained by applying to it the hyperbolic isometry $A = \begin{pmatrix} \cosh r & 0 & 0 & \sinh r \\ 0 & 1 & 0 & 0 \\ 0 & 0 & 1 & 0 \\ \sinh r & 0 & 0 & \cosh r \end{pmatrix} \in \mathbf{SO}_+(3,1)$.

Remark 6.25. As a consequence of Proposition 6.22, any Dupin immersion into hyperbolic space $\mathbf{x} : M^2 \to \mathbf{B}^3$ is, without change, a Dupin immersion into Euclidean space. Conversely, any Dupin immersion into Euclidean space $\mathbf{x} : M^2 \to \mathbf{R}^3$

Fig. 6.5 Isoparametric immersion in $\mathbf{B}^3$ with $a = 1/2$.

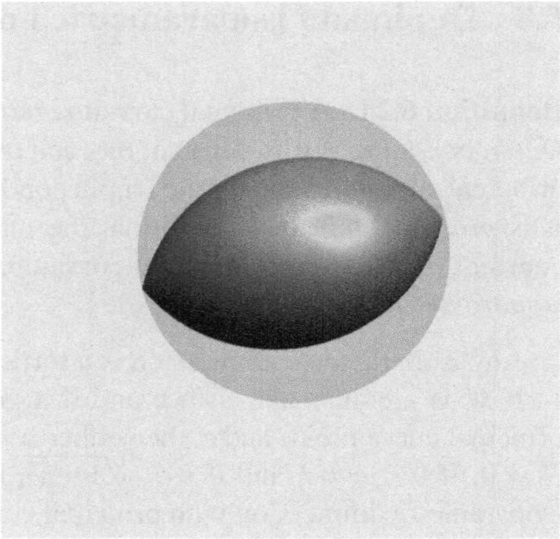

Fig. 6.6 Isoparametric immersion in $\mathbf{B}^3$ with $a = 9/10$.

Fig. 6.7 Hyperbolic congruent copy of $a = 1/2$ case in $\mathbf{B}^3$

whose image is in the unit ball $\mathbf{B}^3$, is, without change, a Dupin immersion into hyperbolic space. Thus, Figures 6.5, 6.6, and 6.7 illustrate examples of proper Dupin immersions in Euclidean space. See Problem 6.54.

6.9 Curves

Exercise 29 (Space curves). Let $\mathbf{f} : J \to \mathbf{H}^3$ be a regular smooth curve, where J is an interval in $\mathbf{R}$. Use the method of moving frames to show that there exists a first order frame field $e = (\mathbf{T}, \mathbf{e}_2, \mathbf{e}_3, \mathbf{f}) : J \to \mathbf{SO}_+(3,1)$, where $\mathbf{T}$ is the unit tangent vector. Find the conditions under which there exists a unique frame field $e = (\mathbf{T}, \mathbf{N}, \mathbf{B}, \mathbf{f})$ along $\mathbf{f}$, called the *Frenet frame*, such that

$$d\mathbf{f} = \mathbf{T}\omega^1, \quad d\mathbf{T} = (\kappa\mathbf{N} + \mathbf{f})\omega^1, \quad d\mathbf{N} = (-\kappa\mathbf{T} + \tau\mathbf{B})\omega^1, \quad d\mathbf{B} = -\tau\mathbf{N}\omega^1,$$

where $\omega^1 = \omega^1_4$ and $\kappa > 0$ and τ are smooth functions on J called the curvature and torsion, respectively, of $\mathbf{f}$. The form $\omega^1 = ds$, for some function s on J, which is the *arclength parameter*. What are the curves that fail to have a unique Frenet frame?

Example 6.26 (Circles). One would expect that a regular smooth curve $\mathbf{f} : J \to \mathbf{H}^3$ with zero torsion and positive constant curvature κ is a circle. This is true, however, only when $\kappa > 1$. If $\kappa = 1$, it is a horocircle with one point at infinity. When $0 < \kappa < 1$, it is an ultra-circle, which goes off to infinity before it can close up. In every case such a curve is the intersection of a totally umbilic surface $S(\kappa\mathbf{y} + \mathbf{v})$ with the plane through $\mathbf{y}$ orthogonal to the unit vector $\mathbf{v} \in T_y\mathbf{H}^3$. The totally geodesic surfaces shown schematically in Figure 6.4 are actually the images by $\mathfrak{s} : \mathbf{H}^3 \to \mathbf{B}^3$ of the circle, horocircle, ultra-circle, and line, respectively, obtained by intersecting these surfaces with any plane through $\boldsymbol{\epsilon}_4$ tangent to $\boldsymbol{\epsilon}_3$ there.

Example 6.27 (Helices). A helix in $\mathbf{H}^3$ is the curve obtained by starting at a point $\mathbf{p}$ a distance $a > 0$ from a line $\boldsymbol{\gamma}$ and rotating it about $\boldsymbol{\gamma}$ at a constant rate as it rises at a constant rate b times the rate of rotation. Here $b \in \mathbf{R}$ is called the *pitch*, and the case $b = 0$ is a circle. Show that if $\boldsymbol{\gamma}(s) = \cosh s\, \boldsymbol{\epsilon}_4 + \sinh s\, \boldsymbol{\epsilon}_3$ and if the initial point $\mathbf{p} = \cosh a\, \boldsymbol{\epsilon}_4 + \sinh a\, \boldsymbol{\epsilon}_1$, then the helix, parametrized by arc-length and with pitch b, is

$$\mathbf{f}(s) = (\cosh a)\boldsymbol{\gamma}(\frac{b}{L}s) + (\sinh a)(\cos\frac{s}{L}\,\boldsymbol{\epsilon}_1 + \sin\frac{s}{L}\,\boldsymbol{\epsilon}_2),$$

where $L = \sqrt{b^2\cosh^2 a + \sinh^2 a}$. See Figure 6.8 for the case $a = 1$ and $b = \frac{1}{8}$, in which case $\kappa \approx 1.2986$ and $\tau \approx 0.0881$.

6.10 Surfaces of revolution

As in Euclidean space, we can take a line l in a plane in $\mathbf{H}^3$, and then rotate about l any smooth *profile curve* lying in an open half-plane of $\boldsymbol{\lambda}$ to obtain a surface of revolution. To study such surfaces in more detail, we choose the plane $S(\boldsymbol{\epsilon}_2)$ through $\boldsymbol{\epsilon}_4$ tangent to $\boldsymbol{\epsilon}_1$ and $\boldsymbol{\epsilon}_3$ there, and we take l to be the line through $\boldsymbol{\epsilon}_4$ tangent to $\boldsymbol{\epsilon}_3$ there. A general profile curve is given by

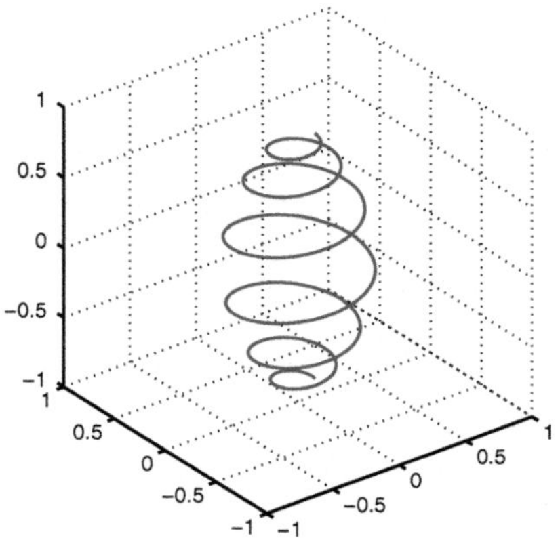

Fig. 6.8 Helix in the Poincaré ball with $a = 1$ and $b = \frac{1}{8}$.

$$\boldsymbol{\sigma} : J \to \mathbf{H}^3, \quad \boldsymbol{\sigma}(s) = f(s)\boldsymbol{\epsilon}_1 + g(s)\boldsymbol{\epsilon}_3 + w(s)\boldsymbol{\epsilon}_4, \tag{6.45}$$

where $J \subset \mathbf{R}$ is an interval, f and g are smooth functions on J with f positive on J, and w is the positive solution of

$$w^2 = 1 + f^2 + g^2$$

on J. The profile curve is parametrized by arclength if and only if

$$1 = \langle \dot{\sigma}, \dot{\sigma} \rangle = \dot{f}^2 + \dot{g}^2 - \dot{w}^2, \tag{6.46}$$

for all $s \in J$. Rotation by $t \in \mathbf{R}$ about l in $\mathbf{H}^3$ is

$$\begin{pmatrix} \cos t & -\sin t & 0 & 0 \\ \sin t & \cos t & 0 & 0 \\ 0 & 0 & 1 & 0 \\ 0 & 0 & 0 & 1 \end{pmatrix},$$

so our surface of revolution $\mathbf{x} : J \times \mathbf{R} \to \mathbf{H}^3$, is give by

$$\mathbf{x}(s,t) = f(s)(\boldsymbol{\epsilon}_1 \cos t + \boldsymbol{\epsilon}_2 \sin t) + g(s)\boldsymbol{\epsilon}_3 + w(s)\boldsymbol{\epsilon}_4. \tag{6.47}$$

The partial derivatives

$$\mathbf{x}_s = \dot{f}(\boldsymbol{\epsilon}_1 \cos t + \boldsymbol{\epsilon}_2 \sin t) + \dot{g}\boldsymbol{\epsilon}_3 + \dot{w}\boldsymbol{\epsilon}_4, \quad \mathbf{x}_t = f(-\boldsymbol{\epsilon}_1 \sin t + \boldsymbol{\epsilon}_2 \cos t)$$

are orthogonal tangent vectors at each point of $\mathbf{x}$. We assume that s is arclength parameter for σ. Then

$$\mathbf{e}_1 = \mathbf{x}_s, \quad \mathbf{e}_2 = -\boldsymbol{\epsilon}_1 \sin t + \boldsymbol{\epsilon}_2 \cos t \tag{6.48}$$

is an orthonormal tangent frame field along $\mathbf{x}$. The unique unit normal vector field $\mathbf{e}_3$ along $\mathbf{x}$ for which the frame field $(\mathbf{e}_1, \mathbf{e}_2, \mathbf{e}_3, \mathbf{x})$ takes values in $\mathbf{SO}_+(3,1)$ at every point of $J \times \mathbf{R}$ is

$$\mathbf{e}_3 = \lambda(\epsilon_1 \cos t + \epsilon_2 \sin t + \epsilon_3 \frac{w\dot{f} - \dot{w}f}{g\dot{w} - w\dot{g}} + \epsilon_4 \frac{g\dot{f} - f\dot{g}}{g\dot{w} - w\dot{g}}), \tag{6.49}$$

where

$$\frac{1}{\lambda^2} = 1 + \frac{(w\dot{f} - f\dot{w})^2 - (g\dot{f} - f\dot{g})^2}{(g\dot{w} - w\dot{g})^2} = 1 + \frac{\dot{f}^2 - f^2}{(g\dot{w} - w\dot{g})^2}.$$

6.11 Tubes and parallel transformations

Exercise 30 (Tubes). Let $\mathbf{f} : J \to \mathbf{H}^3$ be a smooth curve with arclength parameter s and smooth Frenet frame field $e = (\mathbf{T}, \mathbf{N}, \mathbf{B}, \mathbf{f})$. Fix $r > 0$ and define the *tube of radius r about* $\mathbf{f}$ to be the surface $\mathbf{x} : J \times \mathbf{R} \to \mathbf{H}^3$ defined by

$$\mathbf{x}(s,t) = \cosh r\, \mathbf{f}(s) + \sinh r\, (\cos t\, \mathbf{N}(s) + \sin t\, \mathbf{B}(s)). \tag{6.50}$$

Use Subsection 4.8.2 on tubes in Euclidean space as a guide to:

1. Prove that if $\tanh r < 1/\kappa$ on J, then $\mathbf{x}$ is an immersion.
2. Find a second order frame field $(\mathbf{x}, \mathbf{e}_1, \mathbf{e}_2, \mathbf{e}_3)$ along $\mathbf{x}$.
3. Prove that, for some choice of sign of $\mathbf{e}_3$, the principal curvatures of $\mathbf{x}$ are

$$a = \frac{\kappa \cos t}{\cosh r - \kappa \cos t\, \sinh r} > 0, \quad c = -\frac{1}{\sinh r} < 0.$$

4. Find the oriented curvature spheres at $\mathbf{x}(s,t)$ relative to $\mathbf{e}_3(s,t)$.

Definition 6.28. Let $\mathbf{x} : M \to \mathbf{H}^3$ be a smoothly immersed surface with unit normal vector field $\mathbf{e}_3$. For any constant $r \in \mathbf{R}$, the *parallel transformation of* $\mathbf{x}$ *by r in direction* $\mathbf{e}_3$ is the map

$$\tilde{\mathbf{x}} : M \to \mathbf{H}^3, \quad \tilde{\mathbf{x}} = \cosh r\, \mathbf{x} + \sinh r\, \mathbf{e}_3. \tag{6.51}$$

If $\tilde{\mathbf{x}}$ is *singular* at a point $m \in M$, meaning that the rank of $d\tilde{\mathbf{x}}$ at m is less than two, then $\tilde{\mathbf{x}}(m)$ is called a *focal point* of $\mathbf{x}$.

Example 6.29 (Parallel transformations). If

$$e = (\mathbf{e}_1, \mathbf{e}_2, \mathbf{e}_3, \mathbf{x}) : U \subset M \to \mathbf{SO}_+(3,1)$$

is a first order frame field along the immersion $\mathbf{x} : M^2 \to \mathbf{H}^3$, then $\tilde{e} = (\mathbf{e}_1, \mathbf{e}_2, \tilde{\mathbf{e}}_3, \tilde{\mathbf{x}})$, where $\tilde{\mathbf{e}}_3 = \sinh r\, \mathbf{x} + \cosh r\, \mathbf{e}_3$, is a first order frame field along the parallel transformation $\tilde{\mathbf{x}}$ of $\mathbf{x}$ by $r \in \mathbf{R}$ in direction $\mathbf{e}_3$. The principal curvatures $\tilde{a}$ and $\tilde{c}$ of $\tilde{\mathbf{x}}$ relative to its unit normal vector field $\tilde{\mathbf{e}}_3$ are

$$\tilde{a} = \frac{a\cosh r - \sinh r}{\cosh r - a\sinh r}, \quad \tilde{c} = \frac{c\cosh r - \sinh r}{\cosh r - c\sinh r},$$

where a and c are the principal curvatures of $\mathbf{x}$ relative to $\mathbf{e}_3$.

Exercise 31 (Focal loci). Prove that the parallel transformation $\tilde{\mathbf{x}}$ of $\mathbf{x} : M \to \mathbf{H}^3$ by r in direction $\mathbf{e}_3$ is singular at $m \in M$ if and only if $\coth r$ is a principal curvature of $\mathbf{x}$ at m relative to $\mathbf{e}_3(m)$.

Definition 6.30. Let $\mathbf{x} : M \to \mathbf{H}^3$ be an immersion with unit normal $\mathbf{e}_3$. If $a : M \to \mathbf{R}$ is a principal curvature of $\mathbf{x}$ relative to $\mathbf{e}_3$, then a can be taken to be continuous on M, and smooth off the umbilic set. Let $r = \coth^{-1}(a) : M \to \mathbf{R}$, with $0 < r < \pi$. The image of

$$f : M \to \mathbf{H}^3, \quad f(m) = \cosh r\, \mathbf{x} + \sinh r\, \mathbf{e}_3$$

is called a *focal locus* of $\mathbf{x}$.

The parallel transform of a plane is an ultra-sphere, and any ultra-sphere is obtained in this way.

Proposition 6.31 (Parallel transforms of planes). *If* $\mathbf{y} \in \mathbf{H}^3$, *if* $0 \neq r \in \mathbf{R}$, *and if* $\mathbf{v} \in T_{\mathbf{y}}\mathbf{H}^3$ *is a unit vector, then the parallel transform of the plane* $S(\mathbf{v})$ *by* r *in the direction* $\mathbf{v}$ *is the ultra-sphere* $S(a(p\mathbf{y} + q\mathbf{v}) + q\mathbf{y} + p\mathbf{v})$, *where*

$$p = \cosh r, \quad q = \sinh r, \quad a = -\frac{q}{p} = -\tanh r.$$

Proof. It is sufficient to prove this for the case $\mathbf{y} = \epsilon_4$ and $\mathbf{v} = \epsilon_3$, by Problem 6.44. Then $\mathfrak{s}(S(\epsilon_3))$ is the Poincaré plane $\mathbf{B}^2 = \mathbf{B}^3 \cap \epsilon_3^{\perp}$, which is conformally parametrized by the Euclidean coordinates $\{s\epsilon_1 + t\epsilon_2 : s^2 + t^2 < 1\}$. We can then use $\mathfrak{s}^{-1}$ to obtain a conformal parametrization

$$\mathbf{x} : \mathbf{B}^2 \to S(\epsilon_3), \quad \mathbf{x}(s,t) = \frac{1}{1 - s^2 - t^2}((1 + s^2 + t^2)\epsilon_4 + 2s\epsilon_1 + 2t\epsilon_2).$$

Then $\mathbf{x}_s$ and $\mathbf{x}_t$ are orthogonal and we obtain a first order frame field $e : \mathbf{B}^2 \to \mathbf{SO}_+(3,1)$ along $\mathbf{x}$ from

$$\mathbf{e}_1 = \frac{1 - s^2 - t^2}{2}\mathbf{x}_s = s\mathbf{x} + s\epsilon_4 + \epsilon_1,$$
$$\mathbf{e}_2 = \frac{1 - s^2 - t^2}{2}\mathbf{x}_t = t\mathbf{x} + t\epsilon_4 + \epsilon_2,$$
$$\mathbf{e}_3 = \epsilon_3.$$

The parallel transform of $\mathbf{x}$ by r in direction $\mathbf{e}_3$ is then $\tilde{\mathbf{x}} = \cosh r\, \mathbf{x} + \sinh r\, \epsilon_3$, with unit normal vector field $\tilde{\mathbf{e}}_3 = \sinh r\, \mathbf{x} + \cosh r\, \epsilon_3$. Then

$$d\tilde{\mathbf{x}} = \cosh r\, d\mathbf{x}, \quad d\tilde{\mathbf{e}}_3 = \sinh r\, d\mathbf{x} = \frac{\sinh r}{\cosh r} d\tilde{\mathbf{x}},$$

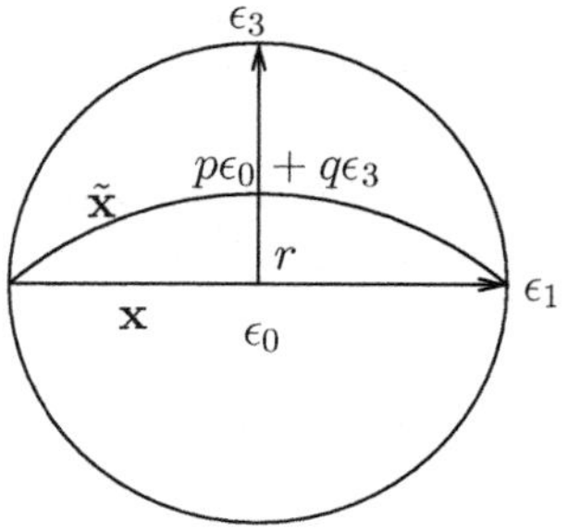

Fig. 6.9 Parallel transform $\tilde{\mathbf{x}}$ of $\mathbf{x} = S(\epsilon_3)$ in $\mathbf{B}^3 \cap \epsilon_2^{\perp}$.

shows that $\tilde{\mathbf{x}}$ is totally umbilic with principal curvature $a = -\tanh r$. It is an elementary exercise to show that

$$\tilde{\mathbf{x}}(\mathbf{B}^2) = S(a(p\epsilon_4 + q\epsilon_3) + q\epsilon_4 + p\epsilon_3),$$

where $p = \cosh r$ and $q = \sinh r$. Figure 6.9 shows the planar slice $\mathbf{B}^3 \cap \epsilon_2^{\perp}$ of the projection of $\mathbf{x}$ and $\tilde{\mathbf{x}}$ into $\mathbf{B}^3$.

□

6.12 Hyperbolic Gauss map

Definition 6.32. The *hyperbolic Gauss map* of a smooth immersion $\mathbf{x} : M^2 \to \mathbf{H}^3 \subset \mathbf{R}^{3,1}$ with unit normal vector field $\mathbf{n} : M \to \mathbf{R}^{3,1}$, so $\mathbf{n}(m) \in \mathbf{x}(m)^{\perp}$ and has unit length at each $m \in M$, is the smooth map

$$g : M \to S^2_{\infty} = N/\mathbf{R}^{\times}, \quad g(m) = [\mathbf{x}(m) + \mathbf{n}(m)]. \tag{6.52}$$

If $e : U \subset M \to \mathbf{O}_+(3,1)$ is a first order frame field along $\mathbf{x}$ such that $\mathbf{e}_3 = \mathbf{n}$ on U, then $g = [\mathbf{e}_4 + \mathbf{e}_3]$. We say that the Gauss map g is *conformal* if g pulls back any metric from the conformal structure on S^2_{∞} to a positive multiple of the metric on M induced by $\mathbf{x}$.

We have the following analog to Theorem 4.29 of Euclidean geometry.

Theorem 6.33. *Let $\mathbf{x} : M \to \mathbf{H}^3$ be a connected immersed surface with unit normal vector field $\mathbf{e}_3$ and mean curvature H determined by $\mathbf{e}_3$. Then its Gauss map $g : M \to S^2_{\infty}$ determined by $\mathbf{e}_3$ is conformal if and only if H is identically equal to 1, or $\mathbf{x}$ is totally umbilic.*

Proof. Let $p \in M$ and let $V \subset S^2_{\infty}$ be an open neighborhood of $g(p)$ on which there is a section $\mathbf{n} : V \to N$ of $N \to N/\mathbf{R}^{\times}$. Let $U = g^{-1}V \subset M$, an open neighborhood of p. Then $\langle d\mathbf{n}, d\mathbf{n}\rangle$ is a Riemannian metric on V in the conformal structure on S^2_{∞} (see Remark 6.6), and

$$g^*\langle d\mathbf{n}, d\mathbf{n}\rangle = \langle d(\mathbf{n}\circ g), d(\mathbf{n}\circ g)\rangle.$$

Shrinking U, if necessary, there exists a first order frame field

$$e = (\mathbf{e}_1, \mathbf{e}_2, \mathbf{e}_3, \mathbf{x}) : U \to \mathbf{SO}_+(3,1)$$

along $\mathbf{x}$ on a neighborhood of p with the given unit normal vector field $\mathbf{e}_3$. Then $g = [\mathbf{e}_4 + \mathbf{e}_3]$ implies that $\mathbf{n} \circ g = t(\mathbf{e}_4 + \mathbf{e}_3)$ for some smooth function $t : U \to \mathbf{R}^\times$, so

$$g^* \langle d\mathbf{n}, d\mathbf{n} \rangle = t^2 \langle d(\mathbf{e}_4 + \mathbf{e}_3), d(\mathbf{e}_4 + \mathbf{e}_3) \rangle,$$

since $\langle \mathbf{e}_4 + \mathbf{e}_3, \mathbf{e}_4 + \mathbf{e}_3 \rangle = 0$, so also $\langle d(\mathbf{e}_4 + \mathbf{e}_3), \mathbf{e}_4 + \mathbf{e}_3 \rangle = 0$. By (6.29),

$$d(\mathbf{e}_4 + \mathbf{e}_3) = ((1 - h_{11})\mathbf{e}_1 - h_{12}\mathbf{e}_2)\,\omega^1 + (-h_{12}\mathbf{e}_1 + (1 - h_{22})\mathbf{e}_2)\,\omega^2,$$

so $\frac{1}{t^2} g^* \langle d\mathbf{n}, d\mathbf{n} \rangle =$

$$\left((1 - h_{11})^2 + h_{21}^2\right) \omega^1 \omega^1 + \left(h_{12}^2 + (1 - h_{22})^2\right) \omega^2 \omega^2 - 2h_{12}(2 - h_{11} - h_{22})\omega^1 \omega^2.$$

Hence, g is conformal on U if and only if

$$(1 - h_{11})^2 + h_{21}^2 = h_{12}^2 + (1 - h_{22})^2 \text{ and } h_{12}(2 - h_{11} - h_{22}) = 0. \tag{6.53}$$

Simplifying, and substituting in the mean curvature $H = (h_{11} + h_{22})/2$, we reduce (6.53) to

$$(h_{11} - h_{22})(H - 1) = 0 \text{ and } h_{12}(H - 1) = 0. \tag{6.54}$$

Hence, g is conformal at a point if and only if either $H = 1$ at the point or $\mathbf{x}$ is umbilic at the point. If H is identically 1 on M, then g is conformal. If $\mathbf{x}$ is totally umbilic, then g is conformal.

Conversely, if g is conformal on M, suppose that H is not identically 1 on M. Let

$$W = \{q \in M : H(q) \neq 1\},$$

a non-empty open subset of M. Let W_0 be a connected component of W. Then g is conformal on W_0 and H is never 1 on W_0, so $\mathbf{x}$ must be totally umbilic on W_0. Therefore, H is constant, not 1, on W_0 by Theorem 6.13. Then the continuous function H must be the same constant on the closure $\overline{W_0}$. But $H = 1$ at any boundary point of W_0, which means that $W_0 = M$ and $\mathbf{x}$ is totally umbilic on M. □

Problems

6.34. Prove that a matrix $A \in \mathbf{O}(3,1)$ if and only if its columns A_1, A_2, A_3, A_4 form an orthonormal basis of $\mathbf{R}^{3,1}$.

6.35. Prove that if $\mathbf{x}, \mathbf{y} \in \mathbf{H}^3$, then $-\langle \mathbf{x}, \mathbf{y} \rangle \geq 1$.

6.36. Prove that $\mathfrak{m}$ is $\operatorname{ad}(\mathbf{O}(3))$-invariant.

6.37. A unit speed curve $\gamma(s)$ in $\mathbf{H}^3$ is a geodesic if $D_{\dot\gamma}\dot\gamma = 0$. Prove that the geodesic $\gamma(s)$ in $\mathbf{H}^3$ with initial position $\gamma(0) = \mathbf{m} \in \mathbf{H}^3$ and initial velocity the unit vector $\dot\gamma(0) = \mathbf{v} \in T_{\mathbf{m}}\mathbf{H}^3$ is

$$\gamma(s) = \cosh s\, \mathbf{m} + \sinh s\, \mathbf{v}.$$

6.38. Prove that the intrinsic distance between points $\mathbf{p}, \mathbf{q} \in \mathbf{H}^3$ is

$$d(\mathbf{p}, \mathbf{q}) = \cosh^{-1}(-\langle \mathbf{p}, \mathbf{q}\rangle) \in (0, \infty).$$

6.39. Prove that $S_r(\mathbf{m})$ is a smoothly embedded surface in $\mathbf{H}^3$ and that $\mathbf{e}_3$ is a smooth unit normal vector field on it.

6.40. Adapt the suggestions given in Problem 5.33 to prove the existence of locally defined smooth unit normal vectors along a surface immersion $\mathbf{x}$. Prove that if m is a point in M, then there exists a smooth first order frame field on some open neighborhood of m.

6.41. Prove that the Gaussian curvature of $S(a\mathbf{y} + \mathbf{v})$ is $K = a^2 - 1$.

6.42. A geodesic in $\mathbf{H}^3$ is called a *line*. Prove that given a line l in $\mathbf{H}^3$ and a point $\mathbf{y} \in \mathbf{H}^3$ not on the line l, then there exist at least two lines through $\mathbf{y}$, in the plane determined by l and $\mathbf{y}$, that never meet l.

6.43. A *plane* in $\mathbf{H}^3$ is a complete, totally geodesic, surface. Prove that for any point $\mathbf{y} \in \mathbf{H}^3$, and any unit vector $\mathbf{v} \in T_{\mathbf{y}}\mathbf{H}^3$, the totally umbilic surface $S(\mathbf{v})$ is a plane with constant unit normal vector $\mathbf{v}$, as claimed in Definition 6.12. Notice that Figure 6.1 actually shows only the plane through $\boldsymbol{\epsilon}_4$ with unit normal $\pm\boldsymbol{\epsilon}_3$ there.

6.44. Given $a \in \mathbf{R}$, $\mathbf{y} \in \mathbf{H}^3$, and unit vector $\mathbf{v} \in T_{\mathbf{y}}\mathbf{H}^3$. Prove the following:

1. $S(a\mathbf{y} + \mathbf{v})$ is a complete surface in $\mathbf{H}^3$.
2. If $A \in \mathbf{SO}_+(3,1)$, then $AS(a\mathbf{y} + \mathbf{v}) = S(aA\mathbf{y} + A\mathbf{v})$.
3. There exists an isometry $A \in \mathbf{SO}_+(3,1)$ such that $S(a\mathbf{y} + \mathbf{v}) = AS(a\boldsymbol{\epsilon}_4 + \boldsymbol{\epsilon}_3)$.
4. Conclude that $\mathbf{SO}_+(3,1)$ acts transitively on the set of oriented planes in $\mathbf{H}^3$.

6.45. Let $a \in \mathbf{R}$, and let $\mathbf{p} = a\boldsymbol{\epsilon}_4 + \boldsymbol{\epsilon}_3 \in \mathbf{R}^{3,1}$. Prove that $S(\mathbf{p}) = \mathbf{H}^3 \cap Q(\mathbf{p})$, where

$$Q(\mathbf{p}) = \{\mathbf{y} \in \mathbf{R}^{3,1} : (a^2 - 1)(y^4)^2 - 2a^2 y^4 + (y^1)^2 + (y^2)^2 = -(1 + a^2)\},$$

which as a subset of $\mathbf{R}^4$ is an ellipsoid if $|a| > 1$, a hyperboloid if $|a| < 1$, and a paraboloid if $|a| = 1$.

6.46. Prove that the geodesics in $\mathbf{B}^3$ trace out circles or lines that are perpendicular to $\partial\mathbf{B}^3$. In particular, prove that $\mathfrak{s}(\boldsymbol{\gamma}(t))$ in (6.43) is a circle or line orthogonal to $\partial\mathbf{B}^3$.

6.47. Prove Proposition 6.18. By Problem 6.46, it is sufficient to prove that through a point $\mathbf{x}' \in \mathbf{B}^3$ and a point $\mathbf{v}' \in \partial\mathbf{B}^3$, there passes a unique circle, or line, orthogonal to $\partial\mathbf{B}^3$ at $\mathbf{v}'$.

6.48. Prove that the oriented curvature spheres of $\mathbf{x}: M \to \mathbf{H}^3$ at $m \in M$ relative to the unit normal $-\mathbf{v}$ at m are the same spheres as those relative to $\mathbf{v}$, but with opposite orientations.

6.49 (Isoparametric immersions). Let M be a connected surface and let $\mathbf{x}: M \to \mathbf{H}^3$ be an immersion with constant principal curvatures a and c. Prove that either $a = c$ and $\mathbf{x}$ is totally umbilic, or $ac = 1$ and $K = 0$. Fix $0 < a < 1$, let $b = \sqrt{1-a^2}$, and integrate the structure equations to show that up to congruence by $\mathbf{SO}_+(3,1)$, the isoparametric immersion with principal curvatures a and $1/a$ is $\mathbf{x}: \mathbf{R}^2 \to \mathbf{H}^3$ given in (6.44).

6.50. Prove that a regular smooth curve $\mathbf{f}: J \to \mathbf{H}^3$ that has a Frenet frame field lies in a plane if and only if its torsion τ is identically zero. Prove that a regular smooth curve is a line if and only if it lies in a plane and its curvature is identically zero.

6.51. Given a constant $\kappa \geq 0$, solve the Frenet-Serret structure equations

$$\dot{\mathbf{f}} = \mathbf{T}, \quad \dot{\mathbf{T}} = \mathbf{f} + \kappa \mathbf{N}, \quad \dot{\mathbf{N}} = -\kappa \mathbf{T}, \quad \dot{\mathbf{B}} = 0$$

with the initial conditions $(\mathbf{T}, \mathbf{N}, \mathbf{B}, \mathbf{f})(0) = (\epsilon_1, \epsilon_2, \epsilon_3, \epsilon_4)$. Prove that the solution curve $\mathbf{f}$ satisfies

$$\mathbf{f}(\mathbf{R}) \subset S(\kappa\epsilon_4 + \epsilon_2) \cap \epsilon_3^{\perp}.$$

6.52. Find the Frenet frame field along the curve $\mathbf{f}$ of Example 6.27 and prove that it has constant curvature

$$\kappa = \frac{(b^2+1)\cosh a \ \sinh a}{L^2}$$

and constant torsion

$$\tau = \frac{b}{L^2},$$

thus proving that $\mathbf{f}$ is a helix. Prove that any helix with constant curvature κ and constant torsion τ is congruent to $\mathbf{f}$ for the appropriate values of a and b.

6.53. Fix constant $0 < a < 1$ and let $b = \sqrt{1-a^2}$. Prove that the isoparametric surface $\mathbf{x}(s,t)$ obtained in Problem 6.49 with principal curvatures a and $1/a$ is the surface of revolution obtained by rotating the profile curve

$$\sigma(s) = \epsilon_1 \frac{a}{b} + \epsilon_3 \frac{\sinh(bs)}{b} + \epsilon_4 \frac{\cosh(bs)}{b}$$

in the plane $S(\epsilon_2)$ about the line through ϵ_4 tangent to ϵ_3 there. Prove that this profile curve has constant curvature $\kappa = a$ and zero torsion. Prove that it is an ultra-circle with the same points at infinity as the axis of rotation.

6.54. Use Proposition 6.22 to prove that if $\mathbf{x} : M^2 \to \mathbf{H}^3$ is a Dupin immersion, then $\mathfrak{s} \circ \mathbf{x} : M \to \mathbf{R}^3$ is a Dupin immersion into Euclidean space. In particular, if $\mathbf{x} : M \to \mathbf{H}^3$ is the isoparametric immersion (6.44), then $\mathfrak{s} \circ \mathbf{x} : M \to \mathbf{R}^3$ is proper Dupin. Show that the coordinate curves are the lines of curvature with Euclidean principal curvatures $\tilde{a} = a$ and $\tilde{c} = \frac{1}{a}(1 + b\cosh(bs))$.

6.55. Use the formulas in Example 6.29 for the principal curvatures of the parallel transformation $\tilde{\mathbf{x}}$ of the immersion $\mathbf{x} : M^2 \to \mathbf{H}^3$ to prove a Hyperbolic Space version of Bonnet's Theorem 4.50: Suppose the mean curvature H of $\mathbf{x}$ is constant with $|H| > 1$. If $r = \frac{1}{2}\coth^{-1} H$, then the Gaussian curvature $\tilde{K}$ of $\tilde{\mathbf{x}}$ is constant, with

$$\tilde{K} = \frac{2\sqrt{H^2 - 1}}{|H| - \sqrt{H^2 - 1}}.$$

If $r = \coth^{-1} H$, then the mean curvature $\tilde{H}$ of $\tilde{\mathbf{x}}$ is constant with $\tilde{H} = -H$.

6.56 (Bonnet's Theorems). State and prove Bonnet's Uniqueness and Existence Theorems. The statement and proofs are nearly identical to those of Bonnet's Theorems 4.18 and 4.19.

6.57. Prove that the Gauss map of the connected immersion $\mathbf{x} : M \to \mathbf{H}^3$ is constant if and only if $\mathbf{x} : M \to \mathbf{H}^3$ is an open subset of a horosphere.

6.58. A horosphere M in $\mathbf{B}^3$ is a Euclidean 2-sphere contained in $\mathbf{B}^3$ except for the point $\mathbf{p}$ where it is tangent to $\partial\mathbf{B}^3$. (See the case $a = 1$ in (6.39)). Prove that any circle or line through $\mathbf{p}$ perpendicular to $\partial\mathbf{B}^3$ is also perpendicular to M at $\mathbf{p}$ and at the second point of its intersection with M. Taking the inward pointing unit normal vector field along M, prove that its Gauss map sends every point of M to $\mathbf{p}$; that is, its Gauss map is constant. Note: a horosphere in $\mathbf{H}^3$ is like a Euclidean plane in that its Gaussian curvature $K = 0$ and its Gauss map is constant, just like a plane in Euclidean space. It is not, however, totally geodesic, a property of Euclidean planes shared by Hyperbolic planes.

Chapter 7
Complex Structure

This chapter reviews complex structures on a manifold, then gives an elementary exposition of the complex structure induced on a surface by a Riemannian metric. In this way a complex structure is induced on any surface immersed into one of the space forms. Surfaces immersed into Möbius space inherit a complex structure. In all cases we use this structure to define a reduction of a moving frame to a unique frame associated to a given complex coordinate. Umbilic points do not hinder the existence of these frames, in contrast to the obstruction they can pose for the existence of second order frame fields. The Hopf invariant and the Hopf quadratic differential play a prominent role in the space forms as well as in Möbius geometry. Using the structure equations of the Hopf invariant h, the conformal factor e^u, and the mean curvature H of such frames, we give an elementary description of the Lawson correspondence between minimal surfaces in Euclidean geometry and constant mean curvature equal to one (CMC 1) surfaces in hyperbolic geometry; and between minimal surfaces in spherical geometry and CMC surfaces in Euclidean geometry.

For supplementary reading see Chern's Lecture Notes [46], Chern's book [48], or the second volume of Kobayashi-Nomizu [101].

7.1 Induced complex structure

We begin with the linear algebraic aspects of complex structures.

Definition 7.1. Let V be a vector space over $\mathbf{R}$ of dimension $2n$. A *complex structure* on V is a linear map $J : V \to V$ such that $J^2 = -\text{id}$.

Such a structure allows us to define a scalar multiplication by $\mathbf{C}$ on V by

$$(a + ib)\mathbf{v} = a\mathbf{v} + bJ\mathbf{v},$$

G.R. Jensen et al., *Surfaces in Classical Geometries*, Universitext,
DOI 10.1007/978-3-319-27076-0_7

for any $a, b \in \mathbf{R}$, which turns V into a vector space over $\mathbf{C}$ with complex dimension n. Check that $J : V \to V$ is then a complex linear map. A dual description of a complex structure on V is given by an $\mathbf{R}$-linear isomorphism

$$\varphi : V \to \mathbf{C}^n,$$

which then defines $J : V \to V$ by $J\mathbf{v} = \varphi^{-1} i \varphi(\mathbf{v})$, where i here means scalar multiplication on $\mathbf{C}^n$ by $i = \sqrt{-1}$. The $\mathbf{R}$-linear isomorphism φ is not uniquely determined by J. If $A \in \mathbf{GL}(n;\mathbf{C})$, then $A \circ \varphi : V \to \mathbf{C}^n$ defines the same J, since

$$(A\varphi)^{-1} i (A\varphi) = \varphi^{-1} A^{-1} i A \varphi = \varphi^{-1} i \varphi,$$

because A commutes with multiplication by i on $\mathbf{C}^n$. Given a complex structure J on V, a dual map $\varphi : V \to \mathbf{C}^n$ can be found as follows. Let $\mathbf{v}_1, \ldots, \mathbf{v}_n$ be a basis of V over $\mathbf{C}$, and then define $\varphi(\mathbf{v}_j) = \boldsymbol{\epsilon}_j$, for $j = 1, \ldots, n$.

Definition 7.2. An *almost complex structure* on a smooth manifold M^{2n} of real dimension $2n$ is a smooth, type $(1,1)$ tensor field J on M such that at each point $m \in M$, the linear map $J_m : T_mM \to T_mM$ is a complex structure on T_mM; that is, $J_m^2 = -\mathrm{id}$.

The local dual description of an almost complex structure on M is given on a neighborhood U of any given point of M by a set of smooth, complex valued 1-forms $\varphi^1, \ldots, \varphi^n$ on U such that, at each point $m \in U$, the map

$$\varphi_m = (\varphi^1_m, \ldots, \varphi^n_m) : T_mM \to \mathbf{C}^n$$

is an $\mathbf{R}$-linear isomorphism. This is equivalent to the condition that the set of all real and imaginary parts of $\varphi^1, \ldots, \varphi^n$ constitute a coframe field on U. Then φ defines a a smooth $(1,1)$ tensor field J on U by $J = \varphi^{-1} i \varphi$ at each point of U. Any complex linear combination $\tilde{\varphi}^j = A^j_k \varphi^k$, where $A = (A^j_k) : U \to \mathbf{GL}(n;\mathbf{C})$ is smooth, defines the same tensor field J.

For the global dual construction on M, we need to cover M with such pairs (U, φ), such that whenever $U \cap \tilde{U} \neq \emptyset$, then $\tilde{\varphi}^1, \ldots, \tilde{\varphi}^n$ must be related to $\varphi^1, \ldots, \varphi^n$ on $U \cap \tilde{U}$ by

$$\tilde{\varphi}^j = \sum_{k=1}^{n} A^j_k \varphi^k,$$

where $A = (A^j_k) : U \cap \tilde{U} \to \mathbf{GL}(n;\mathbf{C})$ is smooth.

Definition 7.3. A *complex structure* on a topological $2n$-manifold M is an atlas of charts $\{(U_a, z_a)\}_{a \in \mathscr{A}}$ such that

1. $z_a : U_a \to \mathbf{C}^n$ is a homeomorphism onto an open subset of $\mathbf{C}^n$.
2. If $U_a \cap U_b \neq \emptyset$, then

$$z_b \circ z_a^{-1} : z_a(U_a \cap U_b) \to z_b(U_a \cap U_b)$$

is holomorphic.

M with such an atlas is called a *complex manifold* of *complex dimension* n. A complex manifold of complex dimension 1 is called a *Riemann surface*.

Technically, we should add the condition that the atlas is maximal in the sense that if any chart (U,z) satisfies the first condition and is compatible with all charts in the atlas in the sense of the second condition, then (U,z) also belongs to the atlas. We do not stress the maximality, because it is easily shown that any atlas that satisfies the two stated conditions, has a unique extension to a maximal atlas. Notice that an atlas that defines a complex n-dimensional structure on M is also an atlas that defines a real analytic structure of real dimension $2n$.

A complex n-manifold induces an almost complex structure on itself. It is defined on each chart (U_a, z_a) by the smooth complex valued 1-forms $dz_a^1, \ldots, dz_1^n$. The required compatibility condition on $U_a \cap U_b \neq \emptyset$ is satisfied because of condition (2), which implies that

$$dz_b^j = \sum_{k=1}^{n} \frac{\partial z_b^j}{\partial z_a^k} dz_a^k,$$

where the map

$$\left(\frac{\partial z_b^j}{\partial z_a^k} \right) : U_a \cap U_b \to \mathbf{GL}(n;\mathbf{C})$$

is holomorphic.

The converse, however, is more complicated. If the smooth manifold M^{2n} has an almost complex structure $\{(U_a, \varphi_a = (\varphi_a^1, \ldots, \varphi_a^n))\}_{a\in\mathscr{A}}$ that is induced by a complex structure on M with atlas $\{(V_b, z_b)\}_{b\in\mathscr{B}}$, then on any nonempty intersection $U_a \cap V_b$,

$$\varphi_a^j = \sum_{k=1}^{n} A_k^j dz_b^k,$$

where $A = (A_k^j) : U_a \cap V_b \to \mathbf{GL}(n;\mathbf{C})$ is smooth. Thus

$$d\varphi_a^j = \sum_{i=1}^{n} \psi_i^j \wedge \varphi_a^i, \tag{7.1}$$

for some smooth, complex valued 1-forms ψ_i^j, which in this case are

$$\psi_i^j = \sum_{k=1}^{n} (A^{-1})_i^k dA_k^j,$$

for $j, i = 1, \ldots, n$. The set of equations (7.1) is called the *integrability condition* on the almost complex structure on M^{2n}. In general, it is not satisfied when $n > 1$. It is a celebrated result of Newlander and Nirenberg [128] that if it is satisfied, then there is a complex structure on M that induces the given almost complex structure.

An almost complex structure on a surface M automatically satisfies the integrability condition. It was first proved by Korn [102] and Lichtenstein [113] that an almost complex structure on a surface is always induced by a complex structure on the surface. In Theorem 7.4 below we will state their result. First, let us see how a Riemannian metric on an oriented surface M induces an almost complex structure on M.

Let (M,I) be a connected, oriented Riemannian surface. An orthonormal coframe field θ^1, θ^2 on $U \subset M$ is oriented, if $\theta^1 \wedge \theta^2 > 0$ at every point of U. For such a coframe field, consider the complex valued 1-form

$$\varphi = \theta^1 + i\theta^2, \qquad i = \sqrt{-1},$$

so that on U we have

$$I = \varphi\bar{\varphi},$$

where $\bar{\varphi} = \theta^1 - i\theta^2$ is the complex conjugate of φ. For each point $m \in U$, this form φ defines an $\mathbf{R}$-linear isomorphism $\varphi_m : T_m M \to \mathbf{C}$, and thus defines an almost complex structure on U. If $\tilde{\theta}^1, \tilde{\theta}^2$ is another oriented, orthonormal coframe field in U, then

$$\tilde{\theta}^1 = A^1_1\theta^1 + A^1_2\theta^2, \quad \tilde{\theta}^2 = A^2_1\theta^1 + A^2_2\theta^2,$$

for some smooth $A = (A^j_k) : U \to \mathbf{SO}(2)$, so

$$\tilde{\varphi} = \tilde{\theta}^1 + i\tilde{\theta}^2 = (A^1_1 + iA^2_1)\theta^1 + (A^1_2 + iA^2_2)\theta^2 = (A^1_1 + iA^2_1)\varphi, \tag{7.2}$$

on U, (see Problem 7.41). Hence, $\tilde{\varphi}$ and φ define the same almost complex structure on U, since $A^1_1 + iA^2_1 : U \to \mathbf{S}^1 \subset \mathbf{C}$ is smooth.

Theorem 7.4 (Korn [102]-Lichtenstein [113]). *If (M,I) is a smooth oriented Riemannian surface, then it possesses a complex structure that induces the same almost complex structure as I. For any complex coordinate chart (U,z) in the atlas of this complex structure, we have*

$$I = e^{2u} dz d\bar{z},$$

where $u : U \to \mathbf{R}$ is a smooth function. In particular, if $z = x + iy$, then $\theta^1 = e^u dx$, $\theta^2 = e^u dy$ is an oriented orthonormal coframe field in U.

For a proof in the smooth category and additional references see Chern [42] or Bers [5].

Proof. Here is a proof, given by Gauss [72], which is valid when M and I are real analytic. Let $m \in M$ and let $(U, (u, v))$ be a chart from the oriented real analytic atlas of M such that $m \in U$. We may assume $u(p) = 0 = v(p)$ and that $(u, v)(U) = D$, an open disk about the origin in $\mathbf{R}^2$. Applying the Gram-Schmidt orthonormalization process to the coframe field du, dv on U, we may construct an oriented real analytic orthonormal coframe field θ^1, θ^2 on U. Then

$$\varphi = \theta^1 + i\theta^2 = P(u,v)du + Q(u,v)dv,$$

where P and Q are complex valued functions, defined and real analytic on $D \subset \mathbf{R}^2$. Thus,

$$P(u,v) = \sum_0^\infty a_{jk} u^j v^k, \quad Q(u,v) = \sum_0^\infty b_{jk} u^j v^k,$$

where these power series have complex coefficients and are convergent on some open neighborhood of the origin of $\mathbf{R}^2$. If we extend u, v to complex variables

$$z = u + i\tilde{u}, \quad w = v + i\tilde{v},$$

then these power series with u and v replaced by z and w, respectively, converge on some open neighborhood O of the origin in $\mathbf{C}^2$. These complex power series then extend $P(u,v)$ and $Q(u,v)$ to holomorphic functions $P(z,w)$ and $Q(z,w)$ on O.

Consider the holomorphic 1-dimensional distribution defined on O by the equation

$$\psi = P(z,w)dz + Q(z,w)dw = 0.$$

It satisfies the Frobenius condition, because $d\psi \wedge \psi = 0$ on O for dimensional reasons. According to the Frobenius Theorem, (whose statement and proof is exactly the same in the holomorphic category as in the smooth real category), there exists a holomorphic chart $(z^1, z^2) : \tilde{O} \to \mathbf{C}^2$ on some open neighborhood $\tilde{O}$ of the origin of $\mathbf{C}^2$ such that the integral submanifolds in $\tilde{O}$ are given by $z^1 = c$, for any constant in the range of z^1. This implies that $\psi = f(z^1, z^2)dz^1$ on $\tilde{O}$, for some holomorphic function $f : \tilde{O} \to \mathbf{C}$.

Consider the smooth map

$$z^1 \circ (u,v) = x^1 + iy^1 : \tilde{U} \to \mathbf{C},$$

defined on an appropriate neighborhood $\tilde{U}$ of m in M. Then

$$\varphi = f(z^1(u,v), z^2(u,v))d(z^1(u,v))$$

has rank equal to 2 at every point of $\tilde{U}$, so z^1 is a smooth diffeomorphism from a neighborhood U_1 of m onto an open subset of $\mathbf{C}$, and f is never zero.

We have now shown how to construct a cover of M by open sets $\{U_a\}_{a \in \mathscr{A}}$ such that on U_a there is an oriented orthonormal coframe field θ^1_a, θ^2_a with

$$\varphi_a = \theta^1_a + i\theta^2_a = f_a dz_a, \tag{7.3}$$

where $f_a : U_a \to \mathbf{C} \setminus \{0\}$ is smooth and $z_a = x_a + iy_a : U_a \to \mathbf{C}$ is a smooth diffeomorphism onto an open subset of $\mathbf{C}$. If $U_a \cap U_b \neq \emptyset$, then on this intersection we have, by (7.2), that $\varphi_b = A_{ba}\varphi_a$ for some smooth function $A_{ba} : U_a \cap U_b \to \mathbf{S}^1 \subset \mathbf{C}$. Thus,

$$dz_b = \frac{1}{f_b}\varphi_b = \frac{A_{ba}}{f_b}\varphi_a = A_{ba}\frac{f_a}{f_b}dz_a = g\,dz_a \tag{7.4}$$

where $g : U_a \cap U_b \to \mathbf{C} \setminus \{0\}$ is smooth. This shows that z_b is a holomorphic function of z_a, by which we mean that $z_b \circ z_a^{-1} : z_a(U_a \cap U_b) \to \mathbf{C}$ is holomorphic. To see this, observe that in terms of the real and imaginary parts of z_a and z_b, (7.4) becomes

$$g(dx_a + idy_a) = dx_b + idy_b = (\frac{\partial x_b}{\partial x_a} + i\frac{\partial y_b}{\partial x_a})dx_a + (\frac{\partial x_b}{\partial y_a} + i\frac{\partial y_b}{\partial y_a})dy_a$$

Comparing the coefficients of dx_a and dy_a, we conclude that x_b, y_b satisfy the Cauchy-Riemann equations

$$\frac{\partial x_b}{\partial x_a} = \frac{\partial y_b}{\partial y_a}, \quad \frac{\partial y_b}{\partial x_a} = -\frac{\partial x_b}{\partial y_a}.$$

Thus, $\{(U_a, z_a)\}_{a\in\mathscr{A}}$ is an atlas of a complex structure on M satisfying (7.3) for each $a \in \mathscr{A}$, which shows that it induces the same almost complex structure as I. □

Definition 7.5. Local *isothermal coordinates* on a Riemannian surface M, I, is any local coordinate patch $U, (x, y)$ on M for which $I = e^{2u}(dx^2 + dy^2)$ on U, for some smooth function $u : U \to \mathbf{R}$.

The Korn-Lichtenstein Theorem 7.4 says that any point of an oriented Riemannian surface (M, I) is contained in some isothermal coordinate patch.

The Frobenius Theorem is an existence result. There is no algorithm for finding the integrating factor f_a in general, although it can be found in special circumstances, as shown in some examples below.

Definition 7.6. On a smooth manifold M, a Riemannian metric $\tilde{I}$ is *conformally related* to a Riemannian metric I if $\tilde{I} = e^{2u}I$, for some smooth function $u : M \to \mathbf{R}$. This is an equivalence relation on the set of all Riemannian metrics on M. An equivalence class is called a *conformal structure* on M.

Lemma 7.7. *Riemannian metrics I and $\tilde{I}$ on an oriented surface M induce the same complex structure on M if and only if they are conformally related.*

Proof. By the proof of the preceding theorem, the metrics I and $\tilde{I}$ induce the same complex structure on M if and only if they induce the same almost complex structures. Suppose this is the case. Then for any point $m \in M$, there exists a neighborhood U of m on which there exist oriented orthonormal coframe fields θ^1, θ^2 for I and $\tilde{\theta}^1, \tilde{\theta}^2$ for $\tilde{I}$, such that

$$\tilde{\varphi} = \tilde{\theta}^1 + i\tilde{\theta}^2 = f(\theta^1 + i\theta^2) = f\varphi$$

for some smooth function $f : U \to \mathbf{C} \setminus \{0\}$. Therefore,

$$\tilde{I} = \tilde{\varphi}\bar{\tilde{\varphi}} = |f|^2 \varphi\bar{\varphi} = |f|^2 I,$$

on U. Since $m \in M$ is arbitrary, if follows that I and $\tilde{I}$ are conformally related on M.

Conversely, suppose $\tilde{I} = e^{2u} I$ on M, where $u : M \to \mathbf{R}$ is some smooth function. Given a point $m \in M$, let U be a neighborhood of m on which there exist oriented orthonormal coframe fields θ^1, θ^2 for I and $\tilde{\theta}^1, \tilde{\theta}^2$ for $\tilde{I}$. Then

$$\tilde{\theta}^j = \sum_{k=1}^{2} A^j_k \theta^k,$$

where $A = (A^j_k) : U \to \mathbf{GL}(2, \mathbf{R})$ is smooth and $\det A > 0$ at every point of U, and

$$\tilde{\varphi} = \tilde{\theta}^1 + i\tilde{\theta}^2 = a\varphi + b\bar{\varphi},$$

where

$$a = \frac{1}{2}(A^1_1 + A^2_2 + i(A^2_1 - A^1_2)), \quad b = \frac{1}{2}(A^1_1 - A^2_2 + i(A^2_1 + A^1_2)).$$

Then

$$\tilde{I} = \tilde{\varphi}\bar{\tilde{\varphi}} = (|a|^2 + |b|^2)\varphi\bar{\varphi} + a\bar{b}\varphi\varphi + \bar{a}b\bar{\varphi}\bar{\varphi} = e^{2u}\varphi\bar{\varphi}$$

implies that $a\bar{b} = 0$ at each point of U. But, if $a = 0$ at a point, then

$$\det A = \det \begin{pmatrix} A^1_1 & A^1_2 \\ A^1_2 & -A^1_1 \end{pmatrix} \leq 0$$

at the point, which cannot happen. Therefore, $b = 0$ at every point of U, so $\tilde{\varphi} = a\varphi$ on U. It follows that I and $\tilde{I}$ induce the same almost complex structure on M. □

One consequence of this lemma is that if Riemannian metrics I and $\tilde{I}$ on a Riemann surface M both induce the given complex structure on M, then they are conformally related. The next lemma shows that any Riemann surface has a Riemannian metric that induces its complex structure.

Lemma 7.8. *If M is a Riemann surface, then there exists a Riemannian metric I on M that induces the complex structure of M.*

Proof. Cover M by complex coordinate charts $\{(U_a, z_a)\}_{a \in \mathscr{A}}$. Let $\{f_a\}_{a \in \mathscr{A}}$ be a smooth partition of unity subordinate to the open cover $\{U_a\}$. Let

$$I = \sum_{a \in \mathscr{A}} f_a |dz_a|^2,$$

which is locally a finite sum of Riemannian metrics, so is a Riemannian metric. To see that I induces the complex structure of M, let $m \in M$ and let U be a neighborhood of m that meets the support of only a finite subset of the f_a, say of $f_1, \dots, f_k$. On U,

$$I = f_1 |dz_1|^2 + \cdots + f_k |dz_k|^2 = (f_1 + f_2 \left| \frac{dz_2}{dz_1} \right|^2 + \cdots + f_k \left| \frac{dz_k}{dz_1} \right|^2) |dz_1|^2,$$

which shows that I is conformally equivalent to $|dz_1|^2$ on a neighborhood of m. Hence, I induces the same almost complex structure as is induced by the given complex structure on M. □

A Riemannian metric I on an oriented surface M defines a complex structure on M, so M is a one-dimensional complex manifold, that is, a Riemann surface. We call this the complex structure induced by I and the given orientation.

The opposite orientation of M would give rise to the almost complex structure $\theta^2 + i\theta^1 = i(\theta^1 - i\theta^2) = i\bar{\varphi}$, which is called the almost complex structure *conjugate to* φ. Its corresponding complex structure would have local complex coordinate $\bar{z}$, and thus defines the conjugate complex structure on M.

We emphasize the following important fact. If (V, w) is any complex coordinate chart in M for the complex structure defined by I, then $I = e^{2v} dw d\bar{w}$, where $v : V \to \mathbf{R}$ is smooth. To see this, let θ^1, θ^2 be an oriented orthonormal coframe field in V and let $\varphi = \theta^1 + i\theta^2$. As seen above, there is a local complex coordinate (U, z) such that $\varphi = f dz$, for some smooth function $f : U \to \mathbf{C} \setminus \{0\}$. Then w must be a holomorphic function of z and $dw = w_z dz$, where $w_z = \frac{dw}{dz}$ is never zero, so

$$I = \varphi\bar{\varphi} = |f|^2 dz d\bar{z} = \frac{|f|^2}{|w_z|^2} dw d\bar{w}$$

as claimed, with $e^{2v} = |f|^2 / |w_z|^2$. In particular, the real and imaginary parts of w are isothermal coordinates for I.

Lemma 7.9 (Adapted coframes). *Let M be a Riemann surface with a conformal metric I. If (U, z) is a complex coordinate chart in M, then there exists a unique, oriented, orthonormal coframe field θ^1, θ^2 on U such that*

$$\theta^1 + i\theta^2 = e^u dz, \tag{7.5}$$

where $u : U \to \mathbf{R}$ is smooth.

Proof. If $z = x + iy$, and if $I = e^{2u} dz d\bar{z}$, for smooth function $u : U \to \mathbf{R}$, then (7.5) implies that

$$\theta^1 = e^u dx, \quad \theta^2 = e^u dy$$

are the unique 1-forms with the desired property. □

7.2 Decomposition of forms into bidegrees

On a Riemann surface M consider a complex coordinate chart $U, z = x + iy$, where the real functions x, y are then isothermal coordinates on U. Any smooth complex valued 1-form α in M is given on U by $\alpha = p\,dx + q\,dy$, for some smooth, complex valued functions p and q on U. Substitution of

$$dx = \frac{1}{2}(dz + d\bar{z}), \quad dy = \frac{1}{2i}(dz - d\bar{z})$$

into the expression for α shows that on U the form α is a linear combination of dz and $d\bar{z}$, namely,

$$\alpha = a\,dz + b\,d\bar{z}, \tag{7.6}$$

for some smooth, complex valued functions $a = \frac{1}{2}(p - iq)$ and $b = \frac{1}{2}(p + iq)$ on U.

Definition 7.10. A complex valued 1-form α has *bidegree* $(1,0)$ if it is a multiple of dz on U. It has *bidegree* $(0,1)$ if it is a multiple of $d\bar{z}$.

These bidegrees are well-defined, since if $\tilde{U}, w$ is another complex coordinate chart, then on $U \cap \tilde{U}$ we have $dw = w'dz$, where $w' = \frac{dw}{dz}$, and $d\bar{w} = \overline{w'}d\bar{z}$. Then (7.6) shows that any complex valued 1 form α decomposes into a sum of a bidegree $(1,0)$ form and a bidegree $(0,1)$ form. We call the summands the bidegree $(1,0)$ and $(0,1)$ parts, respectively, of α, and we call this sum the *decomposition by bidegree* of α.

Notice that if θ^1, θ^2 is an oriented orthonormal coframe field for a conformal metric I on M, then $\varphi = \theta^1 + i\theta^2 = g\,dz$, for some nowhere zero complex function g. Therefore, φ has bidegree $(1,0)$ and any complex valued 1-form can be written as a linear combination of φ and $\bar{\varphi}$, and the part containing φ is the bidegree $(1,0)$ part of the form, while the part containing $\bar{\varphi}$ is the bidegree $(0,1)$ part.

For a complex valued function f we denote the decomposition by bidegree of its differential by

$$df = \partial f + \bar{\partial} f.$$

In terms of the local complex coordinate z, the differential operators ∂ and $\bar{\partial}$ are given by

$$\partial f = f_z dz, \qquad \bar{\partial} f = f_{\bar{z}} d\bar{z}, \tag{7.7}$$

where $z = x + iy$. Then $dz = dx + i dy$ and $d\bar{z} = dx - i dy$ give us

$$\begin{aligned} df &= f_z dz + f_{\bar{z}} d\bar{z} = (f_z + f_{\bar{z}})dx + i(f_z - f_{\bar{z}})dy \\ &= f_x dx + f_y dy. \end{aligned}$$

Therefore,

$$f_x = f_z + f_{\bar{z}}, \quad f_y = i(f_z - f_{\bar{z}}),$$

which can be solved for f_z and $f_{\bar{z}}$ to give

$$f_z = \frac{1}{2}(f_x - if_y), \quad f_{\bar{z}} = \frac{1}{2}(f_x + if_y).$$

This motivates us to define operators

$$\frac{\partial}{\partial z} = \frac{1}{2}(\frac{\partial}{\partial x} - i\frac{\partial}{\partial y}), \quad \frac{\partial}{\partial \bar{z}} = \frac{1}{2}(\frac{\partial}{\partial x} + i\frac{\partial}{\partial y}), \tag{7.8}$$

so that

$$f_z = \frac{\partial f}{\partial z}, \qquad f_{\bar{z}} = \frac{\partial f}{\partial \bar{z}}.$$

For the complex valued 1-form $\alpha = a\,dz + b\,d\bar{z}$, where a and b are complex valued smooth functions, we define

$$\partial\alpha = b_z dz \wedge d\bar{z}, \qquad \bar{\partial}\alpha = a_{\bar{z}} d\bar{z} \wedge dz,$$

so that $d\alpha = \partial\alpha + \bar{\partial}\alpha$. Thus,

$$d = \partial + \bar{\partial}$$

on functions and 1-forms and $d^2 = 0$ implies that

$$\partial^2 = 0 = \bar{\partial}^2, \qquad \partial\bar{\partial} + \bar{\partial}\partial = 0.$$

Then, for the complex valued function f,

$$d\bar{\partial}f = \partial\bar{\partial}f = \frac{1}{4}(f_{xx} + f_{yy})dz \wedge d\bar{z}.$$

A complex valued 1-form $\alpha = f(z)dz$ of bidegree (1,0) is called *holomorphic* or an *abelian differential* if

$$\bar{\partial}\alpha = 0.$$

Observe that this condition is equivalent to the coefficient function $f(z)$ being holomorphic. A function f is harmonic if

$$\partial\bar{\partial}f = 0,$$

which we can see above is equivalent to the usual notion of f being a harmonic function of x and y, namely, that $f_{xx} + f_{yy} = 0$.

Let N be a Riemann surface with a complex structure defined by the form φ_N and the local complex coordinate w, so that $\varphi_N = gdw$, for some smooth function g. A mapping

$$f : M \to N$$

is *holomorphic* if it is given by a holomorphic function $w = w(z)$, where z is a local complex coordinate in M. Equivalently, f is holomorphic if $f^*\varphi_N$ is a multiple of φ_M.

Riemann surfaces M and N are *isomorphic* or *conformally equivalent* if there exists a holomorphic diffeomorphism $f : M \to N$ whose inverse map is also holomorphic. Such a map is called *biholomorphic*. The name *conformally equivalent* comes from the fact that if I_M is a metric on M giving its complex structure and I_N is a metric on N giving its complex structure, then a map $f : M \to N$ is biholomorphic if and only if it is diffeomorphic and $f^*I_N = kI_M$, for some positive smooth function k on M.

7.2.1 Curvature in terms of the complex coordinate

If $z = x + iy$ is a local complex coordinate for M, I, then $I = e^{2u}(dx^2 + dy^2)$ for some real valued, smooth function u. The function e^u is called the *conformal factor* of I relative to z. It follows that

$$\theta^1 = e^u dx, \quad \theta^2 = e^u dy$$

is an orthonormal coframe field in M. Taking the exterior derivative of these equations, one finds that the corresponding Levi-Civita connection form is

$$\omega^1_2 = u_y dx - u_x dy, \tag{7.9}$$

from which it follows that the Gaussian curvature is given by

$$K = -e^{-2u}\Delta u, \tag{7.10}$$

where

$$\Delta = \frac{\partial^2}{\partial x^2} + \frac{\partial^2}{\partial y^2} = 4\frac{\partial^2}{\partial z \partial \bar{z}}$$

is the Euclidean Laplacian.

7.3 Riemann surface examples

Example 7.11 (Complex plane). On $\mathbf{R}^2$ with Euclidean metric $I = dx^2 + dy^2$ an oriented orthonormal coframe field is given by $\theta^1 = dx$, $\theta^2 = dy$. Then

$$\varphi = dx + idy = d(x + iy)$$

so that $z = x + iy$ is the complex structure on $\mathbf{R}^2$ induced by the Euclidean metric. Its conformal factor is 1 with respect to z. The resulting Riemann surface is the complex plane $\mathbf{C}$.

Example 7.12 (Poincaré disk). The open unit disk

$$\mathbf{D} = \{z \in \mathbf{C} : |z| < 1\}$$

is an open complex submanifold of $\mathbf{C}$. Given the conformal Riemannian metric

$$I = \frac{4(dzd\bar{z})}{(1 - |z|^2)^2},$$

it is called the *Poincaré disk*. If we write $z = x + iy$, then $\theta^1 = e^u dx$, $\theta^2 = e^u dy$ is the oriented orthonormal coframe field on $\mathbf{D}$ adapted to z, where $u = \log \frac{2}{1-|z|^2} : \mathbf{D} \to \mathbf{R}$. By (7.9), the Levi-Civita connection form relative to θ^1, θ^2 is

$$\omega_2^1 = 2\frac{ydx - xdy}{1 - |z|^2},$$

and by (7.10) the Gaussian curvature of I is $K = -1$. The Lie group

$$\begin{aligned}\mathbf{SU}(1,1) &= \{A \in \mathbf{GL}(2,\mathbf{C}) : {}^t\!\bar{A} I_{1,1} A = I_{1,1},\ \det A = 1\} \\ &= \left\{A = \begin{pmatrix} a & \bar{b} \\ b & \bar{a} \end{pmatrix} : a, b \in \mathbf{C},\ |a|^2 - |b|^2 = 1\right\}\end{aligned}$$

acts as holomorphic transformations on $\mathbf{D}$ by

$$w = Az = \frac{az + \bar{b}}{bz + \bar{a}},$$

since, if $|z| < 1$, then $|w| < 1$. Here $I_{1,1} = \begin{pmatrix} 1 & 0 \\ 0 & -1 \end{pmatrix}$. Moreover,

$$dw = \frac{dz}{(bz + \bar{a})^2},$$

so

$$\frac{4dwd\bar{w}}{(1-|w|^2)^2} = \frac{4dzd\bar{z}}{(1-|z|^2)^2}$$

shows that A is an isometry. The action of A^{-1} is the inverse transformation. This action of $\mathbf{SU}(1,1)$ on $\mathbf{D}$ is transitive, since if $z \in \mathbf{D}$, then $A0 = z$, when A has the entries

$$a = \frac{1}{\sqrt{1-|z|^2}}, \quad b = a\bar{z}.$$

Its isotropy subgroup at $0 \in \mathbf{D}$ is

$$\left\{ \begin{pmatrix} a & 0 \\ 0 & \bar{a} \end{pmatrix} : |a| = 1 \right\} \cong \mathbf{S}^1 \cong \mathbf{SO}(2).$$

Example 7.13 (Hyperbolic plane: upper half-plane model). The open subset

$$\mathbf{H}^2 = \{z = x + iy \in \mathbf{C} : y > 0\}$$

is an open complex submanifold of $\mathbf{C}$. Given the conformal Riemannian metric

$$I = \frac{dzd\bar{z}}{y^2},$$

$\mathbf{H}^2$ is the *upper half-plane model of the hyperbolic plane.* The conformal factor is $e^u = 1/y$, so by (7.10), the Gaussian curvature is $K = -1$. The *special linear group*

$$\mathbf{SL}(2,\mathbf{R}) = \left\{ A = \begin{pmatrix} a & c \\ b & d \end{pmatrix} : a,b,c,d \in \mathbf{R},\ ad - bc = 1 \right\}$$

acts as holomorphic transformations on $\mathbf{H}^2$ by

$$w = Az = \frac{az+c}{bz+d},$$

since, if $y = \Im z > 0$, then

$$\Im w = \frac{y}{|bz+d|^2} > 0.$$

Moreover, from

$$dw = \frac{dz}{(bz+d)^2},$$

we compute

$$\frac{dwd\bar{w}}{\frac{y^2}{|bz+d|^4}} = \frac{dzd\bar{z}}{y^2}$$

to see that A acts as an isometry. The action is transitive, for if $z = x + iy \in \mathbf{H}^2$, then $Ai = z$, if

$$a = \sqrt{y}, \quad b = 0, \quad c = x\sqrt{y}, \quad d = 1\sqrt{y}.$$

The isotropy subgroup at i is

$$\left\{ \begin{pmatrix} a & -b \\ b & a \end{pmatrix} : a^2 + b^2 = 1 \right\} \cong \mathbf{SO}(2).$$

Example 7.14 (Riemann sphere). Stereographic projections from the north and south poles onto the equatorial plane define complex coordinate charts on $\mathbf{S^2}$ that make it into a Riemann surface called the Riemann sphere. For this it is convenient to identify the $\boldsymbol{\epsilon}_1\boldsymbol{\epsilon}_2$-plane with $\mathbf{C}$, by $x\boldsymbol{\epsilon}_1 + y\boldsymbol{\epsilon}_2 = x + iy$. On the unit sphere $\mathbf{S}^2 = \{\mathbf{x} \in \mathbf{R}^3 : |\mathbf{x}| = 1\}$ let I be the induced metric. That is, at a point $\mathbf{x} \in \mathbf{S}^2$, the tangent space is

$$T_{\mathbf{x}}\mathbf{S^2} = \{\mathbf{y} \in \mathbf{R}^3 : \mathbf{y} \cdot \mathbf{x} = 0\}$$

and I at $\mathbf{x}$ is the Euclidean dot product on this subspace of $\mathbf{R}^3$. Thus, at the point $\mathbf{x} = x\boldsymbol{\epsilon}_1 + y\boldsymbol{\epsilon}_2 + z\boldsymbol{\epsilon}_3 \in \mathbf{S^2}$, so $x^2 + y^2 + z^2 = 1$, we have $I = dx^2 + dy^2 + dz^2$. This metric induces the following complex structure on $\mathbf{S^2}$. If $U = \mathbf{S^2} \setminus \{\boldsymbol{\epsilon}_3\}$, then stereographic projection from $\boldsymbol{\epsilon}_3$ is

$$\mathfrak{s} : U \to \mathbf{C}, \quad \mathfrak{s}(x, y, z) = \frac{x + iy}{1 - z},$$

with inverse transformation

$$\mathfrak{s}^{-1}(u + iv) = \frac{1}{1 + u^2 + v^2}(2u\boldsymbol{\epsilon}_1 + 2v\boldsymbol{\epsilon}_2 + (u^2 + v^2 - 1)\boldsymbol{\epsilon}_3).$$

From the calculation

$$d\mathfrak{s} = \frac{(1 - z)(dx + idy) + (x + iy)dz}{(1 - z)^2},$$

we get

$$dx^2 + dy^2 + dz^2 = I = \frac{4d\mathfrak{s}d\bar{\mathfrak{s}}}{(1 + |\mathfrak{s}|^2)^2}, \tag{7.11}$$

which shows that I is conformal on the chart $(U,\mathfrak{s})$. For a chart covering the south pole, let $V = \mathbf{S^2} \setminus \{-\epsilon_3\}$, and let

$$\mathfrak{t} : V \to \mathbf{C}, \quad \mathfrak{t}(x,y,z) = \frac{x - iy}{1+z}.$$

On $\mathfrak{s}(U \cap V) = \mathbf{C} \setminus \{0\}$ we have

$$\mathfrak{t} \circ \mathfrak{s}^{-1}(u+iv) = \mathfrak{t}(\frac{2u}{u^2+v^2+1}, \frac{2v}{u^2+v^2+1}, \frac{u^2+v^2-1}{u^2+v^2+1}) = \frac{1}{u+iv},$$

which is holomorphic, and thus these two charts define a complex structure on $\mathbf{S^2}$. It is a useful exercise to verify that $\mathfrak{s} \circ \mathfrak{t}^{-1}$ is also holomorphic, and that I is conformal in the chart $(V, \mathfrak{t})$ also.

Example 7.15 (Complex projective space). $\mathbf{CP}^n$ is the set of all one-dimensional complex subspaces of $\mathbf{C}^{n+1}$. Such a subspace is denoted $[\mathbf{x}]$, where $\mathbf{x}$ is any nonzero vector in it. Then $[\mathbf{y}] = [\mathbf{x}]$ if and only if $\mathbf{y} = c\mathbf{x}$, for some nonzero complex number c. Let

$$\pi : \mathbf{C}^{n+1} \setminus \{\mathbf{0}\} \to \mathbf{CP}^n, \quad \pi(\mathbf{x}) = [\mathbf{x}],$$

be the projection mapping any nonzero vector to the one-dimensional subspace containing it. A complex structure is defined on $\mathbf{CP}^n$ by the collection of coordinate charts

$$\{(U_k, \zeta_k) : k = 0, \ldots, n\},$$

where

$$U_k = \{[z^0, \ldots, z^n] : z^k \neq 0\}$$

and

$$\zeta_k : U_k \to \mathbf{C}^n, \quad [z^0, \ldots, z^n] \mapsto \zeta_k = (\frac{z^0}{z^k}, \ldots, \hat{k}, \ldots, \frac{z^n}{z^k}),$$

where $\hat{k}$ means omit that entry. It is an elementary exercise to show that these charts define a complex structure on $\mathbf{CP}^n$ for which $\pi : \mathbf{C}^{n+1} \setminus \{\mathbf{0}\} \to \mathbf{CP}^n$ is holomorphic. There is a special Riemannian metric on $\mathbf{CP}^n$, called the Fubini-Study metric (see Fubini [70] and Study [156]), defined in the chart (U_k, ζ_k) by

$$I = 4\frac{(1+|\zeta_k|^2)(\sum_{j\neq k} d\zeta_k^j d\bar{\zeta}_k^j) - (\sum_{j\neq k} \bar{\zeta}_k^j d\zeta_k^j)(\sum_{j\neq k} \zeta_k^j d\bar{\zeta}_k^j)}{(1+|\zeta_k|^2)^2}.$$

See Kobayashi-Nomizu [101] for details. In this book we will make frequent use of $\mathbf{CP}^1$ and occasional use of $\mathbf{CP}^2$. For the present example we shall restrict our attention to $\mathbf{CP}^1$. Now there are two complex coordinate charts, which we write

$$(U_0, z), \quad z : U_0 \to \mathbf{C}, \quad [z^0, z^1] \mapsto z = z^1/z^0,$$

and

$$(U_1, w), \quad w : U_1 \to \mathbf{C}, \quad [z^0, z^1] \mapsto w = z^0/z^1.$$

On $U_0 \cap U_1$ we have $w = 1/z$. On (U_0, z) the Fubini-Study metric is

$$I = 4\frac{(1+|z|^2)(dzd\bar{z}) - (\bar{z}dz)(zd\bar{z})}{(1+|z|^2)^2} = \frac{4dzd\bar{z}}{(1+|z|^2)^2}, \tag{7.12}$$

and it has an analogous expression on (U_1, w). Thus, I is a conformal metric on the Riemann surface $\mathbf{CP}^1$. The conformal factor is $e^u = 2/(1+|z|^2)$, so its Gaussian curvature by (7.10) is $K = 1$. By continuity of K, its value at the point $[0,1]$ must also be 1. The *special unitary group*

$$\mathbf{SU}(2) = \left\{ A = \begin{pmatrix} a & -\bar{b} \\ b & \bar{a} \end{pmatrix} : a, b \in \mathbf{C},\ |a|^2 + |b|^2 = 1 \right\}$$

acts on $\mathbf{CP}^1$ as holomorphic and isometric transformations. The action comes from the linear action of $\mathbf{SU}(2)$ on $\mathbf{C}^2$,

$$A \begin{bmatrix} z^0 \\ z^1 \end{bmatrix} = \begin{bmatrix} az^0 - \bar{b}z^1 \\ bz^0 + \bar{a}z^1 \end{bmatrix}.$$

Its local expression in the chart (U_0, z) is then

$$\tilde{z} = Az = \frac{az - \bar{b}}{bz + \bar{a}},$$

which is holomorphic. By elementary calculations,

$$d\tilde{z} = \frac{dz}{(bz + \bar{a})^2},$$

and then

$$4\frac{d\tilde{z}d\bar{\tilde{z}}}{(1+|\tilde{z}|^2)^2} = 4\frac{dzd\bar{z}}{(1+|z|^2)^2}$$

shows that A acts as an isometry. The action is transitive on $\mathbf{CP}^1$, for if $\mathbf{z} = z^0\epsilon_1 + z^1\epsilon_2$ represents a point in $\mathbf{CP}^1$, then $|\mathbf{z}| > 0$, so we can let

$$a = z^0/|\mathbf{z}|,\ b = z^1/|\mathbf{z}|, \quad A = \begin{pmatrix} a & -\bar{b} \\ b & \bar{a} \end{pmatrix} \in \mathbf{SU}(2),$$

and then

$$A[\epsilon_1] = [a\epsilon_1 + b\epsilon_2] = [\mathbf{z}].$$

The isotropy subgroup at $[\epsilon_1]$ is

$$\left\{ \begin{pmatrix} a & 0 \\ 0 & \bar{a} \end{pmatrix} : |a| = 1 \right\} \cong \mathbf{S}^1 \cong \mathbf{SO}(2).$$

Example 7.16. [$\mathbf{S^2} \cong \mathbf{CP}^1$] The factorization of $x^2 + y^2 + z^2 = 1$ into

$$(x + iy)(x - iy) = (1 - z)(1 + z)$$

implies

$$\frac{x + iy}{1 - z} = \frac{1 + z}{x - iy}$$

for any point $\mathbf{x} = x\epsilon_1 + y\epsilon_2 + z\epsilon_3 \in \mathbf{S^2}$. This allows us to define the map

$$F : \mathbf{S^2} \to \mathbf{CP}^1, \quad F(\mathbf{x}) = \begin{bmatrix} x + iy \\ 1 - z \end{bmatrix} = \begin{bmatrix} 1 + z \\ x - iy \end{bmatrix}, \tag{7.13}$$

whose inverse is

$$F^{-1} \begin{bmatrix} p \\ q \end{bmatrix} = \frac{1}{|p|^2 + |q|^2} (2\Re(p\bar{q})\epsilon_1 + 2\Im(p\bar{q})\epsilon_2 + (|p|^2 - |q|^2)\epsilon_3).$$

This map is holomorphic, as can be seen from calculations like the following. Let $(U, \mathfrak{s})$ and (U_0, z) be the complex charts on $\mathbf{S^2}$ and $\mathbf{CP}^1$ defined in Examples 7.14 and 7.15, respectively. Then

$$z \circ F \circ \mathfrak{s}^{-1} = \mathrm{id} : \mathbf{C} \to \mathbf{C}.$$

It is an isometry, because the Fubini-Study metric has the same expression (7.12) in the chart (U_0, z) as the standard metric on $\mathbf{S^2}$ has in (7.11) in the chart $(\mathbf{S^2} \setminus \{\epsilon_3\}, \mathfrak{s})$. In Example 7.17 we shall examine how F relates the action of $\mathbf{SO}(3)$ on $\mathbf{S^2}$ to the action of $\mathbf{SU}(2)$ on $\mathbf{CP}^1$.

Example 7.17 (Double cover $\Sigma : \mathbf{SU}(2) \to \mathbf{SO}(3)$). Recall the isometric, biholomorphic map $F : \mathbf{S^2} \to \mathbf{CP}^1$ defined in Example 7.16. For any $A \in \mathbf{SU}(2)$, the composition of isometries

$$\Sigma(A) = F^{-1} \circ A \circ F : \mathbf{S^2} \to \mathbf{S^2}$$

is an isometry. If $A, B \in \mathbf{SU}(2)$, then $\Sigma(AB) = \Sigma(A) \circ \Sigma(B)$, and $\Sigma(I_2)$ is clearly the identity map on $\mathbf{S^2}$. Thus, Σ is a group homomorphism and it is two-to-one since $\Sigma(A) = I_3$, the identity map of $\mathbf{S^2}$, if and only if $A = F \circ I_3 \circ F^{-1}$ is the identity map on $\mathbf{CP}^1$, which requires that $A = \pm I_2$. Any $A \in \mathbf{SU}(2)$ has the form

$$A = \begin{pmatrix} m & n \\ -\bar{n} & \bar{m} \end{pmatrix},$$

for any $m, n \in \mathbf{C}$ such that $|m|^2 + |n|^2 = 1$. By an elementary, but lengthy, calculation we get

$$\Sigma(A) = \begin{pmatrix} \Re(m^2 - n^2) & -\Im(m^2 + n^2) & -2\Re(mn) \\ \Im(m^2 - n^2) & \Re(m^2 + n^2) & -2\Im(mn) \\ 2\Re(m\bar{n}) & -2\Im(m\bar{n}) & |m|^2 - |n|^2 \end{pmatrix} \in \mathbf{SO}(3).$$

So, $\Sigma : \mathbf{SU}(2) \to \mathbf{SO}(3)$ is smooth, and thus is a Lie group homomorphism. Any $X \in \mathfrak{su}(2)$ has the form

$$X = \begin{pmatrix} it & r + is \\ -r + is & -it \end{pmatrix},$$

for any $r, s, t \in \mathbf{R}$. Assuming $A(t)$ is a curve in $\mathbf{SU}(2)$ with $A(0) = I_2$ and $\dot{A}(0) = X$, it is elementary to calculate the induced Lie algebra homomorphism

$$\Sigma_* : \mathfrak{su}(2) \to \mathfrak{o}(3), \quad \Sigma_* X = \begin{pmatrix} 0 & -2t & -2r \\ 2t & 0 & -2s \\ 2r & 2s & 0 \end{pmatrix}.$$

Hence, Σ_* is a Lie algebra isomorphism. Since $\mathbf{SU}(2)$ and $\mathbf{SO}(3)$ are both compact and connected, Σ is a two-to-one Lie group covering homomorphism.

Example 7.18 (Surface of revolution metrics). Let J, K be intervals in $\mathbf{R}$. On a domain $U = J \times K \subset \mathbf{R}^2$ consider a metric of the form

$$I = w(u)^2 du^2 + f(u)^2 dv^2 \tag{7.14}$$

where $f(u)$ and $w(u)$ are smooth, positive functions on J. The induced metric on a surface of revolution has this form, as shown in (4.53) of Example 4.40.

Then $\theta^1 = w(u)du$, $\theta^2 = f(u)dv$ is an oriented orthonormal coframe field in U and $\varphi = w(u)du + if(u)dv$. In general, if $Mdu + Ndv$ is a differential form such that $(M_v - N_u)/N$ depends only on u, then there is an integrating factor depending only on u and its logarithm is given by the integral of this expression. In our case, this expression is $-f'(u)/f(u)$, which depends only on u, and thus an integrating factor is given by $c/f(u)$, for any constant $c \neq 0$. Taking $c = 1$, we see that $\frac{1}{f(u)}\varphi$ is exact,

$$\frac{1}{f(u)}\varphi = \frac{w(u)}{f(u)}du + idv = dz$$

where, up to addition of a complex constant,

$$z = \int \frac{w(u)}{f(u)}du + iv. \tag{7.15}$$

It is the complex coordinate on U induced by this metric, because $I = \varphi\bar{\varphi} = f(u)^2 dz d\bar{z}$.

Example 7.19 (Polar coordinates). On $\mathbf{R}^2$ with its standard coordinates x, y, polar coordinates r, θ are defined on the complement of the negative x-axis by $x = r\cos\theta$, $y = r\sin\theta$, where $r > 0$ and $-\pi < \theta < \pi$. The Euclidean metric is then

$$I = dx^2 + dy^2 = dr^2 + r^2 d\theta^2,$$

which has the form of (7.14). The complex coordinate given in (7.15) is $w = \log r + i\theta$, which is $\log z$, where $z = x + iy = re^{i\theta}$.

Example 7.20 (Tori). The Euclidean plane $\mathbf{R}^2$ with its metric $I = dx^2 + dy^2$ has the complex structure of the complex plane $\mathbf{C}$ with $z = x + iy$. Fix two complex numbers σ and τ which are linearly independent over $\mathbf{R}$, which is the case if and only if $\Im(\sigma/\tau) \neq 0$, where $\Im w$ denotes the imaginary part of the complex number w. Consider the *lattice* in $\mathbf{C}$ generated by σ and τ

$$\Gamma(\sigma, \tau) = \{m\sigma + n\tau : m, n \in \mathbf{Z}\},$$

which is an abelian group with the operation of addition of complex numbers. For example, the square lattice is $\Gamma(1, i) = \mathbf{Z} \times i\mathbf{Z}$. Then Γ acts on $\mathbf{C}$ properly discontinuously by

$$\Gamma \times \mathbf{C} \to \mathbf{C}, \quad (\gamma, z) \mapsto z + \gamma.$$

The quotient manifold $M = \mathbf{C}/\Gamma$ is then a smooth surface, which is topologically a torus. For details see, for example, Boothby [16, Section III.8]. The metric of $\mathbf{C}$ descends to M by the projection map $\pi : \mathbf{C} \to M$, because the metric is invariant under the action of Γ on $\mathbf{C}$. The complex structure induced on M by this metric is such that the projection π is a holomorphic map. A local complex coordinate chart

(U,z) in M is just such a chart in $\mathbf{C}$ such that U is contained within a fundamental domain of the action of Γ. That is, for any point $z \in U$ and any element $\gamma \neq 1 \in \Gamma$, then $\gamma + z \notin U$.

Two different lattices can define tori which are conformally equivalent. This is investigated in Problems 7.44 and 7.45.

Example 7.21 (Compact surfaces). An oriented compact surface is characterized topologically by its *genus*, which is one for a torus, and is three, respectively, four in the surfaces shown in Figure 7.1. The complex structure induced on these figures depends on the induced metric, which depends on the actual immersion. The induced complex structure is not a *visual* concept. A conformal transformation of $\mathbf{R}^3$ preserves the induced complex structure.

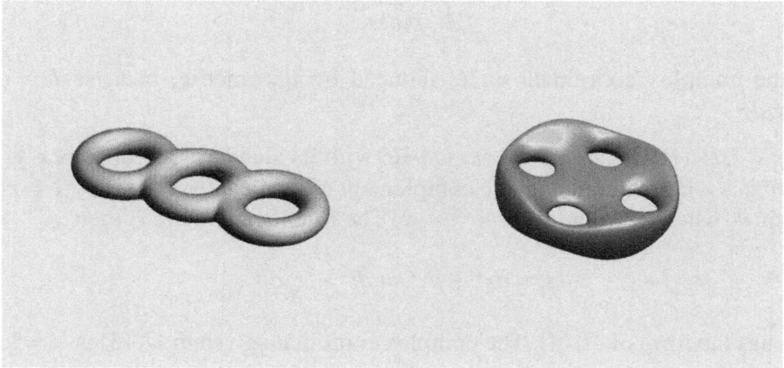

Fig. 7.1 Connected sum of three rotational tori on left, a genus 4 surface on right.

A nonorientable compact surface has a two-to-one covering by an oriented compact surface. It can be immersed into $\mathbf{R}^3$, but not embedded (see, for example, E. Lima [115]). The sphere $\mathbf{S^2}$ is a double cover of the real projective plane $\mathbf{RP}^2$, which is nonorientable. W. Boy [18] found an immersion of $\mathbf{RP}^2$ known as *Boy's surface*. See Hilbert and Cohn-Vossen [89, pp. 317–321] for a detailed explanation of Boy's surface. The left side of Figure 7.2 shows Boy's surface. Figure 14.1 shows a transparent version of Boy's surface to reveal some of its self intersections.

The torus is a double cover of the *Klein bottle*, shown on the right side of Figure 7.2 in a transparent version revealing its self intersections.

7.4 Adapted frames in space forms

Denote the three space forms by $\mathbf{S}_\epsilon$, for $\epsilon \in \{0,+,-\}$, where

$$\mathbf{S}_0 = \mathbf{R}^3, \quad \mathbf{S}_+ = \mathbf{S}^3, \quad \mathbf{S}_- = \mathbf{H}^3, \tag{7.16}$$

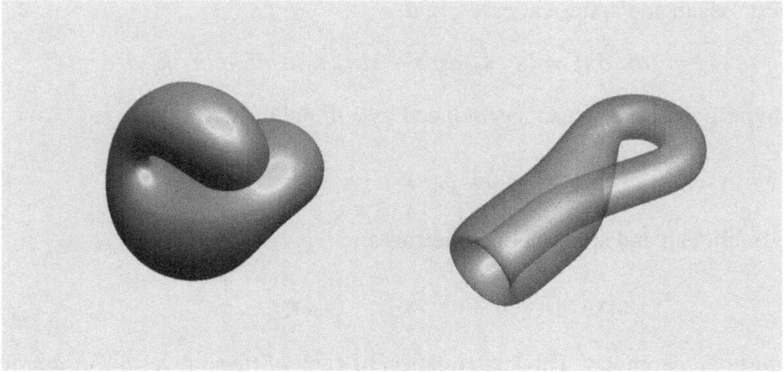

Fig. 7.2 Left: Boy's immersion of $\mathbf{RP}^2$ into $\mathbf{R}^3$. Right: Klein bottle.

with its group of isometries denoted $\mathbf{G}_\epsilon$, where

$$\mathbf{G}_0 = \mathbf{E}(3), \quad \mathbf{G}_+ = \mathbf{O}(4), \quad \mathbf{G}_- = \mathbf{O}_+(3,1).$$

Let $\mathbf{x} : M \to \mathbf{S}_\epsilon$ be a smooth immersion of an oriented surface M. The complex structure on M is that induced by I, the Riemannian metric pulled back by $\mathbf{x}$ to M. When $\epsilon \in \{0,+\}$, then $I = d\mathbf{x} \cdot d\mathbf{x}$, for the standard dot product on $\mathbf{R}^3$ and $\mathbf{R}^4$, and when $\epsilon = -$, then $I = \langle d\mathbf{x}, d\mathbf{x} \rangle$, for the standard Minkowski inner product on $\mathbf{R}^{3,1}$.

Definition 7.22. A first order frame field along $\mathbf{x} : M \to \mathbf{S}_\epsilon$ is *adapted* to the complex coordinate chart (U,z) in M, if it takes values in the connected component of the identity of G_ϵ and

$$\omega^1 + i\omega^2 = e^u dz \tag{7.17}$$

for some smooth real valued function u on U, where $d\mathbf{x} = \omega^1 \mathbf{e}_1 + \omega^2 \mathbf{e}_2$. We call e^u the *conformal factor* of $\mathbf{x}$ relative to z.

Lemma 7.23. *Let (U,z) be a complex coordinate chart in M for the complex structure induced by the first fundamental form I of an immersion $\mathbf{x} : M \to \mathbf{S}_\epsilon$. There exists a unique frame field $(\mathbf{x}, e) : U \to G_\epsilon$ adapted to (U,z).*

Proof. The adapted frame field is determined by the complex coordinate z by taking derivatives as follows. Using the operators in (7.8), we have

$$d\mathbf{x} = \mathbf{x}_z dz + \mathbf{x}_{\bar{z}} d\bar{z}. \tag{7.18}$$

This complex coordinate z is induced by the Riemannian metric I if and only if $I = e^{2u} dz d\bar{z}$, for some smooth $u : U \to \mathbf{R}$. But

$$I = d\mathbf{x} \cdot d\mathbf{x} = \mathbf{x}_z \cdot \mathbf{x}_z dz dz + 2\mathbf{x}_z \cdot \mathbf{x}_{\bar{z}} dz d\bar{z} + \mathbf{x}_{\bar{z}} \cdot \mathbf{x}_{\bar{z}} d\bar{z} d\bar{z}$$

in the Euclidean and spherical cases, and

$$I = \langle d\mathbf{x}, d\mathbf{x}\rangle = \langle \mathbf{x}_z, \mathbf{x}_z\rangle dz dz + 2\langle \mathbf{x}_z, \mathbf{x}_{\bar{z}}\rangle dz d\bar{z} + \langle \mathbf{x}_{\bar{z}}, \mathbf{x}_{\bar{z}}\rangle d\bar{z} d\bar{z}$$

in the hyperbolic case. Thus, z is induced by I if and only if

$$\mathbf{x}_z \cdot \mathbf{x}_z = 0 = \mathbf{x}_{\bar{z}} \cdot \mathbf{x}_{\bar{z}}, \quad \mathbf{x}_z \cdot \mathbf{x}_{\bar{z}} = \frac{1}{2} e^{2u}, \tag{7.19}$$

for the Euclidean and spherical geometries and

$$\langle \mathbf{x}_z, \mathbf{x}_z\rangle = 0 = \langle \mathbf{x}_{\bar{z}}, \mathbf{x}_{\bar{z}}\rangle, \quad \langle \mathbf{x}_z, \mathbf{x}_{\bar{z}}\rangle = \frac{1}{2} e^{2u}, \tag{7.20}$$

for hyperbolic geometry. The last equation of each of these gives a formula for the conformal factor e^u relative to (U, z). If we let

$$\mathbf{e}_1 = e^{-u}(\mathbf{x}_z + \mathbf{x}_{\bar{z}}), \quad \mathbf{e}_2 = i e^{-u}(\mathbf{x}_z - \mathbf{x}_{\bar{z}}), \tag{7.21}$$

then these equations can be solved for $\mathbf{x}_z$ and $\mathbf{x}_{\bar{z}}$ in terms of $\mathbf{e}_1$ and $\mathbf{e}_2$,

$$\mathbf{x}_z = \frac{1}{2} e^u (\mathbf{e}_1 - i\mathbf{e}_2), \quad \mathbf{x}_{\bar{z}} = \overline{\mathbf{x}_z} = \frac{1}{2} e^u (\mathbf{e}_1 + i\mathbf{e}_2) \tag{7.22}$$

which substituted into (7.18) and compared to $d\mathbf{x} = \omega^1 \mathbf{e}_1 + \omega^2 \mathbf{e}_2$ give

$$\omega^1 = \frac{e^u}{2}(dz + d\bar{z}), \quad \omega^2 = -i\frac{e^u}{2}(dz - d\bar{z}).$$

From this we easily verify that equation (7.17) holds. In the Euclidean case, $(\mathbf{x}, \mathbf{e}_1, \mathbf{e}_2, \mathbf{e}_3)$, with $\mathbf{e}_3 = \mathbf{e}_1 \times \mathbf{e}_2$, is the frame field adapted to (U, z). In the spherical case, the orthonormal set of vectors $\mathbf{x}, \mathbf{e}_1, \mathbf{e}_2$ can be completed uniquely to an orthonormal basis of $\mathbf{R}^4$ whose determinant is one at each point of U. The result is the frame field $(\mathbf{x}, \mathbf{e}_1, \mathbf{e}_2, \mathbf{e}_3) : U \to \mathbf{SO}(4)$ adapted to z. Similarly, in the hyperbolic case, the orthonormal set of vectors $\mathbf{e}_1, \mathbf{e}_2, \mathbf{x}$ can be completed uniquely to an orthonormal basis of $\mathbf{R}^{3,1}$ whose determinant is one at each point of U. The result is the frame field $(\mathbf{e}_1, \mathbf{e}_2, \mathbf{e}_3, \mathbf{x}) : U \to \mathbf{SO}_+(3,1)$ adapted to z.

Uniqueness follows from the fact that the conformal factor of a frame field adapted to (U, z) must be given by (7.19) or (7.20). Equations (7.17) and (7.18) combine to show that $\mathbf{e}_1$ and $\mathbf{e}_2$ must be given by (7.21). Then $\mathbf{e}_3$ is uniquely determined by the condition that the frame field must have determinant equal to one. □

For the immersion $\mathbf{x} : M \to \mathbf{S}_\epsilon$, the frame field adapted to a complex coordinate chart (U, z) is always of first order. If we differentiate $\omega^3 = 0$, we find from the structure equations that $\omega^3_i = h_{ij}\omega^j$, where $h_{ij} = h_{ji}$, $i, j = 1, 2$, are smooth functions on U. Then,

$$\omega^3_1 - i\omega^3_2 = h(\omega^1 + i\omega^2) + H(\omega^1 - i\omega^2) = h e^u dz + H e^u d\bar{z}, \tag{7.23}$$

where $H = (h_{11} + h_{22})/2$ is the mean curvature and

$$h = \frac{1}{2}(h_{11} - h_{22}) - ih_{12} \tag{7.24}$$

is the Hopf invariant of Definition 4.2, relative to this first order frame field.

Definition 7.24. For a local complex coordinate chart (U,z) on M, the *Hopf invariant h of* $\mathbf{x}$ *relative to z* is the Hopf invariant of $\mathbf{x}$ relative to the frame field adapted to (U,z).

For the adapted frame field,

$$d\mathbf{e}_3 = -(H\,dz + \bar{h}\,d\bar{z})\mathbf{x}_z - (h\,dz + H\,d\bar{z})\mathbf{x}_{\bar{z}}. \tag{7.25}$$

The second fundamental form has the decomposition into bidegrees

$$\begin{aligned} II &= -d\mathbf{x}\cdot d\mathbf{e}_3 = II^{2,0} + II^{1,1} + II^{0,2}, \text{ for } \epsilon = 0,+, \\ II &= -\langle d\mathbf{x}, d\mathbf{e}_3\rangle = II^{2,0} + II^{1,1} + II^{0,2}, \text{ for } \epsilon = -, \end{aligned} \tag{7.26}$$

independent of local complex coordinate. For the local complex coordinate z we have from equations (7.18) and (7.25) the local expressions

$$II^{2,0} = \frac{1}{2}e^{2u}h\,dzdz = \overline{II^{0,2}}, \quad II^{1,1} = e^{2u}H\,dzd\bar{z}. \tag{7.27}$$

Differentiate (7.17) to obtain

$$\omega_2^1 = i(u_z dz - u_{\bar{z}} d\bar{z}). \tag{7.28}$$

Using equations (7.23) and (7.28), we find

$$\begin{aligned} &d(\mathbf{e}_1 - i\mathbf{e}_2) = \\ &(u_z(\mathbf{e}_1 - i\mathbf{e}_2) + he^u\mathbf{e}_3)dz + (-\epsilon e^u\mathbf{x} - u_{\bar{z}}(\mathbf{e}_1 - i\mathbf{e}_2) + He^u\mathbf{e}_3)d\bar{z} \end{aligned} \tag{7.29}$$

Lemma 7.25. *Let (U,z) be a complex coordinate chart on M for the complex structure induced by the immersion $\mathbf{x}: M \to \mathbf{S}_\epsilon$. Then*

(1) The Hopf invariant h relative to z is real valued on U if and only if the coordinate curves are principal (that is, their tangents are principal directions).
(2) The Hopf invariant h is zero precisely at the umbilic points of $\mathbf{x}$ in U.
(3) The principal vectors of $\mathbf{x}$ are the solutions of

$$\Im(II^{2,0}) = \Im(\frac{1}{2}he^{2u}dzdz) = 0, \tag{7.30}$$

where $\Im(p)$ denotes the imaginary part of the complex number p.

Proof. The Hopf invariant h is zero exactly at the points where the matrix $\mathscr{S} = (h_{ij})$ of the second fundamental form is a multiple of the identity matrix, so these are the umbilic points of $\mathbf{x}$. Furthermore, h is real precisely when $\mathscr{S}$ is in diagonal form, which means that the frame vectors are principal directions. Since

$$(\omega_1^3 - i\omega_2^3)(\omega_0^1 + i\omega_0^2) = \omega_1^3\omega_0^1 + \omega_2^3\omega_0^2 + i(\omega_1^3\omega_0^2 - \omega_2^3\omega_0^1)$$

it follows from Lemma 4.4 that the principal vectors of $\mathbf{x}$ are the solutions of

$$\Im((\omega_1^3 - i\omega_2^3)(\omega_0^1 + i\omega_0^2)) = 0.$$

If $(\mathbf{x}, e)$ is the frame field adapted to the complex chart (U, z), then by (7.23) and (7.17)

$$\Im((\omega_1^3 - i\omega_2^3)(\omega_0^1 + i\omega_0^2)) = \Im(he^{2u}dzdz + He^{2u}dzd\bar{z}),$$

from which (7.30) follows, because $\Im(He^{2u}dzd\bar{z})) = 0$. □

Theorem 7.26 (Complex Structure Equations). *Let (U, z) be a complex coordinate chart in M for the complex structure induced by the immersion $\mathbf{x} : M \to \mathbf{S}_\epsilon$. Let h be the Hopf invariant relative to z, and let H be the mean curvature relative to the adapted frame. Then the Gauss equation is*

$$-4e^{-2u}u_{z\bar{z}} = K = \epsilon 1 + H^2 - |h|^2, \tag{7.31}$$

where K is the Gaussian curvature of the induced metric $I = e^{2u}dzd\bar{z}$, and the Codazzi equations become

$$h_{\bar{z}} + 2hu_{\bar{z}} = H_z. \tag{7.32}$$

Proof. Differentiating (7.28) and using the structure equations of G_ϵ, we have

$$\epsilon\omega^1 \wedge \omega^2 - \omega_3^1 \wedge \omega_2^3 = d\omega_2^1 = -2iu_{z\bar{z}}dz \wedge d\bar{z}.$$

By (7.23) we have

$$\omega_1^3 \wedge \omega_2^3 = \frac{i}{2}e^{2u}(H^2 - |h|^2)dz \wedge d\bar{z}$$

from which (7.31) follows. Since

$$d\omega_2^1 = K\omega^1 \wedge \omega^2 = \frac{i}{2}K(\omega^1 + i\omega^2) \wedge (\omega^1 - i\omega^2) = \frac{i}{2}e^{2u}Kdz \wedge d\bar{z}$$

we see that K equals either side of (7.31). Differentiate (7.23) to obtain

$$-i\omega_2^1 \wedge (\omega_3^1 - i\omega_3^2) = d(\omega_3^1 - i\omega_3^2) = e^u(h_{\bar{z}} + hu_{\bar{z}} - H_z - Hu_z)dz \wedge d\bar{z}. \tag{7.33}$$

Substitute (7.28) and (7.23) into (7.33) to obtain (7.32). □

Theorem 7.27 (Congruence). *Let $\mathbf{x}, \tilde{\mathbf{x}} : M \to \mathbf{S}_\epsilon$ be smooth immersions of a connected oriented surface M, with first fundamental forms $I, \tilde{I}$, unit normal vector fields $\mathbf{n}, \tilde{\mathbf{n}}$ inducing the orientation of M from the standard orientation of $\mathbf{S}_\epsilon$, and mean curvature functions $H, \tilde{H}$, respectively. If $I = \tilde{I}$, $H = \tilde{H}$, and $II^{2,0} = \widetilde{II}^{2,0}$ at every point of M, then there exists an isometry A in the connected component of the identity of $\mathbf{G}_\epsilon$ such that $\tilde{\mathbf{x}} = A\mathbf{x}$.*

Proof. By (7.27) and the hypotheses, we have $I = \tilde{I}$ and

$$II = II^{2,0} + HI + II^{0,2} = \widetilde{II}^{2,0} + \tilde{H}\tilde{I} + \widetilde{II}^{0,2} = \widetilde{II}$$

at every point of M. The result now follows from Bonnet's Congruence Theorem 4.18. □

Theorem 7.28 (Existence). *Let M be a simply connected Riemann surface. Let I be a Riemannian metric on M from its conformal class (if (U,z) is a local complex coordinate chart, then $I = e^{2u} dz d\bar{z}$, for some smooth $u : U \to \mathbf{R}$). Let $II^{2,0}$ be a smooth quadratic differential on M (in (U,z), $II^{2,0} = \frac{1}{2} e^{2u} h dz dz$ for some smooth $h : U \to \mathbf{C}$). Let $H : M \to \mathbf{R}$ be a smooth function. If for each local complex coordinate chart (U,z) in M the functions u, h, and H satisfy the equations* (7.31) *and* (7.32), *then there exists a smooth immersion $\mathbf{x} : M \to \mathbf{S}_\epsilon$, with unit normal vector field $\mathbf{e}_3$, whose first fundamental form $d\mathbf{x} \cdot d\mathbf{x} = I$ and whose second fundamental form*

$$-d\mathbf{e}_3 \cdot d\mathbf{x} = II^{2,0} + \overline{II^{2,0}} + HI, \text{ for } \epsilon = 0, +,$$

$$-\langle d\mathbf{e}_3, d\mathbf{x} \rangle = II^{2,0} + \overline{II^{2,0}} + HI, \text{ for } \epsilon = -.$$

Proof. The Riemannian metric I and the symmetric bilinear form

$$II = II^{2,0} + HI + \overline{II^{2,0}}$$

satisfy the hypotheses of Bonnet's Existence Theorem 4.19, and thus the result follows from that. □

7.5 Constant H and Lawson correspondence

Theorem 7.29 (H. Hopf). *Let $\mathbf{x} : M \to \mathbf{S}_\epsilon$ be an immersion with mean curvature H of an oriented surface M. Then the Hopf quadratic differential $II^{2,0}$ of $\mathbf{x}$ is holomorphic on M if and only if H is constant on M.*

Proof. Let (U,z) be a complex coordinate chart in M. Let e^u be the conformal factor and let h be the Hopf invariant relative to z. By (7.27) $II^{2,0} = \frac{1}{2} e^{2u} h dz dz$ on U so that $II^{2,0}$ is holomorphic means that the coefficient $\frac{1}{2} e^{2u} h$ is a holomorphic function of z, that is, $(e^{2u} h)_{\bar{z}} = 0$ on U. But

$$(e^{2u} h)_{\bar{z}} = e^{2u}(2h u_{\bar{z}} + h_{\bar{z}}) = e^{2u} H_z$$

where the last equation comes from the Codazzi equation (7.32). Since H is a real valued function, it is the case that $H_z = 0$ on U if and only if H is constant on U. As M can be covered by such complex coordinate neighborhoods, the result follows. □

When H is constant, the structure equations (7.31) and (7.32) possess an invariance under certain deformations, which gives rise to the existence of *associate immersions*. This phenomenon persists for isothermic immersions and Willmore immersions into Möbius space.

Definition 7.30. An *associate* of a constant mean curvature immersion $\mathbf{x} : M \to \mathbf{S}_\epsilon$ is a noncongruent immersion $\hat{\mathbf{x}} : M \to \mathbf{S}_\epsilon$ with the same induced metric and mean curvature as $\mathbf{x}$.

Corollary 7.31 (Associate immersions). *Let M be a simply connected Riemann surface with complex coordinate $z : M \to \mathbf{C}$. If $\mathbf{x} : M \to \mathbf{S}_\epsilon$ is an immersion with constant mean curvature and not totally umbilic, then there exists a 1-parameter family of immersions $\mathbf{x}^{(t)} : M \to \mathbf{S}_\epsilon$ for which $\mathbf{x}^{(0)} = \mathbf{x}$ and $\mathbf{x}^{(t)}$ is an associate of $\mathbf{x}$ if $e^{it} \neq 1$. Moreover, $\mathbf{x}^{(t)}$ is congruent to $\mathbf{x}^{(s)}$ if and only if $e^{it} = e^{is}$. Any associate of $\mathbf{x}$ is $\mathbf{x}^{(t)}$ for some $t \in \mathbf{R}$.*

Proof. Let $I = e^{2u} dz d\bar{z}$ be the metric induced by $\mathbf{x}$, let h be the Hopf invariant of $\mathbf{x}$ relative to z, and let H be its constant mean curvature. The structure equations (7.31) and (7.32) for $\mathbf{x}$ become

$$u_{z\bar{z}} + \frac{1}{4} e^{2u}(\epsilon 1 + H^2 - |h|^2) = 0, \quad (he^{2u})_{\bar{z}} = 0. \tag{7.34}$$

Thus, he^{2u} is a holomorphic function on M. If $\mathbf{x}$ is not totally umbilic, then this function is nonzero, so has only isolated zeros on M. For any $t \in \mathbf{R}$, these structure equations are also satisfied by u, $e^{it}h$, and H. By Theorem 7.28, there exists an immersion $\mathbf{x}^{(t)} : M \to \mathbf{S}_\epsilon$ with induced metric I, Hopf invariant $e^{it}h$, and mean curvature H. It is an associate of $\mathbf{x}$ if $e^{it} \neq 1$. Moreover, $\mathbf{x}^{(t)}$ is congruent to $\mathbf{x}^{(s)}$ if and only if $e^{it} = e^{is}$, by Theorem 7.27.

Suppose $\hat{\mathbf{x}} : M \to \mathbf{S}_\epsilon$ is an associate of $\mathbf{x}$, with Hopf invariant $\hat{h}$ relative to z. Then $|\hat{h}| = |h|$ on M, by the first structure equation, and thus $\hat{h}e^{2u}$ and he^{2u} are nonzero holomorphic functions of the same modulus on M. If follows that $\hat{h}/h$ is a holomorphic function of modulus one, off the isolated zero set of h, so it must be constant. Hence, $\hat{h} = e^{it}h$ for some constant $t \in \mathbf{R}$. □

Remark 7.32 (Lawson correspondence). The structure equations (7.34) with $\epsilon = 0$ and $H = 0$ are identical to those with $\epsilon = -$ and $H = 1$. Therefore, there is a local one-to-one correspondence

$$\text{Minimal surfaces in } \mathbf{R}^3 \quad \leftrightarrow \quad \text{CMC1 surfaces in } \mathbf{H}^3.$$

Likewise, these structure equations with $\epsilon = 0$ and $H = 1$ are identical to those with $\epsilon = +$ and $H = 0$, so there is a local one-to-one correspondence

$$\text{CMC1 surfaces in } \mathbf{R}^3 \quad \leftrightarrow \quad \text{Minimal surfaces in } \mathbf{S}^3.$$

7.6 Calculating the invariants

For an immersion $\mathbf{x} : M \to \mathbf{S}_\epsilon$, let (U, z) be a complex coordinate chart in M. We develop formulas for the conformal factor, the Hopf invariant, and the mean curvature relative to z.

Equations (7.19) and (7.20) give us the conformal factor e^u,

$$e^{2u} = 2\mathbf{x}_z \cdot \mathbf{x}_{\bar{z}}, \text{ for } \epsilon = 0, +, \quad e^{2u} = 2\langle \mathbf{x}_z, \mathbf{x}_{\bar{z}} \rangle, \text{ for } \epsilon = -.$$

Differentiation of (7.22) with respect to z and $\bar{z}$ combined with (7.29) gives

$$\mathbf{x}_{zz} = 2u_z \mathbf{x}_z + \frac{1}{2} e^{2u} h \mathbf{e}_3, \quad \mathbf{x}_{z\bar{z}} = -\frac{1}{2} \epsilon e^{2u} \mathbf{x} + \frac{1}{2} e^{2u} H \mathbf{e}_3. \tag{7.35}$$

In the Euclidean case this gives $h = 2e^{-2u} \mathbf{x}_{zz} \cdot \mathbf{e}_3$. Also from (7.22) we see that

$$\mathbf{e}_3 = \mathbf{e}_1 \times \mathbf{e}_2 = -2ie^{-2u} \mathbf{x}_z \times \mathbf{x}_{\bar{z}}, \tag{7.36}$$

so $\mathbf{a} \cdot (\mathbf{b} \times \mathbf{c}) = \det(\mathbf{b}, \mathbf{c}, \mathbf{a})$ for any vectors $\mathbf{a}$, $\mathbf{b}$, and $\mathbf{c}$ in $\mathbf{R}^3$ implies

$$h = -4ie^{-4u} \det(\mathbf{x}_z, \mathbf{x}_{\bar{z}}, \mathbf{x}_{zz}). \tag{7.37}$$

In the same way,

$$H = -4ie^{-4u} \det(\mathbf{x}_z, \mathbf{x}_{\bar{z}}, \mathbf{x}_{z\bar{z}}). \tag{7.38}$$

From (7.27) and (7.37) the Hopf quadratic differential is

$$II^{2,0} = \mathbf{x}_{zz} \cdot \mathbf{e}_3 \, dzdz = -2ie^{-2u} \det(\mathbf{x}_z, \mathbf{x}_{\bar{z}}, \mathbf{x}_{zz}) \, dzdz.$$

In the cases $\epsilon = \pm$, the vectors $\mathbf{x}_z$, $\mathbf{e}_1 - i\mathbf{e}_2$, and so on are in $\mathbf{C}^4$ and

$$\begin{aligned} \det(\mathbf{x}, \mathbf{x}_z, \mathbf{x}_{\bar{z}}, \mathbf{x}_{zz}) &= \det(\mathbf{x}, \frac{1}{2} e^u (\mathbf{e}_1 - i\mathbf{e}_2), \frac{1}{2} e^u (\mathbf{e}_1 + i\mathbf{e}_2), \frac{1}{2} e^{2u} h \mathbf{e}_3) \\ &= \epsilon \frac{i}{4} e^{4u} h \end{aligned}$$

because $\det e = 1$, so the Hopf invariant

$$h = -\epsilon 4ie^{-4u} \det(\mathbf{x}, \mathbf{x}_z, \mathbf{x}_{\bar{z}}, \mathbf{x}_{zz}).$$

In the same way, the mean curvature

$$H = -\epsilon 4ie^{-4u}\det(\mathbf{x}, \mathbf{x}_z, \mathbf{x}_{\bar{z}}, \mathbf{x}_{z\bar{z}}).$$

We record here an important consequence of these simple calculations.

Definition 7.33. A vector $\mathbf{v} \in \mathbf{C}^n$ is *isotropic* if $\mathbf{v} \cdot \mathbf{v} = {}^t\mathbf{v}\mathbf{v} = \sum_1^n v^j v^j = 0$.

Theorem 7.34 (Enneper–Weierstrass characterization). *Let (M^2, I) be an oriented immersion and let M have the induced complex structure. Let $\partial\mathbf{x}$ denote the $\mathbf{C}^3$-valued 1-form on M defined in terms of any local complex coordinate chart (U, z) by*

$$\partial\mathbf{x} = \mathbf{x}_z dz.$$

This form is isotropic by (7.19) *(meaning that its value at any point is an isotropic vector in $\mathbf{C}^3$). Then $\mathbf{x}$ is minimal if and only if $\partial\mathbf{x}$ is holomorphic.*

Proof. The proof follows from the second equation in (7.35). □

Remark 7.35. Given a Riemann surface M, a holomorphic map $f : M \to \mathbf{C}^3$ defines a map $\mathbf{x} = \Re \int f\, dz$ which is an immersion at every point where $f \neq 0$, but the complex structure induced by this immersion will agree with the given complex structure on M if and only if f is isotropic; that is, $f \cdot f = 0$. These ideas, which constitute the Weierstrass representation of a minimal immersion, are considered in detail in Chapter 8.

Corollary 7.36. *Let $\mathbf{x} : M \to \mathbf{R}^3$ be a minimal immersion of an oriented surface M. If $\mathbf{v}$ is any fixed vector in $\mathbf{R}^3$, then the function $\mathbf{x} \cdot \mathbf{v} : M \to \mathbf{R}$ is harmonic on M with respect to the induced metric. Consequently, M cannot be compact (without boundary).*

Proof. Let $z = x + iy$ be a local complex coordinate on M for the induced complex structure. Since $\mathbf{x}$ is minimal and $\mathbf{v}$ is constant, the second equation in (7.35) gives

$$0 = \mathbf{x}_{z\bar{z}} \cdot \mathbf{v} = \frac{1}{4}\left(\frac{\partial^2}{\partial x^2} + \frac{\partial^2}{\partial y^2}\right)(\mathbf{x} \cdot \mathbf{v}),$$

which shows that $\mathbf{x} \cdot \mathbf{v}$ is harmonic. If M were compact, and $\mathbf{v}$ is chosen so that $\mathbf{x} \cdot \mathbf{v}$ is nonconstant, then this function must assume a maximum at some point of M. However, a nonconstant harmonic function cannot have an interior maximum, by the Maximum Principle for harmonic functions (see Ahlfors [1, Theorem 3, p. 179]). □

Example 7.37 (Enneper's Surface). Consider the map

$$\begin{aligned} &\mathbf{x} : \mathbf{R}^2 \to \mathbf{R}^3 \\ &(x, y) \mapsto {}^t\left(x - \frac{1}{3}x^3 + xy^2, -y + \frac{1}{3}y^3 - x^2y, x^2 - y^2\right) \end{aligned} \tag{7.39}$$

After calculation of $d\mathbf{x}$, one finds

$$d\mathbf{x}\cdot d\mathbf{x} = (1+(x^2+y^2))^2(dx^2+dy^2) = e^{2u}dzd\bar{z}$$

where $u = \log(1+|z|^2)$ and $z = x+iy$ is a complex coordinate on $\mathbf{R}^2$ for the induced complex structure. One calculates

$$\mathbf{x}_z = \frac{1}{2}{}^t(1-z^2, i(1+z^2), 2z)$$
$$\mathbf{x}_{zz} = {}^t(-z, iz, 1)$$
$$\mathbf{x}_{z\bar{z}} = 0$$

Therefore, $H = 0$ by (7.38) and $h = -2e^{-2u}$ by (7.37) (which involves some calculation). Thus, $\mathbf{x}$ is a minimal immersion and its Hopf quadratic differential is $II^{2,0} = -dzdz$. This example will be discussed again in Example 8.13 in Chapter 8.

Example 7.38 (Circular Tori of Example 5.12). Consider the immersion $\mathbf{x} = \mathbf{x}_\alpha : \mathbf{R}^2 \to \mathbf{S}^3$ defined in (5.26) of Example 5.12, where $0 < \alpha < \pi/2$, $r = \cos\alpha$, and $s = \sin\alpha$. The induced metric was found to be $d\mathbf{x}\cdot d\mathbf{x} = dx^2+dy^2$, so $z = x+iy$ is a complex coordinate on $\mathbf{R}^2$ for the induced complex structure. Relative to z the conformal factor is $e^u = 1$. By (5.27),

$$\omega_1^3 - i\omega_2^3 = -\frac{1}{2rs}dz + \frac{r^2-s^2}{2rs}d\bar{z},$$

so by (7.23), the Hopf invariant relative to z is $h = -\frac{1}{2rs}$ and the mean curvature is $H = \frac{r^2-s^2}{2rs}$.

7.6.1 Dependence on the complex coordinate

Suppose that w is another complex coordinate in U. Then $w = w(z)$ is a holomorphic function of z and so also is $w' = \frac{dw}{dz}$. Let $e^{\tilde{u}}$ be the conformal factor and $\tilde{h}$ the Hopf invariant relative to w. Of course, the mean curvature H does not depend on the choice of complex coordinate. Comparison of the expressions for the induced metric in the two complex coordinates,

$$e^{2u}dzd\bar{z} = I = e^{2\tilde{u}}dwd\bar{w} = e^{2\tilde{u}}|w'|^2dzd\bar{z},$$

implies that

$$e^u = e^{\tilde{u}}|w'|. \tag{7.40}$$

In the same way,

$$\frac{1}{2}e^{2u}h\,dzdz = II^{2,0} = \frac{1}{2}e^{2\tilde{u}}\tilde{h}\,dwdw$$

implies that

$$\tilde{h} = \frac{\overline{w'}}{w'}h \tag{7.41}$$

Therefore, $|\tilde{h}| = |h|$, which means that $|h|$ is a function defined and smooth on all of M, independent of the choice of local complex coordinate.

Definition 7.39. A function $f : M \to \mathbf{R}$ on a Riemann surface M is *harmonic* if $f_{z\bar{z}} = 0$ for any complex coordinate chart (U,z) in M. A map $\mathbf{y} : M \to \mathbf{R}^n$ is harmonic if each component function is harmonic. If $\mathbf{y}$ takes values in the unit sphere, then it is harmonic as a smooth map $\mathbf{y} : M \to \mathbf{S}^{n-1}$, if the component of $\mathbf{y}_{z\bar{z}}$ tangent to $\mathbf{S}^n$ at $\mathbf{y}$ is zero at each point of M; that is, $\mathbf{y}_{z\bar{z}} - \mathbf{y}\cdot\mathbf{y}_{z\bar{z}}\mathbf{y} = 0$ at each point of M.

Theorem 7.40. *Let M be a connected, oriented surface. Let $\mathbf{x} : M \to \mathbf{R}^3$ be an immersion and let $\mathbf{n}$ be a smooth unit normal vector field along $\mathbf{x}$. Then the Gauss map $\mathbf{n} : M \to \mathbf{S}^2 \subset \mathbf{R}^3$ of x has the following properties.*

1. *The Gauss map is harmonic if and only if the mean curvature H of $\mathbf{x}$ is constant on M. (Ruh-Vilms [146]).*
2. *The Gauss map is conformal and harmonic if and only if $\mathbf{x}$ is minimal or totally umbilic. If $\mathbf{x}$ is minimal, the Gauss map is orientation reversing at the nonumbilic points. If $\mathbf{x}$ is totally umbilic, then $\mathbf{x}(M)$ is an open submanifold of a plane or sphere.*

Proof. Let (U,z) be a complex coordinate chart in M and let $(\mathbf{x},e)$ be the frame field adapted to (U,z). Let e^u be the conformal factor and let h be the invariant with respect to z. Then $\mathbf{n} = \pm\mathbf{e}_3$. In the following we assume that $\mathbf{n} = \mathbf{e}_3$. The case $\mathbf{n} = -\mathbf{e}_3$ requires only obvious changes. To prove (1), use (7.25) to conclude that

$$\mathbf{n}_z = \mathbf{e}_{3z} = -H\mathbf{x}_z - h\mathbf{x}_{\bar{z}} \tag{7.42}$$

and therefore, by (7.35)

$$\begin{aligned}\mathbf{n}_{z\bar{z}} &= -H_{\bar{z}}\mathbf{x}_z - (h_{\bar{z}} + 2u_{\bar{z}}h)\mathbf{x}_{\bar{z}} - \frac{1}{2}e^{2u}(H^2 + |h|^2)\mathbf{e}_3 \\ &= -H_{\bar{z}}\mathbf{x}_z - H_z\mathbf{x}_{\bar{z}} - \frac{1}{2}e^{2u}(H^2 + |h|^2)\mathbf{e}_3\end{aligned}$$

where the last equality was obtained using the Codazzi equation (7.32). Therefore $\mathbf{n} : M \to \mathbf{S}^2$ is harmonic if and only if $H_{\bar{z}} = H_z = 0$ on U, that is, H is constant on U, for every chart (U,z) in M. Since M can be covered by complex coordinate charts, we see that (1) is proved.

To prove (2), use (7.42) to find

$$\mathbf{n}_z \cdot \mathbf{n}_z = \mathbf{e}_{3z} \cdot \mathbf{e}_{3z} = 2Hh\mathbf{x}_z \cdot \mathbf{x}_{\bar{z}} = Hhe^{2u}$$

from which we conclude, by applying (7.19) to $\mathbf{n}$, that $\mathbf{n}$ is conformal if and only if $Hh = 0$ on U. If $\mathbf{n}$ is conformal and harmonic, then H is constant, as was just proved, and $Hh = 0$ on U, so that either the constant $H = 0$ ($\mathbf{x}$ is minimal) or the constant $H \neq 0$ and $h = 0$ on U, for every chart (U,z) ($\mathbf{x}$ is totally umbilic). If $\mathbf{x}$ is minimal, then by (7.42)

$$\mathbf{n}_z \times \mathbf{n}_{\bar{z}} = \mathbf{e}_{3z} \times \mathbf{e}_{3\bar{z}} = -|h|^2 \mathbf{x}_z \times \mathbf{x}_{\bar{z}}$$

which shows that $\mathbf{n}$ is orientation reversing at the nonumbilic points of $\mathbf{x}$. If $\mathbf{x}$ is totally umbilic, then it is an open submanifold of a plane or sphere by Theorem 4.23. □

Problems

7.41. Prove if $A \in \mathbf{SO}(2)$, then $i(A_1^1 + iA_1^2) = A_2^1 + iA_2^2$.

7.42. Prove that for a complex valued function $f = u + iv$, where u and v are real valued smooth functions of x, y, the following are equivalent. If f satisfies any, thus all, of these conditions, then it is called a *holomorphic* function of $z = x + iy$.

1. u, v satisfy the Cauchy-Riemann equations $u_x = v_y$ and $u_y = -v_x$.
2. $\frac{\partial f}{\partial \bar{z}} = 0$.
3. $\bar{\partial} f = 0$.
4. df is a multiple of dz.
5. df is of bidegree $(1,0)$.

7.43. A smooth unit normal vector field on $\mathbf{S}^2$ induces an orientation on $\mathbf{S}^2$ from the standard orientation of $\mathbf{R}^3$. Is the orientation on $\mathbf{S}^2$ coming from its Riemann sphere complex structure in Example 7.14 the one induced by the inward normal or the outward normal?

7.44. Let $\Gamma(\sigma,\tau)$ and $\Gamma(\sigma_1,\tau_1)$ be lattices in $\mathbf{C}$ as defined in Example 7.20. Prove $\mathbf{C}/\Gamma(\sigma,\tau)$ and $\mathbf{C}/\Gamma(\sigma_1,\tau_1)$ are conformally equivalent if and only if there exist integers $\alpha,\beta,\gamma,\delta$ such that $\alpha\delta - \beta\gamma = 1$ and $\sigma_1 = \alpha\sigma + \beta\tau$ and $\tau_1 = \gamma\sigma + \delta\tau$.

7.45. Prove that up to conformal equivalence, any torus is given by $\mathbf{C}/\Gamma(\sigma,\tau)$ where $\sigma = 1$ and $\Im\tau > 0$.

7.46. Prove that the criterion form (see Definition 4.14) of the frame field adapted to the complex chart (U,z) is $\alpha = du$, where e^u is the conformal factor.

7.47. Prove that a cylinder in $\mathbf{R}^3$ over a circle of radius 1/2 (see Problem 4.65) is a Lawson correspondent of the Clifford torus in $\mathbf{S}^3$ (see Example 5.12 for the case $\alpha = \pi/4$).

7.48. For the surface of revolution in Example 4.40, let z be the complex coordinate of the induced structure described in Example 7.18. Find the conformal factor e^u and the Hopf invariant h relative to z. Find the mean curvature H and the Hopf quadratic differential $II^{2,0}$.

7.49. Is there a version of Theorem 7.40 for the hyperbolic Gauss map $g : M \to S^2_\infty$ of an immersion $\mathbf{x} : M \to \mathbf{H}^3$?

Chapter 8
Minimal Immersions in Euclidean Space

This chapter gives a brief history of minimal immersions in Euclidean space. We present the calculation of the first variation of the area functional and we derive the Enneper–Weierstrass representation. Scherk's surface is used to illustrate the problems that arise in integrating the Weierstrass forms. This integration problem is a simpler version of the monodromy problem encountered later in finding examples of CMC 1 immersions in hyperbolic geometry. We present results on complete minimal immersions with finite total curvature, which will be used in Chapter 14 to characterize minimal immersions in Euclidean space that smoothly extend to compact Willmore immersions into Möbius space. The final section on minimal curves applies the method of moving frames to the nonintuitive setting of holomorphic curves in $\mathbf{C}^3$ whose tangent vector is nonzero and isotropic at every point.

8.1 The area functional

The theory of minimal surfaces goes back to Lagrange's 1761 paper, which is regarded as the origin of the Calculus of Variations. James and John Bernoulli had used the ideas of the calculus of variations in their 1697 solutions of the brachistochrone problem, which was discussed first by Galileo (1630) and had been posed as a challenge in 1696 by John Bernoulli (the younger brother). Newton, Leibniz and l'Hospital also solved the problem in response to the challenge.

Let M be an oriented surface and let $\mathbf{x} : M \to \mathbf{R}^3$ be an immersion. Let dA denote the area form on M relative to the induced metric $I = d\mathbf{x} \cdot d\mathbf{x}$. Then the area of the immersed surface is

$$A(\mathbf{x}) = \int_M dA.$$

G.R. Jensen et al., *Surfaces in Classical Geometries*, Universitext,
DOI 10.1007/978-3-319-27076-0_8

If M is not compact, then this integral is improper, even infinite. This area can be thought of as a functional on the set of all immersions $\mathbf{x}$ of M. The immersion is *minimal* if it is a critical point of this functional with respect to admissible variations with compact support, see Definition 8.1.

8.2 A brief history of minimal surfaces

In his 1761 paper [104, Vol.I], Lagrange considered the case where M is a bounded domain in $\mathbf{R}^2$ and the immersions are all graphs of the form

$$\mathbf{x}(x,y) = (x,y,z(x,y)),$$

where $z(x,y)$ is a smooth function on M. In this case the area form is $dA = (1+z_x^2+z_y^2)^{1/2}dx\wedge dy$. We shall use Euler's notation, which for a function $z = z(x,y)$, sets

$$p = z_x, \quad q = z_y, \quad r = z_{xx}, \quad s = z_{xy}, \quad t = z_{yy}.$$

We also set $w = \sqrt{1+p^2+q^2}$. In his 1761 paper Lagrange found that a graph $z = z(x,y)$ is a critical point of the area functional if and only if it satisfies the Euler-Lagrange equation

$$\left(\frac{p}{w}\right)_x + \left(\frac{q}{w}\right)_y = 0. \tag{8.1}$$

The only solution that he found was the plane $z = ax+by+c$, where a, b and c are constants.

In 1776, Meusnier rewrote (8.1) in the form which is now called the minimal surface equation, MSE,

$$(1+q^2)r - 2pqs + (1+p^2)t = 0.$$

The expression on the left is the same as the numerator for the mean curvature of a graph, as expressed in (4.6). Meusnier was the first to observe that the MSE holds if and only if the mean curvature of the graph is zero.

The MSE is a second order, nonlinear, elliptic PDE. Perhaps the term elliptic needs some explanation. A general second order PDE is

$$F(x,y,z,p,q,r,s,t) = 0,$$

where at least one of the partial derivatives F_r, F_s and F_t is nonzero. The *type* of the PDE is the type of the quadratic form

$$F_r\xi^2 + F_s\xi\eta + F_t\eta^2$$

whose type we can read from the sign of its *discriminant* $D = F_rF_t - \frac{1}{4}F_s^2$. The type is:

- *Elliptic* if $D > 0$. E.g., $z_{xx} + z_{yy} = 0$, Laplace equation.
- *Hyperbolic* if $D < 0$. E.g., $z_{xx} - z_{yy} = 0$, wave equation.
- *Parabolic* if $D = 0$. E.g., $z_{xx} - z_y = 0$, heat equation.

The discriminant of the MSE is $w^2 > 0$.

In his paper [119], which wasn't published until 1785, Meusnier observed that any solution z of the two equations

$$r + t = 0 = -q^2 r + 2pqs - p^2 t$$

is a solution of the MSE. For these two equations he found the solution

$$z = b + a \arctan \frac{y}{x}, \quad a, b \text{ constants},$$

whose graph is the *right helicoid*. In the same paper he also found the solution

$$z = \frac{1}{b} \cosh^{-1}\left(\frac{1}{a}\sqrt{x^2 + y^2}\right), \quad a > 0 \text{ and } b > 0 \text{ constants},$$

whose graph is a *catenoid*. Here $M = \{x^2 + y^2 \geq 1\} \subset \mathbf{R}^2$. These examples of Meusnier's were the only minimal surfaces known until 1831, seventy years after the appearance of Lagrange's paper.

In response to a prize offered by the Jablonowsky Society in Leipzig (Society of Sciences of Leipzig) for new solutions to the MSE, Scherk submitted his essay [147], in which he found five new examples and wrote a history of the subject. He won the prize and the Society published his essay. In his paper [148] of 1834, he looked for solutions of the MSE of the form

$$z = f(x) + g(y).$$

Such surfaces are called *surfaces of translation*. He found the solution

$$z = \log\left(\frac{\cos x}{\cos y}\right), \tag{8.2}$$

whose graph is called Scherk's surface. The domain consists of an infinite checkerboard in the x,y-plane of points where $\cos x$ and $\cos y$ have the same sign. It is the union of open squares

$$S_{m,n} = \{(x,y) : |x - m\pi| < \frac{\pi}{2}, \ |y - n\pi| < \frac{\pi}{2}\}, \tag{8.3}$$

where m and n are integers such that $m + n$ is even. Figure 8.1 shows the surface over eight of these squares. It is an infinite array of bedsteads, with the hanging side sheets asymptotically approaching vertical planes parallel to the y-axis along each vertical boundary line, and the head and foot boards asymptotically approaching vertical planes parallel to the x-axis along each horizontal boundary line. As we will see in Example 8.17 below, each bedstead shares an infinite bedpost with each neighbor at each corner so the surface is actually connected. We cannot really see this from our present description, because the whole surface is not a graph.

This must serve here as a tiny introduction to the history of this subject. More will appear below as we study minimal surfaces. In particular, we will see the essential role played by Weierstrass in solving the MSE. A very detailed history to 1890 can be found in H.A. Schwarz's [149][Miscellen, pages 168–198]. Modern comprehensive introductions to minimal surface theory can be found in R.

Fig. 8.1 Scherk's surface as graph over $S_{-1,-1} \cup S_{1,-1} \cup S_{0,0} \cup S_{2,0} \cup S_{-1,1} \cup S_{1,1} \cup S_{0,2} \cup S_{2,2}$.

Osserman's [133], H.B. Lawson's [107], and U. Dierkes, S. Hildebrandt, A. Küster, and O. Wohlrab's [61].

8.3 First variation of the area functional

Let $\mathbf{e}_3$ be a smooth unit normal along the immersion $\mathbf{x} : M \to \mathbf{R}^3$ compatible with the orientation of M.

Definition 8.1. An *admissible variation* of $\mathbf{x}$ is any smooth map

$$X : M \times (-\epsilon, \epsilon) \to \mathbf{R}^3,$$

with compact *support*, such that for each $t \in (-\epsilon, \epsilon)$, the map

$$\mathbf{x}_t : M \to \mathbf{R}^3, \quad \mathbf{x}_t(m) = X(m,t),$$

is an immersion. The support of X is the closure in M of the set of points of M where $\mathbf{x}_t(m) \neq \mathbf{x}(m)$, for some t.

It follows that the *variation vector field* of X,

$$\left.\frac{d}{dt}\right|_{t=0} \mathbf{x}_t : M \to \mathbf{R}^3,$$

has compact support in M.

Example 8.2. If g is a smooth function with compact support $S \subset M$, then there exists $\epsilon > 0$ such that

$$X : M \times (-\epsilon, \epsilon) \to \mathbf{R}^3, \quad X(m,t) = \mathbf{x}(m) + tg(m)\mathbf{e}_3(m)$$

is an admissible variation of $\mathbf{x}$. For example, if a and c are the principal curvatures of $\mathbf{x}$, then $\epsilon = \min_{\text{over} S}\{\frac{1}{|a|}, \frac{1}{|c|}\} > 0$ works.

The *first variation of the area of* $\mathbf{x}$ due to the admissible variation X of $\mathbf{x}$ with support S is

$$\left.\frac{d}{dt}\right|_{t=0} \int_S dA_t = \int_S \left.\frac{d}{dt}\right|_{t=0} dA_t,$$

where dA_t is the area element of the immersion $\mathbf{x}_t : M \to \mathbf{R}^3$. We calculate this first variation as follows.

Let $(\mathbf{x}, e) = (\mathbf{x}, \mathbf{e}_1, \mathbf{e}_2, \mathbf{e}_3)$ be a smooth first order frame field along $\mathbf{x}$ defined on an open set $U \subset M$. The variation vector field of X on U is then

$$\left.\frac{d}{dt}\right|_{t=0} \mathbf{x}_t = \sum_{i=1}^{3} g^i \mathbf{e}_i, \tag{8.4}$$

for some smooth functions $g^1, g^2, g^3 : U \to \mathbf{R}$. Then

$$\left.\frac{d}{dt}\right|_{t=0} d\mathbf{x}_t = d\Big(\left.\frac{d}{dt}\right|_{t=0} \mathbf{x}_t\Big) = d\sum_1^3 g^i \mathbf{e}_i = \sum_i (dg^i + \sum_j g^j \omega^i_j)\mathbf{e}_i.$$

For fixed t, we have on U

$$d\mathbf{x}_t = \omega^1_t \mathbf{e}_1 + \omega^2_t \mathbf{e}_2 + \omega^3_t \mathbf{e}_3, \tag{8.5}$$

where ω^i_t is a smooth curve in $T^*_m M$, for each $m \in U$. Hence

$$\left.\frac{d}{dt}\right|_{t=0} \omega^i_t = dg^i + \sum_{j=1}^{3} g^j \omega^i_j,$$

for $i = 1, 2, 3$.

Lemma 8.3. *About any $m_0 \in U$, there exists a neighborhood $V \subset U$ and a positive number $\delta \leq \epsilon$ for which there exist smooth vector fields*

$$\mathbf{E}_a : V \times (-\delta, \delta) \to \mathbf{R}^3,$$

for $a = 1,2,3$, such that $E_t = (\mathbf{x}_t, \mathbf{E}_1, \mathbf{E}_2, \mathbf{E}_3) : V \to \mathbf{E}(3)$ is a first order frame field along $\mathbf{x}_t$ for each $t \in (-\delta, \delta)$, with the property that

$$\mathbf{E}_1(m,0) = \mathbf{e}_1(m), \quad \mathbf{E}_2(m,0) = \mathbf{e}_2(m), \quad \mathbf{E}_3(m,0) = \mathbf{e}_3(m),$$

for every $m \in V$. Consequently,

$$\frac{\partial \mathbf{E}_j}{\partial t}(m,0) \cdot \mathbf{e}_j(m) = \frac{1}{2} \frac{\partial}{\partial t}\bigg|_{t=0} (\mathbf{E}_j \cdot \mathbf{E}_j) = 0,$$

for every $m \in V$ for $j = 1,2$.

Proof. If u, v are oriented local coordinates on a neighborhood of m_0 in M, then

$$\mathbf{E}_3 = \left(\frac{\partial \mathbf{x}_t}{\partial u} \times \frac{\partial \mathbf{x}_t}{\partial v} \right) \Big/ \left| \frac{\partial \mathbf{x}_t}{\partial u} \times \frac{\partial \mathbf{x}_t}{\partial v} \right|$$

is a unit normal vector field along $\mathbf{x}_t$, for each t. It is smooth in m and t, and when $t = 0$ it is $\mathbf{e}_3$. Then

$$\mathbf{E}_1 = \frac{\mathbf{e}_1 - (\mathbf{e}_1 \cdot \mathbf{E}_3)\mathbf{E}_3}{|\mathbf{e}_1 - (\mathbf{e}_1 \cdot \mathbf{E}_3)\mathbf{E}_3|}, \quad \mathbf{E}_2 = \mathbf{E}_3 \times \mathbf{e}_1$$

completes the desired frame field. □

Let E_t be a frame field constructed in the Lemma. For each t we have

$$d\mathbf{x}_t = \Omega_t^1 \mathbf{E}_1 + \Omega_t^2 \mathbf{E}_2,$$

where Ω_t^1, Ω_t^2 is the orthonormal coframe field for $\mathbf{x}_t$ with respect to E_t. Comparing this to (8.5), we conclude that

$$\Omega_t^j = d\mathbf{x}_t \cdot \mathbf{E}_j = (\mathbf{e}_1 \cdot \mathbf{E}_j)\omega_t^1 + (\mathbf{e}_2 \cdot \mathbf{E}_j)\omega_t^2 + (\mathbf{e}_3 \cdot \mathbf{E}_j)\omega_t^3,$$

for $j = 1,2$. Now the area element of $\mathbf{x}_t$ is

$$dA_t = \Omega_t^1 \wedge \Omega_t^2 = P\omega_t^1 \wedge \omega_t^2 + (Q\omega_t^1 + R\omega_t^2) \wedge \omega_t^3,$$

where P, Q, and R are the smooth functions of m and t

$$P = (\mathbf{e}_1 \cdot \mathbf{E}_1)(\mathbf{e}_2 \cdot \mathbf{E}_2) - (\mathbf{e}_2 \cdot \mathbf{E}_1)(\mathbf{e}_1 \cdot \mathbf{E}_2)$$
$$Q = (\mathbf{e}_1 \cdot \mathbf{E}_1)(\mathbf{e}_3 \cdot \mathbf{E}_2) - (\mathbf{e}_3 \cdot \mathbf{E}_1)(\mathbf{e}_1 \cdot \mathbf{E}_2)$$
$$R = (\mathbf{e}_2 \cdot \mathbf{E}_1)(\mathbf{e}_3 \cdot \mathbf{E}_2) - (\mathbf{e}_3 \cdot \mathbf{E}_1)(\mathbf{e}_2 \cdot \mathbf{E}_2).$$

For any point $m \in V$,

$$P(m,0) = 1, \quad Q(m,0) = 0, \quad R(m,0) = 0, \quad \left.\frac{\partial}{\partial t}\right|_{t=0} P = 0.$$

Then

$$\left.\frac{d}{dt}\right|_{t=0} (\Omega_t^1 \wedge \Omega_t^2) = \left.\frac{d}{dt}\right|_{t=0} (\omega_t^1 \wedge \omega_t^2) = d(g^1\omega^2 - g^2\omega^1) + g^3(\omega_3^1 \wedge \omega^2 + \omega^1 \wedge \omega_3^2),$$

since $\omega_0^3 = 0$ and $Q(m,0) = R(m,0) = 0$.

Exercise 32. Prove:

1. $\omega_3^1 \wedge \omega^2 + \omega^1 \wedge \omega_3^2 = -2HdA$, where H is the mean curvature of $\mathbf{x}$.
2. The 1-form $\alpha = g^1\omega^2 - g^2\omega^1$ is independent of the choice of oriented first order frame field $(\mathbf{x}, e)$ along $\mathbf{x}$, so is globally defined on M.

From all this we conclude that the first variation of the area of $\mathbf{x}$ due to the admissible variation X with variation vector (8.4) is

$$\int_M \left.\frac{d}{dt}\right|_{t=0} dA_t = -2\int_M g^3 H\, dA, \tag{8.6}$$

because $\int_M d\alpha = 0$ by Stokes and the compact support of α.

Remark 8.4. The tangential part of the variation vector never enters into the first variation of area. It is sufficient to consider only variations of the type defined in Example 8.2.

Theorem 8.5 (Lagrange-Meusnier). *An immersion $\mathbf{x} : M \to \mathbf{R}^3$ of an oriented surface M is minimal if and only if the mean curvature is identically zero.*

Proof. We have defined minimal immersion to mean that the first variation of the area is zero for all admissible variations of $\mathbf{x}$. By (8.6), the first variation is zero for all admissible variations, if H is identically zero. To prove the converse, we prove that if the mean curvature is not identically zero then the first variation for some admissible variation is not zero. Suppose $H(m_0) \neq 0$ for some point $m_0 \in M$. By continuity, H is nonzero on some open neighborhood U of m_0. Let $g : M \to \mathbf{R}$ be a smooth function with compact support in U such that $g(m_0) = 1$ and $g \geq 0$ in U. Then

$$\int_M gHdA \neq 0$$

implies that the first variation of the area of $\mathbf{x}$ is nonzero for the admissible variation $X(m,t) = \mathbf{x}(m) + tg(m)\mathbf{e}_3(m)$. □

8.4 Enneper–Weierstrass representation

Nearly a hundred years of research passed before minimal surfaces were finally understood in terms of the induced complex structure on the surface. The Enneper–Weierstrass representation of minimal surfaces comes from the $(1,0)$ part of $d\mathbf{x}$, denoted $\partial\mathbf{x}$ in Theorem 7.34. For Enneper see [65] and for Weierstrass see [169] and [168]. For a more detailed historical background see Darboux [58, Livre II, v.1]. For a modern exposition emphasizing global properties see D. Hoffman and H. Karcher [90].

Theorem 8.6 (Enneper–Weierstrass construction). *If α is a nowhere zero $\mathbf{C}^3$-valued isotropic holomorphic $(1,0)$-form on a connected Riemann surface M, then*

$$\mathbf{x} = 2\Re \int \alpha : \tilde{M} \to \mathbf{R}^3 \tag{8.7}$$

is a conformal minimal immersion of the universal cover $\tilde{M}$ of M. The map $\mathbf{x}$ descends to M if and only if α has no real periods on M.

Here $\Re$ denotes the real part and conformal immersion means that for any local complex coordinate z in $\tilde{M}$, the metric $I = d\mathbf{x}\cdot d\mathbf{x}$ is a positive multiple of $dzd\bar{z}$. The integral $f = \int \alpha$ denotes the holomorphic map $f : \tilde{M} \to \mathbf{C}^3$, determined up to translation, such that $df = \pi^*\alpha$, where $\pi : \tilde{M} \to M$ is the covering projection. More precisely, fixing any point $p_0 \in M$,

$$f(p) = \int_{p_0}^{p} \alpha$$

is the line integral over any path in M from p_0 to p. This line integral is invariant up to homotopy, since α is closed, so if M is not simply connected, then f is well-defined only on $\tilde{M}$. The obstruction to f being well-defined on M is the set of *periods* of f, which is

$$\{\int_\gamma \alpha : \gamma \in \mathscr{A}\},$$

where $\mathscr{A}$ is a set of closed paths whose homotopy classes generate the fundamental group of M. The map $\mathbf{x} : \tilde{M} \to \mathbf{R}^3$ being the real part of f is well-defined on M if the real parts of the periods are all zero. In short, we say that the real periods of α are zero.

Proof. The exterior derivative of equation (8.7) is $d\mathbf{x} = \alpha + \bar{\alpha}$, which is never zero. Its induced metric is $I = d\mathbf{x}\cdot d\mathbf{x} = 2\alpha\cdot\bar{\alpha}$, since α is isotropic, and thus $\mathbf{x}$ is conformal because $\alpha\cdot\bar{\alpha}$ is a metric in the conformal class of the complex structure of $\tilde{M}$. In fact, if z is a local complex coordinate in $\tilde{M}$, then $\alpha = \mathbf{a}(z)dz$ for some holomorphic $\mathbf{C}^3$-valued function $\mathbf{a}(z)$, and $\alpha\cdot\bar{\alpha} = |\mathbf{a}|^2 dz d\bar{z}$. Finally, $\mathbf{x}$ is a minimal immersion by Theorem 7.34, since $\partial\mathbf{x} = \alpha$ is holomorphic. □

8.4.1 Holomorphic isotropic 1-forms

The Enneper–Weierstrass theorems have reformulated the problem of solving the minimal surface equation to that of finding nowhere zero, isotropic $\mathbf{C}^3$-valued holomorphic 1-forms on a Riemann surface. This latter problem has a simple and complete solution in terms of meromorphic functions and ordinary abelian differentials on a Riemann surface.

Definition 8.7. A *meromorphic function* on a Riemann surface M is a holomorphic map

$$F : M \to \mathbf{CP}^1.$$

How does this definition relate to the classical definition of a meromorphic function as a map $g : M \to \mathbf{C}\cup\{\infty\}$ into the Riemann sphere such that each point of M has a neighborhood U on which $g = \mathfrak{p}/\mathfrak{q}$, where $\mathfrak{p}$ and $\mathfrak{q}$ are holomorphic functions on U? The relation comes from the fact that given a point $m \in M$, there exists a neighborhood U of m and a holomorphic map

$${}^t(\mathfrak{p},\mathfrak{q}) : U \to \mathbf{C}^2 \setminus \{\mathbf{0}\}$$

such that $F = {}^t[\mathfrak{p},\mathfrak{q}]$ on U. See Problem 8.53 below. The holomorphic function ${}^t(\mathfrak{p},\mathfrak{q})$ is determined only up to multiplication by a nowhere zero holomorphic function on U, but the ratio

$$g = \mathfrak{p}/\mathfrak{q} : U \to \mathbf{C}\cup\{\infty\}$$

is completely determined by F. This ratio is the local representation of the meromorphic function F as a quotient of holomorphic functions. In this representation of F, the poles of the meromorphic function are the zeros of $\mathfrak{q}$, namely, the points of $F^{-1}\{[\epsilon_1]\}$, a discrete subset of M, except in the case when F is constantly equal to $[\epsilon_1]$. This last case is the only possibility for which $\mathfrak{q}$ can be identically zero.

Moreover, it is important to see that $\mathfrak{p}$ and $\mathfrak{q}$ can be replaced by holomorphic sections of any holomorphic line bundle over M. For example, if μ and ν are holomorphic 1-forms on M, with no simultaneous zeros, then on any complex coordinate chart (U, z) of M, we have $\mu = \mathfrak{p}dz$ and $\nu = \mathfrak{q}dz$, for some holomorphic functions $\mathfrak{p}, \mathfrak{q} : U \to \mathbf{C}$, and the holomorphic map ${}^t[\mathfrak{p}, \mathfrak{q}] : U \to \mathbf{CP}^1$ is independent of the choice of coordinate chart. In this way the expression $F = {}^t[\mu, \nu] : M \to \mathbf{CP}^1$ is a well-defined holomorphic map, that is, a meromorphic function on M.

Theorem 8.8. *Let M be a connected Riemann surface. Any nowhere zero, isotropic, $\mathbf{C}^3$-valued holomorphic 1-form on M is given by*

$$\alpha = \begin{pmatrix} i \\ 1 \\ 0 \end{pmatrix} \eta \quad or \quad \alpha = \begin{pmatrix} \frac{1}{2}(1-g^2) \\ \frac{i}{2}(1+g^2) \\ g \end{pmatrix} \eta, \tag{8.8}$$

where g is a meromorphic function on M and η is a holomorphic 1-form on M, such that a point $p \in M$ is a pole of g of order m if and only if p is a zero of η of order $2m$. There are no other restrictions on g and η.

Proof. If α is defined by (8.8), then the poles and zeros of g and η must balance as stated so that α will be defined at the poles of g and so that α will never be zero.

Conversely, let $\alpha = {}^t(\alpha^1, \alpha^2, \alpha^3)$ be any nowhere zero, $\mathbf{C}^3$-valued isotropic holomorphic 1-form on M. Then each α^j, for $j = 1, 2, 3$, is an ordinary holomorphic 1-form on M and α isotropic implies

$$\alpha^3\alpha^3 = -(\alpha^1\alpha^1 + \alpha^2\alpha^2) = -(\alpha^1 - i\alpha^2)(\alpha^1 + i\alpha^2). \tag{8.9}$$

Being holomorphic, the 1-form $\alpha^1 - i\alpha^2$ is either identically zero or has only isolated zeros.

If $\alpha^1 - i\alpha^2$ is identically zero, then (8.9) implies that $\alpha^3 = 0$ and α must be given by the first formula in (8.8) with $\eta = \alpha^2$.

If $\alpha^1 - i\alpha^2$ has only isolated zeros on M, then

$$F = {}^t[\alpha^3, \alpha^1 - i\alpha^2] = {}^t[\alpha^1 + i\alpha^2, -\alpha^3] : M \to \mathbf{CP}^1 \tag{8.10}$$

is a holomorphic map on all of M, since if $\alpha^1 - i\alpha^2$ is zero at a point $p \in M$, then $\alpha^3(p) = 0$, but then $(\alpha^1 + i\alpha^2)(p) \neq 0$ since $\alpha(p) \neq \mathbf{0}$. Equality of the two expressions for F follows from (8.9). Then F determines the meromorphic function $g = \frac{\alpha^3}{\alpha^1 - i\alpha^2}$ on M and an elementary calculation verifies that α is given by the second expression in (8.8), with $\eta = \alpha^1 - i\alpha^2$. Thus the poles of g balance the zeros of η as required. □

Remark 8.9. The second expression in (8.8) can be replaced by

$$\alpha = \begin{pmatrix} \frac{1}{2}(\frac{1}{g} - g) \\ \frac{i}{2}(\frac{1}{g} + g) \\ 1 \end{pmatrix} \mu$$

where $\mu = g\eta$ is a holomorphic 1-form on M, such that a point $p \in M$ is a pole or zero of g of order m if and only if p is a zero of μ of order m.

8.4.2 Parametrization of isotropic vectors

The set of nonzero isotropic vectors is

$$\mathscr{I} = \{\mathbf{z} \in \mathbf{C}^3 \setminus \{\mathbf{0}\} : \mathbf{z} \cdot \mathbf{z} = 0\}, \tag{8.11}$$

where for $z = \sum_1^3 z^j \epsilon_j \in \mathbf{C}^3$ the dot product is $\mathbf{z} \cdot \mathbf{z} = \sum_1^3 (z^j)^2$. Consider the holomorphic, two-to-one surjective *Enneper–Weierstrass map*

$$\mathfrak{W} : \mathbf{C}^2 \setminus \{\mathbf{0}\} \to \mathscr{I} \subset \mathbf{C}^3, \quad \mathfrak{W}\begin{pmatrix} z \\ w \end{pmatrix} = \begin{pmatrix} \frac{1}{2}(w^2 - z^2) \\ \frac{i}{2}(w^2 + z^2) \\ wz \end{pmatrix}, \tag{8.12}$$

which satisfies

$$\mathfrak{W}(t\mathbf{z}) = t^2 \mathfrak{W}(\mathbf{z}),$$

for any non-zero $\mathbf{z} \in \mathbf{C}^2$ and $t \in \mathbf{C}$. If $\pi : \mathbf{C}^{n+1} \setminus \{\mathbf{0}\} \to \mathbf{CP}^n$ is the projection map, for any natural number n, then $\pi(\mathscr{I})$ is the complex quadric

$$\mathbf{Q}^1 = \{[\mathbf{z}] \in \mathbf{CP}^2 : \mathbf{z} \cdot \mathbf{z} = 0\}.$$

This quadric has a biholomorphic parametrization

$$\mathfrak{w} : \mathbf{CP}^1 \to \mathbf{Q}^1 \subset \mathbf{CP}^2, \quad \mathfrak{w}\begin{bmatrix} z \\ w \end{bmatrix} = \begin{bmatrix} \frac{1}{2}(w^2 - z^2) \\ \frac{i}{2}(w^2 + z^2) \\ wz \end{bmatrix}, \tag{8.13}$$

whose inverse mapping is

$$\mathfrak{w}^{-1}\begin{bmatrix} u \\ v \\ w \end{bmatrix} = \begin{bmatrix} w \\ u - iv \end{bmatrix} = \begin{bmatrix} u + iv \\ -w \end{bmatrix}, \tag{8.14}$$

from the factorization of $u^2 + v^2 + w^2 = 0$ given by $(u + iv)(u - iv) = -w^2$ showing equality of the last two terms in (8.14). The following diagram commutes:

$$\begin{array}{ccc} \mathbf{C}^2 \setminus \{\mathbf{0}\} & \stackrel{\mathscr{W}}{\to} & \mathscr{I} \subset \mathbf{C}^3 \setminus \{\mathbf{0}\} \\ \pi \downarrow & & \downarrow \pi \\ \mathbf{CP}^1 & \stackrel{\mathfrak{w}}{\to} & Q^1 \subset \mathbf{CP}^2 \end{array}$$

Theorem 8.10 (Enneper–Weierstrass). *Let* $\mathbf{x} : M \to \mathbf{R}^3$ *be a minimal immersion of a connected surface. Then* $\partial\mathbf{x}$ *is a nowhere zero holomorphic isotropic 1-form on* M *given by one of the two formulas in* (8.8). *In the case* α *is given by the second of these formulas,*

$$\alpha = \partial \mathbf{x} = \begin{pmatrix} \frac{1}{2}(1-g^2) \\ \frac{i}{2}(1+g^2) \\ g \end{pmatrix} \eta, \tag{8.15}$$

the metric induced by **x** *is*

$$d\mathbf{x} \cdot d\mathbf{x} = I = (1+|g|^2)^2 \eta\bar{\eta}, \tag{8.16}$$

which is in the conformal class of the Riemann surface, the Hopf quadratic differential of **x** *is*

$$II^{2,0} = -\eta\, dg, \tag{8.17}$$

and the Gaussian curvature of I *is*

$$K = \frac{-4}{(1+|g|^2)^4} \left| \frac{dg}{\eta} \right|^2. \tag{8.18}$$

The Gauss map of **x** *is*

$$\mathbf{e}_3 = \mathscr{S}^{-1}(g) = \frac{1}{1+|g|^2} \begin{pmatrix} g+\bar{g} \\ -i(g-\bar{g}) \\ |g|^2 - 1 \end{pmatrix}, \tag{8.19}$$

where

$$\mathscr{S} : \mathbf{S}^2 \setminus \{\boldsymbol{\epsilon}_3\} \to \mathbf{C}$$

is stereographic projection from the north pole $\boldsymbol{\epsilon}_3 = {}^t(0,0,1)$,

$$\mathscr{S}(u,v,w) = \frac{u+iv}{1-w}, \quad \mathscr{S}^{-1}(u+iv) = \frac{1}{u^2+v^2+1} {}^t(2u, 2v, u^2+v^2-1).$$

In the case $\partial\mathbf{x}$ *is given by the first formula in* (8.8)*, then* $\mathbf{x}(M)$ *is a plane with induced metric, Gauss map, and Hopf differential, respectively, given by*

$$d\mathbf{x} \cdot d\mathbf{x} = 4\eta\bar{\eta}, \quad \mathbf{e}_3 = \boldsymbol{\epsilon}_3, \quad II^{2,0} = 0. \tag{8.20}$$

Proof. One calculates (8.16) from $d\mathbf{x} = \alpha + \bar{\alpha}$. Let U, z be a complex coordinate chart in M. Then $\eta = f(z)\, dz$, for some holomorphic function $f(z)$ on U,

$$\mathbf{x}_z\, dz = \alpha = \begin{pmatrix} \frac{1}{2}(1-g^2) \\ \frac{i}{2}(1+g^2) \\ g \end{pmatrix} f\, dz, \tag{8.21}$$

and therefore

$$e^{2u} = 2\mathbf{x}_z \cdot \mathbf{x}_{\bar{z}} = (1 + |g|^2)^2 |f|^2.$$

Then from (7.36) we find formula (8.19) for $\mathbf{e}_3$. Differentiate (8.21) to find

$$\mathbf{x}_{zz} = g_z f \begin{pmatrix} -g \\ ig \\ 1 \end{pmatrix} + \frac{f_z}{f}\mathbf{x}_z,$$

and use this in (7.37) to find the Hopf invariant

$$h = \frac{-2g_z}{(1 + |g|^2)^2 \bar{f}}, \tag{8.22}$$

and substitute this into $II^{2,0} = \frac{1}{2} h e^{2u} dz dz$ to obtain (8.17). By (7.31), $K = H - |h|^2$, which combined with (8.22) gives (8.18).

If $\mathbf{x}$ is given by the first equation in (8.8), then $\mathbf{x}_z = {}^t(i, 1, 0)f$ so that $e^{2u} = 4|f|^2$, from which we obtain the first equation in (8.20). From (7.36) we verify the second equation in (8.20). Now $\mathbf{x}_{zz} = {}^t(i, 1, 0)f_z$, which substituted into (7.37) gives $h = 0$ and that verifies the third equation in (8.20). □

8.4.3 Weierstrass data and associates

We summarize our results for minimal surfaces in $\mathbf{R}^3$. According to Theorem 7.34, an immersion $\mathbf{x} : M \to \mathbf{R}^3$ is minimal if and only if the $\mathbf{C}^3$-valued 1-form $\alpha = \partial \mathbf{x}$ is holomorphic and isotropic. Given any such holomorphic and isotropic 1-form on a Riemann surface M, we know from Theorem 8.6 that $\mathbf{x} = 2\Re \int \alpha : \tilde{M} \to \mathbf{R}^3$ is a conformal minimal immersion, where $\tilde{M}$ is the universal covering space of M. Finally, Theorem 8.8 tells us that, with the exception of one trivial case, any holomorphic isotropic 1-form α on M is given by a meromorphic function g and a holomorphic 1-form η, with poles and zeros balanced in the appropriate way. This data thus determines any nonplanar minimal immersion into $\mathbf{R}^3$.

Definition 8.11. A set of *Weierstrass data* on a Riemann surface M comprises a meromorphic function g and a holomorphic isotropic 1-form η on M, such that a point of M is a pole of g of order m if and only if it is a zero of η of order $2m$.

The set of nonplanar minimal immersions of a simply connected Riemann surface M is in one-to-one correspondence with the set of Weierstrass data on M.

Recall Definition 7.30 of the associates of a constant mean curvature immersion $\mathbf{x} : M \to \mathbf{R}^3$. In terms of a complex coordinate z on M, Corollary 7.31 says that if $\mathbf{x}$ has conformal factor e^u, Hopf invariant h, and constant mean curvature H, then any associate has the same conformal factor and mean curvature, but the Hopf invariant must be $e^{it}h$, for some real constant t for which $e^{it} \neq 0$. If $H \neq 0$, the resulting structure equations must be solved to find this associate. Solving these equations is far simpler when $H = 0$, as then we can use the Weierstrass data.

Suppose the minimal immersion $\mathbf{x} : M \to \mathbf{R}^3$, where M is simply connected, is given by the Weierstrass data (g, η), which determines the holomorphic isotropic 1-form α in (8.21). Let t be any real constant. We see from (8.22) that the Hopf invariant $e^{it}h$ of the associate $\mathbf{x}^{(t)}$ determined by t is the Hopf invariant of the minimal immersion determined by the Weierstrass data $(g, e^{it}\eta)$, which determines the holomorphic isotropic 1-form $\alpha^{(t)} = e^{it}\alpha$. Therefore, by the Enneper–Weierstrass construction of Theorem 8.6,

$$\mathbf{x}^{(t)} = 2\Re \int \alpha^{(t)} = (\cos t)\mathbf{x} - (\sin t)\mathbf{x}^c, \tag{8.23}$$

where

$$\mathbf{x}^c = 2\Im \int \alpha : M \to \mathbf{R}^3,$$

is called the *conjugate surface* to $\mathbf{x}$. It is the associate for which $t = -\pi/2$.

Definition 8.12 (Conformal associate). Let $\mathbf{x} : M \to \mathbf{R}^3$ be a conformal minimal immersion of a Riemann surface M. The *conformal associate* of $\mathbf{x}$ defined by a holomorphic function $\mathfrak{p}$ on M is a minimal immersion $\hat{\mathbf{x}} : M \to \mathbf{R}^3$ such that

$$\partial \hat{\mathbf{x}} = \mathfrak{p}\, \partial \mathbf{x}$$

where ∂ is the operator defined in (7.7).

8.5 Examples

The following examples illustrate features of the Weierstrass representation such as how to integrate α, the existence of real periods, and associate surfaces.

Example 8.13 (Enneper's surface). If $M = \mathbf{C}$, $g(z) = z$ and $\eta = dz$, then

$$\alpha = \frac{1}{2} \begin{pmatrix} 1 - z^2 \\ i(1 + z^2) \\ 2z \end{pmatrix} dz$$

is the nowhere zero, isotropic, $\mathbf{C}^3$-valued holomorphic 1-form on M. Then the *minimal curve* $\boldsymbol{\gamma} : M \to \mathbf{C}^3$ is, up to translation,

$$\boldsymbol{\gamma}(z) = 2 \int_0^z \alpha = \begin{pmatrix} z - \frac{1}{3}z^3 \\ i(z + \frac{1}{3}z^3) \\ z^2 \end{pmatrix} = \mathbf{x}(z) + i\mathbf{x}^c(z).$$

The real part of $\boldsymbol{\gamma}$, gives Enneper's minimal immersion (7.39),

$$\mathbf{x} = \Re \begin{pmatrix} z - \frac{1}{3}z^3 \\ i(z + \frac{1}{3}z^3) \\ z^2 \end{pmatrix} = \begin{pmatrix} x + xy^2 - x^3/3 \\ -y - x^2y + y^3/3 \\ x^2 - y^2 \end{pmatrix},$$

where $z = x + iy$. The imaginary part of $\boldsymbol{\gamma}$ is the Enneper *conjugate surface*.

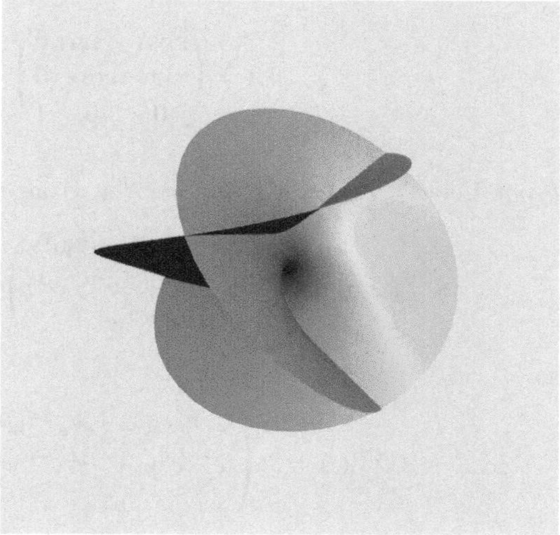

Fig. 8.2 Enneper's surface.

The Enneper associates are given for each real constant t by

$$\mathbf{x}^{(t)} = 2\Re \int e^{it}\alpha = \Re \begin{pmatrix} e^{it}(z - \frac{1}{3}z^3) \\ e^{it}i(z + \frac{1}{3}z^3) \\ e^{it}z^2 \end{pmatrix} = \mathbf{x}\cos t - \mathbf{x}^c \sin t. \tag{8.24}$$

The case $t = 0$, is the Enneper surface itself, shown in Figure 8.2. The case $t = -\pi/2$ is the conjugate surface $\mathbf{x}^c$.

The induced metric of Enneper's surface and all of its associates is

$$I = e^{2u}dzd\bar{z}, \text{ where } e^u = 1 + |z|^2.$$

The Hopf quadratic differential for any t is

$$II_t^{2,0} = -e^{it}dzdz = \frac{1}{2}h_t e^{2u}dzdz,$$

so the Hopf invariant h_t of $\mathbf{x}^{(t)}$ relative to z is

$$h_t = \frac{-2e^{it}}{(1 + |z|^2)^2},$$

thus showing that $|h_t|$ is independent of t. If $e^{it} \neq e^{is}$, then $\mathbf{x}^{(t)}$ is not congruent to $\mathbf{x}^{(s)}$ by a rigid motion, by Corollary 7.31. Nevertheless, the associates are all congruent to $\mathbf{x}$ by a rigid motion after a reparametrization by an internal isometry of (M,I), as we now explain (see [61, pp 147–149]).

Lemma 8.14. *For any real constant t,*

$$A\mathbf{x}^{(2t)}(z) = \mathbf{x} \circ F(z),$$

for all $z \in \mathbf{C}$, *where* $F(z) = e^{it}z$ *and*

$$A = \begin{pmatrix} \cos t & -\sin t & 0 \\ \sin t & \cos t & 0 \\ 0 & 0 & 1 \end{pmatrix} \in \mathbf{SO}(3). \tag{8.25}$$

Proof. Let $w = F(z) = e^{it}z$, so $z = e^{-it}w$. Using (8.24), we get

$$\mathbf{x}^{(2t)}(z) = \Re \begin{pmatrix} e^{i2t}(e^{-it}w - \frac{1}{3}e^{-i3t}w^3) \\ e^{i2t}i(e^{-it}w + \frac{1}{3}e^{-i3t}w^3) \\ e^{i2t}e^{-i2t}w^2 \end{pmatrix} = \Re \begin{pmatrix} e^{it}w - \frac{1}{3}e^{-it}w^3 \\ ie^{it}w + \frac{i}{3}e^{-it}w^3 \\ w^2 \end{pmatrix},$$

so

$$A\mathbf{x}^{(2t)}(z) = \Re \begin{pmatrix} e^{-it}e^{it}w - \frac{1}{3}e^{it}e^{-it}w^3 \\ ie^{-it}e^{it}w + \frac{i}{3}e^{it}e^{-it}w^3 \\ w^2 \end{pmatrix} = \mathbf{x}(w) = \mathbf{x} \circ F(z).$$

Let us see where A and F came from. The map $F : M \to M$ given by $w = F(z) = e^{it}z$ is holomorphic and an isometry, since

$$F^*I = (1+|w|^2)^2 dw d\bar{w} = I.$$

If $II = II^{2,0} + II^{0,2} = -dzdz - d\bar{z}d\bar{z}$ is the second fundamental form of the minimal immersion $\mathbf{x} : M \to \mathbf{R}^3$, then the second fundamental form $\widetilde{II}$ of $\mathbf{x} \circ F$ has the decomposition into bidegrees

$$\widetilde{II} = F^*II = F^*(II^{2,0} + II^{0,2})$$

where

$$F^*II^{2,0} = F^*(-dzdz) = -dwdw = -e^{i2t}dzdz = II^{2,0}_{2t},$$

is the Hopf quadratic differential of $\mathbf{x}^{(2t)}$. Hence, $\mathbf{x} \circ F : M \to \mathbf{R}^3$ must be congruent by rigid motion to $\mathbf{x}^{(2t)} : M \to \mathbf{R}^3$ by Corollary 7.31. Since both immersions send $0 \in M$ to the origin of $\mathbf{R}^3$, there is no translation part to the rigid motion, so there is a rotation $A \in \mathbf{SO}(3)$ such that $A\mathbf{x}^{(2t)}(z) = \mathbf{x} \circ F(z)$, for every $z \in \mathbf{C}$. Comparing $d\mathbf{x}^{(2t)}$ to $d(\mathbf{x} \circ F)$, we find that A must be given by (8.25). □

Example 8.15 (Catenoid). If $M = \mathbf{C}\setminus\{0\}$, $g(z) = 1/z$ and $\eta = dz$, then

$$\alpha = \begin{pmatrix} \frac{1}{2}(1-\frac{1}{z^2}) \\ \frac{i}{2}(1+\frac{1}{z^2}) \\ \frac{1}{z} \end{pmatrix} dz \tag{8.26}$$

is the nowhere zero, isotropic, $\mathbf{C}^3$-valued holomorphic 1-form on M. Then

$$\mathbf{x} = 2\Re\int \alpha = 2\Re \begin{pmatrix} \frac{1}{2}(z+\frac{1}{z}) \\ \frac{i}{2}(z-\frac{1}{z}) \\ \log z \end{pmatrix} = \begin{pmatrix} x+\frac{x}{x^2+y^2} \\ -y-\frac{y}{x^2+y^2} \\ \log(x^2+y^2) \end{pmatrix}, \tag{8.27}$$

which in terms of polar coordinates $x = r\cos\theta$, $y = r\sin\theta$, is

$$\mathbf{x} = \begin{pmatrix} (r+\frac{1}{r})\cos\theta \\ -(r+\frac{1}{r})\sin\theta \\ 2\log r \end{pmatrix}.$$

If we put $s = \log r$, then $r = e^s$, $r + 1/r = 2\cosh s$, and we have

$$\mathbf{x} = \left(2\cosh s\cos\theta, -2\cosh s\sin\theta, 2s\right), \tag{8.28}$$

which can now be seen to be a surface of revolution about the x^3-axis with profile curve $x^1 = 2\cosh\frac{x^3}{2}$, the catenary. See Figure 8.3.

Another approach is to integrate the lift of α to the universal covering

$$\exp : \mathbf{C} \to M, \quad \exp(w) = e^w = z,$$

whose group of deck transformations is

$$\Gamma = \{g_n : g_n(w) = w + 2\pi i n, n \in \mathbf{Z}\} \cong \mathbf{Z}.$$

Then $\tilde{g} = \exp^* g = e^{-w}$, $\tilde{\eta} = \exp^*\eta = e^w dw$, and

$$\tilde{\alpha} = \exp^*\alpha = \begin{pmatrix} \frac{1}{2}(1-e^{-2w}) \\ \frac{i}{2}(1+e^{-2w}) \\ e^{-w} \end{pmatrix} e^w dw = \begin{pmatrix} \sinh w \\ i\cosh w \\ 1 \end{pmatrix} dw,$$

so

$$\tilde{\mathbf{x}} = 2\Re\int_0^w \tilde{\alpha} = 2\Re \begin{pmatrix} \cosh w \\ i\sinh w \\ w \end{pmatrix}.$$

If we set $w = s + i\theta$, then

$$\cosh w = \cosh s\cos\theta + i\sinh s\sin\theta, \quad \sinh w = \sinh s\cos\theta + i\cosh s\sin\theta,$$

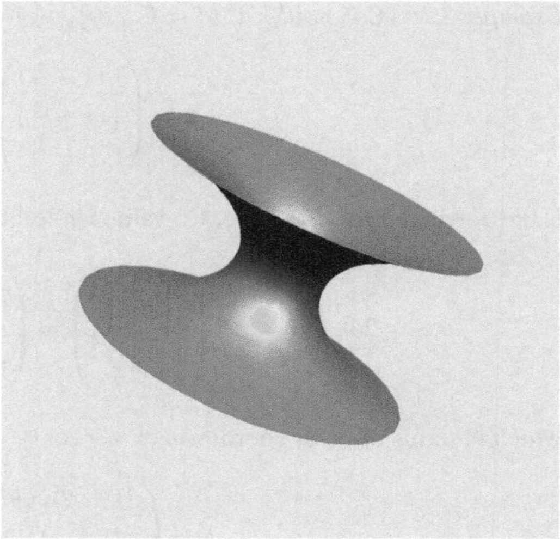

Fig. 8.3 The catenoid

so

$$\tilde{\mathbf{x}} = 2\begin{pmatrix} \cosh s \cos\theta \\ -\cosh s \sin\theta \\ s \end{pmatrix} = \begin{pmatrix} \cos\theta & \sin\theta & 0 \\ -\sin\theta & \cos\theta & 0 \\ 0 & 0 & 1 \end{pmatrix} \begin{pmatrix} 2\cosh s \\ 0 \\ 2s \end{pmatrix},$$

which is invariant under the group of deck transformations Γ because of the periodicity of $\cos\theta$ and $\sin\theta$. Therefore, $\tilde{\mathbf{x}}$ descends to the map $\mathbf{x} : M \to \mathbf{R}^3$ given by the formula (8.28).

The induced metric of $\mathbf{x}$ is

$$I = d\mathbf{x} \cdot d\mathbf{x} = (1 + |g|^2)^2 \eta\bar{\eta} = (1 + \frac{1}{|z|^2})^2 dz d\bar{z}.$$

Its Hopf differential is $II^{2,0} = -\eta dg = \frac{1}{z^2} dz dz$, and its Gauss map (followed by stereographic projection onto $\mathbf{C}$) is $g(z) = 1/z$.

Example 8.16 (Catenoid associates and Helicoid). $M = \mathbf{C} \setminus \{0\}$, $g(z) = 1/z$, $\eta = e^{it} dz$, for a fixed real constant t. The isotropic $\mathbf{C}^3$-valued holomorphic 1-form is $e^{it}\alpha$, where α is the isotropic abelian differential (8.26) defining the Catenoid. If $\mathbf{x}$ denotes the Catenoid solution (8.27), then the present solution is $(\cos t)\mathbf{x} - (\sin t)\mathbf{x}^c$, where $\mathbf{x}^c$ is the *conjugate*

$$\mathbf{x}^c(z) = 2\Im \int \alpha = 2\Im \begin{pmatrix} \frac{1}{2}(z + \frac{1}{z}) \\ \frac{i}{2}(z - \frac{1}{z}) \\ \int \frac{dz}{z} \end{pmatrix},$$

which is multivalued on M, since $\Im \int \frac{dz}{z}$ has real periods. See Figure 8.4.

Fig. 8.4 Catenoid conjugate $t = \frac{\pi}{2}$, the helicoid

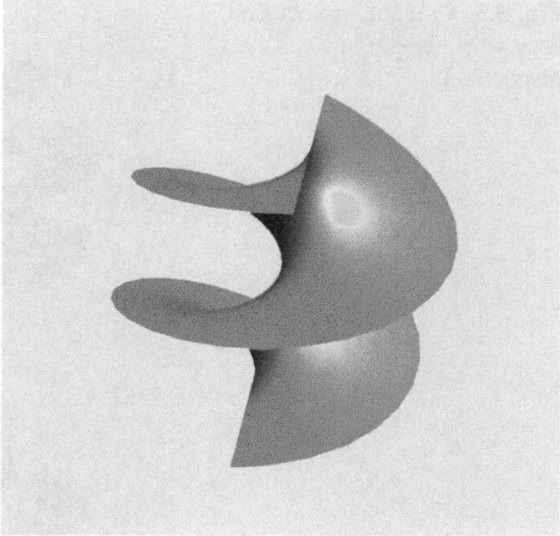

In fact, if we take $0 \neq [\gamma] \in \pi_1(M)$, where $\gamma : [0, 2\pi] \to M$, $\gamma(t) = e^{it}$, then

$$\int_\gamma \frac{dz}{z} = \int_0^{2\pi} e^{-i\theta} de^{i\theta} = 2\pi i.$$

The solution $\mathbf{x}^c$ composed with the universal covering projection $\tilde{M} = \mathbf{C} \to \mathbf{C} \setminus \{0\} = M$, $z = e^w$, where $w = u + iv$ is the complex coordinate on $\tilde{M}$, is

$$\mathbf{x}^c(e^w) = 2\Im \begin{pmatrix} \frac{1}{2}(e^w + e^{-w}) \\ \frac{i}{2}(e^w - e^{-w}) \\ w \end{pmatrix} = 2 \begin{pmatrix} \sinh u \sin v \\ \sinh u \cos v \\ v \end{pmatrix},$$

which is single valued on $\tilde{M}$. It is a helicoid. Figure 8.5 shows the associates when t equals $\frac{\pi}{8}$, $\frac{\pi}{6}$, $\frac{\pi}{4}$, and $\frac{3\pi}{8}$, respectively.

Example 8.17 (Scherk's surface). This example illustrates many important subtleties of the Enneper–Weierstrass representation. On the Riemann surface

$$M = \mathbf{C} \cup \{\infty\} \setminus \{z^4 = 1\} = \mathbf{C} \cup \{\infty\} \setminus \{\pm 1, \pm i\},$$

take the meromorphic function g and the holomorphic differential η,

$$g(z) = z, \quad \eta = 2\frac{dz}{z^4 - 1}.$$

The only pole of g on M is ∞, which is a simple pole. To see the behavior of η at ∞, let $w = 1/z$, a local complex coordinate centered at ∞. Then $dz = -\frac{1}{w^2}dw$ and

$$\eta = 2\frac{-dw/w^2}{1/w^4 - 1} = \frac{2w^2 dw}{w^4 - 1},$$

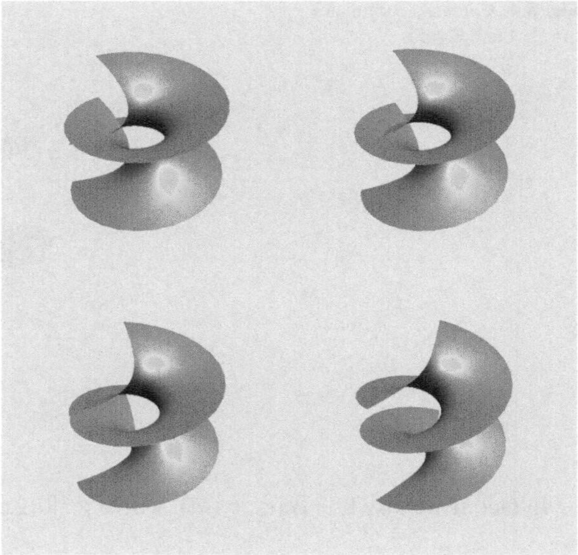

Fig. 8.5 Catenoid associates for $t = \frac{\pi}{8}, \frac{\pi}{6}, \frac{\pi}{4}, \frac{3\pi}{8}$, respectively.

which has a double zero at ∞, that is, at $w = 0$, as required. Then

$$\alpha = \begin{pmatrix} \alpha^1 \\ \alpha^2 \\ \alpha^3 \end{pmatrix} = \begin{pmatrix} \frac{1}{2}(1-z^2) \\ \frac{i}{2}(1+z^2) \\ z \end{pmatrix} \frac{2}{z^4-1}\,dz = \begin{pmatrix} -\frac{i}{2}(\frac{1}{z+i} - \frac{1}{z-i})\,dz \\ -\frac{i}{2}(\frac{1}{z+1} - \frac{1}{z-1})\,dz \\ (\frac{z}{z^2-1} - \frac{z}{z^2+1})\,dz \end{pmatrix} \tag{8.29}$$

is a holomorphic, nowhere zero isotropic differential on M whose integral

$$\int_0^z \alpha = \begin{pmatrix} -\frac{i}{2}\log(\frac{z+i}{z-i}) \\ -\frac{i}{2}\log(\frac{z+1}{z-1}) \\ \frac{1}{2}\log(\frac{z^2-1}{z^2+1}) \end{pmatrix} \tag{8.30}$$

depends on the homotopy class of the piecewise smooth path chosen from 0 to z in M. Thus, to obtain a single valued map, we consider this map as defined on the universal cover $\mu : (\tilde{M}, \tilde{0}) \to (M, 0)$, which does not have an elementary expression like that used above for the Catenoid. We shall use the representation of $\tilde{M}$ as the set of all homotopically equivalent paths in M from the base point 0 (see, for example, A. Hatcher's text [83, §1.3]). For the rest of this example, path will mean piecewise smooth path. A point $[\gamma] \in \tilde{M}$ is represented by a path γ in M with initial point $\gamma_0 = 0$ and terminal point $\gamma_1 = z \in M$. The covering projection is given by $\mu[\gamma] = \gamma_1$. If δ is another path in M from 0 to z, then the homotopy classes $[\gamma] = [\delta]$ if and only if the loop $\gamma\delta^{-1}$ at 0 is homotopically trivial; that is, $[\gamma\delta^{-1}] = 1$ in the fundamental group $G = \pi_1(M, 0)$.

Our minimal immersion is

$$\tilde{\mathbf{x}} : \tilde{M} \to \mathbf{R}^3, \quad \tilde{\mathbf{x}}[\gamma] = 2\Re \int_\gamma \alpha = \begin{pmatrix} \arg_\gamma(\frac{z+i}{z-i}) \\ \arg_\gamma(\frac{z+1}{z-1}) \\ \log|\frac{z^2-1}{z^2+1}| \end{pmatrix} = \begin{pmatrix} x^1_\gamma(z) \\ x^2_\gamma(z) \\ x^3(z) \end{pmatrix}, \tag{8.31}$$

where $\arg_\gamma(\frac{z+i}{z-i})$ denotes the continuous extension of the argument function along the path $\frac{\gamma+i}{\gamma-i}$ from -1 to $\frac{z+i}{z-i}$, with the initial value $\arg(-1) = \pi$. Define $\arg_\gamma(\frac{z+1}{z-1})$ in the same way. Notice that the third component of $\tilde{\mathbf{x}}$ depends only on z, and not on the path from 0 to z. Since $\arg(z)$ is determined by z up to adding an integer multiple of 2π, the functions $\cos x^i_\gamma(z)$ and $\sin x^i_\gamma(z)$, for $i = 1, 2$, depend only on the endpoint z and not on the path γ. For any point $z \in M$,

$$\begin{aligned} w_1 &= \frac{z+i}{z-i} = \frac{|z|^2 - 1 + i(z+\bar{z})}{|z-i|^2} = \left|\frac{z+i}{z-i}\right| (\cos x^1(z) + i \sin x^1(z)), \\ w_2 &= \frac{z+1}{z-1} = \frac{|z|^2 - 1 + \bar{z} - z}{|z-1|^2} = \left|\frac{z+1}{z-1}\right| (\cos x^2(z) + i \sin x^2(z)). \end{aligned} \tag{8.32}$$

Comparing the real and imaginary parts in each line, we find

$$\cos x^1(z) = \frac{|z|^2 - 1}{|z-i|^2} \left|\frac{z-i}{z+i}\right| = \frac{|z|^2 - 1}{|z^2+1|}, \quad \cos x^2(z) = \frac{|z|^2 - 1}{|z^2-1|}.$$

Thus, for any $z \in M$ for which $|z| \neq 1$, we have

$$\frac{\cos x^1(z)}{\cos x^2(z)} = \frac{|z^2-1|}{|z^2+1|} = e^{x^3(z)},$$

by (8.31). This shows that $\tilde{\mathbf{x}}(\tilde{M} \setminus \mu^{-1}\{|z| = 1\})$ is contained in the Scherk surface given as a graph in (8.2). Moreover, from (8.32) we also have

$$\sin x^1(z) = \frac{z+\bar{z}}{|z^2+1|}, \quad \sin x^2(z) = \frac{i(z-\bar{z})}{|z^2-1|}. \tag{8.33}$$

Let us examine $\tilde{\mathbf{x}}$ on the set $\hat{D}$ of all homotopy classes of paths in the open unit disk $D = \{|z| < 1\} \subset M$. Each such class is determined uniquely by its endpoint $z \in D$, since D is contractible, and so we may write $\hat{D} = D$. Then (8.32) shows that both w_1 and w_2 map D biholomorphically onto the left half-plane $\{\Re(w) < 0\}$, with 0 going to -1 in both cases. They are biholomorphic because in each case one can explicitly solve for z in terms of w. Thus,

$$(x^1, x^2)(D) = (\pi/2, 3\pi/2) \times (\pi/2, 3\pi/2) = S_{1,1},$$

in the notation of (8.3). Combining this with (8.33), we get

$$\tilde{\mathbf{x}} : D \to \mathbf{R}^3, \quad \tilde{\mathbf{x}}(z) = \begin{pmatrix} \pi - \arcsin(\frac{z+\bar{z}}{|z^2+1|}) \\ \pi - \arcsin(\frac{i(z-\bar{z})}{|z^2-1|}) \\ \log|\frac{z^2-1}{z^2+1}| \end{pmatrix},$$

which gives the conformal parametrization $\tilde{\mathbf{x}} : D \to \mathbf{R}^3$ of Scherk's surface over $S_{1,1}$ shown in Figure 8.6.

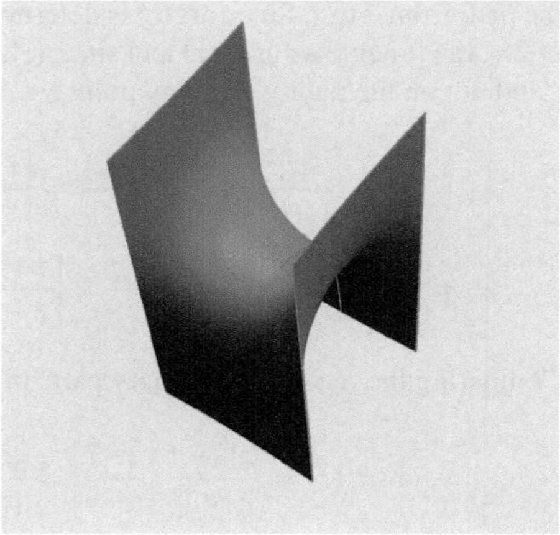

Fig. 8.6 The Weierstrass image of the unit disk D for Scherk's surface.

Next we examine the value of $\tilde{\mathbf{x}}([\gamma_\theta])$, for θ in each of the open intervals $(0,\pi/2)$, $(\pi/2,\pi)$, $(\pi,3\pi/2)$, and $(3\pi/2,2\pi)$, where γ_θ is the the line segment from 0 to $e^{i\theta}$. From (8.32), we get

$$w_1(e^{i\theta}) = i\frac{\cos\theta}{1-\sin\theta}, \quad w_2(e^{i\theta}) = i\frac{-\sin\theta}{1-\cos\theta}.$$

Combining this with the fact that for $z \in D$, along a path inside D, $\arg(w_1(z))$ and $\arg(w_2(z))$ are both in the open interval $(\pi/2, 3\pi/2)$, we conclude that along paths inside D,

$$x^1_{\gamma_\theta}(e^{i\theta}) = \arg_{\gamma_\theta}(w_1(e^{i\theta})) = \begin{cases} \pi/2, & \text{if } 0<\theta<\pi/2 \text{ or } 3\pi/2<\theta<2\pi, \\ 3\pi/2, & \text{if } \pi/2<\theta<\pi \text{ or } \pi<\theta<3\pi/2, \end{cases}$$

$$x^2_{\gamma_\theta}(e^{i\theta}) = \arg_{\gamma_\theta}(w_2(e^{i\theta})) = \begin{cases} 3\pi/2, & \text{if } 0<\theta<\pi/2 \text{ or } \pi/2<\theta<\pi, \\ \pi/2, & \text{if } \pi<\theta<2\pi/2 \text{ or } 3\pi/2<\theta<2\pi. \end{cases} \tag{8.34}$$

See Figure 8.7. Moreover, on each of these four intervals, $x^3(e^{i\theta}) = \log|\tan\theta|$. We conclude that $\tilde{\mathbf{x}}$ maps each of these quarter circles onto a vertical line through a vertex of the square $S_{1,1}$.

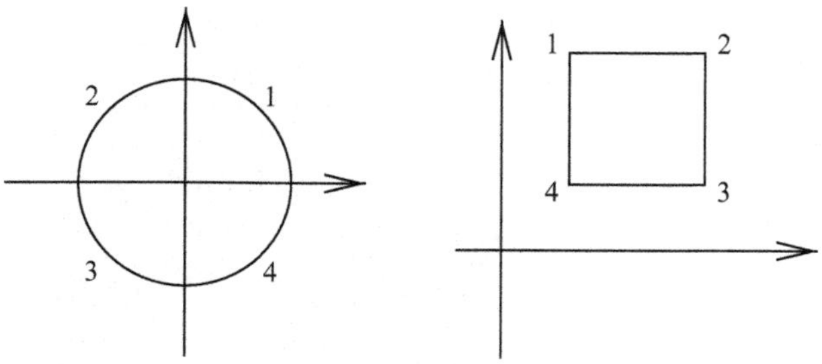

Fig. 8.7 (x^1, x^2) sends numbered arc to same numbered vertex.

Now consider what $\tilde{\mathbf{x}}$ does to $\bar{D}'$, the complement of the closed unit disk in M, on the set of homotopy classes of paths consisting of the line segment from 0 to $1-i$ followed by a path entirely in $\bar{D}'$. Each such class is uniquely determined by its endpoint $z \in \bar{D}'$, since $\bar{D}'$ contains the point ∞ so is topologically a disk. We may thus denote this set of homotopy classes by $\bar{D}'$. By (8.32) we see that both w_1 and w_2 map $\bar{D}'$ biholomorphically onto the right half-plane $\{\Re(w) > 0\}$, and so $\arg(w_1(z))$ and $\arg(w_2(z))$ lie in the open interval $(-\pi/2, \pi/2)$ modulo an additive integer multiple of 2π. But then (8.34) and the required continuity of arg along paths imply that $\mathbf{x}^1(z)$ and $\mathbf{x}^2(z)$ lie in $(-\pi/2, \pi/2)$ for every $z \in \bar{D}'$. This with (8.33) gives us the Enneper–Weierstrass parametrization

$$\tilde{\mathbf{x}} : \bar{D}' \to \mathbf{R}^3, \quad \tilde{\mathbf{x}}(z) = \begin{pmatrix} \arcsin \frac{z+\bar{z}}{|z^2+1|} \\ \arcsin \frac{i(z-\bar{z})}{|z^2-1|} \\ \log |\frac{z^2-1}{z^2+1}|, \end{pmatrix},$$

which is illustrated in Figure 8.8. Putting together the Weierstrass images of D and $\bar{D}'$, we get Figure 8.9.

Finally, we examine $\tilde{\mathbf{x}}$ on all of $\tilde{M}$. If σ is a closed loop in M, then α holomorphic on M implies

$$\int_\sigma \alpha = 2\pi i \sum_{p \in \{\pm 1, \pm i\}} n_\sigma(p) \operatorname{res}_\alpha(p),$$

where $n_\sigma(p)$ is the winding number of σ about p and $\operatorname{res}_\alpha(p)$ denotes the residue of α at p (see [118, Residue Theorem, pp. 217 ff]). Using (8.29), we find

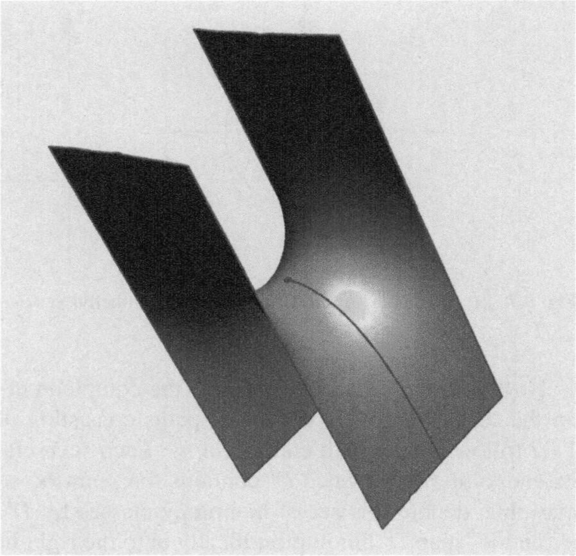

Fig. 8.8 The Weierstrass image of $\bar{D}'$ for Scherk's surface.

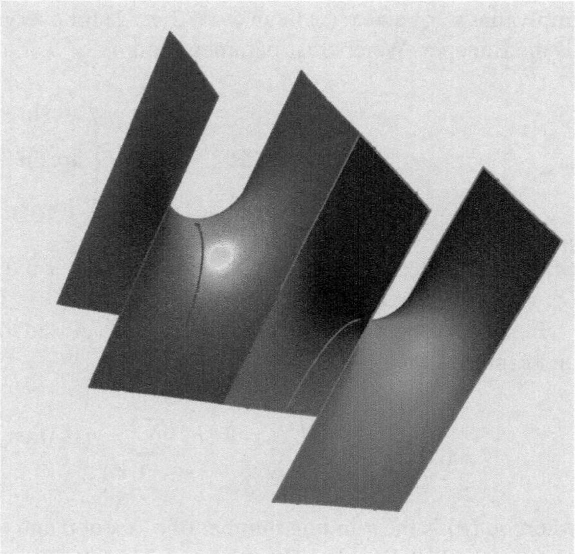

Fig. 8.9 The Weierstrass image of D and $\bar{D}'$ for Scherk's surface.

$$2\pi i\,\mathrm{res}_\alpha(1) = \pi \begin{pmatrix} 0 \\ -1 \\ i \end{pmatrix}, \quad 2\pi i\,\mathrm{res}_\alpha(-1) = \pi \begin{pmatrix} 0 \\ 1 \\ i \end{pmatrix},$$
$$2\pi i\,\mathrm{res}_\alpha(i) = \pi \begin{pmatrix} -1 \\ 0 \\ -i \end{pmatrix}, \quad 2\pi i\,\mathrm{res}_\alpha(-i) = \pi \begin{pmatrix} 1 \\ 0 \\ -i \end{pmatrix}. \tag{8.35}$$

Then the period vectors of the minimal immersion $\tilde{\mathbf{x}} : \tilde{M} \to \mathbf{R}^3$ given in (8.31) are twice the real part of each of the vectors in (8.35). This means that for any closed path σ in M, based at 0,

$$2\Re(\int_\sigma \alpha) = 2\pi((n_\sigma(-i) - n_\sigma(i))\epsilon_1 + (n_\sigma(-1) - n_\sigma(1))\epsilon_2). \tag{8.36}$$

Since the winding number is invariant under homotopy, we have the map

$$F : G \to \mathbf{Z}^2, \quad F[\sigma] = (n_\sigma(-i) - n_\sigma(i))\epsilon_1 + (n_\sigma(-1) - n_\sigma(1))\epsilon_2, \tag{8.37}$$

which is a group homomorphism, since the winding number of a product of closed loops at 0 is the sum of their winding numbers. The fundamental group G of M acts as holomorphic deck transformations on $\tilde{M}$ by

$$G \times \tilde{M} \to \tilde{M}, \quad [\sigma][\gamma] = [\sigma\gamma],$$

for any homotopy class of closed loops $[\sigma]$ at 0 and any homotopy class of paths $[\gamma]$ starting at 0 in M. Under this action, $\tilde{\mathbf{x}}$ satisfies

$$\tilde{\mathbf{x}}([\sigma][\gamma]) = \tilde{\mathbf{x}}([\gamma]) + 2\pi F[\sigma], \tag{8.38}$$

by elementary properties of path integrals and (8.36). If

$$H = \ker(F),$$

is the kernel of F, then H is a normal subgroup of G and

$$\hat{M} = \tilde{M}/H$$

is a Riemann surface covered by $\tilde{M}$ by the holomorphic projection

$$\tilde{\mu} : \tilde{M} \to \hat{M}, \quad \tilde{\mu}[\gamma] = [H\gamma],$$

the H-orbit of $[\gamma]$. Then $H \cong \pi_1(\hat{M}, \hat{0})$, where the base point $\hat{0} = H\tilde{0}$, the H-orbit of the base point of $\tilde{M}$. Moreover,

$$\hat{\mu} : \hat{M} \to M, \quad \hat{\mu}[H\gamma] = \mu[\gamma]$$

is a holomorphic covering space of M whose group of deck transformations is isomorphic to G/H. Then (8.38) implies that $\tilde{\mathbf{x}} : \tilde{M} \to \mathbf{R}^3$ descends to a smooth immersion of $\hat{M}$, which we denote

$$\hat{\mathbf{x}} : \hat{M} \to \mathbf{R}^3, \quad \hat{\mathbf{x}}[H\gamma] = \tilde{\mathbf{x}}[\gamma]. \tag{8.39}$$

Recall how we identified the open unit disk $D \subset M$ with the component of $\tilde{\mu}^{-1}D$ containing the base point $\tilde{0}$, the homotopy class of the constant path at 0 in M. Similarly, we identified the complement of its closure, $\bar{D}'$ with the component of $\tilde{\mu}^{-1}\bar{D}'$ containing the homotopy class of the line segment from 0 to $1-i$ in M. We continue to use the same letters to denote the images $\tilde{\mu}(D)$ and $\tilde{\mu}(\bar{D}')$ in $\hat{M}$, and these sets are mapped biholomorphically onto D and $\bar{D}'$, respectively, by $\hat{\mu}$.

Since the homomorphism F defined in (8.37) is surjective, we know that $G/H \cong \mathbf{Z}^2$. Given $(m,n) \in \mathbf{Z}^2$, let $\sigma_{m,n}$ be a closed loop at 0 in M for which $F[\sigma_{m,n}] = (m,n)$. Then D is evenly covered by $\hat{\mu}$,

$$\hat{\mu}^{-1}D = \cup_{(m,n)\in\mathbf{Z}^2}[\sigma_{m,n}]D$$

is a disjoint union of connected open subsets mapped biholomorphically onto D by $\hat{\mu}$ and (8.38) implies that the restrictions

$$\hat{\mathbf{x}}_{|[\sigma_{m,n}]D} = \hat{\mathbf{x}}_{|D} + 2\pi(m\epsilon_1 + n\epsilon_2),$$

for any $(m,n) \in \mathbf{Z}^2$. Similarly, $\bar{D}'$ is evenly covered by $\hat{\mu}$,

$$\hat{\mu}^{-1}\bar{D}' = \cup_{(m,n)\in\mathbf{Z}^2}[\sigma_{m,n}]\bar{D}'$$

is a disjoint union of connected open subsets mapped biholomorphically onto $\bar{D}'$ by $\hat{\mu}$ and the restrictions

$$\hat{\mathbf{x}}_{|[\sigma_{m,n}]\bar{D}'} = \hat{\mathbf{x}}_{|\bar{D}'} + 2\pi(m\epsilon_1 + n\epsilon_2),$$

for any $(m,n) \in \mathbf{Z}^2$. As observed above, the lifts of the common boundary $\partial D = \partial\bar{D}'$ are mapped to the corresponding vertical lines at the corners of the domains $S_{m,n}$. From all this we conclude that (8.39) is a conformal embedding of $\hat{M}$ onto Scherk's surface. In particular, Scherk's surface is homeomorphic to $\hat{M}$, which is connected and whose fundamental group is isomorphic to H, the group of deck transformations of its universal cover $\tilde{\mu} : \tilde{M} \to \hat{M}$ described above.

Since $\mu : \tilde{M} \to M$ is the universal cover of M, which is homotopically equivalent to $\mathbf{C}$ minus three points, $\tilde{M}$ must be the Poincaré disk (*hyperbolic type*). In fact, the only other possibility for $\tilde{M}$ is that it be the whole complex plane $\mathbf{C}$ (*parabolic type*), in which case the holomorphic covering projection $\mu : \tilde{M} \to M$ would be a non-constant entire function that omits exactly three points, an impossibility (see, for example, [103, Corollary 4, p. 81]). In particular, Scherk's surface is of hyperbolic type.

The image of the Gauss map of $\mathbf{x}$ is $g(M) = M$, which corresponds to the sphere minus four points. Actually, $\hat{g} = g \circ \hat{\mu} : \hat{M} \to \mathbf{C} \cup \{\infty\}$ is the Gauss map on the parameter domain (after stereographic projection), which is infinite to one, so the total curvature of $\mathbf{x}$ is infinite. For more details on this example see Osserman [132], Dierkes et al. [61], and Weber [167].

8.6 The Ricci condition

What characterizes the metrics induced on a Riemann surface by a conformal minimal immersion $\mathbf{x} : M \to \mathbf{R}^3$? If $\mathbf{e}_3 : M \to S^2$ is its Gauss map, then the third fundamental form of $\mathbf{x}$ is $III = d\mathbf{e}_3 \cdot d\mathbf{e}_3$, which is the pull-back to M of the metric of constant curvature 1 on S^2. By Problem 4.69,

$$III = 2H\,II - KI = -KI$$

if the mean curvature $H = 0$. Therefore, on the open set where $K < 0$, the metric $-KI$ on M has constant curvature equal to 1. This necessary condition is also sufficient. This result and its proof are presented by Lawson [108, p. 363].

Theorem 8.18 (Ricci [138]). *Let M be a simply connected Riemann surface with a conformal metric I. Then there exists an isometric minimal immersion of (M, I) into $\mathbf{R}^3$ if the Gaussian curvature K satisfies either*

1. *$K = 0$ on M, or*
2. *$K < 0$ on M and the conformal metric $-KI$ has constant curvature equal to 1.*

Proof. In case 1), M has a conformal minimal immersion as a plane in $\mathbf{R}^3$. To prove case 2), we begin with the following.

Lemma 8.19. *Suppose $K < 0$ on M. The metric*

$$\widetilde{I} = -KI$$

has constant curvature equal to 1 if and only if

$$\Delta \log(-K) = 4K \tag{8.40}$$

on M, where Δ is the Laplace-Beltrami operator of (M, I).

Proof. A simply connected Riemann surface M is either the Riemann sphere or it has a global complex coordinate z. Since the Riemann sphere cannot possess a metric whose Gaussian curvature is always negative, the latter case must hold. Then $I = e^{2u} dz d\bar{z}$, for some smooth function u on M, and

$$\Delta = 4e^{-2u} \frac{\partial^2}{\partial z \partial \bar{z}}, \tag{8.41}$$

so $K = -\Delta u$, by (7.31). Then $\widetilde{I} = e^{2\tilde{u}}dzd\bar{z} = -KI = -Ke^{2u}dzd\bar{z}$, so

$$\tilde{u} = u + \frac{1}{2}\log(-K), \quad \tilde{\Delta} = \frac{1}{-K}\Delta$$

and

$$\tilde{K} = -\tilde{\Delta}\tilde{u} = -1 + \frac{1}{2K}\Delta\log(-K),$$

from which the Lemma follows. □

Returning to the proof of case 2) of the Theorem, we see from (8.40) and (8.41) that

$$\frac{\partial^2}{\partial z \partial \bar{z}}\log(e^{4u}(-K)) = \frac{e^{2u}}{4}(\Delta\log(-K) + 4\Delta u) = e^{2u}(K - K) = 0.$$

Hence $e^{4u}(-K)$ is a positive harmonic function on M. It follows that there exists a holomorphic function $f(z)$ on M such that

$$e^{4u}(-K) = |f|^2.$$

If $h = e^{-2u}f : M \to \mathbf{C}$, then $e^{2u}h = f$ is holomorphic on M, so

$$0 = f_{\bar{z}} = (e^{2u}h)_{\bar{z}} = e^{2u}(2hu_{\bar{z}} + h_{\bar{z}}),$$

which implies (7.32) when $H = 0$. Then Theorem 7.28 implies that there exists an isometric immersion $\mathbf{x} : (M, dzd\bar{z}) \to \mathbf{R}^3$, unique up to rigid motion, whose mean curvature $H \equiv 0$ and whose Hopf invariant is h relative to z. □

8.7 Image of the Gauss map

The image of the Gauss map of a minimal immersion $\mathbf{x} : M \to \mathbf{R}^3$ becomes interesting only if one imposes the condition that the induced metric on M is complete. By the Hopf-Rinow Theorem there are several equivalent formulations of completeness. See, for example, [100, Theorem 4.1, p. 172]. The infinite length of divergent curves is the most useful characterization of completeness in the context of minimal immersions.

Definition 8.20. A *divergent curve* is a smooth curve $\gamma : [0, \infty) \to M$ such that for any compact subset K of M there exists a number T such that $\gamma(t) \notin K$ for all $t > T$.

Definition 8.21. The Riemannian metric I on the manifold M is complete if every divergent curve has infinite length. An immersion is complete if the induced metric is complete.

Theorem 8.22 (Bernstein [4]). *A function $f(x,y)$ is a solution of the minimal surface equation on the whole plane $\mathbf{R}^2$ if and only if $f(x,y) = ax + by + c$, for some real constants a, b, c.*

The proof is elementary and uses properties of the minimal surface equation. We have developed the tools here to sketch a proof of a generalization of Bernstein's Theorem, first conjectured by L. Nirenberg, and subsequently proved by R. Osserman [129] and [133][Theorem 8.1, p. 68] in 1959.

Theorem 8.23 (Osserman). *Let $\mathbf{x} : M \to \mathbf{R}^3$ be a complete minimal immersion. Then either $\mathbf{x}(M)$ is a plane or the image of the Gauss map is dense in the unit sphere S^2.*

Remark 8.24. Bernstein's theorem is an immediate corollary because if a minimal immersion is given by a graph defined on the whole plane, $\mathbf{x}(u,v) = (u, v, f(u,v))$, for all $(u,v) \in \mathbf{R}^2$, then the Gauss map takes values in an open hemisphere of S^2 and the metric $I = d\mathbf{x} \cdot d\mathbf{x}$ induced on $M = \mathbf{R}^2$ is complete because

$$\begin{aligned} I &= (1 + f_u^2)du^2 + 2f_u f_v du dv + (1 + f_v^2)dv^2 \\ &= du^2 + dv^2 + (f_u du + f_v dv)^2 \geq du^2 + dv^2 \end{aligned} \tag{8.42}$$

and $du^2 + dv^2$ on $\mathbf{R}^2$ is complete.

Proof (Proof of Osserman's Theorem.). Let $\pi : \tilde{M} \to M$ denote the universal cover of M. Then $\mathbf{x} \circ \pi$ is a complete minimal immersion whose Gauss map has the same image as the Gauss map of $\mathbf{x}$ and whose image is the same as the image of $\mathbf{x}$. Therefore, it is sufficient to prove the theorem for the case when M is simply connected.

Suppose that the image of the Gauss map is not dense in S^2. Then there is an open set $U \subset S^2$ in the complement of the image of the Gauss map. Applying a rotation to $\mathbf{x}$ if necessary, we may assume that U contains the north pole $(0,0,1)$. Let α be the nowhere zero isotropic $\mathbf{C}^3$-valued holomorphic 1-form on M given by the (1,0) part of $d\mathbf{x}$. If $\alpha^1 - i\alpha^2 = 0$, then $\alpha^3 = 0$ and $\mathbf{x}(M)$ is a plane. Otherwise,

$$\alpha = \begin{pmatrix} \frac{1}{2}(1 - g^2) \\ \frac{i}{2}(1 + g^2) \\ g \end{pmatrix} \eta$$

where g is a meromorphic function on M and η is the abelian differential $\eta = \alpha^1 - i\alpha^2$ on M. We know that g is the composition of stereographic projection from the north pole with the Gauss map. From our assumption on the Gauss map it follows that $|g| < R$ for some positive constant R. Therefore, g has no poles and η has no zeros in M.

Up to biholomorphism, the only non-compact simply connected Riemann surfaces are the complex plane $\mathbf{C}$ and the unit disk $D = \{|z| < 1\}$. If $M = \mathbf{C}$, then g would be a bounded holomorphic function on $\mathbf{C}$, hence constant, which would mean that the Gauss map is constant and therefore that $\mathbf{x}(M)$ is a plane. If M is the unit disk D, then $\eta = f(z)dz$, where f is holomorphic and never zero on D. The induced metric is given in (8.9) as

$$I = (1+|g|^2)^2|f|^2 dz d\bar{z} < (1+R^2)^2|f|^2 dz d\bar{z}.$$

From this we conclude that I is not complete, because the following lemma shows that the metric on the right is not complete. Therefore, M cannot be the unit disk, and we have shown in all possible cases that $\mathbf{x}(M)$ is a plane. □

Lemma 8.25. *Let $f(z)$ be a holomorphic function on the unit disk D such that f has no zeros in D. Then the metric $|f(z)|^2 dz d\bar{z}$ on D is not complete.*

Exercise 33. Prove this lemma. See [133, Lemma 8.5, p. 67].

For nearly forty years mathematicians searched for the optimal version of this theorem. The question is how large is the set O of points on the sphere omitted by the Gauss map of a nonplanar complete minimal immersion. In 1961 Osserman [130] proved that the set of omitted points must have capacity zero. In this paper he gave an example of a complete minimal surface whose Gauss map omits exactly 4 points. In [164], Voss proved the following exercise, thus giving simpler examples where the number of omitted points is precisely k, where k can be 1, 2, 3, or 4.

Exercise 34 ([164]). For $n \geq 1$, let $p_0, p_1, \ldots, p_n$ be distinct points on the sphere S^2. After a possible rotation, we may assume $p_0 = {}^t(0,0,1)$. Let $z_j = \mathfrak{s}(p_j)$, $j = 1, \ldots, n$, be the the stereographic projection from p_0 of these points into the complex plane. Let $M = \mathbf{C} \setminus \{z_1, \ldots, z_n\}$, let $g(z) = z$ and let

$$\eta = \frac{1}{\prod_1^n (z - z_j)} dz.$$

On M, g has no poles and η has no zeros. Prove that this Weierstrass data gives a minimal immersion $\tilde{\mathbf{x}} : \tilde{M} \to \mathbf{R}^3$ whose Gauss image is

$$S^2 \setminus \{p_0, \ldots, p_n\},$$

where $\tilde{\mu} : \tilde{M} \to M$ is a holomorphic covering space of M. The induced metric $\tilde{I}$ on $\tilde{M}$ is the lift of

$$I = (1+|g|^2)^2 \eta\bar{\eta} = \frac{(1+|z|^2)^2}{\prod_1^n |z-z_j|^2} dz d\bar{z}$$

on M. Prove that this metric is complete if and only if $n \leq 3$. Prove that the total curvature is always infinite.

In 1981 F. Xavier [172] proved that O can contain no more than 6 points. In 1988 H. Fujimoto [71] proved that O can contain no more than 4 points, which is the optimal result. Fujimoto's methods use the value distribution theory of R. Nevanlinna. In 1990, X. Mo and R. Osserman [120] proved a stronger version of Fujimoto's result, including that if the Gauss map omits 4 points of the sphere, then it covers every other point infinitely often. In 2002, Ros [140] gave another proof of this result.

Another direction in which Bernstein's Theorem has been generalized is to minimal hypersurfaces in $\mathbf{R}^{n+1}$ – that is, to the case of the minimal surface equation for n independent variables. In 1969 E. Bombieri, E. De Giorgi and E. Giusti [13] proved that the generalization is true for $n \leq 6$, but false for $n = 7$. They made essential use of J. Simon's 1968 paper [153].

8.8 Finite total curvature

Theorem 8.26. *Let (M^2, I) be an oriented, complete Riemannian 2-manifold whose Gaussian curvature K satisfies*

$$K \leq 0 \quad \text{and} \quad \int_M K dA > -\infty.$$

Then there exists a compact Riemann surface $\tilde{M}$ and a finite number of points $p_1, \ldots, p_k$ on $\tilde{M}$, such that the Riemann surface structure induced on M by the metric I is biholomorphic to $\tilde{M} \setminus \{p_1, \ldots, p_k\}$.

Proof. See Osserman [133, Theorem 9.1, pp. 81–82]. □

The following combines the results of Lemma 9.5 and Theorem 9.2 in [133].

Theorem 8.27. *If $\mathbf{x} : M \to \mathbf{R}^3$ is an oriented, complete minimal immersion with finite total curvature,*

$$\int_M K dA > -\infty,$$

then the conclusion of Theorem 8.26 applies, the meromorphic function g of the Enneper–Weierstrass representation extends to a meromorphic function on the Riemann surface $\tilde{M}$, and the total curvature of $\mathbf{x}$ is $-4\pi n$ for some integer $n \geq 0$.

Proof. We know $M = \tilde{M} \setminus \{p_1, \ldots, p_k\}$, where $\tilde{M}$ is a compact Riemann surface. The function g is holomorphic on M. If one of the isolated singularities p_j of g were essential, then by Picard's Great Theorem (see, for example, [103, Theorem 2, pp. 92–93]), g would assume every value infinitely often, with at most two exceptions. This implies that the area of the image of g is infinite, which means that the total curvature must be infinite, since $K < 0$ at all but a discrete set of points and its constant sign means there is no cancelation of areas. Hence, each p_j is a pole or removable singularity of g and g is meromorphic on $\tilde{M}$. Then $\lim_{p\to p_j} g(p)$ exists, possibly equal to ∞, implies that the Gauss map $\mathbf{n} : M \to \mathbf{S}^2$ extends smoothly to the complex surface $\tilde{M}$. By Corollary 4.35,

$$\int_M K\, dA = \int_M \mathbf{n}^*\nu = \int_{\tilde{M}} \mathbf{n}^*\nu = \deg(\mathbf{n})4\pi,$$

where ν is the area form on the unit two-sphere $\mathbf{S}^2$. If K is identically zero, the $\mathbf{n}$ is constant and $\deg(\mathbf{n}) = 0$. Otherwise, $K < 0$ at all but a discrete set of points of M, so $\deg(\mathbf{n})$ is a negative integer. □

Theorem 8.28 (Osserman). *The only complete minimal immersions whose total curvature is -4π are the catenoid and Enneper's surface. These two surfaces are the only complete minimal surfaces whose Gauss map is one-to-one.*

Proof. See [133, Theorem 9.4 and Corollary on p.87]. □

Theorem 8.29 (Osserman). *If $\mathbf{x} : M \to \mathbf{R}^3$ is an oriented, complete minimal immersion with finite total curvature, whose Gauss map $\mathbf{n} : M \to \mathbf{S}^2$ omits more than three points, then $\mathbf{x}(M)$ is a plane.*

Proof. See [131, Theorem 3.3, pp. 359–360]. □

Theorem 8.30 (Osserman). *If $\mathbf{x} : M \to \mathbf{R}^3$ is an oriented, complete, minimal immersion with finite total curvature, whose Gauss map omits exactly three points, then the genus of the compact Riemann surface $\tilde{M}$ of Theorem 8.26 and the total curvature of $\mathbf{x}$ satisfy*

$$\text{genus}(\tilde{M}) \geq 1, \qquad \int_M K\, dA \leq -12\pi.$$

Proof. See [131, Theorem 3.3A, p. 360]. □

C. Costa's 1982 IMPA thesis [54] presents a minimal immersion of genus one and total curvature -12π whose Gauss map omits exactly two points. See Figure 8.10.

Subsequently, D. Hoffman and W. Meeks [91] proved that Costa's surface is an embedding. Their proof uses symmetries in the solution that they found from computer graphics of $\mathbf{x}$. The book [2] by J.L.M. Barbosa and A.G. Colares contains a detailed expositions of how to do calculations on a Riemann surface of arbitrary genus G and of how to work with the Enneper–Weierstrass data on such surfaces. Their book contains many examples, including a detailed description of Costa's surface.

We believe that the following problem, posed in [133, p. 90], remains unsolved.

Osserman's Problem 1. Does there exist a complete minimal surface of finite total curvature whose Gauss map omits exactly three points?

8.9 Goursat transforms

Let M be a connected Riemann surface. A holomorphic curve $\gamma : M \to \mathbf{C}^3$ is *regular* if $\dot{\gamma}(z) \neq 0$ for all z, where dot means the derivative with respect to z, any local complex coordinate in M.

Fig. 8.10 Costa's minimal embedding of a torus with three punctures

Definition 8.31. A nonzero vector $\mathbf{v} \in \mathbf{C}^3$ is *isotropic* if $\mathbf{v} \cdot \mathbf{v} = 0$. Otherwise it is *nonisotropic*. Here the dot product is the symmetric bilinear form

$$^t(x,y,z) \cdot {}^t(u,v,w) = xu + yv + zw$$

for any two vectors in $\mathbf{C}^3$ expressed in terms of the standard basis.

The *complex Euclidean group* is

$$\mathbf{E}(3,\mathbf{C}) = \mathbf{C}^3 \rtimes \mathbf{SO}(3,\mathbf{C}),$$

where

$$\mathbf{SO}(3,\mathbf{C}) = \{A \in \mathbf{GL}(3,\mathbf{C}) : {}^tAA = I, \quad \det A = 1\}$$

is the *complex special orthogonal group*. The complex Euclidean group acts on $\mathbf{C}^3$ by

$$\mathbf{E}(3,\mathbf{C}) \times \mathbf{C}^3 \to \mathbf{C}^3, \quad (\mathbf{v},A)\mathbf{x} = \mathbf{v} + A\mathbf{x}.$$

The theory of regular holomorphic curves $\gamma : M \to \mathbf{C}^3$ under this action is the same as that of regular curves in $\mathbf{R}^3$ under the action of $\mathbf{E}(3) = \mathbf{R}^3 \rtimes \mathbf{SO}(3)$, provided that the tangent vector to the curve is always *nonisotropic*. There is no change in the definition of arclength parameter, first order frame, curvature, second order frame, and torsion for such curves. In this case the curvature and torsion are holomorphic functions on the domain of the curve.

The situation is quite different when the tangents to the curve are isotropic.

Definition 8.32. A *minimal curve in* $\mathbf{C}^3$ is a regular holomorphic curve in $\mathbf{C}^3$ each of whose tangent vectors is isotropic.

If we express the real and imaginary parts of a regular holomorphic curve by $\gamma = \mathbf{x} + i\mathbf{y}$, then $\mathbf{x}, \mathbf{y} : M \to \mathbf{R}^3$ are smooth harmonic maps, since $\gamma_z = \mathbf{x}_z + i\mathbf{y}_z$ is holomorphic and thus $\mathbf{x}_{z\bar{z}} = 0 = \mathbf{y}_{z\bar{z}}$ for any local complex coordinate z in M. They are *harmonic conjugates* in the sense that $\mathbf{x} + i\mathbf{y} : M \to \mathbf{C}^3$ is holomorphic (see [118, p 141]). By the Cauchy-Riemann equations,

$$\dot{\gamma} = 2\mathbf{x}_z = 2i\mathbf{y}_z,$$

so if γ is a minimal curve, then

$$0 = \dot{\gamma} \cdot \dot{\gamma} = 4\mathbf{x}_z \cdot \mathbf{x}_z = -4\mathbf{y}_z \cdot \mathbf{y}_z$$

shows that the immersions $\mathbf{x}, \mathbf{y} : M \to \mathbf{R}^3$ are conformal, by (7.19). That is, the complex structure of M is that induced by these immersions. Hence, the real and imaginary parts of a minimal curve are minimal immersions into $\mathbf{R}^3$.

Conversely, start with a smooth immersion $\mathbf{x} : M^2 \to \mathbf{R}^3$. If (U, z) is a complex coordinate chart for the induced complex structure on M, then $\mathbf{x}$ is conformal so $\mathbf{x}_z : U \to \mathbf{C}^3$ is isotropic at every point of U by (7.19). It is holomorphic if and only if $\mathbf{x}$ is minimal, by Corollary 7.34.

If $\mathbf{x} : M^2 \to \mathbf{R}^3$ is a minimal immersion, then $\mathbf{x}$ is harmonic in the sense that each of its component functions is harmonic, by Corollary 7.36. If U, z is a complex coordinate chart in M for which U is simply connected, then there exists a conjugate harmonic function $\mathbf{y} : U \to \mathbf{R}^3$, determined up to an additive constant vector (see [118, Theorem 32, pp 140–141]), such that

$$\gamma = \mathbf{x} + i\mathbf{y} : U \to \mathbf{C}^3$$

is holomorphic. It is a minimal curve. This is just a recapitulation of the Enneper–Weierstrass representation of a minimal immersion, which is based on the differential $\alpha = d\gamma$ of a minimal curve in $\mathbf{C}^3$.

If $\gamma : M \to \mathbf{C}^3$ is a minimal curve, and if $(\mathbf{v}, T) \in \mathbf{E}(3, \mathbf{C})$, then $(\mathbf{v}, T)\gamma = \mathbf{v} + T\gamma$ is also a minimal curve. How do the real and imaginary parts $\mathbf{x}, \mathbf{y} : M \to \mathbf{R}^3$ of γ transform? If $\mathbf{v} = \mathbf{a} + i\mathbf{b}$, where $\mathbf{a}, \mathbf{b} \in \mathbf{R}^3$, and if $T = A + iB$, where $A, B \in \mathbf{GL}(3, \mathbf{R})$ must satisfy

$$^tAA - {}^tBB = I_3, \quad {}^tAB + {}^tBA = 0, \tag{8.43}$$

then $\tilde{\mathbf{x}} + i\tilde{\mathbf{y}} = (\mathbf{v}, T)\boldsymbol{\gamma}$ satisfies

$$\tilde{\mathbf{x}} = a + A\mathbf{x} - B\mathbf{y}, \quad \tilde{\mathbf{y}} = b + A\mathbf{y} + B\mathbf{x}.$$

We are interested in the cases when $\tilde{\mathbf{x}}$ is not $\mathbf{E}(3)$-congruent to $\mathbf{x}$, so we may assume $\mathbf{v} = \mathbf{0}$ and $B \neq 0$, that is, $T \in \mathbf{SO}(3, \mathbf{C}) \setminus \mathbf{SO}(3)$.

Definition 8.33 (Goursat transform). Let $\mathbf{x} : M \to \mathbf{R}^3$ be a minimal immersion of a simply connected surface M that possesses a global complex coordinate z. Let $\mathbf{y} : M \to \mathbf{R}^3$ be a harmonic conjugate of $\mathbf{x}$. The *Goursat transform* of $\mathbf{x}$ by $T \in \mathbf{SO}(3, \mathbf{C})$ is the minimal immersion $\tilde{\mathbf{x}} : M \to \mathbf{R}^3$ given by the real part of the minimal curve $T(\mathbf{x} + i\mathbf{y}) : M \to \mathbf{C}^3$.

As a tool for understanding the Goursat transform, we consider a double covering of $\mathbf{SO}(3, \mathbf{C})$ by the *complex special linear group*

$$\mathbf{SL}(2, \mathbf{C}) = \{A \in \mathbf{GL}(2, \mathbf{C}) : \det A = 1\}.$$

This is obtained from the following generalization of Example 7.17 (see Goursat [74, 75]). We begin with a parametrization of the set of point pairs in $\mathbf{CP}^1$.

Example 8.34 (Point pairs in $\mathbf{CP}^1$). The set $\mathscr{P}$ of *point pairs in* $\mathbf{CP}^1$,

$$\mathscr{P} = \{(p, q) \in \mathbf{CP}^1 \times \mathbf{CP}^1 : p \neq q\},$$

is an open complex submanifold of $\mathbf{CP}^1 \times \mathbf{CP}^1$ biholomorphically equivalent to the nonsingular complex affine variety

$$V = \{(x, y, z) \in \mathbf{C}^3 : x^2 + y^2 + z^2 = 1\},$$

by the map $F : V \to \mathscr{P}$, where

$$F(x, y, z) = \left(\begin{bmatrix} 1+z \\ x-iy \end{bmatrix}, \begin{bmatrix} x+iy \\ -(1+z) \end{bmatrix} \right) = \left(\begin{bmatrix} x+iy \\ 1-z \end{bmatrix}, \begin{bmatrix} -(1-z) \\ x-iy \end{bmatrix} \right).$$

Equality of these two expressions for F follows from the factorization of $x^2 + y^2 = 1 - z^2$ into $(x+iy)(x-iy) = (1-z)(1+z)$, which gives the ratios

$$u = \frac{x+iy}{1-z} = \frac{1+z}{x-iy}, \quad v = -\frac{x+iy}{1+z} = -\frac{1-z}{x-iy}. \tag{8.44}$$

Thus, u and v are local complex coordinates in $\mathbf{CP}^1$ and

$$F(x, y, z) = (\begin{bmatrix} u \\ 1 \end{bmatrix}, \begin{bmatrix} v \\ 1 \end{bmatrix}) = (u, v).$$

We regard u and v as points on the extended plane $\hat{\mathbf{C}} = \mathbf{C} \cup \{\infty\}$, where $\infty = \begin{bmatrix} 1 \\ 0 \end{bmatrix}$. The equations (8.44) can be inverted, giving

$$x = \frac{1-uv}{u-v}, \quad y = i\frac{1+uv}{u-v}, \quad z = \frac{u+v}{u-v},$$

thus showing that the inverse mapping $F^{-1}(u,v) = (x,y,z)$ exists and is holomorphic.

An element $T \in \mathbf{SL}(2,\mathbf{C})$ acts on $\mathscr{P}$ by

$$T(u,v) = T(\begin{bmatrix} u \\ 1 \end{bmatrix}, \begin{bmatrix} v \\ 1 \end{bmatrix}) = (\left[T\begin{pmatrix} u \\ 1 \end{pmatrix}\right], \left[T\begin{pmatrix} v \\ 1 \end{pmatrix}\right]).$$

An element $T \in \mathbf{SL}(2,\mathbf{C})$ can be written as

$$T = \begin{pmatrix} m & n \\ p & q \end{pmatrix}, \quad m,n,p,q \in \mathbf{C}, \quad mq - np = 1. \tag{8.45}$$

The action of T on $\mathscr{P}$ is then

$$T(u,v) = (\frac{mu+n}{pu+q}, \frac{mv+n}{pv+q}) = (u_1, v_1).$$

Lemma 8.35 (Double cover $\mathbf{SL}(2,\mathbf{C}) \to \mathbf{SO}(3,\mathbf{C})$). *For any $T \in \mathbf{SL}(2,\mathbf{C})$, the map*

$$\Sigma(T) = F^{-1} \circ T \circ F : V \to V \tag{8.46}$$

is the action of a unique element of $\mathbf{SO}(3,\mathbf{C})$. The resulting map

$$\Sigma : \mathbf{SL}(2,\mathbf{C}) \to \mathbf{SO}(3,\mathbf{C})$$

is a 2:1 Lie group homomorphism and a covering map. Its induced isomorphism on the Lie algebras is

$$\Sigma_* : \mathfrak{sl}(2,\mathbf{C}) \to \mathfrak{o}(3,\mathbf{C}),$$

$$\begin{pmatrix} a & b \\ c & -a \end{pmatrix} \mapsto \begin{pmatrix} 0 & 2ia & c-b \\ -2ia & 0 & i(b+c) \\ b-c & -i(b+c) & 0 \end{pmatrix},$$

for all $a,b,c \in \mathbf{C}$.

Thus $\mathbf{SL}(2,\mathbf{C})$ is the universal cover of $\mathbf{SO}(3,\mathbf{C})$, since it is simply connected, being homeomorphic to $\mathbf{S}^3 \times \mathbf{R}^3$ by [84, Lemma 4.3, p 345].

Proof. Let $(x,y,z) \in V$ and let $T = \begin{pmatrix} m & n \\ p & q \end{pmatrix} \in \mathbf{SL}(2,\mathbf{C})$ be given by (8.45). Then an elementary, but long, calculation gives

$$\Sigma(T)(x,y,z) = \begin{pmatrix} \frac{m^2-p^2}{2}(x+iy) + (pq-mn)z + \frac{q^2-n^2}{2}(x-iy) \\ -i\frac{m^2+p^2}{2}(x+iy) + i(mn+pq)z + i\frac{n^2+q^2}{2}(x-iy) \\ -mp(x+iy) + (mq+np)z + nq(x-iy) \end{pmatrix},$$

which is the action on ${}^t(x,y,z) \in V$ by the matrix in $\mathbf{SO}(3,\mathbf{C})$,

$$\begin{pmatrix} \frac{1}{2}(m^2-p^2+q^2-n^2) & \frac{i}{2}(m^2-p^2+n^2-q^2) & pq-mn \\ \frac{i}{2}(n^2+q^2-m^2-p^2) & \frac{1}{2}(m^2+n^2+p^2+q^2) & i(mn+pq) \\ nq-mp & -i(mp+nq) & mq+np \end{pmatrix}.$$

It follows from (8.46) that the map Σ is a Lie group homomorphism. Another direct calculation shows that its derivative map on the Lie algebras is as given above. In particular, Σ is nonsingular at every point of $\mathbf{SL}(2,\mathbf{C})$ and thus it is an open mapping. Since both groups are connected, it follows that Σ is surjective. It is 2:1 because $\Sigma(T) = I_3$ if and only if the action of T on $\mathscr{P}$ is the identity, and this is so if and only if $T = \pm I_2$. □

Exercise 35. Let $T = \begin{pmatrix} m & n \\ p & q \end{pmatrix} \in \mathbf{SL}(2,\mathbf{C})$. Prove that $\Sigma(T) \circ \mathfrak{W} = \mathfrak{W} \circ T$, where $\mathfrak{W}$ is the Enneper–Weierstrass map (8.12) and the matrix $\Sigma(T)$ is given in the proof of Lemma 8.35.

The Goursat transform of a minimal immersion preserves the second fundamental form, as we now show.

Proposition 8.36. *Suppose the surface M is simply connected. If the minimal immersion $\tilde{\mathbf{x}} : M \to \mathbf{R}^3$ is a Goursat transform of a minimal immersion $\mathbf{x} : M \to \mathbf{R}^3$, then $\mathbf{x}$ and $\tilde{\mathbf{x}}$ have the same second fundamental form at each point of M.*

Proof. Suppose that $\mathbf{x}$ is given by the Enneper–Weierstrass data (g,η), so its minimal curve $\boldsymbol{\gamma} = \mathbf{x} + i\mathbf{y}$ satisfies

$$d\boldsymbol{\gamma} = \mathfrak{W}\begin{pmatrix} g \\ 1 \end{pmatrix}\eta = \begin{pmatrix} \frac{1}{2}(1-g^2) \\ \frac{i}{2}(1+g^2) \\ g \end{pmatrix}\eta,$$

by (8.8). The Hopf quadratic differential of $\mathbf{x}$ is $II^{2,0} = -\eta dg$, by (8.17). If $T \in \mathbf{SL}(2,\mathbf{C})$ is given by (8.45), and $\tilde{\mathbf{x}} + i\tilde{\mathbf{y}} = \tilde{\boldsymbol{\gamma}} = \Sigma(T)\boldsymbol{\gamma}$ gives a Goursat transform $\tilde{\mathbf{x}}$ of $\mathbf{x}$, then the commutation formula of Exercise 35 gives

$$d(\Sigma(T)\boldsymbol{\gamma}) = \Sigma(T)d\boldsymbol{\gamma} = \Sigma(T)\mathfrak{W}\begin{pmatrix} g \\ 1 \end{pmatrix}\eta = \mathfrak{W}(T\begin{pmatrix} g \\ 1 \end{pmatrix})\eta = \mathfrak{W}\left(\tilde{g}\right)\tilde{\eta},$$

where

$$\tilde{g} = \frac{mg+n}{pg+q}, \quad \tilde{\eta} = (pg+q)^2\eta$$

are thus the Enneper–Weierstrass data of $\tilde{\mathbf{x}}$. The Hopf quadratic differential of $\tilde{\mathbf{x}}$ is

$$\widetilde{II}^{2,0} = -\tilde{\eta}d\tilde{g} = -\eta((pg+q)mdg - (mg+n)pdg) = -\eta dg = II^{2,0},$$

since $mq - np = 1$. The minimal immersions $\mathbf{x}$ and $\tilde{\mathbf{x}}$ must then have the same second fundamental form at each point of M by (7.26) and (7.27). □

Any element of $\mathbf{SO}(3)$ has a fixed vector in $\mathbf{R}^3$, and is thus rotation through some angle about the line though the origin and this vector. Goursat proved that a similar description holds for any element of $\mathbf{SO}(3,\mathbf{C})$, up to composition with real rotations.

Theorem 8.37 (Goursat). *Let $T \in \mathbf{SL}(2,\mathbf{C})$. The following are equivalent:*

1. $\Sigma(T)\mathbf{S}^2 \subset \mathbf{S}^2$,
2. $T \in \mathbf{SU}(2) = \left\{ \begin{pmatrix} m & n \\ -\bar{n} & \bar{m} \end{pmatrix} : |m|^2 + |n|^2 = 1 \right\}$,
3. $\Sigma(T) \in \mathbf{SO}(3)$.

The restriction of Σ to $\mathbf{SU}(2) \subset \mathbf{SL}(2,\mathbf{C})$ is the double cover $\mathbf{SU}(2) \to \mathbf{SO}(3)$ described in Example 7.17.

If $T \notin \mathbf{SU}(2)$, then there exists a unique pair of antipodal points $\pm\mathbf{x} \in \mathbf{S}^2 \subset \mathbf{R}^3$ such that $\Sigma(T)\mathbf{x} \in \mathbf{S}^2$.

Proof. Let $T \in \mathbf{SL}(2,\mathbf{C})$ be given by (8.45). By Problem 8.63, a point $(x,y,z) \in V$ lies in $\mathbf{S}^2$ if and only if $(u,v) = F(x,y,z)$ satisfies $u\bar{v} = -1$. Suppose this last equation holds for (u,v) and let

$$(u_1, v_1) = T(u,v) = (\frac{mu+n}{pu+q}, \frac{mv+n}{pv+q}).$$

Then $u_1\bar{v}_1 = -1$ if and only if

$$(m\bar{n} + p\bar{q})u^2 + (|n|^2 - |m|^2 + |q|^2 - |p|^2)u - (\bar{m}n + \bar{p}q) = 0.$$

In terms of the columns $\mathbf{T}_1, \mathbf{T}_2 \in \mathbf{C}^2$ of T and the standard hermitian inner product on $\mathbf{C}^2$, this quadratic equation in u can be written

$$\langle \mathbf{T}_1, \mathbf{T}_2 \rangle u^2 + (|\mathbf{T}_2|^2 - |\mathbf{T}_1|^2)u - \overline{\langle \mathbf{T}_1, \mathbf{T}_2 \rangle} = 0. \tag{8.47}$$

The discriminant

$$D = (|\mathbf{T}_2|^2 - |\mathbf{T}_1|^2)^2 + 4|\langle \mathbf{T}_1, \mathbf{T}_2 \rangle|^2 \geq 0,$$

with equality if and only if $T \in \mathbf{SU}(2)$, in which case the coefficients of (8.47) are identically zero, so every $u \in \mathbf{C}$ is a solution and $\Sigma(T)\mathbf{S^2} = \mathbf{S^2}$. If $\Sigma(T) \in \mathbf{SO}(3)$, then $\Sigma(T)\mathbf{S^2} = \mathbf{S^2}$, so $T \in \mathbf{SU}(2)$.

Conversely, if $T \in \mathbf{SU}(2)$, then $\Sigma(T) \in \mathbf{SO}(3)$ by the formula for $\Sigma(T)$ in the proof of Lemma 8.35. When $T \in \mathbf{SU}(2)$, this formula agrees with that of Example 7.17.

In the case $T \notin \mathbf{SU}(2)$, the roots of (8.47),

$$u_{\pm} = \frac{|\mathbf{T}_1|^2 - |\mathbf{T}_2|^2 \pm \sqrt{D}}{2\langle \mathbf{T}_1, \mathbf{T}_2 \rangle},$$

are distinct and $u_+ \bar{u}_- = -1$. Hence $F^{-1}(u_+, u_-)$, $F^{-1}(u_-, u_+)$ is the unique pair of antipodal points in $\mathbf{S^2}$ taken by $\Sigma(T)$ to points in $\mathbf{S^2}$. □

Corollary 8.38. *If $T \in \mathbf{SO}(3,\mathbf{C}) \setminus \mathbf{SO}(3)$, then there exists a vector $\mathbf{e}_3 \in \mathbf{S^2} \subset \mathbf{R}^3$, unique up to sign, such that $T\mathbf{e}_3 \in \mathbf{S^2}$. Complete $\mathbf{e}_3$ to a positively oriented orthonormal basis $\mathbf{e}_3, \mathbf{e}_1, \mathbf{e}_2$ of $\mathbf{R}^3$ and let*

$$\mathbf{e}_3^{\perp} = span_{\mathbf{C}}\{\mathbf{e}_1, \mathbf{e}_2\} \cong \mathbf{C}^2.$$

If $R \in \mathbf{SO}(3)$ rotates $T\mathbf{e}_3$ back to $\mathbf{e}_3$, then $RT\mathbf{e}_3^{\perp} = \mathbf{e}_3^{\perp}$ and

$$RT\mathbf{e}_1 = \mathbf{e}_1 \cos\zeta + \mathbf{e}_2 \sin\zeta, \quad RT\mathbf{e}_2 = -\mathbf{e}_1 \sin\zeta + \mathbf{e}_2 \cos\zeta,$$

for some unique $\zeta \in \mathbf{C}$.

Proof. Relative to the basis $\mathbf{e}_1, \mathbf{e}_2$ of $\mathbf{e}_3^{\perp}$, the matrix of RT is an element of $\mathbf{SO}(2,\mathbf{C})$, whose form is given by Problem 8.64. □

Example 8.39 (Goursat transforms of the catenoid). For $w = u + iv \in \mathbf{C}$, the catenoid is the minimal immersion

$$\mathbf{x} : \mathbf{C} \to \mathbf{R}^3, \quad \mathbf{x}(u,v) = \begin{pmatrix} \cosh u \cos v \\ -\cosh u \sin v \\ u \end{pmatrix},$$

which is the real part of the minimal curve

$$\gamma : \mathbf{C} \to \mathbf{C}^3, \quad \gamma(w) = \begin{pmatrix} \cosh w \\ i \sinh w \\ w \end{pmatrix}.$$

Then

$$d\gamma = \begin{pmatrix} \frac{1}{2}(1 - (e^{-w})^2) \\ \frac{i}{2}(1 + (e^{-w})^2) \\ e^{-w} \end{pmatrix} e^w dw$$

shows that the Enneper–Weierstrass data of $\mathbf{x}$ is $(g = e^{-w},\ \eta = e^{w}dw)$. See Example 8.15. The Goursat transform of $\mathbf{x}$ by

$$T = \begin{pmatrix} \cosh t & -i\sinh t & 0 \\ i\sinh t & \cosh t & 0 \\ 0 & 0 & 1 \end{pmatrix} \in \mathbf{SO}(3,\mathbf{C}) \setminus \mathbf{SO}(3),$$

for any $0 \neq t \in \mathbf{R}$, is the real part of $T\boldsymbol{\gamma} = \tilde{\mathbf{x}} + i\tilde{\mathbf{y}}$, so

$$\tilde{\mathbf{x}} = \begin{pmatrix} \cosh(u+t)\cos v \\ -\cosh(u+t)\sin v \\ u \end{pmatrix} = \mathbf{x}(u+t, v) - t\boldsymbol{\epsilon}_3,$$

which is a translate of a reparametrization of $\mathbf{x}$. In particular, it is $\mathbf{E}(3)$-congruent to $\mathbf{x}$ even though T is not real. The special relation T has with the catenoid is that it fixes the antipodal points of $\mathbf{S}^2$ where the rotation axis of the catenoid intersects $\mathbf{S}^2$.

For a more general Goursat transform consider $T \in \mathbf{SO}(3,\mathbf{C})$ for which the unique antipodal pair of points $\pm\mathbf{e}_3 \in \mathbf{S}^2$ taken by T into $\mathbf{S}^2$ is not $\pm\boldsymbol{\epsilon}_3$. To be specific, suppose $T\boldsymbol{\epsilon}_1 = \boldsymbol{\epsilon}_2$ and let $R = \begin{pmatrix} 0 & 1 & 0 \\ 0 & 0 & 1 \\ 1 & 0 & 0 \end{pmatrix} \in \mathbf{SO}(3)$, so $R\boldsymbol{\epsilon}_1 = \boldsymbol{\epsilon}_3$, $R\boldsymbol{\epsilon}_2 = \boldsymbol{\epsilon}_1$, and $R\boldsymbol{\epsilon}_3 = \boldsymbol{\epsilon}_2$. Then $RT\boldsymbol{\epsilon}_1 = \boldsymbol{\epsilon}_1$, so by Corollary 8.38,

$$RT = \begin{pmatrix} 1 & 0 & 0 \\ 0 & \cos\zeta & -\sin\zeta \\ 0 & \sin\zeta & \cos\zeta \end{pmatrix},$$

for some $\zeta = a + ib \in \mathbf{C}$, where a and $b \neq 0$ are real. Define constants $c = \cos a$, $s = \sin a$, $C = \cosh b$, and $S = \sinh b$. Calculating the real part of $RT\boldsymbol{\gamma}$, we find the Goursat transform of $\mathbf{x}$ by RT to be

$$\tilde{\mathbf{x}}(u,v) = \begin{pmatrix} \cosh u \cos v \\ -cC\cosh u \sin v + sS\sinh u \cos v - sCu + cSv \\ -sC\cosh u \sin v - cS\sinh u \cos v + cCu + sSv \end{pmatrix},$$

which is not $\mathbf{E}(3)$-congruent to $\mathbf{x}$, as is evident in Figures 8.11 and 8.12, which illustrate the cases $a = \pi/2$, and $b = 0, 0.4, 1$, and 20, respectively. One obtains the catenoid when $b = 0$, but as b increases, the Goursat transform of it does not close and it bends.

In his Ph.D. thesis, M. Deutsch [60] uses the Goursat transform to find new CMC-1 immersions in $\mathbf{H}^3$.

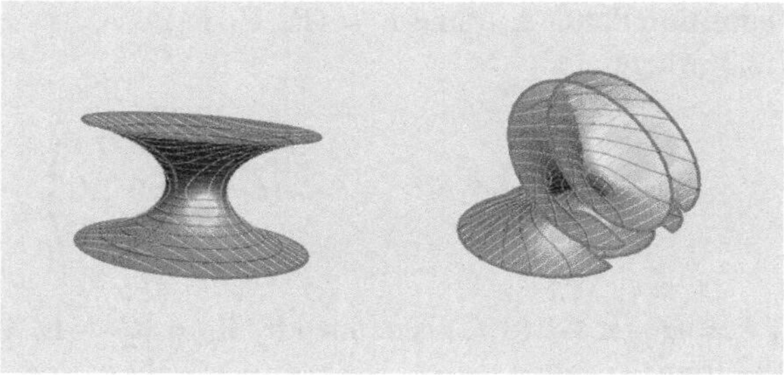

Fig. 8.11 $a = \pi/2$ and $b = 0$ on left, $b = .4$ on right.

8.10 Frames along minimal curves

The action of $(\mathbf{x}, A) \in \mathbf{E}(3,\mathbf{C})$ on $\mathbf{y} \in \mathbf{C}^3$ is

$$(\mathbf{x}, A)\mathbf{y} = \mathbf{x} + A\mathbf{y}$$

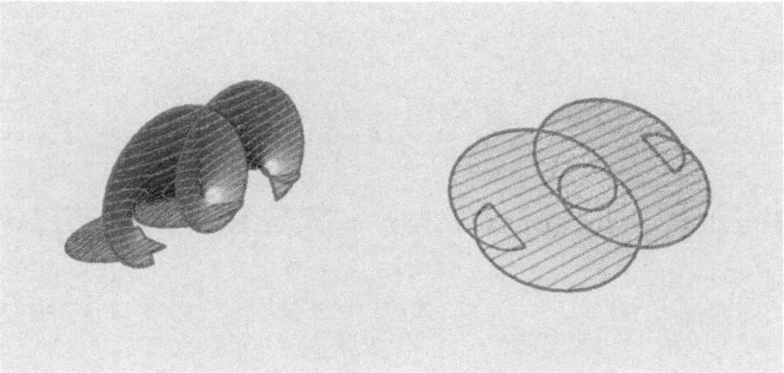

Fig. 8.12 $a = \pi/2$ and $b = 1$ on left, $b = 20$ on right.

This action is transitive. If we choose the zero vector $\mathbf{0}$ as the origin of $\mathbf{C}^3$, then the action defines a principal $\mathbf{SO}(3,\mathbf{C})$-bundle projection

$$\pi : \mathbf{E}(3,\mathbf{C}) \to \mathbf{C}^3, \quad \pi(\mathbf{x}, A) = (\mathbf{x}, A)\mathbf{0} = \mathbf{x}.$$

In the theory of regular curves in Euclidean space, a group element $(\mathbf{x}, A)$ is identified with an orthonormal frame at $\mathbf{x}$ by choosing the standard basis $(\boldsymbol{\epsilon}_1, \boldsymbol{\epsilon}_2, \boldsymbol{\epsilon}_3)$ as the reference frame at $\mathbf{0}$ and then $(\mathbf{x}, A)$ defines the orthonormal frame at $\mathbf{x}$ given by

$$d(\mathbf{x}, A)\boldsymbol{\epsilon}_i = A\boldsymbol{\epsilon}_i = \mathbf{A}_i, \quad i = 1,2,3,$$

where $\mathbf{A}_i$ is column i of A. Consequently, if we use the standard basis of $\mathbf{C}^3$ as reference frame at $\mathbf{0}$, then no frame defined by $\mathbf{E}(3,\mathbf{C})$ contains an isotropic vector, and thus no such frame can have its first vector equal to $\dot{\boldsymbol{\gamma}}(t)$, when $\boldsymbol{\gamma}$ is a minimal curve. The first step in the construction of a Frenet frame along a nonminimal curve makes no sense for a minimal curve.

Following E. Cartan's exposition in his book [32], we begin the frame construction along minimal curves by first changing the choice of reference frame at $\mathbf{0}$.

Definition 8.40. A frame $F = (\mathbf{F}_1, \mathbf{F}_2, \mathbf{F}_3)$ of $\mathbf{C}^3$ is *cyclic* if $\mathbf{F}_i \cdot \mathbf{F}_j = L_{ij}$, for i,j = 1,2,3, where

$$L = (L_{ij}) = \begin{pmatrix} 0 & 0 & 1 \\ 0 & 1 & 0 \\ 1 & 0 & 0 \end{pmatrix}.$$

If $F = (F^j_k) \in \mathbf{GL}(3, \mathbf{C})$ is defined by $\mathbf{F}_k = \sum_1^3 F^j_k \mathbf{E}_j$, then ${}^tFF = L$ and $\det F = \pm i$. The frame is called *direct cyclic*, respectively *indirect cyclic*, according to whether $\det F = i$ or $\det F = -i$.

Exercise 36. Prove that the first and third vectors of a cyclic frame are isotropic and that $L = {}^tL = L^{-1}$.

We choose as our reference frame at $\mathbf{0}$ the direct cyclic frame

$$E = (\mathbf{E}_1, \mathbf{E}_2, \mathbf{E}_3) = (\epsilon_1, \epsilon_2, \epsilon_3) \begin{pmatrix} \frac{1}{2} & 0 & 1 \\ \frac{i}{2} & 0 & -i \\ 0 & 1 & 0 \end{pmatrix},$$

so

$$\mathbf{E}_1 = \frac{1}{2}(\epsilon_1 + i\epsilon_2), \quad \mathbf{E}_2 = \epsilon_3, \quad \mathbf{E}_3 = \epsilon_1 - i\epsilon_2. \tag{8.48}$$

The dot product of vectors $\mathbf{x} = \sum_1^3 x^j \mathbf{E}_j$ and $\mathbf{y} = \sum_1^3 y^j \mathbf{E}_j$ relative to this frame is

$$\mathbf{x} \cdot \mathbf{y} = x^1 y^3 + x^3 y^1 + x^2 y^2 = {}^t\mathbf{x} L \mathbf{y},$$

where now we identify a vector $\mathbf{x} \in \mathbf{C}^3$ with the column matrix of its components relative to E. Relative to E, a frame $F = (\mathbf{F}_1, \mathbf{F}_2, \mathbf{F}_3)$ of $\mathbf{C}^3$ is direct cyclic if and only if ${}^tF_j L F_k = L_{jk}$ and $\det F = 1$, where now $F = (F^j_k) \in \mathbf{GL}(3, \mathbf{C})$ is given by $\mathbf{F}_k = \sum_1^3 F^j_k \mathbf{E}_j$. Let

$$G = \{B \in \mathbf{GL}(3, \mathbf{C}) : {}^tBLB = L, \ \det B = 1\} \cong \mathbf{SO}(3, \mathbf{C}).$$

A matrix $B \in \mathbf{GL}(3, \mathbf{C})$ belongs to G if and only if its columns form a direct cyclic frame (relative to E). The Lie algebra of G is

$$\begin{aligned} \mathfrak{g} &= \{X \in \mathbf{C}^{3\times 3} : {}^tXL + LX = 0, \ \mathrm{trace} X = 0\} \\ &= \left\{ \begin{pmatrix} x & y & 0 \\ z & 0 & -y \\ 0 & -z & -x \end{pmatrix} : x, y, z \in \mathbf{C} \right\}. \end{aligned} \tag{8.49}$$

A matrix F of three column vectors in $\mathbf{C}^3$ is a direct cyclic frame if and only if $F \in G$.

Let M be a connected Riemann surface and let

$$\boldsymbol{\gamma} : M \to \mathbf{C}^3$$

be a regular minimal curve; that is, $d\boldsymbol{\gamma}$ is never zero and $d\boldsymbol{\gamma} \cdot d\boldsymbol{\gamma} = 0$ at every point of U. We emphasize that elements of $\mathbf{C}^3$ are now column matrices relative to E.

Definition 8.41. A *direct cyclic frame field* along $\boldsymbol{\gamma}$ on an open subset $U \subset M$ is a holomorphic map

$$(\boldsymbol{\gamma}, e) : U \to \mathbf{C}^3 \rtimes G \cong \mathbf{E}(3, \mathbf{C}).$$

At each point $m \in U$, $(\boldsymbol{\gamma}, e)$ defines the direct cyclic frame $e(m) = (\mathbf{e}_1, \mathbf{e}_2, \mathbf{e}_3)(m)$ at $\boldsymbol{\gamma}(m)$.

If $(\boldsymbol{\gamma}, e)$ is a direct cyclic frame field along $\boldsymbol{\gamma}$, then we can express $d\boldsymbol{\gamma}$ in terms of the direct cyclic frame e by

$$d\boldsymbol{\gamma} = \sum_{i=1}^{3} \omega^i \mathbf{e}_i \text{ so } e^{-1} d\boldsymbol{\gamma} = {}^t(\omega^1, \omega^2, \omega^3) = {}^t\theta,$$

where the ω^i are $\mathbf{C}$-valued 1-forms on U, and θ is a $\mathbf{C}^3$-valued 1-form. Likewise,

$$d\mathbf{e}_i = \sum_{j=1}^{3} \omega_i^j \mathbf{e}_j,$$

for $i = 1, 2, 3$. In matrix notation this is

$$e^{-1} de = \omega,$$

where ω is a $\mathfrak{g}$-valued 1-form on U, which looks like

$$\omega = \begin{pmatrix} \omega_1^1 & \omega_2^1 & 0 \\ \omega_1^2 & 0 & -\omega_2^1 \\ 0 & -\omega_1^2 & -\omega_1^1 \end{pmatrix},$$

by (8.49). If $(\boldsymbol{\gamma}, e) : U \to \mathbf{C}^3 \rtimes G$ is a direct cyclic frame field along $\boldsymbol{\gamma}$, then any other is given by $(\boldsymbol{\gamma}, \tilde{e})$, where $\tilde{e} = eB$ and

$$B : U \to G$$

is an arbitrary holomorphic map.

Definition 8.42. A direct cyclic frame field (γ, e) along γ is *first order* if $\mathbf{e}_1$ is a non-zero multiple of $d\gamma$ at every point of U. That is,

$$\omega^2 = 0 = \omega^3$$

at every point of U. It follows that ω^1 is a nowhere zero holomorphic 1-form on U, since γ is regular.

Lemma 8.43. *If $(\gamma, e) : U \to \mathbf{C} \rtimes G)$ is a first order direct cyclic frame field along γ, then any other is given by $\tilde{e} = eB$, where*

$$B : U \to G_1$$

is an arbitrary holomorphic map into the complex subgroup

$$G_1 = \{B(r,s) = \begin{pmatrix} r & -rs & -rs^2/2 \\ 0 & 1 & s \\ 0 & 0 & 1/r \end{pmatrix} \in G : r, s \in \mathbf{C},\ r \neq 0\}. \tag{8.50}$$

Proof. On U, $d\gamma$ is a multiple of $\mathbf{e}_1$ and of $\tilde{\mathbf{e}}_1 = \sum_1^3 \mathbf{e}_i B_1^i$ if and only if

$$B_1^2 = B_1^3 = 0 \tag{8.51}$$

on U. The proof is then completed by the following exercise. □

Exercise 37. Prove that $B \in G$ satisfies (8.51) if and only if B is an element of the subgroup G_1 defined in (8.50).

Exercise 38. Prove that the Lie algebra of G_1 is

$$\mathfrak{g}_1 = \left\{ \begin{pmatrix} r & -s & 0 \\ 0 & 0 & s \\ 0 & 0 & -r \end{pmatrix} : r, s \in \mathbf{C} \right\}.$$

For the next step of the frame reduction process, we fix a first order direct cyclic frame field (γ, e), with $\omega = e^{-1}de$, and then examine the entries of ω that do not involve dB when the frame is changed to $\tilde{e} = eB$, for some

$$B = B(r,s) : U \to G_1. \tag{8.52}$$

To do this, we choose a vector subspace $\mathfrak{m}_1 \subset \mathfrak{g}$ complementary to $\mathfrak{g}_1$,

$$\mathfrak{m}_1 = \left\{ \begin{pmatrix} 0 & 0 & 0 \\ z & 0 & 0 \\ 0 & -z & 0 \end{pmatrix} : z \in \mathbf{C} \right\},$$

to arrive at a vector space direct sum

$$\mathfrak{g} = \mathfrak{g}_1 \oplus \mathfrak{m}_1.$$

Writing a subspace as a subscript to denote projection, we have

$$\tilde{\omega} = (eB)^{-1}d(eB) = B^{-1}\omega B + B^{-1}dB = B^{-1}\omega_{\mathfrak{g}_1}B + B^{-1}dB + B^{-1}\omega_{\mathfrak{m}_1}B,$$

which shows, after a calculation, that

$$\tilde{\omega}_{\mathfrak{m}_1} = (B^{-1}\omega_{\mathfrak{m}_1}B)_{\mathfrak{m}_1} = \begin{pmatrix} 0 & 0 & 0 \\ r\omega_1^2 & 0 & 0 \\ 0 & -r\omega_1^2 & 0 \end{pmatrix},$$

so

$$\tilde{\omega}_1^2 = r\omega_1^2.$$

Furthermore,

$$d\boldsymbol{\gamma} = \omega^1\mathbf{e}_1 = \tilde{\omega}^1\tilde{\mathbf{e}}_1 = \tilde{\omega}^1 r\mathbf{e}_1$$

implies that

$$\tilde{\omega}^1 = \frac{1}{r}\omega^1. \tag{8.53}$$

If we set

$$\omega_1^2 = a\omega^1, \quad \tilde{\omega}_1^2 = \tilde{a}\tilde{\omega}^1,$$

where a and $\tilde{a}$ are holomorphic functions on U, then

$$\tilde{a} = r^2 a.$$

There are three cases:

$$\omega_1^2 \text{ never } 0, \quad \omega_1^2 \text{ has isolated zeros}, \quad \omega_1^2 \text{ identically } 0.$$

Example 8.44 (Case ω_1^2 never zero). Now the holomorphic function r above can be chosen so that $\tilde{a} = 1$ on U, that is $\tilde{\omega}_1^2 = \tilde{\omega}^1$.

Definition 8.45. A *second order* direct cyclic frame field along $\boldsymbol{\gamma}$ is a direct cyclic frame field for which

$$\omega^2 = 0 = \omega^3 \qquad \text{(first order), and}$$

$$\omega_1^2 = \omega^1 \qquad \text{(second order).}$$

A second order direct cyclic frame field along $\boldsymbol{\gamma}$ exists on a neighborhood of a point of U, provided that ω_1^2 is never zero on a neighborhood of the point, for some, hence any, first order frame field. If $(\boldsymbol{\gamma}, e)$ is second order, then

$$\tilde{e} = eB(r,s)$$

is also second order if and only if $r = 1$. By (8.53) this condition is equivalent to

$$\tilde{\omega}^1 = \omega^1,$$

so the 1-form ω^1 is invariant under change of second order frames. It is, therefore, a globally defined nowhere zero holomorphic 1-form on M. We shall call it the *element of pseudoarc* on M. We define a *pseudoarc parameter* to be a locally defined holomorphic function z in M such that

$$\omega^1 = dz.$$

At a point of M where $\omega_1^2 \neq 0$, there is a neighborhood of the point on which a pseudoarc parameter exists. It is unique up to additive constant. Let

$$G_2 = \{B(1,s) = \begin{pmatrix} 1 & -s & -s^2/2 \\ 0 & 1 & s \\ 0 & 0 & 1 \end{pmatrix} \in G_1 : s \in \mathbf{C}\},$$

which is a Lie subgroup of G_1 with Lie algebra

$$\mathfrak{g}_2 = \left\{ \begin{pmatrix} 0 & -s & 0 \\ 0 & 0 & s \\ 0 & 0 & 0 \end{pmatrix} : s \in \mathbf{C} \right\}.$$

As a vector subspace complementary to $\mathfrak{g}_2$ in $\mathfrak{g}_1$, we choose

$$\mathfrak{m}_2 = \left\{ \begin{pmatrix} r & 0 & 0 \\ 0 & 0 & 0 \\ 0 & 0 & -r \end{pmatrix} : r \in \mathbf{C} \right\}.$$

For a second order direct cyclic frame field $(\boldsymbol{\gamma}, e)$, we have the decomposition

$$\omega = \omega_{\mathfrak{g}_2} + \omega_{\mathfrak{m}_1 + \mathfrak{m}_2}.$$

Any other second order frame field has

$$\tilde{e} = eB(1,s)$$

from which we calculate that

$$\tilde{\omega}_{\mathfrak{m}_1+\mathfrak{m}_2} = \begin{pmatrix} \omega_1^1 + s\omega^1 & 0 & 0 \\ \omega^1 & 0 & 0 \\ 0 & -\omega^1 & -\omega_1^1 - s\omega^1 \end{pmatrix},$$

which implies that

$$\tilde{\omega}_1^1 = \omega_1^1 + s\omega^1, \quad \tilde{\omega}^1 = \omega^1.$$

We can choose the holomorphic function s so that $\tilde{\omega}_1^1 = 0$ on U.

Definition 8.46. A *third order* direct cyclic frame field (γ, e) is a direct cyclic frame field for which

$$\omega^2 = 0 = \omega^3 \qquad \text{(first order), and}$$
$$\omega_1^2 = \omega^1 \qquad \text{(second order), and}$$
$$\omega_1^1 = 0 \qquad \text{(third order).}$$

on U.

If e is third order, then

$$\tilde{e} = eB(1,s)$$

is also third order if and only if s is identically zero on U; that is, B is the identity matrix and $\tilde{e} = e$. Third order frames are the *Frenet* frames. Under the assumption that for any first order frame field (γ, e) we have ω_1^2 never zero, we have the result that there exists a unique Frenet frame field (γ, e) on M, characterized by

$$\omega = \begin{pmatrix} 0 & \omega_2^1 & 0 \\ \omega^1 & 0 & -\omega_2^1 \\ 0 & -\omega^1 & 0 \end{pmatrix}.$$

The remaining form must be a multiple of ω^1, that is

$$\omega_2^1 = k\omega^1,$$

for some holomorphic function

$$k : M \to \mathbf{C},$$

which is called the *curvature* of the minimal curve γ.

Example 8.47 (From the Weierstrass representation). If $\boldsymbol{\gamma} : M \to \mathbf{C}^3$ is a minimal curve, where M is a connected Riemann surface, then $d\boldsymbol{\gamma}$ is a holomorphic, isotropic, $\mathbf{C}^3$-valued 1-form on M. By the Weierstrass representation, $d\boldsymbol{\gamma}$ has either the degenerate form to be discussed below, or there exists a meromorphic function g on M and a holomorphic 1-form η on M, such that g has a pole of order n at a point if and only if η has a zero of order $2n$ at the point, and

$$d\boldsymbol{\gamma} = (\frac{1}{2}(1-g^2)\boldsymbol{\epsilon}_1 + \frac{i}{2}(1+g^2)\boldsymbol{\epsilon}_2 + g\boldsymbol{\epsilon}_3)\eta = (\mathbf{E}_1 + g\mathbf{E}_2 - \frac{g^2}{2}\mathbf{E}_3)\eta.$$

Relative to E,

$$d\boldsymbol{\gamma} = {}^t(1, g, -\frac{g^2}{2})\eta.$$

A first order direct cyclic frame field $(\boldsymbol{\gamma}, e)$ is given by

$$e = \begin{pmatrix} 1 & 0 & 0 \\ g & 1 & 0 \\ -\frac{g^2}{2} & -g & 1 \end{pmatrix},$$

since then $d\boldsymbol{\gamma} = \eta\mathbf{e}_1$, so

$$\omega^1 = \eta, \quad \omega^2 = 0 = \omega^3.$$

We calculate

$$\omega = e^{-1}de = \begin{pmatrix} 0 & 0 & 0 \\ dg & 0 & 0 \\ 0 & -dg & 0 \end{pmatrix},$$

which shows that

$$\omega_1^2 = dg, \quad \omega_2^1 = 0 = \omega_1^1.$$

Therefore, as can be verified directly,

$$d\mathbf{e}_1 = dg\,\mathbf{e}_2, \quad d\mathbf{e}_2 = -dg\,\mathbf{e}_3, \quad d\mathbf{e}_3 = 0.$$

Any other first order frame field is given by $\tilde{e} = eB(r,s)$, so

$$\tilde{\mathbf{e}}_1 = r\mathbf{e}_1, \quad \tilde{\mathbf{e}}_2 = \mathbf{e}_2 - rs\mathbf{e}_1, \quad \tilde{\mathbf{e}}_3 = \frac{1}{r}\mathbf{e}_3 - \frac{rs^2}{2}\mathbf{e}_1 + s\mathbf{e}_2$$

and

$$\tilde{\omega}^1 = \frac{1}{r}\omega^1 = \frac{1}{r}\eta, \quad \tilde{\omega}_1^2 = r\omega_1^2 = rdg,$$

$$\tilde{\omega}_1^1 = rsdg + \frac{1}{r}dr, \quad \tilde{\omega}_2^1 = -(\frac{s}{r}dr + ds + \frac{s^2 r}{2}dg).$$

We assume that g is not constant, so outside a set of isolated points dg is never zero and we can define a holomorphic function f by

$$\eta = fdg.$$

Then $\tilde{\omega}_1^2 = \tilde{\omega}^1$ if and only if

$$r^2 = f.$$

That is, $\tilde{e}$ is a second order frame if and only if $r = f^{1/2}$. For this value of r, the element of pseudoarc is

$$dz = \tilde{\omega}^1 = \sqrt{f}dg.$$

By (8.17), the Hopf quadratic differential of the minimal immersion defined by the real part of $\boldsymbol{\gamma}$ is

$$II^{2,0} = -\eta dg = -fdgdg = -dzdz. \tag{8.54}$$

This gives another proof that a Goursat transform of a minimal immersion preserves the second fundamental form.

This frame becomes third order if s is chosen so that $\tilde{\omega}_1^1 = 0$, which, by the above expression for $\tilde{\omega}_1^1$, is

$$s = -\frac{1}{r^2}\frac{dr}{dg} = -\frac{1}{2f^{3/2}}\frac{df}{dg} = -\frac{1}{\sqrt{f}}\frac{f'}{2f},$$

where we write $dh = h'dg$ for any holomorphic function h. For these values of r and s, we then calculate that $\tilde{\omega}_2^1 = k\tilde{\omega}^1$, where the holomorphic function k is the curvature,

$$k = \frac{1}{2f}\left((\frac{f'}{f})' - (\frac{f'}{2f})^2\right).$$

Problems

8.48. Prove that the catenoid is the only surface of revolution with mean curvature zero.

8.49. Prove that the plane and the right helicoid are the only minimal surfaces that are ruled surfaces. See Catalan [34].

8.50. Prove that the immersion

$$\mathbf{x} : \cup_{m+n\ \text{even}} S_{m,n} \to \mathbf{R}^3, \quad \mathbf{x}(x,y) = {}^t(x, y, \log(\frac{\cos x}{\cos y}))$$

is not conformal. We invite the reader to try to find a complex coordinate chart for the complex structure induced by $d\mathbf{x} \cdot d\mathbf{x}$, for example by applying the proof of Theorem 7.4. See Weber's geometric solution to this exercise in [167, §5D]. We describe this complex structure in Example 8.17.

8.51. Prove that Scherk's surface is the only surface of translation with mean curvature zero, up to a change of scale $(x,y,z) \mapsto (ax, ay, az)$, for some positive constant a.

8.52. Prove that for an immersion $\mathbf{x} : M \to \mathbf{S}^3$ or $\mathbf{x} : M \to \mathbf{H}^3$, the mean curvature is identically zero if and only if the first variation vanishes for any admissible variation.

8.53. Prove that any point of $\mathbf{CP}^1$ has a neighborhood U on which there is a *holomorphic section* $s : U \to \mathbf{C}^2 \setminus \{\mathbf{0}\}$; that is, $\pi \circ s$ is the identity map of U. Use this to prove that any holomorphic map $F : M \to \mathbf{CP}^1$ locally factors through holomorphic maps $U \subset M \to \mathbf{C}^2 \setminus \{\mathbf{0}\}$.

8.54. Prove that, for the local holomorphic section

$$s : \mathbf{CP}^1 \setminus \{[\epsilon_1]\} \to \mathbf{C}^2, \quad s\begin{bmatrix} z \\ w \end{bmatrix} = \begin{pmatrix} z/w \\ 1 \end{pmatrix},$$

and for $\mathbf{z} = u\epsilon_1 + v\epsilon_2 + w\epsilon_3 \in \mathscr{I}$ for which $u - iv \neq 0$,

$$\mathscr{W} \circ s \circ \mathfrak{w}^{-1} \circ \pi(\mathbf{z}) = \frac{1}{u - iv}\mathbf{z}.$$

This formula lies behind the construction of the map F in (8.10).

8.55. If $\mathbf{x} : M \to \mathbf{R}^3$ is a minimal immersion of a connected surface M, and if $\alpha = \partial\mathbf{x}$ is given by (8.15), derive the formula

$$dA = \frac{i}{2}(1 + |g|^2)^2 \eta \wedge \bar{\eta}$$

for its area element and the formula

$$\int_M K\,dA = -2i \int_M \frac{1}{(1+|g|^2)^2} dg \wedge d\bar{g}$$

for its total curvature. Use (8.18) to prove that $K \leq 0$ on M and that $K(p) = 0$ for some point $p \in M$ if and only if $dg_p = 0$. Conclude that either K is identically zero on M, or the zeros of K are isolated points of M.

8.56. Prove that a minimal immersion $\mathbf{x} : M \to \mathbf{R}^3$ has a conformal associate $\hat{\mathbf{x}}$ defined by any nowhere zero holomorphic function $\mathfrak{p}$ on M. Prove that $d\hat{\mathbf{x}} \cdot d\hat{\mathbf{x}} = |\mathfrak{p}|^2 d\mathbf{x} \cdot d\mathbf{x}$, that $\mathbf{x}$ and $\hat{\mathbf{x}}$ have the same Gauss map, and their Hopf quadratic differentials are related by $\widehat{II}^{2,0} = \mathfrak{p} II^{2,0}$.

8.57. Prove that the induced metric of Scherk's surface $\mathbf{x}$ of Example 8.17 is

$$I = \frac{16(1+|z|^2)^2}{|1-z^4|^2} dz d\bar{z} \tag{8.55}$$

and prove that it is complete.

8.58. Analyze and describe the conjugate to Scherk's surface of Example 8.17. This would be the imaginary part of the complex minimal curve (8.30). See [167] for details.

8.59. Prove that $G = \pi_1(\mathbf{C} \setminus \{\pm 1, \pm i\}, 0) \cong \mathbf{Z} * \mathbf{Z} * \mathbf{Z}$, the free product of three copies of the integers $\mathbf{Z}$. See, for example, [83, §1.2].

8.60. Prove that if $f(x,y)$ is a solution to the minimal surface equation on all of $\mathbf{R}^2$, then the minimal immersion $\mathbf{x} : \mathbf{R}^2 \to \mathbf{R}^3$ given by the graph of f is complete. See (8.42) in Remark 8.24.

8.61. Use Example 8.13 to prove that the total curvature of Enneper's Surface is -4π and its Gauss map omits one point of $\mathbf{S}^2$. Use Example 8.15 to prove the total curvature of the Catenoid is -4π and its Gauss map omits two points of $\mathbf{S}^2$.

8.62. Given $T = A + iB \in \mathbf{SO}(3, \mathbf{C})$, so $A, B \in \mathbf{GL}(n, \mathbf{R})$ satisfy (8.43). Prove that if $A \in \mathbf{SO}(3)$, then $B = 0$. Prove that B cannot be in $\mathbf{SO}(3)$.

8.63. If $(x,y,z) \in V$ and u and v are defined by (8.44), so $F(x,y,z) = (u,v)$, prove that $(x,y,z) \in \mathbf{S}^2 \subset \mathbf{R}^3$ if and only if $u\bar{v} = -1$. Prove also that $-(x,y,z) \in V$ and $F(-x,-y,-z) = (v,u)$. Compare the map F restricted to $\mathbf{S}^2$ with Example 7.16.

8.64. Prove that any element $A \in \mathbf{SO}(2, \mathbf{C})$ is of the form

$$A = \begin{pmatrix} \cos\zeta & -\sin\zeta \\ \sin\zeta & \cos\zeta \end{pmatrix},$$

for some unique $\zeta \in \mathbf{C}$. Writing $\zeta = x + iy$, where $x, y \in \mathbf{R}$, it follows that $A = BC = CB$, where

$$B = \begin{pmatrix} \cos x & -\sin x \\ \sin x & \cos x \end{pmatrix}, \quad C = \begin{pmatrix} \cosh y & -i\sinh y \\ i\sinh y & \cosh y \end{pmatrix}.$$

8.65. Here G is the representation of $\mathbf{SO}(3,\mathbf{C})$ in the basis L as defined in Section 8.10.

8.66. Let $\boldsymbol{\gamma}, \tilde{\boldsymbol{\gamma}} : M \to \mathbf{C}^3$ be minimal curves of the type ω_1^2 never zero, with curvature functions k and $\tilde{k}$, and pseudoarc parameters z and $\tilde{z}$, respectively, on a neighborhood $U \subset M$. Prove that if $k(z) = \tilde{k}(\tilde{z})$ and $dz = d\tilde{z}$ on U, then there exists an element $(\mathbf{a}, A) \in \mathbf{C}^3 \rtimes G$ such that $\tilde{\boldsymbol{\gamma}} = \mathbf{a} + A\boldsymbol{\gamma}$ on U.

8.67. Prove that if $k : M \to \mathbf{C}$ is any holomorphic function on the Riemann surface M, and if $\pi : \tilde{M} \to M$ is the universal cover of M, then there exists a unique, up to congruence, minimal curve $\boldsymbol{\gamma} : \tilde{M} \to \mathbf{C}^3$ whose curvature is $k \circ \pi$.

8.68. Why does formula (8.54) imply the invariance of the second fundamental form under Goursat transform?

8.69 (Enneper's surfaces). Find the curvature of the minimal curve defined by the Enneper surface data $g = z$, $\eta = dz$ on $M = \mathbf{C}$. What is the element of pseudoarc?

8.70 (Catenoid). Find the curvature of the minimal curve defined by the catenoid data on the universal cover $\mathbf{C}$. Here, by Example 8.15, $g = e^{-w}$, $\eta = e^{w}dw$ where w is the standard complex coordinate on $\mathbf{C}$. Find its element of pseudoarc.

8.71 (Scherk's surface). Let $g = z$ and $\eta = \frac{2}{z^4 - 1}dz$ be the Scherk surface data of Example 8.17. Find the curvature and element of pseudoarc of the minimal curve defined by this data.

8.72. What Weierstrass data defines a minimal curve with a given nonzero constant curvature k?

8.73. Find the minimal curve and its curvature defined by the degenerate Weierstrass data, namely, for which $d\boldsymbol{\gamma} = {}^t(0,0,i)\eta$ relative to E.

8.74. Complete the frame reduction for minimal curves $\boldsymbol{\gamma} : M \to \mathbf{C}^3$ for which ω_1^2 has isolated zeros, for any first order frame field along $\boldsymbol{\gamma}$.

8.75. Complete the frame reduction for minimal curves $\boldsymbol{\gamma} : M \to \mathbf{C}^3$ for which ω_1^2 is identically zero, for any first order frame field along $\boldsymbol{\gamma}$.

Chapter 9
Isothermic Immersions

We present here a brief introduction to classical isothermic immersions in Euclidean space, a notion easily extended to immersions of surfaces into each of the space forms. The definition, which is the existence of coordinate charts that are isothermal and whose coordinate curves are lines of curvature, seems more analytic than geometric. We show that CMC immersions are isothermic away from their umbilics, which indicates that isothermic immersions are generalizations of CMC immersions. The Christoffel transform provides geometric content to the concept.

9.1 Background and motivation

Minimal immersions possess two remarkable properties that we now recall in a geometric formulation. Let $\mathbf{x} : M \to \mathbf{R}^3$ be a minimal immersion whose image is not a plane. For ease of exposition we assume that $\mathbf{x}$ has no umbilic points, which would be isolated in any case.

Let the Gauss map of $\mathbf{x}$, denoted $\hat{\mathbf{x}} : M \to \mathbf{S}^2 \subset \mathbf{R}^3$, be given by the unit normal vector field $\mathbf{e}_3$ along $\mathbf{x}$. Then $\mathbf{e}_3$ is also a unit normal vector field along the immersion $\hat{\mathbf{x}}$, which means that $\mathbf{x}$ and $\hat{\mathbf{x}}$ have the same Gauss map, or equivalently, they have parallel tangent planes at each point of M. Of course, this much is true for any immersion $\mathbf{x} : M \to \mathbf{R}^3$. If $\mathbf{x}$ is minimal, however, then by (2) of Theorem 7.40, the immersions $\mathbf{x}$ and $\hat{\mathbf{x}}$ induce conformally related metrics on M and $\mathbf{x}$ and $\hat{\mathbf{x}}$ have opposite orientations. This last property means that for any local coordinate chart $(U, (x, y))$ in M, the parallel vectors $\mathbf{x}_x \times \mathbf{x}_y$ and $\hat{\mathbf{x}}_x \times \hat{\mathbf{x}}_y$ have opposite signs, or equivalently, if $(\mathbf{x}, e)$ is any first order frame field along $\mathbf{x}$ on U, then $(\hat{\mathbf{x}}, e)$ is a first order frame field along $\hat{\mathbf{x}}$ and then $d\mathbf{x} = \omega^1\mathbf{e}_1 + \omega^2\mathbf{e}_2$ and $d\hat{\mathbf{x}} = \hat{\omega}^1\mathbf{e}_1 + \hat{\omega}^2\mathbf{e}_2$ with $\hat{\omega}^1\wedge\hat{\omega}^2 = f\omega^1\wedge\omega^2$ for some negative function f on U. In the language

G.R. Jensen et al., *Surfaces in Classical Geometries*, Universitext,
DOI 10.1007/978-3-319-27076-0_9

of Definition 9.22, the Gauss map $\hat{\mathbf{x}}$ is a Christoffel transform of the minimal immersion $\mathbf{x}$. We shall see that the much wider class of isothermic immersions possess Christoffel transforms.

The second remarkable property of a minimal immersion $\mathbf{x}$ is its possession of a conformal associate $\hat{\mathbf{x}}$ defined by any nowhere zero holomorphic function f on M. By Definition 8.12, this means that $\partial\hat{\mathbf{x}} = f\partial\mathbf{x}$, which shows that $\mathbf{x}$ and $\hat{\mathbf{x}}$ have parallel tangent planes and the same orientation at each point. By Problem 8.56, they have conformally related metrics. We shall see that only minimal immersions have this property.

In his 1867 paper [50], Christoffel considered the problem of which immersions $\mathbf{x} : M \to \mathbf{R}^3$ possess the property that there exists another immersion $\hat{\mathbf{x}} : M \to \mathbf{R}^3$ such that $\mathbf{x}$ and $\hat{\mathbf{x}}$ have the same Gauss map and induce conformally related metrics on M. His answers depend dramatically on whether $\mathbf{x}$ and $\hat{\mathbf{x}}$ are to have the same or opposite orientations. They also require the concept of isothermic immersion. In this chapter we describe this concept and Christoffel's transform.

The classical definition of isothermic immersion in Euclidean space carries over without change to immersions in any space form. As the property is preserved by conformal transformations, we shall profit from studying it in the context of Möbius geometry, which we shall do in Chapter 14. For these chapters we have used the papers by Musso [124] and Bernstein [3] and the monographs of Burstall [24], Hertrich-Jeromin [86] and Kamberov et al. [98].

9.2 Classical isothermic immersions

In his 1871 paper [35], written without apparent knowledge of Christoffel's earlier paper cited above, Cayley posed the following problem, shown graphically in Figure 9.1. Suppose $\mathbf{x} : M \to \mathbf{R}^3$ is an immersion such that at a point $A \in M$ the principal curvatures a and c are independent, meaning that $da \wedge dc \neq 0$ at A. Then a, c are local coordinates on some open neighborhood U about A. Suppose that the coordinates of A are (a_0, c_0) and consider the coordinate curve quadrilateral $ABCD$ in U, where the coordinates of B, C and D are (a, c_0), (a_0, c) and (a, c) respectively. Notice that the coordinate curves are lines of curvature. Consider the infinitesimally nearby lines of curvature, the vertical one given by setting the first curvature to

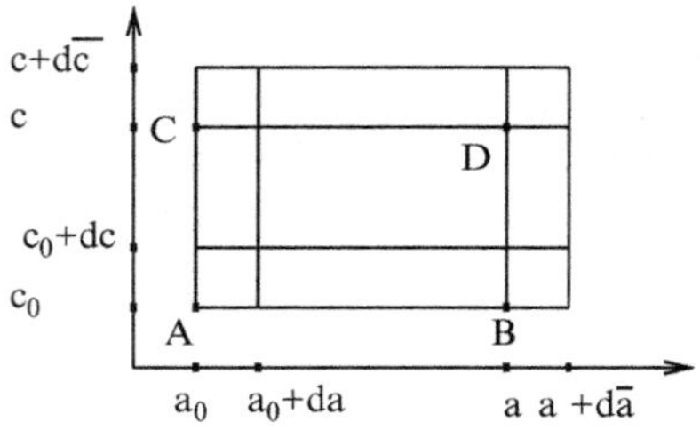

Fig. 9.1 Cayley's problem

$a_0 + da$ and the horizontal one given by setting the second curvature to $c_0 + dc$, with da and dc chosen so that an infinitesimal square is formed at A. Consider the vertical line of curvature given by setting the first curvature to $a + d\bar{a}$ so that an infinitesimal square is formed at B, and the horizontal line of curvature given by setting the second curvature to $c + d\bar{c}$ so that an infinitesimal square is formed at C. Cayley's problem is: When does this construction produce an infinitesimal square at D, for all (a, c)?

His answer goes like this. With these coordinates, the induced metric is $I = E da^2 + G dc^2$, since the lines of curvature are always orthogonal. That the construction produces an infinitesimal square at A means

$$\sqrt{E(a_0, c_0)}\, da = \sqrt{G(a_0, c_0)}\, dc,$$

at B means

$$\sqrt{E(a, c_0)}\, d\bar{a} = \sqrt{G(a, c_0)}\, dc,$$

and at C means

$$\sqrt{E(a_0, c)}\, da = \sqrt{G(a_0, c)}\, d\bar{c}.$$

Consequently, $dc = L da$, where L is a constant, and

$$d\bar{a} = \frac{\sqrt{G(a, c_0)}}{\sqrt{E(a, c_0)}} L da = f(a) da, \quad d\bar{c} = \frac{\sqrt{E(a_0, c)}}{\sqrt{G(a_0, c)}} da = g(c) da.$$

An infinitesimal square is formed at D means $\sqrt{E(a, c)}\, d\bar{a} = \sqrt{G(a, c)}\, d\bar{c}$, which occurs if and only if

$$\frac{E(a, c)}{G(a, c)} = \left(\frac{d\bar{c}}{d\bar{a}}\right)^2 = \frac{g(c)^2}{f(a)^2}$$

in which case the metric takes the form

$$I = \frac{g(c)^2}{f(a)^2} G da^2 + G dc^2 = G g^2 \left(\left(\frac{da}{f(a)}\right)^2 + \left(\frac{dc}{g(c)}\right)^2\right) = e^{2u}(dx^2 + dy^2)$$

where $e^{2u} = G g^2$ and $dx = \frac{da}{f(a)}$ and $dy = \frac{dc}{g(c)}$ are local coordinates whose coordinate curves remain lines of curvature. They are principal isothermal coordinates, a concept defined below.

The only examples given by Cayley of immersions possessing this property are the quadrics

$$\frac{x^2}{p} + \frac{y^2}{q} + \frac{z^2}{r} = 1 \tag{9.1}$$

when the constants p, q and r are distinct. See Problem 9.33.

Using nonstandard analysis, Hertrich-Jeromin [85] has given a modern interpretation of Cayley's geometric characterization of isothermic immersion. We proceed now to the definition of isothermic immersion.

Definition 9.1. A chart $(U,(x,y))$ on a Riemannian surface (M,I) is *isothermal* if it puts the metric in the form

$$I = e^{2u}(dx^2 + dy^2).$$

Definition 9.2. For an immersion $\mathbf{x}: M \to \mathbf{R}^3$, a coordinate chart $(U,(x,y))$ on M is *principal*, if it diagonalizes the second fundamental form of $\mathbf{x}$ as

$$II = L\,dx^2 + N\,dy^2, \tag{9.2}$$

where L and N are smooth functions on U. In this case the coordinate curves are lines of curvature, which are orthogonal at points where $L \neq N$, in which case $I = E\,dx^2 + G\,dy^2$, for some positive smooth functions E and G on U.

Lemma 9.3. *For an immersion $\mathbf{x}: M \to \mathbf{R}^3$, a chart $(U,(x,y))$ is isothermal if and only if*

$$|\mathbf{x}_x| = |\mathbf{x}_y| \ \textit{and}\ \ \mathbf{x}_x \cdot \mathbf{x}_y = 0$$

on U. An isothermal chart is also principal if and only if it also satisfies

$$\mathbf{x}_{xy} \in \textit{span}\ \{\mathbf{x}_x, \mathbf{x}_y\}$$

at each point of U.

Proof. The conditions to be isothermal are clear. If $(U,(x,y))$ is an isothermal chart, let $e^u = |\mathbf{x}_x| = |\mathbf{x}_y|$. Define an orthonormal frame field $(\mathbf{x}, e): U \to \mathbf{E}(3)$ by

$$\mathbf{e}_1 = e^{-u}\mathbf{x}_x, \quad \mathbf{e}_2 = e^{-u}\mathbf{x}_y, \quad \mathbf{e}_3 = \mathbf{e}_1 \times \mathbf{e}_2.$$

Then $\mathbf{x}_x \cdot \mathbf{e}_3 = 0 = \mathbf{x}_y \cdot \mathbf{e}_3$, so

$$\omega_1^3 = e^{-u}(\mathbf{x}_{xx} \cdot \mathbf{e}_3\, dx + \mathbf{x}_{xy} \cdot \mathbf{e}_3\, dy), \quad \omega_2^3 = e^{-u}(\mathbf{x}_{yx} \cdot \mathbf{e}_3\, dx + \mathbf{x}_{yy} \cdot \mathbf{e}_3\, dy).$$

The second fundamental form $II = \omega_1^3\omega^1 + \omega_2^3\omega^2$ has the form (9.2) if and only if $\omega_1^3 = a\omega^1 = ae^u dx$ and $\omega_2^3 = ce^u dy$ if and only if $\mathbf{x}_{xy} \cdot \mathbf{e}_3 = 0$ on U. □

Remark 9.4. For an immersion $\mathbf{x}: M \to \mathbf{R}^3$ of an oriented surface M, an oriented chart (U,x,y) is isothermal if and only if $(U, z = x + iy)$ is a chart for the complex structure induced by $I = d\mathbf{x} \cdot d\mathbf{x}$. A complex chart (U,z) is principal if and only if the Hopf invariant h relative to z is real valued on U. For a principal complex chart (U,z), we have

$$I = e^{2u}dzd\bar{z} = e^{2u}(dx^2 + dy^2), \quad II = e^{2u}(a\,dx^2 + c\,dy^2)$$

where $z = x + iy$, u is a smooth function on U and a, c are the principal curvatures. Then $h = (a - c)/2$.

Definition 9.5. An immersion $\mathbf{x} : M \to \mathbf{R}^3$ of a surface is *isothermic* if M possesses an atlas $\{(U_\alpha, (x_\alpha, y_\alpha))\}_{\alpha\in\mathscr{A}}$ of principal isothermal charts.

Remark 9.6. The definitions of principal chart, principal complex chart, and isothermic immersion remain unchanged for immersions $\mathbf{x}$ of oriented surfaces into $\mathbf{S}^3$ or $\mathbf{H}^3$. In fact, these same definitions apply to an immersion $\mathbf{x}$ of a surface into any three-dimensional Riemannian manifold.

By the Korn-Lichtenstein Theorem, isothermal coordinates always exist about any given point. The existence of principal isothermal coordinates, however, implies special properties of the immersion, as Cayley observed. We begin with some more examples of isothermic immersions.

Example 9.7 (Surfaces of Revolution). A surface of revolution in $\mathbf{R}^3$ is isothermic (provided that it never touches the axis of revolution). Let ${}^t(f(s), 0, g(s)) \in \mathbf{R}^3$ be a smooth curve (the profile curve) immersed in the x^1x^3-plane, for s in some open interval J, with $f(s) > 0$ on J. The surface of revolution obtained by revolving this curve around the x^3-axis is given by the immersion

$$\mathbf{x} : M = J \times \mathbf{S}^1 \to \mathbf{R}^3, \quad \mathbf{x}(s, (\cos t, \sin t)) = {}^t(f(s)\cos t, f(s)\sin t, g(s)) \tag{9.3}$$

Notice that t is a local coordinate on the complement of any point of $\mathbf{S}^1$ and that dt is a 1-form smooth on all of S^1. See Examples 7.18 and 4.40. From

$$d\mathbf{x} = {}^t(\dot{f}\cos t, \dot{f}\sin t, \dot{g})ds + {}^t(-f\sin t, f\cos t, 0)dt,$$

where $\dot{f}$ and $\dot{g}$ denote derivatives with respect to s, the induced metric is

$$d\mathbf{x}\cdot d\mathbf{x} = (\dot{f}^2 + \dot{g}^2)ds^2 + f^2dt^2 = f^2(dr^2 + dt^2), \tag{9.4}$$

so the orthonormal coframe field is $\omega^1 = fdr, \omega^2 = fdt$, where the coordinate function r is defined by

$$r = \int \frac{w}{f}ds \text{ and } w = \sqrt{\dot{f}^2 + \dot{g}^2} > 0,$$

which is the arclength parameter for the profile curve $(f(s(r)), g(s(r))$ in the half-plane $f > 0$ with the hyperbolic metric $\frac{df^2+dg^2}{f^2}$. Therefore r, t are isothermal coordinates with conformal factor $e^u = f$. From the second of equations (4.53) of Example 4.40, we see that these coordinates are also principal, since $w\,ds = f\,dr$ gives

$$II = aw^2ds^2 + cf^2dt^2 = f^2(a\,dr^2 + c\,dt^2) = a\omega^1\omega^1 + c\omega^2\omega^2,$$

where the principal curvatures are

$$a = \frac{\dot{f}\ddot{g} - \dot{g}\ddot{f}}{w^3}, \quad c = \frac{\dot{g}}{wf}.$$

We conclude that any surface of revolution is isothermic. Notice that

$$\mathbf{x}, \quad \mathbf{e}_1 = \frac{1}{w}\begin{pmatrix} \dot{f}\cos t \\ \dot{f}\sin t \\ \dot{g} \end{pmatrix}, \quad \mathbf{e}_2 = \begin{pmatrix} -\sin t \\ \cos t \\ 0 \end{pmatrix}, \quad \mathbf{e}_3 = \frac{1}{w}\begin{pmatrix} -\dot{g}\cos t \\ -\dot{g}\sin t \\ \dot{f} \end{pmatrix}$$

is the frame field adapted to the complex coordinate $z = r + it$, and the Hopf invariant of $\mathbf{x}$ relative to z is real valued,

$$h = \frac{1}{2}(a - c) = \frac{1}{2}\left(\frac{\dot{f}\ddot{g} - \dot{g}\ddot{f}}{w^3} - \frac{\dot{g}}{wf}\right).$$

We want to include the possibility that the profile curve is closed, meaning that it is periodic, in which case its domain is a circle $\mathbf{S}^1(L)$ of radius $L > 0$. In that case s is a local coordinate on $\mathbf{S}^1(L)$ on the complement of any point. For example, the torus of revolution obtained by revolving the circle of radius b in the x^1x^3-plane, center at $(B, 0, 0)$, where $0 < b < B$, would have profile curve parameterized by $f(s) = B + b\cos s$, $g(s) = b\sin s$.

Example 9.8 (Cylinders). Let

$$\mathbf{x} : J \times \mathbf{R} \to \mathbf{R}^3, \quad \mathbf{x}(s,t) = \gamma(s) + t(-\epsilon_3)$$

be the cylinder over the plane curve $\boldsymbol{\gamma} : J \to \mathbf{R}^2$, where $\mathbf{R}^2$ is the span of $\boldsymbol{\epsilon}_1, \boldsymbol{\epsilon}_2$. See Problem 4.65 for details. Here s is arclength parameter and $\kappa : J \to \mathbf{R}$ is the curvature of $\boldsymbol{\gamma}$. We assume $\kappa > 0$ on J. We know $z = s + it$ is a complex coordinate on $J \times \mathbf{R}$, the principal curvatures are $a = \kappa$ and $c = 0$, with $a > c$, and

$$H = \frac{1}{2}(a + c) = \kappa/2 = \frac{1}{2}(a - c) = h \tag{9.5}$$

are the mean curvature of $\mathbf{x}$ and the Hopf invariant of $\mathbf{x}$ relative to z. Thus h is real valued on $J \times \mathbf{R}$, so z is a principal complex coordinate and $\mathbf{x}$ is isothermic.

Example 9.9. Any proper Bonnet immersion is isothermic off its discrete set of umbilic points. See Corollary 10.46.

Example 9.10 (CMC Immersions). Let $\mathbf{x} : M \to \mathbf{R}^3$ be an immersion with constant mean curvature H of an oriented surface M. Let M_c be the set of nonumbilic points of $\mathbf{x}$, an open subset of M, assumed nonempty. Then $\mathbf{x} : M_c \to \mathbf{R}^3$ is isothermic. To see this, we recall that the Hopf quadratic differential $II^{2,0}$ is holomorphic, by Hopf's Theorem 7.29. Given a point in M_c, let (U, z) be a complex coordinate chart in M_c about this point. If e^u is the conformal factor and h is the Hopf invariant relative to z on U, then $II^{2,0} = \frac{1}{2}he^{2u}dzdz$ on U, so $\frac{1}{2}he^{2u}$ is a holomorphic function on U, never zero on U, because h is zero precisely at the umbilic points of $\mathbf{x}$. This nonzero holomorphic function has a holomorphic square root, f, on some neighborhood of the point. If the holomorphic function w is defined by $dw = fdz$, then dw is not zero at the point, so w is a complex coordinate on a neighborhood V of the point and $II^{2,0} = f^2 dzdz = dwdw$ on V. If $e^{\tilde{u}}$ is the conformal factor and $\tilde{h}$ is the Hopf invariant relative to w on V, then $\frac{1}{2}\tilde{h}e^{2\tilde{u}} = 1$ implies that $\tilde{h}$ is real valued on V and therefore (V, w) is a principal complex coordinate chart in M_c about the point. Since the point was arbitrarily chosen, we conclude that $\mathbf{x} : M_c \to \mathbf{R}^3$ is isothermic.

Remark 9.11. If a CMC immersion $\mathbf{x}$ is not totally umbilic, then it is not isothermic at its umbilic points, which are isolated, being the zeros of the nonzero holomorphic quadratic differential $II^{2,0}$. If (U, z) were a connected principal complex coordinate chart about an umbilic point m, then the real, holomorphic function $he^{2u}/2$ must be constant on U, and zero at m, thus contradicting that $II^{2,0}$ has an isolated zero at m. We should not conclude from this, however, that isothermic immersions cannot have umbilic points. Example 4.41 exhibits surfaces of revolution with one or two circles of latitude consisting of umbilic points, and we have seen above that such an immersion is isothermic. See also H. Bernstein's examples in [3].

The criterion form α of Definition 4.14 determines when immersions are isothermic.

Theorem 9.12 (Isothermic criterion). *An immersion $\mathbf{x} : M \to \mathbf{R}^3$ of a surface M is isothermic if and only if each point of M has a neighborhood U on which there is a second order frame field $(\mathbf{x}, e) : U \to \mathbf{E}(3)$ whose criterion form α is closed on U.*

Remark 9.13. If the set of all nonumbilic points of an isothermic immersion $\mathbf{x} : M \to \mathbf{R}^3$ is dense in M, then Theorem 9.12 and Problem 4.63 imply that the unique smooth criterion form on M must be closed.

Proof. If $\mathbf{x}$ is isothermic, then about any point there exists a principal complex coordinate chart $(U, (z = x + iy))$. The frame field $(\mathbf{x}, e) : U \to \mathbf{E}(3)$ adapted to z is second order with $\omega^1 + i\omega^2 = e^u dz$ and $\omega^2_1 = -u_y e^{-u}\omega^1 + u_x e^{-u}\omega^2$, by (7.28), so $\alpha = e^{-u}u_x\omega^1 + e^{-u}u_y\omega^2 = u_x dx + u_y dy = du$, which is closed on U.

Conversely, given arbitrary $m \in M$, let $(\mathbf{x}, e) : U \to \mathbf{E}(3)$ be a second order frame field with closed criterion form α on a neighborhood U of m. The Frobenius Theorem implies the existence of functions x, y, A, and B on a neighborhood $W \subset U$ of m such that $A > 0$, $B > 0$ on W and

$$\omega^1 = Adx, \quad \omega^2 = Bdy.$$

Then, $dx \wedge dy > 0$ on W implies (x, y) is a coordinate system on a neighborhood $V \subset W$ of m, and (see Remark 4.15)

$$\alpha \wedge \omega^1 = d\omega^1 = \frac{A_y}{A} dy \wedge \omega^1, \quad \alpha \wedge \omega^2 = d\omega^2 = \frac{B_x}{B} dx \wedge \omega^2,$$

so

$$\alpha = \frac{B_x}{B} dx + \frac{A_y}{A} dy = (\log B)_x dx + (\log A)_y dy.$$

By the assumption that α is closed on U, we have

$$0 = d\alpha = \big(-(\log B)_{xy} + (\log A)_{yx}\big)\, dx \wedge dy = (\log \frac{A}{B})_{xy} dx \wedge dy$$

on V, which implies that $\log(A/B) = f(x) - g(y)$ for some smooth functions f and g on $x(V) \subset \mathbf{R}$ and $y(V) \subset \mathbf{R}$, respectively. Defining the smooth function

$$u : V \to \mathbf{R}, \quad u = \log A - f(x) = \log B - g(y),$$

we get

$$\omega^1 = A dx = e^u e^{f(x)} dx, \quad \omega^2 = B dy = e^u e^{g(y)} dy.$$

There exist antiderivatives F and G such that $F' = e^f$ and $G' = e^g$. If we let $\tilde{x} = F(x)$ and $\tilde{y} = G(y)$, then

$$d\tilde{x} = e^{f(x)} dx, \quad d\tilde{y} = e^{g(y)} dy.$$

Then $(\tilde{x}, \tilde{y})$ are coordinates on a neighborhood of m, which we continue to call V, on which

$$\omega^1 = e^u d\tilde{x}, \quad \omega^2 = e^u d\tilde{y},$$

so $\omega^1 + i\omega^2 = e^u(d\tilde{x} + i d\tilde{y})$ implies $z = \tilde{x} + i\tilde{y}$ is a complex coordinate of the induced structure on V. The integral curves of the distributions $\omega^1 = 0$ and $\omega^2 = 0$ are lines of curvature, and these coincide with the coordinate curves $d\tilde{x} = 0$ and $d\tilde{y} = 0$, respectively. Thus, z is a principal complex coordinate on V. Hence, $\mathbf{x}$ is isothermic. □

Corollary 9.14. *Suppose the immersion* $\mathbf{x} : M \to \mathbf{R}^3$ *is umbilic free. Let* $h : U \to \mathbf{C}$ *be the Hopf invariant of* $\mathbf{x}$ *relative to a complex chart* (U, z) *of* M, I. *Then* $\mathbf{x} : U \to \mathbf{R}^3$ *is isothermic if and only if*

$$\frac{\partial^2}{\partial z \partial \bar{z}} \Im(\log h) = 0$$

identically on U.

Proof. If $(\mathbf{x}, e) : U \to \mathbf{E}(3)$ is the frame field adapted to z, then its criterion form is du, by Problem 7.46, where e^u is the conformal factor relative to z. Given a point of U, there is a neighborhood $\tilde{U} \subset U$ of this point on which $h = e^{f+ig}$, for smooth functions $f, g : \tilde{U} \to \mathbf{R}$. If

$$B = \begin{pmatrix} \cos t & -\sin t \\ \sin t & \cos t \end{pmatrix} : \tilde{U} \to \mathbf{SO}(2),$$

where $t = -g/2 : \tilde{U} \to \mathbf{R}$, then $B^{-1}dB = \begin{pmatrix} 0 & -dt \\ dt & 0 \end{pmatrix}$. By (4.24), the Hopf invariant $\tilde{h}$ of the frame field

$$(\mathbf{x}, \tilde{e}) = (\mathbf{x}, e\begin{pmatrix} B & 0 \\ 0 & 1 \end{pmatrix}) : \tilde{U} \to \mathbf{E}(3)$$

satisfies $\tilde{h} = e^{2it}h = e^f$, which is positive on $\tilde{U}$. Hence, $(\mathbf{x}, \tilde{e})$ is a second order frame field on $\tilde{U}$. By Exercise 12, its criterion form $\tilde{\alpha}$ on $\tilde{U}$ is $\tilde{\alpha} = du + *dt$. By Exercise 11, this is closed if an only if

$$0 = d * dt = -\frac{1}{2} g_{z\bar{z}}$$

identically on $\tilde{U}$. □

Not all immersions are isothermic, as we now show.

Proposition 9.15. *A tube about a curve in $\mathbf{R}^3$ is isothermic if and only if the curve is (part of) a circle or a line.*

Proof. We will use the notation and calculations for tubes given in Subsection 4.8.2. There $\mathbf{f}(s)$ is a smooth curve in $\mathbf{R}^3$ defined on an interval J and parametrized by arclength s, with curvature $\kappa(s)$ and torsion $\tau(s)$. Its Frenet frame consists of the vector fields $\mathbf{T} = \dot{\mathbf{f}}, \mathbf{N}, \mathbf{B}$, which satisfy the Frenet-Serret equations (4.62). Assume that the curvature is bounded above, that is, there is a positive constant k_M such that $\kappa \leq k_M$ on J. For a constant r satisfying $0 < r < 1/k_M$, the tube about $\mathbf{f}$ of radius r is the immersion (4.63), which is

$$\mathbf{x} : J \times \mathbf{R} \to \mathbf{R}^3, \quad \mathbf{x}(s,t) = \mathbf{f}(s) + r(\cos t \mathbf{N} + \sin t \mathbf{B}).$$

A second order frame field $(\mathbf{x}, e)$ is constructed along $\mathbf{x}$ in (4.64), whose coframe field is

$$\omega^1 = (1 - r\kappa \cos t) ds, \quad \omega^2 = -r(\tau ds + dt),$$

and the principal curvatures are given by

$$\omega_1^3 = \frac{\kappa \cos t}{1 - r\kappa \cos t} \omega^1, \quad \omega_2^3 = -\frac{1}{r} \omega^2.$$

All points are nonumbilic, and the calculations

$$d\omega^1 = \frac{r\kappa \sin t}{1 - r\kappa \cos t} dt \wedge \omega^1, \quad d\omega^2 = 0,$$

imply, by Remark 4.15, that the criterion form is

$$\alpha = \frac{r\kappa \sin t}{1 - r\kappa \cos t} (\tau ds + dt).$$

By Theorem 9.12, $\mathbf{x}$ is isothermic if and only if α is closed. But

$$d\alpha = \frac{r}{(1-r\kappa\cos t)^2}(\dot{\kappa}\sin t - \kappa\tau\cos t + r\kappa^2\tau)\, ds\wedge dt$$

shows that $d\alpha = 0$ if and only if

$$\dot{\kappa}\sin t + \kappa\tau(r\kappa - \cos t) = 0$$

identically for all s and t; that is,

$$\dot{\kappa}\sin t - \kappa\tau\cos t = -r\kappa^2\tau,$$

for all s and t. The right side does not depend on t, so the partial derivative with respect to t of the left side must be identically zero. This gives

$$(\dot{\kappa}, \kappa\tau)\cdot(\cos t, \sin t) = 0$$

for all s and t. This holds if and only if

$$(\dot{\kappa}, \kappa\tau) = (0,0)$$

identically; that is, κ is constant and $\tau = 0$ or $\kappa = 0$. □

9.3 Affine structures

There is a close relationship between isothermic immersions and affine structures on a Riemann surface. For more background on affine structures see Gunning [82, p.167ff].

Definition 9.16. An *affine structure* on a Riemann surface M is an atlas of complex coordinate charts $\{U_\alpha, z_\alpha\}_{\alpha\in\mathscr{A}}$ such that for any $\alpha, \beta \in \mathscr{A}$, the function $c_{\alpha\beta} = \frac{dz_\beta}{dz_\alpha}$ is locally constant on $U_\alpha \cap U_\beta$, meaning that it is constant on each connected component of this set.

Example 9.17. A torus $M = \mathbf{C}/\Gamma$ is a compact Riemann surface with an affine structure. Here Γ is any lattice in $\mathbf{C}$, which we may assume is generated over the integers by 1 and a complex number $\boldsymbol{\tau}$ whose imaginary part is positive. (See Examples 7.44 and 7.45.) Let z be the standard complex coordinate on $\mathbf{C}$ and let $\{U_\alpha\}$ be an open cover of M by open sets which are evenly covered by the projection map $\pi : \mathbf{C} \to M$. Then z descends to a complex coordinate z_α on each U_α such that on $U_\alpha \cap U_\beta$ we must have $z_\alpha = z_\beta + b_{\alpha\beta}$, where $b_{\alpha\beta} \in \Gamma$. In particular, $dz_\alpha = dz_\beta$ on $U_\alpha \cap U_\beta$. Therefore, $\{U_\alpha, z_\alpha\}_{\alpha\in\mathscr{A}}$ is an atlas defining an affine structure on M.

Theorem 9.18. *If* $\mathbf{x} : M \to \mathbf{R}^3$ *is an isothermic immersion of an oriented surface* M*, such that the set of nonumbilic points is open and dense in* M*, then an atlas of principal isothermic charts on* M *defines an affine structure on* M*.*

Proof. Let $\{U_\alpha, z_\alpha\}$ be an atlas of principal complex coordinate charts on M. This atlas exists because $\mathbf{x}$ is isothermic. On $U_\alpha \cap U_\beta$, assumed nonempty, let

$$c_{\alpha\beta} = \frac{dz_\beta}{dz_\alpha}$$

For each chart U_α, z_α in this atlas, its Hopf invariant h_α is real valued on U_α, because it is a principal chart. By the change of coordinate formula (7.41), on $U_\alpha \cap U_\beta$ we have $h_\beta = \frac{\bar{c}_{\alpha\beta}}{c_{\alpha\beta}} h_\alpha$. The nonumbilic points of this intersection form an open dense subset of $U_\alpha \cap U_\beta$, on which $c_{\alpha\beta}^2 = |c_{\alpha\beta}|^2 h_\alpha / h_\beta$ is holomorphic and real valued, therefore locally constant on $U_\alpha \cap U_\beta$. But then $c_{\alpha\beta}$ is locally constant on $U_\alpha \cap U_\beta$, and this atlas defines an affine structure on M. □

Remark 9.19. The hypothesis of open dense nonumbilic set is necessary in the preceding proposition. For example, a sphere of any radius in $\mathbf{R}^3$ is isothermic, since every complex coordinate chart is principal. But as a Riemann surface it has no affine structure, as the next theorem implies, because the genus of the sphere is zero.

Theorem 9.20. *If a compact Riemann surface* M *possesses an affine structure, then its genus is 1, that is, as a Riemann surface* M *is a torus.*

Proof. See Gunning [82, Cor. 3, p. 173]. □

Corollary 9.21. *If* M *is a compact oriented surface and if* $\mathbf{x} : M \to \mathbf{R}^3$ *is an isothermic immersion whose set of nonumbilic points is dense in* M*, then* M *is a torus.*

9.4 Christoffel transforms

Isothermic immersions are characterized geometrically by the existence of a Christoffel transform.

Definition 9.22. A *Christoffel transform* of an immersion $\mathbf{x} : M \to \mathbf{R}^3$ of an oriented surface M is an immersion $\hat{\mathbf{x}} : M \to \mathbf{R}^3$ such that

1. At each point of M, the tangent plane to $\mathbf{x}$ is parallel to the tangent plane to $\hat{\mathbf{x}}$.
2. The metrics induced by $\mathbf{x}$ and $\hat{\mathbf{x}}$ are conformally related.
3. $\mathbf{x}$ and $\hat{\mathbf{x}}$ induce opposite orientations at each point of M.

In more detail: (1) means that $(\mathbf{x}, e) : U \to \mathbf{E}(3)$ is first order along $\mathbf{x}$ if and only if $(\hat{\mathbf{x}}, e) : U \to \mathbf{E}(3)$ is first order along $\hat{\mathbf{x}}$; (2) means $d\hat{\mathbf{x}} \cdot d\hat{\mathbf{x}} = e^{2u} d\mathbf{x} \cdot d\mathbf{x}$, for some smooth $u : M \to \mathbf{R}$; given (1) and (2), (3) means that for any first order frame field $(\mathbf{x}, e) : U \to \mathbf{E}(3)$, so $(\hat{\mathbf{x}}, e) : U \to \mathbf{E}(3)$ is first order along $\hat{\mathbf{x}}$, we have

$$d\mathbf{x} = \omega^1 \mathbf{e}_1 + \omega^2 \mathbf{e}_2, \quad d\hat{\mathbf{x}} = \tau^1 \mathbf{e}_1 + \tau^2 \mathbf{e}_2, \tag{9.6}$$

and $\tau^1 \wedge \tau^2 = -e^{2u} \omega^1 \wedge \omega^2$ on U. All three conditions are equivalent to: for any first order frame field $(\mathbf{x}, e) : U \to \mathbf{E}(3)$ along $\mathbf{x}$, (9.6) holds and

$$\tau^1 - i\tau^2 = f(\omega^1 + i\omega^2), \tag{9.7}$$

for some smooth $f : U \to \mathbf{C} \setminus \{0\}$.

If $\hat{\mathbf{x}}$ is a Christoffel transform of $\mathbf{x}$, then $\mathbf{x}$ is a Christoffel transform of $\hat{\mathbf{x}}$. In addition, for any nonzero real constant r and any constant vector $\mathbf{v} \in \mathbf{R}^3$, the immersion $r\hat{\mathbf{x}} + \mathbf{v}$ is also a Christoffel transform of $\mathbf{x}$.

Example 9.23 (Christoffel transform of a CMC immersion). Recall the statement in Theorem 4.50 of Bonnet's remarkable result that for an immersion $\mathbf{x} : M \to \mathbf{R}^3$ of constant mean curvature $H \neq 0$ and unit normal vector field $\mathbf{e}_3$, the parallel surface $\hat{\mathbf{x}} = \mathbf{x} + \frac{1}{H}\mathbf{e}_3$ has constant mean curvature $\hat{H} = -H$ relative to $\mathbf{e}_3$. This parallel surface is a Christoffel transform of $\mathbf{x}$ on $M_c \subset M$, the open subset of all nonumbilic points of $\mathbf{x}$, assumed nonempty. In fact, let $(\mathbf{x}, e)$ be a first order frame field on $U \subset M_c$, with the given unit normal vector field. Then $d\mathbf{x} = \omega^1 \mathbf{e}_1 + \omega^2 \mathbf{e}_2$ and

$$d\hat{\mathbf{x}} = \omega^1 \mathbf{e}_1 + \omega^2 \mathbf{e}_2 + \frac{1}{H}(\omega^1_3 \mathbf{e}_1 + \omega^2_3 \mathbf{e}_2) = (\omega^1 + \frac{\omega^1_3}{H})\mathbf{e}_1 + (\omega^2 + \frac{\omega^2_3}{H})\mathbf{e}_2,$$

so the tangent plane of $\hat{\mathbf{x}}$ is parallel to that of $\mathbf{x}$, and

$$(\omega^1 + \frac{\omega^1_3}{H}) - i(\omega^2 + \frac{\omega^2_3}{H}) = -\frac{h}{H}(\omega^1 + i\omega^2),$$

where $h = \frac{1}{2}(h_{11} - h_{22}) - ih_{12}$ is the Hopf invariant of $\mathbf{x}$ relative to $(\mathbf{x}, e)$. The function h/H vanishes exactly at the umbilic points of $\mathbf{x}$, so (9.7) holds on a neighborhood of any point of M_c. Hence, $\hat{\mathbf{x}} : M_c \to \mathbf{R}^3$ is a Christoffel transform of $\mathbf{x} : M_c \to \mathbf{R}^3$.

Theorem 9.24 (Christoffel [50]). *Let $\mathbf{x} : M \to \mathbf{R}^3$ be an immersion of a surface M.*

(1) If $\mathbf{x}$ possesses a Christoffel transform, then $\mathbf{x}$ is isothermic.
(2) If M is simply connected and if $\mathbf{x}$ is isothermic and has a dense set of nonumbilic points in M, then $\mathbf{x}$ possesses a Christoffel transform $\hat{\mathbf{x}}$, unique up to homothety and translation, $r\hat{\mathbf{x}} + \mathbf{v}$, where r is a nonzero real constant and $\mathbf{v} \in \mathbf{R}^3$ is constant.

Proof. Suppose that $\mathbf{x}$ possesses the Christoffel transform $\hat{\mathbf{x}} : M \to \mathbf{R}^3$. Choose a point of M and let (U, z) be a complex coordinate chart about this point for the complex structure induced by $\mathbf{x}$. If $(\mathbf{x}, e) : U \to \mathbf{E}(3)$ is the frame field adapted to z, then

$$d\mathbf{x} = \omega^1 \mathbf{e}_1 + \omega^2 \mathbf{e}_2, \quad d\hat{\mathbf{x}} = \tau^1 \mathbf{e}_1 + \tau^2 \mathbf{e}_2, \quad \omega^1 + i\omega^2 = e^u dz,$$

for smooth $u : U \to \mathbf{R}$, and

$$\tau^1 - i\tau^2 = f(\omega^1 + i\omega^2)$$

for some smooth function $f : U \to \mathbf{C} \setminus \{0\}$. From

$$d\mathbf{x} = \mathbf{x}_z dz + \mathbf{x}_{\bar{z}} d\bar{z} = \left(\frac{\mathbf{e}_1 - i\mathbf{e}_2}{2}\right) e^u dz + \left(\frac{\mathbf{e}_1 + i\mathbf{e}_2}{2}\right) e^u d\bar{z},$$

$$d\hat{\mathbf{x}} = \hat{\mathbf{x}}_z dz + \hat{\mathbf{x}}_{\bar{z}} d\bar{z} = \left(\frac{\mathbf{e}_1 + i\mathbf{e}_2}{2}\right) f e^u dz + \left(\frac{\mathbf{e}_1 - i\mathbf{e}_2}{2}\right) \bar{f} e^u d\bar{z}$$

we get

$$\mathbf{x}_z = e^u \left(\frac{\mathbf{e}_1 - i\mathbf{e}_2}{2}\right), \quad \hat{\mathbf{x}}_{\bar{z}} = \bar{f} e^u \left(\frac{\mathbf{e}_1 - i\mathbf{e}_2}{2}\right),$$

so $\hat{\mathbf{x}}_{\bar{z}} = \bar{f}\mathbf{x}_z$ and

$$d\hat{\mathbf{x}} = f\mathbf{x}_{\bar{z}} dz + \bar{f}\mathbf{x}_z d\bar{z} \tag{9.8}$$

on U. Taking the exterior derivative of (9.8) and using the formula $\mathbf{x}_{zz} = 2u_z\mathbf{x}_z + \frac{1}{2}e^{2u}h\mathbf{e}_3$ from (7.35), where h is the Hopf invariant of $\mathbf{x}$ relative to z, we have

$$0 = -(f_{\bar{z}} + 2fu_{\bar{z}})\mathbf{x}_{\bar{z}} + (\bar{f}_z + 2\bar{f}u_z)\mathbf{x}_z + \frac{1}{2}e^{2u}(\bar{h}f - h\bar{f})\mathbf{e}_3.$$

Using the fact that $\mathbf{x}_z, \mathbf{x}_{\bar{z}}, \mathbf{e}_3$ are linearly independent over $\mathbf{C}$ at each point of U, we conclude that

$$f_{\bar{z}} + 2fu_{\bar{z}} = 0 \quad \text{and} \quad \bar{h}f = h\bar{f}. \tag{9.9}$$

The first equation is equivalent to $(e^{2u}f)_{\bar{z}} = 0$, which means that $e^{2u}f$ is a holomorphic function on U. Since $e^{2u}f$ is never zero on U, it has a holomorphic square root g on U. Since g is never zero on U, the holomorphic function w, for which $dw = g\,dz$, is a local complex coordinate on a neighborhood V of the chosen point. By (7.41) the Hopf invariant $\tilde{h}$ of $\mathbf{x}$ relative w satisfies

$$\tilde{h} = \frac{\bar{g}}{g}h = \frac{\bar{g}^2}{|g|^2}h = \frac{e^{2u}\bar{f}h}{|g|^2},$$

which is real valued on V by the second equation in (9.9). Hence, (V,w) is a principal complex coordinate chart for $\mathbf{x}$ containing the chosen point. As the point was arbitrarily chosen, it follows that $\mathbf{x}$ is isothermic.

Remark 9.25. If z itself had been a principal complex coordinate for $\mathbf{x}$, then h would be real valued and (9.9) would imply that $e^{2u}f$ is holomorphic and real valued on U so, supposing U connected, $e^{2u}f = r$, a nonzero real constant on U. Then (9.8) becomes

$$d\hat{\mathbf{x}} = re^{-2u}(\mathbf{x}_{\bar{z}}dz + \mathbf{x}_z d\bar{z}) = re^{-2u}(\omega^1\mathbf{e}_1 - \omega^2\mathbf{e}_2)$$

so

$$\tau^1 - i\tau^2 = re^{-2u}(\omega^1 + i\omega^2) = re^{-u}dz$$

on U, where $(\mathbf{x},e) : U \to \mathbf{E}(3)$ is adapted to z.

For the converse, suppose that M is simply connected and $\mathbf{x} : M \to \mathbf{R}^3$ is isothermic with a dense set of nonumbilic points. To construct a Christoffel transform $\hat{\mathbf{x}} : M \to \mathbf{R}^3$, we try to construct the $\mathbf{R}^3$-valued differential 1-form

$$\boldsymbol{\tau} = d\hat{\mathbf{x}}$$

on M. It is *tangential* in the sense that at any $m \in M$, the value of $\boldsymbol{\tau}$ on any $v \in T_mM$ is in $d\mathbf{x}(T_mM) \subset \mathbf{R}^3$. We first construct $\boldsymbol{\tau}$ on a given principal complex coordinate chart (U,z). The next step will be to see how these locally defined 1-forms are pieced together to give a Christoffel transform of $\mathbf{x}$ on all of M.

9.4.1 Local construction

Let U,z be a principal complex coordinate chart of $\mathbf{x}$. Let $(\mathbf{x},e)$ be the frame adapted to z, so that $\omega^1 + i\omega^2 = e^u dz$, for some smooth $u : U \to \mathbf{R}$, and $h : U \to \mathbf{R}$ is the Hopf invariant of $\mathbf{x}$ relative to z. By Remark 9.25, up to constant multiple the only candidate for an $\mathbf{R}^3$-valued tangential 1-form $\boldsymbol{\tau}$ on U is,

$$\boldsymbol{\tau} = e^{-2u}(\mathbf{x}_{\bar{z}}dz + \mathbf{x}_z d\bar{z}) = e^{-2u}(\omega^1\mathbf{e}_1 - \omega^2\mathbf{e}_2). \tag{9.10}$$

To see when $\boldsymbol{\tau}$ is closed on U, we calculate

$$d\boldsymbol{\tau} = e^{-2u}(2u_{\bar{z}}\mathbf{x}_{\bar{z}} - 2u_z\mathbf{x}_z - \mathbf{x}_{\bar{z}\bar{z}} + \mathbf{x}_{zz})\,dz\wedge d\bar{z}$$

and then substitute in $\mathbf{x}_{zz} = 2u_z\mathbf{x}_z + \frac{1}{2}e^{2u}h\mathbf{e}_3$ from (7.35), to find

$$d\boldsymbol{\tau} = \frac{1}{2}(h - \bar{h})\mathbf{e}_3 dz \wedge d\bar{z}.$$

Hence, $\boldsymbol{\tau}$ is closed if and only if h is real valued on U, which is the case if and only if z is a principal complex coordinate. Thus we see the essential role of the assumption that $\mathbf{x}$ be isothermic. If U is simply connected, then there exists a smooth map $\hat{\mathbf{x}} : U \to \mathbf{R}^3$ such that $d\hat{\mathbf{x}} = \boldsymbol{\tau}$. Then (9.10) shows that $(\hat{\mathbf{x}}, e)$ is first order along $\hat{\mathbf{x}}$ with

$$\tau^1 - i\tau^2 = e^{-2u}\omega^1 + ie^{-2u}\omega^2 = e^{-2u}(\omega^1 + i\omega^2),$$

so $\hat{\mathbf{x}}$ is a Christoffel transform of $\mathbf{x}$ on U, by (9.7). If r is any nonzero real number, then $r\boldsymbol{\tau}$ remains tangentially valued and closed and $d(r\hat{\mathbf{x}}) = r\,d\hat{\mathbf{x}}$. The zeros of h do not affect this local construction. It works even if h is identically zero on U.

9.4.2 Global construction

Let $\{U_\alpha, z_\alpha\}_{\alpha \in \mathscr{A}}$ be an atlas of principal complex coordinate charts on M. We may assume that $U_\alpha \cap U_\beta$ is connected whenever this intersection is non-empty. Let e^{u_α} be the conformal factor and let h_α be the invariant of $\mathbf{x}$ relative to z_α. If $U_\alpha \cap U_\beta \neq \emptyset$, let

$$c_{\alpha\beta} = \frac{dz_\beta}{dz_\alpha}$$

a holomorphic function on this intersection. By (7.41), we know that on this intersection $h_\beta = \frac{\overline{c_{\alpha\beta}}}{c_{\alpha\beta}} h_\alpha$. Since we have principal charts, the functions h_α and h_β are real valued. In addition, we have assumed that the set of nonumbilic points is dense in M. It follows that h_β is nonzero on a dense open subset, V, of $U_\alpha \cap U_\beta$, and therefore $c_{\alpha\beta}^2 = |c_{\alpha\beta}|^2 h_\alpha / h_\beta$ is a real valued holomorphic function on V and so must be locally constant on V. Since V is dense in the connected open set $U_\alpha \cap U_\beta$ where $c_{\alpha\beta}^2$ is continuous, it follows that $c_{\alpha\beta}^2$ is a real valued constant, nonzero, but possibly negative.

Use (9.10) to define $\boldsymbol{\tau}_\alpha$, a tangentially valued closed 1-form on U_α. If $U_\alpha \cap U_\beta \neq \emptyset$, then by (7.40) $e^{-2u_\beta} = e^{-2u_\alpha}|c_{\alpha\beta}|^2$ and

$$\boldsymbol{\tau}_\beta = |c_{\alpha\beta}|^2 e^{-2u_\alpha}\left(\frac{\mathbf{x}_{\bar{z}_\alpha}}{\bar{c}_{\alpha\beta}} c_{\alpha\beta} dz_\alpha + \frac{\mathbf{x}_{z_\alpha}}{c_{\alpha\beta}} \bar{c}_{\alpha\beta} d\bar{z}_\alpha\right) = c_{\alpha\beta}^2 \boldsymbol{\tau}_\alpha, \tag{9.11}$$

because $c_{\alpha\beta}^2$ is real valued. We now use a Čech cohomology argument to piece together the $\boldsymbol{\tau}_\alpha$ into a tangentially valued closed 1-form $\boldsymbol{\tau}$ on all of M. For background on Čech cohomology see Conlon [53, pages 284ff]. We define a Čech 1-cocycle $c = \{c_{\alpha\beta}\}$ by assigning to the ordered pair (U_α, U_β), when $U_\alpha \cap U_\beta \neq \emptyset$, the real number $c^2_{\alpha\beta}$. It satisfies the cocycle condition $c^2_{\alpha\beta}c^2_{\beta\gamma}c^2_{\gamma\alpha} = 1$ whenever $U_\alpha \cap U_\beta \cap U_\gamma \neq \emptyset$. It is proved in [53] that the Čech cohomology equals the de Rham cohomology, $\check{H}^1(M) = H^1(M)$, and this last space is zero by our assumption that M is simply connected. Therefore, there exists a 0-cochain $s = \{s_\alpha\}$ (where s_α is a nonzero real number assigned to U_α) whose coboundary $\delta s = \{s_\beta/s_\alpha\}$ equals c. That is, for the ordered pair (U_α, U_β), we have on $U_\alpha \cap U_\beta$

$$s_\beta/s_\alpha = c^2_{\alpha\beta}$$

Therefore, by (9.11)

$$\boldsymbol{\tau}_\alpha/s_\alpha = \boldsymbol{\tau}_\beta/s_\beta$$

on $U_\alpha \cap U_\beta$ allows us to define $\boldsymbol{\tau}$ on M by defining

$$\boldsymbol{\tau}_{|U_\alpha} = \boldsymbol{\tau}_\alpha/s_\alpha.$$

The resulting 1-form $\boldsymbol{\tau}$ is smooth, tangentially valued and closed on M. Because M is simply connected, there exists a smooth map $\hat{\mathbf{x}} : M \to \mathbf{R}^3$ such that $d\hat{\mathbf{x}} = \boldsymbol{\tau}$. As in the local case, it follows that $\hat{\mathbf{x}}$ is a Christoffel transform of $\mathbf{x}$. If $\tilde{\boldsymbol{\tau}}$ is any smooth, tangentially valued, closed 1-form on M, then for any U_α,

$$\tilde{\boldsymbol{\tau}}|_{U_\alpha} = a_\alpha \boldsymbol{\tau}|_{U_\alpha},$$

for some real constant a_α, by (9.10). But then $a_\alpha = a_\beta$ whenever $U_\alpha \cap U_\beta \neq \emptyset$, so we conclude that $\boldsymbol{\tau}$ is unique up to a constant real multiple. Then $\hat{\mathbf{x}}$ satisfying $d\hat{\mathbf{x}} = \boldsymbol{\tau}$ is unique up to additive constant vector, implies that $\hat{\mathbf{x}}$ is unique up to homothety and translation. □

Corollary 9.26. *If $\hat{\mathbf{x}} : M \to \mathbf{R}^3$ is a Christoffel transform of the isothermic immersion $\mathbf{x}$ and if (U, z) is a principal chart for $\mathbf{x}$ such that $d\hat{\mathbf{x}} = \boldsymbol{\tau}$ is given by* (9.10), *then $(U, \bar{z})$ is a principal chart for $\hat{\mathbf{x}}$, so lines of curvature are preserved, and*

$$e^{\hat{u}} = e^{-u}, \quad \hat{H} = e^{2u}h, \quad \hat{h} = e^{2u}H, \tag{9.12}$$

where e^u and $e^{\hat{u}}$ are the conformal factors and h and $\hat{h}$ are the Hopf invariants relative to z of $\mathbf{x}$ and relative to of $\hat{\mathbf{x}}$, respectively, and H and $\hat{H}$ are the mean curvatures relative to the normal vector $\mathbf{e}_3$. Hence:

(1) If $\mathbf{x}$ is minimal, then $\hat{\mathbf{x}}$ is totally umbilic.
(2) If $\mathbf{x}$ is totally umbilic, then $\hat{\mathbf{x}}$ is minimal.

Proof. The first equation follows immediately from (9.10), where now $\boldsymbol{\tau} = d\hat{\mathbf{x}}$. For the other two equations, use the fact that $\mathbf{e}_3$ is the unit normal vector for both $\mathbf{x}$ and $\hat{\mathbf{x}}$, and thus (7.25) applied to each gives

$$\hat{H}d\hat{\mathbf{x}} + \hat{h}\hat{\mathbf{x}}_z d\bar{z} + \bar{\hat{h}}\hat{\mathbf{x}}_{\bar{z}} dz = -d\mathbf{e}_3 = Hd\mathbf{x} + h\mathbf{x}_{\bar{z}} dz + \bar{h}\mathbf{x}_z d\bar{z}$$

Comparing coefficients of dz, we find that

$$\hat{H}\hat{\mathbf{x}}_z + \bar{\hat{h}}\hat{\mathbf{x}}_{\bar{z}} = H\mathbf{x}_z + h\mathbf{x}_{\bar{z}} \tag{9.13}$$

By (9.10), we have $\hat{\mathbf{x}}_z = e^{-2u}\mathbf{x}_{\bar{z}}$ and $\hat{\mathbf{x}}_{\bar{z}} = e^{-2u}\mathbf{x}_z$, which substituted into (9.13) gives the second and third equations in (9.12). □

Remark 9.27. The above *local construction* works for any principal chart (U,z), regardless of the presence of umbilic points in U. In fact, it works even when $\mathbf{x}$ is totally umbilic on U. The construction depends, however, on the choice of principal chart. As shown in the above *global construction*, if the set of nonumbilic points is dense, then this choice is limited to only one together with its *rotation* by an integer multiple of $\pi/2$. On a totally umbilic set, any complex chart is principal. If a totally umbilic immersion $\mathbf{x} : M \to \mathbf{R}^3$ is nonplanar, then any nonplanar minimal immersion is a Christoffel transform of it for an appropriate choice of complex principal chart.

Corollary 9.28. *If* $\hat{\mathbf{x}} : M \to \mathbf{R}^3$ *is a Christoffel transform of an isothermic immersion* $\mathbf{x} : M \to \mathbf{R}^3$, *then* $\mathcal{Q} = (d\mathbf{x}\cdot d\hat{\mathbf{x}})^{2,0}$ *is a holomorphic, nowhere zero quadratic differential on* M.

Proof. If U,z is any principal complex coordinate chart in M for $\mathbf{x}$, then by (9.10)

$$d\mathbf{x}\cdot d\hat{\mathbf{x}} = ce^{-2u}\mathbf{x}_z\cdot\mathbf{x}_{\bar{z}}(dzdz + d\bar{z}d\bar{z})$$

from which the results follows because $\mathbf{x}_z\cdot\mathbf{x}_{\bar{z}} = 2e^{2u}$. □

Remark 9.29. By the Uniformization Theorem (see Farkas-Kra [66, §IV.4.1, p195 and §IV.5.6, p206], the universal cover $\pi : \tilde{M} \to M$ of a given Riemann surface M must be either the Riemann sphere $\mathbf{S}^2$, the complex plane $\mathbf{C}$ or the unit disk $\mathbf{D} = \{z \in \mathbf{C} : |z| < 1\}$. If $\mathbf{x} : M \to \mathbf{R}^3$ is an isothermic immersion whose induced complex structure is the given structure on M, then the composition $\tilde{\mathbf{x}} = \mathbf{x}\circ\pi : \tilde{M} \to \mathbf{R}^3$ is also isothermic. A point of $\tilde{M}$ is umbilic for $\tilde{\mathbf{x}}$ if and only if it covers an umbilic point of $\mathbf{x}$ in M. It follows that if $\mathbf{x}$ has a dense set of nonumbilic points in M then $\tilde{\mathbf{x}}$ has a dense set of nonumbilic points in $\tilde{M}$, and therefore $\tilde{\mathbf{x}}$ has a Christoffel transform $\hat{\mathbf{x}} : \tilde{M} \to \mathbf{R}^3$. If $U \subset M$ is an open subset evenly covered by the covering projection, and if $\tilde{U} \subset \tilde{M}$ is a connected component of $\pi^{-1}U$, then $\pi : \tilde{U} \to U$ is a diffeomorphism and $\hat{\mathbf{x}}\circ\pi^{-1}_{|\tilde{U}} : U \to \mathbf{R}^3$ is a Christoffel transform of $\mathbf{x}$ restricted to U. But $\hat{\mathbf{x}}$ will be a Christoffel transform of $\mathbf{x}$ on all of M if and only if $\hat{\mathbf{x}}\circ F = \hat{\mathbf{x}}$ for every deck transformation $F : \tilde{M} \to \tilde{M}$ of this covering space. In this case we say that $\hat{\mathbf{x}}$ *descends* to M.

Sometimes the Christoffel transform descends, as for example in every case when $\mathbf{x}$ is of constant mean curvature, because then $\hat{\mathbf{x}} = \mathbf{x} + \mathbf{e}_3/H$ is a parallel surface, as shown in Example 9.23. The next family of examples contains cases where $\hat{\mathbf{x}}$ does not descend.

Example 9.30 (Christoffel transform of a surface of revolution). Consider a surface of revolution $\mathbf{x} : J \times \mathbf{S}^1 \to \mathbf{R}^3$ given in (9.3). See Example 9.7 for the notation we use here. A principal complex coordinate chart is (U,z), where,

$$z = r + it, \quad r = \int \frac{w}{f} ds, \quad w = \sqrt{\dot{f}^2 + \dot{g}^2} > 0. \tag{9.14}$$

Here $U = J \times K$, where K is any open interval of length 2π. The conformal factor of $\mathbf{x}$ relative to z is $e^u = f$, by (9.4). Then

$$dz = dr + idt = \frac{w}{f} ds + idt$$

and

$$\mathbf{x}_z = \frac{1}{2}(\mathbf{x}_r - i\mathbf{x}_t) = \frac{1}{2}(\frac{f}{w}\mathbf{x}_s - i\mathbf{x}_t),$$

from which we find the real part of $\mathbf{x}_{\bar{z}} dz$ to be

$$\Re(\mathbf{x}_{\bar{z}} dz) = \Re(\frac{1}{2}(\frac{f}{w}\mathbf{x}_s + i\mathbf{x}_t)(\frac{w}{f} ds + idt)) = \frac{1}{2}(\mathbf{x}_s ds - \mathbf{x}_t dt),$$

where $\Re(p)$ denotes the real part of the complex number p. By (9.10), a Christoffel transform of $\mathbf{x}$ satisfies, $d\hat{\mathbf{x}} = e^{-2u}(\mathbf{x}_{\bar{z}} dz + \mathbf{x}_z d\bar{z}) = 2e^{-2u}\Re(\mathbf{x}_{\bar{z}} dz)$. Therefore,

$$d\hat{\mathbf{x}} = \frac{1}{f^2}(\mathbf{x}_s ds - \mathbf{x}_t dt) = \begin{pmatrix} \frac{\dot{f}}{f^2}\cos t\, ds + \frac{\sin t}{f} dt \\ \frac{\dot{f}}{f^2}\sin t\, ds - \frac{\cos t}{f} dt \\ \frac{\dot{g}}{f^2} ds \end{pmatrix}.$$

Integrate to find the components of $\hat{\mathbf{x}}$, up to additive constants,

$$\hat{x}^1 = -\frac{\cos t}{f}, \quad \hat{x}^2 = -\frac{\sin t}{f}, \quad \hat{x}^3 = \int \frac{\dot{g}}{f^2} ds. \tag{9.15}$$

Therefore, $\hat{\mathbf{x}}$ is periodic in t, of period 2π, and thus it descends to M. Indeed, $\hat{\mathbf{x}} : M \to \mathbf{R}^3$ is the surface of revolution obtained by revolving the profile curve ${}^t(-1/f, 0, \int(\dot{g}/f^2)\, ds)$ in the x^1x^3-plane about the x^3-axis.

Here the universal cover of M is $\tilde{M} = J \times \mathbf{R}$, which as a Riemann surface is $\mathbf{C}$ or $\mathbf{D}$ according to whether the function r maps J onto $\mathbf{R}$ or not, since the Riemann Mapping Theorem says that a simply connected domain in $\mathbf{C}$, which is not all of $\mathbf{C}$, is biholomorphically equivalent to $\mathbf{D}$.

Consider next the case where $J = \mathbf{S}^1(R)$. The universal cover of M is

$$\varphi : \tilde{M} = \mathbf{R}^2 \to M, \quad \varphi(s,t) = (e^{2\pi i s/R}, e^{it}).$$

Now $f(s)$ and $g(s)$ are R-periodic. Up to translation, the Christoffel transform $\hat{\mathbf{x}} : \tilde{M} \to \mathbf{R}^3$ is given again by (9.15) and it is periodic in t, of period 2π. It is a surface of revolution, but its profile curve ${}^t(-1/f, 0, \int (\dot{g}/f^2)\, ds)$ is R-periodic if and only if $\int (\dot{g}/f^2)\, ds$ is R-periodic. Since the integrand is R-periodic,

$$\frac{d}{dx}\int_x^{x+R} \frac{\dot{g}}{f^2}\, ds = 0,$$

so $\int (\dot{g}/f^2)\, ds$ is R-periodic if and only if

$$\int_0^R (\dot{g}/f^2)\, ds = 0. \tag{9.16}$$

Thus, the Christoffel transform of a torus of revolution is again a torus if and only if (9.16) holds. Figure 9.2 shows part of the Christoffel transform of a circular torus, based on work by J. Hutchings.

Example 9.31 (Torus with torus transform). This example was found by Joseph Hutchings. To find a torus of revolution whose Christoffel transform is a torus, we look for R-periodic functions $f(s)$ and $g(s)$ with $f > 0$ and $\dot{f}^2 + \dot{g}^2 > 0$ such that (9.16) holds. We take $R = 2\pi$ and consider

$$f(s) = 2 + \sin s \cos s, \quad g(s) = (1 + \cos s)\sin s.$$

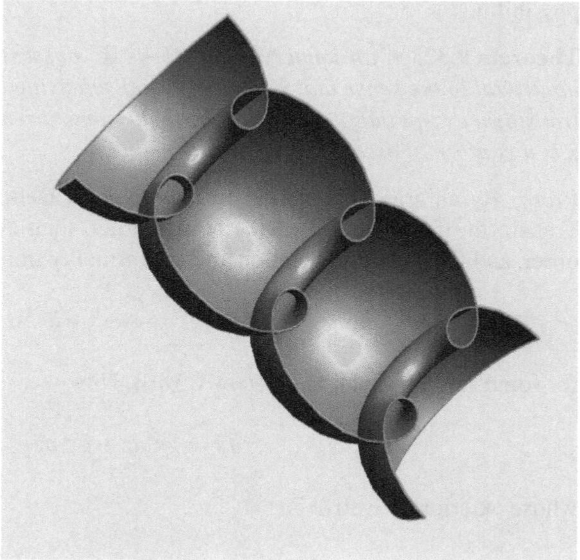

Fig. 9.2 A piece of the Christoffel Transform of a circular torus.

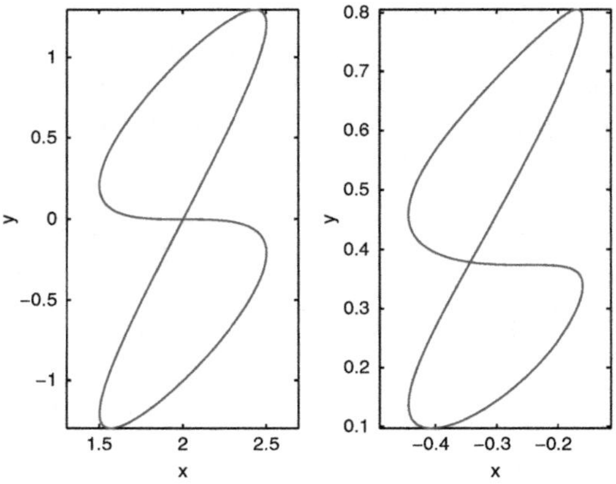

Fig. 9.3 Profile curve of Joey's torus and of its Christoffel transform

Then $\dot{g}(s) = \cos s + \cos 2s, \dot{f}^2 + \dot{g}^2 > 0$, and

$$\int_0^{2\pi} \frac{\dot{g}}{f^2} ds = \int_0^{\pi} \frac{\dot{g}}{f^2} ds + \int_{\pi}^{2\pi} \frac{\dot{g}}{f^2} ds = \int_0^{\pi} \frac{2\cos 2s}{(2 + \frac{1}{2}\sin 2s)^2} ds = 0$$

shows that the Christoffel transform descends. See Figure 9.3 for the profile curves of Joey's torus and of its Christoffel transform, which is also a torus of revolution.

If the definition of Christoffel transform is changed to require that $\mathbf{x}$ and $\hat{\mathbf{x}}$ have the same orientation, then conformal associates of a minimal immersion are the only possibilities.

Theorem 9.32. *If an immersion* $\mathbf{x} : M \to \mathbf{R}^3$ *possesses a transform* $\hat{\mathbf{x}} : M \to \mathbf{R}^3$, *nontrivial in the sense that* $\hat{\mathbf{x}} \neq r\mathbf{x} + \mathbf{v}$, *and satisfying all properties of a Christoffel transform except that* $\mathbf{x}$ *and* $\hat{\mathbf{x}}$ *now have the same orientation, then* $\mathbf{x}$ *is minimal and* $\hat{\mathbf{x}}$ *is a conformal associate of* $\mathbf{x}$.

Proof. By an argument similar to that given in Definition 9.22 of the Christoffel Transform, if $(\mathbf{x}, e) : U \to \mathbf{E}(3)$ is first order, then $(\hat{\mathbf{x}}, e) : U \to \mathbf{E}(3)$ is also first order, and then same orientation and conformality imply

$$\theta^1 + i\theta^2 = g(\omega^1 + i\omega^2),$$

for some smooth function $g : U \to \mathbf{C} \setminus \{0\}$. Now

$$d\hat{\mathbf{x}} = g\mathbf{x}_z dz + \bar{g}\mathbf{x}_{\bar{z}} d\bar{z} \tag{9.17}$$

whose exterior derivative is

$$0 = -g_{\bar{z}}\mathbf{x}_z + \bar{g}_z\mathbf{x}_{\bar{z}} + (\bar{g} - g)\mathbf{x}_{z\bar{z}}$$

Substituting in (7.35), which is $\mathbf{x}_{z\bar{z}} = \frac{1}{2}e^{2u}H\mathbf{e}_3$, we get

$$g_{\bar{z}} = 0 \text{ and } (\bar{g} - g)H = 0 \tag{9.18}$$

at every point of U. Therefore, g a holomorphic function of z on U and, at any point, either g is real or H is zero.

The assumed nontriviality of our transform requires that H be identically zero on M. In fact, suppose that H is not zero at the chosen point of M and assume that U is connected. Then the continuous function H must be nonzero on a nonempty open subset V of U and hence g is a real constant, c, on V. Being a holomorphic function constant on an open subset V, it must be constant on all of the connected open set U. Hence, $g = c$ on U, which substituted into (9.17) gives

$$d\hat{\mathbf{x}} = c d\mathbf{x} \tag{9.19}$$

on U. This must hold on all of M. To see this, let $\tilde{U}, w$ be another complex coordinate chart for $\mathbf{x}$ and suppose that $U \cap \tilde{U} \neq \emptyset$. If $(\mathbf{x}, \tilde{e})$ is the frame field adapted to $(\tilde{U}, w)$, then the above analysis leads to

$$d\hat{\mathbf{x}} = \tilde{g}\mathbf{x}_w dw + \bar{\tilde{g}}\mathbf{x}_{\bar{w}} d\bar{w} \tag{9.20}$$

on $\tilde{U}$, where $\tilde{g}$ must be a holomorphic function of w on $\tilde{U}$ and $(\tilde{g} - \bar{\tilde{g}})H = 0$ at every point of $\tilde{U}$. On $U \cap \tilde{U}$, we have $w = w(z)$ holomorphic and $\mathbf{x}_w dw = \mathbf{x}_z dz$, so that, by (9.17) and (9.20),

$$g\mathbf{x}_z dz + \bar{g}\mathbf{x}_{\bar{z}} d\bar{z} = d\hat{\mathbf{x}} = \tilde{g}\mathbf{x}_z dz + \bar{\tilde{g}}\mathbf{x}_{\bar{z}} d\bar{z},$$

from which we conclude that $\tilde{g} = g = c$ on $U \cap \tilde{U}$. Therefore, the holomorphic function $\tilde{g} = c$ on all of $\tilde{U}$ (assumed connected) and we have $d\hat{\mathbf{x}} = c d\mathbf{x}$ on $U \cup \tilde{U}$. Since M is connected, any point of M lies in U_k of a finite chain of open coordinate neighborhoods $(U_j)_1^k$ with $U_1 = U$ and $U_{j-1} \cap U_j \neq \emptyset$ for $j = 2, \ldots, k$. It follows that (9.19) holds on all of M and this contradicts the nontriviality of the transform.

Therefore, H must be identically zero on M, that is, $\mathbf{x}$ is a minimal immersion. In terms of a complex coordinate z of $\mathbf{x}$ we have, by (9.17),

$$\hat{\mathbf{x}}_z = g\mathbf{x}_z$$

where, by (9.18), the function g is holomorphic in z on U. By Problem 8.56 we see that $\hat{x}$ is a conformal associate of $\mathbf{x}$ on U. □

Problems

9.33. Use Dupin's Theorem, *The surfaces of a triply orthogonal system intersect in the lines of curvature*, (see Struik [155, pages 99–103]), to prove that

$$x^2 = \frac{p(p+a)(p+c)}{(p-q)(p-r)}, \quad y^2 = \frac{q(q+a)(q+c)}{(q-r)(q-p)}, \quad z^2 = \frac{r(r+a)(r+c)}{(r-p)(r-q)}$$

is a parametrization of the quadric (9.1) by its principal curvatures a and c. Use this to show that the induced metric has the form $I = E\,da^2 + G\,dc^2$, where

$$E = \frac{a-c}{4}\frac{a}{(p+a)(q+a)(r+a)}, \quad G = \frac{c-a}{4}\frac{c}{(p+c)(q+c)(r+c)}$$

and therefore $E/G = g(c)/f(a)$, for some functions $f(a)$ and $g(c)$. Figure 9.4 shows a triply orthogonal system of *confocal quadrics*.

9.34 (Cylinder in $\mathbf{S}^3$). Let $\mathbf{y}(s) = {}^t(y^1(s), y^2(s), y^3(s), 0)$ be an immersed curve in the great sphere $\mathbf{S^2} \subset \mathbf{S^3} \subset \mathbf{R}^4$, parameterized by arclength parameter $s \in J$. Let

$$\mathbf{x} : J \times \mathbf{S}^1 \to \mathbf{S^3}, \quad \mathbf{x}(s,t) = \cos t\,\mathbf{y}(s) + \sin t\,\epsilon_4.$$

Is $\mathbf{x}$ isothermic?

9.35 (Cones). Prove that any cone $\mathbf{x} : M \to \mathbf{R}^3$ defined in Example 4.28 is isothermic.

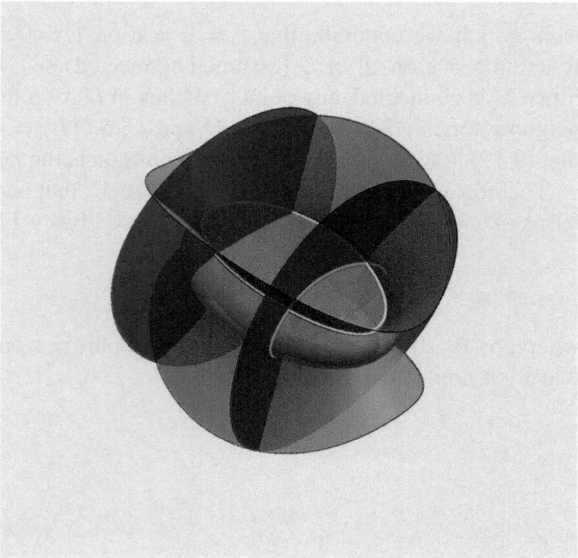

Fig. 9.4 A triply orthogonal system of confocal quadrics.

9.36 (Paraboloid). The immersion

$$\mathbf{x} : M = \mathbf{R}^2 \to \mathbf{R}^3 \quad \mathbf{x}(x,y) = {}^t(x,y,x^2+y^2)$$

is a paraboloid. Prove that $(0,0) \in M$ is an umbilic point, that $\mathbf{x} : M \setminus \{(0,0)\} \to \mathbf{R}^3$ is isothermic, but that there is no principal complex chart (U,z) about $(0,0)$ in M.

9.37. Prove that a Dupin immersion $\mathbf{x} : M \to \mathbf{R}^3$ is isothermic on the open set of nonumbilic points.

9.38. Use (9.10) to prove that a Christoffel transform of the catenoid $\mathbf{x}(s,t) = (\cosh s \cos t, -\cosh s \sin t, s)$ parametrizes the unit sphere with the north and south poles removed. In the same way prove that a Christoffel transform of Enneper's surface $\mathbf{x}(s,t) = (s + st^2 - s^3/3, -t - s^2 t + t^3/3, s^2 - t^2)$ parametrizes a sphere with the north pole removed.

9.39. Prove that if the profile curve of a torus of revolution is embedded, then its Christoffel transform is not a torus. In particular, the Christoffel transform of a circular torus is not a torus, as we have seen in Figure 9.2.

9.40. Find the Christoffel transforms of the cylinders of Example 9.8 and of the cones of Problem 9.35. Do they descend?

Chapter 10
The Bonnet Problem

This chapter presents the Bonnet Problem, which asks whether an immersion of a surface $\mathbf{x} : M \to \mathbf{R}^3$ admits a *Bonnet mate*, which is another noncongruent immersion $\tilde{\mathbf{x}} : M \to \mathbf{R}^3$ with the same induced metric and the same mean curvature at each point. Any immersion with constant mean curvature admits a 1-parameter family of Bonnet mates, all noncongruent to each other. These are its associates. The problem is thus to determine whether an immersion with nonconstant mean curvature has a Bonnet mate. A brief introduction to the notion of G-deformation is used to derive the KPP Bonnet pair construction of Kamberov, Pedit, and Pinkall. We state and prove a new result on proper Bonnet immersions that implies results of Cartan, Bonnet, Chern, and Lawson-Tribuzy. The chapter concludes with a summary of Cartan's classification of proper Bonnet immersions.

10.1 Background

Consider an immersion $\mathbf{x} : M \to \mathbf{R}^3$ of a connected, orientable surface M, with unit normal vector field $\mathbf{e}_3$. Its induced metric $I = d\mathbf{x} \cdot d\mathbf{x}$ and the orientation of M induced by $\mathbf{e}_3$ from the standard orientation of $\mathbf{R}^3$ induce a complex structure on M, which provides a decomposition into bidegrees of the second fundamental form of $\mathbf{x}$ relative to $\mathbf{e}_3$,

$$-d\mathbf{e}_3 \cdot d\mathbf{x} = II = II^{2,0} + HI + II^{0,2}.$$

Here H is the mean curvature of $\mathbf{x}$ relative to $\mathbf{e}_3$ and $II^{2,0} = \overline{II^{0,2}}$ is the Hopf quadratic differential of $\mathbf{x}$. Relative to a complex chart (U, z) in M,

$$I = e^{2u} dz d\bar{z}, \quad II^{2,0} = \frac{1}{2} h e^{2u} dz dz,$$

G.R. Jensen et al., *Surfaces in Classical Geometries*, Universitext,
DOI 10.1007/978-3-319-27076-0_10

where the conformal factor e^u, the Hopf invariant h, and the mean curvature H satisfy the structure equations (7.31) and (7.32) on U:

$$-4e^{-2u}u_{z\bar{z}} = H^2 - |h|^2 \quad \text{Gauss equation}$$

$$(e^{2u}h)_{\bar{z}} = e^{2u}H_z \quad \text{Codazzi equation}$$

and the Gauss curvature is $K = H^2 - |h|^2$. It follows from Bonnet's Congruence Theorem 7.27 and Existence Theorem 7.28 that if functions

$$u, H : U \to \mathbf{R}, \quad h : U \to \mathbf{C}$$

satisfy these structure equations, and if U is simply connected, then there exists an immersion $\mathbf{x} : U \to \mathbf{R}^3$ whose conformal factor is e^u, mean curvature is H, and Hopf invariant is h, and this immersion is unique up to rigid motion.

In his Mémoire [15, pp. 72 ff], Bonnet considered

The Bonnet Problem 1. If two immersions $\mathbf{x}$ and $\tilde{\mathbf{x}}$ have the same induced metric, $I = \tilde{I}$, and the same principal curvatures $a = \tilde{a}$ and $c = \tilde{c}$, then are they congruent?

Bonnet's assumption of equality of the principal curvatures is equivalent to assuming equality of the mean curvatures, $\tilde{H} = H$. In fact, if the principal curvatures agree, then so also do the mean curvatures, since

$$\tilde{H} = (\tilde{a} + \tilde{c})/2 = (a + c)/2 = H.$$

Conversely, equality of the induced metrics implies equality of the Gauss curvatures, $K = \tilde{K}$, and therefore the principal curvatures

$$a = H + \sqrt{H^2 - K}, \quad c = H - \sqrt{H^2 - K}$$

must agree (see (4.25)).

There is an extensive literature on this problem: Bonnet [14, 15], Cartan [33], Roussos [143–145], Bobenko and Eitner [9], and many references within these papers.

Definition 10.1. A *Bonnet immersion* is an immersion $\mathbf{x} : M^2 \to \mathbf{R}^3$ that admits a noncongruent immersion $\tilde{\mathbf{x}} : M \to \mathbf{R}^3$ such that $\tilde{I} = I$ and $\tilde{H} = H$. Call $\tilde{\mathbf{x}}$ a *Bonnet mate* of $\mathbf{x}$ and call $(\mathbf{x}, \tilde{\mathbf{x}})$ a *Bonnet pair*.

This chapter contains a detailed study of Bonnet's Problem. If H is constant, then $\mathbf{x}$ is isothermic and has a 1-parameter family of Bonnet mates, as seen in Example 10.11. If H is nonconstant, then isothermic immersions play an important role in the problem, illustrating again that the isothermic condition is a generalization of constant mean curvature.

Definition 10.2. A *proper Bonnet immersion* is a Bonnet immersion $\mathbf{x} : M \to \mathbf{R}^3$ whose mean curvature is nonconstant and which admits at least two noncongruent Bonnet mates.

We shall see that nonisothermic immersions have a unique Bonnet mate, while proper Bonnet immersions must be isothermic. We believe the first statement is a new result.

During our discussion of the Bonnet problem the following concept of *equivalence* becomes essential.

Definition 10.3. Immersions $\mathbf{x} : M \to \mathbf{R}^3$ and $\hat{\mathbf{x}} : \hat{M} \to \mathbf{R}^3$ are *equivalent* if there exists a diffeomorphism $F : M \to \hat{M}$ such that $\hat{\mathbf{x}} \circ F$ is congruent to $\mathbf{x}$.

Remark 10.4. If $\mathbf{x}, \hat{\mathbf{x}} : M \to \mathbf{R}^3$ are immersions of a simply connected surface M with the same induced metrics $I = \hat{I}$ and if there exists an isometry $F : (M, I) \to (M, I)$ such that $F^*\widehat{II} = II$, then $\hat{\mathbf{x}} \circ F$ and $\mathbf{x}$ are congruent, because then Bonnet's Theorem 4.18 implies there exists $A \in \mathbf{E}(3)$ such that $A \circ \mathbf{x} = \hat{\mathbf{x}} \circ F$.

10.2 The deformation quadratic differential

From the Gauss equation above, the Hopf invariants relative to a complex coordinate z of two immersions with the same induced metric and the same mean curvatures must satisfy

$$|\tilde{h}| = |h|,$$

since $\tilde{u} = u$. Hence, the only possible difference in the invariants of two such immersions must be in the arguments of the complex valued functions h and $\tilde{h}$. Moreover, taking the difference of their Codazzi equations, we get

$$(e^{2u}\tilde{h} - e^{2u}h)_{\bar{z}} = e^{2u}(H_z - H_z) = 0,$$

at every point of the domain U of the complex coordinate z. This means that the function

$$F = e^{2u}(\tilde{h} - h) : U \to \mathbf{C}$$

is holomorphic.

Definition 10.5. If $\mathbf{x}, \tilde{\mathbf{x}} : M \to \mathbf{R}^3$ are immersions that induce the same complex structure on M, then their *deformation quadratic differential* is

$$\mathcal{Q} = \widetilde{II}^{2,0} - II^{2,0}.$$

If $\mathbf{x}$ and $\tilde{\mathbf{x}}$ have the same induced metric and mean curvature, then the expression for $\mathcal{Q}$ relative to a complex coordinate z is

$$\mathcal{Q} = \frac{1}{2}e^{2u}(\tilde{h}-h)dzdz = \frac{1}{2}Fdzdz, \tag{10.1}$$

which shows that $\mathcal{Q}$ is a holomorphic quadratic differential on M, and

$$|F + e^{2u}h| = |e^{2u}\tilde{h}| = |e^{2u}h| \tag{10.2}$$

on U, since $|\tilde{h}| = |h|$. As a convenient shorthand, we will express (10.2) as

$$|\mathcal{Q} + II^{2,0}| = |II^{2,0}|.$$

$\mathcal{Q}$ is identically zero on M if and only if $\tilde{h} = h$ in any complex coordinate system. Therefore, by Bonnet's Congruence Theorem 7.27, $\mathcal{Q} = 0$ if and only if the immersions $\mathbf{x}$ and $\tilde{\mathbf{x}}$ are congruent in the sense that there exists a rigid motion $(\mathbf{y}, A) \in \mathbf{E}(3)$ such that $\tilde{\mathbf{x}} = \mathbf{y} + A\mathbf{x} : M \to \mathbf{R}^3$. Thus, an immersion $\tilde{\mathbf{x}} : M \to \mathbf{R}^3$ is a Bonnet mate of $\mathbf{x} : M \to \mathbf{R}^3$ if it induces the same metric and mean curvature and the deformation quadratic differential is not identically zero.

Proposition 10.6. *If an immersion $\mathbf{x} : M \to \mathbf{R}^3$ possesses a Bonnet mate $\tilde{\mathbf{x}} : M \to \mathbf{R}^3$, then the umbilic points of $\mathbf{x}$ must be isolated.*

Proof. Under the given assumptions, the holomorphic quadratic differential $\mathcal{Q}$ is not identically zero. Therefore, in any complex coordinate chart (U, z), we have $\mathcal{Q} = \frac{1}{2}Fdzdz$, where F is a nonzero holomorphic function of z. Its zeros must be isolated. A point $m \in U$ is umbilic for $\mathbf{x}$ if and only if $h(m) = 0$, in which case $F(m) = 0$ by (10.2). Therefore, the set of umbilic points is a subset of the set of zeros of $\mathcal{Q}$, which is a discrete subset of M. □

Lemma 10.7. *A compact Riemann surface M of genus zero has no nonzero holomorphic quadratic differentials.*

Proof. A compact Riemann surface of genus zero must be the Riemann sphere. This means that there exist points $p, q \in M$ and complex coordinates $z : M \setminus \{q\} \to \mathbf{C}$ and $w : M \setminus \{p\} \to \mathbf{C}$, both onto $\mathbf{C}$, such that $z(p) = 0 = w(q)$, and on $M \setminus \{p, q\}$ we have $w = 1/z$. Let $\mathcal{Q}$ be a holomorphic quadratic differential on M. On $M \setminus \{q\}$, we have $\mathcal{Q} = f(z)dzdz$, and on $M \setminus \{p\}$ we have $\mathcal{Q} = g(w)dwdw$, where f and g are entire holomorphic functions. On $M \setminus \{p, q\}$ we have $dw = -\frac{1}{z^2}dz$, so

$$f(z)dzdz = \mathcal{Q} = g(w)dwdw = g(w)\frac{1}{z^4}dzdz.$$

Therefore, $f(z) = g(w)/z^4$, from which it follows that

$$\lim_{z\to\infty} f(z) = \lim_{z\to\infty} \frac{1}{z^4} \lim_{w\to 0} g(w) = 0,$$

since $g(0)$ is finite. Hence, $f(z)$ must be identically zero, and therefore $\mathcal{Q}$ must be identically zero. □

Proposition 10.8. *Suppose M is homeomorphic to $\mathbf{S}^2$. If two immersions*

$$\mathbf{x}, \tilde{\mathbf{x}} : M \to \mathbf{R}^3$$

have the same induced metrics, $I = \tilde{I}$, and the same mean curvatures, $H = \tilde{H}$, then they are congruent. That is, there exists a rigid motion $(\mathbf{v}, A) \in \mathbf{E}(3)$ such that

$$\tilde{\mathbf{x}} = \mathbf{v} + A\mathbf{x}$$

Proof. In the above discussion we saw that $\mathscr{Q} = \widetilde{II}^{2,0} - II^{2,0}$ is a holomorphic quadratic differential on M. By Lemma 10.7, this must be identically zero on a surface M homeomorphic to $\mathbf{S}^2$. The result now follows from Bonnet's Congruence Theorem 7.27. □

The idea of this proof comes from H. Hopf's proof of the next theorem.

Theorem 10.9 (H. Hopf [92]). *Suppose M is homeomorphic to $\mathbf{S}^2$. If $\mathbf{x} : M \to \mathbf{R}^3$ is an immersion with constant mean curvature, H, then $H \neq 0$ and $\mathbf{x}(M)$ is a sphere of radius $1/|H|$.*

Proof. In any complex coordinate chart (U, z) on M, the Hopf quadratic differential

$$II^{2,0} = \frac{1}{2} e^{2u} h dz dz$$

is holomorphic by the Codazzi equation, which for constant H is

$$(e^{2u} h)_{\bar{z}} = e^{2u} H_z = 0.$$

Therefore, $II^{2,0}$ must be identically zero on M, by Lemma 10.7 above, so $h = 0$ in any complex coordinate chart and $\mathbf{x}$ is totally umbilic. We can now apply Theorem 4.23. If $H = 0$, then the principal curvatures $a = 0 = c$ at every point of M and $\mathbf{x}(M)$ must be an open subset of a plane in $\mathbf{R}^3$. This is impossible for a compact surface M. Hence H is a nonzero constant and the principal curvatures at every point are $a = c = H$ and $\mathbf{x}(M)$ is a sphere of radius $1/|H|$. □

Theorem 10.10. *If $\mathbf{x} : M \to \mathbf{R}^3$ is an immersion admitting a Bonnet mate $\tilde{\mathbf{x}} : M \to \mathbf{R}^3$, then the deformation quadratic differential $\mathscr{Q}$ is nonzero, holomorphic on M (with the complex structure from the induced metric), and satisfies* (10.2).

Conversely, if $\mathbf{x} : M \to \mathbf{R}^3$ is an immersion whose induced complex structure on M admits a nonzero holomorphic quadratic differential $\mathscr{Q}$ satisfying (10.2), *and if M is simply connected, then there exists a Bonnet mate $\tilde{\mathbf{x}} : M \to \mathbf{R}^3$ whose deformation quadratic differential is $\mathscr{Q}$.*

Proof. The first part has been proved in the discussion above. For the converse, suppose given a nonzero holomorphic quadratic differential $\mathscr{Q}$ on the simply connected M satisfying (10.2). Consider the Riemannian metric $I = d\mathbf{x} \cdot d\mathbf{x}$, the mean curvature function $H : M \to \mathbf{R}$ of $\mathbf{x}$, and the smooth quadratic differential

$$\widetilde{II}^{2,0} = II^{2,0} + \mathscr{Q},$$

where $II^{2,0}$ is the Hopf quadratic differential of $\mathbf{x}$. In a complex coordinate chart (U,z) in M, we have $I = e^{2u}dzd\bar{z}$, $II^{2,0} = \frac{1}{2}he^{2u}dzdz$, and $\mathcal{Q} = \frac{1}{2}Fdzdz$, for some holomorphic function F on U, so $\widetilde{II}^{2,0} = \frac{1}{2}\tilde{h}e^{2u}dzdz$, where

$$\tilde{h}e^{2u} = he^{2u} + F.$$

By (10.2), $|II^{2,0} + \mathcal{Q}| = |II^{2,0}|$, which implies

$$|\tilde{h}e^{2u}| = |he^{2u} + F| = |he^{2u}|.$$

on U, so u, H, and $\tilde{h}$ satisfy the Gauss equation (7.31). They also satisfy the Codazzi equation (7.32), because

$$(e^{2u}\tilde{h})_{\bar{z}} = (e^{2u}h + F)_{\bar{z}} = (e^{2u}h)_{\bar{z}} = e^{2u}H_z,$$

since $F_{\bar{z}} = 0$ and since u, H, and h satisfy the Codazzi equation. By Bonnet's Existence Theorem 7.28, there exists an immersion $\tilde{\mathbf{x}} : M \to \mathbf{R}^3$ with induced metric I, mean curvature H, and Hopf quadratic differential $\widetilde{II}^{2,0}$. It is a Bonnet deformation of $\mathbf{x}$ with deformation quadratic differential $\widetilde{II}^{2,0} - II^{2,0} = \mathcal{Q}$. □

Example 10.11 (CMC immersions). Let M be a simply connected surface and let $\mathbf{x} : M \to \mathbf{R}^3$ be an immersion with constant mean curvature H, induced metric I, and Hopf quadratic differential $II^{2,0}$. We exclude the totally geodesic case by assuming that the Hopf quadratic differential $II^{2,0}$ is not identically zero on M. Note that it is holomorphic, since H is constant. For a real constant r, the Riemannian metric I, the constant H, and the smooth quadratic differential $e^{ir}II^{2,0}$ satisfy the Gauss and Codazzi equations in any complex coordinate chart (U,z) in M. Hence, there exists an immersion $\tilde{\mathbf{x}} : M \to \mathbf{R}^3$ with induced metric I, mean curvature H, and Hopf quadratic differential $e^{ir}II^{2,0}$. Its deformation form

$$\mathcal{Q} = e^{ir}II^{2,0} - II^{2,0} = (e^{ir} - 1)II^{2,0}$$

satisfies $|II^{2,0} + \mathcal{Q}| = |II^{2,0}|$ and is nonzero provided that $e^{ir} \neq 1$. This shows that $\mathbf{x}$ possesses a 1-parameter family of Bonnet mates, which we call the *associate immersions* of $\mathbf{x}$. In the case $H = 0$, these are exactly the associates of the minimal immersion $\mathbf{x}$.

10.3 Bonnet versus proper Bonnet

Recall Definition 4.14 of the criterion form α on M of any umbilic free immersion $\mathbf{x} : M \to \mathbf{R}^3$. We know from Theorem 9.12 that $\mathbf{x}$ is isothermic if and only if α is closed on M.

Definition 10.12. An umbilic free immersion $\mathbf{x} : M \to \mathbf{R}^3$ is *totally nonisothermic* if $d\alpha$ is never zero, where α on M is the criterion form of $\mathbf{x}$.

Item (1) of the next theorem is due to Graustein [76]. See Problem 10.52. We believe that item (2) is new.

Theorem 10.13. *Let $\mathbf{x} : M \to \mathbf{R}^3$ be an umbilic free immersion of a simply connected surface M with complex coordinate $z = x + iy : M \to \mathbf{C}$ and nonconstant mean curvature H (i.e., $dH \neq 0$ on a dense open subset of M).*

1. *If $\mathbf{x}$ is isothermic and if it admits a Bonnet mate, then it is proper Bonnet.*
2. *If $\mathbf{x}$ is totally nonisothermic, then it has a unique Bonnet mate (so it cannot be proper Bonnet).*

Proof. Let h and e^u be the Hopf invariant and conformal factor relative to z. Then $h = e^{f+ig}$, for some smooth functions $f, g : M \to \mathbf{R}$. Corollary 9.14 tells us that $\mathbf{x}$ is isothermic if and only if $g_{\bar{z}z} = 0$ identically on M.

By Theorem 10.10, any Bonnet mate of $\mathbf{x}$ must be given by a nonzero holomorphic deformation form $\mathcal{Q} = F dz dz$, where

$$F = he^{2u}(e^{ir} - 1) : M \to \mathbf{C}$$

is holomorphic, for some smooth $r : M \to \mathbf{R}$, which must be nonconstant, since H nonconstant implies he^{2u} is not holomorphic. For convenience we write $f + 2u = G : M \to \mathbf{R}$, so that $he^{2u} = e^{G+ig}$. The Cauchy-Riemann equation $F_{\bar{z}} = 0$ is

$$r_{\bar{z}} = i(G + ig)_{\bar{z}}(1 - e^{-ir}). \tag{10.3}$$

To solve this equation for r, we consider the 2-dimensional distribution defined on $\mathbf{R} \times M$ by the 1-form

$$\rho = dr - r_z dz - r_{\bar{z}} d\bar{z} = dr - r_x dx - r_y dy.$$

This distribution satisfies the Frobenius condition if and only if $\rho \wedge d\rho = 0$ if and only if

$$\Im(r_{\bar{z}z}) = (|G_{\bar{z}} + ig_{\bar{z}}|^2 - G_{\bar{z}z})(\cos r - 1) - g_{\bar{z}z} \sin r = 0 \tag{10.4}$$

on $\mathbf{R} \times M$.

If $\mathbf{x}$ is isothermic, then $g_{\bar{z}z} = 0$ identically on M, so (10.4) holds identically on $\mathbf{R} \times M$ if and only if the invariants f, g, and u of $\mathbf{x}$ satisfy the PDE

$$|G_{\bar{z}} + ig_{\bar{z}}|^2 = G_{\bar{z}z} \tag{10.5}$$

identically on M. If (10.5) is not satisfied, then $\mathbf{x}$ has no Bonnet mates. If (10.5) is satisfied by $\mathbf{x}$, then there is an integrating factor $m : \mathbf{R} \times M \to \mathbf{R}_+$ such that $m\rho = dR$, for some smooth function $R : \mathbf{R} \times M \to \mathbf{R}$, and $R = c$ defines an integrable submanifold of the distribution defined by ρ, for any constant c in the range of R. Since dR never vanishes and is proportional to ρ, it follows from the Implicit Function Theorem that there exists a smooth function $r : M \to \mathbf{R}$ such that $R(r(x,y),x,y) = c$. This function r, which is not identically 0, is a solution of (10.3) and determines a Bonnet mate of $\mathbf{x}$. Distinct values of the constant c determine distinct Bonnet mates. Hence, $\mathbf{x}$ is proper Bonnet in this case.

If $\mathbf{x}$ is totally nonisothermic, then $g_{\bar{z}z}$ is never zero on M and the integrability condition (10.4) holds for a unique value of e^{ir}. In fact,

$$e^{ir} = 1 - \frac{2g_{\bar{z}z}}{D}(g_{\bar{z}z} + i(|G_{\bar{z}} + ig_{\bar{z}}|^2 - G_{\bar{z}z})),$$

where $D = g_{\bar{z}z}^2 + (|G_{\bar{z}} + ig_{\bar{z}}|^2 - G_{\bar{z}z})^2$. This gives a unique Bonnet mate of $\mathbf{x}$. □

We will use Theorem 10.10 to determine all Bonnet cylinders and cones. As these are all umbilic free isothermic immersions, Theorem 10.13 implies that only special cylinders and cones can be Bonnet, and they must be proper Bonnet.

10.3.1 Bonnet cylinders

We consider now the problem of which cylinders (4.76) $\mathbf{x} : J \times \mathbf{R} \to \mathbf{R}^3$, $\mathbf{x}(x,y) = \boldsymbol{\gamma}(x) - y\boldsymbol{\epsilon}_3$, over a plane curve $\boldsymbol{\gamma} : J \to \mathbf{R}^2$, are Bonnet. See Problem 4.65 and Example 9.8 for details. Here x is an arclength parameter for $\boldsymbol{\gamma}$ and κ is its curvature. We assume κ is positive, but nonconstant, so that $H = \kappa/2 = h$, by (9.5), is nonconstant. Then $z = x + iy$ is a complex coordinate determined by the induced metric $d\mathbf{x}\cdot d\mathbf{x} = dx^2 + dy^2$ on $M = J \times \mathbf{R}$. By (10.2), we want to determine conditions on κ that allow the existence of a holomorphic function $F(z)$ satisfying

$$|F + \kappa/2| = |\kappa/2|.$$

Thus, F must be given by

$$F(x,y) = \frac{\kappa(x)}{2}(e^{ir(x,y)} - 1), \tag{10.6}$$

for some smooth real valued function $r : M \to \mathbf{R}$. In order for F to be holomorphic, it must satisfy the Cauchy-Riemann equations $F_y = iF_x$. We compute

$$F_x = \frac{\dot{\kappa}}{2}(e^{ir} - 1) + \frac{\kappa}{2}e^{ir} ir_x, \quad F_y = \frac{\kappa}{2}e^{ir} ir_y,$$

where $\dot{\kappa} = \frac{d\kappa}{dx}$, so

$$F_y - iF_x = \frac{i}{2}\kappa e^{ir}\left(r_y + \frac{\dot{\kappa}}{\kappa}(\cos r - 1) - i(\frac{\dot{\kappa}}{\kappa}\sin r + r_x)\right)$$

is zero if and only if r satisfies

$$r_x = -\frac{\dot{\kappa}}{\kappa}\sin r, \quad r_y = \frac{\dot{\kappa}}{\kappa}(1-\cos r). \tag{10.7}$$

This is solvable for $r(x,y)$ if and only if $r_{xy} = r_{yx}$ if and only if

$$0 = r_{xy} - r_{yx} = (1-\cos r)\left\{\dot{\left(\frac{\dot{\kappa}}{\kappa}\right)} - \left(\frac{\dot{\kappa}}{\kappa}\right)^2\right\}.$$

The case $r = 0$ identically leads to $F = 0$ identically, which gives a trivial deformation. The integrability condition to consider is thus

$$\dot{\left(\frac{\dot{\kappa}}{\kappa}\right)} = \left(\frac{\dot{\kappa}}{\kappa}\right)^2.$$

The general solution for $\dot{\kappa}/\kappa$ is

$$\frac{\dot{\kappa}}{\kappa} = \frac{-1}{x+n_1},$$

where n_1 is an arbitrary real constant. Then $\kappa = n/(x+n_1)$, where n is an arbitrary nonzero real constant making $\kappa > 0$. A change of arclength parameter from x to $x+n_1$, gives the curvature formula

$$\kappa = \frac{n}{x},$$

where $x > 0$ and $n > 0$ or $x < 0$ and $n < 0$. Reversing the orientation of the curve and of the plane, if necessary, we suffer no loss of generality if we assume $n > 0$ and $J = \mathbf{R}^+ = \{x > 0\}$.

For any positive real constant $n > 0$, the curve $\boldsymbol{\gamma}_n : \mathbf{R}^+ \to \mathbf{R}^2$ with curvature $\kappa = n/x$ is, up to rotation and translation in the plane, equal to

$$\begin{aligned}\boldsymbol{\gamma}_n(x) &= \frac{1}{1+n^2}\begin{pmatrix} x\cos(n\log x) - 1 + nx\sin(n\log x) \\ -nx\cos(n\log x) + n + x\sin(n\log x)\end{pmatrix} \\ &= xA_n(x)\mathbf{v}_n - \mathbf{v}_n\end{aligned} \tag{10.8}$$

on $x > 0$, where

$$\mathbf{v}_n = \frac{1}{1+n^2}\begin{pmatrix} 1 \\ -n\end{pmatrix} \in \mathbf{R}^2, \tag{10.9}$$

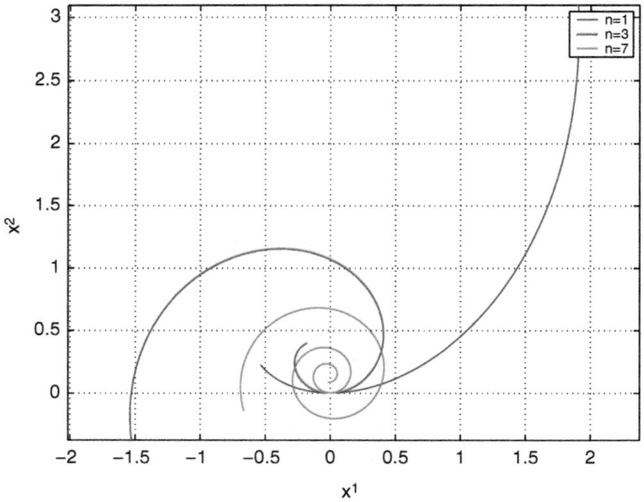

Fig. 10.1 Curves $\boldsymbol{\gamma}_n(x)$ on $.05 \leq x \leq 5$ for $n = 1, 3$, and 7.

and $A_n : \mathbf{R}^+ \to \mathbf{SO}(2)$ is the homomorphism from the multiplicative group of positive real numbers onto $\mathbf{SO}(2)$ given by

$$A_n(x) = \begin{pmatrix} \cos(n \log x) & -\sin(n \log x) \\ \sin(n \log x) & \cos(n \log x) \end{pmatrix}. \tag{10.10}$$

See Figure 10.1.

We summarize our results as follows.

Proposition 10.14 (Bonnet cylinders). *Up to rigid motion and reflection, the only Bonnet cylinders are*

$$\mathbf{x}_n : M \to \mathbf{R}^3, \quad \mathbf{x}_n(x,y) = \boldsymbol{\gamma}_n(x) - y\boldsymbol{\epsilon}_3, \tag{10.11}$$

where $M = \{(x,y) \in \mathbf{R}^2 : x > 0\}$ *and n is any positive constant. The curvature of* $\boldsymbol{\gamma}_n$ *is* $\kappa = n/x$. *Distinct values of n give noncongruent Bonnet cylinders.*

Figure 10.2 shows the cylinder $\mathbf{x}_3(M)$.

We continue our analysis of the cylinder (10.11) to see if we can determine its Bonnet mates. With curvature function $\kappa = n/x$, equations (10.7) become

$$r_x = \frac{\sin r}{x}, \quad r_y = \frac{-1 + \cos r}{x}. \tag{10.12}$$

Integrating the second equation for any fixed x, we find

$$\frac{1 + \cos r}{\sin r} = \int \frac{dr}{-1 + \cos r} = \frac{y}{x} + g(x), \tag{10.13}$$

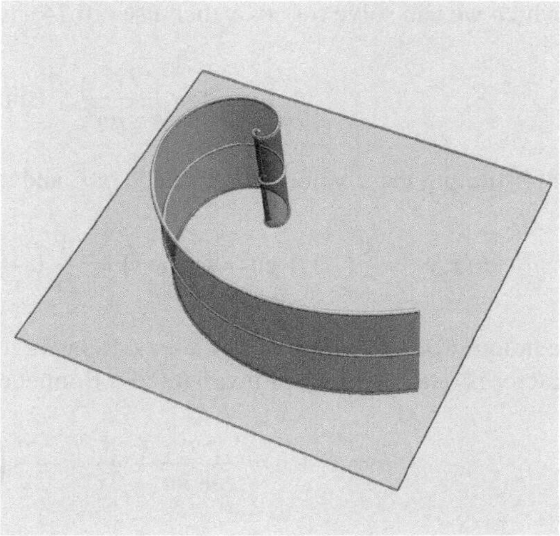

Fig. 10.2 The cylinder through γ_3.

where $g(x)$ is an arbitrary function of x. Take the partial derivative with respect to x, use the first equation in (10.12), and (10.13) again, to get

$$\left(-\frac{y}{x}-g\right)\left(\frac{1}{x}\right)=-\frac{y}{x^2}+\dot{g},$$

where $\dot{g}$ denotes derivative with respect to x. Hence

$$\frac{\dot{g}}{g}=-\frac{1}{x},$$

whose general solution is

$$g(x)=\frac{m}{x},$$

where m is an arbitrary nonzero real constant. Put this into (10.13) to get

$$\frac{1+\cos r}{\sin r}=\frac{y}{x}+\frac{m}{x}=\frac{\tilde{y}}{x}, \tag{10.14}$$

where we let

$$\tilde{y}=y+m.$$

Squaring (10.14), we get

$$\left(\frac{\tilde{y}}{x}\right)^2=\frac{(1+\cos r)^2}{1-\cos^2 r}=\frac{1+\cos r}{1-\cos r},$$

which we can solve for $\cos r$, then use (10.14), to get

$$1+\cos r = \frac{2\tilde{y}^2}{x^2+\tilde{y}^2}, \quad \sin r = \frac{2x\tilde{y}}{x^2+\tilde{y}^2}.$$

Substituting these values of $\sin r$ and $\cos r$, and $\kappa = n/x$, into (10.6), we have

$$F(x,y) = \frac{\kappa}{2}(-1+\cos r + i\sin r) = \frac{n}{2x}\left(\frac{-2x^2}{x^2+\tilde{y}^2} + i\frac{2x\tilde{y}}{x^2+\tilde{y}^2}\right) = \frac{-n}{z+im},$$

a holomorphic function of $z = x+iy$. Relative to z, $h = \frac{n}{2x} = H$, and the conformal factor is one, so the Hopf invariant of a Bonnet mate $\hat{\mathbf{x}}$ relative to z is, by (10.1),

$$\hat{h} = F + h = \frac{-n}{z+im} + \frac{n}{2x} = \frac{-n}{2x}\left(\frac{\bar{z}-im}{z+im}\right). \tag{10.15}$$

We get a mate $\hat{\mathbf{x}}_m$ for each value of the constant $m \in \mathbf{R}$. The frame field $(\hat{\mathbf{x}}_m, \hat{e}_m)$ along $\hat{\mathbf{x}}_m$ and adapted to z pulls back the Maurer–Cartan form of $\mathbf{E}(3)$ to

$$(\hat{e}_m^{-1}d\hat{\mathbf{x}}_m, \hat{e}_m^{-1}d\hat{e}_m) = (\hat{\omega}^i, \hat{\omega}^i_j),$$

where

$$\hat{\omega}^1 = dx, \quad \hat{\omega}^2 = dy, \quad \hat{\omega}^3 = 0, \quad \hat{\omega}^1_2 = 0,$$
$$\hat{\omega}^3_1 - i\hat{\omega}^3_2 = \hat{h}dz + Hd\bar{z} = -n(\frac{y+m}{x} + i)d\theta,$$

where

$$d\theta = d\arctan\frac{y+m}{x} = \frac{x\,d(y+m) - (y+m)\,dx}{x^2+(y+m)^2}.$$

Can we integrate these equations to find $(\hat{\mathbf{x}}_m, \hat{e}_m)$? One suspects that the Bonnet mates of this cylinder should be something like a cylinder. It can't be a cylinder, because we have just proved that these are the only Bonnet cylinders and distinct values of the constant n give distinct values of the mean curvature. It might be a cone.

10.3.2 Bonnet cones

Let

$$\tilde{\mathbf{x}} : J \times \mathbf{R} \to \mathbf{R}^3, \quad \tilde{\mathbf{x}}(s,t) = e^{-t}\sigma(s)$$

be the cone with profile curve $\boldsymbol{\sigma} : J \to \mathbf{R}$. See Example 4.28 for a detailed description of these cones. Assume s is arclength parameter and $\kappa : J \to \mathbf{R}$ is the curvature of $\boldsymbol{\sigma}$. Assume $\kappa > 0$ on J. We know that $w = s + it$ is a complex coordinate on $\tilde{M} = J \times \mathbf{R}$, with conformal factor e^{-t}. The mean curvature and Hopf invariant relative to w are

$$\tilde{H} = \frac{\kappa e^t}{2} = \tilde{h}.$$

Proceeding as we did in the search for Bonnet cylinders above, we must find a holomorphic function $F(w)$ that satisfies

$$F(s,t) = \frac{\kappa}{2} e^{-t}(e^{ir} - 1),$$

for some real valued smooth function $r(s,t)$. Then F is holomorphic if and only if r satisfies

$$r_s = -\frac{\dot{\kappa}}{\kappa} \sin r + \cos r - 1, \quad r_t = \frac{\dot{\kappa}}{\kappa}(1 - \cos r) - \sin r. \tag{10.16}$$

For nonzero F, the integrability condition $r_{st} = r_{ts}$ holds if and only if κ satisfies

$$\dot{\left(\frac{\dot{\kappa}}{\kappa}\right)} - \left(\frac{\dot{\kappa}}{\kappa}\right)^2 - 1 = 0.$$

Translating the arc parameter, if necessary, the general solution is

$$\kappa = n \sec s,$$

for each real constant $n > 0$, on the interval $J = \{-\pi/2 < s < \pi/2\}$. Let

$$\boldsymbol{\sigma}_n : J \to \mathbf{S}^2$$

be the curve with curvature $\kappa = n \sec s$. Its Frenet equations are

$$\dot{\boldsymbol{\sigma}}_n = \mathbf{T}_n, \quad \dot{\mathbf{T}}_n = n \sec s \mathbf{N}_n - \boldsymbol{\sigma}_n, \quad \dot{\mathbf{N}}_n = -n \sec s \mathbf{T}_n.$$

See Figure 10.3 for the solution, up to rigid motion, of the case $n = 3$ with the cone through it.

Proposition 10.15 (Bonnet cones). *Up to rigid motion and reflection, the only Bonnet cones are*

$$\tilde{\mathbf{x}}_n : \tilde{M} \to \mathbf{R}^3, \quad \tilde{\mathbf{x}}_n(s,t) = e^{-t} \boldsymbol{\sigma}_n(s),$$

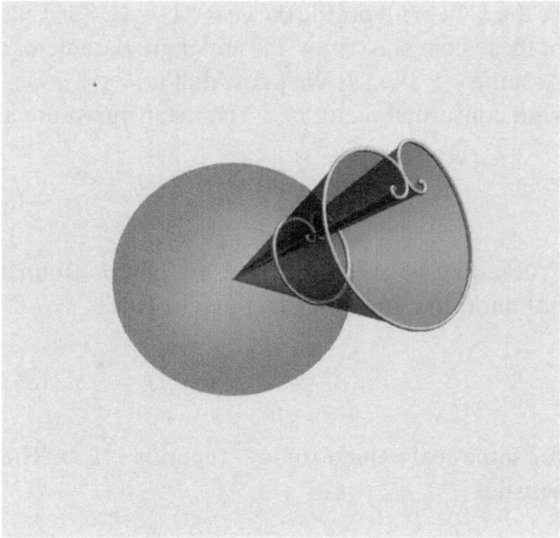

Fig. 10.3 A curve $\boldsymbol{\sigma}_3$ and the cone through it.

where $M = \{(s,t) \in \mathbf{R}^2 : -\pi/2 < s < \pi/2\}$ *and* n *is any positive constant. The curvature of* $\boldsymbol{\sigma}_n$ *is* $\kappa = n \sec s$. *Distinct values of* n *give noncongruent Bonnet cones.*

For $\boldsymbol{\sigma}_n$, the system (10.16) becomes

$$r_s = -\tan s \sin r + \cos r - 1, \quad r_t = \tan s(1 - \cos r) - \sin r.$$

Rather than solve this system for $r(s,t)$ and then use the resulting holomorphic function F to find the Bonnet mate of this Bonnet cone, we will show, for each real constant $n > 0$, that the Bonnet cylinder $\mathbf{x}_n$ and the Bonnet cone $\tilde{\mathbf{x}}_n$ composed with an isometry $\Phi : M \to \tilde{M}$ are Bonnet mates. Define the orientation preserving isometry Φ from the parameter domain M of the Bonnet cylinder $\mathbf{x}_n$ onto the parameter domain $\tilde{M}$ of the Bonnet cone $\tilde{\mathbf{x}}_n$, by

$$\Phi : M \to \tilde{M}, \quad \Phi(x,y) = (s,t),$$

where

$$s = \arctan \frac{y}{x}, \quad t = -\log \sqrt{x^2 + y^2}.$$

Indeed, this is a smooth map with inverse $\Phi^{-1}(s,t) = (e^{-t} \cos s, e^{-t} \sin s)$. It is an isometry, since the metric induced on $\tilde{M}$ by $\tilde{\mathbf{x}}_n$ is $\tilde{I} = e^{-2t}(ds^2 + dt^2)$ and

$$\Phi^* \tilde{I} = I,$$

where $I = dx^2 + dy^2$ is the metric induced on M by $\mathbf{x}_n$. Composing the Bonnet cone $\tilde{\mathbf{x}}(s,t) = e^{-t}\boldsymbol{\sigma}(s)$ with Φ gives an immersion

$$\hat{\mathbf{x}} : M \to \mathbf{R}^3, \quad \hat{\mathbf{x}}(x,y) = \tilde{\mathbf{x}} \circ \Phi(x,y) = \sqrt{x^2 + y^2}\boldsymbol{\sigma}(\arctan \frac{y}{x})$$

whose image is still the given cone $\tilde{\mathbf{x}}_n(\tilde{M})$. The cone $\hat{\mathbf{x}}$ and the cylinder $\mathbf{x}_n$ are both immersions of M into $\mathbf{R}^3$. The metric induced on M by $\hat{\mathbf{x}}$ is

$$\hat{I} = \Phi^* \tilde{I} = I,$$

the metric induced on M by the cylinder $\mathbf{x}_n$. The mean curvature of $\hat{\mathbf{x}}$ is

$$\hat{H}(x,y) = \tilde{H} \circ \Phi(x,y) = \frac{n}{2} \sec(s(x,y)) e^{t(x,y)} = \frac{n}{2x},$$

which is the mean curvature of the Bonnet cylinder $\mathbf{x}_n$. Thus, $\hat{\mathbf{x}}$ is a Bonnet mate of $\mathbf{x}_n$, as they are clearly not congruent. A one parameter family of mates of $\hat{\mathbf{x}}$, and thus also of $\mathbf{x}_n$, is given by

$$\hat{\mathbf{x}}_m = \hat{\mathbf{x}}(z + im),$$

for any constant $m \in \mathbf{R}$, where $z = x + iy$ is the complex coordinate induced on M by I. In fact, the map $z \mapsto z + im$ is a biholomorphic isometry and $\hat{H}(z+im) = H(z)$ on M. If $m \neq m'$, then $\hat{\mathbf{x}}_m$ is not congruent to $\hat{\mathbf{x}}_{m'}$, although they are equivalent, as one is a reparametrization of the other. The Hopf invariant $\hat{h}$ of $\hat{\mathbf{x}}$ relative to z is related to the Hopf invariant $\tilde{h}$ of $\tilde{\mathbf{x}}$ relative to w by (7.41), where now $z = e^{iw}$, so with $z' = \frac{dz}{dw} = iz$,

$$\hat{h}(z) = \frac{\bar{z}'}{z'} \tilde{h}(w(z)) = -\frac{\bar{z}}{z}\frac{n}{2x},$$

which agrees with (10.15) for the case $m = 0$. Applying the same formula to find the Hopf invariant $\hat{h}_m$ of $\hat{\mathbf{x}}_m(z) = \hat{\mathbf{x}}(z+im)$, we get

$$\hat{h}_m(z) = \hat{h}(z+im) = -\frac{n}{2x}\frac{\bar{z}-im}{z+im},$$

again in agreement with (10.15) for any $m \in \mathbf{R}$.

We summarize the results for cylinders and cones as follows.

Proposition 10.16. *A cylinder on a plane curve is Bonnet if and only if the curve is a spiral $\boldsymbol{\gamma}_n(x)$ of curvature n/x on $x > 0$, for some real constant $n > 0$. A cone on a spherical curve is Bonnet if and only if the curve is the spherical spiral $\boldsymbol{\sigma}_n$ of curvature $n \sec s$, on $-\pi/2 < s < \pi/2$, for some real constant $n > 0$. The cylinder on $\boldsymbol{\gamma}_n$ is proper Bonnet and the cone on $\boldsymbol{\sigma}_n$ is a mate, and vice versa. The one parameter family of mates of the cylinder or the cone are pairwise noncongruent, but are all equivalent.*

The cylinder in Figure 10.2 and the cone in Figure 10.3 are thus Bonnet mates. Figure 10.4 shows the Bonnet mates comprising the cone through $\boldsymbol{\sigma}_{2.5}$ and the cylinder through $\boldsymbol{\gamma}_{2.5}$.

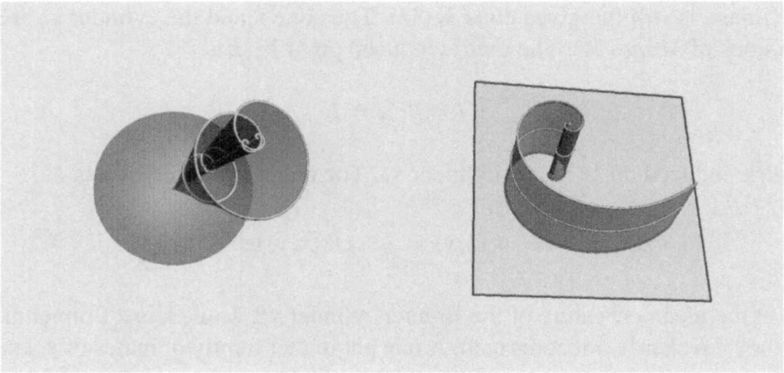

Fig. 10.4 Bonnet mates: Cone through $\sigma_{2.5}$ and cylinder through $\gamma_{2.5}$.

10.4 E(3)-deformations

Proposition 10.17. *Immersions* $\mathbf{x}, \tilde{\mathbf{x}} : M \to \mathbf{R}^3$, *with unit normals* $\mathbf{e}_3$ *and* $\tilde{\mathbf{e}}_3$, *respectively, induce the same orientation and metric on* M *if and only if there exists a smooth map* $A : M \to \mathbf{SO}(3)$ *such that at each point* $m \in M$,

$$A(m)d\mathbf{x}_{(m)} = d\tilde{\mathbf{x}}_{(m)} : T_mM \to \mathbf{R}^3. \tag{10.17}$$

Proof. If A exists so that (10.17) holds, then $\mathbf{x}$ and $\tilde{\mathbf{x}}$ induce the same metric because $A(m) \in \mathbf{SO}(3)$ implies

$$\tilde{I} = d\tilde{\mathbf{x}}_{(m)} \cdot d\tilde{\mathbf{x}}_{(m)} = A(m)d\mathbf{x}_{(m)} \cdot A(m)d\mathbf{x}_{(m)} = d\mathbf{x}_{(m)} \cdot d\mathbf{x}_{(m)} = I.$$

M is oriented by the unit normal vector field $\mathbf{e}_3$ along $\mathbf{x}$. Namely, an orthonormal basis X_1, X_2 of T_mM for the metric induced by $\mathbf{x}$ is positively oriented means that the orthonormal set

$$d\mathbf{x}X_1,\ d\mathbf{x}X_2,\ \mathbf{e}_3(m),$$

is a positively oriented basis of $\mathbf{R}^3$. Then $A(m) \in \mathbf{SO}(3)$ implies that the set

$$d\tilde{\mathbf{x}}X_1 = A(m)d\mathbf{x}X_1,\ d\tilde{\mathbf{x}}X_2 = A(m)d\mathbf{x}X_2,\ A(m)\mathbf{e}_3(m)$$

is again a positively oriented basis of $\mathbf{R}^3$, and therefore

$$\tilde{\mathbf{e}}_3(m) = A(m)\mathbf{e}_3(m).$$

Conversely, suppose $\mathbf{x}$ and $\tilde{\mathbf{x}}$ induce the same metric on M and that the unit normal vector fields $\mathbf{e}_3$ and $\tilde{\mathbf{e}}_3$ along $\mathbf{x}$ and $\tilde{\mathbf{x}}$, respectively, induce the same orientation on M. At a point $m \in M$, if X_1, X_2 is a positively oriented orthonormal basis of T_mM, for the common metric and orientation, then both

$$e = (\mathbf{e}_1 = d\mathbf{x}X_1, \mathbf{e}_2 = d\mathbf{x}X_2, \mathbf{e}_3), \quad \tilde{e} = (\tilde{\mathbf{e}}_1 = d\tilde{\mathbf{x}}X_1, \tilde{\mathbf{e}}_2 = d\tilde{\mathbf{x}}X_2, \tilde{\mathbf{e}}_3)$$

must be positively oriented orthonormal bases of $\mathbf{R}^3$, hence elements of $\mathbf{SO}(3)$. Therefore, the matrix

$$A(m) = \tilde{e}e^{-1} \in \mathbf{SO}(3)$$

and $d\tilde{\mathbf{x}}_{(m)} = A(m)d\mathbf{x}_{(m)}$, since for $i = 1,2$,

$$d\tilde{\mathbf{x}}_{(m)}X_i = \tilde{\mathbf{e}}_i = A(m)\mathbf{e}_i = A(m)d\mathbf{x}_{(m)}X_i.$$

The map $A : M \to \mathbf{SO}(3)$ depends smoothly on m because the vector fields X_1, X_2 can be chosen to be smooth vector fields on a neighborhood U of m, in which case the maps $e, \tilde{e} : U \to \mathbf{SO}(3)$ are smooth and $A = \tilde{e}e^{-1}$ is smooth on U. □

We shall use the following terminology taken from [94].

Definition 10.18. Smooth maps of a surface

$$\mathbf{x}, \tilde{\mathbf{x}} : M \to \mathbf{R}^3$$

agree to first order at a point $m \in M$ if

$$\mathbf{x}(m) = \tilde{\mathbf{x}}(m), \text{ and } d\mathbf{x}_{(m)} = d\tilde{\mathbf{x}}_{(m)} : T_mM \to \mathbf{R}^3.$$

The maps $\mathbf{x}$ and $\tilde{\mathbf{x}}$ are *first order* $\mathbf{E}(3)$*-deformations of each other* if there exists a smooth map

$$(\mathbf{y}, A) : M \to \mathbf{E}(3)$$

such that for each point $m \in M$, the maps $(\mathbf{y}(m), A(m)) \circ \mathbf{x}$ and $\tilde{\mathbf{x}}$ agree to first order at m. The deformation is *trivial* if the map $(\mathbf{y}, A)$ is constant, which means that $\mathbf{x}$ and $\tilde{\mathbf{x}}$ are *congruent*.

Corollary 10.19. *The immersions $\mathbf{x}$ and $\tilde{\mathbf{x}}$ induce the same orientation and metric if and only if they are first order* $\mathbf{E}(3)$*-deformations of each other.*

Proof. Use the map $A : M \to \mathbf{SO}(3)$ of the Proposition, and let $\mathbf{y}(m) = \tilde{\mathbf{x}}(m) - \mathbf{x}(m)$, for any $m \in M$. Then $(\mathbf{y}, A)$ is the desired map. □

10.4.1 The deformation form

Definition 10.20. For an immersion $\mathbf{x} : M \to \mathbf{R}^3$ and a smooth map $A : M \to \mathbf{SO}(3)$, the *deformation form* of A relative to a first order frame field $(\mathbf{x}, e) : U \to \mathbf{E}(3)$ along $\mathbf{x}$ is the $\mathfrak{o}(3)$-valued 1-form

$$\psi = e^{-1}A^{-1}dA\, e.$$

That is, ψ is $A^{-1}dA$ expressed in the frame e at each point.

Suppose that

$$\mathbf{x},\ \tilde{\mathbf{x}} : M \to \mathbf{R}^3$$

are immersions that are first order $\mathbf{E}(3)$-deformations of each other. Then there exists a smooth map

$$A : M \to \mathbf{SO}(3)$$

such that

$$d\tilde{\mathbf{x}} = A d\mathbf{x} \tag{10.18}$$

at every point of M. Therefore,

$$0 = dd\tilde{\mathbf{x}} = dA \wedge d\mathbf{x}$$

on M. If $(\mathbf{x}, e)$ is a first order frame field along $\mathbf{x}$ on $U \subset M$, then $(\tilde{x}, Ae)$ is a first order frame field along $\tilde{\mathbf{x}}$, since

$$\tilde{\theta} = (Ae)^{-1} d\tilde{\mathbf{x}} = e^{-1}A^{-1}Ad\mathbf{x} = e^{-1}d\mathbf{x} = \theta,$$

which shows that $\tilde{\omega}^3 = \omega^3 = 0$. The deformation form of the map A relative to $(\mathbf{x}, e)$ is

$$\psi = e^{-1}A^{-1}dA\, e = (Ae)^{-1}d(Ae) - e^{-1}de = \tilde{\omega} - \omega.$$

In addition, if $(\mathbf{x}, e)$ is adapted to a complex coordinate z, then

$$\tilde{\omega}^1 + i\tilde{\omega}^2 = \omega^1 + i\omega^2 = e^u dz$$

shows that $(\tilde{\mathbf{x}}, \tilde{e})$ also is adapted to z and has the same conformal factor u.

Proposition 10.21 (Deformation criteria). *Consider an immersion $\mathbf{x} : M \to \mathbf{R}^3$ and a smooth map $A : M \to \mathbf{SO}(3)$. If there exists an immersion $\tilde{\mathbf{x}} : M \to \mathbf{R}^3$ such that* (10.18) *holds, then the components of the deformation form ψ of A relative to any first order frame field $(\mathbf{x}, e)$ on $U \subset M$ satisfy the conditions*

$$\begin{aligned} &(i).\ \psi^1_2 = 0 \\ &(ii).\ \psi^3_1 \wedge \omega^1 + \psi^3_2 \wedge \omega^2 = 0. \end{aligned} \tag{10.19}$$

Conversely, if M is simply connected, and if (10.19) *holds for any first order frame $(\mathbf{x}, e)$, then there exists an immersion $\tilde{\mathbf{x}} : M \to \mathbf{R}^3$ such that* (10.18) *holds.*

Let ω^1, ω^2 be the coframe field induced on M by $(\mathbf{x}, e)$, and let $\varphi = \omega^1 + i\omega^2$. Then $\mathbf{x}$ and $\tilde{\mathbf{x}}$ have the same mean curvature if and only if $\psi_1^3 - i\psi_2^3$ is of bidegree $(1,0)$; that is

$$\psi_1^3 - i\psi_2^3 = f\varphi,$$

for some smooth function $f : U \to \mathbf{C}$.

Proof. If there exists an immersion $\tilde{\mathbf{x}} : M \to \mathbf{R}^3$ satisfying (10.18), then

$$0 = d(Ad\mathbf{x}) = dA \wedge d\mathbf{x} = dA\, e \wedge e^{-1}d\mathbf{x}.$$

Multiplying this by $e^{-1}A^{-1}$, we have

$$0 = e^{-1}A^{-1}dAe \wedge e^{-1}d\mathbf{x} = \psi \wedge \theta, \tag{10.20}$$

which is equivalent to (10.19). Conversely, if (10.20) holds for any first order frame field $(\mathbf{x}, e)$, then

$$d(Ad\mathbf{x}) = dAe \wedge e^{-1}d\mathbf{x} = (Ae)\psi \wedge \theta = 0,$$

on U. Therefore, $Ad\mathbf{x}$ is a closed 1-form on M, so if M is simply connected, there exists a smooth map $\tilde{\mathbf{x}} : M \to \mathbf{R}^3$ such that $d\tilde{\mathbf{x}} = Ad\mathbf{x}$, and thus $\tilde{\mathbf{x}}$ is actually an immersion.

If $\tilde{\mathbf{x}}$ satisfies (10.18), then $(\tilde{\mathbf{x}}, Ae)$ is a first order frame field along $\tilde{\mathbf{x}}$ whose dual coframe field is also ω^1, ω^2. The deformation form $\psi = \tilde{\omega} - \omega$ satisfies

$$\psi_1^3 - i\psi_2^3 = \tilde{\omega}_1^3 - i\tilde{\omega}_2^3 - (\omega_1^3 - i\omega_2^3) = (\tilde{h} - h)\varphi + (\tilde{H} - H)\bar{\varphi} \tag{10.21}$$

by (7.23). Hence, $\psi_1^3 - i\psi_2^3$ is of bidegree $(1,0)$ if and only if $\tilde{H} = H$. □

Corollary 10.22. *If immersions $\mathbf{x}, \tilde{\mathbf{x}} : M \to \mathbf{R}^3$ form a Bonnet pair, then their deformation form ψ relative to any first order frame field $(\mathbf{x}, e)$ on $U \subset M$ has isolated zeros in U.*

Proof. By Proposition 10.17, there exists a smooth map $A : M \to \mathbf{SO}(3)$ such that (10.18) holds. Let $(\mathbf{x}, e)$ be a first order frame field on $U \subset M$. Let ω^1, ω^2 be its dual coframe field on U. Then $(\tilde{\mathbf{x}}, \tilde{e}) = (\tilde{\mathbf{x}}, Ae)$ is first order on U with the same dual coframe field. Let $\varphi = \omega^1 + i\omega^2$. The deformation form relative to $(\mathbf{x}, e)$ is

$$\psi = \tilde{\omega} - \omega = \tilde{e}^{-1}d\tilde{e} - e^{-1}de.$$

By (10.21) and the fact that $\tilde{H} = H$,

$$(\psi_1^3 - i\psi_2^3)\varphi = (\tilde{h} - h)\varphi\varphi = \mathcal{Q},$$

the deformation quadratic differential (10.1), which is holomorphic and nonzero on M so has only isolated zeros in U. Hence, ψ can have only isolated zeros in U. □

10.4.2 Specifying the deformation form

For an immersion $\mathbf{x}: M \to \mathbf{R}^3$ with a first order frame field $(\mathbf{x}, e)$, what freedom do we have to specify an $\mathfrak{o}(3)$-valued 1-form on M such that it is the deformation form of a map $A: M \to \mathbf{SO}(3)$ relative to $(\mathbf{x}, e)$?

Proposition 10.23. *Let $\mathbf{x}: M \to \mathbf{R}^3$ be an immersion of a simply connected surface M with a globally defined first order frame field $(\mathbf{x}, e)$. An $\mathfrak{o}(3)$-valued 1-form $\psi = (\psi^i_j)$ on M is the deformation form of a smooth map $A: M \to \mathbf{SO}(3)$ relative to $(\mathbf{x}, e)$ if and only if*

$$\begin{aligned} \psi^3_1 \wedge \omega^3_2 - \psi^3_2 \wedge \omega^3_1 + \psi^3_1 \wedge \psi^3_2 &= 0, \\ d\psi^3_1 + \psi^3_2 \wedge \omega^2_1 &= 0, \\ d\psi^3_2 + \psi^3_1 \wedge \omega^1_2 &= 0. \end{aligned} \tag{10.22}$$

The last two equations are equivalent to

$$d(\psi^3_1 - i\psi^3_2) = -i(\psi^3_1 - i\psi^3_2) \wedge \omega^2_1. \tag{10.23}$$

Proof. If a smooth map $A: M \to \mathbf{SO}(3)$ were to exist such that $\psi = e^{-1}A^{-1}dA\,e$, then

$$A^{-1}dA = e\psi e^{-1}.$$

By the Cartan–Darboux Theorem, such a map A exists if and only if

$$d(e\psi e^{-1}) = -(e\psi e^{-1}) \wedge (e\psi e^{-1}) = -e\psi \wedge \psi e^{-1}. \tag{10.24}$$

Now

$$d(e\psi e^{-1}) = e(d\psi + \omega \wedge \psi + \psi \wedge \omega)e^{-1},$$

so (10.24) is equivalent to

$$d\psi + \psi \wedge \psi + \omega \wedge \psi + \psi \wedge \omega = 0. \tag{10.25}$$

Writing out the components, we find (10.25) is equivalent to (10.22). □

As an application of the idea of deformation forms, we shall find some more examples of Bonnet pairs using a construction of Kamberov, Pedit, and Pinkall [97], which we call the KPP construction. For this we use the quaternions.

10.5 Quaternions

If $\tilde{\mathbf{E}}(3)$ denotes the universal covering group of $\mathbf{E}(3)$, then we shall see that the covering projection $\pi : \tilde{\mathbf{E}}(3) \to \mathbf{E}(3)$ is two-to-one. This property allows us to make a pair of similarity deformations of an isothermic immersion, such that the pair of immersions are Bonnet deformations of each other. We use the quaternions to construct this universal cover.

The normed division algebra of quaternions, $\mathbf{H}$, is a right $\mathbf{C}$-module

$$\mathbf{H} = \mathbf{C} + j\mathbf{C}$$

where $1, j$ is a basis over $\mathbf{C}$, with 1 the identity element and j satisfying

$$j^2 = -1, \quad zj = j\bar{z}, \quad \overline{z + jw} = \bar{z} - jw.$$

for any $z, w \in \mathbf{C}$, where bar denotes quaternion and complex conjugation. Any quaternions $p, q \in \mathbf{H}$ satisfy

$$\overline{pq} = \bar{q}\bar{p}, \quad |p| = \sqrt{p\bar{p}}, \quad p^{-1} = \frac{\bar{p}}{|p|^2}, p \neq 0.$$

If $i = \sqrt{-1} \in \mathbf{C}$, then $1, i, j, ij$ is an orthonormal basis of $\mathbf{H}$ over $\mathbf{R}$ satisfying

$$i^2 = j^2 = (ij)^2 = -1, \quad j(ij) = i, \quad (ij)i = j,$$

which gives the isomorphism over $\mathbf{R}$,

$$\mathbf{H} \cong \mathbf{R}^4, \quad x^0 + x^3 i + (x^1 + ix^2)j \leftrightarrow (x^0, x^1, x^2, x^3).$$

The quaternion conjugation determines a direct sum decomposition

$$\mathbf{H} = \Re\mathbf{H} \oplus \Im\mathbf{H}$$

where

$$\Re\mathbf{H} = \{q \in \mathbf{H} : \bar{q} = q\} = \{x^0 1 : x^0 \in \mathbf{R}\}$$

is the set of *real quaternions*, and

$$\Im\mathbf{H} = \{q \in \mathbf{H} : \bar{q} = -q\} = \{x^3 i + (x^1 + ix^2)j : x^1, x^2, x^3 \in \mathbf{R}\} \tag{10.26}$$

is the set of *imaginary quaternions*. The imaginary quaternions are thus identified with the hyperplane $x^0 = 0$, which we identify with $\mathbf{R}^3$. The standard basis of $\mathbf{R}^3$ corresponds to the following basis of $\Im\mathbf{H}$,

$$\epsilon_1 \leftrightarrow j, \quad \epsilon_2 \leftrightarrow ij, \quad \epsilon_3 \leftrightarrow i.$$

Exercise 39. Prove that if $\mathbf{x},\mathbf{y} \in \Im\mathbf{H}$ are imaginary quaternions, thus vectors in $\mathbf{R}^3$, then their product as quaternions is given in terms of the dot product and vector cross product of $\mathbf{R}^3$ by

$$\mathbf{x}\mathbf{y} = -\mathbf{x}\cdot\mathbf{y} + \mathbf{x}\times\mathbf{y}. \tag{10.27}$$

Exercise 40. Prove that the set of all unit quaternions,

$$S^3 = \{q \in \mathbf{H} : |q| = 1\}$$

under quaternion multiplication is a Lie group whose Lie algebra, identified with its tangent space at the identity element 1, is $\Im\mathbf{H}$. This group is called the *spin group*, often denoted $\mathbf{Spin}(3)$, but we shall denote it by $\mathbf{S}^3$.

The multiplicative group of nonzero quaternions, denoted

$$\mathbf{H}^\times = \mathbf{H}\setminus\{0\},$$

acts on $\mathbf{R}^3$ via the homomorphism

$$\Sigma : \mathbf{H}^\times \to \mathrm{Hom}(\mathbf{R}^3), \quad \Sigma(q)\mathbf{x} = q\mathbf{x}\bar{q}.$$

For $q \in \mathbf{H}^\times$ and $\mathbf{x} \in \Im\mathbf{H} = \mathbf{R}^3$, we use (10.27) to find

$$\Sigma(q)\mathbf{x}\cdot\Sigma(q)\mathbf{x} = -q\mathbf{x}\bar{q}q\mathbf{x}\bar{q} = -|q|^2 q\mathbf{x}\mathbf{x}\bar{q} = |q|^4\mathbf{x}\cdot\mathbf{x},$$

which shows that $\Sigma(q)$ is a *similarity transformation* of $\mathbf{R}^3$ with conformal factor $|q|^2$. In particular, if $|q| = 1$, then $\Sigma(q)$ is an orthogonal transformation.

For each $q \in \mathbf{H}^\times$, we let $A(q)$ denote the matrix of $\Sigma(q)$ relative to the standard basis of $\mathbf{R}^3 \cong \Im\mathbf{H}$. Thus, if A_i denotes column i of the matrix A, then

$$qj\bar{q} = \sum_1^3 A(q)_1^k \epsilon_k, \quad q(-ji)\bar{q} = \sum_1^3 A(q)_2^k \epsilon_k, \quad qi\bar{q} = \sum_1^3 A(q)_3^k \epsilon_k.$$

This defines a group homomorphism

$$A : \mathbf{H}^\times \to \mathbf{GL}(3,\mathbf{R}), \quad q \mapsto A(q).$$

Exercise 41. Verify the statements of the preceding paragraph as follows. For any $q \in \mathbf{H}^\times$, prove:

1. $\Sigma(q)\mathbf{x} \in \Im\mathbf{H}$, for every $\mathbf{x} \in \Im\mathbf{H}$.
2. $\Sigma(q) : \Im\mathbf{H} \to \Im\mathbf{H}$ is a linear transformation.
3. $\Sigma(q)^{-1} = \Sigma(q^{-1})$.
4. $\Sigma(pq) = \Sigma(p)\Sigma(q)$, for every $p,q \in \mathbf{H}^\times$.

5. $\Sigma(q) = I$, the identity transformation, if and only if $q \in \mathbf{R}^\times$.
6. $\Sigma(q)\mathbf{x} \cdot \Sigma(q)\mathbf{y} = |q|^4 \mathbf{x} \cdot \mathbf{y}$, for every $\mathbf{x}, \mathbf{y} \in \Im\mathbf{H}$.
7. If $T \in \mathbf{SO}(3)$ is rotation about the oriented line determined by the unit vector $l \in \mathbf{R}^3 = \Im\mathbf{H}$ through the angle θ, then $T = A(q)$, where q is the unit length quaternion

$$q = \cos(\frac{\theta}{2}) + \sin(\frac{\theta}{2})l.$$

8. The image of A is the *similarity group*,

$$A(\mathbf{H}^\times) = \mathbf{CSO}(3) = \{B \in \mathbf{GL}(3,\mathbf{R}) : {}^tBB = tI, t > 0, \det B > 0\}.$$

9. The restriction of A to $\mathbf{S^3}$ defines a 2:1 covering projection

$$A : \mathbf{S^3} \to \mathbf{SO}(3)$$

whose derivative at the identity is the Lie algebra isomorphism

$$A_* : \Im\mathbf{H} \cong \mathfrak{o}(3)$$

$$ix^3 + j(x^1 - ix^2) \leftrightarrow \begin{pmatrix} 0 & -2x^3 & 2x^2 \\ 2x^3 & 0 & -2x^1 \\ -2x^2 & 2x^1 & 0 \end{pmatrix} \tag{10.28}$$

It follows that the universal covering group of $\mathbf{E}(3)$ is

$$\tilde{\mathbf{E}}(3) = \Im\mathbf{H} \times \mathbf{S^3} \to \mathbf{R}^3 \times \mathbf{SO}(3) = \mathbf{E}(3)$$
$$(\mathbf{x}, p) \mapsto (\mathbf{x}, A(p)).$$

The Lie algebras are isomorphic,

$$\Im\mathbf{H} \times \Im\mathbf{H} = \tilde{\mathscr{E}}(3) \cong \mathscr{E}(3) = \mathbf{R}^3 \times \mathfrak{o}(3)$$
$$(\mathbf{x}, \mathbf{y}) \leftrightarrow (\mathbf{x}, A_*\mathbf{y}).$$

10.5.1 Spin frames

Definition 10.24. A *spin frame* field along an immersed surface $\mathbf{x} : M \to \mathbf{R}^3$ is a smooth map defined on an open subset $U \subset M$,

$$(\mathbf{x}, p) : U \to \tilde{\mathbf{E}}(3).$$

The pull-back of the Maurer–Cartan form of $\mathbf{E}(3)$ by the projected frame field $(\mathbf{x},A(p))$ is

$$(\mathbf{x},A(p))^{-1}d(\mathbf{x},A(p)) = (\theta,\omega)$$

where

$$\theta = \begin{pmatrix}\omega^1\\ \omega^2\\ \omega^3\end{pmatrix}, \quad \omega = \begin{pmatrix}0 & \omega^1_2 & \omega^1_3\\ \omega^2_1 & 0 & \omega^2_3\\ \omega^3_1 & \omega^3_2 & 0\end{pmatrix}.$$

By (10.28) the spin frame field pulls back the Maurer–Cartan form of $\tilde{\mathbf{E}}(3)$ to the $\Im\mathbf{H}+\Im\mathbf{H}$-valued 1-form

$$(\mathbf{x},p)^{-1}d(\mathbf{x},p) = (p^{-1}d\mathbf{x}p, p^{-1}dp)$$

where

$$\begin{aligned} p^{-1}d\mathbf{x}p &= i\omega^3 + j(\omega^1 - i\omega^2)\\ p^{-1}dp &= \frac{1}{2}(i\omega^2_1 + ji(\omega^3_1 - i\omega^3_2)). \end{aligned} \tag{10.29}$$

The spin frame field is *first order* if its projection is a first order Euclidean frame, that is,

$$\omega^3 = 0.$$

It is *adapted* to a complex coordinate chart (U,z) if and only if the Euclidean projection is adapted, that is,

$$p^{-1}d\mathbf{x}p = je^u d\bar{z},$$

for some smooth function $u: U \to \mathbf{R}$.

Exercise 42. Prove that if $(\mathbf{x},p): U \to \tilde{\mathbf{E}}(3)$ is a first order spin frame field along $\mathbf{x}$, then any other on U is given by $(\mathbf{x},pa)$, where $a: U \to \mathbf{C}$ is any smooth map for which $|a| = 1$ at every point of U. If $(\tilde{\theta},\tilde{\omega}) = (\mathbf{x},pa)^{-1}d(\mathbf{x},pa)$, then

$$\tilde{\theta} = \theta a^2, \quad \tilde{\omega} = a^{-1}\omega a + a^{-1}da. \tag{10.30}$$

Writing $\theta = j(\omega^1 - i\omega^2)$ and $\tilde{\theta} = j(\tilde{\omega}^1 - i\tilde{\omega}^2)$, show that the first equation in (10.30) is equivalent to

$$\tilde{\omega}^1 - i\tilde{\omega}^2 = a^2(\omega^1 - i\omega^2).$$

10.5.2 Similarity deformations

Definition 10.25. A *similarity deformation* of an immersion $\mathbf{x} : M \to \mathscr{I}\mathbf{H}$ is an immersion $\tilde{\mathbf{x}} : M \to \Im\mathbf{H}$ for which there is a smooth map $q : M \to \mathbf{H}^\times$ such that

$$d\tilde{\mathbf{x}} = q\, d\mathbf{x}\bar{q} = |q|^2 A(\frac{q}{|q|})d\mathbf{x}$$

at every point of M. Thus, the induced metrics satisfy

$$d\tilde{\mathbf{x}} \cdot d\tilde{\mathbf{x}} = |q|^4 d\mathbf{x} \cdot d\mathbf{x}.$$

If $(\mathbf{x}, p) : U \to \Im\mathbf{H} \times \mathbf{S}^3$ is a spin frame field along $\mathbf{x}$, then

$$(\tilde{\mathbf{x}}, \frac{q}{|q|}p) : U \to \mathscr{I}\mathbf{H} \times \mathbf{S}^3$$

is a spin frame field along the similarity deformation $\tilde{\mathbf{x}}$, and

$$\begin{aligned} i\tilde{\omega}^3 + j(\tilde{\omega}^1 - i\tilde{\omega}^2) &= \left(\frac{q}{|q|}p\right)^{-1} d\tilde{\mathbf{x}}\frac{q}{|q|}p = |q|p^{-1}q^{-1}q\, d\mathbf{x}\bar{q}\frac{q}{|q|}p \\ &= |q|^2 p^{-1} d\mathbf{x} p = |q|^2(i\omega^3 + j(\omega^1 - i\omega^2)) \end{aligned}$$

shows that $\tilde{\omega}^3 = 0$ if and only if $\omega^3 = 0$; that is, the spin frame $(\tilde{\mathbf{x}}, \frac{q}{|q|}p)$ is first order if and only if the spin frame $(\mathbf{x}, p)$ is first order. Assuming that $(\mathbf{x}, p)$ is first order, we set

$$\varphi = \omega^1 + i\omega^2, \quad \tilde{\varphi} = \tilde{\omega}^1 + i\tilde{\omega}^2,$$

so that

$$\tilde{\varphi} = |q|^2 \varphi.$$

In addition, if $(\mathbf{x}, p)$ is adapted to a complex coordinate chart U, z, then

$$\tilde{\varphi} = |q|^2 \varphi = |q|^2 e^u dz$$

implies that $(\tilde{\mathbf{x}}, \frac{q}{|q|}p)$ is adapted as well, and its conformal factor $e^{\tilde{u}}$ is related to that of $(\mathbf{x}, p)$ by

$$e^{\tilde{u}} = |q|^2 e^u. \tag{10.31}$$

Remark 10.26. If $\mathbf{x}, \tilde{\mathbf{x}} : M \to \mathbf{R}^3$ are immersions with the same induced metrics, $|d\mathbf{x}| = |d\tilde{\mathbf{x}}|$, and the same orientation, then by Proposition 10.17 there exists a smooth map $A : M \to \mathbf{SO}(3)$ such that $d\tilde{\mathbf{x}} = Ad\mathbf{x}$. If M is simply connected, then there exists a smooth map

$$q : M \to \mathbf{S}^3 \subset \mathbf{H},$$

such that the projection $A(q) = A$ at every point of M. That is, the map q is a *lift* of A. Moreover,

$$d\tilde{\mathbf{x}} = q\,d\mathbf{x}\,\bar{q}$$

at every point of M.

Definition 10.27. The *deformation form* of a map $q : M \to \mathbf{H}^\times$ relative to a first order spin frame field $(\mathbf{x}, p)$ along an immersion $\mathbf{x}$ is the $\mathbf{H}$-valued 1-form

$$\Psi = p^{-1}q^{-1}dqp = \Psi^1 + j\Psi^2,$$

where Ψ^1 and Ψ^2 are $\mathbf{C}$-valued 1-forms.

Continuing with our similarity deformation $\tilde{\mathbf{x}}$ of $\mathbf{x}$ above, and first order spin frame fields $(\mathbf{x}, p)$ and $(\tilde{\mathbf{x}}, \frac{q}{|q|}p)$, we have

$$\begin{aligned}
\frac{1}{2}(i\tilde{\omega}_1^2 + j(\tilde{\omega}_2^3 + i\tilde{\omega}_1^3)) &= \left(\frac{q}{|q|}p\right)^{-1} d\left(\frac{q}{|q|}p\right) \\
&= -\frac{1}{|q|}d|q| + p^{-1}q^{-1}dqp + p^{-1}dp \\
&= -\frac{1}{|q|}d|q| + \Psi + \frac{1}{2}(i\omega_1^2 + j(\omega_2^3 + i\omega_1^3))
\end{aligned}$$

so the deformation form $\Psi = \Psi^1 + j\Psi^2$ of $q : M \to \mathbf{H}^\times$ relative to $(\mathbf{x}, p)$ is given by

$$\begin{aligned}
\Psi^1 &= \frac{1}{|q|}d|q| + \frac{i}{2}(\tilde{\omega}_1^2 - \omega_1^2) \\
\Psi^2 &= \frac{i}{2}(\tilde{\omega}_1^3 - i\tilde{\omega}_2^3 - (\omega_1^3 - i\omega_2^3))
\end{aligned}$$

Equating the right sides of

$$\begin{aligned}
d\tilde{\varphi} &= -i\tilde{\omega}_1^2 \wedge \tilde{\varphi} = -i|q|^2\tilde{\omega}_1^2 \wedge \varphi, \\
d\tilde{\varphi} &= 2|q|d|q| \wedge \varphi - i|q|^2\omega_1^2 \wedge \varphi,
\end{aligned}$$

we get

$$\left(\frac{2}{|q|}d|q| + i(\tilde{\omega}_1^2 - \omega_1^2)\right) \wedge \varphi = 0,$$

which is

$$\Psi^1 \wedge \varphi = 0.$$

Recalling that for any first order frame field $\omega_1^3 - i\omega_2^3 = h\varphi + H\bar{\varphi}$ (see (7.23)), we get

$$\begin{aligned}\Psi^2 &= \frac{i}{2}\left(\tilde{h}\tilde{\varphi} + \tilde{H}\bar{\tilde{\varphi}} - h\varphi - H\bar{\varphi}\right)\\ &= \frac{i}{2}(|q|^2\tilde{h} - h)\varphi + \frac{i}{2}(|q|^2\tilde{H} - H)\bar{\varphi},\end{aligned} \tag{10.32}$$

which shows that

$$\Psi^2 \wedge \varphi = \frac{i}{2}(|q|^2\tilde{H} - H)\bar{\varphi} \wedge \varphi$$

is real. We look next for a sufficient condition on a smooth map $q : M \to \mathbf{H}^\times$ and an immersion $\mathbf{x} : M \to \mathbf{R}^3$ so that q defines a similarity deformation $\tilde{\mathbf{x}} : M \to \Im\mathbf{H}$. Namely, when is $q\,d\mathbf{x}\,\bar{q}$ a closed $\Im\mathbf{H}$-valued 1-form?

Proposition 10.28. *If $q : M \to \mathbf{H}^\times$ is a smooth map and $\mathbf{x} : M \to \mathbf{R}^3$ is a smooth immersion, then*

$$d(q\,d\mathbf{x}\,\bar{q}) = 0$$

on M if and only if the deformation form of q relative to any first order spin frame field $(\mathbf{x}, p) : U \to \tilde{E}(3)$ satisfies the two conditions

(i) $\Psi^1 \wedge \varphi = 0,$

(ii) $\Psi^2 \wedge \varphi$ *is real.*

Proof. For a first order spin frame, we have $d\mathbf{x} = pj(\omega^1 - i\omega^2)p^{-1}$ and $dq = qp\Psi p^{-1}$, so the form $q\,d\mathbf{x}\,\bar{q}$ is closed on U if and only if

$$\begin{aligned}0 &= dq \wedge d\mathbf{x}\,\bar{q} - q\,d\mathbf{x} \wedge d\bar{q}\\ &= qp\left(\Psi \wedge j\bar{\varphi} - j\bar{\varphi} \wedge \bar{\Psi}\right)p^{-1}\bar{q}\\ &= qp\left((\Psi^1 + j\Psi^2) \wedge j\bar{\varphi} - j\bar{\varphi} \wedge (\bar{\Psi}^1 - j\Psi^2)\right)p^{-1}\bar{q}\\ &= qp\left(-\bar{\Psi}^2 \wedge \bar{\varphi} + \Psi^2 \wedge \varphi + j(2\bar{\Psi}^1 \wedge \bar{\varphi})\right)p^{-1}\bar{q},\end{aligned}$$

from which the result follows. □

Corollary 10.29. *Moreover, if M is simply connected, and if the deformation form Ψ of a map $q : M \to \mathbf{H}^\times$ relative to a first order spin frame satisfies conditions (i) and (ii), then there exists a similarity deformation $\tilde{\mathbf{x}} : M \to \mathbf{R}^3$ such that*

$$d\tilde{\mathbf{x}} = q\,d\mathbf{x}\,\bar{q}$$

on M. The mean curvatures H and $\tilde{H}$ of $\mathbf{x}$ and $\tilde{\mathbf{x}}$, respectively, are related by

$$\frac{i}{2}(|q|^2\tilde{H} - H)\varphi \wedge \bar{\varphi} = \varphi \wedge \Psi^2.$$

Therefore,

$$\tilde{H} = \frac{1}{|q|^2} H$$

if and only if condition (ii) is replaced with the condition

(iii) $$\Psi^2 \wedge \varphi = 0 \text{ on } U.$$

If (iii) holds, then the Hopf invariants $\tilde{h}$ *and* h *relative to the first order spin frames* $(\tilde{\mathbf{x}}, \frac{q}{|q|}p)$ *and* $(\mathbf{x}, p)$*, respectively, satisfy*

$$\Psi^2 = \frac{i}{2}(|q|^2 \tilde{h} - h)\varphi. \tag{10.33}$$

Proof. Conditions (i) and (ii) imply that $d(q\,d\mathbf{x}\,\bar{q}) = 0$ on the simply connected M, so there exists a smooth map $\tilde{\mathbf{x}} : M \to \mathbf{R}^3$ such that $d\tilde{\mathbf{x}} = q\,d\mathbf{x}\,\bar{q}$. Hence, $d\tilde{\mathbf{x}}$ is non-singular and $\tilde{\mathbf{x}}$ is an immersion. The rest of the corollary follows from (10.32). □

10.6 The KPP construction

Recall from Definition 9.22 that a Christoffel transform of an immersion $\mathbf{x} : M \to \mathbf{R}^3$ is an immersion $\hat{\mathbf{x}} : M \to \mathbf{R}^3$ whose induced metric is conformally related to that of $\mathbf{x}$, whose tangent plane at each point is parallel to the tangent plane to $\mathbf{x}$ at that point, and whose orientation induced by a unit normal vector field along it is opposite of that induced by the same unit normal vector field along $\mathbf{x}$.

If $(\mathbf{x}, e)$ is an oriented first order frame field along $\mathbf{x}$ and $d\mathbf{x} = \omega^1 \mathbf{e}_1 + \omega^2 \mathbf{e}_2$, then parallel tangent planes means that

$$d\hat{\mathbf{x}} = \tau^1 \mathbf{e}_1 + \tau^2 \mathbf{e}_2,$$

for real-valued 1-forms τ^1 and τ^2. If $\varphi = \omega^1 + i\omega^2$ and $\tau = \tau^1 + i\tau^2$, then conformality and opposite orientation are equivalent to

$$\bar{\tau} = f\varphi \tag{10.34}$$

for some nowhere zero complex valued function f. By Theorem 9.24, the existence of $\hat{\mathbf{x}}$ implies that $\mathbf{x}$ is isothermic. If (U, z) is a principal complex chart for the isothermic $\mathbf{x}$, then $\varphi = e^u dz$, where $u : U \to \mathbf{R}$ is smooth, and $f = re^{-2u}$ in (10.34) for some nonzero real constant r. (See Remark 9.25).

Relative to any first order spin frame field $(\mathbf{x}, p)$ along $\mathbf{x}$, we have $p^{-1}d\mathbf{x}p \wedge \bar{\varphi} = 0$ on U, by (10.29), so

$$(p^{-1}d\hat{\mathbf{x}}p) \wedge \varphi = 0. \tag{10.35}$$

Remark 10.30. A *Christoffel transform* $\hat{\mathbf{x}}$ of $\mathbf{x}$ is defined up to transformations

$$r(\mathbf{a}+\hat{\mathbf{x}}),$$

where r is a positive constant and $\mathbf{a} \in \mathbf{R}^3$ is any constant vector.

Proposition 10.31 (KPP Construction). *Suppose $\mathbf{x} : M \to \mathbf{R}^3$ is an isothermic immersion of a simply connected surface M. Let $\hat{\mathbf{x}} : M \to \mathbf{R}^3$ be a Christoffel transform of $\mathbf{x}$. Make the identification $\mathbf{R}^3 = \Im\mathbf{H}$ given in* (10.26)*. For any constant $c \in \mathbf{R}$, the smooth map*

$$q = c + \hat{\mathbf{x}} : M \to \mathbf{H} \tag{10.36}$$

satisfies $d(qd\mathbf{x}\bar{q}) = 0$ on M, so determines an immersion $\tilde{\mathbf{x}} : M \to \mathbf{R}^3$, such that $d\tilde{\mathbf{x}} = q\,d\mathbf{x}\,\bar{q}$. If $c > 0$ and if

$$q_\pm = \pm c + \hat{\mathbf{x}},$$

then $|q_+|^2 = |q_-|^2$, and the immersions

$$\mathbf{x}_\pm : M \to \mathbf{R}^3$$

determined by $q_\pm$ form a Bonnet pair.

Proof. For any map $q = c + \hat{\mathbf{x}}$ defined in (10.36),

$$q^{-1} = \frac{1}{|q|^2}\bar{q} = \frac{c}{|q|^2} - \frac{\hat{\mathbf{x}}}{|q|^2}.$$

Let (U, z) be a principal complex coordinate for the isothermic immersion $\mathbf{x}$. If $(\mathbf{x}, p)$ is the first order spin frame field along $\mathbf{x}$ adapted to z, then

$$p^{-1}dqp = p^{-1}d\hat{\mathbf{x}}p = j(\tau^1 - i\tau^2) = jre^{-u}dz,$$

for some nonzero real constant r, by Remark 9.25. By Definition 10.27, the deformation form $\Psi = \Psi^1 + j\Psi^2$ of q relative to $(\mathbf{x}, p)$ is

$$\begin{aligned}\Psi^1 + j\Psi^2 &= p^{-1}q^{-1}dqp = p^{-1}q^{-1}p\,p^{-1}dq\,p = \frac{c - p^{-1}\hat{\mathbf{x}}p}{|q|^2}jre^{-u}dz \\ &= \frac{c}{|q|^2}jre^{-u}dz - \frac{p^{-1}\hat{\mathbf{x}}p}{|q|^2}jre^{-u}dz.\end{aligned}$$

If we let

$$p^{-1}\hat{\mathbf{x}}p = iy^3 + j(y^1 - iy^2) \in \Im\mathbf{H},$$

then

$$\Psi^1 = \frac{1}{|q|^2}(y^1 + iy^2)re^{-u}dz, \quad \Psi^2 = \frac{1}{|q|^2}(c + iy^3)re^{-u}dz, \tag{10.37}$$

so Ψ satisfies (i) and (iii) of Proposition 10.28 and Corollary 10.29:

$$\Psi^1 \wedge e^u dz = 0, \quad \Psi^2 \wedge e^u dz = 0.$$

By Corollary 10.29, there exists an immersion $\tilde{\mathbf{x}} : M \to \mathbf{R}^3$ such that

$$d\tilde{\mathbf{x}} = q\, d\mathbf{x}\, \bar{q}$$

and $(\tilde{\mathbf{x}}, \frac{q}{|q|}p)$ is a first order spin frame field along it with coframe

$$\tilde{\varphi} = |q|^2 e^u dz,$$

thus showing that this frame also is adapted to z. In addition,

$$d\tilde{\mathbf{x}} \cdot d\tilde{\mathbf{x}} = |q|^4 d\mathbf{x} \cdot d\mathbf{x}, \quad \tilde{H} = \frac{1}{|q|^2}H, \quad |q|^2\tilde{h} = h - \frac{2i}{|q|^2}(c + iy^3)re^{-2u},$$

where the last equation comes from comparing Ψ^2 in (10.37) and (10.33). Applying these observations to the two cases

$$q_\pm = \pm c + \hat{\mathbf{x}},$$

we get

$$|q_+|^2 = |c|^2 + |\hat{\mathbf{x}}|^2 = |q_-|^2,$$

so that the immersions $\mathbf{x}_\pm$ determined by $q_\pm$, respectively, have the same induced metric and mean curvature. They are not congruent, because their Hopf invariants relative to z satisfy

$$h_- = \overline{h_+}, \quad \Im(h_+) = \frac{-2cre^{-2u}}{|q|^4},$$

so they are distinct. □

By Remark 10.30, one may assume $\hat{\mathbf{x}}$ chosen in the proof above in such a way that

$$q_\pm = \pm\frac{1}{2} + \hat{\mathbf{x}}.$$

Then $|q_+| = |q_-|$ implies that we have a smooth map

$$q = q_- q_+^{-1} : M \to \mathbf{S}^3 \subset \mathbf{H},$$

and $(1-q)q_+ = q_+ - q_- = 1$ implies that

$$q_+ = (1-q)^{-1}.$$

Essentially all Bonnet pairs are generated by the KPP construction.

Proposition 10.32. *If M is simply connected, and if the immersions*

$$\mathbf{x}_+, \mathbf{x}_- : M \to \mathbf{R}^3$$

form a Bonnet pair, then there exists an isothermic immersion $\mathbf{x} : M \to \mathbf{R}^3$ with Christoffel transform $\hat{\mathbf{x}} : M \to \mathbf{R}^3$ such that the smooth maps

$$q_\pm = \pm\frac{1}{2} + \hat{\mathbf{x}} : M \to \mathbf{H}^\times$$

define similarity deformations of $\mathbf{x}$ to the immersions $\mathbf{x}_\pm$, respectively; that is $d\mathbf{x}_\pm = q_\pm\, d\mathbf{x}\, \bar{q}_\pm$.

Proof. Given a Bonnet pair $\mathbf{x}_\pm : M \to \mathbf{R}^3$, then they have the same induced metric and orientation on M. As observed in Remark 10.26, there exists a smooth map

$$q : M \to \mathbf{S}^3 \subset \mathbf{H}$$

such that

$$d\mathbf{x}_- = q\, d\mathbf{x}_+ \bar{q}. \tag{10.38}$$

Let $(\mathbf{x}_+, p_+)$ be a first order spin frame field along $\mathbf{x}_+$ on M with dual coframe field $\varphi_+ = \omega^1_+ + i\omega^2_+$ on M, and let

$$\Psi_+ = p_+^{-1} q^{-1} dq\, p_+ \tag{10.39}$$

be the deformation form of q relative to this spin frame field. Since $\mathbf{x}_\pm$ have the same induced metric and mean curvature, we know that

$$\Psi_+ \wedge \varphi_+ = 0, \tag{10.40}$$

by Corollary 10.29.

Rotating $\mathbf{x}_+$ if necessary, we may assume that 1 is not in the image of q. In fact, the image of q cannot be all of $\mathbf{S}^3$, so let $b \in \mathbf{S}^3$ be a point not in the image of q. Then $\mathbf{x}_-$ and $b\mathbf{x}_+\bar{b}$ is still a Bonnet pair and

$$d\mathbf{x}_- = q\, d\mathbf{x}_+ \bar{q} = q\bar{b}\, d(b\mathbf{x}_+ \bar{b})(\overline{q\bar{b}})$$

shows that for this pair the map q has been replaced by the map

$$q\bar{b} : M \to \mathbf{S}^3$$

which never takes the value 1 on M. Assuming this done, we have a smooth map

$$(1-q)^{-1} : M \to \mathbf{H}^\times.$$

In order to show that $(1-q)^{-1} - 1/2$ takes all its values in $\Im\mathbf{H}$, write $q = s + \mathbf{y}$, where $s \in \mathbf{R}$, $\mathbf{y} \in \Im\mathbf{H}$ so that $1 = |q|^2 = s^2 + |\mathbf{y}|^2$. Then

$$(1-q)^{-1} = \frac{1-\bar{q}}{|1-q|^2} = \frac{1-s+\mathbf{y}}{1-2s+s^2+|\mathbf{y}|^2} = \frac{1}{2} + \frac{\mathbf{y}}{2(1-s)}.$$

Hence, we have the smooth map

$$\hat{\mathbf{x}} = (1-q)^{-1} - \frac{1}{2} : M \to \Im\mathbf{H},$$

for which, by (10.39),

$$\begin{aligned} d\hat{\mathbf{x}} &= d\left((1-q)^{-1}\right) = -(1-q)^{-1}(-dq)(1-q)^{-1} \\ &= (1-q)^{-1} q p_+ \Psi_+ p_+^{-1} (1-q)^{-1}. \end{aligned}$$

Consider the similarity deformation of $\mathbf{x}_+$ defined by the map $1-q : M \to \mathbf{H}^\times$. Its deformation form relative to $(\mathbf{x}_+, p_+)$ is, by (10.39),

$$\begin{aligned} \Psi &= p_+^{-1}(1-q)^{-1} d(1-q) p_+ = -p_+^{-1}(1-q)^{-1} dq\, p_+ \\ &= -p_+^{-1}(1-q)^{-1} q p_+ \Psi_+. \end{aligned}$$

Therefore, $\Psi \wedge \varphi_+ = 0$ by (10.40), which implies that there exists an immersion $\mathbf{x} : M \to \mathbf{R}^3$ such that

$$d\mathbf{x} = (1-q)\, d\mathbf{x}_+ \,(1-\bar{q}).$$

A first order spin frame field along $\mathbf{x}$ is given by $(\mathbf{x}, \frac{1-q}{|1-q|} p_+)$, whose coframe field is $\varphi = |1-q|^2 \varphi_+$. If we set $p = (1-q)p_+/|1-q|$, then $p^{-1} = \bar{p}_+(1-\bar{q})/|1-q|$, and

$$p^{-1} d\hat{\mathbf{x}} p = \bar{p}_+ \frac{1-\bar{q}}{|1-q|^2}(1-q)^{-1} q p_+ \Psi_+.$$

Therefore, $p^{-1} d\hat{\mathbf{x}} p \wedge \varphi = 0$, so $\hat{\mathbf{x}}$ is a Christoffel transform of $\mathbf{x}$, by (10.35). By our construction, if we let

$$q_+ = \frac{1}{2} + \hat{\mathbf{x}} = (1-q)^{-1},$$

then

$$q_+ d\mathbf{x}\bar{q}_+ = d\mathbf{x}_+.$$

Combining this with (10.38), we get $d\mathbf{x}_- = q_- d\mathbf{x}\bar{q}_-$, where

$$q_- = qq_+ = q(1-q)^{-1} = (1-(1-q))\frac{(1-\bar{q})}{|1-q|^2} = \frac{1-\bar{q}}{|1-q|^2} - 1$$
$$= (1-q)^{-1} - 1 = q_+ - 1 = -\frac{1}{2} + \hat{\mathbf{x}}.$$

□

10.7 KPP construction examples

Cylinders and cones are isothermic immersions simple enough to provide elementary examples of the KPP construction. In each case these pairs are congruent after a reparametrization; that is, they are equivalent.

10.7.1 *The Bonnet pair generated by a cylinder*

Consider again the cylinder over a plane curve described in Subsection 10.3.1. The curve is $\gamma(s) = f(s) + ig(s)$, where s is arclength parameter, and the immersion is

$$\mathbf{x}(s,t) = \gamma(s) - t\epsilon_3 = -it + \gamma(s)j$$

under our identification (10.26) $\mathbf{R}^3 \cong \Im\mathbf{H}$. Then $z = s + it$ is a principal complex coordinate with adapted frame field $(\mathbf{x}, (\mathbf{T}, -\epsilon_3, \mathbf{N}))$, and conformal factor $e^u = 1$. Here $\mathbf{T} = \dot{\gamma}$ and $\mathbf{N} = i\dot{\gamma}$ is the principal normal. The mean curvature H and Hopf invariant h relative to z are both equal to $\kappa/2$, where κ is the curvature of the plane curve γ. By (9.10), up to nonzero constant real multiple, a Christoffel transform $\hat{\mathbf{x}}$ of $\mathbf{x}$ must satisfy

$$d\hat{\mathbf{x}} = e^{-2u}(\mathbf{x}_{\bar{z}}dz + \mathbf{x}_z d\bar{z}) = \mathbf{T}ds + \epsilon_3 dt = idt + \mathbf{T}jds,$$

in $\Im\mathbf{H}$, from which we find the Christoffel transform to be, up to similarity and translation,

$$\hat{\mathbf{x}}(s,t) = ti + \gamma(s)j = \mathbf{x}(s,-t).$$

Choose a real constant $c > 0$ and let

$$q_\pm = \pm c + \hat{\mathbf{x}} = \pm c + it + \gamma j.$$

Applying the KPP construction, we compute

$$d\mathbf{x}_\pm = q_\pm d\mathbf{x}\bar{q}_\pm = (i(|\gamma|^2 - c^2 - t^2) + 2(-t \pm ic))\gamma j)dt +$$
$$(2i(\gamma \cdot (t\mathbf{T} - (\pm c)\mathbf{N})) + ((f\dot{f} + g\dot{g} + i(\dot{f}g - g\dot{g}))\gamma + c^2\mathbf{T} + (-t^2 \pm 2itc)\mathbf{T})j)ds$$

Integrate the coefficient of dt with respect to t to get

$$\mathbf{x}_\pm(s,t) = i\left(-\frac{t^3}{3} + (|\gamma|^2 - c^2)t\right) + \left(2i(\pm ct + i\frac{t^2}{2})\gamma\right)j + \mathbf{a}_\pm(s)$$

where $\mathbf{a}_\pm(s) \in \Im H$ is a constant of integration for each s. Taking the partial derivative with respect to s and equating this to the coefficient of ds in the above expression for $d\mathbf{x}_\pm$, we find

$$\frac{d\mathbf{a}_\pm}{ds} = \dot{\mathbf{a}}_\pm = \begin{pmatrix} (c^2 + f^2 - g^2)\dot{f} + 2fg\dot{g} \\ (c^2 - f^2 + g^2)\dot{g} + 2fg\dot{f} \\ \pm 2c(f\dot{g} - \dot{f}g) \end{pmatrix}. \tag{10.41}$$

In conclusion, the Bonnet pair generated by the cylinder $\mathbf{x} : J \times \mathbf{R} \to \mathbf{R}^3$ comprises $\mathbf{x}_\pm : \tilde{J} \times \mathbf{R} \to \mathbf{R}^3$, where $\tilde{J}$ is the universal cover of J, and

$$\mathbf{x}_\pm(s,t) = \begin{pmatrix} - \pm 2ctg - t^2 f \\ \pm 2ctf - t^2 g \\ -t^3/3 + (f^2 + g^2 - c^2)t \end{pmatrix} + \mathbf{a}_\pm(s). \tag{10.42}$$

Proposition 10.33. *This Bonnet pair generated by the cylinder* $\mathbf{x}$ *satisfy*

$$\mathbf{x}_-(s,t) = A\mathbf{x}_+(s,-t) + \mathbf{b},$$

where $A = \begin{pmatrix} -1 & 0 & 0 \\ 0 & -1 & 0 \\ 0 & 0 & 1 \end{pmatrix}$ *is reflection in the* $\epsilon_1\epsilon_2$*-plane of* $\mathbf{R}^3$ *and* $\mathbf{b} \in \mathbf{R}^3$ *is a constant vector. The immersions* $\mathbf{x}_\pm : \tilde{J} \times \mathbf{R} \to \mathbf{R}^3$ *are equivalent, but not congruent.*

Proof. By (10.41) and (10.42) the components of these vectors satisfy

$$x_-^1(s,t) - x_+^1(s,-t) = a_-^1(s) - a_+^1(s) = b^1,$$
$$x_-^2(s,t) - x_+^2(s,-t) = a_-^2(s) - a_+^2(s) = b^2,$$
$$x_-^3(s,t) + x_+^3(s,-t) = a_-^3(s) + a_+^3(s) = b^3,$$

for some real constants b^1, b^2, b^3. □

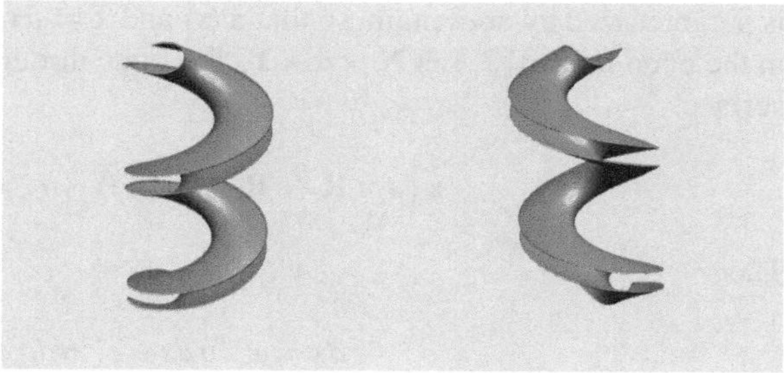

Fig. 10.5 Bonnet pair from KPP applied to circular cylinder.

Example 10.34 (Circular cylinder). The integration of (10.41) can be done explicitly for the case where the cylinder is generated by the circle of radius $R > 0$,

$$\gamma : J \to \mathbf{C} = \mathbf{R}^2, \quad \gamma(s) = (R\cos\frac{s}{R}, R\sin\frac{s}{R}) = Re^{is/R},$$

whose domain is $J = \mathbf{R}/2R\pi$. Integrating (10.41), we get for (10.42)

$$\mathbf{x}_\pm(s,t) = \begin{pmatrix} -2R(\pm ct\sin\frac{s}{R} + \frac{t^2}{2}\cos\frac{s}{R}) \\ 2R(\pm ct\cos\frac{s}{R} - \frac{t^2}{2}\sin\frac{s}{R}) \\ (R^2-c^2)t - t^3/3 \end{pmatrix} + \begin{pmatrix} R(c^2-R^2)\cos\frac{s}{R} \\ R(c^2-R^2)\sin\frac{s}{R} \\ \pm 2cRs \end{pmatrix} + \mathbf{b}_\pm$$

where $\mathbf{b}_\pm \in \mathbf{R}^3$ are arbitrary constant vectors. The domain of $\mathbf{a}_\pm$ is $\mathbf{R}$, the universal cover of J, so the immersions $\mathbf{x}_\pm$ have domain $\mathbf{R}^2$ and are not periodic in s. They are helicoidal immersions $\mathbf{x}_\pm(s,t) = T_\pm(s)\mu_\pm(t)$, with the 1-parameter subgroups

$$T_\pm : \mathbf{R} \to \mathbf{E}_+(3), \quad T_\pm(s) = (\pm 2cRs\epsilon_3, \begin{pmatrix} \cos\frac{s}{R} & -\sin\frac{s}{R} & 0 \\ \sin\frac{s}{R} & \cos\frac{s}{R} & 0 \\ 0 & 0 & 1 \end{pmatrix}) \tag{10.43}$$

and profile curves

$$\mu_\pm : \mathbf{R} \to \mathbf{R}^3, \quad \mu_\pm(t) = {}^t(R(c^2-R^2-t^2), \pm 2Rct, (R^2-c^2)t - t^3/3).$$

The cases $c = 1/2$, $R = 3/2$, and $\mathbf{b}_\pm = \pm{}^t(-10,0,-9)$ over the domain $[0,4R\pi] \times [-2,2]$ are shown in Figure 10.5. These are not embeddings, as can be seen by extending t over $[-4,4]$, for example.

10.7.2 The Bonnet pair generated by a cone

Consider again the cone over a curve in the unit sphere described in Subsection 10.3.2. The curve

$$\sigma : J \to \mathbf{S}^2$$

is parametrized by arc-length, so that $\boldsymbol{\sigma}(s)$ and $\mathbf{T} = \dot{\boldsymbol{\sigma}}(s)$ are unit vectors for every s in the open interval J. Let $\mathbf{N} = \boldsymbol{\sigma} \times \mathbf{T}$. The cone through this curve is the immersed surface

$$\mathbf{x} : J \times \mathbf{R} \to \mathbf{R}^3, \quad \mathbf{x}(s,t) = e^{-t}\boldsymbol{\sigma}(s).$$

Then

$$d\mathbf{x} = e^{-t}\dot{\boldsymbol{\sigma}}\, ds - e^{-t}\boldsymbol{\sigma}\, dt$$

so the induced metric is

$$d\mathbf{x} \cdot d\mathbf{x} = e^{-2t}(ds^2 + dt^2),$$

which shows that $z = s + it$ is a complex coordinate for the induced complex structure, and with respect to it the conformal factor is e^{-t}. The adapted frame field is

$$\mathbf{e}_1 = \dot{\boldsymbol{\sigma}}, \ \mathbf{e}_2 = -\boldsymbol{\sigma}, \ \mathbf{e}_3 = -\dot{\boldsymbol{\sigma}} \times \boldsymbol{\sigma} = \mathbf{N},$$

with coframe field

$$\omega^1 = e^{-t} ds, \quad \omega^2 = e^{-t} dt.$$

Up to nonzero constant real multiple, the Christoffel transform $\hat{\mathbf{x}}$ of $\mathbf{x}$ satisfies

$$d\hat{\mathbf{x}} = e^{2t}(e^{-t} ds \dot{\boldsymbol{\sigma}} + e^{-t} dt \boldsymbol{\sigma}) = d(e^t \boldsymbol{\sigma}),$$

so, up to translation,

$$\hat{\mathbf{x}}(s,t) = e^t \boldsymbol{\sigma}(s) = \mathbf{x}(s,-t).$$

In the quaternionic notation above, we identify $\mathbf{R}^3$ with the imaginary quaternions $\Im \mathbf{H}$. Given any constant $c > 0$, let

$$q_\pm = \pm c + \hat{\mathbf{x}}(s,t) = \pm c + e^t \boldsymbol{\sigma}(s).$$

Then

$$d\mathbf{x}_\pm = q_\pm d\mathbf{x} \bar{q}_\pm = ((c^2 e^{-t} - e^t)\dot{\boldsymbol{\sigma}} \pm 2c\mathbf{N}) ds - (c^2 e^{-t} + e^t)\boldsymbol{\sigma}\, dt.$$

Integrating this in the usual way, we get

$$\mathbf{x}_\pm = (c^2 e^{-t} - e^t)\boldsymbol{\sigma} \pm 2c \int \mathbf{N}(s) ds. \tag{10.44}$$

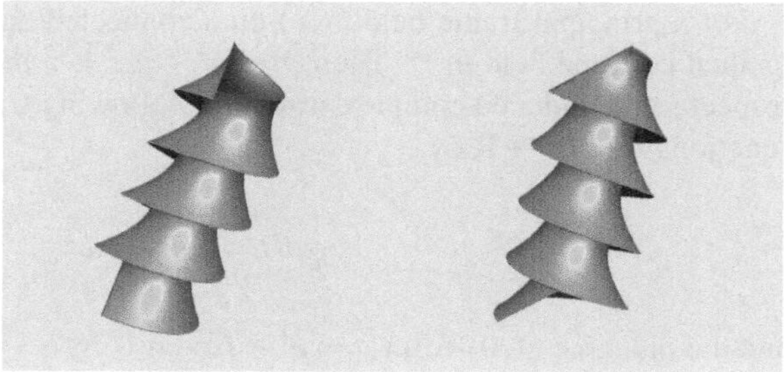

Fig. 10.6 Bonnet pair from KPP applied to circular cone.

In general, the domain of $\mathbf{x}_\pm$ is the universal cover of $J \times \mathbf{R}$, so if $\boldsymbol{\sigma}$ is periodic (so J is a circle), then $\int \mathbf{N}(s)ds$ might not be periodic.

Proposition 10.35. *This Bonnet pair $\mathbf{x}_\pm : \mathbf{R}^2 \to \mathbf{R}^3$ generated by the cone $\mathbf{x}$ satisfies*

$$\mathbf{x}_+(s, -t-2T) + \mathbf{x}_-(s,t) = \mathbf{b},$$

where $\mathbf{b}$ is a constant vector in $\mathbf{R}^3$ and $T = -\log c$. In particular, $\mathbf{x}_+$ and $\mathbf{x}_-$ are equivalent immersions.

Example 10.36 (Circular cone). Fix an angle θ with $0 < \theta < \pi/2$ and let $R = \sin\theta$. Consider the circle on the unit sphere

$$\boldsymbol{\sigma} : \mathbf{R}/2\pi R \to \mathbf{S}^2 \subset \Im H, \quad \boldsymbol{\sigma}(s) = \sqrt{1-R^2}i + Re^{is/R}j,$$

whose principal normal vector is $\mathbf{N}(s) = Ri - \sqrt{1-R^2}e^{is/R}j$. Then $\mathbf{x}_\pm$ is given by (10.44), which is now the helicoidal immersion $\mathbf{x}_\pm(s,t) = T_\pm(s)\nu_\pm(t) + \mathbf{b}_\pm$, where $\mathbf{b}_\pm$ are arbitrary constant vectors in $\mathbf{R}^3$, $T_\pm : \mathbf{R} \to \mathbf{E}(3)$ are the 1-parameter groups (10.43), and the profile curves are lines

$$\nu_\pm(t) = (c^2e^{-t} + e^t)(R\epsilon_1 + \sqrt{1-R^2}\epsilon_3) \pm 2c\sqrt{1-R^2}R\epsilon_2.$$

The domain of $\mathbf{x}_\pm$ is $\mathbf{R}^2$. Figure 10.6 shows $\mathbf{x}_\pm : (-4R\pi, 4R\pi) \times (-\log 3, \log 3) \to \mathbf{R}^3$ for the case $c = 1/2$, $\theta = \pi/4$, and $\mathbf{b}_\pm = \pm 2\epsilon_1$.

10.8 Cartan's Bonnet criterion

Let $\mathbf{x} : M \to \mathbf{R}^3$ be an immersion of an oriented surface M.

Definition 10.37. A *principal frame field* along $\mathbf{x}$, oriented by a unit normal vector field $\mathbf{e}_3$, is a second order frame field $(\mathbf{x}, e) : U \to \mathbf{E}_+(3)$ with the given normal vector such that the principal curvatures a and c satisfy $a - c > 0$; that is, its Hopf invariant $h = (a-c)/2$ is positive on U.

For a principal frame field $(\mathbf{x}, e)$ on a connected open set $U \subset M$, let ω^1, ω^2 be its dual coframe field in U. Then $\varphi = \omega^1 + i\omega^2$ is a bidegree $(1,0)$ form in U with respect to the induced complex structure. Following Cartan [33], we define smooth functions $r, s : U \to \mathbf{R}$ by

$$\frac{1}{h} dH = r\omega^1 + s\omega^2, \tag{10.45}$$

and the bidegree $(1,0)$-form $\rho = \rho^1 + i\rho^2$ on U by

$$\rho = (r + is)(\omega^1 + i\omega^2). \tag{10.46}$$

Remark 10.38. If $e = (\mathbf{e}_1, \mathbf{e}_2, \mathbf{e}_3)$, then the only other principal frame field $(\mathbf{x}, \tilde{e})$ on U is given by $\tilde{e} = (-\mathbf{e}_1, -\mathbf{e}_2, \mathbf{e}_3)$, whose dual coframe field is given by $\tilde{\omega}^1 = -\omega^1$, $\tilde{\omega}^2 = -\omega^2$, so $\tilde{\varphi} = -\varphi$. It follows that $\tilde{h} = h$ and

$$\frac{1}{h} dH = \tilde{r}\tilde{\omega}^1 + \tilde{s}\tilde{\omega}^2$$

where

$$\tilde{r} = -r, \quad \tilde{s} = -s.$$

Thus, relative to $(\mathbf{x}, \tilde{e})$,

$$\tilde{\rho} = (\tilde{r} + i\tilde{s})\tilde{\varphi} = \rho$$

on U. Since each nonumbilic point of M possesses a neighborhood on which there is a principal frame field, it follows that ρ is a bidegree $(1,0)$ form defined on $M \setminus \mathscr{U}$, where $\mathscr{U}$ is the set of all umbilic points of $\mathbf{x}$.

Definition 10.39. The *Cartan form* of an immersion $\mathbf{x} : M \to \mathbf{R}^3$, oriented by the unit normal vector field $\mathbf{e}_3$, is the smooth, bidegree (1,0) form ρ on $M \setminus \mathscr{U}$ defined locally by (10.46). Here $\mathscr{U}$ is the set of all umbilic points of $\mathbf{x}$.

Theorem 10.40 (Cartan's Criterion). *Let $\mathbf{x} : M \to \mathbf{R}^3$ be an immersion without umbilic points, oriented by the unit normal vector field $\mathbf{e}_3$ on M, and with Cartan form $\rho = \rho^1 + i\rho^2$ defined in* (10.46). *If $\mathbf{x}$ is Bonnet, then there exists a nonconstant smooth map*

$$\mathfrak{p} : M \to \mathbf{RP}^1 \tag{10.47}$$

such that

$$\mathscr{Z} = \{m \in M : \mathfrak{p}(m) = [1,0]\} \tag{10.48}$$

is a possibly empty discrete subset of M *and the smooth function*

$$t : \hat{M} = M \setminus \mathscr{Z} \to \mathbf{R}, \quad \mathfrak{p} = [t, 1] \tag{10.49}$$

satisfies

$$dt = t\rho^1 - \rho^2 \tag{10.50}$$

on $\hat{M}$. *Conversely, if* M *is simply connected and if there exists a smooth map* $\mathfrak{p} : M \to \mathbf{RP}^1$ *satisfying* (10.48), (10.49), *and* (10.50), *then* $\mathbf{x}$ *is Bonnet.*

Proof. Suppose $\tilde{\mathbf{x}} : M \to \mathbf{R}^3$ is a Bonnet mate of $\mathbf{x}$. Note that $\tilde{\mathbf{x}}$ has no umbilic points on M, since its principal curvatures are the same as those of $\mathbf{x}$. By Proposition 10.17, there exists a smooth map $A : M \to \mathbf{SO}(3)$, which defines the corresponding $\mathbf{E}(3)$-deformation of $\mathbf{x}$ to $\tilde{\mathbf{x}}$, so that $d\tilde{\mathbf{x}} = A d\mathbf{x}$ on M. Let

$$\psi = e^{-1} A^{-1} dA\, e$$

be its deformation form relative to a principal frame field $(\mathbf{x}, e)$ on an open set $U \subset M$ and let $\theta = {}^t(\omega^1, \omega^2)$ be its dual coframe field on U. This is also the coframe field dual to the first order frame field $(\tilde{\mathbf{x}}, Ae)$ on U. Because the deformation is non-trivial, $(\tilde{\mathbf{x}}, Ae)$ cannot be a principal frame field along $\tilde{\mathbf{x}}$.

Let $(\tilde{\mathbf{x}}, \tilde{e})$ be a principal frame field on U with dual coframe field $\tilde{\theta} = {}^t(\tilde{\omega}^1, \tilde{\omega}^2)$. Because $\tilde{\mathbf{x}}$ is a Bonnet deformation of $\mathbf{x}$, the mean curvature H and Hopf invariant h of $\mathbf{x}$ relative to the principal frame $(\mathbf{x}, e)$ coincide with the mean curvature and Hopf invariant of $\tilde{\mathbf{x}}$ relative to its principal frame $(\tilde{\mathbf{x}}, \tilde{e})$. The two coframes on U are related by

$$\tilde{\theta} = R(\tau)\theta,$$

for some smooth map

$$R(\tau) = \begin{pmatrix} \cos\tau & -\sin\tau \\ \sin\tau & \cos\tau \end{pmatrix} : U \to \mathbf{SO}(2).$$

It follows that the two frame fields $(\tilde{\mathbf{x}}, Ae)$ and $(\tilde{\mathbf{x}}, \tilde{e})$ are related by

$$Ae = \tilde{e} \begin{pmatrix} R(\tau) & 0 \\ 0 & 1 \end{pmatrix}.$$

Therefore, the Hopf invariant $\hat{h}$ of $\tilde{\mathbf{x}}$ relative to the first order frame field $(\tilde{\mathbf{x}}, Ae)$ is

$$\hat{h} = e^{i2\tau} h,$$

(see Exercise 9) so the components of

$$\hat{\omega} = (Ae)^{-1} d(Ae)$$

are given by (see equation (7.23))

$$\hat{\omega}_1^3 - i\hat{\omega}_2^3 = \hat{h}\varphi + H\bar{\varphi},$$

where $\varphi = \omega^1 + i\omega^2$. The deformation form of A relative to $(\mathbf{x}, e)$,

$$\psi_1^3 - i\psi_2^3 = \hat{\omega}_1^3 - i\hat{\omega}_2^3 - (\omega_1^3 - i\omega_2^3) = (\hat{h} - h)\varphi = (e^{i2\tau} - 1)h\varphi, \tag{10.51}$$

has isolated zeros in U, since $(e^{i2\tau} - 1)h\varphi\varphi$ is holomorphic on U by Corollary 10.22. Thus,

$$\mathscr{Z}_U = \{m \in U : e^{i2\tau(m)} = 1\}$$

is a discrete (possibly empty) subset of U. Consider the smooth map

$$\mathfrak{p} = [\cos\tau, \sin\tau] : U \to \mathbf{RP}^1. \tag{10.52}$$

By (10.23) and an elementary calculation we get

$$0 = d(\psi_1^3 - i\psi_2^3) + i(\psi_1^3 - i\psi_2^3) \wedge \omega_1^2 = [(e^{i2\tau} - 1)(dh - 2ih\omega_1^2) + 2ihe^{i2\tau}d\tau] \wedge \varphi.$$

Substituting in the Codazzi equation (7.23), which is

$$(dh - 2ih\omega_1^2) \wedge \varphi + dH \wedge \bar{\varphi} = 0,$$

we get

$$d\tau \wedge \varphi = \frac{e^{i2\tau} - 1}{2ihe^{i2\tau}} dH \wedge \bar{\varphi} = e^{-i\tau} \sin\tau \, \frac{dH}{h} \wedge \bar{\varphi}.$$

Substituting in $\varphi = \omega^1 + i\omega^2$ and (10.45), we get

$$d\tau = -\sin^2\tau \left(\cot\tau(r\omega^1 - s\omega^2) - (s\omega^1 + r\omega^2)\right),$$

on $\hat{U} = U \setminus \mathscr{Z}_U$, which gives (10.50) for

$$t = \cot\tau : \hat{U} \to \mathbf{R}.$$

Any change of the principal frames $(\mathbf{x}, e)$ or $(\tilde{\mathbf{x}}, \tilde{e})$ will at most change the sign of θ or of $\tilde{\theta}$, and therefore at most change the sign of $R(\tau)$, which means $(\cos\tau, \sin\tau)$ changes by $\pm(\cos\tau, \sin\tau)$, and therefore $\mathfrak{p} = [\cos\tau, \sin\tau]$ does not change. Covering M by open sets on each of which there exists a principal frame field along $\mathbf{x}$, we can define the map $\mathfrak{p}$ on all of M such that it satisfies (10.48), (10.49), and (10.50).

Conversely, suppose M is simply connected and that there is a function $\mathfrak{p} : M \to \mathbf{RP}^1$ satisfying (10.48)–(10.50). Let $(\mathbf{x}, e)$ be a principal frame field on

M. Notice that if $\mathfrak{p} = [\cos\tau, \sin\tau]$, then $e^{i2\tau}$ is a well defined smooth function on M. Therefore, we may define the $\mathfrak{o}(3)$-valued 1-form ψ on M by

$$\psi_1^2 = 0, \quad \psi_1^3 - i\psi_2^3 = (e^{i2\tau} - 1)h\varphi.$$

We proceed now to show that ψ_1^3 and ψ_2^3 satisfy equations (10.22). Compare the imaginary parts of

$$(\psi_1^3 - i\psi_2^3) \wedge (\omega_1^3 + i\omega_2^3) = (e^{i2\tau} - 1)h\varphi \wedge (h\bar{\varphi} + H\varphi)$$

to get that

$$\psi_1^3 \wedge \omega_2^3 - \psi_2^3 \wedge \omega_1^3 = -2(\cos 2\tau - 1)h^2\omega^1 \wedge \omega^2.$$

Therefore,

$$\begin{aligned}\psi_1^3 \wedge \omega_2^3 - \psi_2^3 \wedge \omega_1^3 + \psi_1^3 \wedge \psi_2^3 &= \left(-2(\cos 2\tau - 1)h^2 - |e^{i2\tau} - 1|^2 h^2\right)\omega^1 \wedge \omega^2 \\ &= h^2(\cos^2 2\tau - 1 + \sin^2 2\tau)\omega^1 \wedge \omega^2 = 0,\end{aligned}$$

which is the first equation in (10.22). We verify that the remaining two equations in (10.22) hold by showing that (10.23) holds. Now

$$\mathfrak{p} = [\cos\tau, \sin\tau] = [t, 1]$$

on $\hat{M}$, where $t = \cot\tau$ satisfies $dt = t\rho^1 - \rho^2$, implies that

$$d\tau = -\sin\tau(\cos\tau\,\rho^1 - \sin\tau\,\rho^2)$$

on all of M, by continuity. Since $\frac{dH}{h} = r\omega^1 + s\omega^2$ and $(r + is)\varphi = \rho^1 + i\rho^2$, we have

$$\rho^1 \wedge \varphi = -i\rho^2 \wedge \varphi,$$

and

$$\rho^2 \wedge \varphi = (s\omega^1 + r\omega^2) \wedge (\omega^1 + i\omega^2) = (is - r)\omega^1 \wedge \omega^2 = -i\frac{dH}{h} \wedge \bar{\varphi},$$

from all of which we get

$$d\tau \wedge \varphi = e^{-i\tau}\sin\tau \frac{dH}{h} \wedge \bar{\varphi}.$$

Thus, using $d\varphi = i\omega_2^1 \wedge \varphi$, and making use of the Codazzi equation (7.23) in passing from the second line to the third line, we calculate

$$d(\psi_1^3 - i\psi_2^3) + i(\psi_1^3 - i\psi_2^3) \wedge \omega_1^2 = d\left((e^{i2\tau} - 1)h\varphi\right) + i(e^{i2\tau} - 1)h\varphi \wedge \omega_1^2$$
$$= 2ihe^{i2\tau} d\tau \wedge \varphi + (e^{i2\tau} - 1)(dh - 2ih\omega_1^2) \wedge \varphi =$$
$$2ihe^{i\tau} \sin\tau \frac{dH}{h} \wedge \bar{\varphi} - (e^{i2\tau} - 1) dH \wedge \bar{\varphi} = 2ihe^{i\tau}(\sin\tau - \sin\tau)\frac{dH}{h} \wedge \bar{\varphi} = 0.$$

Therefore, equations (10.22) hold on M and the proof is completed by applying Proposition 10.23. □

10.9 Proper Bonnet immersions

A proper Bonnet immersion $\mathbf{x} : M \to \mathbf{R}^3$ induces a nonconstant holomorphic map from M to the *right half-plane* $\mathbf{C}_+ = \{z \in \mathbf{C} : \Re z > 0\}$. To prove this we need the following technical preparation.

Lemma 10.41. *Let M be a connected Riemann surface with smooth bidegree $(1,0)$-form ρ on $\hat{M} = M \setminus \mathscr{U}$, where $\mathscr{U}$ is a discrete subset of M. Let $\mathscr{D}$ be a discrete subset of $\hat{M}$ and let $M' = \hat{M} \setminus \mathscr{D}$. If $u, v : M' \to \mathbf{R}$ are smooth functions such that $u\rho$ is not identically zero on M', and if $w = u + iv : M' \to \mathbf{C}$ satisfies*

$$dw = -u\rho \tag{10.53}$$

on M', then w is defined and holomorphic on all of M and u is never zero on M.

Proof. By (10.53), w is holomorphic and nonconstant on M', since dw is of bidegree $(1,0)$ and not identically zero on M'. Since $-u\rho = dw$ is a holomorphic 1-form on M', the set $\mathscr{Z}$ of zeros of u is a discrete subset of M'.

We first prove that the points of $\mathscr{D}$ are removable singularities of w, so w is defined and holomorphic on $\hat{M}$. For this, consider the trivial real vector bundle $\mathbf{V} = \hat{M} \times \mathbf{R}^2 \to \hat{M}$ on $\hat{M}$, with global smooth frame field $E_1, E_2 : \hat{M} \to \mathbf{V}$ given by

$$E_1(m) = (m, \boldsymbol{\epsilon}_1), \quad E_2(m) = (m, \boldsymbol{\epsilon}_2),$$

where $\boldsymbol{\epsilon}_1, \boldsymbol{\epsilon}_2$ is the standard ordered basis of $\mathbf{R}^2$. Consider the connection D on $\mathbf{V}$ defined by

$$DE_1 = \rho^1 E_1 + \rho^2 E_2, \quad DE_2 = 0,$$

where $\rho = \rho^1 + i\rho^2$ with ρ^1 and ρ^2 smooth, real 1-forms on $\hat{M}$. The section $W : M' \to \mathbf{V}$ defined by

$$W = uE_1 + vE_2$$

is parallel, by (10.53), since

$$\begin{aligned} DW &= duE_1 + uDE_1 + dvE_2 + vDE_2 \\ &= -u\rho^1 E_1 + u(\rho^1 E_1 + \rho^2 E_2) - u\rho^2 E_2 = 0 \end{aligned}$$

on M'. Then W, E_2 are parallel sections on M', linearly independent at each point of $M' \setminus \mathscr{Z}$, so the curvature form of D is zero on $\hat{M} \setminus (\mathscr{D} \cup \mathscr{Z})$, and thus must be zero on all of $\hat{M}$ by continuity. If $m_0 \in \mathscr{D}$, then there must be a connected open neighborhood U of m_0 in $\hat{M}$ on which there exists a parallel frame field F_1, F_2 of $\mathbf{V}$. We may assume that $U \cap \mathscr{D} = \{m_0\}$. Then W, E_2 and F_1, F_2 are parallel frame fields on the connected set $U \setminus \{m_0\}$, so they must be related by

$$W = A_1^1 F_1 + A_1^2 F_2, \quad E_2 = A_2^1 F_1 + A_2^2 F_2,$$

where the coefficients are constants. Thus, W extends smoothly to m_0, so m_0 must be a removable singularity of w. We have thus proved that w is defined and holomorphic on $\hat{M}$.

We next prove that u is never zero on $\hat{M}$. Seeking a contradiction, we suppose that $u(p_0) = 0$ for some point $p_0 \in \hat{M}$. Choose a complex coordinate chart (U, z) around p_0 and write $\rho = R dz$, for some function $R : U \to \mathbf{C}$. In this coordinate the equation $dw = -u\rho$ can be written on U as

$$\frac{\partial w}{\partial z} = -uR. \tag{10.54}$$

By the Cauchy-Riemann equations, $\frac{\partial v}{\partial z} = -i\frac{\partial u}{\partial z}$, so

$$\frac{\partial w}{\partial z} = 2\frac{\partial u}{\partial z}. \tag{10.55}$$

Thus, if $u(p_0) = 0$, then (10.54) and (10.55) imply that

$$\frac{\partial w}{\partial z}(p_0) = 0, \quad \text{and} \quad \frac{\partial u}{\partial z}(p_0) = 0. \tag{10.56}$$

Differentiating (10.54) and (10.55) and using (10.56), we get

$$2\frac{\partial^2 u}{\partial z^2}(p_0) = \frac{\partial^2 w}{\partial z^2}(p_0) = -\frac{\partial u}{\partial z}(p_0)R(p_0) - u(p_0)\frac{\partial R}{\partial z}(p_0) = 0.$$

Proceeding inductively we conclude that

$$\frac{\partial^k w}{\partial z^k}(p_0) = 0, \qquad \text{for all } k \geq 1.$$

Hence, w must be constant on U, and therefore constant on all of $\hat{M}$. This gives the desired contradiction and proves that u is never zero on $\hat{M}$. Since $-w$ also satisfies the theorem and since $\hat{M}$ is connected, we may assume $u > 0$ on $\hat{M}$.

Finally, we will prove that w extends to a holomorphic function on M and that u remains positive on all of M. The Cayley transform

$$F : \mathbf{C}_+ \to \mathbf{D}, \quad F(z) = i\frac{z-1}{z+1},$$

is a biholomorphic map from the right half-plane $\mathbf{C}_+$ onto the unit disk $\mathbf{D} \subset \mathbf{C}$ taking the boundary of $\mathbf{C}_+$ into the boundary of $\mathbf{D}$. Then

$$\tilde{w} = F \circ w : \hat{M} \to \mathbf{D}$$

is a bounded holomorphic function on $\hat{M}$. If $m_0 \in \mathscr{U}$, let (U,z) be a complex coordinate chart of M centered at m_0. We may take U small enough so that $U \cap \mathscr{U} = \{m_0\}$. Then the local representation

$$f = \tilde{w} \circ z^{-1} : z(U \setminus \{m_0\}) \subset \mathbf{C} \to \mathbf{D}$$

is a bounded holomorphic function, so

$$\lim_{z \to 0} zf(z) = 0.$$

Hence, the isolated singularity at $z = 0$ is removable (see Ahlfors [1, Theorem 7, p. 100]). We conclude that w is holomorphic on all of M. Furthermore, if $u(m_0) = 0$, then $|F(w(m_0))| = 1$, which implies that w is constant on U by the maximum principle, and this is not the case. Hence, $u > 0$ on all of M. □

Theorem 10.42. *Let $\mathbf{x} : M \to \mathbf{R}^3$ be an immersion of a connected surface and $\mathbf{e}_3$ a smooth unit normal vector field along $\mathbf{x}$. If $\mathbf{x}$ is proper Bonnet, then there exists a nonconstant holomorphic function*

$$w : M \to \mathbf{C}_+ = \{z \in \mathbf{C} : \Re(z) > 0\},$$

such that

$$dw = -u\rho \tag{10.57}$$

on $M \setminus \mathscr{U}$, where $u = \Re(w) > 0$, $\mathscr{U}$ is the necessarily discrete set of umbilic points of M, and ρ is the Cartan form of $\mathbf{x}$ on $M \setminus \mathscr{U}$ (see Definition 10.39).

Conversely, if M is simply connected, if the immersion $\mathbf{x} : M \to \mathbf{R}^3$ has nonconstant mean curvature, no umbilic points, Cartan form $\rho = \rho^1 + i\rho^2$, and if there exists a smooth function $w : M \to \mathbf{C}$ satisfying (10.57) *on M, where $u = \Re(w)$ and if $u\rho^1$ is nonzero somewhere on M, then w is holomorphic on M and $\mathbf{x}$ is proper Bonnet.*

Remark 10.43. The function w is called a *Cartan holomorphic function* of the proper Bonnet immersion $\mathbf{x}$. It is unique up to a change $lw + in$, where $l > 0$ and n are real constants. The Cartan form ρ was defined in (10.46) only off the set of umbilic points of $\mathbf{x}$. For a proper Bonnet immersion, solving for ρ in (10.57) shows that ρ extends smoothly over the umbilic points.

Proof. By the definition of ρ in (10.46), the assumption that H be non-constant on M is equivalent to ρ being not identically zero on M.

Suppose $\mathbf{x}$ is proper Bonnet with noncongruent Bonnet mates $\tilde{\mathbf{x}}$ and $\hat{\mathbf{x}}$. Let $\mathscr{U}$ be the set of umbilic points of $\mathbf{x}$, a discrete subset (possibly empty) of M, and let $\hat{M} = M \setminus \mathscr{U}$. Then $\mathbf{x} : \hat{M} \to \mathbf{R}^3$ is a proper Bonnet immersion without umbilic points. Let $\tilde{\mathfrak{p}}, \hat{\mathfrak{p}} : \hat{M} \to \mathbf{RP}^1$ be the smooth maps given by $\tilde{\mathbf{x}}$ and $\hat{\mathbf{x}}$, respectively, by Theorem 10.40. We want to prove that

$$\mathscr{Z} = \{m \in \hat{M} : \tilde{\mathfrak{p}}(m) = \hat{\mathfrak{p}}(m)\}$$

is a discrete subset of $\hat{M}$. Writing

$$\tilde{\mathfrak{p}} = [\cos\tilde{\tau}, \sin\tilde{\tau}], \quad \hat{\mathfrak{p}} = [\cos\hat{\tau}, \sin\hat{\tau}],$$

we have $\tilde{\mathfrak{p}}(m) = \hat{\mathfrak{p}}(m)$ if and only if

$$e^{i2\tilde{\tau}(m)} = e^{i2\hat{\tau}(m)}.$$

In (10.51) in the proof of Theorem 10.40 it was shown that relative to any first order frame field $(\mathbf{x}, e)$ on an open set $U \subset \hat{M}$, the quadratic differentials

$$(e^{i2\tilde{\tau}(m)} - 1)h\varphi\varphi, \quad \text{and} \quad (e^{i2\hat{\tau}(m)} - 1)h\varphi\varphi$$

are holomorphic on U. Therefore, their zero sets and the zero set of their difference,

$$\tilde{\mathscr{Z}} = \{m \in \hat{M} : \tilde{\mathfrak{p}}(m) = [1,0]\}, \quad \hat{\mathscr{Z}} = \{m \in \hat{M} : \hat{\mathfrak{p}}(m) = [1,0]\},$$

and $\mathscr{Z}$, are discrete subsets of M. The functions

$$\tilde{t} = \cot\tilde{\tau}, \quad \hat{t} = \cot\hat{\tau},$$

are smooth on the complements of $\tilde{\mathscr{Z}}$ and $\hat{\mathscr{Z}}$, respectively. Let

$$\mathscr{D} = \mathscr{Z} \cup \tilde{\mathscr{Z}} \cup \hat{\mathscr{Z}},$$

a possibly empty discrete subset of $\hat{M}$. On $M' = \hat{M} \setminus \mathscr{D}$ define the smooth complex valued function

$$w = u + iv : M' \to \mathbf{C},$$

by

$$u = \frac{1}{\tilde{t} - \hat{t}}, \quad v = \frac{\tilde{t}}{\tilde{t} - \hat{t}}.$$

Using (10.50), we calculate

$$dw = -\frac{1 + i\tilde{t}}{(\tilde{t} - \hat{t})^2}(d\tilde{t} - d\hat{t}) + \frac{i d\tilde{t}}{\tilde{t} - \hat{t}} = \frac{-\rho^1 - i\rho^2}{\tilde{t} - \hat{t}} = -u\rho$$

on M'. Note that w is not constant because u is never zero on M'. Since ρ is a bidegree $(1,0)$ form, not identically zero, on M', it follows from Lemma 10.41 that w extends to a holomorphic function on M and that we may assume that $u > 0$ at every point of M and that (10.57) holds on M.

Conversely, suppose that M is simply connected, that $\mathbf{x}$ has nonconstant mean curvature and no umbilic points, and that there exists a nonconstant smooth function $w = u + iv : M \to \mathbf{C}$ satisfying $u > 0$ and (10.57) on M. Then w is holomorphic on M by Lemma 10.41. For any real constant c, the function

$$t_c = \frac{v + c}{u}$$

is defined and smooth on M and, by (10.57), satisfies

$$dt_c = \frac{dv}{u} - \frac{v + c}{u^2}du = -\rho^2 - \frac{v + c}{u}(-\rho^1) = t_c\rho^1 - \rho^2.$$

We now apply Theorem 10.40 to conclude that there exists a Bonnet mate $\mathbf{x}_c$ of $\mathbf{x}$ giving rise to the map (10.47). Distinct constants c_1 and c_2 give rise to distinct functions $\mathfrak{p}_{c_1}$ and $\mathfrak{p}_{c_2}$, which in turn define distinct deformation forms relative to a principal frame field, and therefore these define non-congruent Bonnet mates of $\mathbf{x}$.

□

Corollary 10.44. *If $\mathbf{x} : M \to \mathbf{R}^3$ is a proper Bonnet immersion, then its mean curvature function has only isolated critical points.*

Proof. Let $w = u + iv$ be the holomorphic function on M given by the Theorem. Then (10.57) is $dw = -u\rho$, which shows that ρ can be zero only at a zero of the holomorphic 1-form dw, since u is never zero on M, so the zeros of ρ are isolated. By (10.45) and (10.46), the set of zeros of dH is contained in the union of the set of zeros of ρ and the set of umbilic points of $\mathbf{x}$, and this union is a discrete subset of M. □

Corollary 10.45 (Lawson-Tribuzy [109]). *There are no proper Bonnet immersions of a compact surface M.*

Proof. There is no nonconstant holomorphic function on a compact Riemann surface. □

Corollary 10.46 (Bonnet [15]). *If $\mathbf{x}: M \to \mathbf{R}^3$ is a proper Bonnet immersion, and $\mathscr{D}$ is its discrete set of umbilic points, then $\mathbf{x}: M \setminus \mathscr{D} \to \mathbf{R}^3$ is isothermic.*

Proof. Let $w = u + iv : M \to \mathbf{C}_+$ be a holomorphic function satisfying $dw = -u\rho$ and $u > 0$ on M. If $m \in M$ is any nonumbilic point, let $(\mathbf{x}, e)$ be a principal frame field on an open neighborhood U of m. Let $h > 0$ be the Hopf invariant of $(\mathbf{x}, e)$ on U. Then $(hu)^{1/2}\omega^1$ and $(hu)^{1/2}\omega^2$ are closed 1-forms on U. In fact, using (4.36) and (10.45), we get

$$\frac{1}{h}dh = (r - 2q)\omega^1 - (s - 2p)\omega^2, \tag{10.58}$$

and then

$$d\log(uh) = \frac{du}{u} + \frac{dh}{h} = -2q\omega^1 + 2p\omega^2.$$

Thus, again using (4.36), we get

$$d((hu)^{1/2}\omega^1) = (hu)^{1/2}\left(\frac{1}{2}\frac{d(hu)}{hu} \wedge \omega^1 + p\omega^1 \wedge \omega^2\right) = 0,$$

and

$$d((hu)^{1/2}\omega^2) = (hu)^{1/2}\left(\frac{1}{2}\frac{d(hu)}{hu} \wedge \omega^2 + q\omega^1 \wedge \omega^2\right) = 0.$$

Therefore, on a possibly smaller open neighborhood $V \subset U$ of m, there exist smooth functions x and y such that

$$dx = (hu)^{1/2}\omega^1, \quad dy = (hu)^{1/2}\omega^2,$$

and x, y are local coordinates on V. They are principal coordinates, and

$$I = \omega^1\omega^1 + \omega^2\omega^2 = (hu)^{-1}(dx^2 + dy^2)$$

shows that they are isothermal as well. Hence, off the discrete set of umbilic points of $\mathbf{x}$, there exists a principal isothermal coordinate system on a neighborhood of each point, so $\mathbf{x}$ is isothermic by Definition 9.5. □

Corollary 10.47 (Chern [44]). *If $\mathbf{x}: M \to \mathbf{R}^3$ is a proper Bonnet immersion, then the Riemannian metric (singular at the discrete set of zeros of the Cartan form ρ)*

$$\tilde{I} = \frac{\|dH\|^2}{H^2 - K} I$$

on M has Gaussian curvature identically equal to -1.

Proof. Consider a holomorphic function $w : M \to \mathbf{C}_+$ satisfying (10.57). Using this with (10.45) and (10.46), we find that w pulls back the Poincaré metric of constant Gaussian curvature -1 on $\mathbf{C}_+$ to

$$\frac{dw\,d\bar{w}}{u^2} = \rho\bar{\rho} = (r^2+s^2)(\omega^1\omega^1+\omega^2\omega^2) = \|\frac{dH}{h}\|^2 I = \tilde{I},$$

since $H^2 - K = h^2$. Hence, the Gaussian curvature of $\tilde{I}$ is -1. □

10.10 Cartan's Classification

Definition 10.48. A *B-immersion* is a real analytic, conformal, proper Bonnet immersion $\mathbf{x} : D \to \mathbf{R}^3$, where $D \subset \mathbf{C}$ is a simply connected domain for which the standard complex coordinate $z : D \hookrightarrow \mathbf{C}$ is principal (that is, the Hopf invariant h relative to it is positive), and the mean curvature H of $\mathbf{x}$ has no critical points.

Cartan emphasized three classes of B-immersions:

- Class A consists of those B-immersions $\mathbf{x}$ admitting exactly three nonequivalent Bonnet mates, none of which is equivalent to $\mathbf{x}$;
- Class B consists of those B-immersions $\mathbf{x}$ all of whose Bonnet mates are equivalent to $\mathbf{x}$;
- Class C consists of those B-immersions $\mathbf{x}$ all of whose Bonnet mates are equivalent to each other, but not to $\mathbf{x}$.

In Class A, of the four immersions given by $\mathbf{x}$ and its three nonequivalent Bonnet mates, two of them are helicoids, mirror images of each other. Of the other two, one is either a surface of revolution or a cylinder.

Theorem 10.49 (Cartan [33]). *If $\mathbf{x} : M \to \mathbf{R}^3$ is a proper Bonnet immersion of a simply connected Riemann surface M and if dH is never zero on M, then there exists a principal complex coordinate $z = x + iy$ on M such that a Cartan function w of $\mathbf{x}$ has real part $\Re(w)$ equal to one of the following seven cases, which are called the canonical B-immersions.*

1. $\sin x \sinh y$ *and* $M = \{z \in \mathbf{C} : 0 < x < \pi,\ y > 0\}$
2. $\sinh x \sin y$ *and* $M = \{z \in \mathbf{C} : x > 0,\ 0 < y < \pi\}$
3. xy *and* $M = \{z \in \mathbf{C} : x > 0\ y > 0\}$
4. $-xy$ *and* $M = \{z \in \mathbf{C} : x < 0\ y > 0\}$
5. $\cos x \cosh y$ *and* $M = \{z \in \mathbf{C} : -\pi/2 < x < \pi/2\}$
6. $e^y \cos x$ *and* $M = \{z \in \mathbf{C} : -\pi/2 < x < \pi/2\}$
7. x *and* $M = \{z \in \mathbf{C} : x > 0\}$.

Remark 10.50. In cases 1)–4) and 6) the principal complex coordinate z is unique. In case 5), z is unique up to $\pm z$. In case 7), z is defined up to $z + ic$, for any $c \in \mathbf{R}$.

Cartan classifies proper Bonnet immersions into the three Classes, A, B, and C. Class A comprises cases 1)–4), Class B comprises only case 5), and Class C comprises cases 6) and 7).

It would take us too far afield to give a more precise statement of Cartan's classification and to prove it. For all the details see Cartan's 1942 paper [33].

Problems

10.51. Prove that the only Bonnet mates of a CMC immersion $\mathbf{x} : M \to \mathbf{R}^3$ are its associates, as discussed in Example 10.11.

10.52. Prove that if $\mathbf{x} : M \to \mathbf{R}^3$ is isothermic and if the complex coordinate z in Theorem 10.13 is principal, so $h > 0$ on M, then the PDE (10.5) holds if and only if $1/(he^{2u})$ is harmonic. This latter condition is the one used by Graustein [76].

10.53 (Associates of a circular cylinder). For a constant $R > 0$, consider the immersion

$$\mathbf{x} : \mathbf{R}^2 \to \mathbf{R}^3, \quad \mathbf{x}(s,t) = \frac{1}{R}{}^t(\cos Rs, \sin Rs, Rt),$$

whose image is a circular cylinder of radius R. Relative to the outward normal the mean curvature is the constant $H = -R/2$. Let $II^{2,0}$ denote its Hopf quadratic differential. For each constant r with $0 \leq r < 2\pi$, let

$$\omega = R\sqrt{\frac{1+\cos r}{2}}, \quad \nu = R\sqrt{\frac{1-\cos r}{2}}.$$

Then $F : \mathbf{R}^2 \to \mathbf{R}^2$, $F(s,t) = \frac{1}{R}(\omega s - \nu t, \nu s + \omega t)$ is an isometry for the metric $I = ds^2 + dt^2$ induced by $\mathbf{x}$. Prove that, up to congruence by a rigid motion, the associate of $\mathbf{x}$ with Hopf quadratic differential $e^{ir} II^{2,0}$ is

$$\tilde{\mathbf{x}} = \mathbf{x} \circ F : \mathbf{R}^2 \to \mathbf{R}^3,$$

which is a reparametrization of the original cylinder.

10.54. Prove that for any positive real constant $n > 0$, the curve $\boldsymbol{\gamma}_n : \mathbf{R}^+ \to \mathbf{R}^2$ with curvature $\kappa = n/x$ is, up to rotation and translation in the plane, given by (10.8), where $\mathbf{v}_n$ is given by (10.9) and A_n is given by (10.10).

10.55. Given positive real constants $m > 0$ and $n > 0$, let $\tilde{\boldsymbol{\gamma}}(x) = m\boldsymbol{\gamma}_n(x/m)$ on $\mathbf{R}^+$. Prove that x is arclength parameter for $\tilde{\boldsymbol{\gamma}}(x)$ and that this curve has curvature $\tilde{\kappa}(x) = n/x$. Conclude that $\tilde{\boldsymbol{\gamma}}$ is congruent to $\boldsymbol{\gamma}_n$ with congruence given by

$$\tilde{\boldsymbol{\gamma}}(x) = \mathbf{v} + {}^t\!A_n(m)\boldsymbol{\gamma}_n(x),$$

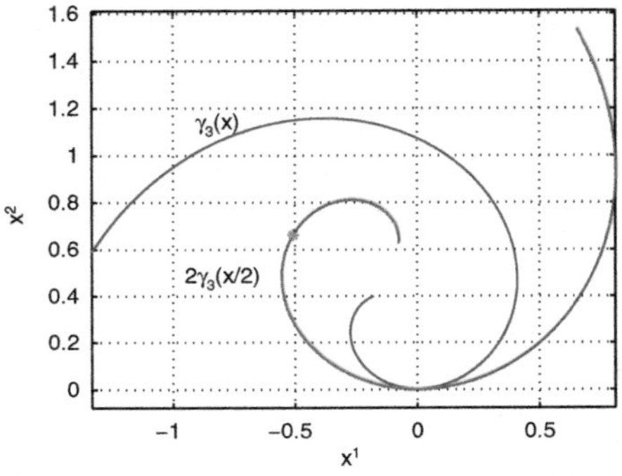

Fig. 10.7 Curve $\boldsymbol{\gamma}_3(x)$ and $2\boldsymbol{\gamma}_3(x/2)$ with $\mathbf{v}$ marked on it.

where the constant vector $\mathbf{v} = \tilde{\boldsymbol{\gamma}}(1) = m\boldsymbol{\gamma}(1/m)$ and where $A_n : \mathbf{R}^+ \to \mathbf{SO}(2)$ is defined in (10.10). See Figure 10.7.

10.56. Prove that $\mathbf{x}_\pm : \tilde{J} \times \mathbf{R} \to \mathbf{R}^3$ defined in (10.42) have the same mean curvatures

$$H_\pm = H/|q_\pm|^2 = \frac{\kappa}{2(c^2 + t^2 + |\boldsymbol{\gamma}|^2)},$$

the same induced metrics

$$I_\pm = (c^2 + t^2 + |\boldsymbol{\gamma}|^2)^2 dz d\bar{z},$$

and their Hopf invariants relative to $z = s + it$ satisfy

$$h_+ - h_- = \frac{-4ci}{(c^2 + t^2 + |\boldsymbol{\gamma}|^2)^2}.$$

10.57. Prove that $F : \mathbf{R}^2 \to \mathbf{R}^2$ given by $F(s,t) = (-s,t)$ is an isometry for the metric induced by the Bonnet pair $\mathbf{x}_\pm$ of Example 10.34 and that $\mathbf{x}_+ \circ F$, $\mathbf{x}_- : \mathbf{R}^2 \to \mathbf{R}^3$ are congruent.

10.58. Find the induced metric and mean curvature H of Bonnet pair $\mathbf{x}_\pm$ in (10.44), and the Hopf invariants $h_\pm$ of $\mathbf{x}_\pm$ relative to $z = s + it$.

10.59. Do Problem 10.57 for the Bonnet pair $\mathbf{x}_\pm$ in (10.44) generated by a cone.

10.60. Prove that if M is simply connected and if the immersion $\mathbf{x} : M \to \mathbf{R}^3$ has no umbilic points, then there exists a principal frame field on all of M. Is it unique?

Chapter 11
CMC 1 Surfaces in $\mathbf{H}^3$

This chapter is an introduction to immersions of surfaces in hyperbolic space with constant mean curvature equal to one (CMC 1 immersions in $\mathbf{H}^3$). Our approach to the subject follows Bryant's fundamental paper [21], in which he replaces the hyperboloid model of $\mathbf{H}^3$ by the set of all 2×2 hermitian matrices with determinant one and positive trace. This model is acted upon isometrically by $\mathbf{SL}(2, \mathbf{C})$, the universal cover of the group of all isometries of hyperbolic space. The method of moving frames is applied to the study of immersed surfaces in this homogeneous space. Departing from Bryant's approach, we use frames adapted to a given complex coordinate to great advantage. A null immersion from a Riemann surface into $\mathbf{SL}(2, \mathbf{C})$ projects to a CMC 1 immersion into hyperbolic space. The null immersions are analogous to minimal curves of the Weierstrass representation of minimal immersions into Euclidean space. A solution of these equations leads to a more complicated monodromy problem, which is described in detail. The chapter ends with some of Bryant's examples as well as more recent examples of Bohle-Peters [12] and Bobenko-Pavlyukevich-Springborn [10]. We have also consulted Umehara and Yamada [162, 163], Rossman [141], and Karcher [99]. There is a vast literature on this active field of research.

In much of the growing literature on this subject a CMC 1 immersion into $\mathbf{H}^3$ is called a *Bryant surface* or *Bryant immersion*. We shall adopt this terminology. As with a minimal surface in $\mathbf{R}^3$, a Bryant surface cannot be compact, but an end can converge to a single point in the sphere at infinity in such a way that the result is a smooth immersion into $\mathbf{H}^3 \cup \mathbf{S}^2_\infty$. These are called *Bryant surfaces with smooth ends*.

G.R. Jensen et al., *Surfaces in Classical Geometries*, Universitext,
DOI 10.1007/978-3-319-27076-0_11

11.1 Hermitian matrix model

The hyperboloid model of hyperbolic space developed in Chapter 6 is given in equation (6.1). It is the homogeneous space

$$\mathbf{H}^3 = \mathbf{SO}_+(3,1)/\mathbf{SO}(3).$$

We want to use the universal covering group $\mathbf{SL}(2,\mathbf{C})$ of $\mathbf{SO}_+(3,1)$. For this purpose, consider the vector space over $\mathbf{R}$ of all 2×2 hermitian matrices

$$\mathbf{Herm} = \{v \in \mathbf{C}^{2\times 2} : {}^t\bar{v} = v\} = \{\begin{pmatrix} r & z \\ \bar{z} & s \end{pmatrix} : r,s \in \mathbf{R},\ z \in \mathbf{C}\},$$

with the signature $(3,1)$ inner product defined by $-\det(v)$,

$$\langle u, v\rangle = \frac{1}{2}(\det(u) + \det(v) - \det(u+v)),$$

for any $u, v \in \mathbf{Herm}$. The map

$$v : \mathbf{R}^{3,1} \to \mathbf{Herm}, \quad v(x^1,x^2,x^3,x^4) = \begin{pmatrix} x^4 + x^3 & x^1 + ix^2 \\ x^1 - ix^2 & x^4 - x^3 \end{pmatrix} \tag{11.1}$$

is a linear isomorphism preserving the inner products.

Exercise 43. Prove that v is an isometry with inverse

$$v^{-1}\begin{pmatrix} r & z \\ \bar{z} & s \end{pmatrix} = \frac{1}{2}{}^t(z+\bar{z}, i(\bar{z}-z), r-s, r+s),$$

for any $r, s \in \mathbf{R}$ and $z \in \mathbf{C}$.

Denote the image of hyperbolic space under this map by

$$\mathbf{H}(3) = v(\mathbf{H}^3) = \{v \in \mathbf{Herm} : \det(v) = 1,\ \mathrm{trace}(v) > 0\}.$$

The origin ϵ_4 of $\mathbf{H}^3$ is mapped by v to the identity matrix in $\mathbf{Herm}$.

11.2 The Universal Cover

For any $A \in \mathbf{SL}(2,\mathbf{C})$ and $v \in \mathbf{Herm}$, define the linear operator

$$\Sigma(A) : \mathbf{Herm} \to \mathbf{Herm} \quad \Sigma(A)v = AvA^*,$$

where $A^* = {}^t\bar{A}$ is the adjoint of A. The operator $\Sigma(A)$ preserves the inner product on **Herm**, since $\det(A) = 1$ implies that

$$\det(\Sigma(A)v) = \det(AvA^*) = \det(A)\det(v)\det(A^*) = \det(v).$$

The map

$$\Sigma : \mathbf{SL}(2,\mathbf{C}) \to \mathbf{GL}(\mathbf{Herm}), \quad A \mapsto \Sigma(A),$$

is a group homomorphism of $\mathbf{SL}(2,\mathbf{C})$ into the group $\mathbf{GL}(\mathbf{Herm})$ of all non-singular linear operators on **Herm**. In fact, if $A, B \in \mathbf{SL}(2,\mathbf{C})$, and $v \in \mathbf{Herm}$, then

$$\Sigma(AB)v = (AB)v(AB)^* = ABvB^*A^* = \Sigma(A)(\Sigma(B)v).$$

The image of this map must lie in the subgroup of $\mathbf{GL}(\mathbf{Herm})$ that preserves the inner product on **Herm**.

The map v sends the standard orthonormal basis $\epsilon_1, \epsilon_2, \epsilon_3, \epsilon_4$ of $\mathbf{R}^{3,1}$ to the orthonormal basis of **Herm** given by the *Pauli matrices*

$$v_4 = v(\epsilon_4) = \begin{pmatrix} 1 & 0 \\ 0 & 1 \end{pmatrix}, \quad v_1 = v(\epsilon_1) = \begin{pmatrix} 0 & 1 \\ 1 & 0 \end{pmatrix},$$

$$v_2 = v(\epsilon_2) = \begin{pmatrix} 0 & i \\ -i & 0 \end{pmatrix}, \quad v_3 = v(\epsilon_3) = \begin{pmatrix} 1 & 0 \\ 0 & -1 \end{pmatrix}.$$

Let $P : \mathbf{GL}(\mathbf{Herm}) \to \mathbf{GL}(4,\mathbf{R})$ be the linear isomorphism that sends an operator on **Herm** to its matrix relative to the Pauli basis of **Herm**.

Lemma 11.1. *The composition*

$$\sigma = P \circ \Sigma : \mathbf{SL}(2,\mathbf{C}) \to \mathbf{GL}(4,\mathbf{R})$$

is a Lie group homomorphism that defines a 2:1 cover of its image,

$$\sigma : \mathbf{SL}(2,\mathbf{C}) \to \mathbf{SO}_+(3,1). \tag{11.2}$$

Proof. Since the image of Σ is contained in the subgroup of $\mathbf{GL}(\mathbf{Herm})$ that preserves the inner product on **Herm**, the map σ must send $\mathbf{SL}(2,\mathbf{C})$ into Σ of this subgroup, which is $\mathbf{O}(3,1)$. These are both Lie groups of dimension 6. It is known that $\mathbf{SL}(2,\mathbf{C})$ is homeomorphic to $\mathbf{SU}(2) \times \mathbf{R}^3$ (see [84, Theorem 2.2, part iii), page 219]), and therefore it is connected and simply connected, because $\mathbf{SU}(2)$ is diffeomorphic to the sphere S^3. It follows that σ sends $\mathbf{SL}(2,\mathbf{C})$ onto $\mathbf{SO}_+(3,1)$, the connected component of the identity of $\mathbf{O}(3,1)$.

To prove that σ is 2:1, it suffices to show that $\sigma(A) = 1$ if and only if $A = \pm 1$. If $\sigma(A) = 1$, then $\Sigma(A)v = v$ for all $v \in \mathbf{Herm}$. In particular, $Av_iA^* = v_i$, for the Pauli matrices v_i, for $i = 4, 1, 2$. From this it follows that $A = \pm 1$. Conversely, $\sigma(\pm 1) = 1$ is clear. □

Remark 11.2. If $v \in \mathbf{Herm}$, and if v_i are the Pauli matrices, then

$$v = \sum_1^4 x^i v_i = \sum_1^4 x^i v(\epsilon_i) = v(\sum_1^4 x^i \epsilon_i) = v(\mathbf{x}),$$

where $\mathbf{x} = \sum_1^4 x^i \epsilon_i \in \mathbf{R}^{3,1}$. If $A \in \mathbf{SL}(2,\mathbf{C})$, then the entries of the matrix $\sigma(A)$ are determined by

$$A v_i A^* = \Sigma(A) v_i = \sum_1^4 \sigma(A)^j_i v_j.$$

The linear action of $\mathbf{SL}(2,\mathbf{C})$ on $\mathbf{Herm}$ preserves $\mathbf{H}(3)$. Its isotropy subgroup at 1 is

$$\mathbf{SU}(2) = \{A \in \mathbf{SL}(2,\mathbf{C}) : AA^* = 1\}.$$

Thus, $\mathbf{H}(3)$ is diffeomorphic to the quotient $\mathbf{SL}(2,\mathbf{C})/\mathbf{SU}(2)$, and we have the principal $\mathbf{SU}(2)$-bundle projection

$$\pi : \mathbf{SL}(2,\mathbf{C}) \to \mathbf{H}(3), \quad \pi(A) = AA^*. \tag{11.3}$$

Lemma 11.3. *The derivative at the identity of the covering* (11.2) *is a Lie algebra isomorphism*

$$d\sigma : \mathfrak{sl}(2,\mathbf{C}) \to \mathfrak{o}(3,1),$$

which sends $X = \begin{pmatrix} X^1_1 & X^1_2 \\ X^2_1 & -X^1_1 \end{pmatrix}$ *to*

$$d\sigma(X) = \begin{pmatrix} 0 & -2\Im(X^1_1) & \Re(-X^1_2 + X^2_1) & \Re(X^1_2 + X^2_1) \\ 2\Im(X^1_1) & 0 & -\Im(X^1_2 + X^2_1) & \Im(X^1_2 - X^2_1) \\ \Re(X^1_2 - X^2_1) & \Im(X^1_2 + X^2_1) & 0 & 2\Re(X^1_1) \\ \Re(X^1_2 + X^2_1) & \Im(X^1_2 - X^2_1) & 2\Re(X^1_1) & 0 \end{pmatrix}. \tag{11.4}$$

If (ω^i_j) *is the* $\mathfrak{o}(3,1)$*-valued Maurer–Cartan form on* $\mathbf{SO}_+(3,1)$ *and* $\alpha = (\alpha^a_b)$ *is the* $\mathfrak{sl}(2,\mathbf{C})$*-valued Maurer–Cartan form on* $\mathbf{SL}(2,\mathbf{C})$*, then*

$$\begin{aligned} \sigma^*\omega^1_4 &= \Re(\alpha^1_2 + \alpha^2_1), & \sigma^*\omega^2_4 &= \Im(\alpha^1_2 - \alpha^2_1), & \sigma^*\omega^3_4 &= 2\Re(\alpha^1_1), \\ \sigma^*\omega^1_2 &= -2\Im(\alpha^1_1), & \sigma^*\omega^3_1 &= \Re(\alpha^1_2 - \alpha^2_1), & \sigma^*\omega^3_2 &= \Im(\alpha^1_2 + \alpha^2_1). \end{aligned} \tag{11.5}$$

If we set

$$\omega = \omega^1_4 + i\omega^2_4, \quad \psi = \omega^3_1 - i\omega^3_2, \tag{11.6}$$

then

$$\sigma^*\omega = \alpha^1_2 + \bar{\alpha}^2_1, \quad \sigma^*\psi = \bar{\alpha}^1_2 - \alpha^2_1,$$

and

$$\alpha = \begin{pmatrix} \alpha^1_1 & \alpha^1_2 \\ \alpha^2_1 & -\alpha^1_1 \end{pmatrix} = \sigma^* \frac{1}{2} \begin{pmatrix} \omega^3_4 - i\omega^1_2 & \omega + \bar{\psi} \\ \bar{\omega} - \psi & -\omega^3_4 + i\omega^1_2 \end{pmatrix}.$$

Proof. For any $X \in \mathfrak{sl}(2, \mathbf{C})$, $d\sigma(X)$ is the matrix of the endomorphism $d\Sigma(X)$: **Herm** $\to$ **Herm**, which is

$$d\Sigma(X)v = Xv + vX^*.$$

In fact, if we let $a(t)$ be a curve in $\mathbf{SL}(2, \mathbf{C})$ such that $a(0) = 1$ and its tangent vector $\dot{a}(0) = X$, then for any $v \in$ **Herm**,

$$d\Sigma(X)v = \left.\frac{d}{dt}\right|_{t=0} \Sigma(a)v = \left.\frac{d}{dt}\right|_{t=0} (ava^*) = Xv + vX^*.$$

One then calculates column j of $d\sigma(X)$ by calculating $d\sigma(X)v_j$ and expanding it in terms of the Pauli basis of **Herm**. This derives (11.4), from which the rest follows. □

Definition 11.4. An $\mathbf{SL}(2, \mathbf{C})$-frame field on an open subset $U \subset \mathbf{H}(3)$ is a smooth map

$$F : U \to \mathbf{SL}(2, \mathbf{C}), \tag{11.7}$$

such that its composition with the projection (11.3) is the identity map on U. That is,

$$\pi \circ F(v) = F(v)F(v)^* = v,$$

for every $v \in U$.

Given a frame field (11.7), any other on U is given by

$$\tilde{F} = FK, \tag{11.8}$$

where

$$K : U \to \mathbf{SU}(2) \tag{11.9}$$

is any smooth map. Their pull-backs $\alpha = F^{-1}dF$ and $\tilde{\alpha} = \tilde{F}^{-1}d\tilde{F}$ of the Maurer–Cartan form of $\mathbf{SL}(2, \mathbf{C})$ are related by

$$\tilde{\alpha} = K^{-1}\alpha K + K^{-1}dK. \tag{11.10}$$

Since any map K of (11.9) is given by

$$K = \begin{pmatrix} a & -\bar{b} \\ b & \bar{a} \end{pmatrix}, \tag{11.11}$$

where $a, b : U \to \mathbf{C}$ are smooth functions satisfying $|a|^2 + |b|^2 = 1$ at every point of U, we can write equations (11.10) as

$$\begin{aligned}
\tilde{\alpha}_1^1 &= (|a|^2 - |b|^2)\alpha_1^1 + \bar{a}b\alpha_2^1 + a\bar{b}\alpha_1^2 + \bar{a}da + \bar{b}db, \\
\tilde{\alpha}_2^1 &= -2\bar{a}\bar{b}\alpha_1^1 + \bar{a}^2\alpha_2^1 - \bar{b}^2\alpha_1^2 - \bar{a}d\bar{b} + \bar{b}d\bar{a}, \\
\tilde{\alpha}_1^2 &= -2ab\alpha_1^1 + a^2\alpha_1^2 - b^2\alpha_2^1 - bda + adb.
\end{aligned}$$

In particular,

$$\begin{aligned}
\tilde{\alpha}_1^1 + \bar{\tilde{\alpha}}_1^1 &= (|a|^2 - |b|^2)(\alpha_1^1 + \bar{\alpha}_1^1) + \bar{a}b(\alpha_2^1 + \bar{\alpha}_1^2) + a\bar{b}(\alpha_1^2 + \bar{\alpha}_2^1), \\
\tilde{\alpha}_2^1 + \bar{\tilde{\alpha}}_1^2 &= -2\bar{a}\bar{b}(\alpha_1^1 + \bar{\alpha}_1^1) + \bar{a}^2(\alpha_2^1 + \bar{\alpha}_1^2) - \bar{b}^2(\alpha_1^2 + \bar{\alpha}_2^1), \\
\tilde{\alpha}_1^2 - \bar{\tilde{\alpha}}_2^1 &= 2ab(\bar{\alpha}_1^1 - \alpha_1^1) + a^2(\alpha_1^2 - \bar{\alpha}_2^1) + b^2(\bar{\alpha}_1^2 - \alpha_2^1) + 2adb - 2bda.
\end{aligned} \tag{11.12}$$

A matrix $X \in \mathfrak{sl}(2, \mathbf{C})$ has a unique decomposition into skew-hermitian and hermitian parts given by

$$X = \frac{1}{2}(X - X^*) + \frac{1}{2}(X + X^*),$$

which gives rise to a vector space direct sum

$$\mathfrak{sl}(2, \mathbf{C}) = \mathfrak{su}(2) + \mathfrak{m},$$

where the Lie algebra of $\mathbf{SU}(2)$ is

$$\mathfrak{su}(2) = \{X \in \mathfrak{sl}(2, \mathbf{C}) : X^* = -X\},$$

and the vector subspace

$$\mathfrak{m} = \{X \in \mathfrak{sl}(2, \mathbf{C}) : X^* = X\}.$$

Then the Maurer–Cartan form α of $\mathbf{SL}(2, \mathbf{C})$ decomposes as

$$\alpha = \alpha_{\mathfrak{su}(2)} + \alpha_{\mathfrak{m}},$$

where

$$\alpha_{\mathfrak{su}(2)} = \frac{1}{2}\begin{pmatrix} \alpha_1^1 - \bar{\alpha}_1^1 & \alpha_2^1 - \bar{\alpha}_1^2 \\ \alpha_1^2 - \bar{\alpha}_2^1 & -\alpha_1^1 + \bar{\alpha}_1^1 \end{pmatrix}, \quad \alpha_{\mathfrak{m}} = \frac{1}{2}\begin{pmatrix} \alpha_1^1 + \bar{\alpha}_1^1 & \alpha_2^1 + \bar{\alpha}_1^2 \\ \alpha_1^2 + \bar{\alpha}_2^1 & -\alpha_1^1 - \bar{\alpha}_1^1 \end{pmatrix}.$$

The subspace $\mathfrak{m}$ is invariant under the adjoint action of $\mathbf{SU}(2)$ on $\mathfrak{sl}(2, \mathbf{C})$; that is, $K\mathfrak{m}K^{-1} \subset \mathfrak{m}$, for any $K \in \mathbf{SU}(2)$.

11.3 Sphere at infinity of $\mathbf{H}(3)$

The sphere at infinity of $\mathbf{H}^3$ is (6.18) $S^2_\infty = N/\mathbf{R}^\times$, where $N \subset \mathbf{R}^{3,1}$ is the light cone (6.17). The *light cone* $N(3) \subset \mathbf{Herm}$ is

$$v(N^3) = N(3) = \{v \in \mathbf{Herm} : \det(v) = 0\}.$$

$\mathbf{R}^\times$ acts on this by scalar multiplication such that

$$v(t\mathbf{n}) = tv(\mathbf{n}),$$

for all $t \in \mathbf{R}^\times$, and $\mathbf{n} \in N^3$. The sphere at infinity of $\mathbf{H}(3)$ is

$$S^2_\infty = N(3)/\mathbf{R}^\times,$$

which we have denoted the same as the sphere at infinity of $\mathbf{H}^3$, because of the natural map induced by v, which we also call v,

$$v : N^3/\mathbf{R}^+ \to N(3)/\mathbf{R}^+, \quad v([\mathbf{n}]) = [v(\mathbf{n})]. \tag{11.13}$$

In Lemma 6.5 one proves that $\mathbf{SO}_+(3,1)$ acts transitively on $N^3/\mathbf{R}^\times$ and one finds the isotropy subgroup at $[\epsilon_3 + \epsilon_4]$, the chosen origin. For an origin of $N(3)$ we choose

$$v[\epsilon_3 + \epsilon_4] = [v_3 + v_4] = \begin{bmatrix} 1 & 0 \\ 0 & 0 \end{bmatrix}.$$

The action of $\mathbf{SL}(2,\mathbf{C})$ on $\mathbf{Herm}$ induces a transitive action on $N(3)/\mathbf{R}^\times$. Its isotropy subgroup at $[v_3 + v_4]$ is

$$\hat{G}_0 = \{\begin{pmatrix} z & w \\ 0 & z^{-1} \end{pmatrix} : z, w \in \mathbf{C}, z \neq 0\}.$$

We have a $\hat{G}_0$-principal bundle

$$\pi_1 : \mathbf{SL}(2,\mathbf{C}) \to N(3)/\mathbf{R}^\times, \quad \pi_1(A) = A\begin{bmatrix} 1 & 0 \\ 0 & 0 \end{bmatrix}A^* = [A_1A_1^*],$$

where A_1 is the first column of A. The standard transitive action of $\mathbf{SL}(2,\mathbf{C})$ on $\mathbf{CP}^1$ has isotropy subgroup at $\begin{bmatrix} 1 \\ 0 \end{bmatrix}$ also equal to $\hat{G}_0$. We have another principal $\hat{G}_0$-bundle

$$\pi_2 : \mathbf{SL}(2,\mathbf{C}) \to \mathbf{C}P^1, \quad \pi_2(A) = A\begin{bmatrix} 1 \\ 0 \end{bmatrix} = [A_1].$$

These bundles are isomorphic.

The light cone $N(3) \subset \mathbf{Herm}$ is

$$N(3) = \{\pm \mathbf{z}\mathbf{z}^* : \mathbf{z} \in \mathbf{C}^2 \setminus \{0\}\}.$$

The map

$$\tau : \mathbf{CP}^1 \to N(3)/\mathbf{R}^\times, \quad \tau[\mathbf{z}] = [\mathbf{z}\mathbf{z}^*]$$

is a diffeomorphism equivariant with the actions of $\mathbf{SL}(2,\mathbf{C})$ on $\mathbf{CP}^1$ and $N(3)/\mathbf{R}^\times$. Namely, if $A \in \mathbf{SL}(2,\mathbf{C})$ and if $[\mathbf{z}] \in \mathbf{CP}^1$, then

$$\tau(A[\mathbf{z}]) = [(A\mathbf{z})(A\mathbf{z})^*] = [A(\mathbf{z}\mathbf{z}^*)A^*] = A(\tau[\mathbf{z}])A^*.$$

In particular, from the case $\mathbf{z} = {}^t[1,0]$ we get

$$\tau \circ \pi_2 = \pi_1.$$

The diagram

$$\begin{array}{ccc} \mathbf{SL}(2,\mathbf{C}) & \xrightarrow{\pi_2} & \mathbf{CP}^1 \\ \downarrow \sigma & & \\ \mathbf{SO}_+(3,1) & \overset{\pi_1}{\searrow} & \downarrow \tau \\ \downarrow \pi & & \\ N^3/\mathbf{R}^\times & \xrightarrow{v} & N(3)/\mathbf{R}^\times \end{array} \tag{11.14}$$

commutes, where $\pi : \mathbf{SO}_+(3,1) \to N^3/\mathbf{R}^\times$ is the G_0-bundle projection in (6.21).

If $\mathbf{x} \in \mathbf{H}^3$ and if $S^2_\mathbf{x}$ is the unit tangent sphere to $\mathbf{H}^3$ at $\mathbf{x}$, then

$$v(S^2_\mathbf{x}) = S^2_{v(\mathbf{x})} = \{w \in \mathbf{Herm} : \langle w, v(\mathbf{x})\rangle = 0, \ \langle w, w\rangle = 1\}$$

is the unit tangent sphere to $\mathbf{H}(3)$ at $v(\mathbf{x})$. Moreover, if v is applied to the map $S^2_\mathbf{x} \to N^3/\mathbf{R}^\times$ given in (6.19), it gives the map

$$S^2_v \to N(3)/\mathbf{R}^\times = S^2_\infty, \quad w \mapsto [v + w].$$

11.4 Surfaces in H(3)

Since $v : \mathbf{H}^3 \to \mathbf{H}(3)$ is an isometry, any immersion

$$f : M^2 \to \mathbf{H}(3) \tag{11.15}$$

comes from an immersed surface $\mathbf{x} : M \to \mathbf{H}^3$ by $f = v(\mathbf{x})$.

Definition 11.5. An $\mathbf{SL}(2,\mathbf{C})$-*frame field* along an immersion (11.15) is a smooth map

$$F : U \subset M \to \mathbf{SL}(2,\mathbf{C}) \tag{11.16}$$

on an open subset U of M such that its projection

$$FF^* = f,$$

at every point of U. Denote the pull-back by F of the Maurer–Cartan form of $\mathbf{SL}(2,\mathbf{C})$ by the $\mathfrak{sl}(2,\mathbf{C})$-valued 1-form on U

$$\alpha = \begin{pmatrix} \alpha_1^1 & \alpha_2^1 \\ \alpha_1^2 & -\alpha_1^1 \end{pmatrix} = F^{-1}dF. \tag{11.17}$$

The reduction of $\mathbf{SL}(2,\mathbf{C})$-frames can be carried out by the general procedure described in Chapter 3, but we want to do this in such a way that the frames of first order, and so on, correspond under the isomorphism $\upsilon : \mathbf{R}^{3,1} \to \mathbf{Herm}$ to frames of first order, etc., as defined in Chapter 6 for surfaces in $\mathbf{H}^3$. In the light of the third equation in (11.5) and our definition of first order frames along immersed surfaces in $\mathbf{H}^3$, we must define first order $\mathbf{SL}(2,\mathbf{C})$-frames as follows.

Definition 11.6. A *first order* $\mathbf{SL}(2,\mathbf{C})$-*frame field* along an immersed surface $f : M \to \mathbf{H}(3)$ is an $\mathbf{SL}(2,\mathbf{C})$-frame field (11.16) along f for which

$$\alpha_1^1 + \bar{\alpha}_1^1 = 0$$

at each point of U.

This is the required definition for the following reasons. If

$$\mathbf{x} : M^2 \to \mathbf{H}^3$$

is an immersion, then

$$f = \upsilon(\mathbf{x}) : M \to \mathbf{H}(3)$$

is an immersion, and they both induce the same metric I on M, since $\upsilon : \mathbf{H}^3 \to \mathbf{H}(3)$ is an isometry. If

$$F : U \subset M \to \mathbf{SL}(2,\mathbf{C})$$

is a frame field along f, then

$$e = \sigma \circ F : U \to \mathbf{SO}_+(3,1)$$

is a frame field along $\mathbf{x}$. We want to define F to be first order along f if and only if $e = \sigma \circ F$ is first order along $\mathbf{x}$. But e is first order along $\mathbf{x}$ if and only if $\omega_4^3 = 0$ if and only if $0 = \sigma^* \omega_4^3 = \alpha_1^1 + \bar{\alpha}_1^1$, by the third equation in (11.5).

For any point of M, there exists a first order $\mathbf{SL}(2, \mathbf{C})$-frame along f on some neighborhood of the point, because this is true for first order frames along $\mathbf{x}$. Namely, if $e : U \to \mathbf{SO}_+(3,1)$ is a first order frame field along $\mathbf{x}$, and if U is simply connected, then there is a lift $F : U \to \mathbf{SL}(2, \mathbf{C})$, since $\sigma : \mathbf{SL}(2, \mathbf{C}) \to \mathbf{SO}_+(3,1)$ is a covering projection, and this lift is a first order frame along $f = \upsilon(\mathbf{x})$.

Lemma 11.7. *If $F : U \to \mathbf{SL}(2, \mathbf{C})$ is a first order frame field on a connected open set $U \subset M$ along an immersion $f : M \to \mathbf{H}(3)$, and if $\omega = \alpha_2^1 + \bar{\alpha}_1^2$, where α is defined in* (11.17)*, then the metric $I = -\det(df)$ induced by f on M satisfies*

$$I = \omega \bar{\omega}.$$

Thus, ω is of bidegree (1,0) *or* (0,1) *relative to the complex structure induced on M, according to whether the area form of F, $\frac{i}{2}\omega \wedge \bar{\omega}$, is positive or negative in the orientation of M. Any other first order frame field along f on $U \subset M$, is given by*

$$\tilde{F} = FK,$$

where $K : U \to G_1$ is a smooth map into the subgroup $G_1 \subset \mathbf{SU}(2)$,

$$G_1 = \{\begin{pmatrix} a & 0 \\ 0 & \bar{a} \end{pmatrix} : |a| = 1\} \cup \{\begin{pmatrix} 0 & -\bar{b} \\ b & 0 \end{pmatrix} : |b| = 1\}.$$

The frame field $\tilde{F}$ defines the same orientation as F if and only if K takes values in the connected component of the identity of G_1; that is, $K = \begin{pmatrix} a & 0 \\ 0 & \bar{a} \end{pmatrix}$, where $a : U \to S^1 \subset \mathbf{C}$ is a smooth map.

Proof. For a first order frame field F we have $\alpha_1^1 + \bar{\alpha}_1^1 = 0$. Since $f = FF^*$ and α is given by (11.17), we have

$$df = d(FF^*) = F(\alpha + \alpha^*)F^*.$$

Knowing $\det(F) = \det(F^*) = 1$, we have

$$I = -\det(df) = -\det(\alpha + \alpha^*) = \omega \bar{\omega}.$$

Frame fields $\tilde{F}$ and F along f must be related by (11.8), where $K : U \to \mathbf{SU}(2)$ has the form (11.11). If $\tilde{F}$ and F are both first order frame fields, then Definition 11.6 and the first two equations of (11.12) imply

$$0 = \tilde{\alpha}_1^1 + \bar{\tilde{\alpha}}_1^1 = \bar{a}b\omega + a\bar{b}\bar{\omega}, \quad \tilde{\omega} = \bar{a}^2\omega - \bar{b}^2\bar{\omega}, \tag{11.18}$$

on U. This implies $\bar{a}b = 0$ on U, since ω and $\bar{\omega}$ are linearly independent over $\mathbf{C}$ at every point of U. With U connected, either a or b is identically zero on U, so either $\tilde{\omega} = \bar{a}^2\omega$ or $\tilde{\omega} = -\bar{b}^2\bar{\omega}$ on U. That the frame fields F and $\tilde{F}$ induce the same orientation on U means $\tilde{\omega}\wedge\bar{\tilde{\omega}} = \omega\wedge\bar{\omega}$ on U, which requires that $b = 0$ identically on U. □

We assume now that M is a connected Riemann surface and that any immersion $f : M \to \mathbf{H}(3)$ is conformal, which means that the complex structure defined by the induced metric agrees with the given complex structure on M.

Definition 11.8. A first order $\mathbf{SL}(2,\mathbf{C})$-frame field (11.16) along an immersion $f : M^2 \to \mathbf{H}(3)$ is *oriented* on U if $\omega = \alpha^1_2 + \bar{\alpha}^2_1$ is of bidegree $(1,0)$; that is, for any complex coordinate z in U,

$$\omega = a\,dz,$$

for some nowhere vanishing function $a : U \to \mathbf{C}$. The frame field is *adapted* to U, z if a is positive on U; that is,

$$\omega = e^u dz,$$

for some smooth function $u : U \to \mathbf{R}$.

Definition 11.9. The *hyperbolic Gauss map* of an immersion $f : M^2 \to \mathbf{H}(3)$ is

$$g_f = v \circ g$$

where $g : M \to S^2_\infty = N^3/\mathbf{R}^+$ is the hyperbolic Gauss map of the immersion $\mathbf{x} : M \to \mathbf{H}^3$ for which $f = v(\mathbf{x})$, and is the map defined in (11.13).

Lemma 11.10. *If $F : U \to \mathbf{SL}(2,\mathbf{C})$ is an oriented first order frame field along the immersion $f : M \to \mathbf{H}(3)$, and $\omega = \alpha^1_2 + \bar{\alpha}^2_1$, then*

$$d\omega = -2\alpha^1_1\wedge\omega, \tag{11.19}$$

$$\psi = \bar{\alpha}^1_2 - \alpha^2_1 = h\omega + H\bar{\omega}, \tag{11.20}$$

where H is the mean curvature of f, and $h : U \to \mathbf{C}$ is the Hopf invariant

$$h = \frac{1}{2}(h_{11} - h_{22}) - ih_{12},$$

where

$$\omega^3_i = \sum_{j=1}^{2} h_{ij}\omega^j,$$

for $i = 1,2$, *are components of* $e^{-1}de \in \mathfrak{o}(3,1)$ *of the corresponding frame* $e = \sigma \circ F : U \to \mathbf{SO}_+(3,1)$. *The Codazzi equations become*

$$(dh - 4h\alpha_1^1)\wedge\omega + dH\wedge\bar{\omega} = 0. \tag{11.21}$$

The hyperbolic Gauss map of f *is given on* U *by*

$$g_f = \tau[F_1] = [F_1F_1^*],$$

where $F_1 : U \to \mathbf{C}^2$ *is the first column of* F. *If, in addition,* F *is adapted to* z, *then* $\omega = e^u dz$ *and*

$$\alpha_1^1 = \frac{1}{2}(u_z dz - u_{\bar{z}} d\bar{z}), \tag{11.22}$$

and the Codazzi equations become

$$(he^{2u})_{\bar{z}} = e^{-2u}H_z.$$

Proof. The structure equations $d\alpha = -\alpha\wedge\alpha$ of $\mathbf{SL}(2,\mathbf{C})$ imply (11.19). Equation (11.20) follows from (11.6). Let $\mathbf{x} : M \to \mathbf{H}^3$ be the immersion for which $f = \upsilon(\mathbf{x})$. Then $e = \sigma \circ F$ is a first order frame field along x and the hyperbolic Gauss map of $\mathbf{x}$ is given on U by $g = \pi \circ e = [\mathbf{e}_3 + \mathbf{e}_4]$, where we use the notation of the commutative diagram (11.14). Then, by Definition 11.9, the hyperbolic Gauss map of f is given on U by

$$g_f = \upsilon \circ g = \upsilon \circ \pi \circ e = \tau \circ \pi_2 \circ F = \tau[F_1],$$

by (11.14). If F is adapted to z, then the exterior derivative of $\omega = e^u dz$ and the structure equations of $\mathbf{SL}(2,\mathbf{C})$ imply that

$$(2\alpha_1^1 + u_{\bar{z}}d\bar{z})\wedge dz = 0,$$

on U. Thus, the $(0,1)$ part of α_1^1 must be $-u_{\bar{z}}d\bar{z}$ and (11.22) follows from the fact that $\overline{\alpha_1^1} = -\alpha_1^1$ when F is first order. □

Theorem 11.11. *If* $f : M^2 \to \mathbf{H}(3)$ *is an immersion whose mean curvature is constant equal to 1, then for any complex coordinate chart* U, z *of* M *for which* U *is simply connected, there exists a frame field*

$$G : U \to \mathbf{SL}(2,\mathbf{C})$$

along f *that is holomorphic, an immersion, and null in the sense that if* $\gamma = G^{-1}dG$, *then* $\det(\gamma) = 0$ *on* U. *If* G *is such a frame field on* U, *then any other is given by* GK, *where* $K \in \mathbf{SU}(2)$ *is constant.*

Proof. Let U, z be a complex coordinate chart for which U is simply connected. Let $F : U \to \mathbf{SL}(2,\mathbf{C})$ be the first order frame field adapted to z. If $\alpha = F^{-1}dF$, and if $H = 1$, then Definition 11.8 and Lemma 11.10 imply that

$$\alpha = \beta + \begin{pmatrix} 0 & e^u dz \\ 0 & 0 \end{pmatrix},$$

where β is the $\mathfrak{su}(2)$-valued 1-form on U,

$$\beta = \begin{pmatrix} \frac{1}{2}(u_z dz - u_{\bar{z}} d\bar{z}) & \frac{1}{2}e^u \bar{h} d\bar{z} \\ -\frac{1}{2}e^u h dz & -\frac{1}{2}(u_z dz - u_{\bar{z}} d\bar{z}) \end{pmatrix}.$$

By the structure equations

$$d\beta + \beta \wedge \beta = \begin{pmatrix} 0 & \frac{1}{2}(d\bar{h} + 4\bar{h}\alpha_1^1)\wedge\bar{\omega} \\ -\frac{1}{2}(dh - 4h\alpha_1^1)\wedge\omega & 0 \end{pmatrix},$$

on U. If H is identically one on M, then the Codazzi equation (11.21) implies that

$$d\beta + \beta\wedge\beta = 0,$$

on U. Therefore, by the Cartan–Darboux Theorem 2.22, there exists a smooth map

$$B : U \to \mathbf{SU}(2),$$

such that

$$B^{-1}dB = \beta,$$

on U. Consider the frame field along f defined by

$$G = FB^{-1} : U \to \mathbf{SL}(2,\mathbf{C}).$$

Then

$$\gamma = G^{-1}dG = B(\alpha - \beta)B^{-1} = B\begin{pmatrix} 0 & e^u dz \\ 0 & 0 \end{pmatrix} B^{-1},$$

which is a bidegree $(1,0)$ form on U. Therefore, G is a holomorphic map on U. It is an immersion, since γ is never zero on U, and it is null because

$$\det(\gamma) = \det\begin{pmatrix} 0 & e^u dz \\ 0 & 0 \end{pmatrix} = 0,$$

on U. Any other frame field along f on U is given by $\tilde{G} = GK$, where

$$K : U \to \mathbf{SU}(2)$$

is a smooth map. If $\tilde{G}$ is holomorphic, then

$$K = G^{-1}\tilde{G} : U \to \mathbf{SL}(2,\mathbf{C})$$

is holomorphic, and takes all its values in $\mathbf{SU}(2)$. Hence, it must be constant (see Problem 11.31). □

This Theorem says that a CMC 1 surface in $\mathbf{H}(3)$ is locally the projection of a holomorphic, null immersion of the surface into $\mathbf{SL}(2,\mathbf{C})$. In the next section we shall see that any such projection is a CMC 1 surface in $\mathbf{H}(3)$.

11.5 Null immersions

Definition 11.12. Let M^2 be a connected Riemann surface. A holomorphic immersion $G : M \to \mathbf{SL}(2,\mathbf{C})$ is *null* if the $\mathfrak{sl}(2,\mathbf{C})$-valued 1-form $\gamma = G^{-1}dG$ satisfies

$$0 = -\det(\gamma) = \gamma_1^1\gamma_1^1 + \gamma_2^1\gamma_1^2$$

at every point of M, where these are symmetric products of 1-forms on M.

The meaning of this definition is enhanced if we view it in a broader context. If X is an element in a Lie algebra $\mathfrak{g}$, then the Jacobi identity implies that the map

$$\mathrm{ad}(X) : \mathfrak{g} \to \mathfrak{g}, \quad \mathrm{ad}(X)Y = [X,Y]$$

is linear and a Lie algebra homomorphism. The *Killing form* of $\mathfrak{g}$ is the symmetric bilinear form $\mathscr{K}$ on $\mathfrak{g}$ given by (see [84, p. 121])

$$\mathscr{K}(X,Y) = \mathrm{trace}(\mathrm{ad}(X)\mathrm{ad}(Y)).$$

On $\mathfrak{sl}(2,\mathbf{C})$ the Killing form is $\mathscr{K}(X,X) = -8\det(X)$, so $G : M^2 \to \mathbf{SL}(2,\mathbf{C})$ is null if and only if

$$G^*\mathscr{K} = \mathscr{K}(G^{-1}dG, G^{-1}dG) = 0.$$

As the parametrization (8.13) of the complex quadric $\mathbf{Q}^1$ led to the Enneper–Weierstrass representation of a minimal immersion into $\mathbf{R}^3$, the following parametrization of V leads to an analogous representation of CMC 1 immersions into $\mathbf{H}^3$.

Lemma 11.13. *If*

$$V = \left\{ \begin{pmatrix} z_1 & z_3 \\ z_2 & -z_1 \end{pmatrix} \in \mathfrak{sl}(2,\mathbf{C}) \setminus \{0\} : z_1 z_1 + z_2 z_3 = 0 \right\},$$

then the map

$$\pi : \mathbf{C}^2 \setminus \{0\} \to V, \quad \pi(p,q) = \begin{pmatrix} -pq & p^2 \\ -q^2 & pq \end{pmatrix} = \begin{pmatrix} p \\ q \end{pmatrix} \begin{pmatrix} -q & p \end{pmatrix} \tag{11.23}$$

is a holomorphic two-to-one covering.

Proof. The component functions of π are polynomials, thus holomorphic. Given $\begin{pmatrix} z_1 & z_3 \\ z_2 & -z_1 \end{pmatrix} \in V$, let

$$q = \pm\sqrt{-z_2}, \quad p = \pm\sqrt{z_3}.$$

Then $z_1^2 = -z_2 z_3 = q^2 p^2$, so $z_1 = \pm pq$. Since z_2 and z_3 are not both zero, p or q is not zero. Reversing the sign on one of these, if necessary, we may assume that

$$z_1 = -pq.$$

Therefore the map is surjective. If

$$\pi(p,q) = \pi(\tilde{p},\tilde{q}),$$

then

$$\tilde{p} = \pm p, \quad \tilde{q} = \pm q, \quad \tilde{p}\tilde{q} = pq,$$

so both signs must be plus or both signs must be minus, and therefore

$$(\tilde{p},\tilde{q}) = \pm(p,q),$$

which shows that π is two-to-one.

It remains to prove that $d\pi$ is non-singular at every point of $\mathbf{C}^2 \setminus \{0\}$, since a two-to-one local diffeomorphism is a covering projection. For a point $(p,q) \in \mathbf{C}^2 \setminus \{0\}$, and for a tangent vector $(z,w) \in \mathbf{C}^2$, suppose

$$0 = d\pi_{(p,q)}(z,w) = \begin{pmatrix} -q & 2p \\ 0 & q \end{pmatrix} z + \begin{pmatrix} -p & 0 \\ -2q & p \end{pmatrix} w = \begin{pmatrix} -qz - pw & 2pz \\ -2qw & qz + pw \end{pmatrix}.$$

If $p \neq 0$, then $z = 0$ so $pw = 0$ and thus $w = 0$ as well. Similarly, if $q \neq 0$, then $z = 0 = w$. Therefore, $d\pi_{(p,q)}$ is non-singular at every point $(p,q) \in V$. □

Lemma 11.14. *If $G : M^2 \to \mathbf{SL}(2, \mathbf{C})$ is a null immersion, then its projection*

$$f = GG^* : M \to \mathbf{H}(3)$$

is an immersion.

Proof. Since

$$df = dGG^* + GdG^* = G(G^{-1}dG + (G^{-1}dG)^*)G^* = G(\gamma + \gamma^*)G^*, \tag{11.24}$$

where $\gamma = G^{-1}dG = (\gamma^i_j) \in \mathfrak{sl}(2, \mathbf{C})$, the metric on M induced by f is

$$\begin{aligned} I &= -\det(df) = -\det(\gamma + \gamma^*) \\ &= (\gamma^1_1 + \bar{\gamma}^1_1)(\gamma^1_1 + \bar{\gamma}^1_1) + (\gamma^1_2 + \bar{\gamma}^2_1)(\gamma^2_1 + \bar{\gamma}^1_2) \\ &= 2\gamma^1_1\bar{\gamma}^1_1 + \gamma^1_2\bar{\gamma}^1_2 + \gamma^2_1\bar{\gamma}^2_1, \end{aligned} \tag{11.25}$$

since $\gamma^1_1\gamma^1_1 + \gamma^1_2\gamma^2_1 = 0$ as a consequence of G being null. Since G is an immersion, the component 1-forms γ^i_j of $G^{-1}dG$ cannot all vanish at any point of M, and therefore the last expression in (11.25) is positive definite and f is an immersion. □

For any immersed surface $f : M^2 \to \mathbf{H}(3)$, if U, z is a complex coordinate chart in M, then there exists a first order $\mathbf{SL}(2, \mathbf{C})$-frame along f in U adapted to z, and it is unique up to multiplication by $-I_2$. When $f = GG^*$ is the projection of a holomorphic null immersion $G : M \to \mathbf{SL}(2, \mathbf{C})$, then this first order frame is determined explicitly in terms of the components of G.

Theorem 11.15. *If $G : M^2 \to \mathbf{SL}(2, \mathbf{C})$ is a holomorphic null immersion, then its projection $f = GG^*$ has constant mean curvature equal to one. Let*

$$G^{-1}dG = \gamma = \begin{pmatrix} \gamma^1_1 & \gamma^1_2 \\ \gamma^2_1 & -\gamma^1_1 \end{pmatrix},$$

where the γ^i_j are holomorphic 1-forms on M, not all zero at any point, and such that the quadratic differential

$$\gamma^1_1\gamma^1_1 + \gamma^1_2\gamma^2_1 = 0,$$

at every point of M (these are symmetric products). If $\tilde{U}, z$ is any complex coordinate chart in M, then

$$\gamma^i_j = g^i_j dz,$$

where the g^i_j are holomorphic functions on $\tilde{U}$. The induced metric of f is given on $\tilde{U}$ by

$$I = e^{2u} dz d\bar{z}, \text{ where } e^u = |g^1_2| + |g^2_1|, \tag{11.26}$$

the Hopf quadratic differential is given on $\tilde{U}$ by

$$II^{2,0} = \frac{1}{2} h e^{2u} dz dz = \frac{g^2_1 dg^1_2 - g^1_2 dg^2_1}{2 g^1_1} dz, \tag{11.27}$$

and the hyperbolic Gauss map

$$g_f : M \to \mathbf{C}P^1,$$

of f, is given on $\tilde{U}$ by

$$g_f = \begin{cases} [\dot{G}_1] \text{ if } \dot{G}_1 \neq 0, \\ [G_1] \text{ if } \dot{G}_1 = 0, \end{cases} \tag{11.28}$$

where G_1 is the first column of G and $\dot{G}_1$ is the derivative of G_1 with respect to z.

Proof. Let $U \subset \tilde{U}$ be any non-empty, open, simply connected subset of $\tilde{U}$. G null implies that

$$g = \begin{pmatrix} g^1_1 & g^1_2 \\ g^2_1 & -g^1_1 \end{pmatrix} : U \to V \subset \mathfrak{sl}(2, \mathbf{C}),$$

where V is defined in Lemma 11.13. Because U is simply connected, the map g has a holomorphic lift

$$(p, q) : U \to \mathbf{C}^2 \setminus \{0\},$$

whose projection (11.23) is

$$\pi \circ (p, q) = g,$$

so that, on U,

$$g^1_1 = -pq, \quad g^1_2 = p^2, \quad g^2_1 = -q^2. \tag{11.29}$$

Hence, on U,

$$G^{-1} dG = \gamma = \begin{pmatrix} -pq & p^2 \\ -q^2 & pq \end{pmatrix} dz. \tag{11.30}$$

If we define the smooth maps $u : U \to \mathbf{R}$ and $a, b : U \to \mathbf{C}$ by

$$e^u = |p|^2 + |q|^2 = |g_2^1| + |g_1^2|, \quad a = e^{-u/2}p, \quad b = e^{-u/2}q,$$

then $|a|^2 + |b|^2 = 1$, so we have a smooth map

$$K = \begin{pmatrix} a & -\bar{b} \\ b & \bar{a} \end{pmatrix} = e^{-u/2}\begin{pmatrix} p & -\bar{q} \\ q & \bar{p} \end{pmatrix} : U \to \mathbf{SU}(2). \tag{11.31}$$

Consider the frame field

$$F = GK : U \to \mathbf{SL}(2, \mathbf{C})$$

along f on U. By an elementary calculation

$$\begin{aligned} \alpha = F^{-1}dF &= K^{-1}G^{-1}(dGK + GdK) = K^{-1}G^{-1}dGK + K^{-1}dK \\ &= \begin{pmatrix} 0 & e^u dz \\ 0 & 0 \end{pmatrix} + \begin{pmatrix} \bar{a}da + \bar{b}db & -\bar{a}d\bar{b} + \bar{b}d\bar{a} \\ -bda + adb & bd\bar{b} + ad\bar{a} \end{pmatrix}. \end{aligned}$$

Therefore,

$$\alpha_1^1 + \bar{\alpha}_1^1 = \bar{a}da + \bar{b}db + ad\bar{a} + bd\bar{b} = 0,$$

because $|a|^2 + |b|^2 = 1$ on U. Hence, F is a first order frame along f on U. Then

$$\alpha_2^1 + \bar{\alpha}_1^2 = e^u dz,$$

shows F is adapted to z, by Definition 11.8, and the induced metric of f is given on U by

$$I = e^{2u}dzd\bar{z} = (|p|^2 + |q|^2)^2 dzd\bar{z},$$

which implies (11.26). In addition,

$$\psi = \bar{\alpha}_2^1 - \alpha_1^2 = 2e^{-u}(qdp - pdq) + e^u d\bar{z} = he^u dz + He^u d\bar{z}$$

shows, since $qdp - pdq$ is a holomorphic 1-form, that $H = 1$ and the Hopf quadratic differential is

$$II^{2,0} = \frac{1}{2}he^{2u}dzdz = (qdp - pdq)dz.$$

Using (11.29), we find by an elementary calculation

$$qdp - pdq = \frac{g_1^2 dg_2^1 - g_2^1 dg_1^2}{2g_1^1},$$

which proves (11.27). As for the hyperbolic Gauss map, by Lemma 11.10 it is given on U as a map into $\mathbf{C}P^1$ by

$$g_f = [F_1],$$

where F_1 is the first column of the frame field F adapted to z. By (11.30), the first column of $\gamma = G^{-1}dG$ is

$$\gamma_1 = -q\begin{pmatrix} p \\ q \end{pmatrix} dz,$$

so that

$$dG_1 = G\gamma_1 = -qG\begin{pmatrix} p \\ q \end{pmatrix} dz.$$

On the other hand, $F = GK$, where K is given by (11.31), so

$$F_1 = e^{-u/2} G\begin{pmatrix} p \\ q \end{pmatrix},$$

and therefore,

$$\dot{G}_1 = \frac{dG_1}{dz} = -qG\begin{pmatrix} p \\ q \end{pmatrix} = -qe^{u/2}F_1 : U \to \mathbf{C}^2.$$

If $q \neq 0$, then $[F_1] = [\dot{G}_1] \in \mathbf{C}P^1$. If $q = 0$ at a point of U, then at this point $p \neq 0$ and

$$F_1 = e^{-u/2} G\begin{pmatrix} p \\ 0 \end{pmatrix} = e^{-u/2} p G_1.$$

The results in (11.28) now follow from these expressions for F_1. □

Corollary 11.16. *Let p and q be holomorphic functions, with no common zeros, on a simply connected domain $U \subset \mathbf{C}$. Consider*

$$\gamma = \begin{pmatrix} -pq & p^2 \\ -q^2 & pq \end{pmatrix} dz, \qquad (11.32)$$

a holomorphic, $\mathfrak{sl}(2, \mathbf{C})$-valued 1-form, never zero, with $\det(\gamma) = 0$, *on U. Then there exist holomorphic maps*

$$G,\ \tilde{G} : U \to \mathbf{SL}(2, \mathbf{C}),$$

such that

$$G^{-1}dG = \gamma = d\tilde{G}\tilde{G}^{-1}.$$

G is unique up to left multiplication by an element $A \in \mathbf{SL}(2,\mathbf{C})$. Its projection $f = GG^* : U \to \mathbf{H}(3)$ is a CMC 1 immersion with induced metric

$$I_f = \left(|p|^2 + |q|^2\right)^2 dzd\bar{z} = \| \begin{pmatrix} p \\ q \end{pmatrix} \|^4 dzd\bar{z},$$

Hopf differential

$$II_f^{2,0} = (qdp - pdq)dz,$$

and hyperbolic Gauss map

$$g_f = G \begin{bmatrix} p \\ q \end{bmatrix} : U \to \mathbf{C}P^1. \tag{11.33}$$

The projection of AG is AfA^*, which is congruent to f, so has the same induced metric and Hopf differential as f, and its hyperbolic Gauss map is $g_{AfA^*} = Ag_f$.

$\tilde{G}$ is unique up to right multiplication by an element $B \in \mathbf{SL}(2,\mathbf{C})$. Its projection $\tilde{f} = \tilde{G}\tilde{G}^* : U \to \mathbf{H}(3)$ is a CMC 1 immersion with induced metric

$$I_{\tilde{f}} = \|\tilde{G}^{-1} \begin{pmatrix} p \\ q \end{pmatrix} \|^4 dzd\bar{z}, \tag{11.34}$$

Hopf differential

$$II_{\tilde{f}}^{2,0} = II_f^{2,0}, \tag{11.35}$$

and hyperbolic Gauss map

$$g_{\tilde{f}} = \begin{bmatrix} p \\ q \end{bmatrix} : U \to \mathbf{C}P^1. \tag{11.36}$$

The projection of $\tilde{G}B$ is

$$\tilde{f}_B = \tilde{G}BB^*\tilde{G}^*,$$

which has induced metric

$$I_{\tilde{f}_B} = \|B^{-1}\tilde{G}^{-1} \begin{pmatrix} p \\ q \end{pmatrix} \|^4 dzd\bar{z},$$

the same Hopf differential, and the same hyperbolic Gauss map as $\tilde{f}$. The projections of $\tilde{G}$ and $\tilde{G}B$ have the same induced metric if and only if $B \in \mathbf{SU}(2)$ if and only if the two projections are the same.

Proof. Since γ is holomorphic, $d\gamma = 0 = \gamma\wedge\gamma$, so the maps G and $\tilde{G}$ exist and are unique up to left (respectively right) multiplication by the Cartan–Darboux Theorem. Moreover,

$$dG = G\gamma, \quad d\tilde{G} = \gamma\tilde{G}$$

are bidegree (1,0) forms, so G and $\tilde{G}$ must be holomorphic. It follows that G is a holomorphic, null immersion. Since

$$\tilde{\gamma} = \tilde{G}^{-1}d\tilde{G} = \tilde{G}^{-1}\gamma\tilde{G}$$

is holomorphic, never zero, and $\det(\tilde{\gamma}) = 0$ on U, the map $\tilde{G}$ is a holomorphic, null immersion. Hence, f and $\tilde{f}$ are CMC 1 immersions by Theorem 11.15. The induced metric, Hopf differential, and hyperbolic Gauss map of f are given in the proof of that Theorem. Write (11.32) as

$$\gamma = \begin{pmatrix} p \\ q \end{pmatrix} \begin{pmatrix} -q & p \end{pmatrix} dz.$$

If we define holomorphic functions $P, Q : U \to \mathbf{C}$ by

$$\begin{pmatrix} P \\ Q \end{pmatrix} = \tilde{G}^{-1} \begin{pmatrix} p \\ q \end{pmatrix},$$

then it follows that

$$\begin{pmatrix} -Q & P \end{pmatrix} = \begin{pmatrix} -q & p \end{pmatrix} \tilde{G}.$$

Therefore,

$$\tilde{G}^{-1}d\tilde{G} = \tilde{\gamma} = \tilde{G}^{-1} \begin{pmatrix} p \\ q \end{pmatrix} \begin{pmatrix} -q & p \end{pmatrix} \tilde{G}\, dz = \begin{pmatrix} P \\ Q \end{pmatrix} \begin{pmatrix} -Q & P \end{pmatrix} dz.$$

This implies (11.34) and (11.36) (using (11.33)). Moreover,

$$\begin{aligned} \begin{pmatrix} -Q & P \end{pmatrix} \begin{pmatrix} dP \\ dQ \end{pmatrix} &= \begin{pmatrix} -q & p \end{pmatrix} \tilde{G} \left(-\tilde{G}^{-1} d\tilde{G}\tilde{G}^{-1} \begin{pmatrix} p \\ q \end{pmatrix} + \tilde{G}^{-1} \begin{pmatrix} dp \\ dq \end{pmatrix} \right) \\ &= -\begin{pmatrix} -q & p \end{pmatrix} \gamma \begin{pmatrix} p \\ q \end{pmatrix} + \begin{pmatrix} -q & p \end{pmatrix} \begin{pmatrix} dp \\ dq \end{pmatrix} = \begin{pmatrix} -q & p \end{pmatrix} \begin{pmatrix} dp \\ dq \end{pmatrix}, \end{aligned}$$

which implies (11.35). The proof is concluded by applying the same argument to $\tilde{G}B$ in place of $\tilde{G}$, for any $B \in \mathbf{SL}(2, \mathbf{C})$. □

Remark 11.17. If γ is a nowhere zero, holomorphic, $\mathfrak{sl}(2,\mathbf{C})$-valued form with $\det(\gamma) = 0$ on the Riemann surface M, and if $\tilde{G} : M \to \mathbf{SL}(2,\mathbf{C})$ satisfies $d\tilde{G}\tilde{G}^{-1} = \gamma$, then the metric of $\tilde{f} = \tilde{G}\tilde{G}^*$ is not determined by γ, but rather by

$$\tilde{\gamma} = \tilde{G}^{-1}d\tilde{G} = \tilde{G}^{-1}d\tilde{G}\tilde{G}^{-1}\tilde{G} = \tilde{G}^{-1}\gamma\tilde{G}.$$

This is to be expected, because the solution $\tilde{G}$ is determined only up to right multiplication by a constant matrix $B \in \mathbf{SL}(2,\mathbf{C})$, and the projection

$$GB(GB)^* = GBB^*G^*$$

is not, in general, isometric to $\tilde{f}$. Nevertheless, the theorem shows that the Hopf differential and the hyperbolic Gauss map do not depend on B, as they are completely determined by γ.

11.6 Solutions

Suppose γ is a holomorphic, $\mathfrak{sl}(2,\mathbf{C})$-valued 1-form with $\det(\gamma) = 0$ on the connected Riemann surface M. Let $\pi : \tilde{M} \to M$ be the universal cover of M. In our applications, $\tilde{M}$ will be either the complex plane $\mathbf{C}$ or the Poincaré disk $\mathbf{D}$. Let Γ be the group of deck transformations, so $\Gamma \cong \pi_1(M)$, and $M = \Gamma\backslash\tilde{M}$, where we regard Γ as acting from the left on $\tilde{M}$. If $g \in \Gamma$, then $\pi \circ g = \pi$. The lift

$$\nu = \pi^*\gamma$$

is a holomorphic, $\mathfrak{sl}(2,\mathbf{C})$-valued 1-form, with $\det(\nu) = 0$ on $\tilde{M}$.

11.6.1 Left-invariant solutions

Let $G : \tilde{M} \to \mathbf{SL}(2,\mathbf{C})$ be a solution of

$$G^{-1}dG = \nu.$$

Any other solution is given by AG, for some constant $A \in \mathbf{SL}(2,\mathbf{C})$. If $g \in \Gamma$, then $G \circ g : \tilde{M} \to \mathbf{SL}(2,\mathbf{C})$ satisfies

$$(G \circ g)^{-1}d(G \circ g) = g^*(G^{-1}dG) = g^*\nu = g^*\pi^*\gamma = (\pi \circ g)^*\gamma = \pi^*\gamma = \nu.$$

Therefore

$$G \circ g = L_G(g)G,$$

for some $L_G(g) \in \mathbf{SL}(2,\mathbf{C})$, thus defining a map

$$L_G : \Gamma \to \mathbf{SL}(2,\mathbf{C}), \quad g \mapsto L_G(g). \tag{11.37}$$

The map L_G is a group homomorphism. The projection

$$f = GG^* : \tilde{M} \to \mathbf{H}(3)$$

is a CMC 1 immersion. If $g \in \Gamma$, then

$$f \circ g = (G \circ g)(G \circ g)^* = (L_G(g)G)(L_G(g)G)^* = L_G(g) f L_G(g)^*,$$

which is the isometric action of $L_G(g)$ on f in $\mathbf{H}(3)$. Then f descends to M if and only if $f \circ g = f$, for all $g \in \Gamma$ if and only if

$$f = L_G(g) f L_G(g)^*, \text{ for all } g \in \Gamma,$$

if and only if the isometric action of $L_G(g)$ on $\mathbf{H}(3)$ fixes every point of $f(\tilde{M})$. Using Problem 11.34, we conclude that f descends to M if and only if $L_G(\Gamma) = \{\pm I_2\}$.

11.6.2 Right-invariant solutions

If $G : \tilde{M} \to \mathbf{SL}(2,\mathbf{C})$ is a solution of $dGG^{-1} = \nu$, then any other solution is given by GB, for some $B \in \mathbf{SL}(2,\mathbf{C})$. If $g \in \Gamma$, then $G \circ g : \tilde{M} \to \mathbf{SL}(2,\mathbf{C})$ satisfies

$$d(G \circ g)(G \circ g)^{-1} = g^*(dGG^{-1}) = g^*\nu = g^*\pi^*\gamma = (\pi \circ g)^*\gamma = \pi^*\gamma = \nu.$$

Therefore

$$G \circ g = GR_G(g),$$

for some $R_G(g) \in \mathbf{SL}(2,\mathbf{C})$, thus defining a map

$$R_G : \Gamma \to \mathbf{SL}(2,\mathbf{C}), \quad g \mapsto R_G(g).$$

The map R_G is a group anti-homomorphism: $R_G(g_1 g_2) = R_G(g_2)R_G(g_1)$, for any $g_1, g_2 \in \Gamma$. The projection

$$f = GG^* : \tilde{M} \to \mathbf{H}(3)$$

is a CMC 1 immersion. If $g \in \Gamma$, then

$$f \circ g = (G \circ g)(G \circ g)^* = (GR_G(g))(GR_G(g))^* = GR_G(g)R_G(g)^*G^*,$$

and this equals $f = GG^*$ if and only if

$$R_G(g)R_G(g)^* = I_2.$$

Therefore, f descends to M if and only if

$$R_G(\Gamma) \subset \mathbf{SU}(2).$$

Exercise 44. Prove that there exists $C \in \mathbf{SL}(2,\mathbf{C})$ such that the CMC 1 immersion

$$(GC)(GC)^*$$

descends to M, if and only if

$$C^{-1}R_G(\Gamma)C \subset \mathbf{SU}(2).$$

Definition 11.18. The map $R_G : \Gamma \to \mathbf{SL}(2,\mathbf{C})$ is *unitary* if $R_G(\Gamma) \subset \mathbf{SU}(2)$. It is *unitarizable* if there exists $C \in \mathbf{SL}(2,\mathbf{C})$ such that R_{GC} is unitary; i.e.,

$$R_{GC}(\Gamma) = C^{-1}R_G(\Gamma)C \subset \mathbf{SU}(2).$$

In conclusion, there exists a solution $G : \tilde{M} \to \mathbf{SL}(2,\mathbf{C})$ of $dG\,G^{-1} = \tilde{\nu}$ such that its projection $f = GG^*$ descends to M if and only if the map $R_G : \Gamma \to \mathbf{SL}(2,\mathbf{C})$ is unitarizable.

11.6.3 How to solve $dG\,G^{-1} = \gamma$

Assume in this subsection that M is simply connected. Let $a, b, c : M \to \mathbf{C}$ be holomorphic functions, with no common zeros and satisfying $a^2 + bc = 0$ identically, so

$$L = \begin{pmatrix} a & b \\ c & -a \end{pmatrix} : M \to \mathfrak{sl}(2,\mathbf{C}),$$

is holomorphic, never zero, and $\det(L) = 0$ on M. To solve the linear system of ODE

$$G' = LG$$

for $G : M \to \mathbf{SL}(2,\mathbf{C})$, let $\begin{pmatrix} x \\ y \end{pmatrix}$ denote either column of G. Then the system to solve is

$$x' = ax + by, \quad y' = cx - ay.$$

If $c \neq 0$, then $x = (y' + ay)/c$ and y must satisfy

$$y'' - \frac{c'}{c}y' + \frac{a'c - ac'}{c}y = 0. \tag{11.38}$$

If $b \neq 0$, then $y = (x' - ax)/b$ and x must satisfy

$$x'' - \frac{b'}{b}x' + \frac{ab' - a'b}{b}x = 0. \tag{11.39}$$

Suppose $y_1, y_2 : M \to \mathbf{C}$ are linearly independent solutions of (11.38). Let

$$\tilde{G} = \begin{pmatrix} \frac{1}{c}(y_1' + ay_1) & \frac{1}{c}(y_2' + ay_2) \\ y_1 & y_2 \end{pmatrix} \tag{11.40}$$

Then $\det\tilde{G}$ is a nonzero constant (see Problem 11.36), so

$$G = \frac{1}{\sqrt{\det\tilde{G}}}\tilde{G} : M \to \mathbf{SL}(2, \mathbf{C})$$

satisfies $G'G^{-1} = L$.

11.7 Spinor data

Recall from Chapter 8 that the data needed on a Riemann surface M to construct a minimal immersion $\mathbf{x} : M \to \mathbf{R}^3$ is the Weierstrass data, which comprises a meromorphic function g and a holomorphic 1-form η, both on M, whose zeros balance the poles of g in the appropriate way. From Corollary 11.16 we see that the analogous data needed to construct a CMC 1 immersion of M into $\mathbf{H}(3)$ is a pair of holomorphic *spinor fields* on M with no common zeros. A spinor field is a section of a certain holomorphic line bundle over the Riemann surface. We begin with a brief description of these. For more details see Conlon [53, 95–99] and Griffiths–Harris [81, pp 66–70].

Definition 11.19. A holomorphic line bundle $\pi : L \to M$ over a Riemann surface M is a complex manifold L of complex dimension 2 and a holomorphic mapping π mapping L onto M such that

1. For each $m \in M$, the *fiber* $L_m = \pi^{-1}\{m\}$ has the structure of a complex vector space of dimension 1.
2. There is an open cover $\{U_\sigma\}_{\sigma \in J}$ of M, together with commutative diagrams

$$\begin{array}{ccc} \pi^{-1}U_\sigma & \stackrel{\varphi_\sigma}{\to} & U_\sigma \times \mathbf{C} \\ \pi \downarrow & & \downarrow \pi_1 \\ U_\sigma & \stackrel{\text{id}}{\to} & U_\sigma \end{array}$$

where φ_σ is biholomorphic, for each $\sigma \in J$, and π_1 is projection onto the first factor.
3. For each $\sigma \in J$ and $m \in U_\sigma$, the restriction of φ_σ to the fiber L_m maps this complex vector space isomorphically onto the complex vector space $\{m\} \times \mathbf{C}$.

The open cover $\{U_\sigma\}$ is called a *trivializing cover* of M for the line bundle and each open set U_σ is called a *trivializing neighborhood* of M.

An example of a holomorphic line bundle is the *trivial line bundle* over M, which is $\pi_1 : M \times \mathbf{C} \to M$, where π_1 is projection onto the first factor.

Holomorphic line bundles $\pi : L \to M$ and $\tilde{\pi} : \tilde{L} \to M$ are *isomorphic*, written $L \cong \tilde{L}$, if there exists a biholomorphic map $\varphi : L \to \tilde{L}$ such that for each $m \in M$, φ maps L_m isomorphically onto $\tilde{L}_m$. A holomorphic line bundle is called *trivial* if it is isomorphic to the trivial line bundle.

A *holomorphic section* of the holomorphic line bundle $\pi : L \to M$ is a holomorphic map $\mathfrak{p} : M \to L$ such that $\pi \circ \mathfrak{p} = \mathrm{id}_M$. A holomorphic section of the trivial bundle is just a holomorphic function $f : M \to \mathbf{C}$.

For a holomorphic line bundle $\pi : L \to M$ with trivializing open cover $\{U_\sigma\}$ as above, if $U_\sigma \cap U_\tau \neq \emptyset$, consider

$$(U_\sigma \cap U_\tau) \times \mathbf{C} \xrightarrow{\varphi_\sigma^{-1}} \pi^{-1}(U_\sigma \cap U_\tau) \xrightarrow{\varphi_\tau} (U_\sigma \cap U_\tau) \times \mathbf{C}.$$

This composition must have the form

$$\varphi_\tau \circ \varphi_\sigma^{-1}(m, \xi) = (m, g_{\tau\sigma}(m)\xi),$$

where the functions

$$g_{\tau\sigma} : U_\sigma \cap U_\tau \to \mathbf{GL}(1, \mathbf{C}) = \mathbf{C} \setminus \{0\} = \mathbf{C}^*$$

are holomorphic, never zero. They are called *the transition functions* of the trivializing cover. They satisfy the *cocycle property*

$$g_{\sigma\tau}(m) g_{\tau\nu}(m) = g_{\sigma\nu}(m),$$

for all $m \in U_\sigma \cap U_\tau \cap U_\nu$. The cocycle property implies $g_{\tau\sigma} = 1/g_{\sigma\tau}$ and $g_{\sigma\sigma} = 1$. The set $\{U_\sigma, g_{\tau\sigma}\}_{\sigma,\tau \in J}$ is called a *structure cocycle* for $\pi : L \to M$.

The structure cocycle gives all the data necessary to construct the line bundle, up to isomorphism. For a detailed explanation of this see the above two references. For our purposes, the structure cocycle is the most workable form of a holomorphic line bundle. For example, if $\mathfrak{f} : M \to L$ is a *holomorphic section* of the line bundle, then for the above structure cocycle we have

$$\varphi_\sigma \circ \mathfrak{f}(m) = (m, f_\sigma(m)),$$

for each $m \in U_\sigma$, where $f_\sigma : U_\sigma \to \mathbf{C}$ must be a holomorphic function. From the definition of the transition functions, we must have

$$f_\sigma(m) = g_{\sigma\tau}(m) f_\tau(m), \tag{11.41}$$

for any $m \in U_\sigma \cap U_\tau$. Thus, in terms of the structure cocycle, a holomorphic section of $\pi : L \to M$ is given by a collection of holomorphic functions

$$f_\sigma : U_\sigma \to \mathbf{C},$$

satisfying (11.41) on $U_\sigma \cap U_\tau$.

Example 11.20. The *canonical bundle* $K \to M$ is given by the covering of complex coordinate charts $\{U_\sigma, z_\sigma\}$ of M, with transition functions on $U_\sigma \cap U_\tau \neq \emptyset$,

$$g_{\sigma\tau} = \frac{dz_\tau}{dz_\sigma}.$$

A holomorphic section of the canonical bundle defines a holomorphic 1-form α on M, and vice versa. In fact, if the section is given by the collection of holomorphic functions

$$f_\sigma : U_\sigma \to \mathbf{C}$$

then on $U_\sigma \cap U_\tau \neq \emptyset$,

$$f_\sigma dz_\sigma = g_{\sigma\tau} f_\tau dz_\sigma = \frac{dz_\tau}{dz_\sigma} f_\tau dz_\sigma = f_\tau dz_\tau$$

shows that

$$\alpha = f_\sigma dz_\sigma$$

is a well-defined holomorphic 1-form on all of M. Conversely, given a holomorphic 1-form α on M, then on U_σ,

$$\alpha = f_\sigma dz_\sigma$$

for some holomorphic function f_σ on U_σ, and this collection of holomorphic functions defines a holomorphic section of the canonical bundle.

Let $L \to M$ be a holomorphic line bundle with structure cocycle $\{U_\sigma, g_{\sigma\tau}\}_{\sigma,\tau\in J}$, and let $\tilde{L} \to M$ be a holomorphic line bundle with structure cocycle $\{U_\sigma, \tilde{g}_{\sigma\tau}\}_{\sigma,\tau\in J}$. Then $L \cong \tilde{L}$ if and only if there exist nowhere zero holomorphic functions

$$f_\sigma : U_\sigma \to \mathbf{C},$$

for each $\sigma \in J$, such that

$$g_{\sigma\tau}(m) f_\tau(m) = f_\sigma(m) \tilde{g}_{\sigma\tau}(m), \tag{11.42}$$

for every $m \in U_\sigma \cap U_\tau$. In fact, if $\varphi : L \to \tilde{L}$ is a bundle isomorphism, then for each $\sigma \in J$,

$$\tilde{\varphi}_\sigma \circ \varphi \circ \varphi_\sigma^{-1} : U_\sigma \times \mathbf{C} \to U_\sigma \times \mathbf{C}$$

must have the form

$$\tilde{\varphi}_\sigma \circ \varphi \circ \varphi_\sigma^{-1}(m, \xi) = (m, f_\sigma(m)\xi), \tag{11.43}$$

for every $(m, \xi) \in U_\sigma \times \mathbf{C}$, where

$$f_\sigma : U_\sigma \to \mathbf{GL}(1, \mathbf{C}) = \mathbf{C}^*$$

is holomorphic. One easily verifies that (11.42) must then hold. Conversely, given the set of holomorphic maps f_σ, the bundle isomorphism φ is defined by (11.43) and is well-defined by (11.42). In particular, $L \to M$ is isomorphic to the trivial line bundle if and only if

$$g_{\sigma\tau}(m) f_\tau(m) = f_\sigma(m),$$

for every $m \in U_\sigma \cap U_\tau$.

The most general isomorphism result allows for the structure cocycles to have different indexing sets. Structure cocycles $\{U_\sigma, g_{\sigma\tau}\}_{\sigma,\tau \in J}$ and $\{V_a, h_{ab}\}_{a,b \in \mathscr{A}}$ *are equivalent* if both are contained in some structure cocycle. This is an equivalence relation on the set of structure cocycles. The holomorphic line bundles defined by two structure cocycles are isomorphic if and only if the structure cocycles are equivalent. We will not need this result here. For a detailed discussion of it see [53, p 99].

The *product of the line bundles* L and $\tilde{L}$ is the holomorphic line bundle, denoted $L\tilde{L}$, given by the structure cocycle

$$\{U_\sigma, g_{\sigma\tau}\tilde{g}_{\sigma\tau}\}_{\sigma,\tau \in J}.$$

Example 11.21. Let $\{U_\sigma, z_\sigma\}$ be an atlas of complex coordinate charts on the Riemann surface M. Suppose that on each U_σ we have a nowhere zero holomorphic 1-form ω_σ. If $U_\sigma \cap U_\tau \neq \emptyset$, there exists a holomorphic function $g_{\sigma\tau}$ on $U_\sigma \cap U_\tau$, such that

$$\omega_\sigma g_{\sigma\tau} = \omega_\tau.$$

It is helpful to write this

$$g_{\sigma\tau} = \omega_\tau / \omega_\sigma.$$

Then $\{U_\sigma, g_{\sigma\tau}\}$ is a structure cocycle defining a holomorphic line bundle isomorphic to the canonical bundle. In fact,

$$\omega_\sigma = f_\sigma dz_\sigma,$$

for some nowhere zero holomorphic function on U_σ, and

$$f_\sigma(m) g_{\sigma\tau}(m) = \frac{dz_\tau}{dz_\sigma}(m) f_\tau(m),$$

for all $m \in U_\sigma \cap U_\tau$.

Definition 11.22. A *spinor bundle* over a Riemann surface M is a holomorphic line bundle $S \to M$ whose square is isomorphic to the canonical bundle; that is

$$SS \cong K.$$

A holomorphic *spinor field* on M is a holomorphic section of a spinor bundle.

If $S \to M$ is a spinor bundle with transition functions $g_{\sigma\tau}$ relative to an atlas $\{U_\sigma, z_\sigma\}$, then $SS \cong K$ implies that

$$g_{\sigma\tau}^2(m) f_\tau(m) = f_\sigma(m) \frac{dz_\tau}{dz_\sigma}(m)$$

for every $m \in U_\sigma \cap U_\tau$, for some nowhere zero holomorphic functions $\{f_\sigma : U_\sigma \to \mathbf{C}\}$. Let

$$\omega_\sigma = \frac{dz_\sigma}{f_\sigma},$$

a nowhere zero holomorphic 1-form on U_σ. Then

$$g_{\sigma\tau}^2 = \frac{\omega_\tau}{\omega_\sigma},$$

where this quotient is explained in Example 11.21. Now $d\omega_\sigma = 0$ on U_σ. If we assume U_σ is contractible, then there exists a holomorphic function $w_\sigma : U_\sigma \to \mathbf{C}$ such that $\omega_\sigma = dw_\sigma$ on U_σ. Then U_σ is a union of open neighborhoods on each of which w_σ is a complex coordinate. In summary, given a spinor bundle $S \to M$, there exists a complex coordinate atlas $\{U_\sigma, w_\sigma\}$ on M relative to which the transition functions $g_{\sigma\tau}$ of S satisfy

$$g_{\sigma\tau}^2 = \frac{dw_\tau}{dw_\sigma}$$

on $U_\sigma \cap U_\tau \neq \emptyset$.

Example 11.23. Let $f : M \to \mathbf{H}(3)$ be a CMC 1 immersion and regard M as the Riemann surface with the complex structure coming from the induced metric. Let $\{U_\sigma, z_\sigma\}$ be an atlas of complex charts on M with each U_σ being simply connected. Let

$$F_\sigma = (F_{\sigma}{}^i_j) : U_\sigma \to \mathbf{SL}(2, \mathbf{C}),$$

for $i,j = 1, 2$, be the first order frame field along f adapted to z_σ. If

$$\alpha_\sigma = F_\sigma^{-1} dF_\sigma$$

on U_σ, then $\alpha_\sigma{}^1_1 + \overline{\alpha}_\sigma{}^1_1 = 0$, by Definition 11.6, and

$$\omega_\sigma = \alpha_\sigma{}^1_2 + \overline{\alpha}_\sigma{}^2_1 = e^{u_\sigma} dz_\sigma,$$

for some smooth $u_\sigma : U_\sigma \to \mathbf{R}$, by Definition 11.8. If $U_\sigma \cap U_\tau \neq \emptyset$, then first order frame fields transform by

$$F_\tau = F_\sigma \begin{pmatrix} a_{\sigma\tau} & 0 \\ 0 & \overline{a}_{\sigma\tau} \end{pmatrix}, \tag{11.44}$$

where $a_{\sigma\tau} : U_\sigma \cap U_\tau \to \mathbf{U}(1)$ is smooth and $|a_{\sigma\tau}| = 1$. Then

$$a_{\sigma\tau}^2 \omega_\tau = \omega_\sigma,$$

by (11.18) (where now $b = 0$), so

$$a_{\sigma\tau}^2 e^{u_\tau} dz_\tau = e^{u_\sigma} dz_\sigma = e^{u_\sigma} \frac{dz_\sigma}{dz_\tau} dz_\tau,$$

which implies that

$$\frac{dz_\sigma}{dz_\tau} = a_{\sigma\tau}^2 e^{u_\tau - u_\sigma}.$$

If we let

$$g_{\tau\sigma} = a_{\sigma\tau} e^{(u_\tau - u_\sigma)/2},$$

then its square is the holomorphic function $\frac{dz_\sigma}{dz_\tau}$, so it is holomorphic. On $U_\sigma \cap U_\tau \cap U_\nu \neq \emptyset$, we have

$$g_{\sigma\tau} g_{\tau\nu} g_{\nu\sigma} = 1,$$

since

$$a_{\tau\sigma} a_{\nu\tau} a_{\sigma\nu} = 1$$

at every point, by (11.44). Therefore, this is a set of holomorphic transition functions defining a holomorphic line bundle $S \to M$ and

$$dz_\sigma = g_{\tau\sigma}^2 dz_\tau$$

shows that S is a spinor bundle. Define functions

$$p_\sigma = e^{u_\sigma/2} F_{\sigma}{}^2_1, \quad q_\sigma = e^{u_\sigma/2} F_{\sigma}{}^1_1.$$

Then $\{p_\sigma\}$ and $\{q_\sigma\}$ define holomorphic sections of $S \to M$, that is, they define holomorphic spinor fields on M, and they have no common zeros. In fact, they are holomorphic functions, because

$$\alpha_{\sigma}{}^1_1 = \frac{1}{2}\left(\frac{\partial u_\sigma}{\partial z_\sigma} dz_\sigma - \frac{\partial u_\sigma}{\partial \bar{z}_\sigma} d\bar{z}_\sigma\right),$$

by (11.22), and

$$\alpha_{\sigma}{}^2_1 = \frac{1}{2}(\bar{\omega}_\sigma - \psi_\sigma) = -\frac{1}{2} h_\sigma e^{u_\sigma} dz_\sigma,$$

by Lemma 11.10, where now $H = 1$, imply that both dp_σ and dq_σ are multiples of dz_σ. From our definition of $g_{\tau\sigma}$ and the transformation rule for the $\{F_\tau\}$, it follows that

$$p_\tau = g_{\tau\sigma} p_\sigma, \quad q_\tau = g_{\tau\sigma} q_\sigma,$$

so these define sections $\mathfrak{p}$ and $\mathfrak{q}$, respectively, of S. They have no common zeros because the first column of F_σ is never zero.

Lemma 11.24. *Let* $\mathfrak{p}$ *and* $\mathfrak{q}$ *be holomorphic spinor fields with no common zeros on a Riemann surface* M*. Let* $\{U_\sigma, z_\sigma\}$ *be an atlas of complex coordinate charts on* M*, relative to which the transition functions* $\{g_{\tau\sigma}\}$ *of the spinor bundle satisfy*

$$g_{\sigma\tau}^2 = \frac{dz_\tau}{dz_\sigma}$$

on $U_\sigma \cap U_\tau \neq \emptyset$*. Then there exists a holomorphic, nowhere zero* $\mathfrak{sl}(2,\mathbf{C})$*-valued 1-form* γ *on* M*, with* $\det(\gamma) = 0$*, whose expression on* U_σ, dz_σ *is*

$$\gamma_\sigma = \begin{pmatrix} -p_\sigma q_\sigma & p_\sigma^2 \\ -q_\sigma^2 & p_\sigma q_\sigma \end{pmatrix} dz_\sigma,$$

where $p_\sigma, q_\sigma : U_\sigma \to \mathbf{C}$ *are the local representations of* $\mathfrak{p}$ *and* $\mathfrak{q}$*.*

Proof. On U_σ, the form γ_σ is holomorphic, nowhere zero, $\mathfrak{sl}(2,\mathbf{C})$-valued, and $\det(\gamma_\sigma) = 0$. It is easily verified that $\gamma_\sigma = \gamma_\tau$ on $U_\sigma \cap U_\tau \neq \emptyset$. □

11.8 Weierstrass and spinor data

If we begin with holomorphic spinor fields $\mathfrak{p}$ and $\mathfrak{q}$ with no common zeros on a Riemann surface M, then they determine the holomorphic, $\mathfrak{sl}(2,\mathbf{C})$-valued 1-form ν on M, with $\det(\nu) = 0$, by

$$\gamma = \begin{pmatrix} -\mathfrak{pq} & \mathfrak{p}^2 \\ -\mathfrak{q}^2 & \mathfrak{pq} \end{pmatrix},$$

which determines CMC 1 immersions of $\tilde{M}$, the universal cover of M, into $\mathbf{H}(3)$, as described in Corollary 11.16. These spinor data determine Weierstrass data as follows. The quotient $g = \mathfrak{p}/\mathfrak{q}$ is a meromorphic function on M and $\eta = -\mathfrak{q}^2$ is a holomorphic 1-form on M whose zeros balance the poles of g as required in Theorem 8.8. These data define the nowhere zero, isotropic, abelian differential α on M,

$$\alpha = \begin{pmatrix} \frac{1}{2}(1-g^2) \\ \frac{i}{2}(1+g^2) \\ g \end{pmatrix} \eta = \begin{pmatrix} \frac{1}{2}(\mathfrak{p}^2-\mathfrak{q}^2) \\ \frac{-i}{2}(\mathfrak{p}^2+\mathfrak{q}^2) \\ -\mathfrak{pq} \end{pmatrix}.$$

The real part of the integral of α (see (8.7)) defines a minimal immersion $\mathbf{x} : \tilde{M} \to \mathbf{R}^3$, with metric, Hopf differential, and Gauss map given by

$$I_{\mathbf{x}} = (1+|g|^2)^2 \eta\bar{\eta}, \quad II^{2,0}_{\mathbf{x}} = -dg\, \eta, \quad g,$$

respectively, by Theorem 8.10.

Conversely, suppose we begin with the Weierstrass data given by a meromorphic function g on M and a holomorphic 1-form η whose zeros balance the poles of g. Suppose there exists a holomorphic spinor field $\mathfrak{q}$ on M such that $\eta = -\mathfrak{q}^2$. Then $\mathfrak{p} = g\mathfrak{q}$ is a holomorphic spinor field on M, since the zeros of $\mathfrak{q}$ remove the poles of g, and we obtain a holomorphic, $\mathfrak{sl}(2,\mathbf{C})$-valued 1-form on M with determinant equal to zero,

$$\gamma = \begin{pmatrix} -\mathfrak{pq} & \mathfrak{p}^2 \\ -\mathfrak{q}^2 & \mathfrak{pq} \end{pmatrix} = \begin{pmatrix} g & -g^2 \\ 1 & -g \end{pmatrix} \eta.$$

Notice that *supposing* the existence of the spinor field $\mathfrak{q}$ led us to this expression for γ, but in fact γ is given in terms of g and η. There exist holomorphic, null solutions $G, \tilde{G} : \tilde{M} \to \mathbf{SL}(2,\mathbf{C})$ of $G^{-1}dG = \gamma = d\tilde{G}\tilde{G}^{-1}$, which give CMC 1 immersions $f = GG^*, \tilde{f} = \tilde{G}\tilde{G}^* : \tilde{M} \to \mathbf{H}(3)$ whose induced metrics, Hopf differentials, and hyperbolic Gauss maps are, by Corollary 11.16,

$$I_f = (1+|g|^2)^2 \eta\bar{\eta}, \quad II^{2,0}_f = -dg\, \eta, \quad g_f = G\begin{bmatrix} g \\ 1 \end{bmatrix} : \tilde{M} \to \mathbf{CP}^1,$$

and

$$I_{\tilde{f}} = ||\tilde{G}^{-1}\begin{pmatrix} g \\ 1 \end{pmatrix}||^4 dz d\bar{z}, \quad II_{\tilde{f}}^{2,0} = II_f^{2,0}, \quad g_{\tilde{f}} = \begin{bmatrix} g \\ 1 \end{bmatrix} : \tilde{M} \to \mathbf{CP}^1,$$

respectively. Thus both have the same Hopf differential as $\mathbf{x}$, while f and $\mathbf{x}$ have the same induced metric while $\tilde{f}$ and $\mathbf{x}$ have, essentially, the same Gauss map. Bryant called the CMC 1 immersions f and $\tilde{f}$ the *cousins* of the minimal immersion $\mathbf{x}$.

11.9 Bryant spheres with smooth ends

A major goal in this subject is to construct complete *Bryant surfaces* with smooth ends. An end is smooth if it extends smoothly as an immersion into $\mathbf{H}^3 \cup \mathbf{S}^2_\infty$, as is the case for any horosphere. In contrast, the isoparametric immersions in Figures 6.5, 6.6, and 6.7 have nonsmooth ends.

Bryant [21] has proved the following analogue to Theorem 8.27 concerning complete minimal surfaces in $\mathbf{R}^3$ with finite total curvature.

Theorem 11.25. *If $f : M \to \mathbf{H}^3$ is a complete Bryant immersion with finite total curvature, then the induced conformal structure on M is that of a compact Riemann surface $\bar{M}$ with a finite set of points $p_1, \ldots, p_k$ removed. Moreover, f extends smoothly to $\bar{M}$ as a map into $\mathbf{H}^3 \cup \mathbf{S}^2_\infty$.*

If $\bar{M}$ is the Riemann sphere, the immersion $f : M \to \mathbf{H}^3$ is called a *Bryant sphere*. Rather than go into these general results, we limit ourselves here to the construction of some explicit examples of Bryant spheres using the theory developed in the preceding sections. We begin with the compact Riemann surface $\mathbf{S}^2$ with two points removed, which is $\mathbf{C}$ with one point removed. We shall also try to expose the relationship between our construction of Bryant immersions in $\mathbf{H}^3$ and the Enneper–Weierstrass construction of minimal immersions in $\mathbf{R}^3$.

Let $M = \mathbf{C} \setminus \{0\}$ with the standard complex coordinate z. From Example 8.15, the Weierstrass data of the catenoid are

$$g = 1/z, \quad \eta = dz.$$

As shown in Section 11.8, these Weierstrass data determine spinor data $\mathfrak{p}$ and $\mathfrak{q}$, which are holomorphic spinor fields, such that

$$\gamma = \begin{pmatrix} g & -g^2 \\ 1 & -g \end{pmatrix} \eta = \begin{pmatrix} -\mathfrak{p}\mathfrak{q} & \mathfrak{p}^2 \\ -\mathfrak{q}^2 & \mathfrak{p}\mathfrak{q} \end{pmatrix}, \tag{11.45}$$

a holomorphic, isotropic, $\mathfrak{sl}(2,\mathbf{C})$-valued 1-form on M. Any nonzero complex constant k defines a new set of spinor data given by $k\mathfrak{p}$ and $k\mathfrak{q}$, which then define the holomorphic, isotropic, $\mathfrak{sl}(2,\mathbf{C})$-valued 1-form

$$\gamma_k = k^2 \gamma$$

on M.

Remark 11.26. These spinor data determine the Weierstrass data of a conformal associate of the catenoid (see Definition 8.12). If $k = re^{it}$, where $r > 0$, and t are real, then these conformal associates $\mathbf{x}^{(k)}$ of the catenoid $\mathbf{x}$ are just the associates of $\mathbf{x}$ rescaled by r,

$$\mathbf{x}^{(k)} = r\mathbf{x}^{(t)},$$

where $\mathbf{x}^{(t)}$ is the associate of $\mathbf{x}$ defined by the Weierstrass data g and $e^{it}\eta$ (see (8.23)). In hyperbolic geometry, however, there is no notion of rescaling (triangles are similar if and only if they are congruent), so one might expect interesting differences in the conformal associates given by real constants $k > 0$. In the examples of this section we shall see that this is the case.

The universal cover of M is

$$\exp : \mathbf{C} \to M, \quad \exp(w) = e^w = z,$$

whose group of deck transformations is

$$\Gamma = \{g_n : n \in \mathbf{Z}\}, \text{ where } g_n(w) = w + 2\pi i n.$$

Let

$$\nu_k = \exp^* \gamma_k = k^2 \begin{pmatrix} 1 & -e^{-w} \\ e^w & -1 \end{pmatrix} dw = \begin{pmatrix} a & b \\ c & -a \end{pmatrix} dw$$

where

$$a = k^2, \quad b = -k^2 e^{-w}, \quad c = k^2 e^w,$$

so c is never zero on $\mathbf{C}$. Now the derivatives with respect to w are $a' = 0$ and $c' = c$, so (11.38) becomes

$$y'' - y' - k^2 y = 0. \tag{11.46}$$

The roots of its auxiliary polynomial are

$$(1 \pm \sqrt{1 + 4k^2})/2.$$

It will be convenient to let

$$m = -(1 + \sqrt{1 + 4k^2})/2,$$

so that $m + 1$ and $-m$ are the roots and $k^2 = m(m+1)$. The roots are distinct if $m \neq -1/2$, which we now assume. Two linearly independent solutions of our ODE are

$$y_1 = e^{(m+1)w}, \quad y_2 = e^{-mw}.$$

From these we construct a basic solution G_m of $dG = \nu_k G$ on $\mathbf{C}$, satisfying

$$\sqrt{\frac{m(m+1)}{2m+1}} \begin{pmatrix} \frac{m+1}{m} e^{mw} & \frac{m}{m+1} e^{-(m+1)w} \\ e^{(m+1)w} & e^{-mw} \end{pmatrix} = G_m \begin{pmatrix} \sqrt{\frac{m+1}{m}} & 0 \\ 0 & \sqrt{\frac{m}{m+1}} \end{pmatrix},$$

where

$$G_m = \frac{1}{\sqrt{2m+1}} \begin{pmatrix} (m+1)e^{mw} & me^{-(m+1)w} \\ me^{(m+1)w} & (m+1)e^{-mw} \end{pmatrix} : \mathbf{C} \to \mathbf{SL}(2,\mathbf{C}). \tag{11.47}$$

Any other solution is given by $G_m B$, for any constant $B \in \mathbf{SL}(2,\mathbf{C})$. From Subsection 11.6.2 we know that the projection $G_m B(G_m B)^* : \mathbf{C} \to \mathbf{H}(3)$ descends to M for some $B \in \mathbf{SL}(2,\mathbf{C})$ if an only if the map $R_{G_m} : \Gamma \to \mathbf{SL}(2,\mathbf{C})$ is unitarizable, as defined in Definition 11.18. For which values of m is R_{G_m} unitarizable and for those m for which values of $B \in \mathbf{SL}(2,\mathbf{C})$ is $R_{G_m B}$ unitary? If $g_n \in \Gamma$, then

$$G_m \circ g_n(w) = \frac{1}{\sqrt{2m+1}} \begin{pmatrix} (m+1)e^{m(w+2\pi in)} & me^{-(m+1)(w+2\pi in)} \\ me^{(m+1)(w+2\pi in)} & (m+1)e^{-m(w+2\pi in)} \end{pmatrix} = G_m(w) D_m^n,$$

where the *monodromy matrix* D_m of G_m is

$$D_m = \begin{pmatrix} e^{2\pi im} & 0 \\ 0 & e^{-2\pi im} \end{pmatrix} \in \mathbf{SU}(2).$$

Lemma 11.27. *Let $m \in \mathbf{C} \setminus \{-1, -1/2, 0\}$, let D_m be the above monodromy matrix, and let $B \in \mathbf{SL}(2,\mathbf{C})$. Then $R_{G_m B}$ is unitary, that is,*

$$B^{-1} D_m B \in \mathbf{SU}(2),$$

if and only if m is real and, either

$$m \in \frac{1}{2}\mathbf{Z}, \text{ and } B \text{ is arbitrary},$$

or

$$m \notin \frac{1}{2}\mathbf{Z}, \text{ and } B = \begin{pmatrix} a & 0 \\ 0 & a^{-1} \end{pmatrix} K,$$

for any $0 \neq a \in \mathbf{R}$ and any $K \in \mathbf{SU}(2)$.

Proof. If $B \in \mathbf{SL}(2,\mathbf{C})$, then it has a unique *polar decomposition* $B = HK$, where H is hermitian positive definite and K is unitary. Then

$$B^{-1} D_m B = K^{-1} H^{-1} D_m H K \in \mathbf{SU}(2)$$

if and only if

$$H^{-1}D_mH \in \mathbf{SU}(2).$$

Therefore, it suffices to prove the Lemma for the case $B = H$ is hermitian positive definite. We leave this to Problem 11.43.

Definition 11.28. Let m be a nonzero real number satisfying $m > -1/2$. A *two-noid* is any CMC 1 immersion $f : M \to \mathbf{H}(3)$ given by $f = G_mB(G_mB)^*$, for any $B \in \mathbf{SL}(2,\mathbf{C})$ for which R_{G_mB} is unitary. The *Catenoid cousins* are the immersions $f_m = G_mG_m^* : M \to \mathbf{H}(3)$.

Suppose that $m \in \mathbf{R} \setminus \{-1,-1/2,0\}$. Since our results depend only on $k^2 = m(m+1)$, it is sufficient to consider only the values $m > -1/2$, $m \neq 0$. For such a value of m, the projection of G_m is

$$\tilde{f}_m = G_mG_m^* : \mathbf{C} \to \mathbf{H}(3),$$

where $(2m+1)\tilde{f}_m(w) =$

$$\begin{pmatrix} (m+1)^2e^{m(w+\bar{w})} + m^2e^{-(m+1)(w+\bar{w})} & (m+1)m(e^{mw+(m+1)\bar{w}} + e^{-(m+1)w-m\bar{w}}) \\ m(m+1)(e^{(m+1)w+m\bar{w}} + e^{-mw-(m+1)\bar{w}}) & m^2e^{(m+1)(w+\bar{w})} + (m+1)^2e^{-m(w+\bar{w})} \end{pmatrix},$$

which descends to

$$f_m : M \to \mathbf{H}(3), \quad f_m \circ \exp = \tilde{f}_m,$$

where

$$f_m(z) = \frac{1}{2m+1}\begin{pmatrix} (m+1)^2(z\bar{z})^m + \frac{m^2}{(z\bar{z})^{m+1}} & m(m+1)((z\bar{z})^m\bar{z} + \frac{1}{(z\bar{z})^m\bar{z}}) \\ m(m+1)((z\bar{z})^m z + \frac{1}{(z\bar{z})^m z}) & m^2(z\bar{z})^{m+1} + \frac{(m+1)^2}{(z\bar{z})^m} \end{pmatrix}.$$

In terms of polar coordinates $z = re^{it}$, with $r > 0$,

$$f_m(z) = R(t)\sigma_m(r)R(t)^*,$$

where

$$R(t) = \begin{pmatrix} e^{-it/2} & 0 \\ 0 & e^{it/2} \end{pmatrix} \in \mathbf{SU}(2)$$

is a 1-parameter family of rotations, and

$$\sigma_m : \mathbf{R}^+ \to \mathbf{H}(3), \quad \sigma_m(r) =$$

$$\frac{1}{2m+1}\begin{pmatrix} (m+1)^2r^{2m} + m^2r^{-2(m+1)} & m(m+1)(r^{2m+1} + r^{-(2m+1)}) \\ m(m+1)(r^{2m+1} + r^{-(2m+1)}) & (m+1)^2r^{-2m} + m^2r^{2(m+1)} \end{pmatrix}$$

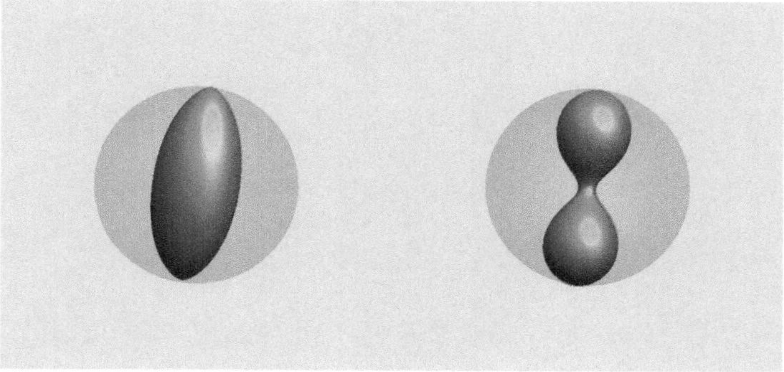

Fig. 11.1 Embedded catenoid cousins with smooth ends. Left: f_m with $m = -0.3$: Right: f_m with $m = -0.08$.

is a planar profile curve. Hence, f_m is a surface of revolution. The profile curve is embedded, so f_m is embedded, if and only if $-1/2 < m < 0$. Figure 11.1 shows two embedded catenoid cousins, $f_m : M \to \mathbf{H}(3)$ with parameter $m = -0.3$ and $m = -0.08$, respectively. Figures in this chapter illustrate $\mathfrak{s} \circ v^{-1} \circ f : M \to \mathbf{B}^3$, where $\mathfrak{s} : \mathbf{H}^3 \to \mathbf{B}^3$ is hyperbolic projection onto the Poincaré ball and $v : \mathbf{H}^3 \to \mathbf{H}(3)$ is the isometry (11.1).

If $m > 0$, then the profile curve is not embedded, so neither is f_m. Figure 11.2 shows the case $m = 0.03$.

If $m \in \frac{1}{2}\mathbf{Z}$ and if $B \in \mathbf{SL}(2, \mathbf{C})$ is non-diagonal, then the immersion $G_m B (G_m B)^* : \mathbf{C} \to \mathbf{H}(3)$ descends to a two-noid $f : M \to \mathbf{H}(3)$ which is neither equivalent to, nor isometric to, a Catenoid cousin. It is not a surface of revolution. Figure 11.3 shows

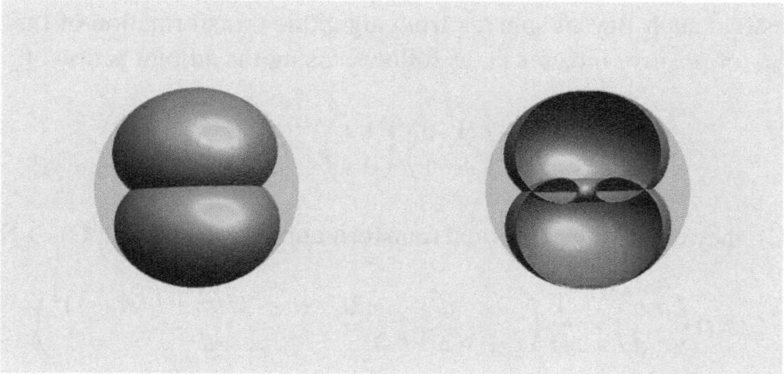

Fig. 11.2 Left: f_m with $m = 0.03$: surface of revolution with two ends. Right: a wedge has been removed to show the self intersection inside.

this Bryant sphere for the case $m = 2$ and $B = \begin{pmatrix} 1 & 3 \\ 0 & 1 \end{pmatrix}$.

Bohle and Peters [12] characterize when a Bryant surface has smooth ends. They prove that there exist Bryant spheres with an arbitrary number of smooth ends.

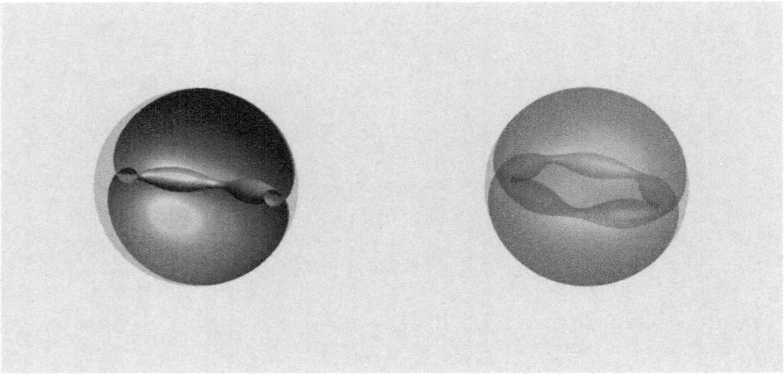

Fig. 11.3 Projection of $G_2B(G_2B)^*$ is a warped 2-noid.

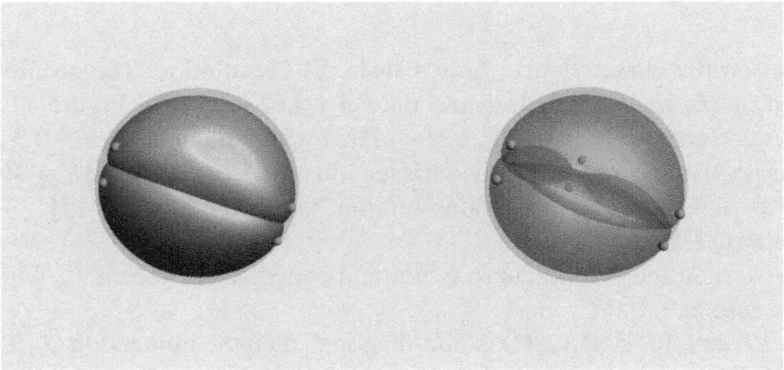

Fig. 11.4 A Bohle-Peters six-noid. The ends are at the blue dots.

They construct such Bryant spheres from algebraic transformation of the catenoid cousins f_m, for positive integers m, as follows. Using the adjoint action of

$$A(s,t) = \begin{pmatrix} 1 & 0 \\ -s & 1 \end{pmatrix} \begin{pmatrix} 1 & t \\ 0 & 1 \end{pmatrix} \in \mathbf{SL}(2,\mathbf{C})$$

on $\mathfrak{sl}(2,\mathbf{C})$, they create the conformal transformation $C(s,t) : \mathbf{SL}(2,\mathbf{C}) \to \mathbf{SL}(2,\mathbf{C})$,

$$C(s,t)\begin{pmatrix} a & b \\ c & d \end{pmatrix} = \frac{1}{D}\begin{pmatrix} a & 2t-(c+2s)t^2+b(st-1)^2 \\ c+s(2-bs) & d \end{pmatrix},$$

where $D = 1 - ct - 2st + bs(st-1)$. They prove:

1. If $st \neq 0$, and if m is a positive integer, then the null immersion $C(s,t)G_m$ projects to a Bryant sphere with $2(m+1)$ smooth ends. Figure 11.4 shows the case $m = 2$ and $s = t = 0.2$, which is a Bryant sphere with six smooth ends.

2. If $t = 0$ and if m is a positive integer, then $C(s,0)G_m$ projects to a Bryant sphere with $m+2$ smooth ends. Figure 11.5 shows the case $m = 9$, $s = 0.4$, and $t = 0$.

So far we have considered only the case when the Riemann surface M is $\mathbf{C}$ with one point removed. If more than one point is removed the situation becomes more complicated on several fronts. The universal cover of such a Riemann surface is no longer given in terms of elementary functions. Secondly, the ODE (11.46) will generally have nonconstant coefficients and its solutions will not be elementary functions. Despite these difficulties Bobenko, Pavlyukevich, and Springborn [10] have constructed trinoids, which are complete Bryant immersions of $M = \mathbf{C} \setminus \{0,1\} \to \mathbf{H}(3)$ with three smooth ends. The spinor data are of the form

$$\mathfrak{p} = (\frac{p_0}{z} + \frac{p_1}{z-1} + p_3)\sqrt{dz}, \quad \mathfrak{q} = (\frac{q_0}{z} + \frac{q_1}{z-1} + q_3)\sqrt{dz},$$

for complex constants p_0, p_1, p_3, q_0, q_1, and q_3. In this case the solution of the ODE will be in terms of hypergeometric functions. Among the solutions found in [10] is a 3-parameter family $\{f_\mu\}_{\mu \in X}$ of embedded trinoids, where the parameter domain is

$$\{(a,b,c) : 0 < a < .4829,\ 0 < b < \frac{1}{2},\ \frac{1}{2} - a - b < c < \min(\frac{1}{2} - a + b, \frac{1}{2} + a - b)\}.$$

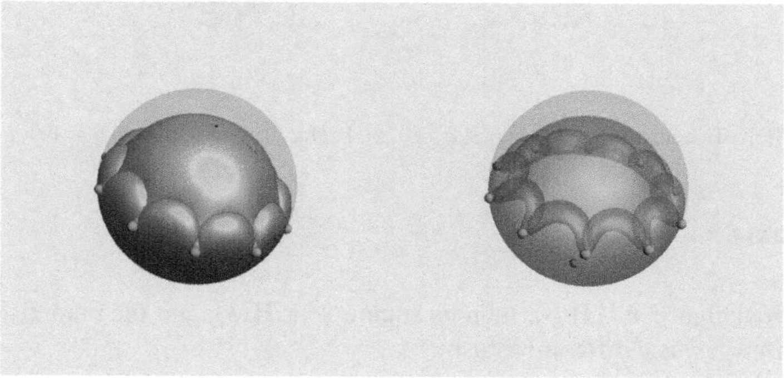

Fig. 11.5 A Bohle-Peters eleven-noid. The ends are at the blue dots.

If the parameters are equal, the trinoid is $\mathbf{Z}_3$-symmetric, otherwise it is not. Figure 11.6 shows the $\mathbf{Z}_3$-symmetric trinoids with parameters all equal to 0.3 and 0.4, respectively.

Figure 11.7 shows nonsymmetric trinoids with parameters (0.3, 0.3, 0.2) and (0.4, 0.4, 0.32), respectively.

All the computations have been implemented in a MATHEMATICA notebook available on line. See [10] for the url and other details. We made the four embedded trinoids here with minor modifications of routines in this notebook.

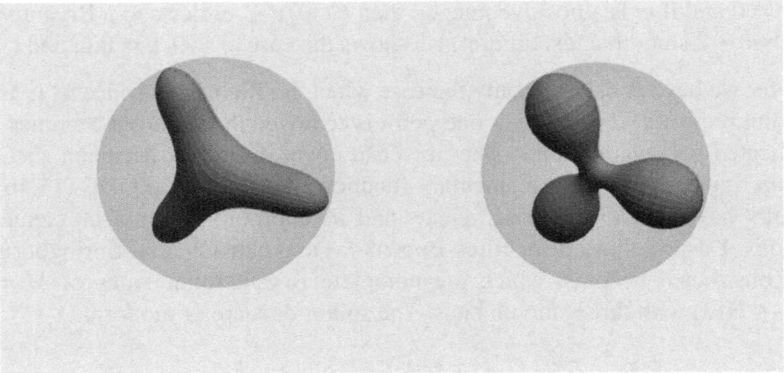

Fig. 11.6 Left: Trinoid with $a = b = c = 0.3$. Right: Trinoid with $a = b = c = 0.4$.

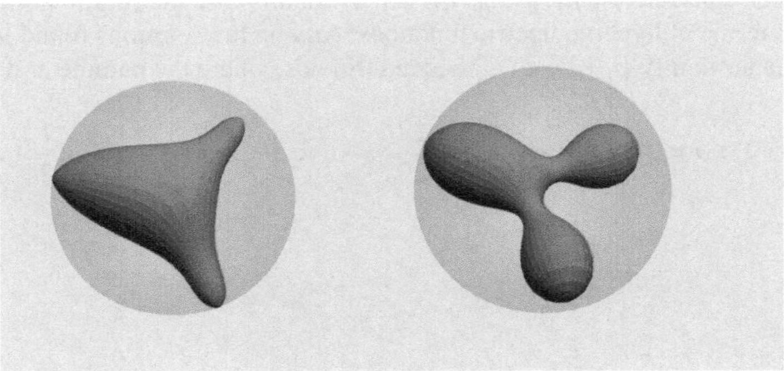

Fig. 11.7 Left: Trinoid with $a = b = 0.3$, $c = 0.2$. Right: Trinoid with $a = b = 0.4$, $c = 0.32$.

Problems

11.29. Prove that if $v \in \mathbf{H}(3)$, then its square $v^2 \in \mathbf{H}(3)$, and the map $T : \mathbf{H}(3) \to \mathbf{H}(3)$, $T(v) = v^2$, is a diffeomorphism.

11.30. Let σ be the homomorphism (11.2).

1. If $A = \begin{pmatrix} x & z \\ y & w \end{pmatrix} \in \mathbf{SL}(2, \mathbf{C})$, find $\sigma(A) \in \mathbf{SO}_+(3, 1)$.
2. If $a = e^{i\theta}$, and if $A = \begin{pmatrix} a & 0 \\ 0 & \bar{a} \end{pmatrix}$, find $\sigma(A)$.
3. If $a = e^{i\theta}$, and if $A = \begin{pmatrix} 0 & -\bar{a} \\ a & 0 \end{pmatrix}$, find $\sigma(A)$.

11.31. Prove that if M is a connected Riemann surface and if $K : M \to \mathbf{SL}(2, \mathbf{C})$ is a holomorphic map for which $K(M) \subset \mathbf{SU}(2)$, then K is constant.

11.32. Prove that the Killing form on $\mathfrak{sl}(n,\mathbf{C})$ is

$$\mathscr{K}(X,Y) = 2n\mathrm{trace}(XY).$$

11.33. Prove that the map L_G defined in (11.37) is a group homomorphism.

11.34. Prove that if $f : \tilde{M} \to \mathbf{H}(3)$ is an immersion and if $A \in \mathbf{SL}(2,\mathbf{C})$ fixes every point of $f(\tilde{M})$, then $A = 1$, the identity transformation (so $A = \pm I_2$).

11.35. Prove that the map R_G is a group anti-homomorphism:

$$R_G(g_1 g_2) = R_G(g_2)R_G(g_1),$$

for all $g_1, g_2 \in \Gamma$.

11.36. Let $\tilde{G}(y_1,y_2)$ be the matrix valued function in (11.40). Prove that $\det\tilde{G}$ is a non-zero constant by showing that its derivative is identically zero.

11.37. Prove that if x_1, x_2 are linearly independent solutions of (11.39), and if

$$\tilde{G} = \begin{pmatrix} x_1 & x_2 \\ \frac{1}{b}(x_1' - ax_1) & \frac{1}{b}(x_2' - ax_2) \end{pmatrix},$$

then $\det\tilde{G}$ is constant, non-zero, and

$$G = \frac{1}{\sqrt{\det\tilde{G}}}\tilde{G} : M \to \mathbf{SL}(2,\mathbf{C})$$

satisfies $G'G^{-1} = L$.

11.38. Derive a method similar to that of Subsection 11.6.3 to find $G : M \to \mathbf{SL}(2,\mathbf{C})$ satisfying $G^{-1}G' = L$.

11.39. Prove that if $\{U_\sigma\}_{\sigma\in J}$ is an open cover of M, and if $g_{\sigma\tau}(m) = 1$, for all $m \in U_\sigma \cap U_\tau$, for all $\sigma,\tau \in J$, then $\{U_\sigma, g_{\sigma\tau}\}$ is a structure cocycle, which determines a holomorphic line bundle isomorphic to the trivial bundle.

11.40. Prove that the product of a line bundle $L \to M$ with the trivial line bundle is isomorphic to L.

11.41. Prove that if the line bundle $L \to M$ has structure cocycle $\{U_\sigma, g_{\sigma\tau}\}$, then $\{U_\sigma, 1/g_{\sigma\tau}\}$ is a structure cocycle defining a line bundle $\tilde{L}$ with the property that $L\tilde{L}$ is isomorphic to the trivial bundle. We write L^{-1} for $\tilde{L}$.

11.42. Prove that any $B \in \mathbf{SL}(2,\mathbf{C})$ has a unique *polar decomposition* $B = HK$, where H is hermitian and positive definite, and $K \in \mathbf{SU}(2)$.

11.43. Prove Lemma 11.27 for the case when B is hermitian positive definite.

11.44. In Example 8.16 we found that the helicoid is the conjugate to the catenoid. Find the cousins of the associates of the catenoid. In particular find the helicoid cousin.

11.45. Find the basic solution $G : \mathbf{C} \to \mathbf{SL}(2,\mathbf{C})$ for the case $m = -1/2$ of equal roots of the auxiliary polynomial of the ODE (11.46). Find its monodromy matrix D. Is G unitarizable?

11.46. If $\mathbf{x} : M \to \mathbf{R}^3$ is the minimal immersion generated by the Weierstrass data g and η on M, and if $k > 0$ is constant, then $k\mathbf{x} : M \to \mathbf{R}^3$ is the minimal immersion generated by the Weierstrass data g and $k\eta$. For the catenoid data $g(z) = 1/z$ and $\eta = dz$ above, find the cousin of the data g and $k\eta$.

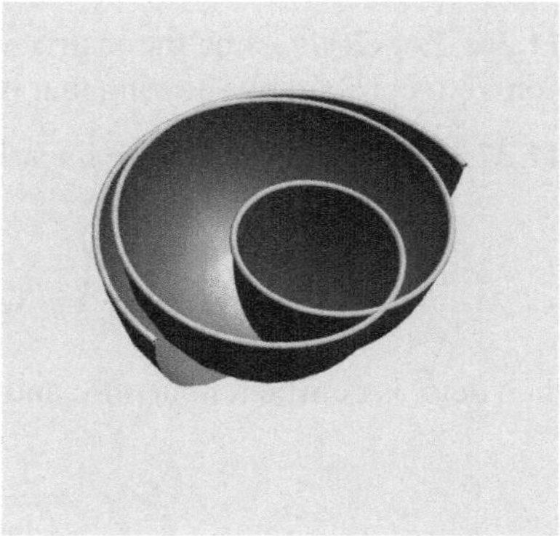

Fig. 11.8 Left Invariant Catenoid cousin: surface of revolution whose rotation trajectories are ultracircles. It has been cut off at the horizontal plane to show the profile curve.

11.47. For γ given by (11.45), find the null immersion $G : \mathbf{C} \to \mathbf{SL}(2,\mathbf{C})$ satisfying $G^{-1}dG = \gamma$ and the resulting Catenoid cousin $f = GG^* : \mathbf{C} \to H(3)$. Prove that it does not descend to $\mathbf{C} \setminus \{0\}$ and that it is a surface of revolution for which the trajectories of the rotation group are ultracircles (planar curves of constant positive curvature less than 1 analogous to ultraspheres of Definition 6.12). Figure 11.8 shows one of the left-invariant solutions. It is a surface of revolution whose rotation trajectories are ultracircles.

Chapter 12
Möbius Geometry

This chapter introduces conformal geometry and Liouville's characterization of conformal transformations of Euclidean space. Through stereographic projection these are all globally defined conformal transformations of the sphere $\mathbf{S}^3$. The Möbius group **Möb** is the group of all conformal transformations of $\mathbf{S}^3$. It is a ten dimensional Lie group containing the group of isometries of each of the space forms as a subgroup. Möbius space $\mathscr{M}$ is the homogeneous space consisting of the sphere $\mathbf{S}^3$ acted upon by **Möb**. $\mathscr{M}$ has a conformal structure invariant under the action of **Möb**. The reduction procedure is applied to Möbius frames. The space forms are each equivariantly embedded into Möbius geometry. Conformally invariant properties, such as Willmore immersion, or isothermic immersion, or Dupin immersion, have characterizations in terms of the Möbius invariants. Oriented spheres in Möbius space provide the appropriate geometric interpretation of the vectors of a frame field.

Throughout this chapter we let $\boldsymbol{\epsilon}_0, \boldsymbol{\epsilon}_1, \boldsymbol{\epsilon}_2, \boldsymbol{\epsilon}_3$ denote the standard orthonormal basis of $\mathbf{R}^4$. Unless otherwise indicated, the indices i,j range through $\{1,2,3\}$.

12.1 Local conformal diffeomorphisms

Definition 12.1 (Conformal diffeomorphism). A smooth map $F : M^m, I \to N^n, \tilde{I}$ of Riemannian manifolds is *conformal* if $F^*\tilde{I} = e^{2u}I$, for some smooth function $u : M \to \mathbf{R}$. The positive function e^u is the *conformal factor*.

Definition 12.2. A *local conformal diffeomorphism* of $\mathbf{R}^3$ is a conformal diffeomorphism $\mathbf{y} = \mathbf{y}(\mathbf{x})$ between open subsets U, V of $\mathbf{R}^3$. Thus,

$$d\mathbf{y} \cdot d\mathbf{y} = e^{2u} d\mathbf{x} \cdot d\mathbf{x} \tag{12.1}$$

for some smooth function $u : U \to \mathbf{R}$ and e^u is the *conformal factor*.

G.R. Jensen et al., *Surfaces in Classical Geometries*, Universitext,
DOI 10.1007/978-3-319-27076-0_12

Geometrically, this condition on $\mathbf{y}$ means that its differential at any point $\mathbf{x}$ of U, $d\mathbf{y}_{\mathbf{x}} : \mathbf{R}^3 \to \mathbf{R}^3$, preserves angles and multiplies lengths by $e^{u(\mathbf{x})}$.

Stereographic projection from a point in the sphere $\mathbf{S}^3$ onto Euclidean space $\mathbf{R}^3$ is conformal. Hyperbolic stereographic projection of hyperbolic space $\mathbf{H}^3$ onto the unit ball in $\mathbf{R}^3$ is conformal. Composing with these projections and their inverses, we obtain any local conformal transformation in Euclidean space or hyperbolic space from a local conformal transformation in the sphere.

Example 12.3 (Isometries). If $(\mathbf{a}, A) \in \mathbf{E}(3)$ is an isometry, then the diffeomorphism $(\mathbf{a}, A) : \mathbf{R}^3 \to \mathbf{R}^3$ given by

$$\mathbf{y} = (\mathbf{a}, A)\mathbf{x} = \mathbf{a} + A\mathbf{x}$$

is a conformal diffeomorphism. Here the conformal factor is 1 at every point. An isometry is a composition of commuting isometries, translation by $\mathbf{a}$, denoted $T_{\mathbf{a}}$, and the orthogonal transformation $A \in \mathbf{O}(4)$.

Example 12.4 (Homotheties). If $0 \neq t \in \mathbf{R}$, then multiplication by t,

$$h_t : \mathbf{R}^3 \to \mathbf{R}^3, \quad h_t(\mathbf{x}) = t\mathbf{x},$$

is a conformal diffeomorphism, called a *homothety*. It is orientation preserving if and only if $t > 0$. If $0 < t < 1$, it is called a *contraction* and it is called a *dilation* if $t > 1$. The conformal factor is $|t|$.

Example 12.5 (Inversions). Inversion in the sphere of radius $r > 0$ with center at the point $\mathbf{m}$ sends a point $\mathbf{x} \neq \mathbf{m}$ to the point $\mathbf{y}$ on the ray from $\mathbf{m}$ through $\mathbf{x}$ such that $|\mathbf{y} - \mathbf{m}||\mathbf{x} - \mathbf{m}| = r^2$. The first condition says that $\mathbf{y} - \mathbf{m} = t(\mathbf{x} - \mathbf{m})$, for some $t > 0$, and then the second condition allows us to solve for t to obtain the formula for the inversion

$$\mathbf{y} = \mathbf{m} + r^2 \frac{\mathbf{x} - \mathbf{m}}{|\mathbf{x} - \mathbf{m}|^2} \tag{12.2}$$

It is a conformal transformation $\mathbf{R}^3 \setminus \{\mathbf{m}\} \to \mathbf{R}^3$ with conformal factor $r^2/|\mathbf{x} - \mathbf{m}|^2$. See Figure 12.1.

As in Example 4.46, let $\mathscr{I}$ denote inversion in the unit sphere centered at the origin, so

$$\mathbf{y} = \mathscr{I}(\mathbf{x}) = \mathbf{x}/|\mathbf{x}|^2.$$

The inversion is not defined at the center of the sphere of inversion. We shall see how to make sense out of the idea that the center is sent to the point at infinity.

Isometries and homotheties send spheres to spheres and planes to planes.

Proposition 12.6. *Inversion in a sphere with center* $\mathbf{m}$ *and radius* $r > 0$ *sends any sphere not passing through* $\mathbf{m}$ *to a sphere, and any sphere passing through* $\mathbf{m}$ *to a* *plane not passing through* **m**. *It sends any plane passing through* **m** *to itself, and any plane not passing through* **m** *to a sphere passing through* **m**. *In summary, inversions map the set of spheres and planes into the set of spheres and planes.*

Proof. By Problem 12.54, it is sufficient to prove this proposition for the case of inversion in the unit sphere centered at the origin. Let us see what happens to the sphere with center **m** and radius $r > 0$, whose equation is

Fig. 12.1 **x** and **y** are inversions of each other

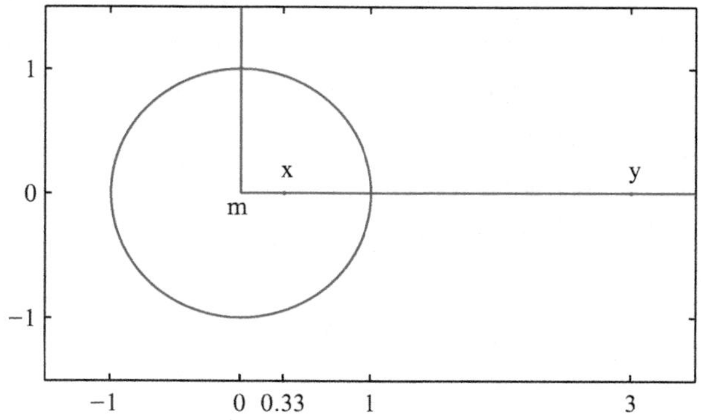

$$|\mathbf{x}-\mathbf{m}|^2 = r^2, \tag{12.3}$$

when it is inverted in the unit sphere. A point **x** on this sphere is mapped to $\mathbf{y} = \frac{\mathbf{x}}{|\mathbf{x}|^2}$, which is never the origin, and inverting again sends **y** back to **x**, so

$$\mathbf{x} = \frac{\mathbf{y}}{|\mathbf{y}|^2}.$$

Substituting this into (12.3), we find that a point **y** in the image of the inverted sphere must satisfy the equation

$$\left|\frac{\mathbf{y}}{|\mathbf{y}|^2} - \mathbf{m}\right|^2 = r^2,$$

which expands to

$$\left|\frac{\mathbf{y}}{|\mathbf{y}|^2}\right|^2 - 2\frac{\mathbf{y}}{|\mathbf{y}|^2}\cdot\mathbf{m} + |\mathbf{m}|^2 = r^2.$$

Multiplying through by $|\mathbf{y}|^2$, which is never zero, gives

$$1 - 2\mathbf{y}\cdot\mathbf{m} + |\mathbf{y}|^2(|\mathbf{m}|^2 - r^2) = 0.$$

If $|\mathbf{m}|^2 - r^2 \neq 0$, that is, if the given sphere does not pass through the origin, then we can complete the square to obtain the equation

$$\left| \mathbf{y} - \frac{\mathbf{m}}{|\mathbf{m}|^2 - r^2} \right|^2 - \frac{r^2}{(|\mathbf{m}|^2 - r^2)^2} = 0,$$

which is the sphere of radius $\frac{r}{|\mathbf{m}|^2 \text{Ũ} r^2}$ and center $\frac{\mathbf{m}}{|\mathbf{m}|^2 \text{Ũ} r^2}$. This is illustrated in the 2-dimensional case in Figure 12.2. The rest of the proof will be left for Problem 12.56. □

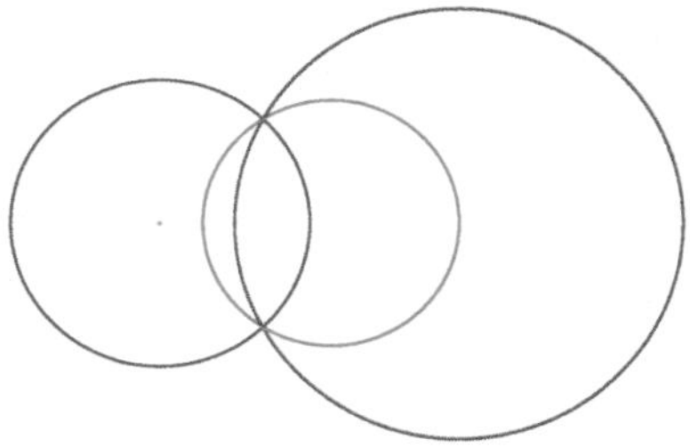

Fig. 12.2 The red and green circles are inversions of each other through the blue circle.

Theorem 12.7 (J. Liouville, Note VI in [121]). *Any local conformal transformation of* $\mathbf{R}^3$ *is either*

1. *a homothety followed by an isometry, or*
2. *an inversion in a unit sphere followed by a homothety followed by an isometry.*

For a proof see do Carmo [31, pp 170–176] or Dubrovin, Novikov, Fomenko [62, pp 138–141].

Corollary 12.8. *A local conformal diffeomorphism of* $\mathbf{R}^3$ *sends the set of all spheres and planes onto the set of all spheres and planes.*

Definition 12.9. Let $\mathrm{Conf}(\mathbf{S}^3)$ denote the set of all conformal diffeomorphisms of $\mathbf{S}^3$ to itself. It is a group under composition of maps.

Corollary 12.10. *Let* $\mathscr{S} = \mathscr{S}_{\mathbf{p}}$ *be stereographic projection from a point* $\mathbf{p} \in \mathbf{S}^3$. *Any map* $f \in \mathrm{Conf}(\mathbf{S}^3)$ *is a composition of a finite number of maps of the form* $\mathscr{S}^{-1} \circ T \circ \mathscr{S}$, *where T is a translation or an orthogonal transformation or a homothety or inversion in the unit sphere centered at the origin, of* $\mathbf{R}^3$.

Proof. For any $f \in \mathrm{Conf}(\mathbf{S}^3)$, the map $\mathscr{S} \circ f \circ \mathscr{S}^{-1}$ is a local conformal transformation of $\mathbf{R}^3$. The result now follows from Liouville's Theorem 12.7, Example 12.3 and Problem 12.54. □

Our next goal is to find a linear representation of the group $\mathrm{Conf}(\mathbf{S}^3)$. This is done in Theorem 12.17 below.

12.2 Möbius space

Let $\mathbf{R}^{4,1}$ denote $\mathbf{R}^5$ with a Lorentzian inner product. Let $\epsilon_0,\dots,\epsilon_4$ denote the standard orthonormal basis of $\mathbf{R}^{4,1}$ given by the standard orthonormal basis $\epsilon_0,\dots,\epsilon_3$ of the Euclidean space $\mathbf{R}^4$ and with $\langle\epsilon_4,\epsilon_4\rangle = -1$. The Lorentzian inner products $\langle\epsilon_a,\epsilon_b\rangle$, for $a,b = 0,\dots,4$, are the entries of the matrix

$$I_{4,1} = \begin{pmatrix} I_4 & 0 \\ 0 & -1 \end{pmatrix}. \tag{12.4}$$

Write elements of $\mathbf{R}^{4,1} = \mathbf{R}^4 \oplus \mathbf{R}\epsilon_4$ as $\mathbf{x} + t\epsilon_4$, where $\mathbf{x} \in \mathbf{R}^4$ and $t \in \mathbf{R}$. The Lorentzian inner product is then

$$\langle \mathbf{x} + s\epsilon_4, \mathbf{y} + t\epsilon_4\rangle = \mathbf{x}\cdot\mathbf{y} - st.$$

Write $\mathbf{P}(\mathbf{R}^{4,1})$ for the projective space on $\mathbf{R}^{4,1}$. We use the notation that if $\mathbf{u}$ is any nonzero vector in $\mathbf{R}^{4,1}$, then $[\mathbf{u}]$ denotes the point in $\mathbf{P}(\mathbf{R}^{4,1})$ determined by it. That is, $[\mathbf{u}]$ is the 1-dimensional subspace of $\mathbf{R}^{4,1}$ spanned by $\mathbf{u}$.

Definition 12.11. *Möbius space* is the smooth quadric hypersurface of $\mathbf{P}(\mathbf{R}^{4,1})$ defined by

$$\mathscr{M} = \{[\mathbf{u}] \in \mathbf{P}(\mathbf{R}^{4,1}) : \langle\mathbf{u},\mathbf{u}\rangle = 0\}.$$

Points $[\mathbf{u}], [\mathbf{v}] \in \mathbf{P}(\mathbf{R}^{4,1})$ determine the vectors $\mathbf{u},\mathbf{v} \in \mathbf{R}^{4,1}$ only up to nonzero multiples, but the condition that $\langle\mathbf{u},\mathbf{u}\rangle$ is zero, or positive, or $\langle\mathbf{u},\mathbf{v}\rangle = 0$, is independent of these multiples.

Proposition 12.12. *The map*

$$f_+ : \mathbf{S}^3 \to \mathscr{M}, \quad f_+(\mathbf{x}) = [\mathbf{x} + \epsilon_4] \tag{12.5}$$

with inverse

$$f_+^{-1}[\mathbf{x} + t\epsilon_4] = \frac{1}{t}\mathbf{x}$$

is a smooth diffeomorphism.

Proof. Consider the standard coordinate chart (U_4, φ_4) on $\mathbf{P}(\mathbf{R}^{4,1})$, where the open set

$$U_4 = \{[\mathbf{x} + t\epsilon_4] \in \mathbf{P}(\mathbf{R}^{4,1}) : t \neq 0\}$$

contains $\mathscr{M}$ and has the coordinate map

$$\varphi_4 : U_4 \to \mathbf{R}^4, \quad \varphi_4[\mathbf{x} + t\epsilon_4] = \frac{1}{t}\mathbf{x}.$$

Then the local representation

$$\varphi_4 \circ f_+(\mathbf{x}) = \mathbf{x}$$

is smooth and f_+^{-1} is just the restriction of φ_4 to $\mathscr{M}$, so it also is smooth. Notice that $\varphi_4(\mathscr{M}) = \mathbf{S}^3$. □

Definition 12.13. The *light cone* of $\mathbf{R}^{4,1}$ is the smooth hypersurface

$$\mathscr{L} = \{\mathbf{u} \in \mathbf{R}^{4,1} \setminus \{0\} : \langle \mathbf{u}, \mathbf{u} \rangle = 0\}.$$

The projection $\pi : \mathbf{R}^{4,1} \setminus \{0\} \to \mathbf{P}(\mathbf{R}^{4,1})$ restricted to the light cone defines a principal $\mathbf{R}^\times$-bundle

$$\pi : \mathscr{L} \to \mathscr{M}, \tag{12.6}$$

where $\mathbf{R}^\times = \mathbf{R} \setminus \{0\}$ is the multiplicative group of real numbers.

Definition 12.14. A *conformal structure* on a manifold M is an open covering $\{(U_a, I_a)\}_{a\in\mathscr{A}}$, where I_a is a Riemannian metric on U_a, such that on any non-empty intersection $U_a \cap U_b$, we have $I_b = f_{ab} I_a$, for some smooth positive function f_{ab} on $U_a \cap U_b$. Such a covering can be made maximal with respect to this property.

Proposition 12.15. *Local sections of the principal $\mathbf{R}^\times$-bundle* (12.6) *pull back the Lorentzian inner product on $\mathbf{R}^{4,1}$ to a conformal structure on $\mathscr{M}$.*

Proof. A local section $\sigma : U \subset \mathscr{M} \to \mathscr{L}$ pulls back the Lorentzian inner product on $\mathbf{R}^{4,1}$ to a smooth symmetric bilinear form on U. Any other local section on U is given by $\tilde{\sigma} = t\sigma$, for some smooth function $t : U \to \mathbf{R}^\times$, and then

$$d\tilde{\sigma} = \sigma\, dt + t\, d\sigma,$$

so the induced metrics are related by

$$\langle d\tilde{\sigma}, d\tilde{\sigma} \rangle = \langle \sigma\, dt + t\, d\sigma, \sigma\, dt + t\, d\sigma \rangle = t^2 \langle d\sigma, d\sigma \rangle,$$

since $\langle \sigma, \sigma \rangle = 0$ on U, and thus also $\langle d\sigma, \sigma \rangle = 0$ on U. Since these induced metrics are all related by a positive factor, they are all positive definite if the induced metric of a global section is positive definite. We have the global smooth section

$$\sigma_4 : \mathscr{M} \to \mathscr{L}, \quad \sigma_4[\mathbf{x} + t\epsilon_4] = \frac{1}{t}\mathbf{x} + \epsilon_4 = \varphi_4[\mathbf{x} + t\epsilon_4] + \epsilon_4, \tag{12.7}$$

which pulls back the Lorentzian inner product to

$$\langle d\sigma_4, d\sigma_4 \rangle = d\varphi_4 \cdot d\varphi_4.$$

This is positive definite since the dot product of $\mathbf{R}^4$ is positive definite and $d\varphi_4$ is a linear isomorphism at any point of U_4. □

Proposition 12.16. *The map f_+ of* (12.5) *is a conformal diffeomorphism.*

Proof. Consider the composition

$$\sigma_4 \circ f_+ : \mathbf{S}^3 \to \mathscr{L}, \quad \sigma_4 \circ f_+(\mathbf{x}) = \mathbf{x} + \epsilon_4,$$

where σ_4 is the section (12.7). Then $\sigma_4^* \langle\ ,\ \rangle$ is a Riemannian metric in the conformal structure on $\mathscr{M}$, and

$$f_+^* \sigma_4^* \langle\ ,\ \rangle = \langle (d\mathbf{x}, 0), (d\mathbf{x}, 0) \rangle = d\mathbf{x} \cdot d\mathbf{x},$$

which is the standard Riemannian metric on $\mathbf{S}^3$ defining the conformal structure on $\mathbf{S}^3$. □

Orient $\mathbf{R}^{4,1} = \mathbf{R}^4 \oplus \mathbf{R}\epsilon_4$ by

$$\epsilon_0 \wedge \epsilon_1 \wedge \epsilon_2 \wedge \epsilon_3 \wedge \epsilon_4 > 0. \tag{12.8}$$

Time orientation of $\mathbf{R}^{4,1}$ is defined by the *positive light cone*

$$\mathscr{L}^+ = \{\mathbf{u} = \sum_0^4 u^a \epsilon_a \in \mathscr{L} : u^4 > 0\}.$$

The orthogonal group for the Lorentzian inner product is

$$\mathbf{O}(\mathbf{R}^{4,1}) = \{T \in \mathrm{Hom}(\mathbf{R}^{4,1}, \mathbf{R}^{4,1}) : \langle T\mathbf{u}, T\mathbf{v} \rangle = \langle \mathbf{u}, \mathbf{v} \rangle,\ \forall\ \mathbf{u}, \mathbf{v} \in \mathbf{R}^{4,1}\}.$$

Its representation in the standard basis $\epsilon_0, \epsilon_1, \ldots, \epsilon_4$ of $\mathbf{R}^{4,1}$ is

$$\mathbf{O}(4,1) = \{A \in \mathbf{GL}(5, \mathbf{R}) : {}^t\!A I_{4,1} A = I_{4,1}\},$$

where $I_{4,1}$ is defined in (12.4). The group $\mathbf{O}(\mathbf{R}^{4,1})$ has four connected components. The subgroup that preserves the orientation (12.8) of $\mathbf{R}^{4,1}$ is

$$\mathbf{SO}(\mathbf{R}^{4,1}) = \{T \in \mathbf{O}(\mathbf{R}^{4,1}) : \det T = 1\},$$

which has two connected components (see [84, p. 346]). Then $\mathbf{SO}(4,1) = \{A \in \mathbf{O}(4,1) : \det A = 1\}$ is its representation in the standard basis. Its connected component of the identity is the subgroup preserving time orientation,

$$\mathbf{SO}_+(\mathbf{R}^{4,1}) = \{T \in \mathbf{SO}(\mathbf{R}^{4,1}) : T\mathscr{L}^+ = \mathscr{L}^+\},$$

and $\mathbf{SO}_+(4,1)$ is its representation in the standard basis. The action of $\mathbf{SO}(\mathbf{R}^{4,1})$ on $\mathbf{P}(\mathbf{R}^{4,1})$ is

$$T[\mathbf{u}] = [T\mathbf{u}],$$

for any $T \in \mathbf{SO}(\mathbf{R}^{4,1})$ and $[\mathbf{u}] \in \mathbf{P}(\mathbf{R}^{4,1})$. This action sends $\mathscr{M}$ to itself, since it preserves the inner product on $\mathbf{R}^{4,1}$.

Theorem 12.17. *The map*

$$F : \mathbf{SO}(\mathbf{R}^{4,1}) \to \mathrm{Conf}(\mathbf{S}^3), \quad F(T) = f_+^{-1} \circ T \circ f_+$$

is a group isomorphism. Here $f_+ : \mathbf{S}^3 \to \mathscr{M}$ *is the conformal diffeomorphism* (12.5).

Proof. It is elementary to prove that F is a group monomorphism. It remains to prove that any conformal diffeomorphism $f \in \mathrm{Conf}(\mathbf{S}^3)$ is given by $F(T)$ for some $T \in \mathbf{SO}(\mathbf{R}^{4,1})$. By Corollary 12.10 of Liouville's Theorem 12.7, it suffices to show that if $L : \mathbf{R}^3 \to \mathbf{R}^3$ is a local conformal transformation equal to a translation $L_{\mathbf{a}}$, or orthogonal transformation $A \in \mathbf{O}(3)$, or homothety h_t, or inversion $\mathscr{I}$ in the unit sphere centered at the origin, then

$$\mathscr{S}^{-1} \circ L \circ \mathscr{S} = f_+^{-1} \circ T \circ f_+ : \mathbf{S}^3 \to \mathbf{S}^3, \tag{12.9}$$

for some $T \in \mathbf{SO}(\mathbf{R}^{4,1})$, where $\mathscr{S} = \mathscr{S}_{\mathbf{p}}$ is stereographic projection from some point $\mathbf{p} \in \mathbf{S}^3$. In particular, this shows that the local conformal diffeomorphism on the left side of (12.9) is defined and smooth on all of $\mathbf{S}^3$ and hence is an element of $\mathrm{Conf}(\mathbf{S}^3)$.

In the following let $\mathscr{S} = \mathscr{S}_{-\epsilon_0}$ be stereographic projection from the point $-\epsilon_0 \in \mathbf{S}^3$, as described in Definition 5.22. Then $\mathbf{R}^3$ is the span of $\epsilon_1, \epsilon_2, \epsilon_3$ in $\mathbf{R}^4 \subset \mathbf{R}^{4,1}$. The proof is completed by verifying the following Exercises.

Exercise 45 (Translation $L_{\mathbf{a}}$ by $\mathbf{a} \in \mathbf{R}^3$). Prove that

$$\mathscr{S}^{-1} \circ L_{\mathbf{a}} \circ \mathscr{S} = f_+^{-1} \circ T_{\mathbf{a}} \circ f_+ : \mathbf{S}^3 \to \mathbf{S}^3,$$

where the matrix of $T_{\mathbf{a}}$ in the standard basis is

$$\begin{pmatrix} \frac{2-|\mathbf{a}|^2}{2} & -{}^t\mathbf{a} & -\frac{|\mathbf{a}|^2}{2} \\ \mathbf{a} & I_3 & \mathbf{a} \\ \frac{|\mathbf{a}|^2}{2} & {}^t\mathbf{a} & \frac{2+|\mathbf{a}|^2}{2} \end{pmatrix} \in \mathbf{SO}_+(4,1). \tag{12.10}$$

Exercise 46 (Orthogonal transformation $A \in \mathbf{O}(3)$). Prove that

$$\mathscr{S}^{-1} \circ A \circ \mathscr{S} = f_+^{-1} \circ T_A \circ f_+ : \mathbf{S}^3 \to \mathbf{S}^3,$$

where the matrix of T_A in the standard basis is

$$(\det A) \begin{pmatrix} 1 & 0 & 0 \\ 0 & A & 0 \\ 0 & 0 & 1 \end{pmatrix} \in \mathbf{SO}(4,1). \tag{12.11}$$

Exercise 47 (Homothety h_t, for $t \in \mathbf{R}^\times$). Prove that

$$\mathscr{S}^{-1} \circ h_t \circ \mathscr{S} = f_+^{-1} \circ T_t \circ f_+,$$

where the matrix of T_t in the standard basis is

$$\begin{pmatrix} \frac{1+t^2}{2t} & 0 & \frac{1-t^2}{2t} \\ 0 & I_3 & 0 \\ \frac{1-t^2}{2t} & 0 & \frac{1+t^2}{2t} \end{pmatrix} \in \mathbf{SO}_+(4,1). \tag{12.12}$$

Exercise 48 (Inversion $\mathscr{I}$ in the unit sphere). Prove that

$$\mathscr{S}^{-1} \circ \mathscr{I} \circ \mathscr{S} = f_+^{-1} \circ T_{\mathscr{I}} \circ f_+,$$

where the matrix of $T_{\mathscr{I}}$ in the standard basis is

$$\begin{pmatrix} 1 & 0 & 0 \\ 0 & -I_3 & 0 \\ 0 & 0 & -1 \end{pmatrix} \in \mathbf{SO}(4,1). \tag{12.13}$$

□

Corollary 12.18. *The Lie group* $\mathbf{SO}(\mathbf{R}^{4,1})$ *is generated by elements of the four types described in equations* (12.10) *through* (12.13).

12.3 Möbius frames

The group Conf($\mathbf{S}^3$) acts transitively on $\mathbf{S}^3$, so the action of $\mathbf{SO}(\mathbf{R}^{4,1})$ on $\mathscr{M}$ is transitive, by Theorem 12.17. For any chosen origin $p \in \mathscr{M}$, we have a principal bundle map

$$\pi : \mathbf{SO}(\mathbf{R}^{4,1}) \to \mathscr{M}, \ \pi(T) = Tp. \tag{12.14}$$

In order to do calculations for this action, we want to represent the group in some convenient basis of $\mathbf{R}^{4,1}$. One problem with using the standard basis for this purpose is that $[\epsilon_a] \notin \mathscr{M}$, for any of the standard basis vectors ϵ_a. This complicates the expression for the map (12.14). We consider bases of $\mathbf{R}^{4,1}$ that contain null vectors.

Definition 12.19. A *Möbius frame* of $\mathbf{R}^{4,1}$ is an oriented basis $\mathbf{Y}_0, \mathbf{Y}_1, \dots, \mathbf{Y}_4$ (so $\mathbf{Y}_0 \wedge \dots \wedge \mathbf{Y}_4 > 0$), such that

$$\langle \mathbf{Y}_a, \mathbf{Y}_b \rangle = g_{ab},$$

where the g_{ab}, for $a, b = 0, \dots, 4$, are the entries of

$$g = \begin{pmatrix} 0 & 0 & -1 \\ 0 & I_3 & 0 \\ -1 & 0 & 0 \end{pmatrix}.$$

In particular, $\mathbf{Y}_0, \mathbf{Y}_4 \in \mathscr{L}$. The frame is *time oriented* if, in addition,

$$\mathbf{Y}_0, \mathbf{Y}_4 \in \mathscr{L}^+.$$

We choose and fix the time oriented Möbius frame of $\mathbf{R}^{4,1}$

$$\boldsymbol{\delta}_0 = \frac{1}{2}(\boldsymbol{\epsilon}_4 + \boldsymbol{\epsilon}_0), \;\; \boldsymbol{\delta}_4 = \boldsymbol{\epsilon}_4 - \boldsymbol{\epsilon}_0, \;\; \boldsymbol{\delta}_i = \boldsymbol{\epsilon}_i, \; i = 1,2,3. \tag{12.15}$$

The Lorentzian inner product of $\mathbf{u} = \sum_0^4 u^a \boldsymbol{\delta}_a$ with $\mathbf{v} = \sum_0^4 v^b \boldsymbol{\delta}_b$ is then

$$\langle \mathbf{u}, \mathbf{v} \rangle = -(u^0 v^4 + u^4 v^0) + \sum_1^3 u^i v^i.$$

The positive light cone in this Möbius frame is

$$\mathscr{L}^+ = \{\mathbf{u} = \sum_0^4 u^a \boldsymbol{\delta}_a : -2u^0 u^4 + \sum_1^3 (u^i)^2 = 0, \; u^0 + 2u^4 > 0\}.$$

The group $\mathbf{SO}(\mathbf{R}^{4,1})$ is represented in the standard frame by $\mathbf{SO}(4,1)$, and in this Möbius frame by

$$\mathbf{M\ddot{o}b} = \{Y \in \mathbf{SL}(5, \mathbf{R}) : {}^tYgY = g\}. \tag{12.16}$$

The isomorphism between these two representations is given as follows. Let

$$L = \begin{pmatrix} 1/2 & 0 & -1 \\ 0 & I_3 & 0 \\ 1/2 & 0 & 1 \end{pmatrix}, \quad L^{-1} = \begin{pmatrix} 1 & 0 & 1 \\ 0 & I_3 & 0 \\ -1/2 & 0 & 1/2 \end{pmatrix}, \tag{12.17}$$

be the matrix in the standard basis of $\mathbf{R}^{4,1}$ of the change of basis transformation $L\boldsymbol{\epsilon}_a = \boldsymbol{\delta}_a$, for $a = 0, \ldots, 4$. Then

$$\mathscr{F} : \mathbf{SO}(4,1) \to \mathbf{M\ddot{o}b}, \quad \mathscr{F}(A) = L^{-1}AL \tag{12.18}$$

is the desired isomorphism. An element $T \in \mathbf{SO}(\mathbf{R}^{4,1})$ is represented by the matrix $Y \in \mathbf{M\ddot{o}b}$ that satisfies

$$T\boldsymbol{\delta}_a = \sum_0^4 Y_a^b \boldsymbol{\delta}_b,$$

for $a = 0, \ldots, 4$. A matrix Y is in **Möb** if and only if its columns,

$$\mathbf{Y}_a = \sum_0^4 Y_a^b \boldsymbol{\delta}_b,$$

for $a = 0,\dots,4$, constitute a Möbius frame of $\mathbf{R}^{4,1}$. Conversely, if $\mathbf{Y}_0,\dots,\mathbf{Y}_4$ is any Möbius frame, and if $\mathbf{Y}_a = \sum Y_a^b \boldsymbol{\delta}_b$, then the matrix $Y = (Y_a^b) \in \mathbf{Möb}$. The connected component of the identity of this matrix group is

$$\mathbf{Möb}_+ = \{Y \in \mathbf{Möb} : \mathbf{Y}_0,\ \mathbf{Y}_4 \in \mathscr{L}^+\} \cong \mathbf{SO}_+(4,1),$$

which is the representation in the fixed Möbius frame of the group $\mathbf{SO}_+(\mathbf{R}^{4,1})$. A matrix Y is in $\mathbf{Möb}_+$ if and only if its columns, in the Möbius frame (12.15), constitute a time oriented Möbius frame of $\mathbf{R}^{4,1}$, and conversely.

Remark 12.20. If $T \in \mathbf{SO}(\mathbf{R}^{4,1})$, then $T\mathscr{L} = \mathscr{L}$ and $\mathscr{L}^+$ connected imply that either $T\mathscr{L}^+ = \mathscr{L}^+$ or $T\mathscr{L}^+ = -\mathscr{L}^+ = \mathscr{L}^-$. From this it follows that if $Y \in \mathbf{Möb}$, then either both $\mathbf{Y}_0$ and $\mathbf{Y}_4$ are in $\mathscr{L}^+$, or neither is in $\mathscr{L}^+$.

In what follows, whenever a 5×1 column vector is identified with an element of $\mathbf{R}^{4,1}$, it is understood that this is done with the fixed Möbius frame (12.15). The Lie algebra of $\mathbf{Möb}$ is identified with the Lie algebra of its connected component $\mathbf{Möb}_+$, which is

$$\mathfrak{m\ddot{o}b} = \{X \in \mathfrak{sl}(5,\mathbf{R}) : {}^t\!Xg + gX = 0\} \cong \mathfrak{o}(4,1).$$

Explicitly, a 5×5 matrix X is in $\mathfrak{m\ddot{o}b}$ iff for $i,j = 1,2,3$

$$X_j^i = -X_i^j,\ X_4^0 = 0 = X_0^4,\ X_4^i = X_i^0,\ X_j^4 = X_0^j,\ X_4^4 = -X_0^0. \tag{12.19}$$

The Maurer–Cartan form of $\mathbf{Möb}$ is the $\mathfrak{m\ddot{o}b}$-valued 1-form on $\mathbf{Möb}$,

$$\omega = Y^{-1}dY = \begin{pmatrix} \omega_0^0 & \omega_j^0 & 0 \\ \omega_0^i & \omega_j^i & \omega_i^0 \\ 0 & \omega_0^j & -\omega_0^0 \end{pmatrix}, \tag{12.20}$$

where $i,j = 1,2,3$, and $\omega_j^i = -\omega_i^j$. In terms of the columns of $Y \in \mathbf{Möb}$, equation (12.20) says

$$d\mathbf{Y}_a = \sum_0^4 \mathbf{Y}_b \omega_a^b, \quad a,b = 0,1,\dots,4. \tag{12.21}$$

It expresses the derivative of column a of Y, which is the map $\mathbf{Y}_a : \mathbf{Möb} \to \mathbf{R}^{4,1}$, as a linear combination of the Möbius frame of $\mathbf{R}^{4,1}$ given by all five columns, where the coefficients are 1-forms on $\mathbf{Möb}$. The exterior differential of equation (12.20) gives the structure equations of $\mathbf{Möb}$,

$$d\omega_b^a = -\sum_{c=0}^4 \omega_c^a \wedge \omega_b^c, \tag{12.22}$$

for $a,b = 0,1,\dots,4$.

In terms of our fixed Möbius frame, $\boldsymbol{\delta}_0, \ldots, \boldsymbol{\delta}_4$ defined in (12.15), Möbius space $\mathscr{M}$ of Definition 12.11 is

$$\mathscr{M} = \left\{ [\mathbf{u}] = [\sum_0^4 u^a \boldsymbol{\delta}_a] \in \mathbf{P}(\mathbf{R}^{4,1}) : 0 = -2u^0u^4 + \sum_1^3 (u^i)^2 \right\}.$$

In this Möbius frame, the diffeomorphism $f_+ : \mathbf{S}^3 \to \mathscr{M}$ defined in (12.5) has the expression

$$f_+(\sum_0^3 x^i \boldsymbol{\epsilon}_i) = \left[(1+x^0)\boldsymbol{\delta}_0 + \sum_1^3 x^i \boldsymbol{\delta}_i + \frac{1-x^0}{2}\boldsymbol{\delta}_4 \right], \tag{12.23}$$

and its inverse mapping $f_+^{-1} : \mathscr{M} \to \mathbf{S}^3$ is

$$f_+^{-1}\left[\sum_0^4 u^a \boldsymbol{\delta}_a \right] = \frac{1}{u^0 + 2u^4} \left((u^0 - 2u^4)\boldsymbol{\epsilon}_0 + 2\sum_1^3 u^i \boldsymbol{\epsilon}_i \right).$$

Note that if $[\sum_0^4 u^a \boldsymbol{\delta}_a] \in \mathscr{M}$, then $2u^0u^4 = \sum (u^i)^2 > 0$, so $u^0 + 2u^4 \neq 0$.

The projective action of $\mathbf{SO}(\mathbf{R}^{4,1})$ on $\mathscr{M}$ becomes, in our Möbius frame representation, the matrix group action

$$\mathbf{M\ddot{o}b} \times \mathscr{M} \to \mathscr{M}, \quad (Y, [\mathbf{u}]) \mapsto Y[\mathbf{u}] = [Y\mathbf{u}],$$

where $Y\mathbf{u}$ is matrix multiplication of $Y \in \mathbf{M\ddot{o}b}$ with the 5×1 coefficient matrix of the vector $\mathbf{u} = \sum_0^4 u^a \boldsymbol{\delta}_a$. By Theorem 12.17, this action is transitive, which means that the orbit of any given point of $\mathscr{M}$ is all of $\mathscr{M}$. Choose $[\boldsymbol{\delta}_0]$ to be the origin of $\mathscr{M}$. If $Y \in \mathbf{M\ddot{o}b}_+$, then $Y\boldsymbol{\delta}_0 = \mathbf{Y}_0 \in \mathscr{L}^+$ is the first column of Y relative to the basis (12.15). Transitivity of the action of $\mathbf{M\ddot{o}b}_+$ on $\mathscr{M}$ means that any vector $\mathbf{Y}_0 \in \mathscr{L}^+$ can be completed to a time oriented Möbius frame $\mathbf{Y}_0, \mathbf{Y}_1, \mathbf{Y}_2, \mathbf{Y}_3, \mathbf{Y}_4$ of $\mathbf{R}^{4,1}$. In a vector space with positive definite inner product, this completion is done by using the Gram-Schmidt orthogonalization process. A similar process exists in the present case. See Problem 12.59.

The transitivity of the action of $\mathbf{M\ddot{o}b}_+$ on $\mathscr{M}$ implies transitivity of the action of $\mathbf{M\ddot{o}b}$ on $\mathscr{M}$, which means that the map

$$\pi : \mathbf{M\ddot{o}b} \to \mathscr{M}, \quad \pi(Y) = Y[\boldsymbol{\delta}_0] = [\mathbf{Y}_0] \tag{12.24}$$

is surjective and its restriction to $\mathbf{M\ddot{o}b}_+$ is surjective. The set of elements of $\mathbf{M\ddot{o}b}$ that fix $[\boldsymbol{\delta}_0]$ is called the *isotropy subgroup* of $\mathbf{M\ddot{o}b}$ at $[\boldsymbol{\delta}_0]$. We denote it

$$G_0 = \{K(r, A, \mathbf{y}) : r \neq 0,\ \mathbf{y} \in \mathbf{R}^3,\ A \in \mathbf{SO}(3)\}, \tag{12.25}$$

where we introduce the notation

$$K(r,A,\mathbf{y}) = \begin{pmatrix} \frac{1}{r} & {}^t\mathbf{y}A & \frac{1}{2}r|\mathbf{y}|^2 \\ 0 & A & r\mathbf{y} \\ 0 & 0 & r \end{pmatrix} \in \mathbf{M\ddot{o}b}, \tag{12.26}$$

for any real number $r \neq 0$, vector $\mathbf{y} \in \mathbf{R}^3$, and matrix $A \in \mathbf{SO}(3)$. If $K(r,A,\mathbf{y})$ and $K(s,B,\mathbf{z})$ in G_0, then

$$K(r,A,\mathbf{y})K(s,B,\mathbf{z}) = K(rs,AB,\mathbf{y}+r^{-1}A\mathbf{z}), \tag{12.27}$$

from which it follows that

$$K(r,A,\mathbf{y})^{-1} = K(r^{-1},A^{-1},-rA^{-1}\mathbf{y}). \tag{12.28}$$

Then $\mathscr{M} \cong \mathbf{M\ddot{o}b}/G_0$ and (12.24) is a principal G_0-bundle over $\mathscr{M}$. The isotropy subgroup of $\mathbf{M\ddot{o}b}_+$ at $[\boldsymbol{\delta}_0]$ is the connected component of the identity of G_0,

$$G_{0+} = \{K(r,A,\mathbf{y}) \in G_0 : r > 0\},$$

and we also have $\mathscr{M} \cong \mathbf{M\ddot{o}b}_+/G_{0+}$.

The Lie algebra $\mathfrak{m\ddot{o}b}$ has a direct sum decomposition as vector spaces

$$\mathfrak{m\ddot{o}b} = \mathfrak{g}_0 + \mathfrak{m}_0, \tag{12.29}$$

(see (12.19) for the structure of $\mathfrak{m\ddot{o}b}$) where $\mathfrak{g}_0$ is the Lie algebra of G_0, and of G_{0+},

$$\mathfrak{g}_0 = \{\begin{pmatrix} X_0^0 & X_j^0 & 0 \\ 0 & X_j^i & X_i^0 \\ 0 & 0 & -X_0^0 \end{pmatrix} : X_j^i + X_i^j = 0, i,j = 1,2,3\},$$

and $\mathfrak{m}_0$ is the complementary subspace

$$\mathfrak{m}_0 = \{X = \begin{pmatrix} 0 & 0 & 0 \\ \mathbf{x} & 0 & 0 \\ 0 & {}^t\mathbf{x} & 0 \end{pmatrix} : \mathbf{x} \in \mathbf{R}^3\}.$$

We identify $\mathfrak{m}_0$ with $\mathbf{R}^3$ by

$$\mathfrak{m}_0 \ni X \leftrightarrow \mathbf{x} \in \mathbf{R}^3. \tag{12.30}$$

Unlike the situation with Euclidean, spherical or hyperbolic geometry, however, the subspace $\mathfrak{m}_0$ cannot be chosen to be invariant under the adjoint action of G_0 or of G_{0+}. That is, in general $K^{-1}XK \notin \mathfrak{m}_0$. We have

$$K\begin{pmatrix}0 & 0 & 0\\ \mathbf{x} & 0 & 0\\ 0 & {}^t\mathbf{x} & 0\end{pmatrix}K^{-1} = \begin{pmatrix}0 & 0 & 0\\ rA\mathbf{x} & 0 & 0\\ 0 & r{}^t\mathbf{x}{}^tA & 0\end{pmatrix} \quad \text{mod } \mathfrak{g}_0.$$

In terms of the identification (12.30), the adjoint action of G_0 on $\mathfrak{m\ddot{o}b}/\mathfrak{g}_0 \cong \mathfrak{m}_0$ is

$$G_0 \times \mathbf{R}^3 \to \mathbf{R}^3, \quad K(r,A,\mathbf{y})\mathbf{x} = rA\mathbf{x}. \tag{12.31}$$

Definition 12.21. A *Möbius frame field on an open subset $U \subset \mathscr{M}$* is a smooth local section

$$Y : U \to \mathbf{M\ddot{o}b} \tag{12.32}$$

of the principal G_0-bundle (12.24). This means that the columns $\mathbf{Y}_a$ of Y are smooth maps

$$\mathbf{Y}_a : U \to \mathbf{R}^{4,1}, \quad a = 0,1,2,3,4,$$

which form a Möbius frame at each point of U and such that $\pi \circ \mathbf{Y}_0 = [\mathbf{Y}_0]$ is the identity map of U. A *time oriented Möbius frame field on U* is a smooth local section $Y : U \to \mathbf{M\ddot{o}b}_+$.

A Möbius frame field (12.32) pulls back the Maurer–Cartan form (12.20) to U. We continue to denote these pulled back forms by the same letters, $Y^{-1}dY = \omega = (\omega^a_b)$, so

$$d\mathbf{Y}_a = \sum_0^4 \mathbf{Y}_b \omega^b_a,$$

where $a,b = 0,1,2,3,4$, and the ω^b_a are now 1-forms on U satisfying the structure equations (12.22). Let

$$\theta = {}^t(\omega^1_0, \omega^2_0, \omega^3_0)$$

denote the $\mathfrak{m}_0$-component of ω (using the identification (12.30)). Any other Möbius frame field on U is given by

$$\tilde{Y} = YK(r,A,\mathbf{y}), \tag{12.33}$$

where

$$K(r,A,\mathbf{y}) : U \to G_0$$

is a smooth map. Let $\tilde{\omega} = \tilde{Y}^{-1}d\tilde{Y}$ and let $\tilde{\theta}$ denote its $\mathfrak{m}_0$-component. Then

$$\tilde{\omega} = K^{-1}Y^{-1}(dYK + YdK) = K^{-1}\omega K + K^{-1}dK$$

shows that $\tilde{\theta}$ is the $\mathfrak{m}_0$-component of the adjoint action of K^{-1} on θ. Then

$$\tilde{\theta} = \frac{1}{r}A^{-1}\theta, \tag{12.34}$$

by (12.31). From

$$^t\tilde{\theta}\tilde{\theta} = \frac{1}{r^2}{}^t\theta\theta, \tag{12.35}$$

it follows that the collection of symmetric bilinear forms ${}^t\theta\theta$ on U, for all open sets $U \subset \mathscr{M}$ on which there is a Möbius frame field $Y : U \to \mathbf{M\ddot{o}b}$, defines a conformal structure on $\mathscr{M}$. Actually, this is the conformal structure induced by local sections of $\pi : \mathscr{L} \to \mathscr{M}$ introduced in Proposition 12.15. For, if $\mathbf{Y}_0$ is the first column of the Möbius frame field $Y : U \to \mathbf{M\ddot{o}b}$, then $\mathbf{Y}_0 : U \to \mathscr{L}$ is a local section and

$$\langle d\mathbf{Y}_0, d\mathbf{Y}_0\rangle = \langle \omega^0_0\mathbf{Y}_0 + \sum_1^3 \omega^i_0\mathbf{Y}_i, \omega^0_0\mathbf{Y}_0 + \sum_1^3 \omega^j_0\mathbf{Y}_j\rangle = \sum_1^3 \omega^i_0\omega^i_0 = {}^t\theta\theta. \tag{12.36}$$

For any change of frame field (12.33), the 3-forms

$$\tilde{\omega}^1_0\wedge\tilde{\omega}^2_0\wedge\tilde{\omega}^3_0 = \frac{1}{r^3}\omega^1_0\wedge\omega^2_0\wedge\omega^3_0,$$

so they define an orientation on $\mathscr{M}$ only if we restrict to local time oriented Möbius frame fields on $\mathscr{M}$.

12.4 Möbius frames along a surface

An immersion $f : M \to \mathscr{M}$ of a surface pulls back the conformal structure (12.35) of Möbius space to a conformal structure on M, which in turn induces a complex structure on M, by Problem 12.57.

Definition 12.22. An immersion

$$f : M \to \mathscr{M}$$

of a Riemann surface M is *conformal* if it induces the given complex structure on M. In more detail, the conformality of $f : M \to \mathscr{M}$ means that if (V, I) belongs to the conformal structure of $\mathscr{M}$, and if (U, z) is a complex coordinate chart of M with $U \subset f^{-1}V$, then $f^*I = e^{2u}dzd\bar{z}$, for some smooth function $u : U \to \mathbf{R}$.

Definition 12.23. A *Möbius frame field along* f on an open subset $U \subset M$ is a smooth map

$$Y : U \to \mathbf{M\ddot{o}b},$$

such that $\pi \circ Y = [\mathbf{Y}_0] = f$ on U. In other words, if the columns of Y are $\mathbf{Y}_a$, for $a = 0, \dots, 4$, then the smooth maps

$$\mathbf{Y}_0, \mathbf{Y}_1, \mathbf{Y}_2, \mathbf{Y}_3, \mathbf{Y}_4 : U \to \mathbf{R}^{4,1} \tag{12.37}$$

constitute a Möbius frame at each point of U, and $f = [\mathbf{Y}_0]$. A *lift of* f is a smooth map $\mathbf{F} : U \to \mathscr{L} \subset \mathbf{R}^{4,1}$ such that $f = [\mathbf{F}]$. Conversely, a set of smooth maps (12.37) forming a Möbius frame at each point of U, such that $\mathbf{Y}_0$ is a lift of f, determines a Möbius frame field

$$Y = (\mathbf{Y}_0, \dots, \mathbf{Y}_4) : U \to \mathbf{M\ddot{o}b}$$

along f. If $Y : U \to \mathbf{M\ddot{o}b}_+$, then Y is a *time oriented Möbius frame field along* f on U. A lift $\mathbf{F} : U \to \mathscr{L}^+$ of f is a *time oriented lift of* f.

We now apply the reduction scheme of Chapter 3 to time oriented Möbius frames along a conformal immersion $f : M \to \mathscr{M}$. If $Y : U \subset M \to \mathbf{M\ddot{o}b}_+$ is a time oriented Möbius frame field along f, then any other on U is given by $\tilde{Y} = YK(r, A, \mathbf{y})$, where $K(r, A, \mathbf{y}) : U \to G_{0+} \subset \mathbf{M\ddot{o}b}_+$ is a smooth map into

$$G_{0+} = \{ \begin{pmatrix} \frac{1}{r} & {}^t\mathbf{y}A & \frac{r}{2}|\mathbf{y}|^2 \\ 0 & A & r\mathbf{y} \\ 0 & 0 & r \end{pmatrix} : r > 0,\ A \in \mathbf{SO}(3),\ \mathbf{y} \in \mathbf{R}^3 \}.$$

The notation $K(r, A, \mathbf{y})$ is defined in (12.26). The vector space direct sum $\mathfrak{m\ddot{o}b} = \mathfrak{g}_0 + \mathfrak{m}_0$ of (12.29) decomposes the Maurer–Cartan form $\omega = Y^{-1}dY$ into $\omega = \omega_{\mathfrak{m}_0} + \omega_{\mathfrak{g}_0}$, where

$$\omega_{\mathfrak{m}_0} = \begin{pmatrix} 0 & 0 & 0 \\ \theta & 0 & 0 \\ 0 & {}^t\theta & 0 \end{pmatrix}, \quad \theta = \begin{pmatrix} \omega_0^1 \\ \omega_0^2 \\ \omega_0^3 \end{pmatrix}.$$

If $\tilde{\omega} = \tilde{Y}^{-1}d\tilde{Y}$, then as shown in (12.34) we have $\tilde{\theta} = \frac{1}{r}A^{-1}\theta$.

Definition 12.24. A *first order* Möbius frame field along f is a time oriented Möbius frame field $Y : U \subset M \to \mathbf{M\ddot{o}b}_+$ along f for which

$$\omega_0^3 = 0 \quad \text{and} \quad \omega_0^1 \wedge \omega_0^2 > 0, \tag{12.38}$$

where positivity of the 2-form is relative to the orientation of the Riemann surface M.

Lemma 12.25. *Let $f : M \to \mathscr{M}$ be a conformal immersion of a Riemann surface M. Given a point $m \in M$, there exists a first order Möbius frame field along f on some neighborhood U of m.*

Proof. There exists a lift $\mathbf{F} : U \to \mathscr{L}^+$ on some neighborhood U of m. Using the standard orthonormal basis $\boldsymbol{\epsilon}_0, \ldots, \boldsymbol{\epsilon}_4$ of $\mathbf{R}^{4,1}$, we have $\mathbf{F} = \sum_0^4 F^a \boldsymbol{\epsilon}_a$ for smooth functions $F^a : U \to \mathbf{R}$ for which $F^4 > 0$ on U. Since $\frac{1}{F^4}\mathbf{F}$ is also a time oriented lift of f, we may assume $F^4 = 1$ on U, in which case $\mathbf{F} = \mathbf{x} + \boldsymbol{\epsilon}_4$, where $\mathbf{x} : U \to \mathbf{S}^3 \subset \mathbf{R}^4$ is a smooth immersion. We know there exists a first order frame field $e : U \to \mathbf{SO}(4)$ along $\mathbf{x}$, possibly after shrinking U. Then $\mathbf{x} = \mathbf{e}_0$ and $d\mathbf{x} = \theta^1 \mathbf{e}_1 + \theta^2 \mathbf{e}_2$, for smooth 1-forms θ^1 and θ^2 on U. If we let $\mathbf{Y}_4 = \frac{1}{2}(-\mathbf{x} + \boldsymbol{\epsilon}_4)$, then

$$Y = (\mathbf{F}, \mathbf{e}_1, \mathbf{e}_2, \mathbf{e}_3, \mathbf{Y}_4) : U \to \mathbf{M\ddot{o}b}_+$$

is a smooth time oriented first order Möbius frame field along f. □

Let $Y : U \to \mathbf{M\ddot{o}b}_+$ be first order along f. Taking the exterior differential of $\omega_0^3 = 0$ and using the structure equations (12.22) of **Möb**, we find that

$$\omega_1^3 \wedge \omega_0^1 + \omega_2^3 \wedge \omega_0^2 = 0,$$

which implies, by Cartan's Lemma, that,

$$\omega_i^3 = \sum_{j=1}^{2} h_{ij} \omega_0^j, \quad h_{ij} = h_{ji}, \quad i,j = 1,2,$$

for some smooth functions $h_{ij} : U \to \mathbf{R}$. Let $S = (h_{ij}) : U \to \mathscr{S}$, where $\mathscr{S}$ is the space of real 2×2 symmetric matrices, and let

$$h = \frac{1}{2}(h_{11} - h_{22}) - i h_{12} = L(S), \quad H = \frac{1}{2}(h_{11} + h_{22}) = \frac{1}{2}\mathrm{trace} S, \tag{12.39}$$

where L is the Hopf transform defined in (4.15) of Exercise 9. Then

$$\omega_1^3 - i\omega_2^3 = h(\omega_0^1 + i\omega_0^2) + H(\omega_0^1 - i\omega_0^2) \tag{12.40}$$

and this equation determines $h : U \to \mathbf{C}$ and $H : U \to \mathbf{R}$.

Exercise 49. Prove that, if $Y : U \to \mathbf{M\ddot{o}b}_+$ is first order along f, then any other first order frame field along f on U is given by

$$\tilde{Y} = YK(r, A, \mathbf{y}), \tag{12.41}$$

where $K(r, A, \mathbf{y}) : U \to G_1 \subset G_{0+}$ is any smooth map into the subgroup

$$G_1 = \{ \begin{pmatrix} 1/r & {}^t\mathbf{y}A & r|\mathbf{y}|^2/2 \\ 0 & A & r\mathbf{y} \\ 0 & 0 & r \end{pmatrix} : r > 0,\ A = \begin{pmatrix} a & 0 \\ 0 & 1 \end{pmatrix},\ a \in \mathbf{SO}(2),\ \mathbf{y} \in \mathbf{R}^3 \}.$$

Hint: Be sure to apply both conditions in (12.38).

As a shorthand we shall write

$$G_1 = \{K(r, e^{it}, \mathbf{y}) : r > 0,\ t \in \mathbf{R},\ \mathbf{y} \in \mathbf{R}^3\},$$

where $a = \begin{pmatrix} \cos t & -\sin t \\ \sin t & \cos t \end{pmatrix}$. Its Lie algebra is

$$\mathfrak{g}_1 = \{\begin{pmatrix} s & {}^t\mathbf{x} & 0 \\ 0 & X & \mathbf{x} \\ 0 & 0 & -s \end{pmatrix} : \mathbf{x} \in \mathbf{R}^3,\ X = \begin{pmatrix} 0 & t & 0 \\ -t & 0 & 0 \\ 0 & 0 & 0 \end{pmatrix},\ s, t \in \mathbf{R}\}.$$

A complementary subspace is

$$\mathfrak{m}_1 = \{\begin{pmatrix} 0 & 0 & 0 & 0 & 0 \\ 0 & 0 & 0 & s & 0 \\ 0 & 0 & 0 & t & 0 \\ 0 & -s & -t & 0 & 0 \\ 0 & 0 & 0 & 0 & 0 \end{pmatrix} : s, t \in \mathbf{R}\}.$$

Write (12.40) for $\tilde{Y}$ of (12.41) as

$$\tilde{\omega}_1^3 - i\tilde{\omega}_2^3 = \tilde{h}(\tilde{\omega}_0^1 + i\tilde{\omega}_0^2) + \tilde{H}(\tilde{\omega}_0^1 - i\tilde{\omega}_0^2).$$

Then

$$\tilde{\omega}_0^1 + i\tilde{\omega}_0^2 = \frac{1}{r} e^{-it}(\omega_0^1 + i\omega_0^2),$$
$$\tilde{\omega}_1^3 - i\tilde{\omega}_2^3 = re^{2it} h(\tilde{\omega}_0^1 + i\tilde{\omega}_0^2) + r(H - y^3)(\tilde{\omega}_0^1 - i\tilde{\omega}_0^2),$$

which shows that the coefficients transform by (see Problem 12.63)

$$\tilde{h} = re^{2it} h, \quad \tilde{H} = r(H - y^3). \tag{12.42}$$

Definition 12.26. The *conformal area element* of a conformal immersion $f : M \to \mathcal{M}$ is the smooth 2-form Ω on M, defined by $\Omega = \Omega_Y = |h|^2 \omega_0^1 \wedge \omega_0^2$ on U, for any first order frame field $Y : U \to \mathbf{M\ddot{o}b}_+$. See Problem 12.64.

Definition 12.27. An *umbilic point* of a conformal immersion $f : M \to \mathcal{M}$ is a point $m \in M$ for which $\Omega_{(m)} = 0$.

The next frame reduction involves the action of G_1 on $\mathbf{C} \times \mathbf{R}$ coming from (12.42), which for $K(r, e^{it}, \mathbf{y}) \in G_1$ is

$$K(r, e^{it}, \mathbf{y})(h, H) = (re^{2it} h, r(H - y^3)). \tag{12.43}$$

This action on the $\mathbf{C}$ factor has two orbits of distinct types, $\{0\}$ and $\mathbf{C}\setminus\{0\}$, while the action on the $\mathbf{R}$ factor is transitive. Since $h = 0$ exactly at the umbilic points of f in U, by Lemma 12.64, the orbit type of the action on $\mathbf{C}$ depends on whether a point is umbilic. The action on $\mathbf{R}$ requires no assumptions on f. Following Bryant in [20], at this stage we make an intermediate reduction by choosing $y^3 = H$ in (12.42).

Definition 12.28. A *γ-frame along f* is a first order Möbius frame field $Y : U \to$ $\mathbf{Möb}_+$ along f for which the coefficient H in (12.40) is identically zero on U. It is characterized by

$$\omega_0^3 = 0, \quad \omega_1^3 - i\omega_2^3 = h(\omega_0^1 + i\omega_0^2), \quad \omega_0^1\wedge\omega_0^2 > 0,$$

for some smooth function $h : U \to \mathbf{C}$.

Exercise 50. Prove that if $Y : U \to \mathbf{Möb}_+$ is a γ-frame along f, then any other γ-frame on U is given by $\tilde{Y} = YK(r,A,\mathbf{y})$, where $K : U \to G_\gamma$,

$$G_\gamma = \{K(r,e^{it},\mathbf{y}) \in G_1 : y^3 = 0\}.$$

We have just observed that any first order frame field can be reduced to a γ-frame. If $Y : U \to \mathbf{Möb}_+$ is a γ-frame field, we take the exterior derivative of $\omega_1^3 - i\omega_2^3 = h(\omega_0^1 + i\omega_0^2)$ and apply the structure equations to get

$$(dh + h(\omega_0^0 - 2i\omega_1^2))\wedge\varphi + \omega_3^0\wedge\bar{\varphi} = 0,$$

where $\varphi = \omega_0^1 + i\omega_0^2$. It follows that

$$dh + h(\omega_0^0 - 2i\omega_1^2) = h_1\varphi + h_2\bar{\varphi}, \quad \omega_3^0 = h_2\varphi + \bar{h}_2\bar{\varphi}, \tag{12.44}$$

for some smooth functions $h_1, h_2 : U \to \mathbf{C}$. The exterior differential of $\omega_3^0 = h_2\varphi + \bar{h}_2\bar{\varphi}$, combined with the structure equations, leads to

$$dh_2 + h_2(2\omega_0^0 + i\omega_2^1) - \frac{h}{2}(\omega_1^0 + i\omega_2^0) = P\varphi + W\bar{\varphi}, \tag{12.45}$$

for some smooth functions, called the *γ-invariants* of Y,

$$P : U \to \mathbf{C}, \quad W : U \to \mathbf{R}.$$

Definition 12.29. The *Willmore function of $f : M \to \mathcal{M}$ relative to the γ-frame* $Y : U \to \mathbf{Möb}$ is the real valued γ-invariant $W : U \to \mathbf{R}$ defined in (12.45). As we shall see in Section 13.6, this could also be called the *conformal mean curvature* of f relative to Y, or the *Bryant function* of f relative to Y, as he was the first to identify its importance in [20]

Lemma 12.30. *If $Y : U \to \mathbf{Möb}$ is a γ-frame along the conformal immersion $f :$ $M \to \mathcal{M}$, with Willmore function $W : U \to \mathbf{R}$, and if $\tilde{Y} = YK(r,e^{it},\mathbf{y})$ is any other*

γ-frame on U, where $K(r, e^{it}, \mathbf{y}) : U \to G_\gamma$, then its Willmore function

$$\tilde{W} = r^3 W : U \to \mathbf{R}.$$

Hence $W^{1/3}\mathbf{Y}_0 : U \to \mathscr{L} \cup \{\mathbf{0}\} \subset \mathbf{R}^{4,1}$ is independent of the choice of γ-frame Y, so defines a smooth map on all of M, which we call the conformal mean curvature vector field of f.

Proof. This is a long, elementary calculation of the type we have done several times. It is simplified considerably by carrying out an arbitrary change of γ-frame in two steps, using

$$K(r, e^{it}, \mathbf{y}) = K(1, 1, \mathbf{y})K(r, e^{it}, \mathbf{0}),$$

for any $r > 0$, $t \in \mathbf{R}$, and $\mathbf{y} \in \mathbf{R}^2$. Then

$$\tilde{Y} = \hat{Y}K(r, e^{it}, \mathbf{0}), \quad \hat{Y} = YK(1, 1, \mathbf{y})$$

are γ-frames with Willmore functions

$$\tilde{W} = r^3 \hat{W}, \quad \hat{W} = W,$$

by much simpler calculations. □

12.4.1 Second order frame fields

We now consider the umbilic free case, which allows additional frame reductions.

Definition 12.31. A *second order* Möbius frame field along f is a first order frame field $Y : U \to \mathbf{M\ddot{o}b}_+$ along f for which $h = 1$ and $H = 0$, that is

$$\begin{aligned} &\omega_0^3 = 0, \quad \omega_0^1 \wedge \omega_0^2 > 0 \quad \text{(first order)}, \\ &\omega_1^3 - i\omega_2^3 = \omega_0^1 + i\omega_0^2 \quad \text{(second order)}. \end{aligned} \tag{12.46}$$

Lemma 12.32. *Given any nonumbilic point m_0 of f in M, there exists a neighborhood U_0 of m_0 on which there is a second order Möbius frame field $\tilde{Y}$ along f.*

Proof. There exists a γ-frame field $Y : U \to \mathbf{M\ddot{o}b}_+$ on some neighborhood U of m_0. Intersecting U with the open set of nonumbilic points, we may assume that U contains no umbilic points. Then the function $h : U \to \mathbf{C}$ in (12.40) is never zero, since the umbilic points in U are precisely the zeros of h, and we have a polar representation $h = re^{it}$ on U, where $r = |h| > 0$. If $U_0 \subset U$ is the connected component containing m_0 of

$$h^{-1}(\mathbf{C} \setminus \{sh(m_0) : s \leq 0\}),$$

then there is a smooth function $t : U_0 \to \mathbf{R}$ such that $h = re^{it}$ on U_0. Using the shorthand introduced in Exercise 49, we let $\tilde{Y} = YK(1/r, e^{-it/2}, \mathbf{0}) : U_0 \to \mathbf{M\ddot{o}b}_+$. By (12.42) we have $\tilde{h} = 1$ and $\tilde{H} = 0$, which means that $\tilde{Y}$ is a second order Möbius frame field along f on U_0. □

Let $Y : U \to \mathbf{M\ddot{o}b}_+$ be second order along f. Differentiating the second equation of (12.46), we get (12.44) for the case $h = 1$, which is

$$\omega_0^0 - 2i\omega_1^2 = h_1\varphi + h_2\bar{\varphi}, \quad \omega_3^0 = h_2\varphi + \bar{h}_2\bar{\varphi}, \tag{12.47}$$

where $h_1, h_2 : U \to \mathbf{C}$ are smooth functions. Any other second order frame field on U is given by

$$\tilde{Y} = YK : U \to \mathbf{M\ddot{o}b}_+, \tag{12.48}$$

where $K : U \to G_2$ is a smooth map into the subgroup G_2 of G_γ,

$$G_2 = \left\{ \begin{pmatrix} 1 & \epsilon{}^t\mathbf{y} & 0 & \frac{1}{2}|\mathbf{y}|^2 \\ 0 & \epsilon I_2 & 0 & \mathbf{y} \\ 0 & 0 & 1 & 0 \\ 0 & 0 & 0 & 1 \end{pmatrix} : \epsilon = \pm 1,\ \mathbf{y} = {}^t(y^1, y^2) \in \mathbf{R}^2 \right\},$$

which is the isotropy subgroup of G_1 of the action (12.43) at $(1, 0) \in \mathbf{C} \times \mathbf{R}$. Its Lie algebra is

$$\mathfrak{g}_2 = \{ \begin{pmatrix} 0 & {}^t\mathbf{x} & 0 \\ 0 & 0 & \mathbf{x} \\ 0 & 0 & 0 \end{pmatrix} : \mathbf{x} = \begin{pmatrix} s \\ t \\ 0 \end{pmatrix} \in \mathbf{R}^2 \subset \mathbf{R}^3 \}.$$

A complementary subspace is

$$\mathfrak{m}_2 = \{ \begin{pmatrix} r & 0 & 0 & s & 0 \\ 0 & 0 & t & 0 & 0 \\ 0 & -t & 0 & 0 & 0 \\ 0 & 0 & 0 & 0 & s \\ 0 & 0 & 0 & 0 & -r \end{pmatrix} : r, s, t \in \mathbf{R} \}.$$

Under the change of second order frame field (12.48), we have

$$\tilde{\varphi} = \epsilon\varphi, \quad \tilde{\omega}_3^0 = \omega_3^0 + \frac{1}{2}(y\varphi + \bar{y}\bar{\varphi}),$$

$$\tilde{\omega}_0^0 = \omega_0^0 - \frac{1}{2}(\bar{y}\varphi + y\bar{\varphi}), \quad \tilde{\omega}_1^2 = \omega_1^2 - \frac{i}{2}(\bar{y}\varphi - y\bar{\varphi}),$$

where $y = y^1 + iy^2$ and $\varphi = \omega_0^1 + i\omega_0^2$. Then

$$\tilde{h}_1 = \epsilon(h_1 - \frac{3}{2}\bar{y}), \quad \tilde{h}_2 = \epsilon(h_2 + \frac{1}{2}y).$$

Following Bryant [20], we choose to make $\tilde{\omega}_3^0 = 0$ for this final reduction. This is accomplished by setting $y = -2h_2$.

Definition 12.33 (Central frames). A *third order* Möbius frame field along f is a second order frame field $Y : M \to \mathbf{M\ddot{o}b}_+$ along f for which

$$\omega_0^3 = 0, \ \omega_0^1 \wedge \omega_0^2 > 0 \qquad \text{(first order)}$$

$$\omega_1^3 - i\omega_2^3 = \omega_0^1 + i\omega_0^2 \qquad \text{(second order)}$$

$$\omega_3^0 = 0 \qquad \text{(third order).}$$

These are called *central frame* fields along f.

Lemma 12.34. *Given a nonumbilic point of f in M, there is a neighborhood U of this point on which there exists a central frame field $Y = (\mathbf{Y}_0, \dots, \mathbf{Y}_4)$. If U is connected, the only central frame fields on U are*

$$Y = (\mathbf{Y}_0, \epsilon\mathbf{Y}_1, \epsilon\mathbf{Y}_2, \mathbf{Y}_3, \mathbf{Y}_4), \tag{12.49}$$

where $\epsilon = \pm 1$.

Proof. Existence was verified above. If $Y : U \to \mathbf{M\ddot{o}b}_+$ is third order, then a change (12.48) is also third order if and only if $\mathbf{y} = \mathbf{0}$. If U is connected, such a frame change is given by (12.49). □

The isotropy subgroup of G_2 at $h_2 = 0$ is

$$G_3 = \{\begin{pmatrix} 1 & 0 & 0 \\ 0 & \epsilon I_2 & 0 \\ 0 & 0 & I_2 \end{pmatrix} \} : \epsilon = \pm 1\},$$

whose Lie algebra $\mathfrak{g}_3 = 0$. The frame reduction ends here.

Let $Y : U \to \mathbf{M\ddot{o}b}_+$ be a central frame field along $f : M \to \mathscr{M}$. Then $\omega_3^0 = 0$ in (12.47) implies $\omega_0^0 - 2i\omega_1^2$ is a multiple of φ, which we write as

$$\omega_0^0 - 2i\omega_1^2 = -2i(q_1 - iq_2)\varphi \tag{12.50}$$

for some smooth functions $q_1, q_2 : U \to \mathbf{R}$. This is equivalent to

$$\omega_1^2 = q_1\omega_0^1 + q_2\omega_0^2, \quad \omega_0^0 = -2(q_2\omega_0^1 - q_1\omega_0^2), \tag{12.51}$$

which is Bryant's notation in [20]. Taking the exterior differential of $\omega_3^0 = 0$, we get (12.45) for the case $h = 1$ and $h_2 = 0$ on U, which we rewrite in the form

$$\omega_1^0 = p_1\omega_0^1 + p_2\omega_0^2, \quad \omega_2^0 = -p_2\omega_0^1 + p_3\omega_0^2, \tag{12.52}$$

for smooth functions $p_1, p_2, p_3 : U \to \mathbf{R}$. In the form of (12.45) this is,

$$\omega_1^0 + i\omega_2^0 = \frac{1}{2}(p_1 + p_3 - 2ip_2)\varphi + \frac{1}{2}(p_1 - p_3)\bar{\varphi} = -2P\varphi - 2W\bar{\varphi}. \tag{12.53}$$

The functions q_1, q_2, p_1, p_2, and p_3 are the *third order Möbius invariants* of f as defined in [20]. Differentiating (12.50) and (12.53), and using

$$d\varphi = (\omega_0^0 - i\omega_1^2) \wedge \varphi = -\frac{i}{2}(q_1 + iq_2)\varphi \wedge \bar{\varphi},$$

we get the structure equations of a central frame field,

$$d(q_2 + iq_1) \wedge \varphi = -\frac{1}{2}(p_1 + p_3 + 1 + q_1^2 + q_2^2 + ip_2)\varphi \wedge \bar{\varphi}, \tag{12.54}$$

and

$$\begin{aligned} &d(p_1 + p_3 - 2ip_2) \wedge \varphi + d(p_1 - p_3) \wedge \bar{\varphi} \\ &\quad = (4p_2q_1 - 3p_3q_2 - p_1q_2 + i(4p_2q_2 + 3p_1q_1 + p_3q_1))\varphi \wedge \bar{\varphi}. \end{aligned} \tag{12.55}$$

Remark 12.35. A central frame field $Y : U \to \mathbf{M\ddot{o}b}_+$ is a γ-frame with γ-invariant $P = -\frac{1}{4}(p_1 + p_3 - 2ip_2)$ and Willmore function

$$W = \frac{1}{4}(p_3 - p_1),$$

by (12.53).

Exercise 51. Prove that under the frame change (12.49), the third order invariants transform by

$$\tilde{q}_1 = \epsilon q_1, \quad \tilde{q}_2 = \epsilon q_2, \quad \tilde{p}_1 = p_1, \quad \tilde{p}_2 = p_2, \quad \tilde{p}_3 = p_3.$$

Hint: Differentiate $\tilde{\mathbf{Y}}_0 = \mathbf{Y}_0$ and $\tilde{\mathbf{Y}}_4 = \mathbf{Y}_4$ and then compare the coefficients of $\mathbf{Y}_0$, $\mathbf{Y}_1$, and $\mathbf{Y}_2$.

Let $\mathscr{U}_f$ be the set of umbilic points of a conformal immersion $f : M \to \mathscr{M}$, which we assume is not totally umbilic. Each point of $M \setminus \mathscr{U}_f$ has a neighborhood on which there is a central frame field. At any point, the two possible central frame fields have the same vector $\mathbf{Y}_4$.

Definition 12.36. The *conformal dual map* of a conformal immersion $f : M \to \mathscr{M}$, assumed not totally umbilic, is the smooth map

$$\hat{f} : M \setminus \mathscr{U}_f \to \mathscr{M}, \quad \hat{f} = [\mathbf{Y}_4],$$

where $\mathbf{Y}_4 : M \setminus \mathscr{U}_f \to \mathscr{L}^+$ is the last vector in any central frame field.

12.5 Space forms in Möbius geometry

Möbius geometry consists of the sphere $\mathbf{S}^3$ with its group of orientation preserving conformal diffeomorphisms. Since any orientation preserving isometry of $\mathbf{S}^3$ is conformal, we see that spherical geometry is naturally a subgeometry of Möbius geometry. Stereographic projection $\mathscr{S} : \mathbf{S}^3 \setminus \{\mathbf{p}\} \to \mathbf{R}^3$ is a conformal diffeomorphism that lifts any Euclidean isometry to a conformal transformation that extends to all of $\mathbf{S}^3$. In this way Euclidean geometry is a subgeometry of Möbius geometry. Hyperbolic geometry realized as the Poincaré Ball is conformally diffeomorphic to an open subset of Euclidean space and $\mathscr{S}$ lifts any hyperbolic isometry to a conformal transformation that extends to all of $\mathbf{S}^3$. In this way Hyperbolic geometry is a subgeometry of Möbius geometry.

We need explicit expressions for these inclusions in order to do calculations. For this we choose stereographic projection $\mathscr{S} = \mathscr{S}_{-\epsilon_0} : \mathbf{S}^3 \setminus \{-\epsilon_0\} \to \mathbf{R}^3$ of Definition 5.22,

$$\mathscr{S}(x^0\epsilon_0 + \sum_1^3 x^i\epsilon_i) = \frac{\sum_1^3 x^i\epsilon_i}{1+x^0}, \quad \mathscr{S}^{-1}(\mathbf{y}) = \frac{1}{1+|\mathbf{y}|^2}((1-|\mathbf{y}|^2)\epsilon_0 + 2\mathbf{y}),$$

where $\mathbf{y} = \sum_1^3 y^i\epsilon_i \in \mathbf{R}^3$. Using the standard orthonormal basis $\epsilon_0, \dots, \epsilon_4$ of $\mathbf{R}^{4,1}$, where $\langle \epsilon_4, \epsilon_4 \rangle = -1$, we identify the subspaces with the spans

$$\mathbf{R}^3 = \text{span}\{\epsilon_1, \epsilon_2, \epsilon_3\} \subset \mathbf{R}^4 = \text{span}\{\epsilon_0, \epsilon_1, \epsilon_2, \epsilon_3\}, \quad \mathbf{R}^{3,1} = \text{span}\{\epsilon_1, \epsilon_2, \epsilon_3, \epsilon_4\}.$$

As in (7.16), let

$$\mathbf{S}_0 = \mathbf{R}^3, \quad \mathbf{S}_+ = \mathbf{S}^3 \subset \mathbf{R}^4, \quad \mathbf{S}_- = \mathbf{H}^3 \subset \mathbf{R}^{3,1},$$

with their groups of isometries denoted by

$$\mathbf{G}_0 = \mathbf{E}(3), \quad \mathbf{G}_+ = \mathbf{O}(4), \quad \mathbf{G}_- = \mathbf{O}_+(3,1),$$

and Lie algebras

$$\mathfrak{g}_0 = \mathscr{E}(3), \quad \mathfrak{g}_+ = \mathfrak{o}(4), \quad \mathfrak{g}_- = \mathfrak{o}(3,1),$$

respectively. For each $\epsilon \in \{0,+,-\}$, we shall define a conformal embedding

$$f_\epsilon : \mathbf{S}_\epsilon \to \mathscr{M},$$

onto an open subset of $\mathscr{M}$, where $f_+ : \mathbf{S}^3 \to \mathscr{M}$ is the diffeomorphism defined in (12.5), and a Lie group monomorphism

$$F_\epsilon : \mathbf{G}_\epsilon \to \mathbf{M\ddot{o}b},$$

which is equivariant with f_ϵ in the sense that

$$F_\epsilon(A) f_\epsilon(\mathbf{x}) = f_\epsilon(A\mathbf{x}),$$

for any $\mathbf{x} \in \mathbf{S}_\epsilon$ and any $A \in \mathbf{G}_\epsilon$. Actually, the maps f_ϵ determine the monomorphisms F_ϵ, since for any $T \in \mathbf{G}_\epsilon$, the locally defined

$$F_\epsilon(T) = f_\epsilon \circ T \circ f_\epsilon^{-1}$$

extends to a conformal diffeomorphism on all of $\mathbf{S}^3$. The images of the Lie algebras are the Lie subalgebras

$$dF_\epsilon \mathfrak{g}_\epsilon = \{X \in \mathfrak{m\ddot{o}b} : X^0_0 = 0,\ X^0_i = -\epsilon X^i_0,\ i = 1,2,3\}.$$

12.5.1 *Spherical geometry*

The diffeomorphism $f_+ : \mathbf{S}^3 \to \mathscr{M}$ of (12.23) is, for $\mathbf{x} = \sum_0^3 x^i \boldsymbol{\epsilon}_i \in \mathbf{S}^3$,

$$f_+(\mathbf{x}) = [\mathbf{x} + \boldsymbol{\epsilon}_4] = [(1+x^0)\boldsymbol{\delta}_0 + \sum_1^3 x^i \boldsymbol{\delta}_i + \frac{(1-x^0)}{2}\boldsymbol{\delta}_4], \tag{12.56}$$

where $\boldsymbol{\delta}_0 = \frac{1}{2}(\boldsymbol{\epsilon}_4 + \boldsymbol{\epsilon}_0), \boldsymbol{\delta}_i = \boldsymbol{\epsilon}_i, \boldsymbol{\delta}_4 = \boldsymbol{\epsilon}_4 - \boldsymbol{\epsilon}_0$ is the Möbius frame (12.15). Writing elements of $\mathbf{R}^4$ as 4×1 column vectors whose entries are the coefficients relative to the standard basis, and elements of $\mathbf{R}^{4,1}$ as 5×1 column vectors whose entries are the coefficients relative to the Möbius frame, we have

$$f_+(\mathbf{x}) = L^{-1} \begin{bmatrix} \mathbf{x} \\ 1 \end{bmatrix},$$

where the matrix L is defined in (12.17).

The group of isometries of $\mathbf{S}^3$ is $\mathbf{O}(4)$ acting on $\mathbf{S}^3$ by its standard linear action on $\mathbf{R}^4$, as described in Chapter 5. It is naturally included in $\mathbf{SO}(4,1)$ by

$$\mathbf{O}(4) \hookrightarrow \mathbf{SO}(4,1), \quad e \mapsto \det(e)\begin{pmatrix} e & 0 \\ 0 & 1 \end{pmatrix}.$$

Following this with the group isomorphism

$$\mathscr{F} : \mathbf{SO}(4,1) \to \mathbf{Möb}, \quad \mathscr{F}(A) = L^{-1}AL$$

as defined in (12.18), we can define

$$F_+ : \mathbf{O}(4) \to \mathbf{Möb},$$

$$F_+(e) = \det(e)L^{-1}\begin{pmatrix} e & 0 \\ 0 & 1 \end{pmatrix}L = \det(e)\begin{pmatrix} \frac{1+e_0^0}{2} & e_j^0 & 1-e_0^0 \\ \frac{e_0^i}{2} & e_j^i & -e_0^i \\ \frac{1-e_0^0}{4} & -\frac{e_j^0}{2} & \frac{1+e_0^0}{2} \end{pmatrix}, \tag{12.57}$$

a Lie group monomorphism equivariant with the embedding f_+ and the actions of $\mathbf{O}(4)$ on $\mathbf{S}^3$ and of $\mathbf{Möb}$ on $\mathscr{M}$. Equivariance means that, if $e \in \mathbf{O}(4)$ and if $\mathbf{x} \in \mathbf{S}^3$, then $f_+(e\mathbf{x}) = F_+(e)f_+(\mathbf{x})$, which follows from

$$F_+(e)f_+(\mathbf{x}) = \det(e)L^{-1}\begin{pmatrix} e & 0 \\ 0 & 1 \end{pmatrix}LL^{-1}\begin{bmatrix} \mathbf{x} \\ 1 \end{bmatrix} = L^{-1}\begin{bmatrix} e\mathbf{x} \\ 1 \end{bmatrix} = f_+(e\mathbf{x}).$$

The derivative at the identity of the Lie group monomorphism F_+ induces the Lie algebra monomorphism

$$dF_+ : \mathfrak{o}(4) \to \mathfrak{möb}, \quad dF_+X = \begin{pmatrix} 0 & X_j^0 & 0 \\ \frac{X_0^i}{2} & X_j^i & X_i^0 \\ 0 & \frac{X_0^j}{2} & 0 \end{pmatrix} \tag{12.58}$$

where, because X is skew-symmetric, $X_j^0 = -X_0^j$ and $X_j^i = -X_i^j$.

A frame field $e : U \subset \mathbf{S}^3 \to \mathbf{O}(4)$ followed by the monomorphism (12.57) defines a map $F_+ \circ e : U \to \mathbf{Möb}$. Because f_+ is an embedding, there exists a frame field $Y : f_+(U) \to \mathbf{Möb}$ such that $Y \circ f_+ = F_+ \circ e$. If $\omega = Y^{-1}dY$, then the Riemannian metric $\sum_1^3(\omega_0^i)^2$ on $f_+(U)$ is in the conformal class of $\mathscr{M}$. If $\eta = e^{-1}de$, then

$$f_+^*\omega = dF_+\eta,$$

so $f_+^*\omega_0^i = dF_+\eta_0^i = \frac{\eta_0^i}{2}$ and

$$f_+^*\sum_1^3(\omega_0^i)^2 = \frac{1}{4}\sum_1^3(\eta_0^i)^2$$

shows that f_+ is conformal, since $\sum_1^3(\eta_0^i)^2$ is the standard Riemannian metric on $U \subset \mathbf{S}^3$ and $\sum_1^3(\omega_0^i)^2$ is a metric in the conformal structure of $\mathscr{M}$, as shown in (12.35) and (12.36).

Exercise 52. Prove that if $e : U \to \mathbf{SO}(4)$ is a first order frame field on an open set $U \subset M$ along an immersion $\mathbf{x} : M^2 \to \mathbf{S}^3$, then $Y = F_+ \circ e : U \to \mathbf{M\ddot{o}b}_+$ is a first order frame field along $f = f_+ \circ \mathbf{x} : M \to \mathscr{M}$. Prove that the umbilic points of $\mathbf{x}$ coincide with the umbilic points of f.

12.5.2 Euclidean geometry

Using the Möbius frame (12.15) $\boldsymbol{\delta}_0 = \frac{\boldsymbol{\epsilon}_4+\boldsymbol{\epsilon}_0}{2}$, $\boldsymbol{\delta}_i = \boldsymbol{\epsilon}_i$, $\boldsymbol{\delta}_4 = \boldsymbol{\epsilon}_4 - \boldsymbol{\epsilon}_0$ of $\mathbf{R}^{4,1}$, we embed Euclidean space into Möbius space $\mathscr{M}$ with the map

$$f_0 = f_+ \circ \mathscr{S}^{-1} : \mathbf{R}^3 \to \mathscr{M}, \quad f_0(\mathbf{x}) = [\boldsymbol{\delta}_0 + \mathbf{x} + \frac{|\mathbf{x}|^2}{2}\boldsymbol{\delta}_4], \tag{12.59}$$

where $\mathscr{S} = \mathscr{S}_{-\boldsymbol{\epsilon}_0}$ is stereographic projection. The image of this map is $\mathscr{M} \setminus \{[\boldsymbol{\delta}_4]\}$. Being the composition of conformal maps, this map is conformal, which means that it pulls back any metric of the conformal structure of $\mathscr{M}$ to a positive multiple of the Euclidean metric on $\mathbf{R}^3$.

We map the Euclidean group $\mathbf{E}(3)$ into the Möbius group $\mathbf{M\ddot{o}b}$, defined in (12.16) relative to the fixed Möbius frame (12.15), by applying the isomorphism $\mathscr{F}$ in (12.18) to the conformal transformation $\mathscr{S}^{-1} \circ (\mathbf{x}, e) \circ \mathscr{S}$ of $\mathbf{S}^3$. Using (12.10) and (12.11), we calculate this composition to be

$$F_0 : \mathbf{E}(3) \to \mathbf{M\ddot{o}b}, \quad F_0(\mathbf{x}, e) = \det(e) \begin{pmatrix} 1 & 0 & 0 \\ \mathbf{x} & e & 0 \\ \frac{|\mathbf{x}|^2}{2} & {}^t\mathbf{x}e & 1 \end{pmatrix}, \tag{12.60}$$

where $e = (\mathbf{e}_1, \mathbf{e}_2, \mathbf{e}_3)$ and ${}^t\mathbf{x}e$ denotes the row vector $(\mathbf{x} \cdot \mathbf{e}_1, \mathbf{x} \cdot \mathbf{e}_2, \mathbf{x} \cdot \mathbf{e}_3)$.

The Lie group monomorphism F_0 induces a Lie algebra monomorphism of the Lie algebras

$$dF_0 : \mathscr{E}(3) \to \mathfrak{m\ddot{o}b}, \quad dF_0(\mathbf{x}, X) = \begin{pmatrix} 0 & 0 & 0 \\ \mathbf{x} & X & 0 \\ 0 & {}^t\mathbf{x} & 0 \end{pmatrix} \tag{12.61}$$

Exercise 53. Prove that if $(\mathbf{x}, e) : U \to \mathbf{E}_+(3)$ is a first order frame field on an open set $U \subset M$ along the immersion $\mathbf{x} : M^2 \to \mathbf{R}^3$, then $Y = F_0 \circ (\mathbf{x}, e) : U \to \mathbf{M\ddot{o}b}_+$ is a first order frame field along $f = f_0 \circ \mathbf{x} : M \to \mathscr{M}$. Prove that the umbilic points of $\mathbf{x}$ coincide with the umbilic points of f.

12.5.3 Hyperbolic geometry

Our inclusion $\mathbf{R}^{3,1} \subset \mathbf{R}^{4,1}$ gives a natural embedding

$$f_- : \mathbf{H}^3 \to \mathscr{M}, \quad f_-(\mathbf{x}) = [\boldsymbol{\epsilon}_0 + \mathbf{x}],$$

where

$$\mathbf{H}^3 = \{\mathbf{x} = \sum_1^4 x^i \boldsymbol{\epsilon}_i \in \mathbf{R}^{3,1} : \langle \mathbf{x}, \mathbf{x} \rangle = \sum_1^3 (x^i)^2 - (x^4)^2 = -1,\ x^4 \geq 1\}.$$

In the Möbius frame (12.15) this embedding is

$$f_-(\sum_1^4 x^i \boldsymbol{\epsilon}_i) = [\boldsymbol{\epsilon}_0 + \sum_1^4 x^i \boldsymbol{\epsilon}_i] = [(x^4+1)\boldsymbol{\delta}_0 + \sum_1^3 x^i \boldsymbol{\delta}_i + \frac{x^4-1}{2}\boldsymbol{\delta}_4]. \tag{12.62}$$

That is, for any $\mathbf{x} \in \mathbf{H}^3$, using the matrix L defined in (12.17), we have

$$f_-(\mathbf{x}) = L^{-1} \begin{bmatrix} 1 \\ \mathbf{x} \end{bmatrix}.$$

Recall Definition 6.15 of the hyperbolic stereographic projection of hyperbolic space onto the open ball $\mathbf{B}^3 \subset \mathbf{R}^3$,

$$\mathfrak{s} : \mathbf{H}^3 \to \mathbf{B}^3 \subset \mathbf{R}^3, \quad \mathfrak{s}(\mathbf{x}) = \sum_1^3 \frac{x^i}{1+x^4} \boldsymbol{\epsilon}_i, \quad \mathbf{x} = \sum_1^4 x^i \boldsymbol{\epsilon}_i.$$

We have

$$f_- = f_0 \circ \mathfrak{s} = f_+ \circ \mathscr{S}^{-1} \circ \mathfrak{s}, \tag{12.63}$$

(see Problem 12.68). Thus, f_- is a composition of conformal maps, so is itself conformal. Its image is an open hemisphere in $\mathscr{M}$ described as follows.

Exercise 54. Consider the open subset

$$\begin{aligned} \mathscr{M}_- &= \{[\sum_0^4 x^a \boldsymbol{\epsilon}_a] \in \mathscr{M} : x^0 \neq 0,\ \frac{x^4}{x^0} \geq 1\} \\ &= \{[\sum_0^4 u^a \boldsymbol{\delta}_a] \in \mathscr{M} : u^0 - 2u^4 \neq 0,\ \frac{u^0 + 2u^4}{u^0 - 2u^4} \geq 1\}. \end{aligned}$$

Prove that $f_-(\mathbf{H}^3) = \mathscr{M}_-$ and the inverse mapping $f_-^{-1} : \mathscr{M}_- \to \mathbf{H}^3$ is

$$f_-^{-1}[\sum_0^4 u^a \delta_a] = \frac{1}{u^0 - 2u^4}(2\sum_1^3 u^i \epsilon_i + (u^0 + 2u^4)\epsilon_4)$$

in the Möbius frame, and

$$f_-^{-1}[\sum_0^4 x^a \epsilon_a] = \frac{1}{x^0}\sum_1^4 x^i \epsilon_i$$

in the standard frame. The sphere at infinity (6.18) S^2_∞ of $\mathbf{H}^3$ is

$$\{[\sum_1^4 n^i \epsilon_i] \in P(\mathbf{R}^{3,1}) : \sum_1^3 (n^i)^2 - (n^4)^2 = 0\} = \{[\sum_0^4 u^a \epsilon_a] \in \mathscr{M} : u^0 = 0\}.$$

Using the conformal diffeomorphism $\mathfrak{s}_\infty : S^2_\infty \to \partial\mathbf{B}^3$ defined in (6.41), prove that the map

$$f_-^\infty = f_0 \circ \mathfrak{s}_\infty : S^2_\infty \to \mathscr{M}$$

is the inclusion $f_-^\infty[\sum_1^4 n^i \epsilon_i] = [\sum_1^4 n^i \epsilon_i] \in \mathscr{M}$, which is a conformal immersion with image

$$f_-^\infty(S^2_\infty) = \partial\mathscr{M}_-.$$

For any unit speed geodesic $\gamma(t)$ in $\mathbf{H}^3$, prove that

$$\lim_{t\to\infty} f_- \circ \gamma(t) = f_-^\infty[\gamma(0) + \dot\gamma(0)]. \tag{12.64}$$

Remark 12.37. The preceding exercise displays nicely a difference between the conformal geometry of Euclidean space and the conformal geometry of Hyperbolic space. If $\gamma(t)$ is any smooth divergent curve in $\mathbf{R}^3$, such as any geodesic, then $f_0 \circ \gamma(t) \to [\delta_4] \in \mathscr{M}$. In this sense there is only one point at infinity of $\mathbf{R}^3$. Contrast this with the situation in Hyperbolic space where, for any divergent curve $\gamma(t)$, the image curve $f_- \circ \gamma(t)$ converges to a point in $\partial\mathscr{M}_-$, as $t \to \infty$. If γ and σ are unit speed geodesics in $\mathbf{H}^3$ such that $[\gamma(0) + \dot\gamma(0)] \neq [\sigma(0) + \dot\sigma(0)] \in S^2_\infty$, then $f_- \circ \gamma(t)$ converges to a point in $\partial\mathscr{M}_-$ distinct from the point to which $f_- \circ \sigma(t)$ converges as $t \to \infty$. In this sense there is a whole $\mathbf{S}^2$ of points at infinity of $\mathbf{H}^3$, a fact illustrated quite clearly by the Poincaré ball model.

The group of isometries of $\mathbf{H}^3$ is the Lie group $\mathbf{O}_+(3,1)$ defined in (6.4), acting on $\mathbf{H}^3$ by its standard linear action on $\mathbf{R}^{3,1}$, as described in Section 6.1. There is a natural inclusion

$$\mathbf{O}_+(3,1) \subset \mathbf{SO}(4,1), \quad e \mapsto \det(e)\begin{pmatrix} 1 & 0 \\ 0 & e \end{pmatrix}.$$

Following this with the group isomorphism $\mathscr{F} : \mathbf{SO}(4,1) \to \mathbf{Möb}$ of (12.18), we can define the Lie group monomorphism

$$F_- : \mathbf{O}_+(3,1) \to \mathbf{Möb},$$

$$F_-(e) = L^{-1}\det(e)\begin{pmatrix} 1 & 0 \\ 0 & e \end{pmatrix} L = \det(e)\begin{pmatrix} \frac{e_4^4+1}{2} & e_j^4 & e_4^4 - 1 \\ \frac{e_4^i}{2} & e_j^i & e_4^i \\ \frac{e_4^4-1}{4} & \frac{e_j^4}{2} & \frac{e_4^4+1}{2} \end{pmatrix}. \tag{12.65}$$

The Lie group monomorphism (12.65) induces the Lie algebra monomorphism

$$dF_- : \mathfrak{o}(3,1) \to \mathfrak{möb}, \quad dF_-X = \begin{pmatrix} 0 & X_4^j & 0 \\ \frac{X_4^i}{2} & X_j^i & X_4^i \\ 0 & \frac{X_j^4}{2} & 0 \end{pmatrix}, \tag{12.66}$$

where $X_j^4 = X_4^j$ and $X_j^i = -X_i^j$, for $i,j = 1,2,3$.

Remark 12.38. The Poincaré ball model of hyperbolic space shows geometrically how any isometry of hyperbolic space gives rise to a conformal transformation of the sphere. Indeed, the isometries of the Poincaré ball model $\mathbf{B}^3 \subset \mathbf{R}^3$ are generated by inversions in spheres that intersect the unit sphere $\partial\mathbf{B}^3$ orthogonally. This includes also reflection in any plane passing through the origin. See, for example, Greenberg [78] for the two-dimensional case.

Exercise 55. Prove that if $e : U \to \mathbf{SO}_+(3,1)$ is a first order frame field on an open set $U \subset M$ along an immersion $\mathbf{x} : M^2 \to \mathbf{H}^3$, then $Y = F_- \circ e : U \to \mathbf{Möb}_+$ is a first order frame field along $f = f_- \circ \mathbf{x} : M \to \mathscr{M}$. Prove that the umbilic points of $\mathbf{x}$ coincide with the umbilic points of f.

12.6 Spheres in Möbius space

Recall Definition 5.4 of the oriented sphere $S_r(\mathbf{m}) = \{\mathbf{y} \in \mathbf{S}^3 : \mathbf{y} \cdot \mathbf{m} = \cos r\}$ in $\mathbf{S}^3$ with center $\mathbf{m} \in \mathbf{S}^3$, signed radius $r \in \mathbf{R}$, and unit normal $\mathbf{n}(\mathbf{y}) = \frac{\mathbf{m} - \cos r\, \mathbf{y}}{\sin r}$. Then Exercise 19 parametrizes the set $\tilde{\Sigma}$ of all nonpoint oriented spheres in $\mathbf{S}^3$ by the identification

$$\tilde{\Sigma} = \{S_r(\mathbf{m}) : \mathbf{m} \in \mathbf{S}^3,\ 0 < r < \pi\} \cong \mathbf{S}^3 \times (0,\pi).$$

Definition 12.39. The *space of oriented spheres* in $\mathscr{M}$ is

$$\mathbf{S}^{3,1} = \{S \in \mathbf{R}^{4,1} : \langle S, S \rangle = 1\}. \tag{12.67}$$

Fig. 12.3 The light cone $\mathscr{L}$ and set of oriented spheres $\mathbf{S}^{3,1}$ in the $\epsilon_0\epsilon_4$-subspace of $\mathbf{R}^{4,1}$.

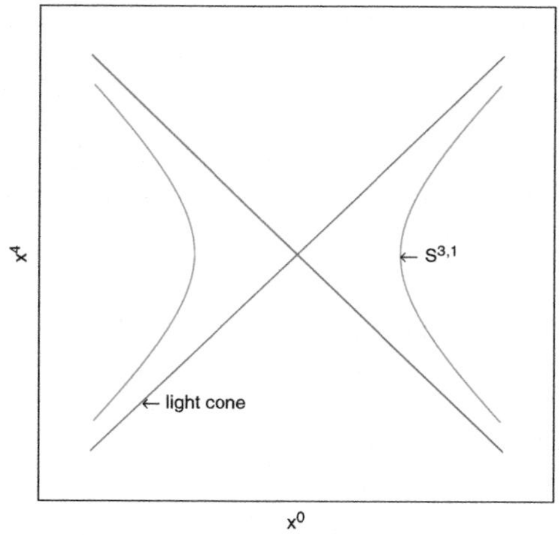

Figure 12.3 illustrates the light cone $\mathscr{L}$ with $\mathbf{S}^{3,1}$ in span$\{\epsilon_0, \epsilon_4\} \subset \mathbf{R}^{4,1}$. A point $S = \mathbf{x} + t\epsilon_4 \in \mathbf{S}^{3,1} \subset \mathbf{R}^4 \oplus \mathbf{R}\epsilon_4$ determines the subset

$$S^{\perp} = \{[\mathbf{u}] \in \mathscr{M} : \langle u, S \rangle = 0\} = \{[\mathbf{y} + \epsilon_4] \in \mathscr{M} : \mathbf{y} \in \mathbf{S}^3, \ \mathbf{y} \cdot \mathbf{x} = t\},$$

so $S^{\perp} = f_+ S_r(\mathbf{m})$, where $r = \cot^{-1} t \in (0, \pi)$ and $\mathbf{m} = \sin r \, \mathbf{x} \in \mathbf{S}^3$. Our name for $\mathbf{S}^{3,1}$ comes from identifying it with $\tilde{\Sigma}$ by the following diffeomorphism $\Psi : \tilde{\Sigma} \to \mathbf{S}^{3,1}$.

Exercise 56. Prove that the map

$$\Psi : \tilde{\Sigma} = \mathbf{S}^3 \times (0, \pi) \to \mathbf{S}^{3,1}, \quad \Psi(S_r(\mathbf{m})) = \frac{1}{\sin r}\mathbf{m} + \cot r \, \epsilon_4$$

is smooth with smooth inverse

$$\Upsilon : \mathbf{S}^{3,1} \to \tilde{\Sigma}, \quad \Upsilon(\mathbf{x} + t\epsilon_4) = S_r(\sin r \, \mathbf{x}),$$

where $r = \cot^{-1} t \in (0, \pi)$. Prove, in addition, that

$$f_+(S_r(\mathbf{m})) = (\frac{1}{\sin r}\mathbf{m} + \cot r \, \epsilon_4)^{\perp}.$$

The action of $\mathbf{SO}(\mathbf{R}^{4,1})$ on $\mathbf{R}^{4,1}$ sends $\mathbf{S}^{3,1}$ to itself, so induces a group of isometries on $\mathbf{S}^{3,1}$ with its induced Lorentzian metric. This action sends oriented spheres of $\mathscr{M}$ to oriented spheres of $\mathscr{M}$, since if $S \in \mathbf{S}^{3,1}$ and $T \in \mathbf{SO}(\mathbf{R}^{4,1})$, then

$$T(S^{\perp}) = \{T[\mathbf{u}] \in \mathscr{M} : 0 = \langle \mathbf{u}, S \rangle = \langle T\mathbf{u}, TS \rangle\} = (TS)^{\perp}.$$

Proposition 12.40. *If $S \in \mathbf{S}^{3,1}$, then the oriented sphere $S^\perp \subset \mathcal{M}$ is a totally umbilic immersed surface. Conversely, if $f : M \to \mathcal{M}$ is a totally umbilic immersion of a connected surface M, then $f(M)$ is an open subset of $S^\perp$, for some $S \in \mathbf{S}^{3,1}$.*

Proof. By the transitivity of the action of $\mathbf{M\ddot{o}b}_+$ on $\mathbf{S}^{3,1}$, it is sufficient to consider the case of $\boldsymbol{\delta}_3 \in \mathbf{S}^{3,1}$. Then

$$\boldsymbol{\delta}_3^\perp = \{[\mathbf{x} + \boldsymbol{\epsilon}_4] : \mathbf{x} \in \mathbf{S}^2 = \mathbf{S}^3 \cap \boldsymbol{\epsilon}_3^\perp \subset \mathbf{R}^4\}.$$

For a point in $\mathbf{S}^2$, there is a first order frame field $e = (\mathbf{x}, \mathbf{e}_1, \mathbf{e}_2, \boldsymbol{\epsilon}_3) : U \to \mathbf{SO}(4)$ on some neighborhood U of the point, so $d\mathbf{x} = \theta^1 \mathbf{e}_1 + \theta^2 \mathbf{e}_2$ and $d\boldsymbol{\epsilon}_3 = 0$. Then

$$\mathbf{Y}_0 = \mathbf{x} + \boldsymbol{\epsilon}_4,\ \mathbf{Y}_1 = \mathbf{e}_1,\ \mathbf{Y}_2 = \mathbf{e}_2,\ \mathbf{Y}_3 = \boldsymbol{\delta}_3,\ \mathbf{Y}_4 = \frac{1}{2}(-\mathbf{x} + \boldsymbol{\epsilon}_4)$$

is a time oriented first order Möbius frame field on U for which $\omega_1^3 = 0 = \omega_2^3$ at every point of U. Hence, this is a γ-frame and $\boldsymbol{\delta}_3^\perp$ is totally umbilic. Conversely, if $f : M \to \mathcal{M}$ is totally umbilic, then for any γ-frame $Y : U \subset M \to \mathbf{M\ddot{o}b}_+$, we have $\omega_1^3 = 0 = \omega_2^3$, and thus $\omega_3^0 = 0$ as well, by the structure equations. With respect to this frame field,

$$\omega_0^3 = 0,\ \omega_1^3 = 0,\ \omega_2^3 = 0,\ \omega_3^0 = 0$$

on U. Regard these as left invariant forms defining a distribution on $\mathbf{M\ddot{o}b}_+$. It satisfies the Frobenius condition, and therefore defines a Lie subalgebra, $\mathfrak{h} \subset \mathfrak{m\ddot{o}b}$. The maximal integral submanifold through the identity is the Lie subgroup $H = \{Y \in \mathbf{M\ddot{o}b}_+ : Y\boldsymbol{\delta}_3 = \boldsymbol{\delta}_3\}$. Every other integral submanifold is a left coset of H. The set of all γ frames along f must lie in one of these integral submanifolds. Thus, for any point $m_0 \in U$, the coset $Y(m_0)H$ must contain all γ frames on f, so $f(M) \subset Y(m_0)H[\boldsymbol{\delta}_0]$. But $H[\boldsymbol{\delta}_0] = \boldsymbol{\delta}_3^\perp$. □

A point $\mathbf{x} \in \mathbf{S}^3$ and a unit tangent vector $\mathbf{n} \in T_\mathbf{x}\mathbf{S}^3 = \mathbf{x}^\perp \subset \mathbf{R}^4$ determine the pencil of oriented tangent spheres

$$\{S_r(\cos r\, \mathbf{x} + \sin r\, \mathbf{n}) : 0 < r < \pi\}.$$

Then Ψ of this pencil is the curve in $\mathbf{S}^{3,1}$,

$$\{\mathbf{n} + \cot r(\mathbf{x} + \boldsymbol{\epsilon}_4) : 0 < r < \pi\}.$$

These observations lead to the following definition.

Definition 12.41. The *pencil of oriented spheres* through a point $[\mathbf{u}] \in \mathcal{M}$ determined by an oriented sphere $S \in \mathbf{S}^{3,1}$ passing through $[\mathbf{u}]$ is the curve

$$\{S + t\mathbf{u} : t \in \mathbf{R}\} \subset \mathbf{S}^{3,1}. \tag{12.68}$$

An oriented sphere $S \in \mathbf{S}^{3,1}$ passes through a point $[\mathbf{u}] \in \mathscr{M}$ if and only if $[\mathbf{u}] \in S^{\perp}$ if and only if $\langle S, \mathbf{u} \rangle = 0$. Thus, if S passes through $[\mathbf{u}] \in \mathscr{M}$, then every oriented sphere in the pencil (12.68) passes through $[\mathbf{u}]$.

For an immersion $\mathbf{x} : M^2 \to \mathbf{S}^3$ with unit normal vector field $\mathbf{n}$, the oriented tangent sphere at a point $p \in M$, oriented by $\mathbf{n}(p)$, of radius $r \in (0, \pi)$, is $S_r(\cos r\, \mathbf{x}(p) + \sin r\, \mathbf{n}(p))$. Its characterizing features are that it passes through $\mathbf{x}(p)$, is oriented by the unit normal $\mathbf{n}(p)$ at $\mathbf{x}(p)$, and is tangent to $\mathbf{x}(M)$ at $\mathbf{x}(p)$, that is $d\mathbf{x}_p \cdot \mathbf{n}(p) = 0$. Then

$$\Psi S_r(\cos r\, \mathbf{x}(p) + \sin r\, \mathbf{n}(p)) = \mathbf{n}(p) + \cot r\, (\mathbf{x}(p) + \boldsymbol{\epsilon}_4)$$

is an oriented sphere in $\mathbf{S}^{3,1}$. It passes through $f_+ \circ \mathbf{x}(p) = [\mathbf{x}(p) + \boldsymbol{\epsilon}_4]$, since

$$\langle \mathbf{n}(p) + \cot r\, (\mathbf{x}(p) + \boldsymbol{\epsilon}_4), \mathbf{x}(p) + \boldsymbol{\epsilon}_4 \rangle = 0,$$

it lies in the pencil of oriented spheres through $[\mathbf{x}(p) + \boldsymbol{\epsilon}_4]$ determined by the oriented sphere $\mathbf{n}(p) \in \mathbf{S}^{3,1}$, and it has the property that

$$\langle d(\mathbf{x} + \boldsymbol{\epsilon}_4)_p, \mathbf{n}(p) + \cot r\, (\mathbf{x}(p) + \boldsymbol{\epsilon}_4) \rangle = 0.$$

These observations lead to the following definitions.

Definition 12.42. A *tangent sphere* of a conformal immersion $f : M \to \mathscr{M}$ at $p \in M$ is an oriented sphere $S \in \mathbf{S}^{3,1}$ such that, if $F : U \subset M \to \mathscr{L}$ is any lift of f about p, then

$$\langle F(p), S \rangle = 0, \text{ and } \langle dF_p, S \rangle = 0.$$

The *pencil of oriented tangent spheres* of f at $p \in M$ determined by the oriented tangent sphere S at p is

$$\{S + tF(p) : t \in \mathbf{R}\}.$$

A *tangent sphere map* along f on $U \subset M$ is a smooth map $S : U \to \mathbf{S}^{3,1}$ such that $S(p)$ is an oriented tangent sphere of f at each point $p \in U$. A *smooth pencil of oriented tangent spheres* along f is the set of pencils determined by a smooth tangent sphere map along f.

Exercise 57. Prove that, if $Y : U \to \mathbf{M\ddot{o}b}$ is first order along f, and if $t : U \to \mathbf{R}$, then $\mathbf{Y}_3 + t\mathbf{Y}_0 : U \to \mathbf{S}^{3,1}$ is an oriented tangent sphere map along f. Prove that any two first order frame fields $Y, \tilde{Y} : U \to \mathbf{S}^{3,1}$ along f induce the same pencil of oriented tangent spheres along f on U.

Remark 12.43. The pencil of oriented tangent spheres of f at p determined by the oriented tangent sphere $-S \in \mathbf{S}^{3,1}$ is disjoint from the pencil determined by S. The following Lemma shows that these are the only two pencils of oriented tangent spheres of f at p. Thus, if there exists a smooth pencil of oriented tangent spheres along f, determined by the smooth tangent sphere map $S : M \to \mathbf{S}^{3,1}$, and if M is connected, then the only other smooth pencil is the one determined by $-S : M \to \mathbf{S}^{3,1}$.

Lemma 12.44. *If $F : U \to \mathscr{L}^+$ is a lift of the immersion $f : M \to \mathscr{M}$ on a neighborhood U of $p \in M$, and if $S, T \in \mathbf{S}^{3,1}$ are tangent spheres of f at p, then $T = \pm S + tF(p)$, for some $t \in \mathbf{R}$.*

Proof. $F = \sum_0^4 F^a \epsilon_a : U \to \mathscr{L}^+$ has $F^4 > 0$ on U. Dividing by F^4, we may assume $F^4 = 1$ on U. Then $F = \mathbf{x} + \epsilon_4$, where $\mathbf{x} : U \to \mathbf{S}^3 \subset \mathbf{R}^4$, so $d\mathbf{x} \in \mathbf{x}^\perp \subset \mathbf{R}^4$. There exists an orthonormal basis $\mathbf{x}(p), \mathbf{e}_1, \mathbf{e}_2, \mathbf{e}_3$ of $\mathbf{R}^4$ such that $d\mathbf{x}_p = \theta^1 \mathbf{e}_1 + \theta^2 \mathbf{e}_2$. Let $Y = \frac{1}{2}(-\mathbf{x}(p) + \epsilon_4)$, so $\langle Y, Y \rangle = 0$ and $\langle Y, F \rangle = -1$. Then $F(p), \mathbf{e}_1, \mathbf{e}_2, \mathbf{e}_3, Y$ is a basis of $\mathbf{R}^{4,1}$ for which the null vectors $F(p)$ and Y are orthogonal to $\mathbf{e}_1, \mathbf{e}_2, \mathbf{e}_3$.

If $S = s^0 F(p) + s^1 \mathbf{e}_1 + s^2 \mathbf{e}_2 + s^3 \mathbf{e}_3 + s^4 Y \in \mathbf{S}^{3,1}$ is a tangent sphere of f at p, then $s^1 = s^2 = s^4 = 0$, and $s^3 = \pm 1$. In the same way, any other tangent sphere $T \in \mathbf{S}^{3,1}$ has the expansion $T = t^0 F + t^3 \mathbf{e}_3$, where $t^3 = \pm 1$. Substituting $\mathbf{e}_3 = \frac{1}{s^3}(S - s^0 F(p))$ into the expansion of T, we get $T = \pm S + tF(p)$, for some $t \in \mathbf{R}$. □

Remark 12.45. If follows from the Lemma that there are exactly two pencils of oriented tangent spheres at any point of an immersion $f : M \to \mathscr{M}$: the one determined by $\mathbf{Y}_3(m)$ of any first order frame field at the point, at its opposite, $-\mathbf{Y}_3(m)$.

Definition 12.46. A smooth tangent sphere map $S : M \to \mathbf{S}^{3,1}$ along a smooth immersion $f : M \to \mathscr{M}$ *induces the orientation* of the Riemann surface M, if the pencil of oriented tangent spheres it determines at each point is the same as that determined by the first order frame fields.

Definition 12.47. An *oriented curvature sphere* of an immersion $f : M \to \mathscr{M}$ at a point $p \in M$ is an oriented tangent sphere $\hat{S}$ to f at p such that, if $S : U \to \mathbf{S}^{3,1}$ is an oriented tangent sphere map on a neighborhood U of p and $S(p) = \hat{S}$, then $dS_p \mod f(p)$ has rank less than two. Here the expression $\mod f(p)$ means modulo the 1-dimensional subspace $f(p)$ of $\mathbf{R}^{4,1}$.

If $dS_p \mod f(p)$ has rank less than two for some tangent sphere map S, then any other such map has the same property. In fact, it is given by $T = S + tF$, where F is a local lift of f to $\mathscr{L}^+$ and $t : U \to \mathbf{R}$ satisfies $t(p) = 0$, so $dT_p = dS_p + dt_p F(p) \equiv dS_p \mod F(p)$.

This definition is consistent with our notion of curvature sphere of an immersion $\mathbf{x} : M \to \mathbf{S}^3$.

Proposition 12.48. *If $\mathbf{x} : M \to \mathbf{S}^3$ is a smooth immersion oriented by the unit normal vector field $\mathbf{n}$, then any oriented curvature sphere of the conformal immersion $f = f_+ \circ \mathbf{x} : M \to \mathscr{M}$ at a point $p \in M$ is the image under $f_+ : \mathbf{S}^3 \to \mathscr{M}$ of an oriented curvature sphere of $\mathbf{x}$ at p.*

Proof. For fixed $r \in (0,\pi)$, an oriented tangent sphere map along $\mathbf{x}$ is $S_r(\cos r\ \mathbf{x} + \sin r\ \mathbf{n}) : M \to \tilde{\Sigma}$. Then

$$S = \Psi S_r(\cos r\ \mathbf{x} + \sin r\ \mathbf{n}) = \mathbf{n} + \cot r\ (\mathbf{x} + \epsilon_4) : M \to \mathbf{S}^{3,1} \subset \mathbf{R}^{4,1}$$

is an oriented tangent sphere map along $f = f_+ \circ \mathbf{x}$ and

$$dS = d\mathbf{n} + \cot r\ d\mathbf{x} \mod \mathbf{x} + \epsilon_4$$

has rank less than two at $p \in M$ precisely when $d\mathbf{n} + \cot r\ d\mathbf{x}$ has rank less than two at p, which is the condition that $S_r(\cos r\ \mathbf{x} + \sin r\ \mathbf{n})$ at p is an oriented curvature sphere of $\mathbf{x}$ at p. □

12.7 Canal and Dupin immersions

Definition 12.49. An immersion $f : M^2 \to \mathscr{M}$ is *canal* if one of its curvature sphere maps is smooth and has rank less than two at every point of M. The immersion is *Dupin* if both curvature sphere maps are smooth and each has rank less than two at every point of M.

Lemma 12.50 (Canal and Dupin criteria). *An umbilic free immersion $f : M^2 \to \mathscr{M}$ is canal if and only if its third order invariants satisfy $q_1 q_2 = 0$ on M. It is Dupin if and only if $q_1 = 0 = q_2$ on M.*

Proof. Let $Y : U \to \mathbf{M\ddot{o}b}_+$ be a central frame field along f. Then the curvature sphere maps are $\mathbf{Y}_3 + \epsilon \mathbf{Y}_0$, where $\epsilon = \pm 1$, and

$$d(\mathbf{Y}_3 + \epsilon \mathbf{Y}_0) = \epsilon \omega_0^0 \mathbf{Y}_0 + (-\omega_0^1 + \epsilon \omega_0^1)\mathbf{Y}_1 + (\omega_0^2 + \epsilon \omega_0^2)\mathbf{Y}_2.$$

For $\epsilon = 1$, this has rank less than two if and only if $\omega_0^0 \wedge \omega_0^2 = 0$ if and only if $q_2 = 0$ on M, by (12.51). For $\epsilon = -1$, it has rank less than two if and only if $q_1 = 0$. □

Here are some results about Dupin immersions. For more results about canal immersions see Musso and Nicolodi [126].

Theorem 12.51 (Dupin immersions). *If $f : M^2 \to \mathscr{M}$ is an umbilic free Dupin immersion of a connected Riemann surface M, then, up to Möbius transformation, $f(M)$ is an open submanifold of the standard embedding into $\mathscr{M}$ of an isoparametric immersion in $\mathbf{R}^3$, or $\mathbf{S}^3$, or $\mathbf{H}^3$.*

Proof. This is a special case of Proposition 3.11. Let $Y : U \to \mathbf{M\ddot{o}b}_+$ be a central frame field along f on a connected open subset U of M. Then $q_1 = 0 = q_2$ on U, by Lemma 12.50, so $\omega_0^0 = 0 = \omega_1^2$ on U. Then $p_1 + p_3 + 1 = 0 = p_2$ on U, by (12.54), so $p_1 - p_3$ is constant on U, by (12.55). Set $p_1 - p_3 = 2C$, a real constant. Then $p_1 = -\frac{1}{2} + C$ and $p_3 = -\frac{1}{2} - C$ are constant on U, so $Y : U \to \mathbf{M\ddot{o}b}_+$ is an integral submanifold of the 2-dimensional distribution on $\mathbf{M\ddot{o}b}_+$ defined by the equations

$$\omega_0^0 = \omega_1^2 = \omega_0^3 = \omega_3^0 = 0, \quad \omega_1^3 = \omega_0^1, \quad \omega_2^3 = -\omega_0^2,$$

$$\omega_1^0 = (-\frac{1}{2} + C)\omega_0^1, \quad \omega_2^0 = (-\frac{1}{2} - C)\omega_0^2.$$

For any constant $C \in \mathbf{R}$, these equations define a 2-dimensional, abelian, Lie subalgebra $\mathfrak{h}^C$ of $\mathfrak{möb}$. If H^C is its connected Lie subgroup of **Möb**, then the maximal integral submanifolds of this distribution are the right cosets of H^C. Hence, $Y(U) \subset Y(m_0)H^C$, for any given point $m_0 \in U$, so $f(U) \subset Y(m_0)H^C[\boldsymbol{\delta}_0]$. Any point $m \in M$ can be reached by a chain of open subsets U_i, for $i = 1, \ldots, k$, such that $U_1 = U$, $U_i \cap U_{i+1} \neq \emptyset$, $m \in U_k$, and there exists a central frame field $Y^{(i)} : U_i \to$ **Möb**. By Lemma 12.34, the invariants of two frame fields with overlapping domains must be the same so their images must be in the same coset of H^C. In particular, $f(U_k) \subset Y(m_0)H^C[\boldsymbol{\delta}_0]$. We conclude that $f(M) \subset Y(m_0)H^C[\boldsymbol{\delta}_0]$, which up to the Möbius transformation $Y(m_0)$ is $H^C[\boldsymbol{\delta}_0]$. This one-parameter family of immersions comes from the isoparametric immersions in space forms. See Problems 12.74, 12.75, and 12.76 for details. □

Remark 12.52. If $Y : M \to$ **Möb** is a central frame field along a Dupin conformal immersion $f : M \to \mathcal{M}$, with invariant C, then the Dupin conformal immersion with invariant $-C$ is the same map $f : \bar{M} \to \mathcal{M}$ with the Riemann surface M replaced by its complex conjugate $\bar{M}$. A central frame field along it is $\tilde{Y} = (\mathbf{Y}_0, \mathbf{Y}_2, \mathbf{Y}_1, -\mathbf{Y}_3, \mathbf{Y}_4)$. The central frame field along a Dupin immersion has been used by Musso and Nicolodi [127] as a tool for studying the Darboux transform of Dupin immersions. This transform has been used also by H. Bernstein [3] to find many new isothermic immersions of tori. Figure 12.4 shows one of her non-special, non-canal, isothermic tori, opened to exhibit the self-intersections. See Section 14.4 for the meaning of special isothermic immersion.

One consequence of the preceding proof is that the Möbius invariants of a Dupin immersion are all constant. Are there any other immersions of surfaces into $\mathcal{M}$ whose Möbius invariants are all constant? It turns out that there are and we have already encountered them as the Bonnet cylinders of Proposition 10.14. These immersions are known as *spiral cylinders*. See Sulanke [158] and additional references in that paper.

Corollary 12.53 (Spiral cylinders). *Let $f : M \to \mathcal{M}$ be an umbilic free conformal immersion of a connected Riemann surface. If the conformal invariants of f are all constant, then, up to Möbius transformation, either f is a Dupin immersion or $f(M)$ is contained in the image of*

$$f_0 \circ \mathbf{x}_n : \mathbf{R}^+ \times \mathbf{R} \to \mathcal{M}, \tag{12.69}$$

for some real constant $n > 0$, where $\mathbf{x}_n : \mathbf{R}^+ \times \mathbf{R} \to \mathbf{R}^3$ is a Bonnet cylinder of Proposition 10.14, shown in Figure 10.2.

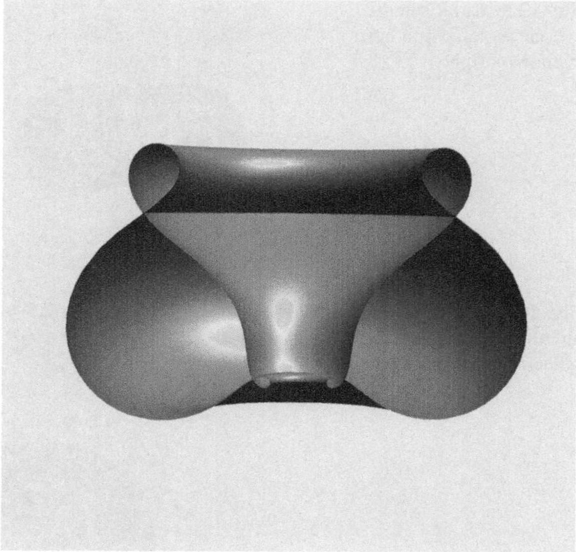

Fig. 12.4 A non-special, non-canal, isothermic torus of Bernstein.

Proof. In the preceding proof we saw that if the Möbius invariants q_1 and q_2 are zero, then the remaining three invariants are constant and then we classified this case. Suppose then that all invariants are constant and that q_1 or q_2 is not zero, say $q_2 \neq 0$. We may assume $q_2 > 0$, by Exercise 51. Then

$$p_2 = 0 = q_1, \quad p_1 = -3p_3, \quad p_3 = \frac{1}{2}(1 + q_2^2), \tag{12.70}$$

by (12.54) and (12.55). Then (12.69), for $n = \frac{2}{q_2}$, is the unique conformal immersion, up to Möbius transformation, whose invariants satisfy (12.70). See Problem 12.77 for more details. The case $q_1 > 0$ is considered in Problem 12.78. □

In the standard classification of Dupin immersions in $\mathbf{R}^3$ – see, for example, [40] – a *limit spindle cyclide* is a circular cylinder inverted in any sphere whose center lies inside the cylinder. Inversion of the cylinder $x^2 + y^2 = 1$ in the unit sphere is shown on the left in Figure 12.5. It has been opened to show how all the circles forming it meet at one point.

The *limit horn cyclide* is a circular cylinder inverted in a sphere whose center lies outside of the cylinder. Inversion of the cylinder $(x-2)^2 + y^2 = 1$ in the unit sphere is shown on the right side in Figure 12.5. See Problem 12.75.

In the standard classification, a *spindle cyclide* is the projection into the Poincaré ball of a circular hyperboloid whose axis passes through the center, together with its inversion in the unit sphere, as shown in Figure 12.6. A *horn cyclide* is a circular hyperboloid whose axis does not pass through the center of the Poincaré ball, together with its reflection in the unit boundary sphere, as shown in Figure 12.7. See Problem 12.76.

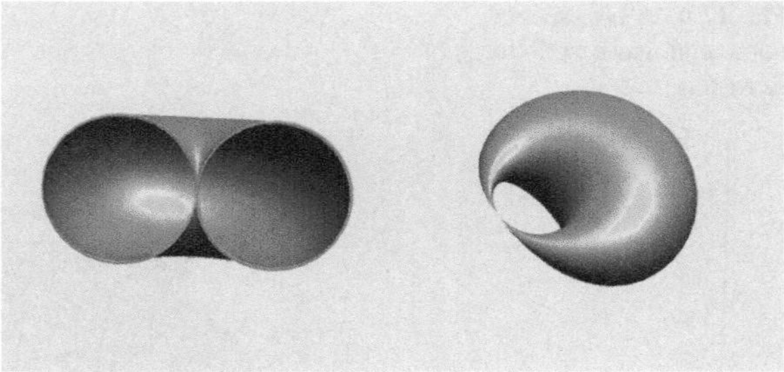

Fig. 12.5 Limit spindle cyclide on left, limit horn cyclide on right.

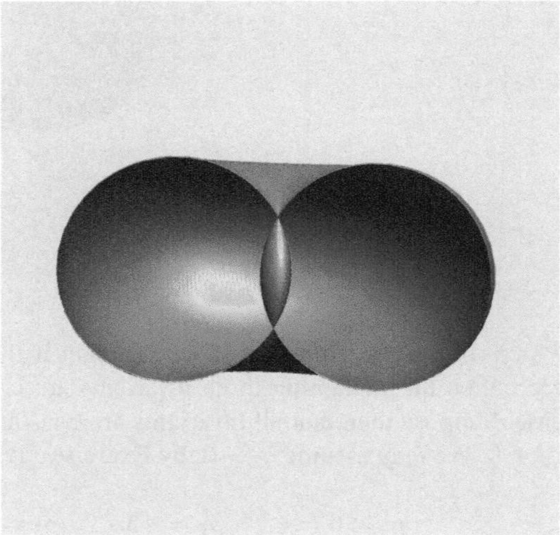

Fig. 12.6 Spindle cyclide: circular hyperboloid and its reflection in unit sphere

Problems

12.54. Let $\mathscr{I}_{\mathbf{m}}$ denote inversion in the unit sphere centered at the point $\mathbf{m} \in \mathbf{R}^3$, so $\mathscr{I} = \mathscr{I}_0$. Prove that inversion in the sphere with center $\mathbf{m}$ and radius $r > 0$ given by the formula (12.2) is the composition of a translation, a homothety, and an inversion.

12.55. Prove that inversion in the unit sphere centered at $\mathbf{m}$ is a composition of translations and inversion in the unit sphere centered at the origin.

12.56. Prove the following.

1. If $|\mathbf{m}| = r$, then the inversion of the sphere $|\mathbf{x} - \mathbf{m}|^2 = r^2$ in the unit sphere is a plane. Find a normal vector and a point of the plane.
2. Prove that the inversion in the unit sphere of a plane not containing the origin is a sphere that passes through the origin. Find the center and radius of the sphere in terms of the data of the plane (point on it and a normal vector). It should be clear that the inversion in the unit sphere of a plane passing through the origin is the same plane with the origin removed.

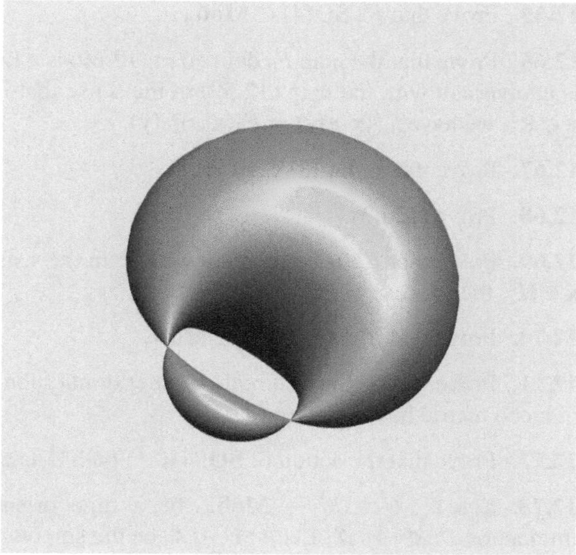

Fig. 12.7 Horn cyclide: Off-center circular hyperboloid and its reflection

12.57. Prove that a conformal structure on a surface M induces a complex structure on M such that for any complex coordinate chart (U,z) in M, the pair $(U,dzd\bar{z})$ belongs to the conformal structure. Prove also that a complex structure on a surface defines a conformal structure on it.

12.58. Prove (12.18) and prove that ${}^tLI_{4,1}L = g$, where L is defined in (12.17). Calculate the matrices in **Möb** of translations $T_{\mathbf{a}} \in \mathbf{SO}(4,1)$ in (12.10), orthogonal transformations $T_A \in \mathbf{SO}(4,1)$ in (12.11), homotheties $T_t \in \mathbf{SO}(4,1)$ in (12.12), and the inversion $T_{\mathscr{I}} \in \mathbf{SO}(4,1)$ in (12.13).

12.59. Prove that a vector $\mathbf{Y}_0 = \sum_0^4 u^a \boldsymbol{\delta}_a \in \mathscr{L}^+$ can be completed to a time oriented Möbius frame $\mathbf{Y}_0, \mathbf{Y}_1, \dots, \mathbf{Y}_4$ of $\mathbf{R}^{4,1}$.

12.60. Prove (12.27) and (12.28).

12.61. Prove that $\mathfrak{m}_0$ is not invariant under the adjoint action of G_0.

12.62. Follow the procedure of Section 3.1 to prove Lemma 12.25.

12.63. Prove (12.42).

12.64. For a first order frame field $Y : U \to \mathbf{Möb}_+$ along the immersion $f : M \to \mathscr{M}$, define the 2-form

$$\Omega_Y = |h|^2 \omega_0^1 \wedge \omega_0^2$$

on U, where h is defined in (12.39). Prove that Ω_Y is independent of the first order frame field Y along f.

12.65. Prove that $F_+\mathbf{SO}(4) \subset \mathbf{M\ddot{o}b}_+$.

12.66. Prove that the map F_0 defined in (12.60) is a Lie group monomorphism that is equivariant with the map (12.59) in the sense that for any $(\mathbf{x}, e) \in \mathbf{E}(3)$ and any $\mathbf{y} \in \mathbf{R}^3$, we have $f_0((\mathbf{x}, e)\mathbf{y}) = F_0(\mathbf{x}, e)f_0(\mathbf{y})$.

12.67. Prove that $F_0\mathbf{E}_+(3) \subset \mathbf{M\ddot{o}b}_+$.

12.68. Prove (12.63).

12.69. Prove that F_- is equivariant with f_- in the sense that, if $e \in \mathbf{O}_+(3,1)$ and if $\mathbf{x} \in \mathbf{H}^3$, then $f_-(e\mathbf{x}) = F_-(e)f_-(\mathbf{x})$.

12.70. Prove that $F_-\mathbf{SO}_+(3,1) \subset \mathbf{M\ddot{o}b}_+$.

12.71. Prove that $\mathbf{S}^{3,1}$ is a smooth 4-dimensional submanifold of $\mathbf{R}^{4,1}$ for which the induced metric has signature $(3,1)$.

12.72. Prove that the action of $\mathbf{SO}_+(\mathbf{R}^{4,1})$ on $\mathbf{S}^{3,1}$ is transitive.

12.73. Let $Y : U \subset M \to \mathbf{M\ddot{o}b}_+$ be a time oriented γ-frame field along an immersion $f : M \to \mathcal{M}$. Let $h : U \to \mathbf{C}$ be the smooth function of Definition 12.28. Prove that $\mathbf{Y}_3 \pm |h|\mathbf{Y}_0 : U \to \mathbf{S}^{3,1}$ are independent of the choice of γ-frame field along f. Prove that these globally defined maps $\mathbf{Y}_3 \pm |h|\mathbf{Y}_0 : M \to \mathbf{S}^{3,1}$ are curvature sphere maps of f. Note that these maps are smooth away from the umbilic points of f, and on any open set of umbilic points.

12.74 (Circular tori). The isoparametric immersions in $\mathbf{S}^3$ were determined in Theorem 5.15 to be the circular tori $\mathbf{S}^1(r) \times \mathbf{S}^1(s) \subset \mathbf{S}^3$, where $r = \cos\alpha$ and $s = \sin\alpha$ for any constant $0 < \alpha < \pi/2$. As discussed in Example 5.12, these are given by immersions

$$\mathbf{x} : \mathbf{R}^2 \to \mathbf{S}^3, \quad \mathbf{x}(x,y) = {}^t(r\cos\frac{x}{r}, r\sin\frac{x}{r}, s\cos\frac{y}{s}, s\sin\frac{y}{s})$$

Prove that $f = f_+ \circ \mathbf{x} : \mathbf{R}^2 \to \mathcal{M}$ is a conformal Dupin immersion with invariant $C = \frac{p_1 - p_3}{2} = -\cos 2\alpha$.

12.75 (Circular cylinder). The isoparametric immersions in $\mathbf{R}^3$ were determined in Proposition 4.39 to be, up to congruence, the circular cylinders of radius $R > 0$,

$$\mathbf{x} : \mathbf{R}^2 \to \mathbf{R}^3, \quad \mathbf{x}(x,y) = {}^t(R\cos x, R\sin x, y)$$

with second order frame field $(\mathbf{x}, e)$, where $\mathbf{e}_1 = \frac{1}{R}\mathbf{x}_x$, $\mathbf{e}_2 = \mathbf{x}_y = \epsilon_3$, and $\mathbf{e}_3 = \mathbf{e}_1 \times \mathbf{e}_2$. Prove that $f = f_0 \circ \mathbf{x} : \mathbf{R}^2 \to \mathcal{M}$ is a conformal Dupin immersion with invariant $C = 1$.

12.76 (Circular hyperboloid). The isoparametric immersions in $\mathbf{H}^3$ were found in Problem 6.49 to be, up to congruence, the circular hyperboloids $\mathbf{S}^1(\frac{a}{s}) \times \mathbf{H}^1(\frac{1}{s}) \subset \mathbf{H}^3 \subset \mathbf{R}^2 \times \mathbf{R}^{1,1}$, where a is any constant $0 < a < 1$ and $s = \sqrt{1-a^2}$. A parametrization with principal curvatures a and $1/a$ is given by $\mathbf{x} : \mathbf{R}^2 \to \mathbf{H}^3$,

where

$$\mathbf{x}(x,y) = \frac{1}{s}(a\cos(\frac{sy}{a})\epsilon_1 + a\sin(\frac{sy}{a})\epsilon_2 + \sinh(sx)\epsilon_3 + \cosh(sx)\epsilon_4),$$

with unit normal vector field

$$\mathbf{e}_3 = -\frac{1}{s}(\cos(\frac{sy}{a})\epsilon_1 + \sin(\frac{sy}{a})\epsilon_2 + a\sinh(sx)\epsilon_3 + a\cosh(sx)\epsilon_4).$$

Prove that $f = f_- \circ \mathbf{x} : \mathbf{R}^2 \to \mathscr{M}$ is a conformal Dupin immersion with invariant $C = \frac{p_1 - p_3}{2} = \frac{a+a^{-1}}{a-a^{-1}} \in (-\infty, -1)$.

12.77. Verify that the Möbius invariants of the immersion in (12.69) are given by (12.70) with $q_2 = \frac{2}{n}$. See also Problem 13.53 in the next chapter.

12.78. Classify the umbilic free conformal immersions $f : M \to \mathscr{M}$ of a connected Riemann surface M for which the conformal invariants are all constant and $q_1 > 0$.

Chapter 13
Complex Structure and Möbius Geometry

This chapter takes up the Möbius invariant conformal structure on Möbius space. It induces a conformal structure on any immersed surface, which in turn induces a complex structure on the surface. Möbius geometry is the study of properties of conformal immersions of Riemann surfaces into Möbius space $\mathscr{M}$ that remain invariant under the action of **Möb**. Each complex coordinate chart on an immersed surface has a unique Möbius frame field adapted to it, whose first order invariant we call k and whose second order invariant we call b. These are smooth, complex valued functions on the domain of the frame field. These frames are used to derive the structure equations for k and b, the conformal area, the conformal Gauss map, and the conformal area element. The equivariant embeddings of the space forms into Möbius space are conformal. Relative to a complex coordinate, the Hopf invariant, conformal factor, and mean curvature of an immersed surface in a space form determine the Möbius invariants k and b of the immersion into $\mathscr{M}$ obtained by applying the embedding of the space form into $\mathscr{M}$ to the given immersed surface. This gives a formula for the conformal area element showing that the Willmore energy is conformally invariant.

13.1 Conformal immersions

In this chapter we exploit the complex structure induced on a surface M by an immersion $f : M \to \mathscr{M}$, as described at the beginning of Section 12.4. If M is a given Riemann surface, then the immersion is conformal if the given complex structure is the same as that induced by the immersion, by Definition 12.22. The inner product on $\mathbf{R}^{4,1}$ is extended to be bilinear over $\mathbf{C}$ on the complexification $\mathbf{R}^{4,1} \otimes \mathbf{C}$. The following exercise gives an important criterion for when the immersion of a Riemann surface is conformal.

G.R. Jensen et al., *Surfaces in Classical Geometries*, Universitext,
DOI 10.1007/978-3-319-27076-0_13

Let $f : M \to \mathscr{M}$ be an immersion of a Riemann surface M. The map f is conformal if and only if any point of M is contained in some complex coordinate chart $(U, z = x + iy)$ of M for which there exists a lift $F : U \to \mathscr{L}$ of f satisfying

$$\langle dF, dF \rangle = e^{2u} dz d\bar{z}, \tag{13.1}$$

for some smooth $u : U \to \mathbf{R}$. This condition is equivalent to

$$\langle F_z, F_{\bar{z}} \rangle > 0 \text{ and } \langle F_z, F_z \rangle = 0,$$

where $F_z = \frac{1}{2}(\frac{\partial F}{\partial x} - i\frac{\partial F}{\partial y})$ and $F_{\bar{z}} = \frac{1}{2}(\frac{\partial F}{\partial x} + i\frac{\partial F}{\partial y})$.

Example 13.1. Let $M = \mathbf{C}$, the complex plane with its complex coordinate $z = x + iy$. Let $f : \mathbf{C} \to \mathscr{M}$ be the map defined by the time oriented lift

$$F : \mathbf{C} \to \mathscr{L}^+, \quad F(z) = \mathbf{Y}_0 = \boldsymbol{\delta}_0 + x\boldsymbol{\delta}_1 + y\boldsymbol{\delta}_2 + \frac{|z|^2}{2}\boldsymbol{\delta}_4. \tag{13.2}$$

Then

$$F_z = \frac{1}{2}(\boldsymbol{\delta}_1 - i\boldsymbol{\delta}_2 + \bar{z}\boldsymbol{\delta}_4), \quad F_{\bar{z}} = \frac{1}{2}(\boldsymbol{\delta}_1 + i\boldsymbol{\delta}_2 + z\boldsymbol{\delta}_4),$$

and then

$$\langle dF, dF \rangle = 2\langle F_z, F_{\bar{z}} \rangle dz d\bar{z} = dz d\bar{z}$$

shows that f is conformal. If $z \neq 0$, then

$$f(z) = [\frac{1}{|z|^2}\boldsymbol{\delta}_0 + \frac{x}{|z|^2}\boldsymbol{\delta}_1 + \frac{y}{|z|^2}\boldsymbol{\delta}_2 + \frac{1}{2}\boldsymbol{\delta}_4] \to [\boldsymbol{\delta}_4] \text{ as } z \to \infty$$

shows that f can be extended to a continuous map on the Riemann sphere $\hat{\mathbf{C}} = \mathbf{C} \cup \{\infty\}$. Using the complex coordinate $w = 1/z$ on $\hat{\mathbf{C}} \setminus \{0\}$, where $w(\infty) = 0$, one can show that f is a conformal immersion $\hat{\mathbf{C}} \to \mathscr{M}$.

A time oriented Möbius frame field Y along f on $\mathbf{C}$ is given by

$$\mathbf{Y}_0 = F, \quad \mathbf{Y}_1 = \boldsymbol{\delta}_1 + x\boldsymbol{\delta}_4, \quad \mathbf{Y}_2 = \boldsymbol{\delta}_2 + y\boldsymbol{\delta}_4, \quad \mathbf{Y}_3 = \boldsymbol{\delta}_3, \quad \mathbf{Y}_4 = \boldsymbol{\delta}_4,$$

in the Möbius frame (12.15). In general, $d\mathbf{Y}_a = \sum_0^4 \mathbf{Y}_b \omega_a^b$, for $a = 0, \dots, 4$, where the ω_a^b are the 1-forms on $M = \mathbf{C}$ obtained by pulling back the Maurer–Cartan forms of **Möb** by the frame field $Y : \mathbf{C} \to$ **Möb**. We calculate

$$d\mathbf{Y}_0 = \mathbf{Y}_1 dx + \mathbf{Y}_2 dy = \sum_0^4 \omega_0^a \mathbf{Y}_a,$$

which implies that

$$\omega_0^1 = dx, \quad \omega_0^2 = dy, \quad \omega_0^3 = 0 = \omega_0^0, \tag{13.3}$$

and we calculate

$$d\mathbf{Y}_1 = \boldsymbol{\delta}_4 dx, \quad d\mathbf{Y}_2 = \boldsymbol{\delta}_4 dy, \quad d\mathbf{Y}_3 = 0 = d\mathbf{Y}_4, \tag{13.4}$$

which imply that

$$\omega_1^0 = \omega_2^0 = \omega_3^0 = \omega_3^1 = \omega_3^2 = 0. \tag{13.5}$$

The forms in (13.5) can be read from (13.4) by inspection, or be found by

$$\omega_a^0 = -\langle d\mathbf{Y}_a, \mathbf{Y}_4\rangle, \quad \omega_a^i = \langle d\mathbf{Y}_a, \mathbf{Y}_i\rangle, \quad \omega_a^4 = -\langle d\mathbf{Y}_a, \mathbf{Y}_0\rangle,$$

where $a = 0, 1, \dots, 4$ and $i = 1, 2, 3$. Observe that $\omega_0^1 + i\omega_0^2 = dz$ in this example.

13.2 Adapted frames

For a conformal immersion $f : M \to \mathscr{M}$ (see Definition 12.22) we want to make frame reductions relative to a fixed local complex coordinate in the Riemann surface M. After the second reduction, which is unique for time oriented frames, we will then see how the second order frame and the invariants change with a change of local complex coordinate. Adapting the frame to a given coordinate system is a major innovation in the subject introduced by F. Burstall, F. Pedit, and U. Pinkall in [26].

Let $(U, z = x + iy)$ be a complex coordinate chart of M. Let

$$Y = (\mathbf{Y}_0, \dots, \mathbf{Y}_4) : U \to \mathbf{M\ddot{o}b}$$

be a Möbius frame field along f on U. Then

$$d\mathbf{Y}_a = \sum_0^4 \omega_a^b \mathbf{Y}_b,$$

for $a = 0, \dots, 4$, where ω_a^b denote the pull backs of the Maurer–Cartan forms by Y. Under the decomposition $\mathfrak{m\ddot{o}b} = \mathfrak{g}_0 + \mathfrak{m}_0$ of (12.29), the entries of the $\mathfrak{m}_0$-component of ω are ω_0^i, for $i = 1, 2, 3$. We set

$$\omega_0^i = p^i dz + \bar{p}^i d\bar{z}, \quad i = 1, 2, 3, \tag{13.6}$$

where $p_i : U \to \mathbf{C}$ are smooth functions. Let

$$\mathbf{p} = {}^t(p^1, p^2, p^3) : U \to \mathbf{C}^3.$$

Then the $\mathfrak{m}_0$-component of ω is

$$\theta = {}^t(\omega_0^1, \omega_0^2, \omega_0^3) = \mathbf{p}dz + \bar{\mathbf{p}}d\bar{z},$$

so

$${}^t\theta\theta = {}^t\mathbf{p}\mathbf{p}\, dz^2 + {}^t\bar{\mathbf{p}}\bar{\mathbf{p}}\, d\bar{z}^2 + 2\,{}^t\mathbf{p}\bar{\mathbf{p}}\, dzd\bar{z}.$$

The condition that f be a conformal immersion requires that ${}^t\theta\theta$ be a smooth positive multiple of $dzd\bar{z}$ on U, which requires

$${}^t\mathbf{p}\mathbf{p} = 0 = {}^t\bar{\mathbf{p}}\bar{\mathbf{p}}, \quad {}^t\mathbf{p}\bar{\mathbf{p}} > 0.$$

Therefore, $\mathbf{p}$ takes values in the set $\mathscr{I}$ of *isotropic vectors* in $\mathbf{C}^3$, which is the complex quadric cone

$$\mathscr{I} = \{\mathbf{0} \neq \mathbf{z} = (z^j) \in \mathbf{C}^3 : {}^t\mathbf{z}\mathbf{z} = \sum_1^3 (z^j)^2 = 0\}.$$

Notice that ${}^t\mathbf{z}\mathbf{z} = \mathbf{z} \cdot \mathbf{z}$ is the dot product of $\mathbf{R}^3$ extended to $\mathbf{C}^3$ to be bilinear over $\mathbf{C}$. Any other frame field along f on U is given by $\tilde{Y} = YK(r, A, \mathbf{y})$, for

$$K(r, A, \mathbf{y}) = \begin{pmatrix} \frac{1}{r} & {}^t\mathbf{y}A & \frac{r}{2}|\mathbf{y}|^2 \\ 0 & A & r\mathbf{y} \\ 0 & 0 & r \end{pmatrix} : U \to G_0, \tag{13.7}$$

$r \neq 0$, $A \in \mathbf{SO}(3)$, and $\mathbf{y} \in \mathbf{R}^3$, defined in (12.25) and (12.26). Set $\tilde{Y}^{-1}d\tilde{Y} = \tilde{\omega}$. Denote the quantities in (13.6) for $\tilde{Y}$ with the same letters marked with a tilde. From the change of frame formula (12.34), it follows that

$$\tilde{\mathbf{p}} = \frac{1}{r}A^{-1}\mathbf{p}. \tag{13.8}$$

Exercise 58. Prove that the action of $\mathbf{R}^\times \times \mathbf{SO}(3)$ on $\mathscr{I}$ given by $(r, A)\mathbf{p} = rA\mathbf{p}$ is transitive. To be specific, prove that this action sends the point ${}^t(1, -i, 0) \in \mathscr{I}$ to any point $\mathbf{p} \in \mathscr{I}$. Prove that the isotropy subgroup at ${}^t(1, -i, 0)$ is

$$\{(\epsilon, \begin{pmatrix} \epsilon I_2 & 0 \\ 0 & 1 \end{pmatrix}) : \epsilon = \pm 1\}.$$

Hint: If $\mathbf{p} \in \mathbf{C}^3$, then $\mathbf{p} = \mathbf{u}_1 + i\mathbf{u}_2$, where $\mathbf{u}_1, \mathbf{u}_2 \in \mathbf{R}^3$ are the real and imaginary parts of $\mathbf{p}$. Then $\mathbf{p} \in \mathscr{I}$ if and only if $\mathbf{u}_1$ and $\mathbf{u}_2$ are orthogonal and of the same length $r > 0$. If

$$\mathbf{e}_1 = \frac{1}{r}\mathbf{u}_1,\ \mathbf{e}_2 = -\frac{1}{r}\mathbf{u}_2,\ \mathbf{e}_3 = \mathbf{e}_1 \times \mathbf{e}_2 \in \mathbf{R}^3,$$

then $A = (\mathbf{e}_1, \mathbf{e}_2, \mathbf{e}_3) \in \mathbf{SO}(3)$ and $rA^t(1, -i, 0) = \mathbf{p}$.

Exercise 59. Prove that if $\mathbf{p} : U \to \mathscr{I} \subset \mathbf{C}^3$ is a smooth map, then there exists a unique smooth map $(r, A) : U \to \mathbf{R}^+ \times \mathbf{SO}(3)$ such that

$$\frac{1}{r}A^{-1}\mathbf{p} = \frac{1}{2}(\boldsymbol{\epsilon}_1 - i\boldsymbol{\epsilon}_2),$$

at every point of U, where $\boldsymbol{\epsilon}_1, \boldsymbol{\epsilon}_2, \boldsymbol{\epsilon}_3$ is the standard basis of $\mathbf{C}^3$.

It follows from (13.8) and Exercise 59 that there exists a smooth map (13.7) such that $\tilde{\mathbf{p}} = \frac{1}{2}(\boldsymbol{\epsilon}_1 - i\boldsymbol{\epsilon}_2)$ at every point of U. That is,

$$\tilde{\omega}_0^1 = \frac{1}{2}(dz + d\bar{z}) = dx, \quad \tilde{\omega}_0^2 = -\frac{i}{2}(dz - d\bar{z}) = dy, \quad \tilde{\omega}_0^3 = 0.$$

Definition 13.2. A Möbius frame field $Y : U \to \mathbf{Möb}$ is of *first order relative to the local complex coordinate z in U* if

$$\omega_0^1 + i\omega_0^2 = dz \text{ and } \omega_0^3 = 0 \tag{13.9}$$

hold on U.

Remark 13.3. If $Y : U \to \mathbf{Möb}$ is first order relative to a complex coordinate z on U, then Y is a first order Möbius frame field along f according to Definition 12.24.

By Exercise 59, for any complex coordinate chart (U, z) in M, there exists a Möbius frame field $Y : U \to \mathbf{Möb}$ which is first order relative to z. Then (13.9) holds and its differential, combined with the structure equations (12.22), gives

$$(\omega_0^0 + i\omega_2^1) \wedge dz = 0, \quad (\omega_1^3 - i\omega_2^3) \wedge dz + (\omega_1^3 + i\omega_2^3) \wedge d\bar{z} = 0.$$

From this we conclude that

$$\begin{aligned} &\omega_0^0 + i\omega_2^1 = l dz, \text{ some smooth } l : U \to \mathbf{C}, \\ &\omega_1^3 - i\omega_2^3 = k dz + s d\bar{z}, \text{ some smooth } k : U \to \mathbf{C} \text{ and } s : U \to \mathbf{R}. \end{aligned} \tag{13.10}$$

The next step is to see how the functions l, k, and s transform under a change of first order Möbius frame field.

Let $Y : U \to \mathbf{M\ddot{o}b}$ be a first order Möbius frame field along f relative to the complex coordinate chart U, z. By Exercise 58, any other first order frame field $\tilde{Y} : U \to \mathbf{M\ddot{o}b}$ along f is given by

$$\tilde{Y} = YK,$$

where $K : U \to G_1$ is a smooth map into the subgroup of G_0,

$$G_1 = \{K(\epsilon, E_\epsilon, \mathbf{y}) \in G_0 : \epsilon = \pm 1,\ E_\epsilon = \begin{pmatrix} \epsilon & 0 & 0 \\ 0 & \epsilon & 0 \\ 0 & 0 & 1 \end{pmatrix},\ \mathbf{y} \in \mathbf{R}^3\}, \tag{13.11}$$

whose Lie algebra is the subalgebra $\mathfrak{g}_1$ of $\mathfrak{g}_0$,

$$\mathfrak{g}_1 = \{\begin{pmatrix} 0 & {}^tX & 0 \\ 0 & 0 & X \\ 0 & 0 & 0 \end{pmatrix} : X \in \mathbf{R}^3\}.$$

To calculate how the functions l, k, and s in (13.10) transform under this change of frame, we will use the complementary subspace $\mathfrak{m}_1$ to $\mathfrak{g}_1$ in $\mathfrak{g}_0$,

$$\mathfrak{m}_1 = \{\begin{pmatrix} X_0^0 & 0 & 0 \\ 0 & X_j^i & 0 \\ 0 & 0 & -X_0^0 \end{pmatrix} : X_j^i + X_i^j = 0,\ i,j = 1,2,3\}.$$

This gives vector space direct sums $\mathfrak{g}_0 = \mathfrak{g}_1 + \mathfrak{m}_1$ and

$$\mathfrak{m\ddot{o}b} = \mathfrak{g}_0 + \mathfrak{m}_0 = \mathfrak{g}_1 + \mathfrak{m}_1 + \mathfrak{m}_0.$$

The $(\mathfrak{m}_0 + \mathfrak{m}_1)$-component of $\omega = Y^{-1}dY$ is

$$\omega_{\mathfrak{m}_0+\mathfrak{m}_1} = \begin{pmatrix} \omega_0^0 & 0 & 0 \\ \theta & \eta & 0 \\ 0 & {}^t\theta & -\omega_0^0 \end{pmatrix},$$

where we set, for this calculation,

$$\theta = {}^t(\omega_0^1,\ \omega_0^2,\ 0), \quad \eta = (\omega_j^i),$$

where $i,j = 1,2,3$. We calculate

$$\begin{aligned} \tilde{\omega}_{\mathfrak{m}_0+\mathfrak{m}_1} &= (K^{-1}\omega_{\mathfrak{m}_0+\mathfrak{m}_1}K)_{\mathfrak{m}_0+\mathfrak{m}_1} \\ &= \begin{pmatrix} \omega_0^0 - {}^t\mathbf{y}\theta & 0 & 0 \\ \epsilon E_\epsilon \theta & E_\epsilon(\eta + \theta{}^t\mathbf{y} - \mathbf{y}{}^t\theta)E_\epsilon & 0 \\ 0 & \epsilon{}^t\theta E_\epsilon & -\omega_0^0 + {}^t\theta\mathbf{y} \end{pmatrix}, \end{aligned}$$

from which, writing ${}^t\mathbf{y} = (y^1, y^2, y^3)$, we find

$$\begin{aligned} \tilde{\omega}_0^0 + i\tilde{\omega}_2^1 &= \omega_0^0 + i\omega_2^1 - (y^1 - iy^2)dz, \\ \tilde{\omega}_1^3 - i\tilde{\omega}_2^3 &= \epsilon(\omega_1^3 - i\omega_2^3 - y^3 d\bar{z}). \end{aligned} \tag{13.12}$$

Substituting (13.10) into (13.12), with quantities relative to $\tilde{Y}$ denoted by the same letters with a tilde, we get

$$\tilde{l} = l - (y^1 - iy^2), \quad \tilde{k} = \epsilon k, \quad \tilde{s} = \epsilon(s - y^3). \tag{13.13}$$

We draw two important conclusions from these calculations. The first of these is that if we allow only time oriented Möbius frame fields along f, then $\epsilon = 1$ in (13.13) and

$$k : U \to \mathbf{C} \tag{13.14}$$

is independent of the choice of first order time oriented Möbius frame field relative to z. The second conclusion is that there exists a unique time oriented frame field $\tilde{Y} : U \to \mathbf{M\ddot{o}b}_+$ along f for which $\tilde{l}$ and $\tilde{s}$ are identically zero on U.

Definition 13.4. The *first order Möbius invariant* relative to a complex coordinate (U, z) of a conformal immersion $f : M \to \mathcal{M}$ is the smooth function $k : U \to \mathbf{C}$ defined in (13.10) for any time oriented Möbius frame field that is first order relative to z.

Definition 13.5 (Adapted frames). A Möbius frame field $Y : U \to \mathbf{M\ddot{o}b}$ along a conformal immersion $f : M \to \mathcal{M}$ is *adapted* to a complex coordinate chart (U, z) if it is time oriented and $\omega = Y^{-1}dY$ satisfies

$$\omega_0^1 + i\omega_0^2 = dz, \quad \omega_0^3 = 0, \quad \omega_0^0 = 0 = \omega_2^1, \quad \omega_1^3 - i\omega_2^3 = kdz, \tag{13.15}$$

for some smooth $k : U \to \mathbf{C}$, which must be the first order invariant of f relative to z. Differentiating equations (13.15) and using (12.22), we get

$$\omega_3^0 = k_{\bar{z}}dz + \overline{k_{\bar{z}}}d\bar{z}, \quad \omega_1^0 - i\omega_2^0 = -bdz - \frac{|k|^2}{2}d\bar{z}, \tag{13.16}$$

for some smooth function

$$b : U \to \mathbf{C},$$

which is the *second order Möbius invariant of f relative to z.*

Theorem 13.6. *Let $f : M \to \mathcal{M}$ be a conformal immersion of a Riemann surface M and let U, z be a complex coordinate chart in M. There exists a unique Möbius* *frame field $Y : U \to \mathbf{M\ddot{o}b}_+$ along f adapted to z. The invariants k and b relative to z satisfy the* structure equations

$$b_{\bar{z}} = \frac{1}{2}(|k|^2)_z + k\overline{k_{\bar{z}}}, \tag{13.17}$$

$$\Im(k_{\bar{z}\bar{z}} + \frac{1}{2}\bar{b}k) = 0, \tag{13.18}$$

on U, where $\Im(p)$ indicates the imaginary part of a complex number p.

Proof. By the calculations above, there exists a unique frame field $Y : U \to \mathbf{M\ddot{o}b}_+$ adapted to z. All the entries of $\omega = Y^{-1}dY$ are given in (13.15) and (13.16). Then $d\omega = -\omega\wedge\omega$ implies (13.17) and (13.18). □

Remark 13.7. If $Y : U \to \mathbf{M\ddot{o}b}_+$ is the frame field adapted to z on U, then $\tilde{Y} = YK(-1, E_{-1}, \mathbf{0}) = (-\mathbf{Y}_0, -\mathbf{Y}_1, -\mathbf{Y}_2, \mathbf{Y}_3, -\mathbf{Y}_4)$, where E_{-1} is defined in (13.11), is a time reversing frame field on U satisfying (13.15) and (13.16) for the functions $\tilde{k} = -k$ and $\tilde{b} = b$, which also satisfy the structure equations (13.17) and (13.18).

Theorem 13.8 (Existence and Uniqueness). *Let M be a Riemann surface and let U, z be a local complex chart for which $U \subset M$ is simply connected. If $k : U \to \mathbf{C}$ and $b : U \to \mathbf{C}$ are smooth functions satisfying the structure equations* (13.17) *and* (13.18)*, then there exists a conformal immersion $f : U \to \mathscr{M}$, unique up to time oriented Möbius transformation, whose time oriented first order invariant is k and whose second order invariant is b.*

Theorem 13.9. *Let $f : M \to \mathscr{M}$ be a conformal immersion of a Riemann surface M. If $Y : U \to \mathbf{M\ddot{o}b}_+$ is the frame field adapted to a complex coordinate chart (U, z) of M, then Y is a γ-frame field along f with Willmore function*

$$W = k_{\bar{z}\bar{z}} + \frac{1}{2}k\bar{b} : U \to \mathbf{R}, \tag{13.19}$$

where k and b are the invariants of f relative to z.

Proof. If $Y : U \to \mathbf{M\ddot{o}b}_+$ is the frame field adapted to z, then (see Definition 12.28) Y is a γ-frame along f, with $h = k$, the first order invariant relative to z, $\varphi = \omega_0^1 + i\omega_0^2 = dz$, and

$$\omega_0^0 = 0 = \omega_2^1, \quad \omega_1^3 - i\omega_2^3 = kdz,$$

$$\omega_3^0 = k_{\bar{z}}dz + \overline{k_{\bar{z}}}d\bar{z}, \quad \omega_1^0 - i\omega_2^0 = -bdz - \frac{|k|^2}{2}d\bar{z},$$

where $b : U \to \mathbf{C}$ is the second order invariant of f relative to z. Thus, $h_2 = k_{\bar{z}}$ in (12.44), and so (12.45) becomes

$$dk_{\bar{z}} + \frac{k}{2}(\bar{b}d\bar{z} + \frac{|k|^2}{2}dz) = Pdz + Wd\bar{z},$$

which implies that (13.19) gives the Willmore function relative to Y, and

$$P = k_{\bar{z}\bar{z}} + \frac{1}{4}k|k|^2 : U \to \mathbf{C}$$

is the other γ-invariant of Y. □

Note that W in (13.19) is real valued because of the structure equation (13.18).

13.3 Dependence on z

We want to determine transformation rules relating the invariants relative to two different complex coordinates.

Theorem 13.10. *For a conformal immersion $f : M \to \mathcal{M}$, let z and w be complex coordinates on a domain $U \subset M$. Let k and b be the invariants of f relative to z and let $\hat{k}$ and $\hat{b}$ be the invariants relative to w. Then w is a holomorphic function of z on U, $w' = \frac{dw}{dz}$ is a nowhere zero holomorphic function on U, and*

$$\hat{k} = \frac{k|w'|}{(w')^2}, \tag{13.20}$$

$$\hat{b} = \frac{b - \mathcal{S}_z(w)}{(w')^2}, \tag{13.21}$$

where

$$\mathcal{S}_z(w) = \left(\frac{w''}{w'}\right)' - \frac{1}{2}\left(\frac{w''}{w'}\right)^2$$

is the Schwarzian derivative of w with respect to z on U.

Proof. The nowhere zero holomorphic function $w' = \frac{dw}{dz}$ on U has a well-defined polar representation

$$w' = re^{it}, \tag{13.22}$$

where $r : U \to \mathbf{R}^+$ and $e^{it} : U \to \mathbf{S}^1 \subset \mathbf{C}$ are smooth. Although the argument function $t : U \to \mathbf{R}$ is multivalued, defined only up to adding integer multiples of 2π, the functions $\cos t$, $\sin t$ and the real 1-form dt are well-defined and smooth on U. Then

$$A = \begin{pmatrix} \cos t & \sin t & 0 \\ -\sin t & \cos t & 0 \\ 0 & 0 & 1 \end{pmatrix} : U \to \mathbf{SO}(3)$$

is a smooth map, and we use the holomorphic function

$$y = y^1 - iy^2 = \frac{w''}{(w')^2} \tag{13.23}$$

to define the smooth map

$$\mathbf{y} = {}^t(y^1, y^2, 0) : U \to \mathbf{R}^3.$$

Let $Y = (\mathbf{Y}_0, \ldots, \mathbf{Y}_4) : U \to \mathbf{M\ddot{o}b}_+$ be the time oriented Möbius frame field along f adapted to z. We claim that

$$\tilde{Y} = YK(r^{-1}, A, rA\mathbf{y}) = Y \begin{pmatrix} r & r^t\mathbf{y} & \frac{1}{2}r|\mathbf{y}|^2 \\ 0 & A & A\mathbf{y} \\ 0 & 0 & \frac{1}{r} \end{pmatrix} : U \to \mathbf{M\ddot{o}b}_+ \tag{13.24}$$

is the time oriented Möbius frame field along f adapted to w. The columns of $\tilde{Y}$ are

$$\begin{aligned}
\tilde{\mathbf{Y}}_0 &= r\mathbf{Y}_0, \\
\tilde{\mathbf{Y}}_1 - i\tilde{\mathbf{Y}}_2 &= e^{-it}(\mathbf{Y}_1 - i\mathbf{Y}_2) + ry\mathbf{Y}_0, \\
\tilde{\mathbf{Y}}_3 &= \mathbf{Y}_3, \\
\tilde{\mathbf{Y}}_4 &= \frac{1}{r}\mathbf{Y}_4 + \frac{r}{2}|y|^2\mathbf{Y}_0 + \frac{\bar{y}}{2}e^{-it}(\mathbf{Y}_1 - i\mathbf{Y}_2) + \frac{y}{2}e^{it}(\mathbf{Y}_1 + i\mathbf{Y}_2).
\end{aligned} \tag{13.25}$$

To establish this claim, we make the long, but elementary, calculation

$$\alpha = \tilde{Y}^{-1}d\tilde{Y} = K^{-1}\omega K + K^{-1}dK.$$

Using Definition 13.5 and this calculation, we get

$$\begin{aligned}
\alpha_0^1 + i\alpha_0^2 &= re^{it}(\omega_0^1 + i\omega_0^2) = w'dz = dw, \\
\alpha_0^3 &= \omega_0^3 = 0, \\
\alpha_0^0 &= d\log r - y^1\alpha_0^1 - y^2\alpha_0^2 = ((\log r)_w - \frac{1}{2}y)dw + ((\log r)_{\bar{w}} - \frac{1}{2}\bar{y})d\bar{w}, \\
\alpha_1^0 - i\alpha_2^0 &= \frac{1}{w'}(\omega_1^0 - i\omega_2^0) + \frac{1}{2}(y)^2dw + dy \\
\alpha_1^3 - i\alpha_2^3 &= e^{-it}(\omega_1^3 - i\omega_2^3) = e^{-it}k\,dz = \frac{rk}{(w')^2}dw, \\
\alpha_3^0 &= \frac{1}{r}\omega_3^0 + \frac{y}{2}e^{it}(\omega_1^3 + i\omega_2^3) + \frac{\bar{y}}{2}e^{-it}(\omega_1^3 - i\omega_2^3).
\end{aligned}$$

Then $\alpha_0^1 + i\alpha_0^2 = dw$ and $\alpha_0^3 = 0$ imply that $\tilde{Y}$ is a first order frame field relative to w. The expression for $\alpha_1^3 - i\alpha_2^3$ then verifies (13.20). We claim $\alpha_0^0 = 0$, because $w' = re^{it}$ holomorphic implies

$$0 = (w')_{\bar{z}} = ((\log r)_{\bar{z}} + it_{\bar{z}})w',$$

so w' never zero and $\log r$ and dt real imply

$$0 = \overline{((\log r)_{\bar{z}} + it_{\bar{z}})} = (\log r)_z - it_z.$$

Thus

$$\begin{aligned} 0 &= \frac{1}{2}((\log r)_z - it_z)dz = ((\log r)_z - \frac{1}{2}(\log r + it)_z)dz \\ &= ((\log r)_z - \frac{1}{2}(\log w')_z)dz = ((\log r)_z - \frac{1}{2}\frac{w''}{w'})dz \\ &= (\frac{(\log r)_z}{w'} - \frac{1}{2}\frac{w''}{(w')^2})dw = ((\log r)_w - \frac{1}{2}y)dw \end{aligned}$$

shows that $\alpha_0^0 = 0$ on U. It follows that $\alpha_2^1 = 0$ as well, since we know already that $\tilde{Y}$ is first order relative to w, which implies that $\alpha_0^0 + i\alpha_2^1 = l dw$, for some smooth function $l : U \to \mathbf{C}$, and therefore $l = 0$ if $\alpha_0^0 = 0$. By Definition 13.5, $\tilde{Y}$ is the time oriented frame field adapted to w on U. Substituting (13.16) into the formula for $\alpha_1^0 - i\alpha_2^0$, we get

$$\begin{aligned} \alpha_1^0 - i\alpha_2^0 &= -\frac{1}{w'}(bdz + \frac{|k|^2}{2}d\bar{z}) + \frac{1}{2}(y)^2 dw + \frac{y'}{w'}dw \\ &= (-\frac{b}{(w')^2} + \frac{1}{2}(y)^2 + \frac{y'}{w'})dw - \frac{|k|^2}{2(w')^2}d\bar{w}. \end{aligned}$$

Then $-\hat{b}$ is the coefficient of dw, by (13.16) for $\tilde{Y}$, so

$$\hat{b} = \frac{b}{(w')^2} - \frac{1}{2}(\frac{w''}{(w')^2})^2 - \frac{1}{w'}(\frac{1}{w'}(\frac{w''}{w'}))' = \frac{1}{(w')^2}(b + \frac{1}{2}(\frac{w''}{w'})^2 - (\frac{w''}{w'})'),$$

which verifies (13.21). □

Remark 13.11. The frame change (13.24) is found in two steps. The frame

$$\hat{Y} = YK(r^{-1}, A, 0)$$

is a first order frame field relative to w, provided that (13.22) is used to determine r and t. Then

$$\tilde{Y} = \hat{Y}K(1, I_3, \mathbf{y})$$

is the frame field adapted to w, provided that $\mathbf{y}$ is given by (13.23). The frame change (13.24) follows from the product formula (12.27).

Corollary 13.12. *The smooth map* $\mathbf{Y}_3 : U \to \mathbf{R}^{4,1}$ *and the smooth 2-form*

$$\Omega = \frac{i}{2}|k|^2 dz \wedge d\bar{z}$$

are independent of the choice of local complex coordinate chart (U,z) *in* M, *and thus are smooth and well-defined on all of* M.

Proof. These statements follow from (13.25) and (13.20), respectively. □

13.3.1 Calculating the Invariants

Let $Y : U \to \mathbf{M\ddot{o}b}_+$ be the time oriented Möbius frame field adapted to the complex coordinate chart (U,z) in M. From (12.21) and (13.16) we have

$$\begin{aligned} d\mathbf{Y}_0 &= \frac{1}{2}(\mathbf{Y}_1 - i\mathbf{Y}_2)dz + \frac{1}{2}(\mathbf{Y}_1 + i\mathbf{Y}_2)d\bar{z}, \\ d(\mathbf{Y}_1 - i\mathbf{Y}_2) &= (-b\mathbf{Y}_0 + k\mathbf{Y}_3)dz + (-\frac{|k|^2}{2}\mathbf{Y}_0 + \mathbf{Y}_4)d\bar{z}, \\ d\mathbf{Y}_3 &= (k_{\bar{z}}\mathbf{Y}_0 - \frac{k}{2}(\mathbf{Y}_1 + i\mathbf{Y}_2))dz + (\overline{k_{\bar{z}}}\mathbf{Y}_0 - \frac{\bar{k}}{2}(\mathbf{Y}_1 - i\mathbf{Y}_2))d\bar{z}, \\ d\mathbf{Y}_4 &= (-\frac{b}{2}(\mathbf{Y}_1 + i\mathbf{Y}_2) - \frac{|k|^2}{4}(\mathbf{Y}_1 - i\mathbf{Y}_2) + k_{\bar{z}}\mathbf{Y}_3)dz \\ &\quad + (-\frac{\bar{b}}{2}(\mathbf{Y}_1 - i\mathbf{Y}_2) - \frac{|k|^2}{4}(\mathbf{Y}_1 + i\mathbf{Y}_2) + \overline{k_{\bar{z}}}\mathbf{Y}_3)d\bar{z}. \end{aligned} \tag{13.26}$$

Comparing the first equation in (13.26) with

$$d\mathbf{Y}_0 = \mathbf{Y}_{0z}dz + \mathbf{Y}_{0\bar{z}}d\bar{z},$$

we conclude that

$$\mathbf{Y}_{0z} = \frac{1}{2}(\mathbf{Y}_1 - i\mathbf{Y}_2), \quad \mathbf{Y}_{0\bar{z}} = \frac{1}{2}(\mathbf{Y}_1 + i\mathbf{Y}_2). \tag{13.27}$$

Therefore,

$$\langle d\mathbf{Y}_0, d\mathbf{Y}_0 \rangle = dz d\bar{z},$$

which is equivalent to

$$\langle \mathbf{Y}_{0z}, \mathbf{Y}_{0z} \rangle = 0 \text{ and } \langle \mathbf{Y}_{0z}, \mathbf{Y}_{0\bar{z}} \rangle = \frac{1}{2}.$$

Differentiating the first equation in (13.27), and using the second equation in (13.26), we conclude that

$$\mathbf{Y}_{0zz} = \frac{1}{2}(-b\mathbf{Y}_0 + k\mathbf{Y}_3), \quad \mathbf{Y}_{0z\bar{z}} = \frac{1}{2}(-\frac{|k|^2}{2}\mathbf{Y}_0 + \mathbf{Y}_4), \tag{13.28}$$

from which we can calculate

$$b = 4\langle \mathbf{Y}_{0zz}, \mathbf{Y}_{0z\bar{z}} \rangle.$$

Combining this last equation with (13.28) determines $k\mathbf{Y}_3 = 2\mathbf{Y}_{0zz} + b\mathbf{Y}_0$. The invariant k is determined by

$$\det(\mathbf{Y}_0, \mathbf{Y}_{0z}, \mathbf{Y}_{0\bar{z}}, \mathbf{Y}_{0zz}, \mathbf{Y}_{0z\bar{z}}) = \frac{ik}{8}\det(\mathbf{Y}_0, \mathbf{Y}_1, \mathbf{Y}_2, \mathbf{Y}_3, \mathbf{Y}_4) = \frac{ik}{8}.$$

We summarize these calculations as follows.

Theorem 13.13 (Calculation of the invariants). *Let $f : U \to \mathcal{M}$ be a conformal immersion and let z be a complex coordinate on U. If $F : U \to \mathcal{L}^+ \subset \mathbf{R}^{4,1}$ is a lift of f for which*

$$\langle F_z, F_z \rangle = 0 \text{ and } \langle F_z, F_{\bar{z}} \rangle = \frac{1}{2},$$

then the Möbius frame field $Y : U \to \mathbf{M\ddot{o}b}_+$ adapted to z has $\mathbf{Y}_0 = F$ and $\mathbf{Y}_3$ and the Möbius invariants k and b of f relative to z are determined by the equations

$$b = 4\langle F_{zz}, F_{z\bar{z}} \rangle, \quad k\mathbf{Y}_3 = 2F_{zz} + bF, \tag{13.29}$$

and

$$k = -8i\det(F, F_z, F_{\bar{z}}, F_{zz}, F_{z\bar{z}}). \tag{13.30}$$

Let us try these calculations on some examples.

Example 13.14 (Round Riemann sphere). This is Example 13.1 revisited. A time oriented lift of $f : \mathbf{C} \to \mathcal{M}$ is given by $F = \delta_0 + x\delta_1 + y\delta_2 + \frac{1}{2}|z|^2\delta_4$ in (13.2), where $z = x + iy$. Then

$$F_z = \frac{1}{2}(F_x - iF_y) = \frac{1}{2}(\delta_1 - i\delta_2 + \bar{z}\delta_4)$$

and

$$\langle F_z, F_z \rangle = 0, \quad \langle F_z, F_{\bar{z}} \rangle = \frac{1}{2},$$

so $F : \mathbf{C} \to \mathscr{L}^+$ is the desired time oriented lift. We calculate

$$F_{zz} = 0, \quad F_{z\bar{z}} = \frac{1}{2}\boldsymbol{\delta}_4$$

to conclude that $b = 0 = k$ by (13.29). We can also see from equations (13.3) and (13.5) that the time oriented Möbius frame field Y given in Example 13.1 is adapted to z and $b = k = 0$.

Example 13.15 (Cylinders). Consider the regular plane curve

$$\sigma : J \subset \mathbf{R} \to \mathbf{C} = \mathbf{R}^2, \quad \sigma(x) = \sigma^1(x)\boldsymbol{\delta}_1 + \sigma^2(x)\boldsymbol{\delta}_2,$$

where $\mathbf{R}^2 \subset \mathbf{R}^{4,1}$ is the $\boldsymbol{\delta}_1\boldsymbol{\delta}_2$-plane, J is a connected interval containing 0, and x is arclength parameter. Then

$$\dot{\sigma} = \mathbf{T}, \quad \ddot{\sigma} = \dot{\mathbf{T}} = \kappa\mathbf{N},$$

where $\mathbf{T}$ and $\mathbf{N}$ are the oriented unit tangent and principal normal vectors and $\kappa : J \to \mathbf{R}$ is the curvature. Let $f : J \times \mathbf{R} \to \mathscr{M}$ be given by the lift

$$F : J \times \mathbf{R} \to \mathscr{L}^+, \quad F(x,y) = \boldsymbol{\delta}_0 + \sigma(x) - y\boldsymbol{\delta}_3 + \frac{|\sigma|^2 + y^2}{2}\boldsymbol{\delta}_4,$$

where $z = x + iy$ is the complex coordinate on $J \times \mathbf{R}$. Then

$$F_z = \frac{1}{2}(F_x - iF_y) = \frac{1}{2}(\mathbf{T} + i\boldsymbol{\delta}_3 + (\sigma \cdot \mathbf{T} - iy)\boldsymbol{\delta}_4),$$

from which we calculate $\langle F_z, F_z\rangle = 0$ and $\langle F_z, F_{\bar{z}}\rangle = 1/2$, so f is a conformal immersion. From the calculation

$$F_{zz} = \frac{\kappa}{4}(\mathbf{N} + (\sigma \cdot \mathbf{N})\boldsymbol{\delta}_4), \quad F_{z\bar{z}} = \frac{\kappa}{4}(\mathbf{N} + (\sigma \cdot \mathbf{N})\boldsymbol{\delta}_4) + \frac{1}{2}\boldsymbol{\delta}_4,$$

we conclude from Theorem 13.13 that

$$b = 4\langle F_{zz}, F_{z\bar{z}}\rangle = \kappa^2/4,$$

$$k\mathbf{Y}_3 = 2F_{zz} + \frac{1}{4}F = \frac{1}{4}(\boldsymbol{\delta}_0 + 2\kappa\mathbf{N} + \sigma - y\boldsymbol{\delta}_3 + (2\kappa\sigma \cdot \mathbf{N} + \frac{|\sigma|^2 + y^2}{2})\boldsymbol{\delta}_4).$$

Calculate $\langle k\mathbf{Y}_3, k\mathbf{Y}_3\rangle = \kappa^2/4$ to conclude that $k = \pm\kappa/2$. We use (13.30) to establish that $k(x,y) = \kappa(x)/2$ at every point $(x,y) \in J \times \mathbf{R}$.

Example 13.16 (Cones). Recall the cones in Euclidean space described in Example 4.28, parametrized by the smooth immersion

$$\mathbf{x} : J \times \mathbf{R} \to \mathbf{R}^3, \quad \mathbf{x}(x,y) = e^{-y}\sigma(x),$$

where $\boldsymbol{\sigma} : J \to \mathbf{S}^2$ is the profile curve with arclength parameter x in the interval $J \subset \mathbf{R}$. Then $z = x + iy$ is a complex coordinate for the induced complex structure. Using the conformal embedding (12.59), we consider the conformal immersion

$$f = f_0 \circ \mathbf{x} : J \times \mathbf{R} \to \mathscr{M}, \quad f(x,y) = [\boldsymbol{\delta}_0 + \mathbf{x} + \frac{|\mathbf{x}|^2}{2} \boldsymbol{\delta}_4],$$

expressed in the Möbius frame (12.15), with lift

$$F : J \times \mathbf{R} \to \mathscr{L}^+, \quad F(x,y) = e^y \boldsymbol{\delta}_0 + \boldsymbol{\sigma}(x) + \frac{e^{-y}}{2} \boldsymbol{\delta}_4,$$

which satisfies the conditions of Theorem 13.13, as can be verified from the calculation

$$F_z = \frac{1}{2}(\dot{\boldsymbol{\sigma}} - ie^y \boldsymbol{\delta}_0 + i\frac{e^{-y}}{2} \boldsymbol{\delta}_4).$$

Using the Frenet-Serret equation (4.46) $\ddot{\boldsymbol{\sigma}} = -\boldsymbol{\sigma} + \kappa \boldsymbol{\sigma} \times \dot{\boldsymbol{\sigma}}$, we find

$$F_{zz} = \frac{1}{4}(-\boldsymbol{\sigma} + \kappa \boldsymbol{\sigma} \times \dot{\boldsymbol{\sigma}} - e^y \boldsymbol{\delta}_0 - \frac{e^{-y}}{2} \boldsymbol{\delta}_4)$$
$$F_{z\bar{z}} = \frac{1}{4}(-\boldsymbol{\sigma} + \kappa \boldsymbol{\sigma} \times \dot{\boldsymbol{\sigma}} + e^y \boldsymbol{\delta}_0 + \frac{e^{-y}}{2} \boldsymbol{\delta}_4).$$

By (13.29) we find

$$\begin{aligned} b = 4\langle F_{zz}, F_{z\bar{z}} \rangle &= \frac{2+\kappa^2}{4} \\ k\mathbf{Y}_3 = 2F_{zz} + bF &= \frac{\kappa^2 e^y}{4} \boldsymbol{\delta}_0 + \frac{\kappa^2}{4} \boldsymbol{\sigma} + \frac{\kappa}{2} \boldsymbol{\sigma} \times \dot{\boldsymbol{\sigma}} + \frac{e^{-y}\kappa^2}{8} \boldsymbol{\delta}_4, \end{aligned} \tag{13.31}$$

so

$$k^2 = \langle k\mathbf{Y}_3, k\mathbf{Y}_3 \rangle = \frac{\kappa^2}{4}. \tag{13.32}$$

13.4 Curvature spheres

Proposition 13.17. *If (U,z) is a complex coordinate chart on the Riemann surface M, then the curvature spheres at each point of U for the conformal immersion $f : M \to \mathscr{M}$ are*

$$\mathbf{Y}_3 \pm |k| \mathbf{Y}_0 : U \to \mathbf{S}^{3,1},$$

where $Y : U \to \mathbf{M\ddot{o}b}_+$ is the Möbius frame field adapted to z and $k : U \to \mathbf{C}$ is the first order Möbius invariant of f relative to z.

Remark 13.18. If $F : U \to \mathscr{L}^+$ is a lift of f satisfying $\langle F_z, F_z\rangle = 0$ and $\langle F_z, F_{\bar z}\rangle = 1/2$, then Theorem 13.13 tells us that $F = \mathbf{Y}_0$. It also tells us how to compute $\mathbf{Y}_3$ and $|k|$ in terms of F.

Proof. For any smooth function $r : U \to \mathbf{R}$, the smooth map

$$S = \mathbf{Y}_3 + r\mathbf{Y}_0 : U \to \mathbf{S}^{3,1} \subset \mathbf{R}^{4,1}$$

is an oriented tangent sphere map, since $d\mathbf{Y}_0 = \omega_0^1\mathbf{Y}_1 + \omega_0^2\mathbf{Y}_2$, so $\langle \mathbf{Y}_0, S\rangle = 0 = \langle d\mathbf{Y}_0, S\rangle$ on U, as required by Definition 12.42. Then

$$dS = (\omega_3^0 + dr)\mathbf{Y}_0 + \sum_1^2(\omega_3^i + r\omega_0^i)\mathbf{Y}_i \equiv \sum_1^2(\omega_3^i + r\omega_0^i)\mathbf{Y}_i \mod \mathbf{Y}_0$$

on U, by (13.15), so $dS \mod \mathbf{Y}_0$ has rank less than two at a point $p \in U$ if and only if

$$0 = (\omega_3^1 + r\omega_0^1)\wedge(\omega_3^2 + r\omega_0^2) = \omega_3^1\wedge\omega_3^2 + r(\omega_3^1\wedge\omega_0^2 + \omega_0^1\wedge\omega_3^2) + r^2\omega_0^1\wedge\omega_0^2.$$

By (13.15), $\omega_3^1 - i\omega_3^2 = -k(\omega_0^1 + i\omega_0^2)$ on U, so $\omega_3^1\wedge\omega_3^2 = -|k|^2\omega_0^1\wedge\omega_0^2$ and

$$0 = (\omega_3^1 - i\omega_3^2)\wedge(\omega_0^1 + i\omega_0^2) = \omega_3^1\wedge\omega_0^1 + \omega_3^2\wedge\omega_0^2 + i(\omega_3^1\wedge\omega_0^2 - \omega_3^2\wedge\omega_0^1).$$

As the real and imaginary parts must be zero, we get

$$(\omega_3^1 + r\omega_0^1)\wedge(\omega_3^2 + r\omega_0^2) = (-|k|^2 + r^2)\omega_0^1\wedge\omega_0^2,$$

which shows that $dS \mod \mathbf{Y}_0$ has rank less than two at a point in U if and only if $r^2 = |k|^2$ at this point. The result follows from Definition 12.47. □

Exercise 60 (Independence from z). The curvature spheres at a point $p \in M$ for a conformal immersion $f : M \to \mathscr{M}$ are independent of the choice of complex coordinate about the point. Verify that, if U, z and U, w are complex coordinate charts about p, if $Y, \hat{Y} : U \to$ **Möb** are the Möbius frame fields adapted to z and w, respectively, and if k and $\hat{k}$ are the respective first order invariants relative to z and w, then

$$\mathbf{Y}_3 \pm |k|\mathbf{Y}_0 = \hat{\mathbf{Y}}_3 \pm |\hat{k}|\hat{\mathbf{Y}}_0.$$

Example 13.19 (Circular Tori of Example 12.74 revisited). Fix α to satisfy $0 < \alpha < \pi/2$ and let $r = \cos\alpha$, $s = \sin\alpha$. If $C(R)$ denotes the circle in $\mathbf{R}^2$ of radius $R > 0$, center at the origin, then the torus $C(r)\times C(s) \subset \mathbf{S}^3$ is parameterized by

$$\mathbf{x}(x,y) = {}^t(r\cos\frac{x}{r}, r\sin\frac{x}{r}, s\cos\frac{y}{s}, s\sin\frac{y}{s}),$$

where $\mathbf{x}(x,y)$ is defined on $\mathbf{R}^2$ and is periodic of period $2\pi r$ in x and of period $2\pi s$ in y. Then $d\mathbf{x}\cdot d\mathbf{x} = dx^2+dy^2$ shows that $z = x+iy$ is a complex coordinate in $\mathbf{R}^2$ for the complex structure induced by $\mathbf{x}$. Composing $\mathbf{x}$ with the diffeomorphism $f_+ : \mathbf{S}^3 \to \mathscr{M}$ of (12.23) defines an immersion $f : \mathbf{R}^2 \to \mathscr{M}$ for which a lift $F : \mathbf{R}^2 \to \mathscr{L}^+$ in the Möbius frame (12.15) is

$$F(x,y) = {}^t(1+r\cos\frac{x}{r}, r\sin\frac{x}{r}, s\cos\frac{y}{s}, s\sin\frac{y}{s}, \frac{1-r\cos\frac{x}{r}}{2}). \tag{13.33}$$

Then

$$F_z = \frac{1}{2}{}^t(-\sin\frac{x}{r}, \cos\frac{x}{r}, i\sin\frac{y}{s}, -i\cos\frac{y}{s}, \frac{1}{2}\sin\frac{x}{r}),$$

from which we see that $\langle F_z, F_z\rangle = 0$ and $\langle F_z, F_{\bar{z}}\rangle = 1/2$, so that F is the lift of f required by Theorem 13.13. From

$$F_{zz} = \frac{1}{4}{}^t(-\frac{1}{r}\cos\frac{x}{r}, -\frac{1}{r}\sin\frac{x}{r}, \frac{1}{s}\cos\frac{y}{s}, \frac{1}{s}\sin\frac{y}{s}, \frac{1}{2r}\cos\frac{x}{r}),$$
$$F_{z\bar{z}} = \frac{1}{4}{}^t(-\frac{1}{r}\cos\frac{x}{r}, -\frac{1}{r}\sin\frac{x}{r}, -\frac{1}{s}\cos\frac{y}{s}, -\frac{1}{s}\sin\frac{y}{s}, \frac{1}{2r}\cos\frac{x}{r}),$$

we use (13.29) to calculate

$$b = 4\langle F_{zz}, F_{\bar{z}\bar{z}}\rangle = \frac{1}{4}\frac{s^2-r^2}{r^2s^2} = -\frac{\cos 2\alpha}{\sin^2 2\alpha},$$

which is constant on $\mathbf{R}^2$, and $k\mathbf{Y}_3 = 2F_{zz} + bF =$

$$\frac{1}{4r^2s^2}{}^t(-r\cos\frac{x}{r} + s^2 - r^2, -r\sin\frac{x}{r}, s\cos\frac{y}{s}, s\sin\frac{y}{s}, \frac{r\cos\frac{x}{r} + s^2 - r^2}{2}),$$

so

$$k^2 = \langle k\mathbf{Y}_3, k\mathbf{Y}_3\rangle = \frac{1}{4r^2s^2}.$$

Using (13.30) at $z = 0$, we get $k = -\frac{1}{2rs} = -\csc 2\alpha$, also constant on $\mathbf{R}^2$. As α goes through the range $0 < \alpha < \pi/2$ the constant Möbius invariants go through the ranges $-\infty < k \leq -1$ and $-\infty < b < \infty$. For $\alpha = \pi/4$, the Clifford torus, $k = -1$ and $b = 0$ relative to z.

Relative to the complex coordinate z, the curvature spheres of f are

$$S_\pm = \mathbf{Y}_3 \pm \frac{1}{2rs}F,$$

by Proposition 13.17 and Remark 13.18. These are

$$S_+ = \frac{r+\cos\frac{x}{r}}{s}\boldsymbol{\varepsilon}_0 + \frac{\sin\frac{x}{r}}{s}\boldsymbol{\varepsilon}_1 + \frac{r-\cos\frac{x}{r}}{2s}\boldsymbol{\varepsilon}_4$$

and

$$S_- = -\frac{s}{r}\boldsymbol{\varepsilon}_0 - \frac{\cos\frac{y}{s}}{r}\boldsymbol{\varepsilon}_2 - \frac{\sin\frac{y}{s}}{r}\boldsymbol{\varepsilon}_3 - \frac{s}{2r}\boldsymbol{\varepsilon}_4.$$

Definition 13.20 (Lines of curvature). A *principal vector* for a conformal immersion $f : M \to \mathscr{M}$ is a tangent vector on which the symmetric bilinear form $\Im(k\,dzdz)$ is zero, where k is the first order Möbius invariant relative to a complex chart U, z in M. A smooth curve $\gamma : J \to M$ is a *line of curvature* if $\gamma^*\Im(k\,dzdz) = 0$ on J.

The transformation formula (13.20) shows that for complex coordinate charts (U,z) and (U,w) on M, we have

$$\hat{k}dwdw = |w'|kdzdz,$$

which shows that the zeros of the symmetric bilinear form $\Im(k\,dzdz)$ are independent of the choice of complex coordinate chart (U,z).

For the cylinder $f : J \times \mathbf{R} \to \mathscr{M}$ in Example 13.15,

$$kdzdz == \frac{-\kappa(x)}{2}(dx^2 - dy^2 + 2idxdy),$$

whose imaginary part is $-\kappa(x)dxdy$. The (x,y) coordinate curves are the lines of curvature on $M = J \times \mathbf{R}$.

Lemma 13.21 (Principal vectors). *If $f : M \to \mathscr{M}$ is a conformal immersion with Möbius frame field $Y : U \subset M \to$ **Möb**$_+$ satisfying $\omega_0^3 = 0$ on U, then the principal vectors on U are the zeros of the symmetric bilinear form*

$$\Im((\omega_0^1 + i\omega_0^2)(\omega_1^3 - i\omega_2^3)) = \omega_0^2\omega_1^3 - \omega_0^1\omega_2^3.$$

Proof. Let z be a complex coordinate on U. By Problem 13.43, the entries of $Y^{-1}dY$ satisfy $\omega_0^1 + i\omega_0^2 = e^{u+iv}dz$ and $\omega_1^3 - i\omega_2^3 = pdz + qd\bar{z}$, for some smooth functions $u, v : U \to \mathbf{R}$ and $p, q : U \to \mathbf{C}$. The first order invariant relative to z is $k = e^{iv}p$, and $e^{iv}q$ is real valued. Thus,

$$\begin{aligned}\Im((\omega_0^1 + i\omega_0^2)(\omega_1^3 - i\omega_2^3)) &= \Im(e^u e^{iv}(pdz + qd\bar{z})) = e^u\Im(kdzdz + e^{iv}qdzd\bar{z})\\ &= e^u\Im(kdzdz),\end{aligned}$$

since $e^{iv}qdzd\bar{z}$ is real valued. □

13.5 Complex structure and space forms

Recall the notation introduced in Section 12.5 for the simply connected space forms $\mathbf{S}_\epsilon$, their groups of isometries G_ϵ, the conformal embeddings $f_\epsilon : S_\epsilon \to \mathcal{M}$, and the group monomorphisms $F_\epsilon : G_\epsilon \to$ **Möb**.

Theorem 13.22 (Möbius invariants from space form invariants). *If $\mathbf{x} : M \to \mathbf{S}_\epsilon$ is an immersion of an oriented surface M, with mean curvature H and Gaussian curvature K, and if (U, z) is a complex coordinate chart in M relative to which the conformal factor is e^u and the Hopf invariant is h, then the composition*

$$f_\epsilon \circ \mathbf{x} : M \to \mathcal{M}$$

is a conformal immersion whose first and second order Möbius invariants relative to z are

$$k = he^u \quad \text{and} \quad b = 2u_{zz} - 2u_z^2 + Hhe^{2u}. \tag{13.34}$$

The proof will be given for each case of $\epsilon \in \{0, +, -\}$ in the next three subsections.

13.5.1 Surfaces in Euclidean space

Proof. Let

$$\mathbf{x} : M \to \mathbf{R}^3$$

be an immersion of an oriented surface M. Let (U, z) be a complex coordinate chart for the complex structure induced on M by the metric $d\mathbf{x} \cdot d\mathbf{x}$. Let $(\mathbf{x}, e) : U \to \mathbf{E}_+(3)$ be the frame field along $\mathbf{x}$ adapted to z and let

$$(\theta, \eta) = (\mathbf{x}, e)^{-1} d(\mathbf{x}, e) = (e^{-1} d\mathbf{x}, e^{-1} de)$$

denote the pull back to M by $(\mathbf{x}, e)$ of the Maurer–Cartan form of $\mathbf{E}_+(3)$. Then $\theta^3 = 0$ and

$$\theta^1 + i\theta^2 = e^u dz, \tag{13.35}$$

for some smooth function $u : U \to \mathbf{R}$. Then

$$\tilde{\mathbf{Y}} = F_0 \circ (\mathbf{x}, e) : U \to \mathbf{Möb}$$

is an oriented Möbius frame field along the conformal immersion $f_0 \circ \mathbf{x} : M \to \mathscr{M}$, where $f_0 : \mathbf{R}^3 \to \mathscr{M}$ is the conformal embedding (12.59). By (12.61), $\tilde{Y}$ pulls back the Maurer–Cartan form of $\mathbf{M\ddot{o}b}_+$ to

$$\beta = \tilde{Y}^{-1} d\tilde{Y} = dF_0(\theta, \eta) = \begin{pmatrix} 0 & 0 & 0 \\ \theta & \eta & 0 \\ 0 & {}^t\theta & 0 \end{pmatrix}, \tag{13.36}$$

so

$$\beta^3_0 = \theta^3 = 0, \quad \beta^1_0 + i\beta^2_0 = \theta^1 + i\theta^2 = e^u dz.$$

If $K(e^u, I, 0)$ is defined as in (12.26), then

$$\tilde{Y} K(e^u, I, 0) = (e^{-u}\tilde{\mathbf{Y}}_0, \tilde{\mathbf{Y}}_i, e^u \tilde{\mathbf{Y}}_4)$$

is a first order Möbius frame field along f relative to z (see Definition 13.2), as we verify by computing

$$\omega = (\tilde{Y}K)^{-1} d(\tilde{Y}K) = \begin{pmatrix} -du & 0 & 0 \\ e^{-u}\theta & \eta & 0 \\ 0 & e^{-u}\,{}^t\theta & du \end{pmatrix},$$

from which we see that (13.9) holds: $\omega^3_0 = 0$ and $\omega^1_0 + i\omega^2_0 = dz$. In addition, we see that $\omega^0_0 = -du$ and $\omega^1_2 = \eta^1_2$. From (13.35) (see (7.28)) we have

$$\eta^1_2 = iu_z dz - iu_{\bar{z}} d\bar{z},$$

and

$$\omega^0_0 + i\omega^1_2 = -du - u_z dz + u_{\bar{z}} d\bar{z} = -2u_z dz, \tag{13.37}$$

so in (13.10) $l = -2u_z : U \to \mathbf{C}$, and

$$\omega^3_1 - i\omega^3_2 = \eta^3_1 - i\eta^3_2 = he^u dz + He^u d\bar{z}, \tag{13.38}$$

where the last equation comes from (7.23), H is the mean curvature of $\mathbf{x}$, and h is the Hopf invariant of $\mathbf{x}$ relative to z (see Definition 7.24). Comparing this with (13.10), we see that the first order Möbius invariant of f relative to z is

$$k = he^u.$$

Comparing (13.37) and (13.38) with (13.10), we see from (13.13) that if we let

$$y^1 - iy^2 = -2u_z, \quad y^3 = He^u, \quad \mathbf{y} = {}^t(y^1, y^2, y^3), \tag{13.39}$$

then the Möbius frame field

$$Y = \tilde{Y}K(e^u, I, 0)K(1, I, \mathbf{y}) = \tilde{Y}K(e^u, I, e^{-u}\mathbf{y}) : U \to \mathbf{M\ddot{o}b}_+$$

is adapted to z. We calculate $\alpha = Y^{-1}dY$ to verify that $\alpha_0^0 + i\alpha_2^1 = 0$ and that $\alpha_1^3 - i\alpha_2^3 = he^u dz$ as expected, and also that

$$\alpha_1^0 = dy^1 - y^1 du + \frac{1}{2}|\mathbf{y}|^2 \omega_0^1 + \eta_2^1 y^2 + \eta_3^1 y^3 - y^1(y^1\omega_0^1 + y^2\omega_0^2),$$
$$\alpha_2^0 = dy^2 - y^2 du + \frac{1}{2}|\mathbf{y}|^2 \omega_0^2 + \eta_1^2 y^1 + \eta_3^2 y^3 - y^2(y^1\omega_0^1 + y^2\omega_0^2).$$

From this we find

$$\alpha_1^0 - i\alpha_2^0 = (-2u_{zz} + 2u_z^2 - Hhe^{2u})dz - \frac{1}{2}(4u_{z\bar{z}} + H^2 e^{2u})d\bar{z}.$$

Comparing this with (13.16), we see that the second order Möbius invariant b of f is the coefficient of $-dz$, thus confirming (13.34) and completing the proof of Theorem 13.22 when $\epsilon = 0$. □

Remark 13.23 (Hopf quadratic differential and lines of curvature). The Hopf quadratic differential of $\mathbf{x} : M \to \mathbf{R}^3$ relative to the complex coordinate z on $U \subset M$ is $II^{2,0} = \frac{1}{2}he^{2u}\,dzdz$, which by (13.34) is expressed in terms of the first order Möbius invariant k of $f_0 \circ \mathbf{x} : M \to \mathscr{M}$ by

$$II^{2,0} = \frac{1}{2}ke^u\,dzdz. \tag{13.40}$$

By Lemma 7.25, the principal vectors of $\mathbf{x} : M \to \mathbf{R}^3$ are the solutions of $\Im(II^{2,0}) = 0$. Compare this with Definition 13.20, which defines the principal vectors of $f : M \to \mathscr{M}$ to be the solutions of $\Im(k\,dzdz) = 0$, for any complex chart U, z. It follows then from (13.40) that the principal vectors of $\mathbf{x} : M \to \mathbf{R}^3$ coincide with those of $f_0 \circ \mathbf{x} : M \to \mathscr{M}$. It follows that principal vectors and lines of curvature are preserved by conformal transformations of S_0.

Example 13.24 (Cones revisited). Let us compare the values of k and b relative to z calculated for the cones in Example 13.16 to the values given by Theorem 13.22. In Example 4.28 we began with the cones $\mathbf{x}(x,y) = e^{-y}\boldsymbol{\sigma}(x)$, where x is arclength parameter and $\kappa(x)$ is the curvature of the curve $\boldsymbol{\sigma} : J \to \mathbf{S}^2$. The frame field $(\mathbf{x}, e) : M \times \mathbf{R} \to \mathbf{E}_+(3)$ given by

$$e = (\dot{\boldsymbol{\sigma}}, -\boldsymbol{\sigma}, \boldsymbol{\sigma} \times \dot{\boldsymbol{\sigma}})$$

is adapted to the complex coordinate $z = x + iy$. The Hopf invariant h and mean curvature H are given by $h = \kappa(x)e^y/2 = H$. The conformal factor is $e^u = e^{-y}$, so $u = -y$ and $u_z = i/2$. For the conformal immersion $f = f_0 \circ \mathbf{x} : J \times \mathbf{R} \to \mathscr{M}$, Theorem 13.22 gives the invariants relative to z to be

$$k = he^u = \frac{\kappa}{2}, \quad b = 2u_{zz} - 2u_z^2 + Hhe^{2u} = \frac{2+\kappa^2}{4},$$

which agree with the calculations in Example 13.16. Moreover,

$$\mathbf{Y}_3 = \frac{\kappa e^y}{2}\boldsymbol{\delta}_0 + \frac{\kappa}{2}\boldsymbol{\sigma} + \boldsymbol{\sigma} \times \dot{\boldsymbol{\sigma}} + \frac{\kappa e^{-y}}{4}\boldsymbol{\delta}_4,$$

by (13.31), where we used the lift $F(x,y) = e^y\boldsymbol{\delta}_0 + \boldsymbol{\sigma}(x) + \frac{e^{-y}}{2}\boldsymbol{\delta}_4$ of f required by Theorem 13.13. The oriented curvature spheres of $f : \mathbf{R}^2 \to \mathscr{M}$ (see Definition 12.47) are given by Proposition 13.17 and Remark 13.18 to be $S_\pm = \mathbf{Y}_3 \pm \frac{\kappa}{2}F$, which are (in the Möbius frame (12.15))

$$S_+ = \kappa e^y\boldsymbol{\delta}_0 + \kappa\boldsymbol{\sigma} + \boldsymbol{\sigma} \times \dot{\boldsymbol{\sigma}} + \frac{\kappa e^{-y}}{2}\boldsymbol{\delta}_4, \quad S_- = \boldsymbol{\sigma} \times \dot{\boldsymbol{\sigma}} \in \mathbf{S}^{3,1}.$$

Using Problem 13.48, one easily checks that these are the images under f_0 of the curvature spheres (4.48) found in Example 4.28.

13.5.2 Surfaces in $\mathbf{S}^3$

Proof. Let $\mathbf{x} : M \to \mathbf{S}^3$ be an immersion of an oriented surface M. Let

$$e = (\mathbf{e}_0, \mathbf{e}_1, \mathbf{e}_2, \mathbf{e}_3) : U \subset M \to \mathbf{SO}_+(4)$$

be an oriented frame field along $\mathbf{x}$. Denote the pull-back by e of the Maurer–Cartan form of $\mathbf{SO}(4)$ by

$$\eta = e^{-1}de = \begin{pmatrix} 0 & \eta^0_j \\ \eta^i_0 & \eta^i_j \end{pmatrix},$$

where $\eta^0_j = -\eta^j_0$ and $\eta^i_j = -\eta^j_i$, as in (5.5) and $i,j = 1,2,3$. Then

$$d\mathbf{e}_0 = \sum_1^3 \eta^i_0\mathbf{e}_i, \quad d\mathbf{e}_i = \eta^0_i\mathbf{e}_0 + \sum_1^3 \eta^j_i\mathbf{e}_j, \tag{13.41}$$

Let z be a local complex coordinate in $U \subset M$ for the complex structure induced by the metric $I = d\mathbf{x} \cdot d\mathbf{x}$ on M. Suppose that the frame field $e : U \to \mathbf{SO}_+(4)$ is adapted to z, (see Definition 7.22). Then

$$\eta^3_0 = 0, \quad \eta^1_0 + i\eta^2_0 = e^u dz,$$

for some smooth function $u : U \to \mathbf{R}$, so that e^u is the conformal factor of $\mathbf{x}$ relative to z. A time oriented Möbius frame field $\tilde{Y}$ along

$$f = f_+ \circ \mathbf{x} : M \to \mathscr{M}, \quad f = [(1+x^0)\boldsymbol{\delta}_0 + x^i\boldsymbol{\delta}_i + \frac{(1-x^0)}{2}\boldsymbol{\delta}_4]$$

is given by

$$\tilde{Y} = F_+ \circ e : U \to \mathbf{M\ddot{o}b}_+.$$

Then, by (12.58)

$$\beta = \tilde{Y}^{-1}d\tilde{Y} = dF_+(e^{-1}de) = \begin{pmatrix} 0 & \eta^0_j & 0 \\ \frac{\eta^i_0}{2} & \eta^i_j & -\eta^i_0 \\ 0 & -\frac{\eta^0_j}{2} & 0 \end{pmatrix}.$$

Taking into account equations (13.41), and using $K(\frac{e^u}{2}, I, 0)$ defined in (12.26), we find that the time oriented Möbius frame field along f,

$$\hat{Y} = \tilde{Y}K(\frac{e^u}{2}, I, 0) = (2e^{-u}\tilde{\mathbf{Y}}_0, \tilde{\mathbf{Y}}_i, \frac{e^u}{2}\tilde{\mathbf{Y}}_4),$$

is of first order relative to z, by the computation

$$\omega = (\tilde{Y}K)^{-1}d(\tilde{Y}K) = \begin{pmatrix} -du & \frac{e^u}{2}\eta^0_j & 0 \\ e^{-u}\eta^i_0 & \eta^i_j & -\frac{e^u}{2}\eta^i_0 \\ 0 & -e^{-u}\eta^0_j & du \end{pmatrix}, \tag{13.42}$$

from which we see that $\omega^3_0 = 0$ and $\omega^1_0 + i\omega^2_0 = dz$, (see equations (13.9) of Definition 13.2). In addition, we see that $\omega^0_0 = -du$ and $\omega^1_2 = \eta^1_2$. Recalling from equation (7.28) that $\eta^1_2 = i(u_z dz - u_{\bar{z}} d\bar{z})$, we have

$$\omega^0_0 + i\omega^1_2 = -2u_z dz \tag{13.43}$$

and, by (7.23),

$$\omega^3_1 - i\omega^3_2 = \eta^3_1 - i\eta^3_2 = he^u\, dz + He^u\, d\bar{z}, \tag{13.44}$$

where H is the mean curvature of $\mathbf{x}$ and h is the Hopf invariant of $\mathbf{x}$ relative to z, defined in (7.24). It follows from (13.10) and (13.14) that the first order Möbius invariant of f relative to z is

$$k = he^u,$$

and the coefficients $l = -2u_z$ and $s = He^u$. Comparing (13.43) and (13.44) with (13.10), we see from (13.13) that if we let

$$y^1 - iy^2 = -2u_z, \quad y^3 = He^u, \quad \mathbf{y} = {}^t(y^1, y^2, y^3),$$

then the time oriented Möbius frame field

$$Y = \tilde{Y}K(\frac{e^u}{2}, I, 0)K(1, I, \mathbf{y}) = \tilde{Y}K(\frac{e^u}{2}, I, 2e^{-u}\mathbf{y})$$

is the Möbius frame field adapted to z. Using ω in (13.42), we calculate

$$\alpha = Y^{-1}dY = K(1,I,\mathbf{y})^{-1}\omega K(1,I,\mathbf{y}) + K(1,I,\mathbf{y})^{-1}dK(1,I,\mathbf{y})$$

$$= \begin{pmatrix} 1 & -{}^t\mathbf{y} & |\mathbf{y}|^2/2 \\ 0 & I & -\mathbf{y} \\ 0 & 0 & 1 \end{pmatrix} \begin{pmatrix} -du & e^u\eta^0_j/2 & 0 \\ e^{-u}\eta^i_0 & \eta^i_j & -e^u\eta^i_0/2 \\ 0 & -e^{-u}\eta^0_j & du \end{pmatrix} \begin{pmatrix} 1 & {}^t\mathbf{y} & |\mathbf{y}|^2/2 \\ 0 & I & \mathbf{y} \\ 0 & 0 & 1 \end{pmatrix}$$

$$+ \begin{pmatrix} 0 & d{}^t\mathbf{y} & 0 \\ 0 & 0 & d\mathbf{y} \\ 0 & 0 & 0 \end{pmatrix} =$$

$$\begin{pmatrix} -du - e^{-u}y^i\eta^i_0 & d{}^t\mathbf{y} - (du + e^{-u}y^i\eta^i_0){}^t\mathbf{y} - y^i\eta^i_j + \frac{1}{2}(e^u - e^{-u}|\mathbf{y}|^2)\eta^0_j & 0 \\ e^{-u}\eta^i_0 & \eta^i_j + e^{-u}(y^j\eta^i_0 + y^i\eta^0_j) & \star \\ 0 & \star & \star \end{pmatrix},$$

where the omitted entries are determined by the fact that α is $\mathfrak{m\ddot{o}b}$-valued. Since $\eta^3_0 = 0$, $\eta^1_0 + i\eta^2_0 = e^u dz$, and $\eta^1_2 = i(u_z dz - u_{\bar{z}} d\bar{z})$, we have

$$y^i\eta^i_0 = -e^u du, \quad y^2\eta^1_0 - y^1\eta^2_0 = -ie^u(u_z dz - u_{\bar{z}} d\bar{z}).$$

It is now easily verified that $\alpha^0_0 + i\alpha^1_2 = 0$ and $\alpha^3_1 - i\alpha^3_2 = he^u dz$ and thus Y is the Möbius frame field adapted to z, as expected. Finally, we calculate

$$\alpha^0_1 - i\alpha^0_2 = -(2u_{zz} - 2u_z^2 + Hhe^{2u})dz - \frac{1}{2}(4u_{z\bar{z}} + (1 + H^2)e^{2u})d\bar{z}.$$

Comparing this with (13.16), we see that the second order Möbius invariant b of f is the coefficient of $-dz$, thus confirming (13.34) and completing the proof of Theorem 13.22 when $\epsilon = +$. □

The Hopf quadratic differential of $\mathbf{x} : M \to \mathbf{S}^3$ in terms of the complex coordinate z is $II^{2,0} = \frac{1}{2}he^{2u}\, dzdz$. In terms of the first order Möbius invariant k it is

$$II^{2,0} = \frac{1}{2}ke^u\, dzdz$$

just as in the Euclidean case. Lines of curvature are the solutions of the differential equation $\Im(II^{2,0}) = 0$.

Example 13.25 (Circular Tori of Example 13.19 revisited). The conformal immersions $f : \mathbf{R}^2 \to \mathscr{M}$ of Example 13.19 are the compositions $f_+ \circ \mathbf{x}$, where $\mathbf{x} = \mathbf{x}^{(\alpha)} : \mathbf{R}^2 \to \mathbf{S}^3$, for fixed $0 < \alpha < \pi/2$, are the circular tori of Example 5.12. In Example 7.38 we found that relative to the induced complex coordinate $z = x + iy$ on $\mathbf{R}^2$, the conformal factor is $e^u = 1$ and the Hopf invariant and mean curvature of $\mathbf{x}$ are

$$h = -\frac{1}{2rs}, \quad H = \frac{r^2 - s^2}{2rs},$$

where $r = \cos\alpha$ and $s = \sin\alpha$. Using this data in Theorem 13.22, we find the conformal invariants of f relative to z are

$$k = he^u = -\frac{1}{2rs}, \quad b = 2u_{zz} - 2u_z^2 + Hhe^{2u} = \frac{s^2 - r^2}{4r^2s^2},$$

in agreement with the values found in Example 13.19.

13.5.3 *Surfaces in* $\mathbf{H}^3$

Proof. Let $\mathbf{x} : M \to \mathbf{H}^3$ be an immersion of an oriented surface M. Let

$$e = (\mathbf{e}_1, \ldots, \mathbf{e}_4) : U \subset M \to \mathbf{SO}_+(3,1)$$

be an oriented frame field along $\mathbf{x} = \mathbf{e}_4$. Denote the pull-back of the Maurer–Cartan form (6.10) of $\mathbf{SO}_+(3,1)$ by

$$\eta = e^{-1}de = \begin{pmatrix} \eta^i_j & \eta^i_4 \\ \eta^4_j & 0 \end{pmatrix} \in \mathfrak{o}(3,1),$$

where $\eta^4_j = \eta^j_4$ and $\eta^i_j = -\eta^j_i$, $i,j = 1,2,3$. Then

$$d\mathbf{x} = d\mathbf{e}_4 = \sum_1^3 \eta^i_4 \mathbf{e}_i, \quad d\mathbf{e}_i = \eta^4_i \mathbf{e}_4 + \sum_1^3 \eta^j_i \mathbf{e}_j.$$

Let z be a local complex coordinate in $U \subset M$ for the complex structure induced by the metric $I = \langle d\mathbf{x}, d\mathbf{x} \rangle$ on M. Suppose that $e : U \to \mathbf{SO}_+(1,3)$ is the frame field adapted to z, (see Definition 7.22). Then

$$\eta^3_4 = 0, \quad \eta^1_4 + i\eta^2_4 = e^u dz,$$

for some smooth function $u : U \to \mathbf{R}$, so that e^u is the conformal factor of $\mathbf{x}$ relative to z. Let

$$f = f_- \circ \mathbf{x} : M \to \mathscr{M}, \quad f = [(x^4+1)\boldsymbol{\delta}_0 + \sum_1^3 x^i \boldsymbol{\delta}_i + \frac{(x^4-1)}{2}\boldsymbol{\delta}_4]$$

be the conformal immersion obtained by composing the map f_- in (12.62) with $\mathbf{x}$. A time oriented Möbius frame field $\tilde{Y}$ along f is obtained by composing the map F_- of (12.65) with e,

$$\tilde{Y} = F_- \circ e : U \to \mathbf{M\ddot{o}b}_+.$$

By (12.66),

$$\beta = \tilde{Y}^{-1} d\tilde{Y} = dF_+(e^{-1}de) = \begin{pmatrix} 0 & \eta^4_j & 0 \\ \frac{\eta^i_4}{2} & \eta^i_j & \eta^i_4 \\ 0 & \frac{\eta^4_j}{2} & 0 \end{pmatrix}.$$

The rest of the proof of the $\epsilon = -$ case is almost identical to the spherical geometry case. Its completion is left to the following Exercise. □

Exercise 61. Using the proof for the $\epsilon = +$ case as a guide, complete the present proof.

The Hopf quadratic differential of $\mathbf{x} : M \to \mathbf{H}^3$ in terms of the complex coordinate z is $II^{2,0} = \frac{1}{2} h e^{2u}\, dzdz$. In terms of the first order Möbius invariant k it is

$$II^{2,0} = \frac{1}{2} k e^u\, dzdz$$

just as in the Euclidean and spherical cases. Lines of curvature are the solutions of the differential equation $\Im(II^{2,0}) = 0$.

13.6 Conformal area and Willmore functionals

Let $f : M \to \mathscr{M}$ be a conformal immersion of a Riemann surface M. Recall the conformal area element Ω of Definition 12.26. It is defined locally by Ω_Y, for any first order frame $Y : U \to \mathbf{M\ddot{o}b}$ along f. If (U, z) is any complex coordinate chart in M, then the Möbius frame field $Y : U \to \mathbf{M\ddot{o}b}$ adapted to z is first order with $h = k$ and $\omega^1_0 \wedge \omega^2_0 = \frac{i}{2} dz \wedge d\bar{z}$, so by Problem 12.64,

$$\Omega_Y = \frac{i}{2} |k|^2 dz \wedge d\bar{z}. \tag{13.45}$$

Corollary 13.26 (Conformal area element). *If $\mathbf{x} : M \to \mathbf{S}_\epsilon$ is an immersion of an oriented surface, and if (U,z) is any complex coordinate chart in M for the complex structure induced on M by $\mathbf{x}$, then the conformal area element of $f_\epsilon \circ \mathbf{x} : M \to \mathscr{M}$ is*

$$\Omega = \frac{i}{2}|k|^2 dz \wedge d\bar{z} = (H^2 - K + \epsilon 1)dA, \tag{13.46}$$

where $dA = \frac{i}{2}e^{2u}dz \wedge d\bar{z}$ is the area element, H is the mean curvature, and K is the Gaussian curvature of $\mathbf{x}$.

Proof. Since f_ϵ is conformal, the complex structure induced on M by $\mathbf{x}$ is the same as the complex structure induced on M by $f_\epsilon \circ \mathbf{x}$. If U,z is a complex chart in M, and $Y : U \to \mathbf{M\ddot{o}b}_+$ is the frame field along $f_\epsilon \circ \mathbf{x}$ adapted to z, then the conformal area element Ω is given on U by (13.45). The first order Möbius invariant k of $f_\epsilon \circ \mathbf{x}$ relative to z satisfies $k = he^u$, by (13.34) of Theorem 13.22, where h is the Hopf invariant and e^u is the conformal factor of $\mathbf{x}$ relative to z. This, with the Gauss equation (7.31),

$$|h|^2 = \epsilon 1 + H^2 - K,$$

implies (13.46). □

Remark 13.27 (Willmore functional in each space form). The integrand of the Willmore functional (4.74) of an immersion $\mathbf{x} : M \to \mathbf{R}^3$ is the right side of (13.46) for the case $\epsilon = 0$. This equation indicates that the Willmore functional for immersions $\mathbf{x} : M \to \mathbf{S}_\epsilon$ should be

$$\int_M (H^2 - K + \epsilon 1)dA,$$

where H is the mean curvature, K is the Gaussian curvature, and dA is the area element of $\mathbf{x}$. These functionals are then invariant under conformal transformations from $\mathbf{S}_\epsilon$ to $\mathbf{S}_\delta$, for any $\epsilon, \delta \in \{0, +, -\}$. For example, if H, K, and dA are the mean curvature, Gaussian curvature, and area element on M, for an immersion $\mathbf{x} : M \to \mathbf{S}^3 \setminus \{-\boldsymbol{\epsilon}_0\}$, and if $\tilde{H}$, $\tilde{K}$, and $d\tilde{A}$ denote the same quantities for the immersion $\mathscr{S} \circ \mathbf{x} : M \to \mathbf{R}^3$, where $\mathscr{S} : \mathbf{S}^3 \setminus \{-\boldsymbol{\epsilon}_0\} \to \mathbf{R}^3$ is stereographic projection from $-\boldsymbol{\epsilon}_0$ (see Definition 5.22), then

$$(H^2 - K + 1)dA = (\tilde{H}^2 - \tilde{K})d\tilde{A},$$

since both sides of this equation equal $\frac{i}{2}|k|^2 dz \wedge d\bar{z}$ for any complex coordinate chart (U,z) for the immersion $f_+ \circ \mathbf{x} = f_0 \circ \mathscr{S} \circ \mathbf{x} : M \to \mathscr{M}$.

Definition 13.28. An *admissible variation* of an immersion $f : M^2 \to \mathscr{M}$ is any smooth map $\mathscr{F} : M \times (-\epsilon, \epsilon) \to \mathscr{M}$, for some $\epsilon > 0$, with compact support, such that for each $t \in (-\epsilon, \epsilon)$, the map

$$f_t : M \to \mathscr{M}, \quad f_t(m) = \mathscr{F}(m,t)$$

is an immersion. The *support* of $\mathscr{F}$ is the closure in M of the set of points $m \in M$ where $f_t(m) \neq f(m)$, for some t. The *variation vector field* of an admissible variation is the vector field along f given by

$$\frac{d}{dt}\bigg|_0 f_t : M \to T\mathscr{M}.$$

Definition 13.29. A conformal immersion $f : M \to \mathscr{M}$ of a connected Riemann surface M is *Willmore* if f is a *critical point* of the *conformal area functional*; that is, the *first variation* of the conformal area

$$\frac{d}{dt}\bigg|_0 \int_S \Omega_t = \int_S \frac{d}{dt}\bigg|_0 \Omega_t$$

is zero for any admissible variation of f with compact support $S \subset M$.

Proposition 13.30. *A conformal immersion $f : M \to \mathscr{M}$ of a connected Riemann surface M is Willmore if and only if its Willmore function W of Definition 12.29 (see also Theorem 13.9) is identically zero for any γ-frame along f.*

Proof. See Bryant [20]. □

Example 13.31. Recall the circular tori discussed in Example 13.19. Since

$$\hat{k}_{\bar{w}\bar{w}} + \frac{\hat{k}}{2}\hat{\bar{b}} = \frac{1}{2}(r^2 - s^2) = \cos 2\alpha$$

is zero if and only if $\alpha = \pi/4$, the Clifford torus is the only circular torus that is conformally minimal.

Proposition 13.32 ([20], Theorem C). *If $f : M \to \mathscr{M}$ is a Willmore immersion of a connected Riemann surface M, then either f is totally umbilic or the set $\mathscr{U}_f$ of umbilic points of f is a closed subset of M without interior points.*

Proof. Let k and b be the invariants of f relative to a complex coordinate chart U, z. Then the Willmore condition (13.19) can be written as

$$\begin{pmatrix} k \\ k_{\bar{z}} \end{pmatrix}_{\bar{z}} = \begin{pmatrix} 0 & 1 \\ -\frac{1}{2}\bar{b} & 0 \end{pmatrix} \begin{pmatrix} k \\ k_{\bar{z}} \end{pmatrix}.$$

This implies that the map

$$\mathfrak{k} = \begin{pmatrix} k \\ k_{\bar{z}} \end{pmatrix} : U \to \mathbf{C}^2$$

is of *analytic type* (Chern [43, p.32]), which means that either $\mathfrak{k}$ is identically zero on U, in which case we say $n(m) = \infty$ for every $m \in U$, or for each point $m \in U$, there exists a unique whole number $n(m)$ such that

$$\mathfrak{k} = (z - z(m))^{n(m)}\mathbf{v},$$

where $\mathbf{v} : U \to \mathbf{C}^2$ is a smooth map with $\mathbf{v}(m) \neq \mathbf{0}$. In this latter case, the zeros of $\mathfrak{k}$ must be isolated in U, and each zero $m \in U$ has a well-defined multiplicity $n(m)$. If $\mathfrak{k}$ is identically zero on any open subset of U, then it must be identically zero on U, and this will be true for this function relative to any complex coordinate chart, since M is connected. Consequently, f is totally umbilic in this case. If f is not totally umbilic, then for any chart (U,z), the map $\mathfrak{k}$ must have isolated zeros. If there were an open set V of umbilic points in U, then k would be identically zero on V, which would imply that $k_{\bar{z}}$ is also identically zero on V, and this is impossible. □

13.7 Conformal Gauss map

By Corollary 13.12 the vector $\mathbf{Y}_3$ of the Möbius frame field along f adapted to z is independent of the choice of local complex coordinate z in M. This vector field is therefore globally defined on M.

Definition 13.33. The *conformal Gauss map* of the conformal immersion $f : M \to \mathcal{M}$ is the smooth map

$$\mathfrak{n} : M \to \mathbf{S}^{3,1} = \{v \in \mathbf{R}^{4,1} : \langle v, v \rangle = 1\}$$

defined on any complex chart (U,z) by $\mathfrak{n} = \mathbf{Y}_3$, where $(\mathbf{Y}_0, \mathbf{Y}_1, \mathbf{Y}_2, \mathbf{Y}_3, \mathbf{Y}_4)$ is the Möbius frame field adapted to (U,z).

Remark 13.34. The conformal Gauss map was introduced and studied by Bryant in [20], who proved the following theorem. The study and result was extended to surfaces in n-dimensional Möbius space $\mathcal{M}^n$, for any $n \geq 3$, by M. Rigoli in [139]. Further results were obtained for surfaces in $\mathcal{M}^4$ by E. Musso in [123].

Theorem 13.35. *The conformal Gauss map $\mathfrak{n} : M \to \mathbf{S}^{3,1}$ of the conformal immersion $f : M \to \mathcal{M}$ has rank less than two at each umbilic point of f and is a conformal immersion on the complement of the set of umbilic points. It is harmonic if and only if f is Willmore.*

Proof. If Y is the Möbius frame field adapted to a complex chart (U,z) in M, then $\mathfrak{n} = \mathbf{Y}_3$ on U. By (13.26),

$$\mathbf{Y}_{3_z} = k_{\bar{z}} \mathbf{Y}_0 - \frac{k}{2}(\mathbf{Y}_1 + i\mathbf{Y}_2)$$

is a null vector at every point, and

$$\langle d\mathbf{Y}_3, d\mathbf{Y}_3 \rangle = 2\langle \mathbf{Y}_{3_z}, \mathbf{Y}_{3_{\bar{z}}} \rangle dz d\bar{z} = |k|^2 dz d\bar{z}$$

shows that the Gauss map is singular where $k = 0$ and conformal at any point where $k \neq 0$. The umbilic points in U are the zeros of k. The Laplace-Beltrami operator of the metric $dz d\bar{z}$, applied to $\mathbf{Y}_3 : U \to \mathbf{R}^{4,1}$, is $4\mathbf{Y}_{3_{z\bar{z}}}$. By (13.26),

$$\mathbf{Y}_{3z\bar{z}} = W\mathbf{Y}_0 - \frac{1}{2}|k|^2\mathbf{Y}_3,$$

where $W : U \to \mathbf{R}$ is the Willmore function (13.19) relative to Y. The map $\mathbf{Y}_3 : M \to \mathbf{S}^{3,1} \subset \mathbf{R}^{4,1}$ is harmonic, by definition, if the tangential component of the Laplace-Beltrami operator (of the metric $dzd\bar{z}$) of it,

$$4\mathbf{Y}_{3z\bar{z}} - \langle 4\mathbf{Y}_{3z\bar{z}}, \mathbf{Y}_3\rangle = 4W\mathbf{Y}_0,$$

vanishes. Thus, the conformal Gauss map is harmonic if and only if f is Willmore. □

Theorem 13.36 (Willmore associates). *Let M be a simply connected Riemann surface with complex coordinate z. Let b and k be smooth, complex valued functions on M satisfying*

$$b_{\bar{z}} = \frac{1}{2}(|k|^2)_z + k\overline{k_{\bar{z}}}, \quad 2k_{\bar{z}\bar{z}} + \bar{b}k = 0. \tag{13.47}$$

Up to Möbius transformation, there exists a unique Willmore immersion $f : M \to \mathscr{M}$ whose Möbius invariants relative to z are b and k. Given solutions b and k of these equations, let t be any real constant and let

$$\hat{k} = e^{it}k.$$

Then b and $\hat{k}$ are solutions to (13.47). *The corresponding conformal immersions $f_t : M \to \mathscr{M}$ constitute a family, parameterized by the circle $\mathbf{S}^1$, of distinct (that is, not Möbius congruent) conformal Willmore immersions called the associates of $f = f_0$.*

Proof. If k and b satisfy (13.47), then so also do $e^{it}k$ and b, for any real constant t. Now everything follows from Theorem 13.8. □

13.8 Relating the invariants

To gain insight into to how the Möbius invariants behave near umbilic points, we will express them in terms of the invariants relative to a complex coordinate chart (U,z) in M.

Proposition 13.37 (Möbius invariants related to k and b). *Let $Y : U \to$ $\mathbf{M\ddot{o}b}_+$ be the frame field adapted to the complex coordinate chart (U,z) in M for a conformal immersion $f : M \to \mathscr{M}$. If $m_0 \in U$ is a nonumbilic point of f, then there is a neighborhood $U' \subset U$ of m_0 on which there is a central frame field $\tilde{Y}$, whose third order Möbius invariants defined in* (12.50) *and* (12.53) *on U' are related to the* *invariants k and b relative to z by*

$$q_1 - iq_2 = \frac{i}{2\sqrt{k|k|}}(k^{-1}k_z + 3\overline{k^{-1}k_{\bar z}}),$$
$$p_1 + p_3 - 2ip_2 = -\frac{4}{|k|^2k^2}Q, \quad p_1 - p_3 = -\frac{4}{|k|^3}W, \tag{13.48}$$

where

$$Q = kk_{\bar z\bar z} + \frac{1}{4}k^2|k|^2 - k_z k_{\bar z} : U \to \mathbf{C} \tag{13.49}$$

and

$$W = k_{\overline{zz}} + \frac{1}{2}k\bar b : U \to \mathbf{R},$$

is the Willmore function of Definition 12.29.

Proof. The choice of square root in the first equation determines the right side up to sign, which corresponds to the choice of time oriented central frame determining the left side up to sign, by Exercise 51. The frame field Y adapted to z is a time oriented γ-frame along f. If $\omega = Y^{-1}dY$, then

$$\omega_0^0 = 0 = \omega_2^1, \quad \omega_0^3 = 0, \quad \omega_0^1 + i\omega_0^2 = dz,$$
$$\omega_1^3 - i\omega_2^3 = k dz, \quad \omega_3^0 = k_{\bar z}dz + \overline{k_{\bar z}}d\bar z, \quad \omega_1^0 - i\omega_2^0 = -b dz - \frac{|k|^2}{2}d\bar z.$$

For the following calculation it is convenient to let

$${}^t\theta = (\omega_0^1, \omega_0^2), \quad {}^t\mu = (\omega_1^3, \omega_2^3), \quad {}^t\eta = (\omega_1^0, \omega_2^0).$$

Since $k : U \to \mathbf{C}$ is never zero on $U \subset M \setminus \mathscr{U}_f$, we know $k(m_0) = r_0 e^{it_0}$, where $r_0 > 0$ and $0 \le t_0 < 2\pi$. Let U' be the connected component containing m_0 of the set

$$k^{-1}(\mathbf{C} \setminus \{re^{it_0} : r \le 0\}) \subset U.$$

Then $\log(k)$ is defined on U', and there are well defined smooth functions $r, t : U' \to \mathbf{R}$ such that $r > 0$ and $-\log k = \log r + 2it$; that is,

$$re^{2it} = 1/k$$

on U'. Thus $|k| = 1/r > 0$ and

$$\sqrt{k} = r^{-1/2}e^{-it} : U' \to \mathbf{C} \tag{13.50}$$

is smooth. Define a smooth map $K = K(1, a, \mathbf{y}) : U' \to G_\gamma$, where

$$\mathbf{y} = \begin{pmatrix} y^1 \\ y^2 \end{pmatrix} \in \mathbf{R}^2, \quad y^1 + iy^2 = -2k^{-1}k_{\bar{z}}, \quad a = \begin{pmatrix} \cos t & -\sin t \\ \sin t & \cos t \end{pmatrix}. \tag{13.51}$$

Then

$$\tilde{Y} = YK = Y \begin{pmatrix} r^{-1} & {}^t\mathbf{y}a & 0 & \frac{r}{2}|\mathbf{y}|^2 \\ 0 & a & 0 & r\mathbf{y} \\ 0 & 0 & 1 & 0 \\ 0 & 0 & 0 & r \end{pmatrix} : U' \to \mathbf{M\ddot{o}b}_+, \tag{13.52}$$

is a central frame field along f on U'. To verify this, we compute:

$$K^{-1}dK = \begin{pmatrix} r & -r{}^t\mathbf{y} & 0 & \frac{r}{2}|\mathbf{y}|^2 \\ 0 & a^{-1} & 0 & -a^{-1}\mathbf{y} \\ 0 & 0 & 1 & 0 \\ 0 & 0 & 0 & r^{-1} \end{pmatrix} d \begin{pmatrix} r^{-1} & {}^t\mathbf{y}a & 0 & \frac{r}{2}|\mathbf{y}|^2 \\ 0 & a & 0 & r\mathbf{y} \\ 0 & 0 & 1 & 0 \\ 0 & 0 & 0 & r \end{pmatrix}$$

$$= \begin{pmatrix} -r^{-1}dr & r(d{}^t\mathbf{y})a & 0 & 0 \\ 0 & a^{-1}da & 0 & ra^{-1}d\mathbf{y} \\ 0 & 0 & 0 & 0 \\ 0 & 0 & 0 & r^{-1}dr \end{pmatrix}$$

and

$$K^{-1}\omega K = K^{-1} \begin{pmatrix} 0 & {}^t\eta & \omega_3^0 & 0 \\ \theta & 0 & -\mu & \eta \\ 0 & {}^t\mu & 0 & \omega_3^0 \\ 0 & {}^t\theta & 0 & 0 \end{pmatrix} K.$$

Thus

$$\tilde{\omega} = \tilde{Y}^{-1}d\tilde{Y} = K^{-1}\omega K + K^{-1}dK = \begin{pmatrix} \tilde{\omega}_0^0 & {}^t\tilde{\eta} & \tilde{\omega}_3^0 & 0 \\ \tilde{\theta} & \tilde{\Omega} & -\tilde{\mu} & \tilde{\eta} \\ 0 & {}^t\tilde{\mu} & 0 & \tilde{\omega}_3^0 \\ 0 & {}^t\tilde{\theta} & 0 & -\tilde{\omega}_0^0 \end{pmatrix},$$

where

$$\begin{gathered} \tilde{\omega}_0^0 = -r^{-1}dr - {}^t\mathbf{y}\theta, \quad \tilde{\theta} = r^{-1}a^{-1}\theta, \quad \tilde{\omega}_3^0 = r\omega_3^0 + r{}^t\mathbf{y}\mu, \\ {}^t\tilde{\mu} = {}^t\mu a, \quad {}^t\tilde{\eta} = r({}^t\eta - ({}^t\mathbf{y}a){}^t\mathbf{y} + \frac{1}{2}|\mathbf{y}|^{2t}\theta + d\,{}^t\mathbf{y})a, \\ \begin{pmatrix} 0 & \tilde{\omega}_2^1 \\ \tilde{\omega}_1^2 & 0 \end{pmatrix} = \tilde{\Omega} = a^{-1}(\theta{}^t\mathbf{y} - \mathbf{y}{}^t\theta)a + \begin{pmatrix} 0 & -1 \\ 1 & 0 \end{pmatrix} dt. \end{gathered} \tag{13.53}$$

Then

$$\tilde{\varphi} = \tilde{\omega}_0^1 + i\tilde{\omega}_0^2 = r^{-1}e^{-it}(\omega_0^1 + i\omega_0^2) = r^{-1}e^{-it}dz = \sqrt{k|k|}dz,$$

since $r^{-1}e^{-it} = \sqrt{k|k|}$ by (13.50), and

$$\tilde{\omega}_1^3 - i\tilde{\omega}_2^3 = e^{it}(\omega_1^3 - i\omega_2^3) = e^{it}kdz = r^{-1}e^{-it}dz = \tilde{\varphi},$$

so $\tilde{Y}$ is a second order frame field. Moreover,

$$\begin{aligned}{}^t\mathbf{y}\mu &= y^1\omega_1^3 + y^2\omega_2^3 = \frac{1}{2}((y^1 + iy^2)(\omega_1^3 - i\omega_2^3) + (y^1 - iy^2)(\omega_1^3 + i\omega_2^3)) \\ &= -(k_{\bar{z}}dz + \overline{k_{\bar{z}}}d\bar{z}),\end{aligned}$$

if $y^1 + iy^2$ is given by (13.51), and thus

$$\tilde{\omega}_3^0 = r(\omega_3^0 + {}^t\mathbf{y}\mu) = 0,$$

so $\tilde{Y}$ is a third order frame field. From (13.53),

$${}^t\tilde{\eta} = r(\omega_1^0 - {}^t\mathbf{y}\theta y^1 + \frac{1}{2}|\mathbf{y}|^2\omega_0^1 + dy^1,\ \omega_2^0 - {}^t\mathbf{y}\theta y^2 + \frac{1}{2}|\mathbf{y}|^2\omega_0^2 + dy^2)a.$$

Using (13.51) in this, we get

$$\begin{aligned}&r^{-1}e^{it}(\tilde{\omega}_1^0 + i\tilde{\omega}_2^0) \\ &= \omega_1^0 + i\omega_2^0 - {}^t\mathbf{y}\theta(y^1 + iy^2) + \frac{1}{2}|\mathbf{y}|^2(\omega_0^1 + i\omega_0^2) + d(y^1 + iy^2) \\ &= -\bar{b}d\bar{z} - \frac{|k|^2}{2}dz - 2\frac{k_{\bar{z}}}{k}(\frac{\overline{k_{\bar{z}}}}{\bar{k}}dz + \frac{k_{\bar{z}}}{k}d\bar{z}) + 2|\frac{k_{\bar{z}}}{k}|^2dz - 2d(\frac{k_{\bar{z}}}{k}) \\ &= -\frac{2}{k^2}(kk_{z\bar{z}} + \frac{1}{4}k^2|k|^2 - k_zk_{\bar{z}})dz - \frac{2}{k}(\frac{1}{2}k\bar{b} + k_{\bar{z}\bar{z}})d\bar{z},\end{aligned}$$

so

$$\tilde{\omega}_1^0 + i\tilde{\omega}_2^0 = \frac{-2}{k^2|k|^2}Q\tilde{\varphi} + \frac{-2}{|k|^3}W\bar{\tilde{\varphi}}. \tag{13.54}$$

This with (12.53) confirms the last two equations in (13.48). The last equation in (13.53) is

$$\begin{pmatrix}0 & -1\\ 1 & 0\end{pmatrix}\tilde{\omega}_1^2 = a^{-1}\begin{pmatrix}0 & y^2\omega_0^1 - y^1\omega_0^2\\ y^1\omega_0^2 - y^2\omega_0^1 & 0\end{pmatrix}a + \begin{pmatrix}0 & -1\\ 1 & 0\end{pmatrix}dt,$$

so, since **SO**(2) is abelian,

$$\tilde{\omega}_1^2 = y^1\omega_0^2 - y^2\omega_0^1 + dt.$$

Thus, using $-r^{-1}dr - 2idt = k^{-1}dk$, (13.53), and finally (13.51), we get

$$\begin{aligned}
\tilde{\omega}_0^0 - 2i\tilde{\omega}_1^2 &= -r^{-1}dr - y^1\omega_0^1 - y^2\omega_0^2 - 2i(y^1\omega_0^2 - y^2\omega_0^1) - 2idt \\
&= -r^{-1}dr - 2idt - (y^1 - 2iy^2)\omega_0^1 - (y^2 + 2iy^1)\omega_0^2 \\
&= k^{-1}dk - (y^1 - 2iy^2)\frac{dz + d\bar{z}}{2} - (y^2 + 2iy^1)\frac{dz - d\bar{z}}{2i} \\
&= k^{-1}dk - \frac{3}{2}(y^1 - iy^2)dz + \frac{1}{2}(y^1 + iy^2)d\bar{z} = k^{-1}dk + 3\overline{k^{-1}k_{\bar{z}}}dz - k^{-1}k_{\bar{z}}d\bar{z} \\
&= (k^{-1}k_z + 3\overline{k^{-1}k_{\bar{z}}})dz = \frac{1}{\sqrt{k|k|}}(k^{-1}k_z + 3\overline{k^{-1}k_{\bar{z}}})\varphi,
\end{aligned}$$

so

$$\tilde{\omega}_0^0 - 2i\tilde{\omega}_1^2 = \frac{1}{\sqrt{k|k|}}(k^{-1}k_z + 3\overline{k^{-1}k_{\bar{z}}})\tilde{\varphi}.$$

This with (12.50) confirms the first equation in (13.48). □

If w is another complex coordinate on U', and if $\hat{k}$ is the first order invariant relative to w, then (13.20) says $\hat{k} = k|w'|/(w')^2$, where $w' = \frac{dw}{dz}$. By (13.48),

$$\frac{\hat{Q}}{|\hat{k}|^2\hat{k}^2} = -\frac{1}{4}(p_1 + p_3 - 2ip_2) = \frac{Q}{|k|^2k^2},$$

where $\hat{Q}$ is defined by (13.49) for $\hat{k}$. Hence, the quartic differentials

$$\hat{Q}dwdwdwdw = Qdzdzdzdz$$

are independent of complex coordinate, so define a global quartic differential.

Definition 13.38. The *Bryant quartic differential* of a conformal immersion $f : M \to \mathcal{M}$ is the smooth quartic differential $\mathcal{Q}$ on the Riemann surface M defined in any complex coordinate chart (U, z) by

$$\mathcal{Q} = (k(k_{z\bar{z}} + \frac{1}{4}k|k|^2) - k_zk_{\bar{z}})dzdzdzdz,$$

where k is the first order Möbius invariant of f relative to z.

Remark 13.39. For the central frame field $\tilde{Y}: U' \to \mathbf{M\ddot{o}b}_+$ defined in (13.52),

$$\tilde{\mathbf{Y}}_4 = r(\mathbf{Y}_4 + \frac{1}{2}|\mathbf{y}|^2\mathbf{Y}_0 + (\mathbf{Y}_1, \mathbf{Y}_2)\mathbf{y})$$
$$= \frac{1}{|k|^3}(|k|^2\mathbf{Y}_4 + 2|k_{\bar{z}}|^2\mathbf{Y}_0 - \bar{k}k_{\bar{z}}(\mathbf{Y}_1 - i\mathbf{Y}_2) - k\overline{k_{\bar{z}}}(\mathbf{Y}_1 + i\mathbf{Y}_2)),$$

by (13.51). Thus, the dual map $\hat{f}: M \setminus \mathcal{U}_f \to \mathcal{M}$ of Definition 12.36 is given on U' by

$$\begin{aligned}\hat{f} &= [\tilde{\mathbf{Y}}_4] = [|k|^3\tilde{\mathbf{Y}}_4]\\ &= [|k|^2\mathbf{Y}_4 + 2|k_{\bar{z}}|^2\mathbf{Y}_0 - \bar{k}k_{\bar{z}}(\mathbf{Y}_1 - i\mathbf{Y}_2) - k\overline{k_{\bar{z}}}(\mathbf{Y}_1 + i\mathbf{Y}_2)].\end{aligned} \tag{13.55}$$

Then

$$d\tilde{\mathbf{Y}}_4 = \tilde{\omega}_1^0\tilde{\mathbf{Y}}_1 + \tilde{\omega}_2^0\tilde{\mathbf{Y}}_2, \tag{13.56}$$

so by (13.54)

$$\langle d\tilde{\mathbf{Y}}_4, d\tilde{\mathbf{Y}}_4\rangle = (\tilde{\omega}_1^0 + i\tilde{\omega}_2^0)(\tilde{\omega}_1^0 - i\tilde{\omega}_2^0)$$
$$= 4|k|^2\left((\frac{|Q|^2}{|k|^4} + \frac{|W|^2}{|k|^2})\varphi\bar{\varphi} + \frac{Q\bar{W}}{k^2|k|}\varphi\varphi + \frac{\bar{Q}W}{\bar{k}^2|k|}\bar{\varphi}\bar{\varphi}\right).$$

Theorem 13.40 (Bryant [20]). *Let $f: M \to \mathcal{M}$ be a Willmore immersion of a connected Riemann surface M. Let $U_f \subset M$ be its set of umbilic points. If f is not totally umbilic, then its conformal dual map $\hat{f}: M \setminus \mathcal{U}_f \to \mathcal{M}$ extends smoothly to all of M. If the quartic differential $\mathcal{Q}$ of f is identically zero on M, then $\hat{f}$ is constant on M.*

Proof. If (U, z) is any complex coordinate chart in M, then $\hat{f}$ is given on $U' = U \setminus U_f$ by (13.55). If f is not totally umbilic, then by Proposition 13.32, $\mathcal{U}_f$ is a closed subset of M without interior points, and for any point $m \in U$, there is a neighborhood $V \subset U$ of m on which

$$k = (z - z(m))^n g_1, \quad k_{\bar{z}} = (z - z(m))^n g_2,$$

where $g_1, g_2 : V \to \mathbf{C}$ are smooth functions, not both zero at any point of V, and n is a nonnegative integer. We may replace z by $z - z(m)$ and thereby assume that $z(m) = 0$. On V,

$$|k|^2 = |z|^{2n}|g_1|^2, \quad |k_{\bar{z}}|^2 = |z|^{2n}|g_2|^2, \quad \bar{k}k_{\bar{z}} = |z|^{2n}\bar{g}_1 g_2,$$

so after factoring out $|z|^{2n}$ in (13.55), $\hat{f}$ is given on $V\setminus\{m\}$ by

$$[\tilde{\mathbf{Y}}_4] = [g_1^2\mathbf{Y}_4 + 2|g_2|^2\mathbf{Y}_0 - \bar{g}_1 g_2(\mathbf{Y}_1 - i\mathbf{Y}_2) - g_1\bar{g}_2(\mathbf{Y}_1 + i\mathbf{Y}_2)],$$

which is smooth on all of V. Hence $\hat{f}$ extends smoothly to U, for any chart (U,z), and thus it extends smoothly to M.

If Q and W are identically zero on U, for any complex coordinate chart (U,z), then by (13.56) and (13.54) the map $\tilde{\mathbf{Y}}_4$ is constant on U. Hence, $\mathscr{Q}$ identically zero on M implies $\hat{f}$ is constant on M. □

Problems

13.41. Prove that the sum $\mathbf{z}+\mathbf{w}$ of isotropic vectors is again isotropic if and only if they are orthogonal, meaning ${}^t\mathbf{z}\mathbf{w} = 0$.

13.42. Suppose $f : M \to \mathscr{M}$ is a conformal immersion and that (U,z) is a complex coordinate chart in M. Prove that if $Y : U \to \mathbf{M\ddot{o}b}_+$ is a time oriented frame satisfying

$$\omega_0^1 + i\omega_0^2 = e^u dz \text{ and } \omega_0^3 = 0$$

on U, where $u : U \to \mathbf{R}$, then

$$\omega_1^3 - i\omega_2^3 = kdz + sd\bar{z},$$

where k is the first order invariant of f relative to z and $s : U \to \mathbf{R}$.

13.43. Suppose $f : M \to \mathscr{M}$ is a conformal immersion and that (U,z) is a complex coordinate chart in M. Prove that if $Y : U \to \mathbf{M\ddot{o}b}_+$ is a time oriented frame field satisfying $\omega_0^3 = 0$ on U, then

$$\omega_0^1 + i\omega_0^2 = e^{u+iv}dz, \quad \omega_1^3 - i\omega_2^3 = pdz + qd\bar{z},$$

for some smooth functions $u, v : U \to \mathbf{R}$ and $p, q : U \to \mathbf{C}$. Prove that the first order invariant of f relative to z is

$$k = e^{iv}p,$$

and $e^{iv}q$ is real valued on U.

13.44. Prove that if k and b are the conformal invariants relative to a complex chart (U,z) of a conformal immersion $f : M \to \mathscr{M}$, and if $\hat{k}$ and $\hat{b}$ are the conformal invariants relative to a complex chart (U,w), then

$$\hat{k}_{\bar{w}\bar{w}} + \frac{1}{2}\hat{k}\bar{\hat{b}} = \frac{1}{|w'|^3}(k_{\bar{z}\bar{z}} + \frac{1}{2}k\bar{b}),$$

where $w' = \frac{dw}{dz}$. This shows that the Willmore function (13.19) relative to (U,z) transforms under a change of complex coordinate as expected from Lemma 12.30.

13.45. Find the Möbius invariants k and b relative to $z = x+iy$ for the cylinder of Example 13.15 over the circle of radius $R > 0$, $\boldsymbol{\sigma}(x) = R(\cos\frac{x}{R}\boldsymbol{\delta}_1 + \sin\frac{x}{R}\boldsymbol{\delta}_2)$.

13.46. Fine the Möbius invariants k and b relative to $z = x+iy$ for the cone of Example ex:17:cone in the case when the profile curve is the circle on $\mathbf{S}^2$

$$\boldsymbol{\sigma}(x) = (l\cos(\frac{x}{l}), l\sin(\frac{x}{l}), \sqrt{1-l^2}),$$

where $l = \cos\alpha$, for some fixed angle $0 < \alpha < \pi/2$.

13.47. The circular torus for given α has constant Möbius invariants k and b relative to z calculated in Example 13.19. Change the complex coordinate to $w = ikz$, and denote the invariants relative to w by $\hat{k}$ and $\hat{b}$. Use Theorem 13.10 to find $\hat{k}$ and $\hat{b}$.

13.48. Using the decomposition $f_0 = f_+ \circ \mathscr{S}^{-1}$ together with Proposition 5.23 and Exercise 56, prove that the map from oriented spheres and planes in $\mathbf{R}^3$ to $\mathbf{S}^{3,1}$ defined by $f_0 : \mathbf{R}^3 \to \mathscr{M}$ is

$$f_0(S_R(\mathbf{c})) = \frac{1}{R}(\boldsymbol{\delta}_0 + \mathbf{c} + \frac{|\mathbf{c}|^2 - R^2}{2}\boldsymbol{\delta}_4), \quad f_0(\Pi_h(\mathbf{n})) = \mathbf{n} + h\boldsymbol{\delta}_4 \in \mathbf{S}^{3,1}.$$

13.49. Prove that the oriented curvature spheres of the circular torus $\mathbf{x}^{(\alpha)}$ relative to $\mathbf{e}_3$ found in Example 5.12 correspond, under the conformal diffeomorphism $f_+ : \mathbf{S}^3 \to \mathscr{M}$ to the points $S_\pm \in \mathbf{S}^{3,1}$ found above in Example 13.19.

13.50. If H, K, and dA are the mean curvature, Gaussian curvature, and area element on M for an immersion $\mathbf{x} : M \to \mathbf{H}^3$, and if $\tilde{H}$, $\tilde{K}$, and $d\tilde{A}$ denote the same quantities for the immersion $\mathfrak{s} \circ \mathbf{x} : M \to \mathbf{R}^3$, where $\mathfrak{s} : \mathbf{H}^3 \to \mathbf{B}^3 \subset \mathbf{R}^3$ is hyperbolic stereographic projection (see Definition 6.15), prove that

$$(H^2 - K - 1)dA = (\tilde{H}^2 - \tilde{K})d\tilde{A}.$$

13.51 (Clifford torus associates). Find the associates of the Clifford torus $f : \mathbf{R}^2 \to \mathscr{M}$ defined by the lift (13.33) for the case $\theta = \pi/4$. Relative to $z = x+iy$ its invariants are $k = -1$ and $b = 0$. For any real constant t, its associate f_t has invariants $\hat{k} = -e^{it}$ and $\hat{b} = 0$, relative to z.

13.52. Use the change of coordinate formula (13.20) to prove directly that the Bryant quartic differential $\mathscr{Q}$ of Definition 13.38 is independent of the choice of complex coordinate chart in M, and thus is smooth and well-defined on all of M.

13.53 (Spiral cylinder). Use Theorem 13.22 and Proposition 13.37 to verify that the Möbius invariants of the immersion in (12.69) are given by (12.70) with $q_2 = \frac{2}{n}$.

Chapter 14
Isothermic Immersions into Möbius Space

The theory of isothermic surfaces in conformal geometry was intensely studied at the turn of the 20th century by L. Bianchi [6, 7], P. Calapso [27, 28], Darboux [55–57], and R. Rothe [142]. The modern theory of integrable systems and loop groups has renewed interest in this theory. The methods of integrable systems theory made their first appearance in the study of isothermic surfaces with Cieśliński, Goldstein, and Sym's zero-curvature formulation of the Gauss–Codazzi equations of an isothermic surface [51, 52]. This work was taken up by Burstall, Hertrich-Jeromin, Pedit, and Pinkall [25], who described the integrable system of isothermic surfaces in the context of Möbius geometry as an example of the curved flat system of Ferus and Pedit [69]. An equivalent description was given by Brück, Park, and Terng in [19]. This approach provided a coherent framework for discussing the classical transformations of isothermic surfaces. For recent accounts of the transformation theory of isothermic surfaces, we refer the reader to Hertrich-Jeromin's book [86] and Burstall's monograph [24]. In the latter reference, the Darboux transformation is described using the loop group formulation according to the general theory of Bäcklund transformations due to Terng and Uhlenbeck [159].

This chapter presents the classical theory in terms of the Möbius invariants k and b relative to a complex coordinate.

14.1 Isothermic immersions

Let $f : M \to \mathcal{M}$ be a conformal immersion of a Riemann surface M. Let (U,z) be a complex coordinate chart in M. Recall Definition 13.5 of the Möbius frame field $Y : U \to \mathbf{M\ddot{o}b}_+$ adapted to z together with the first order invariant $k : U \to \mathbf{C}$ and the second order invariant $b : U \to \mathbf{C}$ of f relative to z.

G.R. Jensen et al., *Surfaces in Classical Geometries*, Universitext,
DOI 10.1007/978-3-319-27076-0_14

Definition 14.1. A complex coordinate chart (U, z) in M is called *principal* for f if the first order invariant k of f relative to z is real valued on U. In this case we call k the *Calapso potential* for f relative to z.

Definition 14.2. The conformal immersion $f : M \to \mathcal{M}$ is *isothermic* if there exists an atlas $\{U_\alpha, z_\alpha\}_{\alpha \in \mathcal{A}}$ on M of principal complex coordinate charts for f.

Remark 14.3. By this definition a totally umbilic immersion $f : M \to \mathcal{M}$ is isothermic because k is identically zero, therefore real valued, for any complex chart (U, z) in M.

Recall the conformal embeddings $f_\epsilon : \mathbf{S}_\epsilon \to \mathcal{M}$, $\epsilon \in \{0, +, -\}$, defined in Section 12.5.

Lemma 14.4. *If $\mathbf{x} : M \to \mathbf{S}_\epsilon$ is an isothermic immersion, then the composition $f = f_\epsilon \circ \mathbf{x} : M \to \mathcal{M}$ is a conformal isothermic immersion.*

Proof. Let (U, z) be a principal chart in M for the complex structure coming from the metric I induced on M by the isothermic immersion $\mathbf{x}$. Let e^u be the conformal factor and let h be the Hopf invariant of $\mathbf{x}$ relative to z. Then the first order Möbius invariant k of $f = f_\epsilon \circ \mathbf{x}$ is $k = he^u$, by (13.34) in Theorem 13.22. Since h is real valued, k is real valued. An atlas of principal charts in M for $\mathbf{x}$ is therefore also an atlas of principal charts for f. □

Proposition 14.5 (Isothermic criterion). *A conformal immersion $f : M \to \mathcal{M}$ is isothermic if and only if for any complex coordinate chart (U, z) in M, the first order Möbius invariant k relative to z has the property that for any point in U there is a neighborhood $V \subset U$ about the point such that $k = rg^2$ on V, where $r : V \to \mathbf{R}$ is smooth and $g : V \to \mathbf{C} \setminus \{0\}$ is holomorphic.*

Proof. Suppose f is isothermic and suppose (U, z) is a complex coordinate chart of M. About a given point in U, let (V, w) be a principal chart for f and let $\hat{k}$ be the Calapso potential relative to w. By the change of coordinate formula (13.20),

$$k = \hat{k}\left(\frac{dw}{dz}\right)^2 \Big/ \left|\frac{dw}{dz}\right|, \tag{14.1}$$

which is a real valued smooth function $\hat{k} / \left|\frac{dw}{dz}\right|$ times the square of the nowhere zero holomorphic function $\frac{dw}{dz}$.

Conversely, if $k = rg^2$ on V, where g is holomorphic and never zero on V, and r is real, then there is a holomorphic function w on a possibly smaller neighborhood of the point in V such that $dw = g dz$ and w is a complex coordinate on this neighborhood. By (14.1), the first order invariant $\hat{k}$ of f relative to w is

$$\hat{k} = \frac{|g| k}{g^2} = |g| r,$$

which is real valued. □

Recall Definition 9.16 of an affine structure on a Riemann surface M.

Theorem 14.6 (Affine structure). *If $\{U_\alpha, z_\alpha\}_{\alpha\in\mathscr{A}}$ is an atlas of principal charts for an isothermic conformal immersion $f : M \to \mathscr{M}$ whose set of nonumbilic points is dense in M, then this atlas defines an affine structure on M.*

Proof. Let $\{U_\alpha, z_\alpha\}_{\alpha\in\mathscr{A}}$ be an atlas of principal charts on M and let k_α be the first order Möbius invariant of f relative to z_α. If $U_\alpha \cap U_\beta \neq \emptyset$, then by (13.20)

$$k_\beta \left(\frac{dz_\beta}{dz_\alpha}\right)^2 = k_\alpha \left|\frac{dz_\beta}{dz_\alpha}\right|$$

Since the set of nonumbilic points of f is assumed dense in M, its intersection with $U_\alpha \cap U_\beta$ is dense in this set, and on this dense subset of $U_\alpha \cap U_\beta$ we have

$$\left(\frac{dz_\beta}{dz_\alpha}\right)^2 = \left|\frac{dz_\beta}{dz_\alpha}\right| k_\alpha / k_\beta$$

is holomorphic and real valued, thus locally constant. It follows that $\frac{dz_\beta}{dz_\alpha}$ is locally constant on $U_\alpha \cap U_\beta$. Therefore, the given atlas defines an affine structure on M. □

Corollary 14.7. *If M is a compact Riemann surface and if $f : M \to \mathscr{M}$ is a conformal isothermic immersion whose set of nonumbilic points is dense in M, then M is a torus.*

Proof. By Theorem 9.20, a compact Riemann surface which possesses an affine structure must have genus equal to one. □

Theorem 14.8. *If M is a simply connected Riemann surface with a conformal isothermic immersion $f : M \to \mathscr{M}$ whose set of nonumbilic points is dense in M, then there exists an atlas $\{U_\alpha, z_\alpha\}_{\alpha\in\mathscr{A}}$ of principal charts for f such that, if k_α and b_α are the first and second order Möbius invariants of f relative to z_α, then*

$$z_\alpha = z_\beta, \quad k_\alpha = k_\beta, \quad b_\alpha = b_\beta$$

for any $\alpha, \beta \in \mathscr{A}$ with $U_\alpha \cap U_\beta \neq \emptyset$. Therefore, there exist a holomorphic function $z : M \to \mathbf{C}$ and smooth functions $k : M \to \mathbf{R}$ and $b : M \to \mathbf{C}$ such that

$$z_\alpha = z|_{U_\alpha}, \quad k_\alpha = k|_{U_\alpha}, \quad b_\alpha = b|_{U_\alpha},$$

for any $\alpha \in \mathscr{A}$.

Proof. Let $\{U_\alpha, z_\alpha\}_{\alpha\in\mathscr{A}}$ be an atlas of principal charts on M. We may assume that the open cover $\mathscr{U} = \{U_\alpha\}_{\alpha\in\mathscr{A}}$ is a *simple cover*, which implies that any nonempty intersection $U_\alpha \cap U_\beta$ is contractible (see [53, Theorem 8.5.7, p. 268]). By the proof of Theorem 14.6, if $U_\alpha \cap U_\beta \neq \emptyset$, then

$$c_{\alpha\beta} = \frac{dz_\beta}{dz_\alpha}$$

is locally constant, hence constant, on $U_\alpha \cap U_\beta$, and $c^2_{\alpha\beta}$ is real valued and nonzero. We now apply a Čech cohomology argument of the kind used in the global construction part of the proof of Theorem 9.24.

It is known (see [53, Lemma 8.9.8, p. 285]) that for any abelian group G, the Čech cohomology of a simple cover, $\check{H}^1(\mathscr{U}, G)$, is equal to $H^1(M,G)$, which is just the trivial group consisting of the identity element of G when M is simply connected. That said, we now go through two cohomology arguments to arrive at the desired atlas.

For the first argument, we let $G = \mathbf{R}_\times \cup i\mathbf{R}_\times$ be the multiplicative group of nonzero real and pure imaginary numbers. Thus $c_{\alpha\beta} \in G$ whenever $U_\alpha \cap U_\beta \neq \emptyset$. Define a 1-cochain c on M by assigning to the pair U_α, U_β, if $U_\alpha \cap U_\beta \neq \emptyset$, the element $c_{\alpha\beta} \in G$. Then, whenever $U_\alpha \cap U_\beta \cap U_\gamma \neq \emptyset$, the cocycle condition $c_{\alpha\beta} c_{\beta\gamma} c_{\gamma\alpha} = 1$ is satisfied, which shows that c is a Čech cocycle. Since the Čech cohomology group $\check{H}^1(\mathscr{U}, G) = \{1\}$, there exists a 0-cochain t whose coboundary $\delta t = c$. That is, t assigns to the open set U_α the number $t_\alpha \in G$, such that for any pair U_α, U_β with nonempty intersection,

$$(\delta t)_{\alpha\beta} = t_\beta / t_\alpha = c_{\alpha\beta}. \tag{14.2}$$

Let $w_\alpha : U_\alpha \to \mathbf{C}$ be defined by $w_\alpha = z_\alpha / t_\alpha$. Then U_α, w_α is a complex chart in M such that $\frac{dw_\alpha}{dz_\alpha} = 1/t_\alpha$ on U_α and, by (13.20), its first order Möbius invariant

$$\tilde{k}_\alpha = \frac{k_\alpha |\frac{dw_\alpha}{dz_\alpha}|}{(\frac{dw_\alpha}{dz_\alpha})^2} = \frac{k_\alpha t_\alpha^2}{|t_\alpha|},$$

which is real valued on U_α. Hence, U_α, w_α is a principal chart for f, for any $\alpha \in \mathscr{A}$, and whenever $U_\alpha \cap U_\beta \neq \emptyset$,

$$\frac{dw_\beta}{dw_\alpha} = \frac{dz_\beta}{t_\beta} \frac{t_\alpha}{dz_\alpha} = \frac{t_\alpha}{t_\beta} c_{\alpha\beta} = 1.$$

We may thus assume that our original atlas of principal charts $\{U_\alpha, z_\alpha\}_{\alpha \in \mathscr{A}}$ is such that $\mathscr{U}$ is a simple cover of M and that $\frac{dz_\beta}{dz_\alpha} = 1$ on any nonempty intersection $U_\alpha \cap U_\beta$. It follows that on this intersection, which is connected, we have $z_\beta = z_\alpha + a_{\alpha\beta}$, for some constant $a_{\alpha\beta} \in \mathbf{C}$.

Now consider Čech cohomology with coefficients in the additive group of complex numbers $\mathbf{C}$. Let a denote the 1-cochain which assigns to a pair U_α, U_β, with $U_\alpha \cap U_\beta \neq \emptyset$, the constant $a_{\alpha\beta} \in \mathbf{C}$. Then a is a cocycle because $a_{\alpha\beta} + a_{\beta\gamma} + a_{\gamma\alpha} = 0$. Hence, there exists a 0-cochain $s = (s_\alpha)$ such that $\delta s = a$, which means that s assigns to U_α the constant $s_\alpha \in \mathbf{C}$ in such a way that on $U_\alpha \cap U_\beta \neq \emptyset$ we have $s_\beta - s_\alpha = (\delta s)_{\alpha\beta} = a_{\alpha\beta}$. Let $w_\alpha : U_\alpha \to \mathbf{C}$ be defined by $w_\alpha = z_\alpha - s_\alpha$. Then U_α, w_α remains a principal complex chart for each $\alpha \in \mathscr{A}$ and on any nonempty intersection

$$w_\beta - w_\alpha = z_\beta - s_\beta - z_\alpha + s_\alpha = a_{\alpha\beta} - a_{\alpha\beta} = 0$$

We have shown that there exists an atlas $\{U_\alpha, z_\alpha\}_{\alpha\in\mathscr{A}}$ on M of principal complex charts for f such that $z_\alpha = z_\beta$ on $U_\alpha \cap U_\beta$. Then $\frac{dz_\beta}{dz_\alpha} = 1$ and the Schwarzian derivative $\mathscr{S}_{z_\alpha}(z_\beta) = 0$, so by (13.20) and (13.21), we have $k_\beta = k_\alpha$ and $b_\beta = b_\alpha$ on $U_\alpha \cap U_\beta$. □

Remark 14.9. In the notation of the preceding theorem, the function $z : M \to \mathbf{C}$ is not necessarily a complex coordinate on all of M, because it need not be one-to-one, but for each U_α in this atlas, the function $z|_{U_\alpha} = z_\alpha$ is a principal complex coordinate on U_α for f. For this reason there is a well defined operator ∂_z on functions and 1-forms, which on U_α is ∂_{z_α}. The function $k : M \to \mathbf{R}$ is the Calapso potential for f relative to z, in the sense that $k|_{U_\alpha} = k_\alpha$ is the Calapso potential relative to z_α. In the same way, the function $b : M \to \mathbf{C}$ is the second order Möbius invariant of f relative to z in the sense that $b|_{U_\alpha} = b_\alpha$ is this invariant for f relative to z_α. We summarize this in the following definition.

Definition 14.10. A *principal function* for an isothermic immersion $f : M \to \mathscr{M}$ is a holomorphic function $z : M \to \mathbf{C}$ together with *Calapso potential* $k : M \to \mathbf{R}$ and *second order Möbius invariant* $b : M \to \mathbf{C}$ such that there is an atlas $\{U_\alpha, z_\alpha\}_{\alpha\in\mathscr{A}}$ of principal charts for f with Calapso potential k_α and second order Möbius invariant b_α relative to z_α such that $z|_{U_\alpha} = z_\alpha$, $k|_{U_\alpha} = k_\alpha$ and $b|_{U_\alpha} = b_\alpha$, for each $\alpha \in \mathscr{A}$.

This definition allows the possibility that $\mathscr{A}$ consists of only one element so that (M, z) is a principal chart with Calapso potential k and second order Möbius invariant b. According to Theorem 14.8, if M is simply connected and if $f : M \to \mathscr{M}$ is a conformal isothermic immersion with dense set of nonumbilic points, then there exists a principal function z on M with Calapso potential $k : M \to \mathbf{R}$ and second order Möbius invariant $b : M \to \mathbf{C}$ relative to z.

Example 14.11. The following example has a principal function that is not a global complex coordinate on M. Let $M = \mathbf{C}$ with its standard complex structure $z = x + iy$. Let $f : M \to \mathscr{M}$ be the conformal immersion defined by the lift $F : M \to \mathscr{L}^+$ defined by

$$F(x,y) = \frac{1}{c}\begin{pmatrix} c + \cos(ce^x \cos y) \\ \sin(ce^x \cos y) \\ \cos(ce^x \sin y) \\ \sin(ce^x \sin y) \\ \frac{1}{2}(c - \cos(ce^x \cos y)) \end{pmatrix}$$

where $c = \sqrt{2}$. One calculates $\langle F_z, F_z\rangle = 0$ and $\langle F_z, F_{\bar{z}}\rangle = e^{2x}/2$ to verify that $e^{-x}F$ is the normalized lift required in Theorem 13.13 to calculate the invariants k and b of f relative to z. As an exercise, verify that

$$k = -e^{2z}e^{-x}, \quad b = -1/2.$$

By Proposition 14.5, f is an isothermic immersion, because k is a real function times a nowhere zero holomorphic function. But the complex coordinate z is not principal, because k is not real valued. The holomorphic map $w = e^z : \mathbf{C} \to \mathbf{C}$ is a principal function for f. In fact, on a sufficiently small neighborhood of any point of $\mathbf{C}$, w is a local complex coordinate, because $w' = \frac{dw}{dz} = e^z$ is never zero. It is not a global complex coordinate because it is not one-to-one. According to the change of coordinate formulas (13.20) and (13.21), the invariants $\hat{k}$ and $\hat{b}$ relative to w are

$$\hat{k} = \frac{k|w'|}{(w')^2} = -1, \quad \hat{b} = \frac{b - \mathscr{S}_z(w)}{(w')^2} = 0.$$

Actually, the normalized lift $\hat{F} : M \to \mathscr{L}^+$ of f relative to w is the map defined in (13.33) with $r = s = 1/\sqrt{2}$, which is the minimal Clifford torus in $\mathbf{S}^3$ composed with the conformal diffeomorphism (12.23) of $\mathbf{S}^3$ with $\mathscr{M}$.

14.2 T-Transforms and Calapso's equation

Definition 14.12. A *T-transform* of an isothermic immersion $f : M \to \mathscr{M}$ with principal function $z : M \to \mathbf{C}$ is an isothermic immersion $\hat{f} : M \to \mathscr{M}$ with the same Calapso potential as f.

T-transforms are analogs of associates of CMC immersions in the space forms.

Theorem 14.13 (T-Transforms of Isothermic Immersions). *Suppose $z : M \to \mathbf{C}$ is a principal function for an isothermic immersion $f : M \to \mathscr{M}$ of a simply connected Riemann surface M. Let $k : U \to \mathbf{R}$ be the Calapso potential and let $b : U \to \mathbf{C}$ be the second order invariant of f relative to z. Then the structure equations* (13.17) *and* (13.18) *for f become*

$$b_{\bar{z}} = (k^2)_z, \quad \Im(k_{zz} + \frac{1}{2} bk) = 0, \tag{14.3}$$

where $\Im(p)$ denotes the imaginary part of the complex number p. For each constant $t \in \mathbf{R}$, there exists an isothermic immersion $f_t : M \to \mathscr{M}$, called the T-transform of f, such that $z : M \to \mathbf{C}$ remains a local principal complex coordinate for f_t, the Calapso potential for f_t relative to z is $k_t = k$ and the second order Möbius invariant of f_t relative to z is $b_t = b + t$. If $t \neq s$, then f_t is not Möbius congruent to f_s. If $\hat{f} : M \to \mathscr{M}$ is a T-transform of f, then $\hat{f} = f_t$, for some constant t.

Proof. Let $\{U_\alpha, z_\alpha\}_{\alpha \in \mathscr{A}}$ be an atlas of principal charts on M such that $z|_{U_\alpha} = z_\alpha$ for each $\alpha \in \mathscr{A}$. Let k_α be the Calapso potential and b_α the second order Möbius invariant of f relative to z_α. Since k_α is real valued on U_α, the structure equations (13.17) and (13.18) become

$$b_{\alpha \bar{z}_\alpha} = (k_\alpha^2)_{z_\alpha} \quad \Im(k_{\alpha zz} + \frac{1}{2} b_\alpha k_\alpha) = 0.$$

Then (14.3) follows from this because $k|_{U_\alpha} = k_\alpha$, $b|_{U_\alpha} = b_\alpha$ and $z|_{U_\alpha} = z_\alpha$.

For any constant $t \in \mathbf{R}$, the smooth functions $k_t = k : M \to \mathbf{R}$ and $b_t = b + t : M \to \mathbf{C}$ satisfy the structure equations (14.3). By Theorem 13.8, there exists an immersion $f_t : M \to \mathcal{M}$ with Calapso potential $k_t = k$ and second order Möbius invariant $b_t = b + t$. As the invariants b_t are distinct for distinct values of t, the immersions f_t are noncongruent for distinct values of t.

If $\hat{f} : M \to \mathcal{M}$ is a T-transform of f, and if $\hat{b}$ is the second order invariant of $\hat{f}$ relative to z, then $\hat{b}_{\bar{z}} = b_{\bar{z}}$ and $\Im(\hat{b}) = \Im(b)$, by (14.3). That a real holomorphic function must be a real constant then implies that $\hat{b} = b + t$, for some real constant t, so $\hat{f} = f_t$ up to Möbius transformation. □

Proposition 14.14 (Calapso's equation [27]). *Let M be a simply connected Riemann surface with a complex coordinate $z : M \to \mathbf{C}$.*

If $f : M \to \mathcal{M}$ is a conformal isothermic immersion for which (M, z) is a principal chart with first order invariant $k : M \to \mathbf{R} \setminus \{0\}$, then k satisfies Calapso's equation

$$\Delta\left(\frac{k_{xy}}{k}\right) + 2(k^2)_{xy} = 0, \tag{14.4}$$

where $z = x + iy$ and $\Delta = 4\frac{\partial^2}{\partial z \partial \bar{z}} = \frac{\partial^2}{\partial x^2} + \frac{\partial^2}{\partial y^2}$ is the Laplace operator.

Conversely, if $k : M \to \mathbf{R} \setminus \{0\}$ is a smooth function satisfying Calapso's equation (14.4)*, then there exists a conformal isothermic immersion $f : M \to \mathcal{M}$ for which (M, z) is a principal chart whose first order invariant is k.*

Proof. For a real valued function g we have $\Im(g_{zz}) = -\frac{1}{2} g_{xy}$ since $g_{zz} = \frac{1}{4}(g_{xx} - g_{yy} - 2i g_{xy})$. Let $b : M \to \mathbf{C}$ be the second order invariant of f relative to z. Write the second structure equation in (14.3) as $\Im(\frac{k_{zz}}{k} + b/2) = 0$. To it apply the Laplace operator Δ, which commutes with $\Im$, and then use the first structure equation in (14.3), to get

$$\begin{aligned}
0 &= \Delta \Im\left(\frac{k_{zz}}{k} + \frac{1}{2} b\right) = \Delta\left(-\frac{1}{2}\frac{k_{xy}}{k}\right) + \frac{1}{2}\Im(\Delta b) \\
&= -\frac{1}{2}\Delta\left(\frac{k_{xy}}{k}\right) + \frac{1}{2}\Im(4(b_{\bar{z}})_z) = -\frac{1}{2}\Delta\left(\frac{k_{xy}}{k}\right) + 2\Im((k^2)_{zz}) \\
&= -\frac{1}{2}\Delta\left(\frac{k_{xy}}{k}\right) - (k^2)_{xy},
\end{aligned}$$

from which (14.4) follows.

To prove the converse, we construct a smooth function $b : M \to \mathbf{C}$ such that k and b satisfy the structure equations (14.3). The result will then follow from Theorem 13.8. If we write $b = b_1 + i b_2$, where b_1 and b_2 are real valued functions on M, then the second structure equation in (14.3) holds if and only if

$$b_2 = -2\Im\left(\frac{k_{zz}}{k}\right) = \frac{k_{xy}}{k}.$$

The first structure equation holds if and only if

$$b_{1x} = 2kk_x + b_{2y}, \quad b_{1y} = -2kk_y - b_{2x}.$$

A solution b_1 of these two equations exists if and only if

$$(2kk_x + b_{2y})_y + (2kk_y + b_{2x})_x = 0,$$

which is true by Calapso's equation (14.4). □

Definition 14.15. A *principal* frame field along a conformal immersion $f : M \to \mathscr{M}$ is a Möbius frame field $Y : U \subset M \to \mathbf{M\ddot{o}b}_+$ along f for which

$$\omega_0^3 = 0, \quad \omega_1^3 \wedge \omega_0^1 = 0 = \omega_2^3 \wedge \omega_0^2.$$

Then $\omega_1^2 = p\omega_0^1 + q\omega_0^2$, for smooth functions $p, q : U \to \mathbf{R}$, and the *criterion form* of Y is

$$\alpha = \omega_0^0 + q\omega_0^1 - p\omega_0^2.$$

The conditions $\omega_1^3 \wedge \omega_0^1 = 0 = \omega_2^3 \wedge \omega_0^2$ on U are equivalent to the conditions $\omega_1^3 = a\omega_0^1$ and $\omega_2^3 = c\omega_0^2$, for smooth $a, c : U \to \mathbf{R}$. The principal vectors are the zeros of $\omega_0^2\omega_1^3 - \omega_0^1\omega_2^3 = (a-c)\omega_0^1\omega_0^2$, by Lemma 13.21. If f is umbilic free on U, then $a - c$ is never zero on U, so the integral curves of $\omega_0^1 = 0$ and $\omega_0^2 = 0$ are the lines of curvature of f in U.

Any second order Möbius frame field along f is principal, and these exist on a neighborhood of any nonumbilic point. We have an isothermic criterion analogous to Theorem 9.12 given in terms of a principal frame field.

Theorem 14.16 (Isothermic criterion form). *Let $f : M \to \mathscr{M}$ be a conformal immersion of a Riemann surface M.*

If f is isothermic, then each point of M has a neighborhood on which there is a principal frame field along f. If the nonumbilic points of f are dense in M, then for any principal frame field along f the criterion form is closed.

Conversely, if each point of M has a neighborhood on which there is a principal frame field with closed criterion form, then f is isothermic.

Proof. If f is isothermic, then for any $m \in M$ there exists a principal complex coordinate chart about m. The Möbius frame field adapted to this complex coordinate is principal with criterion form $\alpha = 0$.

Suppose $Y : U \to \mathbf{M\ddot{o}b}_+$ is any principal frame field along isothermic f with criterion form α. If we show that $d\alpha = 0$ on some neighborhood of any nonumbilic point $m \in U$, then $d\alpha = 0$ on U under the further assumption that the nonumbilic points of f are dense in M. Let $z = x + iy$ be a principal complex coordinate on a connected, umbilic free neighborhood $V \subset U$ of m. The integral curves of $\omega_0^1 = 0$ and of $\omega_0^2 = 0$ are the lines of curvature of f, as are the coordinate curves $dx = 0$ and

$dy = 0$. Thus $\omega_0^1 = ldx$, $\omega_0^2 = ndy$ or $\omega_0^1 = ldy$, $\omega_0^2 = ndx$, for smooth $l, n : U \to \mathbf{R}$. Since $\omega_0^1 \wedge \omega_0^2$ and $dx \wedge dy$ are positive, we must have $ln > 0$, respectively, $ln < 0$ on V. By conformality, $\omega_0^1\omega_0^1 + \omega_0^2\omega_0^2$ is a multiple of $dx^2 + dy^2$, so we must have $l^2 = n^2$. Thus, either $\omega_0^1 = ldx$, $\omega_0^2 = ldy$ or $\omega_0^1 = ldy$, $\omega_0^2 = -ldx$. In the latter case, replace z by the principal coordinate iz, to get the former case. Finally, if $l < 0$, replace z by $-z$. Hence, we may assume that on some open neighborhood $V \subset U$ of m,

$$\omega_0^1 + i\omega_0^2 = e^u dz,$$

for some smooth $u : V \to \mathbf{R}$. With $\omega_1^2 = p\omega_0^1 + q\omega_0^2$ and $\omega_0^0 = v\omega_0^1 + w\omega_0^2$, for smooth $p, q, v, w : U \to \mathbf{R}$, the criterion form is

$$\alpha = \omega_0^0 + q\omega_0^1 - p\omega_0^2 = \bar{Q}dz + Qd\bar{z}$$

on V, where

$$Q = \frac{1}{2}e^u(v + q + i(w - p)) : V \to \mathbf{C}$$

is smooth, and $\omega_0^0 - i\omega_1^2 = Pdz + Qd\bar{z}$, for some smooth $P : V \to \mathbf{C}$. By the structure equations,

$$-e^u u_{\bar{z}} dz \wedge d\bar{z} = d(\omega_0^1 + i\omega_0^2) = (\omega_0^0 - i\omega_1^2) \wedge e^u dz = -e^u Q dz \wedge d\bar{z},$$

so $Q = u_{\bar{z}}$ and $\alpha = u_z dz + u_{\bar{z}} d\bar{z} = du$ is closed on V.

Conversely, given a point $m \in M$, let $Y : U \to \mathbf{M\ddot{o}b}_+$ be a principal frame field on a neighborhood U of m with closed criterion form α on U. By an argument identical to that given in the proof of the converse of Theorem 9.12, there exists a complex coordinate z on a neighborhood $V \subset U$ of m such that $\omega_0^1 + i\omega_0^2 = e^u dz$ on V. Now Y principal implies $\omega_1^3 = a\omega_0^1$, $\omega_2^3 = c\omega_0^2$, for smooth $a, c : U \to \mathbf{R}$, so

$$\omega_1^3 - i\omega_2^3 = \frac{a-c}{2}(\omega_0^1 + i\omega_0^2) + \frac{a+c}{2}(\omega_0^1 - i\omega_0^2) = \frac{a-c}{2}e^u dz + \frac{a+c}{2}e^u d\bar{z}.$$

Problem 13.42 now implies that the first order invariant of f relative to z is $k = \frac{a-c}{2}e^u$, which is real valued on V, so z is a principal complex coordinate about m. Hence, f is isothermic. □

Corollary 14.17. *A conformal immersion $f : M \to \mathscr{M}$ with dense set of nonumbilic points M' is isothermic if and only if its third order Möbius invariant $p_2 = 0$ on M'.*

Proof. If f is isothermic, then each point of M' is contained in a principal complex coordinate neighborhood (U, z) for which the first order Möbius invariant $k > 0$ on U. Then Q, the coefficient of the quartic form $\mathscr{Q}_f$ relative to z must be real, by its formula in Proposition 13.37, and thus $p_2 = 0$ on U by (13.48).

Conversely, suppose $p_2 = 0$ on M'. Any point of M' has a neighborhood U on which there is a Bryant central frame $Y : U \to \mathbf{M\ddot{o}b}_+$, by Lemma 12.34. This frame is principal, and by (12.51) $\omega_1^2 = q_1\omega_0^1 + q_2\omega_0^2$, and $\omega_0^0 = -2(q_2\omega_0^1 - q_1\omega_0^2)$, so

$$\alpha = \omega_0^0 + q_2\omega_0^1 - q_1\omega_0^2 = \frac{1}{2}\omega_0^0.$$

By (12.52) and the structure equations of **Möb**, we get

$$d\alpha = -\frac{1}{2}(\omega_1^0 \wedge \omega_0^1 + \omega_2^0 \wedge \omega_0^2) = p_2\omega_0^1 \wedge \omega_0^2 = 0$$

on U. Hence, f is isothermic by Theorem 14.16. □

Corollary 14.18. *If $f : M \to \mathscr{M}$ is a Dupin immersion without umbilics, then it is isothermic and relative to any principal complex coordinate the Calapso potential is constant.*

Proof. By Theorem 12.51, the Möbius invariants $q_1 = q_2 = p_2 = 0$, so f is isothermic. Relative to a principal complex coordinate (U, z) in M, the Calapso potential $k : U \to \mathbf{R}$ satisfies $k_z = 0$ on U, by (13.48) in Proposition 13.37, so it is constant. □

Definition 14.19. An immersion $f : M \to \mathscr{M}$ is *densely nonisothermic* if for any complex chart (U, z) in M the first order Möbius invariant k of f relative to z has nonzero imaginary part on a dense subset of U. In other words, f is not isothermic on any open subset of M.

Theorem 14.20 (Burstall, Pedit, Pinkall [26]). *Suppose that the Riemann surface M has a global complex coordinate z. If $f, \hat{f} : M \to \mathscr{M}$ are conformal immersions for which their first order Möbius invariants k and $\hat{k}$ relative to z are equal at every point of M, $\hat{k} = k$ on M, and if there is no open subset M' of M for which $f : M' \to \mathscr{M}$ is isothermic, then f and $\hat{f}$ are Möbius congruent; that is, there exists an element $T \in$ **Möb** such that $\hat{f} = T \circ f$.*

Proof. Let b and $\hat{b}$ be the second order Möbius invariants of f and $\hat{f}$ relative to z. Then the assumption $\hat{k} = k$ and the structure equation (13.17) imply that $(\hat{b} - b)_{\bar{z}} = 0$. Hence, $\hat{b} - b$ is holomorphic on M. Seeking a contradiction, suppose $\hat{b} - b$ is not zero at some point $m \in M$. Then a square root of it is defined and nonzero on a neighborhood U of m. Let $\hat{b} - b = u + iv$ and $k = p + iq$, where u, v, p, q are smooth, real valued functions on U. Then the structure equation (13.18) gives

$$0 = \Im(\overline{(\hat{b} - b)}k) = uq - vp,$$

which implies that $(p, q) = r(u, v)$, for some smooth real valued function r on U. Hence,

$$k = r(u + iv) = r(\hat{b} - b)$$

is a real valued function times the square of a holomorphic function on U, which implies that $f : U \to \mathscr{M}$ is isothermic, contrary to our assumption about f. Thus, $\hat{b} = b$ at every point of M, and our result now follows from Theorem 13.8. □

14.3 Möbius deformation

Let $f, \hat{f} : M \to \mathscr{M}$ be smooth maps of a 2-dimensional manifold into Möbius space. For a given point $m_0 \in M$, let $\{U; x^1, x^2\}$ be a local coordinate system of M about m_0. Since $\mathbf{M\ddot{o}b}_+$ acts transitively on $\mathscr{M}$, we may assume that

$$f(m_0) = \hat{f}(m_0) = [\boldsymbol{\delta}_0].$$

Let $\mathbf{F}, \hat{\mathbf{F}} : U \to \mathscr{L} \subset \mathbf{R}^{4,1}$ be arbitrary lifts of f and $\hat{f}$, where $\mathbf{F} = F^0\boldsymbol{\delta}_0 + \sum_1^3 F^i\boldsymbol{\epsilon}_i + F^4\boldsymbol{\delta}_4$, $\hat{\mathbf{F}} = \hat{F}^0\boldsymbol{\delta}_0 + \sum_1^3 \hat{F}^i\boldsymbol{\epsilon}_i + \hat{F}^4\boldsymbol{\delta}_4$. Let $\xi^a = F^a/F^0$, $\zeta^a = \hat{F}^a/\hat{F}^0$, $a = 1,2,3,4$, and define $\xi = \boldsymbol{\delta}_0 + \sum_1^3 \xi^i\boldsymbol{\epsilon}_i + \xi^4\boldsymbol{\delta}_4$, $\zeta = \boldsymbol{\delta}_0 + \sum_1^3 \zeta^i\boldsymbol{\epsilon}_i + \zeta^4\boldsymbol{\delta}_4$.

Definition 14.21. We say that $f, \hat{f} : M \to \mathscr{M}$ *agree to order k at $m_0 \in M$* if $\xi(m_0) = \zeta(m_0)$ and for any integer r, $1 \le r \le k$,

$$\frac{\partial^r \xi^a}{\partial x^{i_1} \cdots \partial x^{i_r}}(m_0) = \frac{\partial^r \zeta^a}{\partial x^{i_1} \cdots \partial x^{i_r}}(m_0),$$

for all $i_1, \ldots, i_r$ and $a = 1,2,3,4$. This amounts to saying that $f, \hat{f}$ have the same kth order Taylor polynomials at m_0.

We adopt the following notation. If $\lambda = \lambda_{i_1 \cdots i_h} dx^{i_1} \cdots dx^{i_h}$ is a homogeneous symmetric differential form of degree h on U, by $j(\lambda)$ we mean the homogeneous symmetric differential form of degree $h+1$ defined by

$$j(\lambda) := \frac{\partial \lambda_{i_1 \cdots i_h}}{\partial x^{i_{h+1}}} dx^{i_1} \cdots dx^{i_h} dx^{i_{h+1}}.$$

Lemma 14.22. *The smooth maps $f, \hat{f} : M \to \mathscr{M}$ agree to order k at $m_0 \in M$ if and only if*

$$j^r(\hat{\mathbf{F}})_{|m_0} = \sum_{h=0}^{r} \binom{r}{h} j^{r-h}(\hat{F}^0/F^0)_{|m_0} j^h(\mathbf{F})_{|m_0} \quad (r = 0, \ldots, k), \tag{14.5}$$

for arbitrary lifts $\hat{\mathbf{F}}, \mathbf{F} : U \to \mathscr{L} \subset \mathbf{R}^{4,1}$ such that $f = [\mathbf{F}]$, $\hat{f} = [\hat{\mathbf{F}}]$.

Proof. If f and $\hat{f}$ agree to order k at m_0, since $\mathbf{F} = F^0\xi$, $\hat{\mathbf{F}} = \hat{F}^0\zeta$ and $j^r(\xi)(m_0) = j^r(\zeta)(m_0)$, $r = 0, \ldots, k$, we compute

$$
\begin{aligned}
j^r(\hat{\mathbf{F}})_{|m_0} &= \sum_{h=0}^{r}\binom{r}{h} j^h(\hat{F}^0)_{|m_0} j^{r-h}(\zeta)_{|m_0} = \sum_{h=0}^{r}\binom{r}{h} j^h(\hat{F}^0)_{|m_0} j^{r-h}(\xi)_{|m_0}\\
&= \sum_{h=0}^{r}\binom{r}{h} j^h(\hat{F}^0)_{|m_0} j^{r-h}(\frac{\mathbf{F}}{F_0})_{|m_0}\\
&= \sum_{h=0}^{r}\sum_{m=0}^{r-h}\binom{r}{h}\binom{r-h}{m} j^h(\hat{F}^0)_{|m_0} j^{r-h-m}(\frac{1}{F_0})_{|m_0} j^m(\mathbf{F})_{|m_0}\\
&= \sum_{m=0}^{r}\sum_{h=0}^{r-m}\binom{r}{m}\binom{r-m}{h} j^h(\hat{F}^0)_{|m_0} j^{r-h-m}(\frac{1}{F_0})_{|m_0} j^m(\mathbf{F})_{|m_0}\\
&= \sum_{m=0}^{r}\binom{r}{m} j^{r-m}(\hat{F}^0/F^0)_{|m_0} j^m(\mathbf{F})_{|m_0},
\end{aligned}
$$

and hence condition (14.5).

Conversely, if conditions (14.5) hold true for arbitrary lifts $\mathbf{F}$ and $\hat{\mathbf{F}}$, by choosing $\mathbf{F} = \boldsymbol{\delta}_0 + \sum_1^3 \xi^i \boldsymbol{\epsilon}_i + \xi^4 \boldsymbol{\delta}_4$ and $\hat{\mathbf{F}} = \boldsymbol{\delta}_0 + \sum_1^3 \zeta^i \boldsymbol{\epsilon}_i + \zeta^4 \boldsymbol{\delta}_4$, then f and $\hat{f}$ agree to order k. □

Corollary 14.23. *In particular, we have:*

- *$f, \hat{f}$ agree at first order if and only if*

$$
\hat{\mathbf{F}}_{|m_0} = \rho_{0|m_0}\mathbf{F}_{|m_0}, \quad j^1(\hat{\mathbf{F}})_{|m_0} = \rho_{1|m_0}\mathbf{F}_{|m_0} + \rho_{0|m_0} j^1(\mathbf{F})_{|m_0}, \tag{14.6}
$$

  where $\rho_0 = \hat{F}^0/F^0$ and $\rho_1 = j^1(\hat{F}^0/F^0)$.
- *$f, \hat{f}$ agree at second order if and only if* (14.6) *holds and*

$$
j^2(\hat{\mathbf{F}})_{|m_0} = \rho_{2|m_0}\mathbf{F}_{|m_0} + 2\rho_{1|m_0} j^1(\mathbf{F})_{|m_0} + \rho_{0|m_0} j^2(\mathbf{F})_{|m_0}, \tag{14.7}
$$

  where $\rho_2 = j^2(\hat{F}^0/F^0)$.
- *$f, \hat{f}$ agree at third order if and only if* (14.6) *and* (14.7) *hold and*

$$
\begin{aligned}
j^3(\hat{\mathbf{F}})_{|m_0} &= \rho_{3|m_0}\mathbf{F}_{|m_0} + 3\rho_{2|m_0} j^1(\mathbf{F})_{|m_0}\\
&\quad + 3\rho_{1|m_0} j^2(\mathbf{F})_{|m_0} + \rho_{0|m_0} j^3(\mathbf{F})_{|m_0},
\end{aligned} \tag{14.8}
$$

  where $\rho_3 = j^3(\hat{F}^0/F^0)$.

Definition 14.24. Two smooth maps $f, \hat{f} : M \to \mathscr{M}$ are kth *order Möbius deformations of each other* if there exists a smooth map $D : M \to \mathbf{M\ddot{o}b}_+$ such that for each point $m \in M$, the maps $\hat{f}$ and $D(m)f$ *agree to order k at m*. The map D is called the *infinitesimal displacement* of the deformation. If $D(m)$ does not depend on $m \in M$, the deformation is called *trivial*, in which case $\hat{f} = Df$ is Möbius congruent to f.

A given map $f : M \to \mathscr{M}$ is *rigid* to kth order deformation if there is no nontrivial kth order deformation of it. It is *Möbius deformable of order k* if it admits a nontrivial kth order deformation.

Proposition 14.25. *Let f, $\hat{f} : M \to \mathscr{M}$ be smooth immersions of the oriented 2-dimensional manifold M into Möbius apace $\mathscr{M}$, viewed as homogeneous space of the group $\mathbf{M\ddot{o}b}_+$. Then the following statements hold true:*

1. *The immersions f and $\hat{f}$ are first order Möbius deformations of each other if and only if there exist first order frame fields $Y, \hat{Y} : M \to \mathbf{M\ddot{o}b}_+$ along f and $\hat{f}$, respectively, such that*

$$\hat{Y}^* \omega_0^1 = Y^* \omega_0^1, \quad \hat{Y}^* \omega_0^2 = Y^* \omega_0^2, \tag{14.9}$$

where ω is the Maurer–Cartan form of $\mathbf{M\ddot{o}b}_+$. In particular, any smooth immersion $f : M \to \mathscr{M}$ is Möbius deformable of first order.

2. *The immersions f and $\hat{f}$ are second order Möbius deformations of each other if and only if there exist second order frame fields Y and $\hat{Y}$ along f and $\hat{f}$, respectively, such that*

$$\begin{aligned} \hat{Y}^* \omega_0^1 &= Y^* \omega_0^1, & \hat{Y}^* \omega_0^2 &= Y^* \omega_0^2, & \hat{Y}^* \omega_1^2 &= Y^* \omega_1^2, \\ \hat{Y}^* \omega_0^0 &= Y^* \omega_0^0, & \hat{Y}^* \omega_1^3 &= Y^* \omega_1^3, & \hat{Y}^* \omega_2^3 &= Y^* \omega_2^3. \end{aligned} \tag{14.10}$$

3. *The immersions f and $\hat{f}$ are third order Möbius deformations of each other if and only if there exist second order frame fields Y and $\hat{Y}$ along f and $\hat{f}$, respectively, such that*

$$\hat{Y}^* \omega = Y^* \omega. \tag{14.11}$$

Thus, any smooth immersion $f : M \to \mathscr{M}$ is rigid to third order.

Proof. (1) Suppose f and $\hat{f}$ are first order Möbius deformations of each other. Then $D : M \to \mathbf{M\ddot{o}b}_+$ exists so that $\hat{f}$ and $D(m)f$ agree to order one at m, for each $m \in M$. Let Y be a first order Möbius frame field along f and define $\hat{Y} : M \to \mathbf{M\ddot{o}b}_+$ by $\hat{Y}(m) = D(m)Y(m)$, for each $m \in M$. Then $\hat{Y}$ is a frame field along $\hat{f}$ and $Y' = D(m_0)Y : M \to \mathbf{M\ddot{o}b}_+$ is a frame field along $D(m_0)f$, for each $m_0 \in M$. According to (14.6) in Corollary 14.23, we have

$$\hat{\mathbf{Y}}_0(m_0) = \rho_0(m_0)\mathbf{Y}'_0(m_0) \tag{14.12}$$

$$d\hat{\mathbf{Y}}_{0|m_0} = \rho_0(m_0)d\mathbf{Y}'_{0|m_0} + \rho_1(m_0)\mathbf{Y}'_0(m_0). \tag{14.13}$$

Equation (14.12) yields

$$\rho_0(m_0) = 1, \tag{14.14}$$

since $\hat{Y}$ and Y' agree at m_0. Now, the structure equations of $\mathbf{M\ddot{o}b}_+$ imply

$$d\hat{\mathbf{Y}}_0 = \sum_{j=0}^{5} \hat{\alpha}_0^j \hat{\mathbf{Y}}_j, \quad d\mathbf{Y}'_0 = \sum_{j=0}^{5} \alpha_0^j \mathbf{Y}'_j, \tag{14.15}$$

where we have set $\alpha = Y^*\omega$ and $\hat{\alpha} = \hat{Y}^*\omega$. Substitution of (14.15) into (14.13) yields

$$\rho_1 = (\hat{\alpha}_0^0 - \alpha_0^0), \quad \hat{\alpha}_0^i = \alpha_0^i, \quad \text{at } m_0, \text{ for } i = 1,2,3. \tag{14.16}$$

Since m_0 was chosen arbitrarily, equations (14.14) and (14.16) hold on all M, which implies that $\hat{Y}$ is a first order frame along $\hat{f}$ with the required properties.

Conversely, suppose (14.9) hold for first order frames $Y, \hat{Y} : M \to \mathbf{M\ddot{o}b}_+$ along f and $\hat{f}$, respectively. Define $D : M \to \mathbf{M\ddot{o}b}_+$ by

$$D(m) = \hat{Y}(m)Y(m)^{-1}, \quad m \in M.$$

It follows from (14.14), (14.15), and (14.16) that (14.12) and (14.13) hold, which proves that D induces a first order Möbius deformation.

(2) We retain the notation of part (1) and suppose that f and $\hat{f}$ are second order Möbius deformations of each other. Then $\hat{f}$ and $D(m)f$ agree to second order at m, for each $m \in M$. Let $Y : M \to \mathbf{M\ddot{o}b}_+$ be a second order frame field along f and $\hat{Y}$ and Y' be as in part (1). We have to show that $\hat{Y}$ defines a second order frame field along $\hat{f}$ such that

$$\hat{\alpha}_0^1 = \alpha_0^1, \quad \hat{\alpha}_0^2 = \alpha_0^2, \quad \hat{\alpha}_1^2 = \alpha_1^2, \quad \hat{\alpha}_0^0 = \alpha_0^0, \quad \hat{\alpha}_1^3 = \alpha_1^3, \quad \hat{\alpha}_2^3 = \alpha_2^3. \tag{14.17}$$

By Corollary 14.23 and the discussion in part (1), the frame fields $\hat{Y}$ and Y' must satisfy (14.12), (14.13), and

$$j^2(\hat{\mathbf{Y}}_0)_{|m_0} = \rho_2(m_0)\mathbf{Y}'_{0|m_0} + 2(\hat{\alpha}_0^0 - \alpha_0^0)_{|m_0} d\mathbf{Y}'_{0|m_0} + j^2(\mathbf{Y}'_0)_{|m_0}. \tag{14.18}$$

Writing out (14.18), using the structure equations (14.15), equations $\hat{\alpha}_0^i = \alpha_0^i$, $i = 1,2$, $\hat{\alpha}_0^3 = \alpha_0^3$, and the fact that $\hat{\mathbf{Y}}_0(m_0) = \mathbf{Y}'_0(m_0)$, we find

$$\begin{aligned}
\rho_2 &= j(\hat{\alpha}_0^0 - \alpha_0^0) + (\hat{\alpha}_0^0 - \alpha_0^0)^2 + \alpha_0^1(\hat{\alpha}_1^0 - \alpha_1^0) + \alpha_0^2(\hat{\alpha}_2^0 - \alpha_2^0), \\
0 &= \alpha_0^2(\hat{\alpha}_2^1 - \alpha_2^1) - \alpha_0^1(\hat{\alpha}_0^0 - \alpha_0^0), \\
0 &= \alpha_0^1(\hat{\alpha}_1^2 - \alpha_1^2) - \alpha_0^2(\hat{\alpha}_0^0 - \alpha_0^0), \\
0 &= \alpha_0^1(\hat{\alpha}_1^3 - \alpha_1^3) + \alpha_0^2(\hat{\alpha}_2^3 - \alpha_2^3).
\end{aligned}$$

From these equation it follows that

$$\hat{\alpha}_2^1 = \alpha_2^1, \quad \hat{\alpha}_0^0 = \alpha_0^0, \quad \hat{\alpha}_i^3 = \alpha_i^3 \quad (i = 1, 2). \tag{14.19}$$

Thus $\hat{Y}$ is a second order frame field along $\hat{f}$ and the conditions (14.17) are satisfied.

Conversely, suppose (14.17) hold for second order frame fields $Y, \hat{Y} : M \to \mathbf{M\ddot{o}b}_+$ along f and $\hat{f}$, respectively. As above, let $D : M \to \mathbf{M\ddot{o}b}_+$ be given by

$$D(m) = \hat{Y}(m)Y(m)^{-1}, \quad m \in M.$$

By reversing the arguments above, we see that (14.12), (14.13), and (14.18) are satisfied, so that D induces a second order Möbius conformal deformation of f and $\hat{f}$.

As for (3), writing out (14.8), after some lengthy but straightforward computations one can prove that $\alpha = Y^{-1}dY = \hat{Y}^{-1}d\hat{Y} = \hat{\alpha}$. By the Cartan–Darboux theorem, we then have that $dD_{|m} = 0$, for every $m \in M$. □

The next result characterizes isothermic surfaces as the surfaces which are Möbius deformable of second order.

Proposition 14.26. *An umbilic free immersion $f : M \to \mathcal{M}$ of an oriented surface into Möbius space is Möbius deformable of order two if and only if it is isothermic.*

Proof. Assume that $f : M \to \mathcal{M}$ and $\hat{f} : M \to \mathcal{M}$ are second order Möbius deformation of each other. Let Y and $\hat{Y}$ be central frame fields along f and $\hat{f}$, respectively, and write $\alpha = Y^*\omega$, $\hat{\alpha} = \hat{Y}^*\omega$. According to Proposition 14.25, we then have

$$\hat{\alpha}_0^1 = \alpha_0^1, \quad \hat{\alpha}_0^2 = \alpha_0^2, \tag{14.20}$$

$$\hat{\alpha}_0^0 = \alpha_0^0, \quad \hat{\alpha}_1^2 = \alpha_1^2. \tag{14.21}$$

Differentiating the two equations in (14.21) and using the structure equations of the group $\mathbf{M\ddot{o}b}_+$, we get

$$\alpha_1^0 - \hat{\alpha}_1^0 = r\alpha_0^1, \quad \alpha_2^0 - \hat{\alpha}_2^0 = -r\alpha_0^2, \tag{14.22}$$

for some smooth function $r : M \to \mathbf{R}$. We have $r \neq 0$ on an open, dense subset of M, since f and $\hat{f}$ are not Möbius congruent on any open subset of M. Next, differentiating (14.22), taking into account (14.21), and using again the structure equations, we obtain

$$(dr + 2rq_1\alpha_0^2) \wedge \alpha_0^1 = 0,$$
$$(dr - 2rq_2\alpha_0^1) \wedge \alpha_0^2 = 0,$$

which imply

$$dr - r\alpha_0^0 = 0, \tag{14.23}$$

since $\alpha_0^0 = 2(q_2\alpha_0^1 - q_1\alpha_0^2)$. By the structure equations,

$$d\alpha_0^0 = 2p_2\alpha_0^1 \wedge \alpha_0^2, \tag{14.24}$$

which combined with (14.23) implies that the third order invariant $p_2 = 0$, and hence f is isothermic.

Conversely, if $p_2 = 0$, by (14.24), α_0^0 is a closed 1-form. Given $m_0 \in M$, let U be a simply connected open neighborhood of m_0 and let $u : U \to \mathbf{R}$ be a smooth function such that $\alpha_0^0 = du$. Let $a = e^u$ and set

$$\hat{\alpha}_1^0 = \alpha_1^0 - a\alpha_0^1, \quad \hat{\alpha}_2^0 = \alpha_2^0 = a\alpha_0^2. \tag{14.25}$$

Now, the $\mathfrak{möb}$-valued 1-form

$$\hat{\alpha} = \begin{pmatrix} \alpha_0^0 & \hat{\alpha}_1^0 & \hat{\alpha}_2^0 & 0 & 0 \\ \alpha_0^1 & 0 & -\alpha_1^2 & -\alpha_0^1 & \hat{\alpha}_1^0 \\ \alpha_0^2 & \alpha_1^2 & 0 & \alpha_0^2 & \hat{\alpha}_2^0 \\ 0 & \alpha_0^1 & -\alpha_0^2 & 0 & 0 \\ 0 & \alpha_0^1 & \alpha_0^2 & 0 & -\alpha_0^0 \end{pmatrix}$$

satisfies the Maurer–Cartan structure equations, which implies the existence of $\hat{Y} : U \to \mathbf{Möb}_+$, unique up to left translations, such that $\hat{\alpha} = \hat{Y}^*\omega$. Let $\hat{f} : U \to \mathscr{M}$ be defined by $f(m) := [\hat{\mathbf{Y}}_0(m_0)]$, for each $m \in U$. Note that $\hat{Y}$ is a central frame field along $\hat{f}$ and that, in particular, Y and $\hat{Y}$ are second order frame fields satisfying the conditions of Proposition 14.25. □

14.4 Special isothermic immersions

In the following,

$$\mathbf{S}_0 = \mathbf{R}^3, \quad \mathbf{S}_+ = \mathbf{S}^3, \quad \mathbf{S}_- = \mathbf{H}^3,$$

$\epsilon \in \{0, +, -\}$, and $f_\epsilon : \mathbf{S}_\epsilon \to \mathscr{M}$ are the conformal embeddings (12.59), (12.56), and (12.62) of the space forms.

Definition 14.27. A *special isothermic immersion* is an umbilic free isothermic immersion $f : M \to \mathscr{M}$ of a connected Riemann surface M such that for some $\epsilon \in \{0, +, -\}$ there exists a constant mean curvature immersion $\mathbf{x} : M \to \mathbf{S}_\epsilon$ and a $T \in \mathbf{Möb}_+$ for which $T \circ f_\epsilon \circ \mathbf{x} = f : U \to \mathscr{M}$.

This defines a subclass of immersions studied by L. Bianchi [7]. The essence of the following result is in L. Bianchi's [7].

Theorem 14.28. *An umbilic free isothermic immersion $f : M \to \mathscr{M}$ is special isothermic if and only if for any connected principal complex chart (U,z) in M, the Calapso potential k relative to z satisfies*

$$4(\log k)_{z\bar{z}} = sk^{-2} - k^2, \tag{14.26}$$

on U, for some constant $s \in \mathbf{R}$ called the character of f relative to z.

Proof. Suppose f is special isothermic. Let $\mathbf{x} : M \to \mathbf{S}_\epsilon$ be a CMC H immersion for which $f_\epsilon \circ \mathbf{x} : M \to \mathscr{M}$ is Möbius congruent to f. Let (U,z) be a connected principal chart for f. The conformal factor e^u and Hopf invariant h of $\mathbf{x}$ relative to z are related to the Calapso potential k and second order invariant b of f by $k = he^u$ and $b = 2u_{zz} - 2u_z^2 + Hhe^{2u}$. Replacing z by iz, if necessary, we may assume $k > 0$ on U, and thus $h > 0$ on U. Since H is constant, the Codazzi equation (7.32), $(e^{2u}h)_{\bar{z}} = e^{2u}H_z = 0$ on U, implies that he^{2u} is a real valued holomorphic function on U, hence equal to a positive constant

$$t = he^{2u} = ke^u. \tag{14.27}$$

Then $0 = (ke^u)_{\bar{z}} = e^u(k_{\bar{z}} + ku_{\bar{z}})$ implies

$$k_{\bar{z}} = -ku_{\bar{z}}, \tag{14.28}$$

on U. Differentiating this with respect to $\bar{z}$ to find $k_{\bar{z}\bar{z}}$, using the fact that u is real valued, and using the formula above for b, we get

$$k_{\bar{z}\bar{z}} + \frac{1}{2}\bar{b}k = k\frac{H}{2}he^{2u} \tag{14.29}$$

on U. Note that the function $W = k_{\bar{z}\bar{z}} + \frac{1}{2}\bar{b}k$ is always real valued, by the structure equation (13.18) of k and b. Combining (14.27) and (14.29), we get

$$W/k = tH/2, \tag{14.30}$$

on U. By (14.28),

$$k_{\bar{z}z} = -(ku_{\bar{z}})_z = -ku_{\bar{z}z} - k_z u_{\bar{z}}.$$

Taking log of both sides of (14.27) and using the Gauss equation (7.31)

$$-u_{z\bar{z}} = \frac{1}{4}e^{2u}(\epsilon 1 + H^2 - |h|^2),$$

for $\mathbf{x}$, we obtain (14.26) on U for the constant

$$s = t^2(\epsilon 1 + H^2). \tag{14.31}$$

Conversely, suppose that (14.26) holds for any connected principal coordinate chart for f in M. Suppose first that M is simply connected. Then by Theorem 14.8, there exists a principal complex function $z : M \to \mathbf{C}$ with associated Calapso potential $k : M \to \mathbf{R}^+$ and second order Möbius invariant $b : M \to \mathbf{C}$ (see Definition 14.10). We may assume $k > 0$ on M in which case the structure equation (13.17) becomes

$$b_{\bar{z}} = 2kk_z. \tag{14.32}$$

Each point of M has a connected neighborhood on which z is a principal coordinate of f with invariants k and b restricted to this neighborhood. Thus, (14.26) holds on M for some constant s, since it holds on a neighborhood of any point of M. Now $W = k_{\bar{z}\bar{z}} + \frac{1}{2}\bar{b}k : M \to \mathbf{R}$ is a globally defined smooth function. Using (14.32), and solving for $s/4$ in (14.26), we get

$$\left(\frac{W}{k}\right)_z = \frac{1}{k^2}\left(\frac{s}{4}\right)_{\bar{z}} = 0 \tag{14.33}$$

on M, so W/k is a real constant on M. By the preceding calculations, if there exists a CMC H immersion $\mathbf{x} : M \to \mathbf{S}_\epsilon$, then (14.31) and (14.30) determine $\epsilon \in \{0, -, +\}$ and the real constants $t > 0$ and H. We then use (14.27) to define $e^u = t/k$ and $h = k^2/t$. The functions e^u, h, and H satisfy the Codazzi equation (7.32) and the Gauss equation (7.31) follows from (14.26).

The details depend on whether $s < 0$, $s = 0$, or $s > 0$.

If $s < 0$, then $\epsilon = -$ and $t^2 = 4W^2/k^2 - s > 0$ determines $t > 0$, and $H^2 = (s + t^2)/t^2 \geq 0$ determines H^2. Note $H = 0$ if and only if $W = 0$.

If $s = 0$, then (14.30) implies that H is zero or not depending on whether W is zero or not. If W is zero on M, then $f : M \to \mathscr{M}$ is Willmore, $\epsilon = 0$, and $H = 0$ by (14.31). If W is not identically zero, then $\epsilon = -$ and $H = \pm 1$.

If $s > 0$, then $tH = 2W/k$, so $\epsilon t^2 = s - 4(W/k)^2$. If this is not zero, it determines ϵ and $t > 0$, and then $H = 2(W/k)/t$. If $s - 4(W/k)^2 = 0$, then $\epsilon = 0$ and $tH = \sqrt{s}$. If $\mathbf{x} : M \to \mathbf{R}^3$ is the solution for $t = 1$, so $H = \sqrt{s}$, then the solution for any constant $t > 0$ is $\frac{1}{t}\mathbf{x} : M \to \mathbf{R}^3$, and f_0 composed with any of these is Möbius congruent to f.

To conclude the proof, we must prove the converse without the assumption that M be simply connected. In this general case, let $\mu : \tilde{M} \to M$ be the universal cover of M. Here $\tilde{M}$ is a Riemann surface and the projection map μ is holomorphic. For background on covering spaces, see, for example, [83] or [110]. Then $f \circ \mu : \tilde{M} \to \mathscr{M}$ satisfies the hypotheses of the theorem, including (14.26), so we have already proved that there exists a CMC H immersion $\tilde{\mathbf{x}} : \tilde{M} \to \mathbf{S}_\epsilon$, for some $\epsilon \in \{0, -, +\}$ and a $T \in \mathbf{M\ddot{o}b}_+$ such that

$$T \circ f_\epsilon \circ \tilde{\mathbf{x}} = f \circ \mu : \tilde{M} \to \mathscr{M}.$$

If $m,n \in \tilde{M}$ and if $\mu(m) = \mu(n)$, then

$$T \circ f_\epsilon \circ \tilde{\mathbf{x}}(m) = f \circ \mu(m) = f \circ \mu(n) = T \circ f_\epsilon \circ \tilde{\mathbf{x}}(n)$$

implies $\tilde{\mathbf{x}}(m) = \tilde{\mathbf{x}}(n)$. Hence, $\tilde{\mathbf{x}}$ descends to M in the sense that there exists a smooth map $\mathbf{x} : M \to \mathbf{S}_\epsilon$ such that $\mathbf{x} \circ \mu = \tilde{\mathbf{x}}$. Then $\mathbf{x}$ is a CMC H immersion satisfying $T \circ f_\epsilon \circ \mathbf{x} = f$ on M. □

Proposition 14.29 (CMC is T-transform of minimal). *If $\mathbf{x} : M \to \mathbf{S}_\epsilon$ is an umbilic free, constant mean curvature $H \neq 0$ immersion of a simply connected surface M, then there exists an $\hat{\epsilon} \in \{0,+,-\}$ and a minimal immersion $\hat{\mathbf{x}} : M \to \mathbf{S}_{\hat{\epsilon}}$ such that $f = f_\epsilon \circ \mathbf{x}$ is a T-transform of $\hat{f} = f_{\hat{\epsilon}} \circ \hat{\mathbf{x}}$.*

Proof. Let $z : M \to \mathbf{C}$ be a principal function of $f = f_\epsilon \circ \mathbf{x} : M \to \mathscr{M}$, with Calapso potential $k : M \to \mathbf{R}^+$ and second order Möbius invariant $b : M \to \mathbf{C}$. Use the notation of the above proof. Then $e^{2u}h = t$ is a nonzero constant on M that we may assume positive. The invariants relative to z of f are

$$k = te^{-u}, \quad b = 2u_{zz} - 2u_z^2 + Ht.$$

Let $\hat{f} : M \to \mathscr{M}$ be the T-transform of f obtained from the invariants

$$\hat{k} = k, \quad \hat{b} = b - Ht.$$

We want to find an immersion $\hat{\mathbf{x}} : M \to \mathbf{S}_{\hat{\epsilon}}$ with mean curvature $\hat{H} = 0$ such that $f_{\hat{\epsilon}} \circ \hat{\mathbf{x}}$ is Möbius congruent to $\hat{f}$. The conformal factor $e^{\hat{u}}$ and Hopf invariant $\hat{h}$ of $\hat{\mathbf{x}}$ must satisfy

$$\hat{h} = \hat{t}e^{-2\hat{u}}, \quad \hat{t}e^{-\hat{u}} = \hat{k} = k = e^{-u},$$

for some constant $\hat{t} > 0$. Then $\hat{u} = u + \log(\hat{t}/t)$. To satisfy the Gauss equation for $\hat{\mathbf{x}}$ requires

$$\hat{\epsilon}\hat{t}^2 = t^2(H^2 + \epsilon 1).$$

If $H^2 + \epsilon 1 \neq 0$, this equation has a unique solution $\hat{\epsilon} \in \{-,+\}$ and $\hat{t} > 0$. If $H^2 + \epsilon 1 = 0$, so $\mathbf{x}$ is a CMC ± 1 immersion in $\mathbf{H}^3$, then $\hat{\epsilon} = 0$ and $\hat{t}$ is any positive constant. This 1-parameter family of solutions consists of the dilations $\frac{\hat{t}}{t}\hat{\mathbf{x}}$ of the solution $\hat{\mathbf{x}} : M \to \mathbf{R}^3$ given by $\hat{t} = t$, which is the Lawson correspondent of the given immersion $\mathbf{x} : M \to \mathbf{H}^3$. □

Bryant [20] proved that if $f : M \to \mathscr{M}$ is Willmore, then its quartic differential $\mathscr{Q}$ of Definition 13.38 is holomorphic. In Bohle and Peters [12] and Bohle [11], the following more general result is attributed to K. Voss.

Theorem 14.30 (K. Voss [165]). *The Bryant quartic differential $\mathcal{Q}$ of an umbilic free immersion $f : M \to \mathcal{M}$ is holomorphic if and only if f is Willmore or $f : M' \to \mathcal{M}$ is special isothermic, where M' is the complement of a discrete set of points of M.*

Proof. Let $k, b : U \to \mathbf{C}$ be the invariants of f relative to a complex chart (U, z) in M. The umbilic free assumption means k is never zero on U. Then $\mathcal{Q} = Q(dz)^4$ on U, where

$$Q = kk_{z\bar{z}} + \frac{1}{4}k^2|k|^2 - k_z k_{\bar{z}}, \tag{14.34}$$

by Definition 13.38. By (13.19) in Theorem 13.9, the Willmore function of f relative to z is $W = k_{\bar{z}\bar{z}} + \frac{1}{2}\bar{b}k : U \to \mathbf{R}$. One calculates

$$Q_{\bar{z}} = kW_z - k_z W = k^2 (W/k)_z. \tag{14.35}$$

Suppose $\mathcal{Q}$ is holomorphic on M. Then Q is holomorphic on U, so $W = k\bar{g}$, for some holomorphic function g on U. There are two cases possible. Either W is identically zero on U, or it has only isolated zeros in U. By the transformation formula in Problem 13.44, whichever case holds for W on U, must be the case that holds for the Willmore function relative to any complex chart in M. Thus, in the former case, f is a Willmore immersion of M.

In the latter case, let $\mathcal{Z} \subset M$ be the discrete set of zeros of the Willmore function relative to any complex chart in M. Let $M' = M \setminus \mathcal{Z}$. Then g is never zero on $U' = U \cap M'$, so $k = W/\bar{g} = \frac{W}{|g|^2}g$, satisfies the criterion of Proposition 14.5 on U', which implies $f : U' \to \mathcal{M}$ is isothermic. Covering M with complex charts, we conclude that $f : M' \to \mathcal{M}$ is isothermic.

Given any point of M', we may assume z is principal on a connected neighborhood $U'' \subset M'$ of the point, so k is positive on U''. Then Q is holomorphic and real, hence constant on U'', and by (14.34),

$$4(\log k)_{z\bar{z}} = 4Qk^{-2} - k^2.$$

Thus, $f : M' \to \mathcal{M}$ is special isothermic.

Conversely, if $f : M \to \mathcal{M}$ is Willmore, then relative to any complex chart (U, z) we have $W = 0$ on U and thus Q is holomorphic on U by (14.35). If $f : M' \to \mathcal{M}$ is special isothermic, where M' is the complement of a discrete subset of M, then W/k relative to any principal coordinate is constant, by (14.30) (also by (14.33)), and hence Q is holomorphic by (14.35). □

Remark 14.31. By the preceding two propositions we see that any special isothermic immersion is a T-transform of a Willmore isothermic immersion. For the relation between the T-transform and the Lawson correspondence in $\mathbf{H}^3$ see Hertrich-Jeromin, Musso, and Nicolodi [88].

14.5 Thomsen's Theorem

Definition 14.32. A point p in a surface M is an *end* of an immersion $\mathbf{x} : M \setminus \{p\} \to \mathbf{S}_\epsilon$ if either $\epsilon = 0$ and $\lim_{m\to p} f_0 \circ \mathbf{x}(m) = [\boldsymbol{\delta}_4] \in \mathscr{M}$, or $\epsilon = -$ and $\lim_{m\to p} f_- \circ \mathbf{x}(m) \in \partial\mathscr{M}_-$, the boundary of $\mathscr{M}_- = f_-(\mathbf{H}^3)$.

Theorem 14.33 (Thomsen [160]). *If $f : M \to \mathscr{M}$ is a Willmore immersion of a connected Riemann surface M, if the set $\mathscr{D}$ of umbilic points of f is discrete, and if $f : M \setminus \mathscr{D} \to \mathscr{M}$ is isothermic, then there exists a minimal immersion $\mathbf{x} : M \setminus \mathscr{D} \to S_\epsilon$, for some $\epsilon \in \{+, -, 0\}$, and a $T \in$ **Möb**, such that*

$$f_\epsilon \circ \mathbf{x} = T \circ f : M \setminus \mathscr{D} \to \mathscr{M}. \tag{14.36}$$

If $p \in \mathscr{D}$, then either $\mathbf{x}$ extends smoothly to p and p is an umbilic point of $\mathbf{x}$, or p is an embedded end of $\mathbf{x}$.

Proof. From the proof of Theorem 14.30, f Willmore implies that the quartic form $\mathscr{Q}$ is holomorphic, and then $f : M \setminus \mathscr{D} \to \mathscr{M}$ isothermic implies it is special isothermic and the coefficient Q of $\mathscr{Q}$ relative to any principal coordinate z is constant. By Theorem 14.26, there exists a CMC H immersion $\mathbf{x} : M \setminus \mathscr{D} \to S_\epsilon$, for which H must be zero because W is identically zero, and there exists $T \in \mathbf{Möb}_+$ such that (14.36) holds. If $Q < 0$, then $\epsilon = -$, if $Q > 0$, then $\epsilon = +$, and if $Q = 0$, then $\epsilon = 0$, by the proof of Theorem 14.26.

For a point $p \in \mathscr{D}$, if

$$T(f(p)) \in f_\epsilon(\mathbf{S}_\epsilon),$$

(always the case if $\epsilon = +$), then $\mathbf{x}$ extends smoothly to p by $\mathbf{x} = f_\epsilon^{-1} \circ T \circ f$ on a neighborhood of p, and p is an umbilic point of $\mathbf{x}$. There exists a neighborhood U of p in M on which the immersion $T \circ f : U \to \mathscr{M}$ is an embedding. If

$$T(f(p)) \notin f_\epsilon(\mathbf{S}_\epsilon),$$

which is possible only if $\epsilon \neq +$, then $f_\epsilon \circ \mathbf{x} = T \circ f : U \setminus \{p\} \to f_\epsilon(\mathbf{S}_\epsilon)$ is an embedding, and thus $\mathbf{x} : U \setminus \{p\} \to \mathbf{S}_\epsilon$ is an embedding. In the Euclidean case, we must have $T(f(p)) = [\boldsymbol{\delta}_4]$, the only point not in $f_0(\mathbf{R}^3)$, and so $\lim_{m\to p} f_0 \circ \mathbf{x} = [\boldsymbol{\delta}_4]$ and p is an embedded end of $\mathbf{x} : M \setminus \mathscr{D} \to \mathbf{R}^3$, by Definition 14.32. In the hyperbolic case,

$$\lim_{m\to p} f_- \circ \mathbf{x}(m) = \lim_{m\to p} T \circ f(m) = T(f(p)) \in \partial\mathscr{M}_-$$

implies that p is an embedded end of $\mathbf{x} : M \setminus \mathscr{D} \to \mathbf{H}^3$. □

Corollary 14.34 (Bryant [20]). *If M is compact and if the quartic form $\mathscr{Q}$ of f is identically zero on M, then for some nonempty subset $\mathscr{E} \subset \mathscr{D}$, the minimal immersion $\mathbf{x} : M \setminus \mathscr{E} \to \mathbf{R}^3$ is complete, has finite total curvature, and each point of $\mathscr{E}$ is an embedded planar end.*

Proof. The discrete subset $\mathscr{D}$ of M must be finite when M is compact. By the Theorem, up to Möbius transformation, $f = f_0 \circ \mathbf{x}$ on $M \setminus \mathscr{D}$, where $\mathbf{x} : M \setminus \mathscr{D} \to \mathbf{R}^3$ is conformal and minimal. Since $\mathbf{x}$ cannot be a minimal immersion of the compact M, the set $\mathscr{E} \subset \mathscr{D}$ of ends of $\mathbf{x}$ must be nonempty. The minimal immersion $\mathbf{x} : M \setminus \mathscr{E} \to \mathbf{R}^3$ is complete, since if $\gamma(t)$ is a curve in $M \setminus \mathscr{E}$ that approaches a point of $\mathscr{E}$, then $\mathbf{x} \circ \gamma(t)$ is divergent. By the Theorem, these ends are all embedded. For the proof that the ends are planar – that is, they have zero logarithmic growth – we refer to Bryant [20]. The total curvature of $\mathbf{x}$ is

$$-\int_{M \setminus \mathscr{E}} K dA = \int_{M \setminus \mathscr{E}} (H^2 - K) dA = \int_{M \setminus \mathscr{E}} \Omega_f = \int_M \Omega_f,$$

which is finite since M is compact and the conformal area element Ω_f is smoothly defined on all of M. □

Bryant and R. Kusner [23] have parametrized Boy's surface to give a Willmore immersion $\mathbf{x} : \mathbf{RP}^2 \to \mathbf{R}^3$ shown in Figure 14.1. The figure is partially transparent in order to show self-intersections.

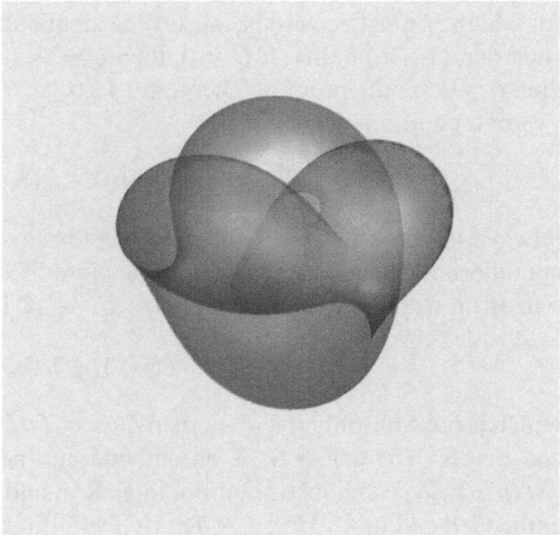

Fig. 14.1 Boy's surface as Willmore immersion.

14.6 Hopf cylinders are not generically isothermic

Recall Hopf cylinders of Definition 5.29.

Proposition 14.35. *A Hopf cylinder $f_\gamma : N \times \mathbf{S}^1 \to \mathbf{S}^3$, $f_\gamma(s, \theta) = e^{i\theta} \Gamma_1(s)$ over a unit speed curve $\gamma : N \to \mathbf{S}^2$ with curvature $\kappa : N \to \mathbf{R}$ is isothermic if and only if $\kappa(s) = \tan(cs)$ for some constant $c \in \mathbf{R}$.*

Proof. By (5.52), $\alpha^1 = \frac{1}{2}ds$, $\alpha^2 = d\theta + \frac{\kappa}{2}ds$ is an orthonormal coframe field for the metric induced by f_γ and

$$\alpha^1 + i\alpha^2 = \frac{1}{2}(ds + id\varphi),$$

where

$$\varphi = 2\theta + \int_0^s \kappa(u)du.$$

Thus $z = s + i\varphi$ is a complex coordinate and $e^u = 1/2$ is the conformal factor for the induced metric. The calculations done in (5.53) show that the Hopf invariant of f_γ relative to z is

$$h = \kappa + i,$$

and the Hopf quadratic form is

$$II^{2,0} = \frac{1}{2}he^{2u}dzdz = \frac{1}{8}(\kappa + i)dzdz.$$

By Lemma 14.4 and Proposition 14.5, f_γ is Möbius congruent to an isothermic immersion into $\mathbf{S}^3$ if and only if its Hopf quadratic form is given by

$$II^{2,0} = \frac{1}{8}rg\, dzdz,$$

where $r : N \times \mathbf{R} \to \mathbf{R}$ is smooth and $g : N \times \mathbf{R} \to \mathbf{C}$ is holomorphic. Thus, r and g are never zero and f_γ is Möbius congruent to an isothermic immersion if and only if

$$0 = \frac{\partial}{\partial \bar{z}}\left(\frac{\kappa + i}{r}\right) = \frac{1}{2r}\left(\dot{\kappa} - (\kappa + i)\frac{r_s + ir_\varphi}{r}\right),$$

which holds if and only if

$$\frac{(\kappa - i)\dot{\kappa}}{\kappa^2 + 1} = \frac{\dot{\kappa}}{\kappa + i} = \frac{r_s}{r} + i\frac{r_\varphi}{r},$$

which is equivalent to the pair of equations

$$\frac{\kappa\dot{\kappa}}{\kappa^2 + 1} = \frac{r_s}{r}, \quad -\frac{\dot{\kappa}}{\kappa^2 + 1} = \frac{r_\varphi}{r}, \tag{14.37}$$

which implies

$$0 = \left(\frac{\kappa\dot{\kappa}}{\kappa^2 + 1}\right)_\varphi = \left(\frac{r_s}{r}\right)_\varphi = \left(\frac{r_\varphi}{r}\right)_s = -\left(\frac{\dot{\kappa}}{\kappa^2 + 1}\right)_s.$$

These equations imply

$$\frac{\dot{\kappa}}{\kappa^2+1} = n, \tag{14.38}$$

for some real constant n. Integrating, and using the fact that the arclength parameter s is defined up to an additive constant, we can conclude that $\kappa(s) = \tan(ns)$, for some real constant n. Conversely, if this is the expression for $\kappa(s)$, for any real constant n, then (14.38) holds and implies that

$$\frac{\kappa\dot{\kappa}}{\kappa^2+1}ds - \frac{\dot{\kappa}}{\kappa^2+1}d\varphi = n(\kappa ds - d\varphi)$$

is a closed 1-form on $N \times \mathbf{R}$ and thus there exists a positive real function $r : N \times \mathbf{R} \to \mathbf{R}$ satisfying (14.37) and the above argument is reversible. □

Corollary 14.36. *Pinkall's Willmore tori in $\mathbf{S}^3$ constructed in Section 5.8 are not Möbius congruent to any minimal torus in $\mathbf{S}^3$ except when the curvature κ of the base curve is identically zero, in which case the Clifford torus is generated.*

Proof. The function $\kappa(s) = \tan(ns)$ satisfies (5.54) if and only if the constant $n = 0$. □

Problems

14.37. Let (U,z) be a connected principal chart for an umbilic free isothermic immersion $f : M \to \mathscr{M}$. Any other principal complex coordinate on U is given by $w = lz + \zeta$, where l and ζ are any complex constants such that l^2 is nonzero and real. Let $k : U \to \mathbf{R} \setminus \{0\}$ be the Calapso potential, $b : U \to \mathbf{C}$ the second order invariant, and $W = k_{\bar{z}\bar{z}} + \frac{1}{2}\bar{b}k : U \to \mathbf{R}$, relative to z. Denote these functions relative to w by the same letters with tildes. Use Theorem 13.10 to prove

$$\hat{k} = k|l|/l^2, \quad \hat{b} = b/l^2, \quad \hat{W}/\hat{k} = l^2 W/k,$$

so both k and $\hat{k}$ are positive if and only if l is real valued on U. Prove also that if (14.26) holds for both, then $\hat{s} = s/l^4$.

14.38. Let M be a Riemann surface with complex coordinate $z = x + iy$. Prove that if a smooth positive function $k : M \to \mathbf{R}$ satisfies the special isothermic equation (14.26), then it satisfies Calapso's equation.

14.39. For a Hopf cylinder over a curve in $\mathbf{S}^2$ with curvature κ and arclength parameter s, use Theorem 13.22 to find the Möbius invariants k and b relative to $z = s + i\varphi$. Use the notation of Proposition 13.37 to find Q and W. Use Proposition 13.37 to find the Möbius invariants p_1, p_2, p_3, q_1, and q_2 of the Hopf cylinder.

Chapter 15
Lie Sphere Geometry

Some properties of oriented surfaces immersed in a space form – for example, the Dupin condition – are preserved by Möbius transformations and by parallel translations (see §4.8.1, Problem 5.48, and Example 6.29). These two sets of transformations generate the Lie sphere group G.

The set Q of all oriented spheres in $\mathbf{S}^3$ is the disjoint union of $\mathbf{S}^{3,1}$, the set of all oriented spheres of nonzero radius, with $\mathscr{M}$, the set of all point spheres. Q is realized as the Lie sphere quadric in the projective space $\mathbf{P}(\mathbf{R}^{4,2})$. The lines in Q are called pencils of oriented spheres. A pencil is determined by the unique point sphere and unique oriented great sphere in it. The set of all lines in Q is denoted Λ, a five-dimensional manifold. The spherical projection $\pi : \Lambda \to \mathscr{M}$ sends a line to the unique point sphere in it. Λ possesses a natural contact structure. An immersion $\mathbf{x} : M^2 \to \mathbf{S}^3$ with unit normal vector field $\mathbf{e}_3$ has a Legendre lift $\lambda : M \to \Lambda$ given by $\lambda(m)$ is the line determined by the point sphere $\mathbf{x}(m)$ and the oriented great sphere through $\mathbf{x}(m)$ with unit normal $\mathbf{e}_3(m)$.

The Lie sphere group G is the group of all diffeomorphisms of Q that send pencils to pencils. G acts transitively on Λ and preserves the contact structure on λ. It is isomorphic to $\mathbf{O}(4,2)/\{\pm I\}$. Lie sphere geometry is the study of properties of Legendre immersions into the homogeneous space Λ that are invariant under the action of G. We use the method of moving frames to carry out this study. In conclusion we prove that any Dupin immersion is Lie sphere congruent to an open subset of the Legendre lift of a great circle.

15.1 Oriented spheres in $\mathbf{S}^3$

Recall Definition 5.4 of the sphere $S_r(\mathbf{m})$ of radius $r \in \mathbf{R}$ and center $\mathbf{m} \in \mathbf{S}^3$ and Definition 5.6 of the oriented sphere of signed radius r and center $\mathbf{m}$. For convenience we repeat this latter definition here.

G.R. Jensen et al., *Surfaces in Classical Geometries*, Universitext,
DOI 10.1007/978-3-319-27076-0_15

Definition 15.1. An *oriented sphere* in $\mathbf{S}^3$ is a sphere $S_r(\mathbf{m})$, together with a choice of continuous unit normal vector field on it.

We can use the radius to define an orientation of $S_r(\mathbf{m})$ as follows. Recall that $\mathbf{x} \in S_r(\mathbf{m})$ if and only if $\mathbf{x} \cdot \mathbf{m} = \cos r$. Let $\mathbf{n} \in T_{\mathbf{x}}\mathbf{S}^3$ be the unit vector satisfying the equation

$$\mathbf{m} = \cos r\, \mathbf{x} + \sin r\, \mathbf{n}. \tag{15.1}$$

Except for the cases $r = a\pi$ for any integer a, this uniquely determines $\mathbf{n}$ for each point x on the locus. The cases $r = a\pi$, a any integer, are the point spheres $\mathbf{m}$ and $-\mathbf{m}$, and there is no condition on $\mathbf{n}$ in (5.24). We consider point spheres to be in the set of all oriented spheres, but without orientation. Thus, the oriented spheres with center $\mathbf{m}$ are parametrized by their radius r satisfying $0 \le r < 2\pi$.

We illustrate the oriented nonpoint spheres in Figures 15.1, 15.2 and 15.3. Picturing $\mathbf{m}$ as the north pole, the diagrams show the plane determined by $\mathbf{x}$ and $\mathbf{m}$, which will also contain $\mathbf{n}$, the unit normal at $\mathbf{x}$ determining the orientation by (5.24). The diagram on the left in Figure 15.1 shows a point $\mathbf{x}$ and the normal $\mathbf{n}$ at $\mathbf{x}$ of the oriented sphere with center $\mathbf{m}$ and radius r in the range $0 < r < \pi/2$. This sphere is in the northern hemisphere with normal pointing towards $\mathbf{m}$.

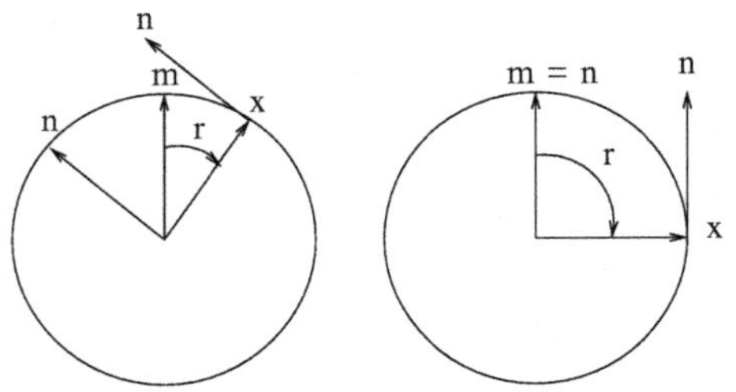

Fig. 15.1 Center $\mathbf{m}$, normal n at point x. Left: $0 < r < \pi/2$. Right: great sphere.

The case $r = \pi/2$ in the diagram on the right in Figure 15.1 is the oriented great sphere with unit normal $\mathbf{n} = \mathbf{m}$.

When $\pi/2 < r < \pi$, we have the oriented spheres in the southern hemisphere with normal pointing towards $\mathbf{m}$, as shown in the diagram on the left in Figure 15.2.

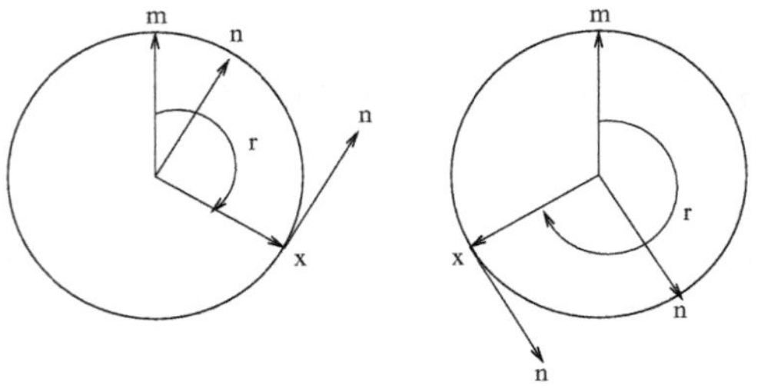

Fig. 15.2 Center $\mathbf{m}$, normal $\mathbf{n}$ at point x. Left: $\pi/2 < r < \pi$. Right: $\pi < r < 3\pi/2$.

The case $r = -\pi$ is the point sphere at $-\mathbf{m}$. When $\pi < r < 3\pi/2$, we have the oriented spheres in the southern hemisphere with normal pointing away from $\mathbf{m}$, as shown in the diagram on the right in Figure 15.2.

The case $r = 3\pi/2$ is the great sphere with unit normal $\mathbf{n} = -\mathbf{m}$ shown in the left diagram of Figure 15.3.

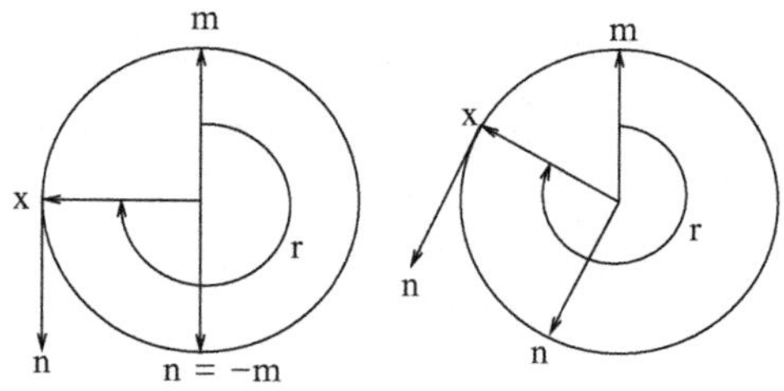

Fig. 15.3 Left: Center $\mathbf{m}$, normal $\mathbf{n} = -\mathbf{m}$ at $\mathbf{x}$, radius $r = 3\pi/2$. Right: Center $\mathbf{m}$, normal $\mathbf{n}$ at $\mathbf{x}$, radius $3\pi/2 < r < 2\pi$

Finally, when $3\pi/2 < r < 2\pi$, we have the oriented spheres in the northern hemisphere with normal pointing away from $\mathbf{m}$, as shown in the right diagram in Figure 15.3.

Proposition 15.2. *The set of all oriented spheres in* $\mathbf{S}^3$ *is in 1:1 correspondence with*

$$(\mathbf{S}^3 \times \mathbf{S}^1)/\{\pm 1\} = \{(\mathbf{m}, \cos r, \sin r) : \mathbf{m} \in \mathbf{S}^3,\ r \in \mathbf{R}\}/\{\pm 1\},$$

which is compact.

Proof. The oriented sphere with center $\mathbf{m} \in \mathbf{S}^3$ and signed radius $r \in \mathbf{R}$ is

$$S_r(\mathbf{m}) = \{\mathbf{x} \in \mathbf{S}^3 : \mathbf{x} \cdot \mathbf{m} = \cos r\},$$

which is the point sphere $\mathbf{m}$ if $r = 0$, and otherwise is oriented by the unit normal vector field $\mathbf{n}(\mathbf{x}) = \frac{\mathbf{m} - \cos r\,\mathbf{x}}{\sin r}$. Then

$$S_r(\mathbf{m}) \leftrightarrow (\mathbf{m}, \cos r, \sin r) \in \mathbf{S}^3 \times \mathbf{S}^1.$$

By the discussion above, the same oriented sphere $S_r(\mathbf{m})$ also corresponds to $-(\mathbf{m}, \cos r, \sin r)$. It is clear that oriented spheres with centers $\mathbf{m}$ and $\tilde{\mathbf{m}}$ cannot coincide if $\tilde{\mathbf{m}} \neq \pm\mathbf{m}$. If their centers are the same, then they coincide if and only if their radii differ by an integer multiple of 2π. By (2) and (3) of Problem 15.57, if $\tilde{\mathbf{m}} = -\mathbf{m}$, then $S_{\tilde{r}}(-\mathbf{m})$ and $S_r(\mathbf{m})$ coincide as oriented spheres if and only if $\tilde{r} = r + (2a+1)\pi$, for some integer a, in which case

$$(\tilde{\mathbf{m}}, \cos \tilde{r}, \sin \tilde{r}) = -(\mathbf{m}, \cos r, \sin r).$$

□

The set of all oriented spheres in $\mathbf{S}^3$ has the following important projective description. Let $\mathbf{R}^{4,2}$ denote $\mathbf{R}^6 = \mathbf{R}^4 \oplus \mathbf{R}^2$ with the inner product of signature $(4,2)$,

$$\langle \mathbf{u}, \mathbf{v} \rangle = \sum_0^3 u^i v^i - u^4 v^4 - u^5 v^5, \tag{15.2}$$

where we label the components of vectors in $\mathbf{R}^{4,2}$ from 0 to 5 relative to the standard basis $\epsilon_0, \dots, \epsilon_5$ of $\mathbf{R}^6$, which is an orthonormal basis of signature $(++++--)$. Denote the projective space $\mathbf{RP}^5$ by $\mathbf{P}(\mathbf{R}^{4,2})$ in order to emphasize the inner product of $\mathbf{R}^{4,2}$.

The following inner product spaces are naturally embedded in $\mathbf{R}^{4,2}$ as

$$\begin{aligned} \mathbf{R}^3 &= \operatorname{span}\{\epsilon_1, \epsilon_2, \epsilon_3\}, \\ \mathbf{R}^4 &= \operatorname{span}\{\epsilon_0, \epsilon_1, \epsilon_2, \epsilon_3\}, \\ \mathbf{R}^{3,1} &= \operatorname{span}\{\epsilon_1, \epsilon_2, \epsilon_3, \epsilon_4\}, \\ \mathbf{R}^{4,1} &= \operatorname{span}\{\epsilon_0, \epsilon_1, \epsilon_2, \epsilon_3, \epsilon_4\}. \end{aligned} \tag{15.3}$$

Definition 15.3. The *Lie quadric* $Q \subset \mathbf{P}(\mathbf{R}^{4,2})$ is the smooth quadric hypersurface

$$Q = \{[\mathbf{u}] \in \mathbf{P}(\mathbf{R}^{4,2}) : \langle \mathbf{u}, \mathbf{u} \rangle = 0\}.$$

Exercise 62. Verify that there is a natural isomorphism

$$(\mathbf{S}^3 \times \mathbf{S}^1)/\{\pm 1\} \to Q, \quad [\mathbf{m}, (\cos r, \sin r)] \mapsto [\mathbf{m} + \cos r\, \epsilon_4 + \sin r \epsilon_5],$$

with inverse

$$Q \to (\mathbf{S}^3 \times \mathbf{S}^1)/\{\pm 1\}, \quad [q] \mapsto [\sum_0^3 \frac{q^i}{D} \epsilon_i, (\frac{q^4}{D}, \frac{q^5}{D})],$$

where $D^2 = (q^4)^2 + (q^5)^2$.

Recall the set $\mathbf{S}^{3,1}$ of oriented nonpoint spheres in $\mathbf{S}^3$ defined in (12.67) by,

$$\mathbf{S}^{3,1} = \{\mathbf{u} = (u^0, \dots, u^4) \in \mathbf{R}^{4,1} : \langle \mathbf{u}, \mathbf{u} \rangle = \sum_0^3 (u^i)^2 - (u^4)^2 = 1\}.$$

As shown in Exercise 56, $\mathbf{S}^{3,1}$ is identified with $\tilde{\Sigma}$, the set of nonpoint oriented spheres in $\mathbf{S}^3$. The set of point spheres is identified with the Möbius space $\mathscr{M}$ defined in (12.56).

Proposition 15.4. *The Lie quadric Q is a compactification of $\mathbf{S}^{3,1}$ obtained by attaching the set of point spheres $\mathscr{M}$ to $\mathbf{S}^{3,1}$. Namely,*

$$Q = \mathbf{S}^{3,1} \cup \mathscr{M} \text{ and } \mathbf{S}^{3,1} \cap \mathscr{M} = \emptyset,$$

where

$$\mathbf{S}^{3,1} = \{[\mathbf{q}] \in Q : \langle \mathbf{q}, \boldsymbol{\epsilon}_5 \rangle \neq 0\}, \tag{15.4}$$

and

$$\mathscr{M} = \{[\mathbf{q}] \in Q : \langle \mathbf{q}, \boldsymbol{\epsilon}_5 \rangle = 0\}. \tag{15.5}$$

Proof. The natural inclusions (15.3) give rise to additional inclusions

$$\mathbf{S}^{3,1} = \{\mathbf{u} \in \mathbf{R}^{4,1} : \langle \mathbf{u}, \mathbf{u} \rangle = 1\} \subset Q, \quad \mathbf{u} \mapsto \mathbf{u} + \boldsymbol{\epsilon}_5,$$

whose image is (15.4), and

$$\mathscr{M} = \{[\mathbf{u}] \in \mathbf{P}(\mathbf{R}^{4,1}) : \langle \mathbf{u}, \mathbf{u} \rangle = 0\} \subset Q, \quad [\mathbf{u}] \mapsto [\mathbf{u} + 0\boldsymbol{\epsilon}_5],$$

whose image is (15.5). □

Proposition 15.5. *In the identification of oriented spheres in $\mathbf{S}^3$ with points in Q given in Proposition 15.2 and Exercise 62, the oriented sphere of center $\mathbf{m} \in \mathbf{S}^3$ and signed radius $r \in \mathbf{R}$,*

$$S_r(\mathbf{m}) = \{\mathbf{x} \in \mathbf{S}^3 : \mathbf{x} \cdot \mathbf{m} = \cos r\}, \quad \mathbf{n}(\mathbf{x}) = \frac{\mathbf{m} - \cos r\, \mathbf{x}}{\sin r},$$

is identified with

$$[\mathbf{q}] = [\mathbf{m} + \cos r\, \boldsymbol{\epsilon}_4 + \sin r\, \boldsymbol{\epsilon}_5] \in Q. \tag{15.6}$$

An oriented sphere in $\mathbf{R}^3$ of center $\mathbf{p}$ and signed radius $r \in \mathbf{R}$,

$$S_r(\mathbf{p}) = \{\mathbf{x} \in \mathbf{R}^3 : |\mathbf{x} - \mathbf{p}|^2 = r^2\}, \quad \mathbf{n}(\mathbf{x}) = \frac{\mathbf{p} - \mathbf{x}}{r},$$

is identified with

$$[\mathbf{q}] = [\frac{1 - |\mathbf{p}|^2 + r^2}{2}\boldsymbol{\epsilon}_0 + \mathbf{p} + \frac{1 + |\mathbf{p}|^2 - r^2}{2}\boldsymbol{\epsilon}_4 + r\boldsymbol{\epsilon}_5] \in Q. \tag{15.7}$$

An oriented plane in $\mathbf{R}^3$ with unit normal $\mathbf{n}$ and signed height $h \in \mathbf{R}$,

$$\Pi_h(\mathbf{n}) = \{\mathbf{x} \in \mathbf{R}^3 : \mathbf{x} \cdot \mathbf{n} = h\},$$

is identified with

$$[\mathbf{q}] = [-h\epsilon_0 + \mathbf{n} + h\epsilon_4 + \epsilon_5] \in Q. \tag{15.8}$$

The oriented sphere in $\mathbf{H}^3$ *of center* $\mathbf{m} \in \mathbf{H}^3 \subset \mathbf{R}^{3,1}$ *and signed radius* r,

$$S_r(\mathbf{m}) = \{\mathbf{x} \in \mathbf{H}^3 : \langle \mathbf{x}, \mathbf{m} \rangle = -\cosh r\},$$

is identified with

$$[\mathbf{q}] = [\cosh r\, \epsilon_0 + \mathbf{m} + \sinh r\, \epsilon_5] \in Q. \tag{15.9}$$

More generally, the oriented totally umbilic surface through $\mathbf{y} \in \mathbf{H}^3$ *with unit normal* $\mathbf{v} \in T_\mathbf{y}\mathbf{H}^3 = \mathbf{y}^\perp$ *and principal curvature* $a \in \mathbf{R}$, *given in Lemma 6.9 as*

$$S(a\mathbf{y} + \mathbf{v}) = \{\mathbf{x} \in \mathbf{H}^3 : \langle \mathbf{x}, a\mathbf{y} + \mathbf{v} \rangle = -a\},$$

is identified with

$$[\mathbf{q}] = [a\epsilon_0 + a\mathbf{y} + \mathbf{v} + \epsilon_5] \in Q. \tag{15.10}$$

Proof. (15.6) is characterized by

$$f_+ S_r(\mathbf{m}) = [\mathbf{q}]^\perp = \{[\mathbf{x} + \epsilon_4] \in \mathscr{M} : \langle \mathbf{x} + \epsilon_4, \mathbf{q} \rangle = 0\}.$$

The stereographic projections $\mathscr{S}^{-1} : \mathbf{R}^3 \to \mathbf{S}^3$ and $\mathscr{S}^{-1} \circ \mathfrak{s} : \mathbf{H}^3 \to \mathbf{S}^3$ send oriented spheres, planes, horospheres, and ultraspheres onto open connected subsets of oriented spheres in $\mathbf{S}^3$, so these correspond to points of Q. To obtain explicit expressions of the correspondences we use the conformal embeddings $f_0 : \mathbf{R}^3 \to \mathscr{M}$ of (12.59) and $f_- : \mathbf{H}^3 \to \mathscr{M}$ of (12.62).

$[\mathbf{q}]$ in (15.7) is characterized by $f_0 S_r(\mathbf{p}) = [\mathbf{q}]^\perp =$

$$\{[\frac{1-|\mathbf{x}|^2}{2}\epsilon_0 + \mathbf{x} + \frac{1+|\mathbf{x}|^2}{2}\epsilon_4] : \langle \frac{1-|\mathbf{x}|^2}{2}\epsilon_0 + \mathbf{x} + \frac{1+|\mathbf{x}|^2}{2}\epsilon_4, \mathbf{q} \rangle = 0\}.$$

$[\mathbf{q}]$ of (15.8) is characterized by $f_0 \Pi_h(\mathbf{n}) = [\mathbf{q}]^\perp =$

$$\{[\frac{1-|\mathbf{x}|^2}{2}\epsilon_0 + \mathbf{x} + \frac{1+|\mathbf{x}|^2}{2}\epsilon_4] : \langle \frac{1-|\mathbf{x}|^2}{2}\epsilon_0 + \mathbf{x} + \frac{1+|\mathbf{x}|^2}{2}\epsilon_4, \mathbf{q} \rangle = 0\}.$$

$[\mathbf{q}]$ in (15.9) is characterized by $f_- S_r(\mathbf{m}) = [\mathbf{q}]^\perp =$

$$\{[\mathbf{x} + \epsilon_4] \in \mathscr{M} : \langle \mathbf{x} + \epsilon_4, \mathbf{q} \rangle = 0\}.$$

$[\mathbf{q}]$ in (15.10) is characterized by $f\_S(a\mathbf{y}+\mathbf{v}) = [\mathbf{q}]^\perp =$

$$\{[\mathbf{x}+\epsilon_4] \in \mathscr{M} : \langle \mathbf{x}+\epsilon_4, a\mathbf{y}+\mathbf{v}\rangle = 0\}.$$

In each of these five cases the elementary calculation $\langle \mathbf{q},\mathbf{q}\rangle = 0 = \langle \mathbf{q}, f_\epsilon(\mathbf{x})\rangle$ is easily verified. By Definition 6.12, the last case is an oriented sphere if $|a| > 1$, an oriented horosphere if $|a| = 1$, an oriented ultrasphere if $0 < |a| < 1$, and an oriented plane if $a = 0$. □

15.2 Pencils of oriented spheres

A line in $\mathbf{RP}^n$ is the projection of a 2-dimensional subspace of $\mathbf{R}^{n+1}$. It is determined by any two distinct points $[\mathbf{x}]$ and $[\mathbf{y}]$ on it, for if $[\mathbf{x}] \neq [\mathbf{y}]$, then $\mathbf{x}$ and $\mathbf{y}$ are linearly independent vectors in $\mathbf{R}^{n+1}$ and so span the 2-dimensional subspace containing them. For distinct points $[\mathbf{x}]$ and $[\mathbf{y}]$ in $\mathbf{RP}^n$, we let $[\mathbf{x},\mathbf{y}]$ denote the line in $\mathbf{RP}^n$ spanned by them.

Lemma 15.6. *If* $[\mathbf{u}]$ *and* $[\mathbf{v}]$ *are distinct points in the Lie quadric* Q*, then the line they span lies in* Q *if and only if*

$$\langle \mathbf{u},\mathbf{v}\rangle = 0.$$

Proof. The line spanned by $[\mathbf{u}]$ and $[\mathbf{v}]$,

$$L = \{[r\mathbf{u}+s\mathbf{v}] : r,s \in \mathbf{R}\} \subset \mathbf{P}(\mathbf{R}^{4,2})$$

lies in Q if and only if

$$0 = \langle r\mathbf{u}+s\mathbf{v}, r\mathbf{u}+s\mathbf{v}\rangle = 2rs\langle \mathbf{u},\mathbf{v}\rangle,$$

for all $r,s \in \mathbf{R}$. □

Definition 15.7. Let Λ denote the set of all lines in the Lie quadric Q.

Lemma 15.8. *If* λ *is a line in* Q *spanned by points* $q = [\mathbf{m}+a\epsilon_4+b\epsilon_5]$ *and* $\tilde{q} = [\tilde{\mathbf{m}}+\tilde{a}\epsilon_4+\tilde{b}\epsilon_5]$ *in* Q*, where* $a,b,\tilde{a},\tilde{b} \in \mathbf{R}$ *and* $\mathbf{m},\tilde{\mathbf{m}} \in \mathbf{S}^3$*, then*

$$a\tilde{b} - \tilde{a}b \neq 0.$$

Proof. By definition of Q and by Lemma 15.6,

$$a^2+b^2 = |\mathbf{m}|^2, \quad \tilde{a}^2+\tilde{b}^2 = |\tilde{\mathbf{m}}|^2, \quad a\tilde{a}+b\tilde{b} = \mathbf{m}\cdot\tilde{\mathbf{m}}. \tag{15.11}$$

Seeking a contradiction, we suppose that $a\tilde{b} - \tilde{a}b = \det\begin{pmatrix} a & b \\ \tilde{a} & \tilde{b} \end{pmatrix} = 0$. Then $(\tilde{a}, \tilde{b}) = t(a,b)$, for some $t \in \mathbf{R}$, so $|\tilde{\mathbf{m}}|^2 = t^2|\mathbf{m}|^2$ and $t|\mathbf{m}|^2 = \mathbf{m}\cdot\tilde{\mathbf{m}}$, which imply

$$|\mathbf{m}\cdot\tilde{\mathbf{m}}| = |t||\mathbf{m}|^2 = |\mathbf{m}||\tilde{\mathbf{m}}|.$$

By the Cauchy-Schwarz inequality, $\tilde{\mathbf{m}} = s\mathbf{m}$, for some $s \in \mathbf{R}$, so then $t|\mathbf{m}|^2 = \mathbf{m}\cdot\tilde{\mathbf{m}} = \mathbf{m}\cdot s\mathbf{m} = s|\mathbf{m}|^2$ implies that $s = t$. This leads to the contradiction

$$\tilde{q} = [t\mathbf{m} + ta\epsilon_4 + tb\epsilon_5] = q.$$

□

Proposition 15.9. *Any line $\lambda \in \Lambda$ contains a unique point sphere $[\mathbf{x} + \epsilon_4] \in Q$, for some $\mathbf{x} \in \mathbf{S}^3$, and a unique great sphere $[\mathbf{n} + \epsilon_5] \in Q$, for some $\mathbf{n} \in \mathbf{S}^3 \cap \mathbf{x}^\perp$. Thus,*

$$\lambda = [\mathbf{x} + \epsilon_4, \mathbf{n} + \epsilon_5]. \tag{15.12}$$

Proof. Let λ be a line in Q. It is spanned by any two distinct points q and $\tilde{q}$ in it, which must be represented as

$$q = [\mathbf{m} + a\epsilon_4 + b\epsilon_5], \quad \tilde{q} = [\tilde{\mathbf{m}} + \tilde{a}\epsilon_4 + \tilde{b}\epsilon_5], \tag{15.13}$$

for some $\mathbf{m}, \tilde{\mathbf{m}} \in \mathbf{S}^3$ and $a, b, \tilde{a}, \tilde{b} \in \mathbf{R}$ satisfying (15.11). Then

$$\lambda = \{[r\mathbf{m} + s\tilde{\mathbf{m}} + (ra + s\tilde{a})\epsilon_4 + (rb + s\tilde{b})\epsilon_5] : r, s \in \mathbf{R}\}.$$

A point in λ is a point sphere if and only if $rb + s\tilde{b} = 0$, which is true if and only if

$$(r, s) = k(-\tilde{b}, b),$$

for some non-zero $k \in \mathbf{R}$. For any such value of k, the resulting point in λ is

$$[b\tilde{\mathbf{m}} - \tilde{b}\mathbf{m} + (b\tilde{a} - a\tilde{b})\epsilon_4], \tag{15.14}$$

which is independent of k. Hence, this is the unique point sphere in λ. In the same way, the unique oriented great sphere in λ is

$$[a\tilde{\mathbf{m}} - \tilde{a}\mathbf{m} + (a\tilde{b} - \tilde{a}b)\epsilon_5].$$

By Lemma 15.8, $D = a\tilde{b} - \tilde{a}b \neq 0$. Thus, λ is given by (15.12), where $\mathbf{x} = \frac{1}{D}(\tilde{b}\mathbf{m} - b\tilde{\mathbf{m}})$ and $\mathbf{n} = \frac{1}{D}(a\tilde{\mathbf{m}} - \tilde{a}\mathbf{m})$. □

Definition 15.10. Two oriented spheres in $\mathbf{S}^3$ are in *oriented contact* if they are tangent at a point $\mathbf{x} \in \mathbf{S}^3$ and they have the same orientation defining unit normal vector $\mathbf{n}$ at $\mathbf{x}$.

Definition 15.11. A *pencil* of oriented spheres in $\mathbf{S}^3$ is the set of all oriented spheres that contain a given point $\mathbf{x} \in \mathbf{S}^3$ and have the same orientation defining unit normal vector $\mathbf{n} \in T_{\mathbf{x}}\mathbf{S}^3$.

Figure 15.4 shows the pencil determined by the point $\mathbf{x} = \epsilon_2 \in \mathbf{S}^3$ with unit normal $\mathbf{n} = \epsilon_3 \in T_{\mathbf{x}}\mathbf{S}^3$, as it appears in the hyperplane $x^0 = 0$ of $\mathbf{R}^4$. The dotted circle is the locus of centers $\mathbf{m}$ of the oriented spheres of the pencil. It is the great circle through $\mathbf{x}$ tangent to $\mathbf{n}$.

Proposition 15.12. *There is a one-to-one correspondence between the set of pencils of oriented spheres in $\mathbf{S}^3$ and Λ, the set of lines in the Lie quadric Q.*

Proof. The pencil of oriented spheres determined by $\mathbf{x} \in \mathbf{S}^3$ and unit vector $\mathbf{n} \in T_{\mathbf{x}}\mathbf{S}^3$ is the set of oriented spheres

$$\{[\mathbf{m} + \cos r\epsilon_4 + \sin r\epsilon_5] : \mathbf{m} = \cos r\mathbf{x} + \sin r\mathbf{n},\ r \in \mathbf{R}\},$$

which is precisely the line in Q determined by the point sphere $[\mathbf{x} + \epsilon_4]$ and the great sphere $[\mathbf{n} + \epsilon_5]$. Conversely, any line $\lambda \in \Lambda$ is spanned by a unique point sphere $[\mathbf{x} + \epsilon_4]$ and a unique great sphere $[\mathbf{n} + \epsilon_5]$, and thus

$$\lambda = \{[a\mathbf{x} + b\mathbf{n} + a\epsilon_4 + b\epsilon_5] \in Q : a^2 + b^2 = 1\},$$

which is the set of oriented spheres in the pencil determined by $\mathbf{x} \in \mathbf{S}^3$ and unit normal $\mathbf{n} \in T_{\mathbf{x}}\mathbf{S}^3$. □

Lemma 15.13. *The set Λ is a smooth, five-dimensional submanifold of the Grassmannian $G(2,6)$.*

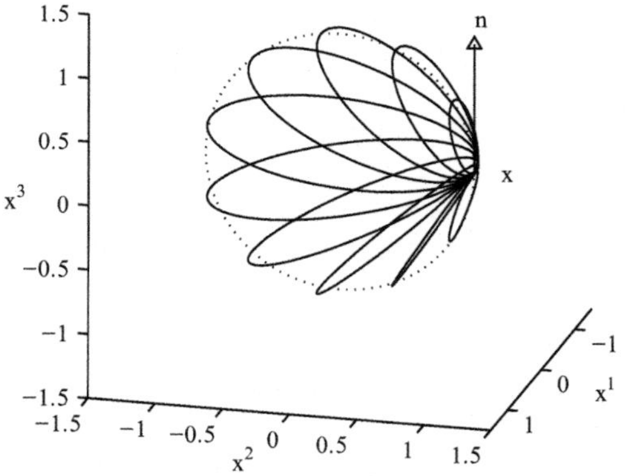

Fig. 15.4 The pencil of oriented spheres determined by $\mathbf{x} = \epsilon_2$, $n = \epsilon_3$.

Proof. Recall that the Grassmannian $G(2,6)$ is the set of all two-dimensional subspaces of $\mathbf{R}^6 = \mathbf{R}^{4,2}$. It is naturally represented by the quotient of the space $\mathbf{R}^{6\times 2^*}$ of all 6×2 matrices of rank two by the action of right multiplication by $\mathbf{GL}(2,\mathbf{R})$. If X is such a matrix, let $[X]$ denote its equivalence class in $G(2,6)$. If we write vectors in $\mathbf{R}^{4,2}$ as columns, then

$$\Lambda = \{[\mathbf{u},\mathbf{v}] \in G(2,6) : 0 = \langle \mathbf{u},\mathbf{u}\rangle = \langle \mathbf{v},\mathbf{v}\rangle = \langle \mathbf{u},\mathbf{v}\rangle\}.$$

Consider the open coordinate neighborhood of $G(2,6)$ given by

$$U = \{[\mathbf{u},\mathbf{v}] \in G(2,6) : \det\begin{pmatrix} u^4 & v^4 \\ u^5 & v^5 \end{pmatrix} \neq 0\},$$

with coordinate map

$$F : U \to \mathbf{R}^{4\times 2}, \quad F[\mathbf{u},\mathbf{v}] = \begin{pmatrix} u^0 & v^0 \\ \vdots & \vdots \\ u^3 & v^3 \end{pmatrix} \begin{pmatrix} u^4 & v^4 \\ u^5 & v^5 \end{pmatrix}^{-1}.$$

Then $\Lambda \subset U$, by Lemma 15.8, and

$$F(\Lambda) = \{(\mathbf{x},\mathbf{y}) \in \mathbf{R}^{4\times 2} : |\mathbf{x}| = |\mathbf{y}| = 1,\ \mathbf{x}\cdot\mathbf{y} = 0\}, \tag{15.15}$$

which is a smooth, five-dimensional submanifold of $\mathbf{R}^{4\times 2}$. □

Exercise 63. Verify (15.15).

Definition 15.14. The *spherical projection* is the map

$$\pi : \Lambda \to \mathscr{M},$$

that sends any line $\lambda \in \Lambda$ to the unique point sphere in it. If $\lambda = [q,\tilde{q}]$, where $q = [\mathbf{m} + a\epsilon_4 + b\epsilon_5]$ and $\tilde{q} = [\tilde{\mathbf{m}} + \tilde{a}\epsilon_4 + \tilde{b}\epsilon_5]$ are points in Q given by (15.13), then

$$\pi(\lambda) = [\tilde{b}\mathbf{m} - b\tilde{\mathbf{m}} + (a\tilde{b} - b\tilde{a})\epsilon_4] \in \mathscr{M},$$

by (15.14).

15.3 Lie sphere transformations

Definition 15.15. A *Lie sphere transformation* $T : Q \to Q$ is a diffeomorphism on the space of oriented spheres that sends lines to lines, or, equivalently, preserves oriented contact. Thus, it defines a diffeomorphism $T : \Lambda \to \Lambda$. The set $\mathrm{Lie}(\mathbf{S}^3)$ of all Lie sphere transformations is a group under composition of mappings.

Definition 15.16. Let $\mathbf{O}(4,2)$ denote the set of all linear transformations T of $\mathbf{R}^{4,2}$ that preserve the signature $(4,2)$ inner product (15.2); that is

$$\langle T\mathbf{x}, T\mathbf{y}\rangle = \langle \mathbf{x}, \mathbf{y}\rangle,$$

for any $\mathbf{x}, \mathbf{y} \in \mathbf{R}^{4,2}$. This is a Lie group of dimension 15.

In the standard orthonormal basis $\epsilon_0, \dots, \epsilon_5$ of $\mathbf{R}^{4,2}$ used in (15.2), the group $\mathbf{O}(4,2)$ is represented by

$$\mathbf{O}(4,2) = \{T \in \mathbf{GL}(6,\mathbf{R}) : {}^tTI_{4,2}T = I_{4,2}\},$$

where

$$I_{4,2} = \begin{pmatrix} I_4 & 0 \\ 0 & -I_2 \end{pmatrix}.$$

Exercise 64. Prove that the Lie algebra of $\mathbf{O}(4,2)$ is

$$\mathfrak{o}(4,2) = \{X \in \mathfrak{gl}(6,\mathbf{R}) : {}^tXI_{4,2} + I_{4,2}X = 0\}. \tag{15.16}$$

Find its dimension.

Example 15.17. The linear action of $\mathbf{O}(4,2)$ on $\mathbf{R}^{4,2}$ maps linear subspaces to linear subspaces, so it induces a smooth action on $\mathbf{P}(\mathbf{R}^{4,2})$, which sends Q to Q and lines in Q to lines in Q. Hence, the elements of $\mathbf{O}(4,2)$ induce Lie sphere transformations. Moreover, this action induces a smooth action of $\mathbf{O}(4,2)$ on Λ, the space of lines in Q. An element $T \in \mathbf{O}(4,2)$ induces the identity transformation on Q if and only if $T = \pm I_6$. Thus $\mathbf{O}(4,2)/\{\pm I\}$ is a subgroup of $\mathrm{Lie}(\mathbf{S}^3)$.

Exercise 65. Prove that the Lie sphere transformation defined by the element

$$I_{5,1} = \begin{pmatrix} I_5 & 0 \\ 0 & -1 \end{pmatrix} \in \mathbf{O}(4,2)$$

fixes each point sphere and sends any oriented non-point sphere to the same sphere with the opposite orientation.

Consider the Lie group inclusion

$$\mathbf{O}(4,1) \subset \mathbf{O}(4,2), \quad A \hookrightarrow \begin{pmatrix} A & 0 \\ 0 & 1 \end{pmatrix}.$$

If $T \in \mathbf{O}(4,2)$ sends $\mathscr{M} = [\epsilon_5]^\perp \subset Q$ to itself, then, up to sign, $T = \begin{pmatrix} A & 0 \\ 0 & 1 \end{pmatrix}$, where $A \in \mathbf{O}(4,1)$. If $\det A = -1$, then $-A \in \mathbf{SO}(4,1)$ and

$$-\begin{pmatrix} A & 0 \\ 0 & 1 \end{pmatrix} = I_{5,1} \begin{pmatrix} -A & 0 \\ 0 & 1 \end{pmatrix}.$$

Thus, up to sign, $T \in \mathbf{SO}(4,1)$ or $T \in I_{5,1}\mathbf{SO}(4,1)$. In words, a Lie sphere transformation in $\mathbf{O}(4,2)/\{\pm I\}$ that sends point spheres to point spheres is either a Möbius transformation or a Möbius transformation followed by the orientation reversal mapping of Exercise 65.

Example 15.18. A conformal diffeomorphism $f : \mathbf{S}^3 \to \mathbf{S}^3$ (see Definition 12.9) sends spheres to spheres. If $\mathbf{n}$ is a unit normal vector field orienting a sphere, then df sends $\mathbf{n}$ to a normal vector field along the image sphere, thus orienting the image sphere. A pencil of spheres through $\mathbf{x} \in \mathbf{S}^3$ with unit normal $\mathbf{n}$ at $\mathbf{x}$ is sent by f to the set of oriented spheres through $f(\mathbf{x})$ with unit normal $\frac{df_{\mathbf{x}}\mathbf{n}}{|df_{\mathbf{x}}\mathbf{n}|}$, which is again a pencil. Hence, f induces a map $Q \to Q$, which sends lines to lines. To see that this induced map is a diffeomorphism of Q, we recall the group isomorphism

$$F : \mathbf{SO}(4,1) \to \mathrm{Conf}(\mathbf{S}^3), \quad F(T) = f_+^{-1} \circ T \circ f_+$$

introduced in Theorem 12.17. Thus, the map induced by $f \in \mathrm{Conf}(\mathbf{S}^3)$ on Q is that of $T = F^{-1}(f) \in \mathbf{SO}(4,1)$, whose projective action on Q is smooth with smooth inverse given by T^{-1}.

Example 15.19. Lie parallel transformations by $t \in \mathbf{R}$ are the Lie transformations induced by the elements $T_t^\epsilon \in \mathbf{SO}(4,2)$, where $\epsilon \in \{+,0,-\}$, that send the oriented sphere with given center in $\mathbf{S}^3$, $\mathbf{R}^3$, or $\mathbf{H}^3$, respectively, and signed radius r, to the oriented sphere with the same center and signed radius $r+t$. These are:

$$T_t^+ = \begin{pmatrix} I_4 & 0 \\ 0 & \begin{pmatrix} \cos t & -\sin t \\ \sin t & \cos t \end{pmatrix} \end{pmatrix} \in \mathbf{SO}(4,2), \tag{15.17}$$

$$T_t^0 = \begin{pmatrix} 1+\frac{t^2}{2} & 0 & \frac{t^2}{2} & t \\ 0 & I_3 & 0 & 0 \\ -\frac{t^2}{2} & 0 & 1-\frac{t^2}{2} & -t \\ t & 0 & t & 1 \end{pmatrix} \in \mathbf{SO}(4,2), \tag{15.18}$$

and

$$T_t^- = \begin{pmatrix} \cosh t & 0 & \sinh t \\ 0 & I_4 & 0 \\ \sinh t & 0 & \cosh t \end{pmatrix} \in \mathbf{SO}(4,2). \tag{15.19}$$

The following is called the fundamental theorem of Lie sphere geometry.

Theorem 15.20 ([114]). *The group* $\mathrm{Lie}(\mathbf{S}^3)$, *of all Lie sphere transformations of* $\mathbf{S}^3$, *is* $\mathbf{O}(4,2)/\{\pm I\}$.

Proof. This is in Lie's thesis [114, p. 186]. Pinkall [135, p. 431] proved this theorem for the group of all Lie sphere transformations of $\mathbf{S}^n$, for any $n \geq 3$. An excellent exposition of the proof, with all details provided, is in Cecil [36, Theorem 3.5, pp. 28–30]. □

Theorem 15.21 (Cecil-Chern [37]). *Any Lie sphere transformation decomposes into a Möbius transformation followed by a parallel transformation followed by a Möbius transformation.*

Proof. See Cecil [36, Theorem 3.18, p. 49]. □

15.4 Lie sphere frames

The basic space of Lie sphere geometry is Λ, the space of lines in the Lie quadric Q. The standard action of the Lie sphere group $\mathbf{O}(4,2)/\{\pm I\}$ on $\mathbf{R}^{4,2}$ induces an action on $\mathbf{P}(\mathbf{R}^{4,2})$, which preserves Q and sends lines to lines. In this way the Lie sphere group acts on the set Λ of lines in Q. In the notation developed above, for any $T \in \mathbf{O}(4,2)$ and $\lambda = [\mathbf{u},\mathbf{v}] \in \Lambda$, the action is $T\lambda = [T\mathbf{u}, T\mathbf{v}]$.

Lemma 15.22. *The Lie sphere group acts transitively on Λ.*

Proof. By Lemma 15.12 and its proof, any point in $\lambda \in \Lambda$ is given by $[\mathbf{x}+\boldsymbol{\epsilon}_4, \mathbf{n}+\boldsymbol{\epsilon}_5]$, where $\mathbf{x},\mathbf{n} \in \mathbf{S}^3 \subset \mathbf{R}^4$ and $\mathbf{x}\cdot\mathbf{n} = 0$. Fix the point $o = [\boldsymbol{\epsilon}_0+\boldsymbol{\epsilon}_4, \boldsymbol{\epsilon}_1+\boldsymbol{\epsilon}_5] \in \Lambda$. Applying Gram-Schmidt to $\mathbf{R}^4$, we see that there exists $A \in \mathbf{SO}(4)$ such that $A\boldsymbol{\epsilon}_0 = \mathbf{x}$ and $A\boldsymbol{\epsilon}_1 = \mathbf{n}$. Then $T = \begin{pmatrix} A & 0 \\ 0 & I_2 \end{pmatrix} \in \mathbf{SO}(4,2)$ and $To = \lambda$. □

As in the case of Möbius geometry, the orthonormal basis of $\mathbf{R}^{4,2}$ is not convenient for the study of this action. What we need are Lie frames.

Definition 15.23. A *Lie frame* is a basis $\mathbf{v}_0,\dots,\mathbf{v}_5$ of $\mathbf{R}^{4,2}$ that satisfies

$$\langle \mathbf{v}_a, \mathbf{v}_b \rangle = \hat{g}_{ab},$$

for $a,b = 0,\dots,5$, where the $\hat{g}_{ab}$ are the entries of the matrix

$$\hat{g} = \begin{pmatrix} 0 & 0 & -L \\ 0 & I_2 & 0 \\ -L & 0 & 0 \end{pmatrix}, \quad L = \begin{pmatrix} 0 & 1 \\ 1 & 0 \end{pmatrix}.$$

To be specific, we fix the Lie frame

$$\begin{aligned} \boldsymbol{\lambda}_0 &= \frac{\boldsymbol{\epsilon}_4+\boldsymbol{\epsilon}_0}{2}, \quad \boldsymbol{\lambda}_5 = \boldsymbol{\epsilon}_4-\boldsymbol{\epsilon}_0, \quad \boldsymbol{\lambda}_2 = \boldsymbol{\epsilon}_2, \quad \boldsymbol{\lambda}_3 = \boldsymbol{\epsilon}_3, \\ \boldsymbol{\lambda}_1 &= \frac{\boldsymbol{\epsilon}_5+\boldsymbol{\epsilon}_1}{2}, \quad \boldsymbol{\lambda}_4 = \boldsymbol{\epsilon}_5-\boldsymbol{\epsilon}_1, \end{aligned} \tag{15.20}$$

where $\boldsymbol{\epsilon}_0,\dots,\boldsymbol{\epsilon}_5$ is the standard orthonormal basis of $\mathbf{R}^{4,2}$. In this Lie frame, the inner product of vectors $\mathbf{x} = \sum_0^5 x^a \boldsymbol{\lambda}_a$ and $\mathbf{y} = \sum_0^5 y^b \boldsymbol{\lambda}_b$ is

$$\langle \mathbf{x},\mathbf{y}\rangle = -(x^0y^5 + x^5y^0) - (x^1y^4 + x^4y^1) + x^2y^2 + x^3y^3.$$

The representation of $\mathbf{O}(4,2)$ in this Lie frame is

$$G = \{T \in \mathbf{GL}(6,\mathbf{R}) : {}^tT\hat{g}T = \hat{g}\},$$

which we also call the *Lie sphere group*. Its Lie algebra is

$$\mathfrak{g} = \{\mathscr{T} \in \mathfrak{gl}(6,\mathbf{R}) : {}^t\mathscr{T}\hat{g} + \hat{g}\mathscr{T} = 0\}.$$

Thus, $\mathscr{T} \in \mathfrak{g}$ if and only if $\hat{g}\mathscr{T}$ is skew-symmetric. These linear relations on the entries of elements of $\mathfrak{g}$ can be expressed with the block form

$$\mathfrak{g} = \left\{ \begin{pmatrix} X & {}^tY & rI_{1,1} \\ V & W & YL \\ sI_{1,1} & L^tV & -L^tXL \end{pmatrix} : X, Y, V \in \mathbf{R}^{2\times 2},\ W \in \mathfrak{o}(2),\ r,s \in \mathbf{R} \right\},$$

where $I_{1,1} = \begin{pmatrix} 1 & 0 \\ 0 & -1 \end{pmatrix}$. The Maurer–Cartan form of G is the left-invariant $\mathfrak{g}$-valued 1-form $\omega = T^{-1}dT = (\omega^a_b)$ on G. We express it in the block form used for $\mathfrak{g}$ as

$$\omega = (\omega^a_b) = \begin{pmatrix} \eta & {}^t\Sigma & \beta I_{1,1} \\ \theta & \Omega & \Sigma L \\ \alpha I_{1,1} & L^t\theta & -L^t\eta L \end{pmatrix},$$

where $a,b,c = 0,\dots,5$. These imply the following relations:

$$\begin{gathered}
\begin{pmatrix} \omega^0_0 & \omega^0_1 \\ \omega^1_0 & \omega^1_1 \end{pmatrix} = \eta, \quad \begin{pmatrix} \omega^4_4 & \omega^4_5 \\ \omega^5_4 & \omega^5_5 \end{pmatrix} = -L^t\eta L = -\begin{pmatrix} \omega^1_1 & \omega^0_1 \\ \omega^1_0 & \omega^0_0 \end{pmatrix} \\
\begin{pmatrix} \omega^2_0 & \omega^2_1 \\ \omega^3_0 & \omega^3_1 \end{pmatrix} = \theta, \quad \begin{pmatrix} \omega^4_2 & \omega^4_3 \\ \omega^5_2 & \omega^5_3 \end{pmatrix} = L^t\theta = \begin{pmatrix} \omega^2_1 & \omega^3_1 \\ \omega^2_0 & \omega^3_0 \end{pmatrix} \\
\begin{pmatrix} \omega^0_2 & \omega^0_3 \\ \omega^1_2 & \omega^1_3 \end{pmatrix} = {}^t\Sigma, \quad \begin{pmatrix} \omega^2_4 & \omega^2_5 \\ \omega^3_4 & \omega^3_5 \end{pmatrix} = \Sigma L = \begin{pmatrix} \omega^1_2 & \omega^0_2 \\ \omega^1_3 & \omega^0_3 \end{pmatrix} \\
\begin{pmatrix} 0 & \omega^2_3 \\ \omega^3_2 & 0 \end{pmatrix} = \Omega = -{}^t\Omega, \\
\alpha = \omega^4_0 = -\omega^5_1, \quad \omega^4_1 = 0 = \omega^5_0, \quad \beta = \omega^0_4 = -\omega^1_5, \quad \omega^0_5 = 0 = \omega^1_4.
\end{gathered} \tag{15.21}$$

Now $dT = T\omega$ means

$$d\mathbf{T}_a = \sum_0^5 \mathbf{T}_b \omega_a^b,$$

where $\mathbf{T}_a = \sum_0^5 T_a^b \boldsymbol{\lambda}_b$ is column a of T in the Lie frame (15.20), for $a,b = 0,\ldots,5$. We can calculate the forms ω_b^a from the inner products

$$\langle d\mathbf{T}_a, \mathbf{T}_c \rangle = \sum_0^5 \langle \mathbf{T}_b, \mathbf{T}_c \rangle \omega_a^b.$$

For example,

$$\alpha = \omega_0^4 = -\langle d\mathbf{T}_0, \mathbf{T}_1 \rangle, \quad \beta = \omega_4^0 = -\langle d\mathbf{T}_4, \mathbf{T}_5 \rangle.$$

The structure equations of G are

$$d\omega_b^a = -\sum_0^5 \omega_c^a \wedge \omega_b^c.$$

Applying this to $\alpha = \omega_0^4$ and using (15.21), we get

$$d\alpha = -\alpha \wedge (\omega_0^0 + \omega_1^1) + \omega_1^2 \wedge \omega_0^2 + \omega_1^3 \wedge \omega_0^3. \tag{15.22}$$

Choose $o = [\boldsymbol{\lambda}_0, \boldsymbol{\lambda}_1] \in \Lambda$ as the origin of the space of lines Λ in the Lie quadric Q. The transitive action of G on Λ has the projection map

$$\pi : G \to \Lambda, \quad \pi(T) = To = T[\boldsymbol{\lambda}_0, \boldsymbol{\lambda}_1] = [T\boldsymbol{\lambda}_0, T\boldsymbol{\lambda}_1] = [\mathbf{T}_0, \mathbf{T}_1]. \tag{15.23}$$

The isotropy subgroup of G at $o = [\boldsymbol{\lambda}_0, \boldsymbol{\lambda}_1]$ is

$$G_0 = \{k(c,B,Z,b) : bL^t c + cL^t b = {}^t ZZ\}, \tag{15.24}$$

where we introduce the notation

$$k(c,B,Z,b) = \begin{pmatrix} c & {}^t Z & b \\ 0 & B & BZ^t c^{-1} L \\ 0 & 0 & L^t c^{-1} L \end{pmatrix}, \tag{15.25}$$

for

$$c \in \mathbf{GL}(2,\mathbf{R}),\ b, Z \in \mathbf{R}^{2\times 2},\ B \in \mathbf{O}(2).$$

Exercise 66. Prove that

$$k(c,B,Z,b)^{-1} = k(c^{-1},B^{-1},-BZ^tc^{-1},L^tbL).$$

The Lie algebra of G_0 is the subalgebra $s = 0 = V$ of $\mathfrak{g}$,

$$\mathfrak{g}_0 = \left\{ \begin{pmatrix} X & {}^tY & rI_{1,1} \\ 0 & W & YL \\ 0 & 0 & -L^tXL \end{pmatrix} : X, Y \in \mathbf{R}^{2\times 2},\ W \in \mathfrak{o}(2),\ r \in \mathbf{R} \right\}.$$

We choose the complementary subspace

$$\mathfrak{m}_0 = \left\{ \begin{pmatrix} 0 & 0 & 0 \\ V & 0 & 0 \\ sI_{1,1} & L^tV & 0 \end{pmatrix} : V \in \mathbf{R}^{2\times 2}, s \in \mathbf{R} \right\}$$

of $\mathfrak{g}_0$ in $\mathfrak{g}$, so that $\mathfrak{g} = \mathfrak{g}_0 + \mathfrak{m}_0$ is a vector space direct sum. We shall need the adjoint action of G_0 on $\mathfrak{m}_0$.

Exercise 67. Prove that if $k = k(c,B,Z,b) \in G_0$ and if

$$X_0 = \begin{pmatrix} 0 & 0 & 0 \\ V & 0 & 0 \\ 0 & L^tV & 0 \end{pmatrix} \in \mathfrak{m}_0,$$

then the $\mathfrak{m}_0$-component of $\mathrm{Ad}(k^{-1})X_0$ (relative to $\mathfrak{g} = \mathfrak{g}_0 + \mathfrak{m}_0$) is

$$\mathrm{Ad}(k^{-1})_{\mathfrak{m}_0} X_0 = \begin{pmatrix} 0 & 0 & 0 \\ (B^{-1}V - sZLI_{1,1})c & 0 & 0 \\ s\det(c) & L^tc({}^tVB - sI_{1,1}L^tZ) & 0 \end{pmatrix}.$$

Definition 15.24. A *local Lie frame field* on an open subset $U \subset \Lambda$ is a smooth map $T : U \to G$ such that $\pi \circ T$ is the identity map of U.

If $\mathbf{T}_a$ is column $a = 0,\dots,5$ of a local Lie frame field $T : U \to G$, then each $\mathbf{T}_a : U \to \mathbf{R}^{4,2}$ is a smooth map such that $\mathbf{T}_0,\dots,\mathbf{T}_5$ is a Lie frame of $\mathbf{R}^{4,2}$, at each point of U, and

$$\pi \circ T = [\mathbf{T}_0, \mathbf{T}_1] : U \to U$$

is the identity map. If we denote the $\mathfrak{m}_0$-component of ω by

$$\omega_0 = \begin{pmatrix} 0 & 0 & 0 \\ \theta & 0 & 0 \\ \alpha I_{1,1} & L^t\theta & 0 \end{pmatrix}, \tag{15.26}$$

where θ is defined in (15.21), and if we denote the entries of $T^*\omega_0$ with the same symbols, which are now 1-forms on U, then

$$\alpha, \omega_0^2, \omega_1^2, \omega_0^3, \omega_1^3 \tag{15.27}$$

is a coframe field on U, by Proposition 3.3.

Given any Lie frame field on U, any other is given by

$$\tilde{T} = Tk(c,B,Z,b),$$

for some smooth map $k(c,B,Z,b) : U \to G_0$. Denoting $\tilde{T}^*\omega_0$ as in (15.26) with $\tilde{\theta}$ and $\tilde{\alpha}$, then

$$\tilde{\theta} = (B^{-1}\theta - \alpha Z L I_{1,1})c, \quad \tilde{\alpha} = \det(c)\alpha, \tag{15.28}$$

by Exercise 67.

Definition 15.25. A *contact structure* on a smooth, $2n+1$-dimensional manifold M is a collection $\mathscr{A}$ of pairs (U,α), where U is an open subset of M and α is a smooth 1-form on U such that $\alpha \wedge (d\alpha)^n \neq 0$ at every point of U, with the following properties.

1. The open sets U of all pairs in $\mathscr{A}$ cover M.
2. If (U,α) and $(\tilde{U},\tilde{\alpha})$ are elements of $\mathscr{A}$ such that $U \cap \tilde{U} \neq \emptyset$, then

$$\tilde{\alpha} = l\alpha$$

on $U \cap \tilde{U}$ for some nowhere zero smooth function $l : U \cap \tilde{U} \to \mathbf{R}^\times$.
3. $\mathscr{A}$ is maximal for properties 1) and 2).

For the pair $(U,\alpha) \in \mathscr{A}$, the form α is called the *contact form* on U. Generally we take property 3) for granted, since any collection $\mathscr{A}$ satisfying the first two properties can always be augmented to satisfy the third property. In particular, a 1-form α on M satisfying $\alpha \wedge (d\alpha)^n$ never zero on M defines a contact structure on M.

Example 15.26. Consider the 1-form on $\mathbf{R}^4 \times \mathbf{R}^4$ defined by

$$d\mathbf{x} \cdot \mathbf{n} = \sum_1^4 n^i dx^i,$$

where ϵ_i, $i = 1, \ldots, 4$ is the standard basis of $\mathbf{R}^4$, and $(\mathbf{x},\mathbf{n}) \in \mathbf{R}^4 \times \mathbf{R}^4$ is given by $\mathbf{x} = \sum_1^4 x^i \epsilon_i$ and $\mathbf{n} = \sum_1^4 n^i \epsilon_i$. Let β be this 1-form pulled back to the unit tangent bundle $U\mathbf{S}^3$ of $\mathbf{S}^3$, which is the submanifold $U\mathbf{S}^3 \subset \mathbf{S}^3 \times \mathbf{S}^3 \subset \mathbf{R}^4 \times \mathbf{R}^4$ defined in Problem 15.62. Then β is a *contact form* on $U\mathbf{S}^3$, which means that the 5-form

$$\beta \wedge d\beta \wedge d\beta$$

is never zero on $U\mathbf{S}^3$.

Example 15.27 (Contact structure on Λ). Let $\mathscr{A}$ be the set of all pairs (U,α), where U is an open subset of Λ on which there is a local section $T : U \to G$, and $\alpha = T^*\omega_0^4$. By (15.22),

$$\alpha\wedge d\alpha\wedge d\alpha = 2\alpha\wedge\omega_1^2\wedge\omega_0^2\wedge\omega_1^3\wedge\omega_0^3,$$

which is non-zero at every point of U, by (15.27). If $T : U \to G$ and $\tilde{T} : \tilde{U} \to G$ are Lie frame fields and if $U \cap \tilde{U} \neq \emptyset$, then $\tilde{T} = Tk(c,B,Z,b)$ for some smooth map $k(c,B,Z,b) : U \cap \tilde{U} \to G_0$ and $\tilde{\alpha} = \det(c)\alpha$ on $U \cap \tilde{U}$. Thus, $\mathscr{A}$ is a contact structure on Λ.

Example 15.28. The diagram

$$\begin{array}{ccc} U\mathbf{S}^3 & \stackrel{F}{\to} & \Lambda \\ \pi\downarrow & & \downarrow\pi \\ \mathbf{S}^3 & \stackrel{f_+}{\to} & \mathscr{M} \end{array},$$

where F, the diffeomorphism of Problem 15.62, is a fiber bundle isomorphism that preserves the contact structures. That is, F maps the fiber of $U\mathbf{S}^3$ over $\mathbf{x} \in \mathbf{S}^3$ onto the fiber of Λ over $f_+(\mathbf{x})$, and F^* of a contact form on Λ is a nonzero multiple of the contact form $\beta = d\mathbf{x}\cdot\mathbf{n}$ on $U\mathbf{S}^3$. To see this last point in more detail, let $T : U \to G$ be a local Lie frame field on an open subset $U \subset \Lambda$. Then $T^*\omega_0^4 = -\langle d\mathbf{T}_0, \mathbf{T}_1\rangle$ is the contact form on U, where $T = (\mathbf{T}_0, \mathbf{T}_1, \dots, \mathbf{T}_5)$. On the other hand, if $(\mathbf{x},\mathbf{n}) \in F^{-1}U \subset \mathbf{US}^3$, then

$$F(\mathbf{x},\mathbf{n}) = [\mathbf{x}+\epsilon_4, \mathbf{n}+\epsilon_5] = [\mathbf{T}_0\circ F, \mathbf{T}_1\circ F]$$

implies that $\mathbf{T}_0\circ F = c_0^0(\mathbf{x}+\epsilon_4) + c_0^1(\mathbf{n}+\epsilon_5)$ and $\mathbf{T}_1\circ F = c_1^0(\mathbf{x}+\epsilon_4) + c_1^1(\mathbf{n}+\epsilon_5)$ for some smooth functions $c_j^i : U \to \mathbf{R}$, $i,j = 1,2$, such that $c_0^0c_1^1 - c_0^1c_1^0$ is never zero on U. Thus,

$$F^*T^*\omega_0^4 = -\langle d(\mathbf{T}_0\circ F), \mathbf{T}_1\circ F\rangle = -(c_0^0c_1^1 - c_0^1c_1^0)d\mathbf{x}\cdot\mathbf{n}.$$

15.5 Legendre submanifolds

Definition 15.29. A *Legendre map* into a manifold N with contact structure $\mathscr{A}$ is a smooth map $\lambda : M \to N$ such that

$$\lambda^*\alpha = 0,$$

for every local contact form $(U,\alpha) \in \mathscr{A}$.

Definition 15.30. A *Legendre lift* of a conformal immersion $f : M \to \mathscr{M}$ of a Riemann surface M is a Legendre immersion $\lambda : M \to \Lambda$ such that $\pi \circ \lambda = f$, where $\pi : \Lambda \to \mathscr{M}$ is the spherical projection map of Definition 15.14.

Remark 15.31. A lift $\lambda : M \to \Lambda$ of an immersion $f : M \to \mathscr{M}$ is a smooth section along f of the fibre bundle $\pi : \Lambda \to \mathscr{M}$. Let $F : U \to \mathscr{L}$ be a local representation of f on an open subset U of M. At a point $m \in M$, the lift $\lambda(m)$ is a pencil of oriented tangent spheres to f at m. Thus, $\lambda(m) = [F(m), S(m) + \epsilon_5]$, where $S : U \to \mathbf{S}^{3,1}$ is a smooth map such that $S(m)$ is an oriented tangent sphere of f at m. That is,

$$\langle F(m), S(m)\rangle = 0 = \langle dF(m), S(m)\rangle,$$

for each $m \in U$, by Definition 12.42. The next proposition shows that this lift is Legendrian.

Proposition 15.32. *If M is a Riemann surface with a smooth immersion $\lambda : M \to \Lambda$ given by $\lambda = [S_0, S_1]$, where $S_0, S_1 : M \to \mathbf{R}^{4,2} \setminus \{0\}$ are smooth maps satisfying $0 = \langle S_i, S_j\rangle$, for all $i,j = 0,1$, then λ pulls back any contact form of the contact structure of Λ to a multiple of $\langle dS_0, S_1\rangle$. In particular, λ is a Legendre map if $\langle dS_0, S_1\rangle = 0$ on M.*

Proof. Given a point $m \in M$, let $T = (\mathbf{T}_0, \dots, \mathbf{T}_4) : U \to G$ be a Lie frame field on a neighborhood $U \subset \Lambda$ of $\lambda(m)$. Then

$$T \circ \lambda = [\mathbf{T}_0 \circ \lambda, \mathbf{T}_1 \circ \lambda] = [c_0^0 S_0 + c_0^1 S_1, c_1^0 S_0 + c_1^1 S_1]$$

on $\lambda^{-1}U \subset M$, where $c = (c_j^i) : \lambda^{-1}U \to \mathbf{GL}(2, \mathbf{R})$ is a smooth map. Thus,

$$\lambda^* T^* \omega_0^4 = -\lambda^* \langle d\mathbf{T}_0, \mathbf{T}_1\rangle = -\det c \langle dS_0, S_1\rangle.$$

□

Corollary 15.33. *Legendre lifts along surface immersions into various spaces are constructed as follows.*

$\mathscr{M}$ *A smooth oriented tangent sphere map $S : M \to \mathbf{S}^{3,1}$ along an immersion $f : M^2 \to \mathscr{M}$ determines the Legendre lift $\lambda = [F, S + \epsilon_5] : M \to \Lambda$ of f, where $F : U \subset M \to \mathscr{L}$ is a local lift of f.*

$\mathbf{S}^3$ *A unit normal vector field $\mathbf{n} : M \to \mathbf{S}^3$ along an immersion $\mathbf{x} : M^2 \to \mathbf{S}^3$ determines the Legendre lift $\lambda = [\mathbf{x} + \epsilon_4, \mathbf{n} + \epsilon_5] : M \to \Lambda$ of the conformal immersion $f = f_+ \circ \mathbf{x} : M \to \mathscr{M}$.*

$\mathbf{R}^3$ *A unit normal vector field $\mathbf{n} : M \to \mathbf{S}^2 \subset \mathbf{R}^3$ along an immersion $\mathbf{x} : M \to \mathbf{R}^3$ determines the Legendre lift*

$$\lambda = [(1 - |\mathbf{x}|^2)\epsilon_0 + 2\mathbf{x} + (1 + |\mathbf{x}|^2)\epsilon_4, -(\mathbf{x} \cdot \mathbf{n})\epsilon_0 + \mathbf{n} + (\mathbf{x} \cdot \mathbf{n})\epsilon_4 + \epsilon_5] : M \to \Lambda$$

of the conformal immersion $f_0 \circ \mathbf{x} : M \to \mathscr{M}$.

$\mathbf{H}^3$ *A unit normal vector field* $\mathbf{n} : M \to \mathbf{R}^{3,1}$ *along an immersion* $\mathbf{x} : M^2 \to \mathbf{H}^3 \subset \mathbf{R}^{3,1}$ *determines the Legendre lift*

$$\lambda = [\epsilon_0 + \mathbf{x}, \mathbf{n} + \epsilon_5] : M \to \Lambda \tag{15.29}$$

of the conformal immersion $f_- \circ \mathbf{x} : M \to \mathscr{M}$.

This construction works even in cases when the surface has singularities. Such cases cannot be avoided because in general the spherical projection of a Legendre immersion $\lambda : M^2 \to \Lambda$ will have singularities.

Example 15.34 (Legendre lift of surfaces with singularities). Let a and b be positive constants. Consider the surface obtained by rotating the circle in the $\epsilon_1\epsilon_3$-plane with center $(a,0,0) \in \mathbf{R}^3$ and radius b about the ϵ_3-axis. It is parametrized by

$$\mathbf{x} : \mathbf{R}^2 \to \mathbf{R}^3, \quad \mathbf{x}(s,t) = (a + b\cos s)\cos t\epsilon_1 + (a + b\cos s)\sin t\epsilon_2 + b\sin s\epsilon_3.$$

This is an immersion if $0 < b < a$, but if $0 < a \leq b$, then it has singularities, as shown in Figure 15.5, where some of the outer surface is cut away to show the singularities inside.

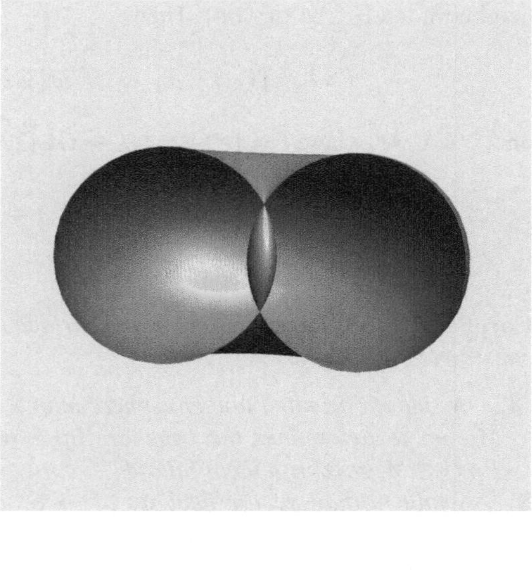

Fig. 15.5 Circle in $\epsilon_2^\perp$, center $(a,0,0)$, radius $b > a > 0$, rotated about ϵ_3-axis

The map $(\mathbf{x}, \mathbf{e}_1, \mathbf{e}_2, \mathbf{e}_3) : \mathbf{R}^2 \to \mathbf{E}(3)$, where

$$\begin{aligned}
\mathbf{e}_1 &= -\sin s\cos t\epsilon_1 - \sin s\sin t\epsilon_2 + \cos s\epsilon_3, \\
\mathbf{e}_2 &= -\sin t\epsilon_1 + \cos t\epsilon_2, \\
\mathbf{e}_3 &= -\cos s\cos t\epsilon_1 - \cos s\sin t\epsilon_2 - \sin s\epsilon_3,
\end{aligned}$$

is a frame field along $\mathbf{x}$ for which $\mathbf{e}_1, \mathbf{e}_2$ are tangent at the nonsingular points, and $\mathbf{e}_2, \mathbf{e}_3$ are normal at the singular points. The map $\lambda : \mathbf{R}^2 \to \Lambda$,

$$\lambda(s,t) = [(1 - |\mathbf{x}|^2)\epsilon_0 + 2\mathbf{x} + (1 + |\mathbf{x}|^2)\epsilon_4, (a+b)\epsilon_0 + \mathbf{e}_3 - (a+b)\epsilon_4 + \epsilon_5],$$

is a Legendre lift of $f_+ \circ \mathscr{S}^{-1} \circ \mathbf{x} : \mathbf{R}^2 \to \mathscr{M}$.

Example 15.35. A smooth curve in $\mathbf{S}^3$ has a Legendre lift. This is important because the spherical projection of a Legendre immersion of a surface can be a curve, as we now show. The unit normal bundle $\mathbf{N}$ of a smooth curve $\mathbf{f} : J \to \mathbf{S}^3$, where $J \subset \mathbf{R}$ is an interval, is

$$\mathbf{N} = \{(s, \mathbf{v}) \in J \times \mathbf{S}^3 : \mathbf{f}(s) \cdot \mathbf{v} = 0 = \dot{\mathbf{f}}(s) \cdot \mathbf{v}\} \subset T_{\mathbf{f}(s)}\mathbf{S}^3.$$

It has a Legendre immersion

$$\lambda : \mathbf{N} \to \Lambda, \quad \lambda(s, \mathbf{v}) = [\mathbf{f}(s) + \epsilon_4, \mathbf{v} + \epsilon_5].$$

Its spherical projection is the curve $\pi \circ \lambda = f_+ \circ \mathbf{f}$ in $\mathscr{M}$.

Lie sphere geometry is the study of properties of Legendre immersions of surfaces into Λ that remain invariant under the action of the Lie sphere group G. It is an extension of Möbius geometry in the following sense. If $f : M \to \mathscr{M}$ is a conformal immersion of a Riemann surface into Möbius space, then its Legendre lift $\lambda : M \to \Lambda$ is a Legendre immersion. The Möbius group is a subgroup of G that sends point spheres to point spheres. Any property of the Legendre lift invariant under the Lie sphere group, is also invariant under the Möbius subgroup. To see the geometric effects of a Lie sphere transformation on the Legendre lift, we must project it back into Möbius space via the spherical projection of Definition 15.14. As we just observed in the preceding two examples, this projection can fail to be an immersion.

Let us examine this last statement in more detail. Suppose $f : M \to \mathscr{M}$ is a conformal immersion with Legendre lift $\lambda : M \to \Lambda$, which we know is a Legendre immersion. If $\pi : \Lambda \to \mathscr{M}$ is the spherical projection (each line goes to the unique point sphere in it), then $\pi \circ \lambda = f$. If $T \in G$ is any Lie sphere transformation, then $T \circ \lambda : M \to \Lambda$ is again a Legendre immersion, since T acts on Λ as a contact diffeomorphism. The spherical projection

$$\pi \circ T \circ \lambda : M \to \mathscr{M},$$

however, need not be an immersion. It could have discrete points where its differential has rank less than two, or the whole map can have rank less than two. The Lie parallel transformations are the culprits.

Proposition 15.36. *Let* $\mathbf{x} : M \to \mathbf{R}^3$ *be an immersion with unit normal vector field* $\mathbf{n}$ *and Legendre lift* $\lambda : M \to \Lambda$ *given in Corollary 15.33. For any* $r \in \mathbf{R}$, *let $T_r^0 \in \mathbf{SO}(4,2)$ be the Lie parallel transformation* (15.18). *Then the spherical projection of $T_r^0 \circ \lambda : M \to \Lambda$ is $f_0 \circ \tilde{\mathbf{x}}$, where $\tilde{\mathbf{x}} = \mathbf{x} - r\mathbf{n}$ is the parallel transformation of $\mathbf{x}$ by $-r$.*

Thus, the spherical projection of the smooth Legendre immersion $T_r^0 \circ \lambda : M \to \Lambda$ is singular at any point of M where $-1/r$ is a principal curvature of $\mathbf{x}$ relative to $\mathbf{n}$. In general, these points will be discrete, but consider an immersion such as a circular torus obtained by revolving a circle of radius $r > 0$ about a line disjoint from it. With the inward pointing normal, the parallel transformation $\tilde{\mathbf{x}}$ by r will have rank equal to one at every point. The parallel transformation of an oriented sphere of radius r by an amount r along the inward pointing normal is a constant map.

Proof. This is an elementary calculation. Use Definition 15.14 to calculate the spherical projection. □

15.6 Tangent and curvature spheres

Following the general theory of Chapter 3, we now apply the method of moving frames to Legendre immersions $\lambda : M \to \Lambda$, where M is a surface, and the Lie sphere group G acts transitively on Λ. The geometry of Lie frames is based on the concepts of tangent and curvature spheres. Recall that the Lie quadric Q is identified with the set of all oriented spheres in $\mathbf{S}^3$ in Proposition 15.2 and Exercise 62.

Definition 15.37. An *oriented tangent sphere* at a point $m \in M$ of a Legendre immersion $\lambda : M^2 \to \Lambda$ is a point on the line $\lambda(m) \subset Q$. An *oriented tangent sphere map* is a smooth map $S : U \subset M \to Q$ such that $S(m) \in \lambda(m)$, for every $m \in U$. An *oriented curvature sphere* at a point $m \in M$ of λ is an oriented tangent sphere $\hat{S}$ of λ at m for which, if $S : U \to Q$ is an oriented tangent sphere map of λ on a neighborhood U of m with $S(m) = \hat{S}$, then $dS_m \mod \lambda(m)$ has rank less than 2.

Example 15.38. The oriented curvature spheres of the Legendre lift of a conformal immersion $f : M \to \mathscr{M}$ with tangent sphere map $S : M \to \mathbf{S}^{3,1}$ correspond to the oriented curvature spheres of f via the inclusion $\mathbf{S}^{3,1} \subset Q$, $S \mapsto [S + \epsilon_5]$. In fact, if $F : U \subset M \to \mathscr{L}$ is a lift of f, then any oriented tangent sphere of f at a point on U is given by $S + rF \in \mathbf{S}^{3,1}$, for any $r \in \mathbf{R}$. This is an oriented curvature sphere if and only if $d(S + rF) \mod F$ has rank less than two. The Legendre lift of f on U is $\lambda = [F, S + \epsilon_5] : U \to \Lambda$. All of its oriented tangent spheres are given on U by $[rF + s(S + \epsilon_5)] \in Q$, for ${}^t[r,s] \in \mathbf{RP}^1$. This is a curvature sphere of λ if and only if $d(rF + s(S + \epsilon_5)) \mod \{F, S + \epsilon_5\}$ has rank less than two if and only if $s \neq 0$ and $d(S + \frac{r}{s}F) \mod F$ has rank less than two if and only if $S + \frac{r}{s}F$ is a curvature sphere of f.

Definition 15.39. A *Lie frame field along a Legendre immersion* $\lambda : M \to \Lambda$ is a smooth map $T : U \to G$ defined on an open subset $U \subset M$ such that $\pi \circ T = \lambda$, where $\pi : G \to \Lambda$ is the projection (15.23).

Let $T = (\mathbf{T}_0, \dots, \mathbf{T}_5) : U \to G$ be a Lie frame field along the Legendre immersion $\lambda : M \to \Lambda$. Then $\lambda = [\mathbf{T}_0, \mathbf{T}_1]$, which is a line in the Lie quadric $Q \subset \mathbf{P}(\mathbf{R}^{4,2})$. Any oriented tangent sphere S of λ at a point of U is given by

$$S = [r\mathbf{T}_0 + s\mathbf{T}_1],$$

for any ${}^t[r,s] \in \mathbf{RP}^1$, which are called the *projective coordinates of S relative to T*. Since λ is a Legendre map, it satisfies

$$T^*\omega_0^4 = 0$$

on U. We shall omit writing T^* in the entries of $T^*\omega$. Taking the exterior derivative of this equation, and using the structure equations of G, we get

$$\omega_1^2 \wedge \omega_0^2 + \omega_1^3 \wedge \omega_0^3 = 0 \tag{15.30}$$

on U. By (15.26), the linearly independent entries of $T^*\omega_{\mathfrak{m}_0}$ lie in

$$\theta = \begin{pmatrix} \omega_0^2 & \omega_1^2 \\ \omega_0^3 & \omega_1^3 \end{pmatrix}.$$

The smooth map $\mathbf{T}_0 \wedge \mathbf{T}_1 : U \to \Lambda_2\mathbf{R}^{4,2}$ is a lift of $\lambda : U \to \Lambda$ and

$$d(\mathbf{T}_0 \wedge \mathbf{T}_1) \equiv \sum_2^3 \omega_0^i \mathbf{T}_i \wedge \mathbf{T}_1 + \sum_2^3 \omega_1^i \mathbf{T}_0 \wedge \mathbf{T}_i \mod \mathbf{T}_0 \wedge \mathbf{T}_1.$$

Thus, $d\lambda$ has rank two if and only if the dimension of $\operatorname{span}\{\omega_0^2, \omega_0^3, \omega_1^2, \omega_1^3\}$ is two at every point of U. This is equivalent to the condition that for any point $m \in U$, there exists $\mathfrak{p} = {}^t[r,s] \in \mathbf{RP}^1$, such that

$$(r\omega_0^2 + s\omega_1^2) \wedge (r\omega_0^3 + s\omega_1^3) \neq 0 \tag{15.31}$$

on a neighborhood of m in U. In fact, if $\omega_0^2 \wedge \omega_0^3 \neq 0$ at m, then $\mathfrak{p} = {}^t[1,0]$ works. Otherwise, suppose that $\omega_0^2 \wedge \omega_0^3 = 0$ at m. Now, if for every $r \in \mathbf{R}$ we have at m,

$$0 = (r\omega_0^2 + \omega_1^2) \wedge (r\omega_0^3 \wedge \omega_1^3) = r(\omega_0^2 \wedge \omega_1^3 + \omega_1^2 \wedge \omega_0^3) + \omega_1^2 \wedge \omega_1^3,$$

then taking $r = 0$ implies $\omega_1^2 \wedge \omega_1^3 = 0$ at m, and thus also $\omega_0^2 \wedge \omega_1^3 + \omega_1^2 \wedge \omega_0^3 = 0$ at m. These equations combined with (15.30) imply that the rank of $\{\omega_0^2, \omega_0^3, \omega_1^2, \omega_1^3\}$ is less than two at m, a contradiction.

Lemma 15.40. *A curvature sphere of a Legendre immersion* $\lambda : M \to \Lambda$ *at a point* $m \in M$ *is a tangent sphere* S *at* m *with the property that for any Lie frame field* $T : U \to G$*, with* $m \in U$*, the projective coordinates* ${}^t[r,s]$ *of* S *relative to* T *and the Maurer–Cartan forms* $(\omega^b_a) = T^{-1}dT$ *satisfy*

$$(r\omega^2_0 + s\omega^2_1) \wedge (r\omega^3_0 + s\omega^3_1) = 0 \tag{15.32}$$

at m.

Proof. Let $T : U \to G$ be a Lie frame field along λ on a neighborhood U of m in M. Let ${}^t[s,t] \in \mathbf{RP}^1$ be the projective coordinates relative to T of a curvature sphere S at m. The tangent sphere map $[r\mathbf{T}_0 + s\mathbf{T}_1] : U \to Q$ has value S at m and the rank of $d[r\mathbf{T}_0 + s\mathbf{T}_1]_m \mod \lambda(m)$ is less than 2 if and only if

$$d(r\mathbf{T}_0 + s\mathbf{T}_1) \equiv (r\omega^2_0 + s\omega^2_1)\mathbf{T}_2 + (r\omega^3_0 + s\omega^3_1)\mathbf{T}_3, \quad \mod \{\mathbf{T}_0, \mathbf{T}_1\}$$

has rank less than two at m if and only if (15.32) holds at m. □

How do the projective coordinates of a tangent sphere depend on the Lie frame? Let S be a curvature sphere of λ at $m \in M$. Let $T : U \to G$ be a Lie frame field about $m \in M$ such that $S = [s\mathbf{T}_0 + r\mathbf{T}_1](m) \in Q$. Any other Lie frame field on U is given by

$$\tilde{T} = Tk(c,B,Z,b),$$

where $k(c,B,Z,b) : U \to G_0$ is defined in (15.25). Here $c : U \to \mathbf{GL}(2,\mathbf{R})$, $B : U \to \mathbf{O}(2)$, and $Z, b : U \to \mathbf{R}^{2\times 2}$ are smooth maps. Then

$$(\tilde{\mathbf{T}}_0, \tilde{\mathbf{T}}_1) = (\mathbf{T}_0, \mathbf{T}_1)c \text{ and } \tilde{\theta} = B^{-1}\theta c,$$

by (15.28). The projective coordinates of S relative to T and $\tilde{T}$ are related by

$$\begin{bmatrix} \tilde{r} \\ \tilde{s} \end{bmatrix} = c^{-1} \begin{bmatrix} r \\ s \end{bmatrix}, \tag{15.33}$$

so

$$\tilde{\theta} \begin{bmatrix} \tilde{r} \\ \tilde{s} \end{bmatrix} = B^{-1}\theta c c^{-1} \begin{bmatrix} r \\ s \end{bmatrix} = B^{-1}\theta \begin{bmatrix} r \\ s \end{bmatrix},$$

which implies that

$$(\tilde{r}\tilde{\omega}^2_0 + \tilde{s}\tilde{\omega}^2_1) \wedge (\tilde{r}\tilde{\omega}^3_0 + \tilde{s}\tilde{\omega}^3_1) = (\det B^{-1})(r\omega^2_0 + s\omega^2_1) \wedge (r\omega^3_0 + s\omega^3_1).$$

A change of Lie frame changes the projective coordinates of a tangent sphere S by (15.33). A basic property of this action of $\mathbf{GL}(2,\mathbf{R})$ on $\mathbf{RP}^1$ is its three point transitivity. See Problem 15.76.

The discussion around (15.31) showed that at any point $m \in M$, there exists a neighborhood U of m on which there is a smooth tangent sphere $S : U \to Q$, which is not a curvature sphere at any point of U. By Problem 15.76, there exists a Lie frame field $T : U \to G$ such that $S = [\mathbf{T}_0]$ on U. Now (15.32) becomes $\omega_0^2 \wedge \omega_0^3 \neq 0$ at every point of U. Then Cartan's Lemma applied to (15.30) implies that $\omega_1^i = h_j^i \omega_0^j$ for smooth functions on U satisfying $h_j^i = h_i^j$, for all $i,j \in \{2,3\}$. Any tangent sphere not equal to S is given by $[r\mathbf{T}_0 + \mathbf{T}_1]$, where $r : U \to \mathbf{R}$. Then

$$(r\omega_0^2 + \omega_1^2) \wedge (r\omega_0^3 + \omega_1^3) = (r^2 + r(h_2^2 + h_3^3) + h_2^2 h_3^3 - h_3^2 h_2^3)\omega_0^2 \wedge \omega_0^3$$

shows that $[r\mathbf{T}_0 + \mathbf{T}_1]$ is a curvature sphere at a point $m \in U$ if and only if $-r$ is an eigenvalue of the symmetric matrix $(h_j^i(m))$. In particular, there are precisely two curvature spheres, counting multiplicities, at each point of M. Distinct solutions depend smoothly on m, but the dependence is only continuous at points where the solutions coincide. Thus, if $r_0, r_1 : U \to \mathbf{R}$ are the solutions, then the curvature spheres

$$[r_i\mathbf{T}_0 + \mathbf{T}_1] : U \to Q$$

are continuous maps and smooth where r_0 and r_1 are distinct. Thus, we have shown that if

$$S_0, S_1 : M \to Q$$

denote the curvature spheres at each point of M, then these are continuous maps, smooth when they are distinct, and $\langle S_0, S_1 \rangle = 0$.

Definition 15.41. A point $m \in M$ is an *umbilic point* of the Legendre immersion $\lambda : M \to \Lambda$, if the curvature sphere at m has multiplicity two; that is, $S_0(m) = S_1(m)$. Otherwise, the point is nonumbilic.

15.7 Frame reductions

The first step of the reduction procedure for Lie frame fields along a Legendre immersion $\lambda : M \to \Lambda$ requires the absence of umbilic points.

Definition 15.42. A *first order* Lie frame field $T : U \to G$ along a Legendre immersion $\lambda : M^2 \to \Lambda$ for which there are distinct smooth curvature spheres $S_0, S_1 : U \to Q$ is characterized by

1. $[\mathbf{T}_0] = S_0$ and $[\mathbf{T}_1] = S_1$ at each point of U, and
2. $\omega_0^2 = 0 = \omega_1^3$ and $\omega_1^2 \wedge \omega_0^3 \neq 0$ at every point of U.

Lemma 15.43. *If $\lambda : M \to \Lambda$ is a Legendre immersion, then for any nonumbilic point $m \in M$, there exists a neighborhood U of m in M on which there exists a first order Lie frame field.*

Proof. Let $T : U \to G$ be any smooth Lie frame field on a neighborhood U of m. By the three point transitivity of the action of $\mathbf{GL}(2,\mathbf{R})$ on $\mathbf{RP}^1$, as proved in Problem 15.76, there exists a smooth map $c : U \to \mathbf{GL}(2,\mathbf{R})$ such that relative to the Lie frame field $\tilde{T} = Tk(c, I_2, 0, 0) : U \to G$ the curvature spheres are $[\tilde{\mathbf{T}}_0]$ and $[\tilde{\mathbf{T}}_1]$ at each point of U. Change the notation to assume this is the case for the original Lie frame field T. Then $[\mathbf{T}_0]$ and $[\mathbf{T}_1]$ being curvature spheres on U implies that

$$\omega_0^2 \wedge \omega_0^3 = 0 = \omega_1^2 \wedge \omega_1^3, \tag{15.34}$$

at each point of U, by (15.32). In particular, ω_0^2 and ω_0^3 are linearly dependent at each point of U, and they are not both zero at any point, because if $r, s \in \mathbf{R} \setminus \{0\}$, then $[r\mathbf{T}_0 + s\mathbf{T}_1]$ is not an oriented curvature sphere at any point of U, so by (15.32) and (15.34),

$$0 \neq (r\omega_0^2 + s\omega_1^2) \wedge (r\omega_0^3 + s\omega_1^3) = rs(\omega_0^2 \wedge \omega_1^3 + \omega_1^2 \wedge \omega_0^3), \tag{15.35}$$

at every point of U. There exists a smooth map

$$(a,b) : U \to \mathbf{S}^1 \subset \mathbf{R}^2$$

such that $a\omega_0^2 + b\omega_0^3 = 0$ at each point of U, so

$$B = \begin{pmatrix} a & -b \\ b & a \end{pmatrix} : U \to \mathbf{SO}(2)$$

is a smooth map. The Lie frame field $\tilde{T} = Tk(I_2, B, 0, 0) : U \to G$ has $\tilde{\mathbf{T}}_0 = \mathbf{T}_0$ and $\tilde{\mathbf{T}}_1 = \mathbf{T}_1$ and

$$\tilde{\theta} = B^{-1}\theta = \begin{pmatrix} 0 & \tilde{\omega}_1^2 \\ \tilde{\omega}_0^3 & \tilde{\omega}_1^3 \end{pmatrix},$$

by (15.28). Now $\tilde{\omega}_0^2 = 0$ on U implies $\tilde{\omega}_1^2 \wedge \tilde{\omega}_0^3 \neq 0$ at every point of U, by (15.35), which holds for $\tilde{\theta}$ as well. Thus, $\tilde{\omega}_1^2, \tilde{\omega}_0^3$ is a coframe field on U, so $\tilde{\omega}_1^3 = 0$ on U, since $\tilde{\omega}_1^3 \wedge \tilde{\omega}_1^2 = 0$ on U, by (15.34), and $\tilde{\omega}_1^3 \wedge \tilde{\omega}_0^3 = 0$ on U by the Legendre condition (15.30). Hence, $\tilde{T} : U \to G$ is of first order. □

Let $T : U \to G$ be a first order Lie frame field along λ. Let

$$\theta^2 = \omega_1^2, \quad \theta^3 = \omega_0^3. \tag{15.36}$$

By (2) of Definition 15.42, θ^1, θ^2 is a coframe field on U and

$$0 = d\omega_0^2 = \omega_0^1 \wedge \theta^2 - \omega_3^2 \wedge \theta^3, \quad 0 = d\omega_1^3 = \omega_1^0 \wedge \theta^3 - \omega_2^3 \wedge \theta^2, \tag{15.37}$$

by the structure equations of G. If we let

$$\omega_0^1 = A_2\theta^2 + A_3\theta^3, \quad \omega_1^0 = B_2\theta^2 + B_3\theta^3, \tag{15.38}$$

for some smooth functions $A_2, A_3, B_2, B_3 : U \to \mathbf{R}$, then (15.37) implies

$$\omega_3^2 = -A_3\theta^2 + B_2\theta^3 = -\omega_2^3. \tag{15.39}$$

In the notation of (15.25), any other first order Lie frame field on U is

$$\tilde{T} = Tk(c, B, Z, b) : U \to G,$$

where $k(c, B, Z, b) : U \to G_1$ is a smooth map into the closed subgroup

$$G_1 = \{k(c, B, Z, b) \in G_0 : c = \begin{pmatrix} r & 0 \\ 0 & s \end{pmatrix}, \, Z, b \in \mathbf{R}^{2\times 2}, \, B = \begin{pmatrix} \epsilon & 0 \\ 0 & \delta \end{pmatrix}\},$$

where $rs \neq 0$ and $\epsilon, \delta \in \{\pm 1\}$. Its Lie algebra is

$$\mathfrak{g}_1 = \left\{ \begin{pmatrix} X & {}^tY & X_4^0 I_{1,1} \\ 0 & 0 & YL \\ 0 & 0 & -L{}^tXL \end{pmatrix} \in \mathfrak{g}_0 : X = \begin{pmatrix} X_0^0 & 0 \\ 0 & X_1^1 \end{pmatrix}, Y \in \mathbf{R}^{2\times 2}, \, X_4^0 \in \mathbf{R} \right\}.$$

Equations (15.38) suggest that for a vector space complement of $\mathfrak{g}_1$ in $\mathfrak{g}_0$ we choose

$$\mathfrak{m}_1 = \left\{ \begin{pmatrix} X_0 & 0 & 0 \\ 0 & W & 0 \\ 0 & 0 & -L{}^tX_0L \end{pmatrix} : X_0 = \begin{pmatrix} 0 & X_1^0 \\ X_0^1 & 0 \end{pmatrix}, \, W \in \mathfrak{o}(2) \right\}.$$

The $(\mathfrak{m}_0 + \mathfrak{m}_1)$-component of the Maurer–Cartan form ω is

$$\omega_{\mathfrak{m}_0+\mathfrak{m}_1} = \begin{pmatrix} \eta_0 & 0 & 0 \\ \theta & \Omega & 0 \\ 0 & L{}^t\theta & -L{}^t\eta_0 L \end{pmatrix},$$

where

$$\eta_0 = \begin{pmatrix} 0 & \omega_1^0 \\ \omega_0^1 & 0 \end{pmatrix}, \quad \theta = \begin{pmatrix} 0 & \theta^2 \\ \theta^3 & 0 \end{pmatrix}, \quad \Omega = \begin{pmatrix} 0 & \omega_3^2 \\ \omega_2^3 & 0 \end{pmatrix}.$$

Let $k = k(c,B,Z,b) : U \to G_1$. Denote the $(\mathfrak{m}_0 + \mathfrak{m}_1)$-component of $\mathrm{Ad}(k^{-1})\omega_{\mathfrak{m}_0+\mathfrak{m}_1}$ with the same letters covered by a tilde. Then

$$\tilde\theta^2 = \epsilon s \theta^2, \quad \tilde\theta^3 = \delta r \theta^3,$$

$$\tilde\omega_0^1 = \frac{r}{s}(\omega_0^1 - \delta Z_3^1 \theta^3) = \frac{r\epsilon A_2}{s^2}\tilde\theta^2 + \frac{\delta A_3 - Z_3^1}{s}\tilde\theta^3,$$

$$\tilde\omega_1^0 = \frac{s}{r}(\omega_1^0 - \epsilon Z_2^0 \theta^2) = \frac{\epsilon B_2 - Z_2^0}{r}\tilde\theta^2 + \frac{s\delta B_3}{r^2}\tilde\theta^3,$$

$$\tilde\omega_2^3 = \epsilon\delta\omega_2^3 + \delta Z_2^0\theta^3 - \epsilon Z_3^1\theta^2 = -\tilde\omega_3^2.$$

Hence, the coefficients in (15.38) relative to the new frame satisfy

$$\tilde A_2 = \frac{\epsilon r}{s^2}A_2, \quad \tilde A_3 = \frac{\delta A_3 - Z_3^1}{s}, \quad \tilde B_2 = \frac{\epsilon B_2 - Z_2^0}{r}, \quad \tilde B_3 = \frac{\delta s}{r^2}B_3, \tag{15.40}$$

from which we see that we can always choose $k : U \to G_1$ to make $\tilde A_3 = 0 = \tilde B_2$ on U, but the orbit structure of the action of G_1 on A_2 and B_3 depends on whether these functions are zero or not. There are four orbit types.

Definition 15.44. A Legendre immersion $\lambda : M \to \Lambda$ is of the following *type*, if for every first order Lie frame field $T : U \to G$,

A). $\omega_0^1 \wedge \theta^3 \neq 0$ and $\theta^2 \wedge \omega_1^0 \neq 0$ at every point of U;
B). $\omega_0^1 \wedge \theta^3 \neq 0$ and $\theta^2 \wedge \omega_1^0 = 0$ at every point of U;
C). $\omega_0^1 \wedge \theta^3 = 0$ and $\theta^2 \wedge \omega_1^0 \neq 0$ at every point of U;
D). $\omega_0^1 \wedge \theta^3 = 0$ and $\theta^2 \wedge \omega_1^0 = 0$ at every point of U.

Type A is the *non-degenerate* case, immersions of Types B or C are called *canal*, and immersions of Type D are called *Dupin*.

15.8 Frame reductions for generic immersions

Suppose now that $\lambda : M \to \Lambda$ is a Legendre immersion of Type A.

Definition 15.45. A *second order Lie frame field* $T : U \to G$ along λ is a first order Lie frame field for which

$$\omega_0^1 = \theta^2, \quad \omega_1^0 = \theta^3, \quad \omega_3^2 = 0 \tag{15.41}$$

on U, where θ^2, θ^3 is the coframe field defined in (15.36).

It is evident from (15.40) that smooth second order Lie frame fields exist on some neighborhood of any point of M. These same equations also show that if T and $\tilde T$ are both second order Lie frame fields on U, then

$$Z_2^0 = 0 = Z_3^1, \quad r = \epsilon, \quad s = \delta.$$

Let $T : U \to G$ be a second order Lie frame field. Taking the exterior differential of the forms in (15.41) and using the structure equations, we get

$$\begin{aligned}(\omega_0^0 - 2\omega_1^1) \wedge \theta^2 - \omega_3^1 \wedge \theta^3 &= 0,\\ \omega_2^0 \wedge \theta^2 - (\omega_1^1 - 2\omega_0^0) \wedge \theta^3 &= 0,\\ \omega_3^1 \wedge \theta^2 - \omega_2^0 \wedge \theta^3 &= 0\end{aligned}$$

on U. These equations imply that

$$\begin{aligned}\omega_2^0 = D_2\theta^2 + D_3\theta^3, &\quad \omega_3^1 = E_2\theta^2 - D_2\theta^3,\\ \omega_0^0 - 2\omega_1^1 = C_2\theta^2 - E_2\theta^3, &\quad \omega_1^1 - 2\omega_0^0 = -D_3\theta^2 + C_3\theta^3,\end{aligned} \tag{15.42}$$

for some smooth functions $D_2, D_3, E_2, C_2, C_3 : U \to \mathbf{R}$. Any other second order Lie frame field on U is given by

$$\tilde{T} = Tk(c, c, Z, b) : U \to G,$$

where $k(c, c, Z, b) : U \to G_2$ is a smooth map into the closed subgroup

$$G_2 = \{k(c, c, Z, b) \in G_1 : c = \begin{pmatrix} \epsilon & 0 \\ 0 & \delta \end{pmatrix}, \ \epsilon, \delta = \pm 1, \ {}^tZ = \begin{pmatrix} 0 & s \\ r & 0 \end{pmatrix}, \ r, s \in \mathbf{R}\},$$

where we use the notation of (15.25). We note that the condition imposed on $b \in \mathbf{R}^{2\times 2}$ in (15.24) implies the entries b_j^i, $i = 0, 2$, $j = 4, 5$, of b satisfy

$$b_5^0 = \frac{\epsilon}{2}s^2, \quad b_4^1 = \frac{\delta}{2}r^2, \quad \epsilon b_5^1 + \delta b_4^0 = 0.$$

Note r, s, and b_4^0 can be chosen arbitrarily. The Lie algebra of G_2 is

$$\mathfrak{g}_2 = \left\{ \begin{pmatrix} 0 & {}^tY & tI_{1,1} \\ 0 & 0 & YL \\ 0 & 0 & 0 \end{pmatrix} : Y = \begin{pmatrix} 0 & Y_3^0 \\ Y_2^1 & 0 \end{pmatrix}, \ t \in \mathbf{R} \right\}.$$

Equations (15.42) suggest that for the vector space complement of $\mathfrak{g}_2$ in $\mathfrak{g}_1$ we should choose

$$\mathfrak{m}_2 = \left\{ \begin{pmatrix} X & {}^tY & 0 \\ 0 & 0 & YL \\ 0 & 0 & -L{}^tXL \end{pmatrix} : {}^tY = \begin{pmatrix} Y_2^0 & 0 \\ 0 & Y_3^1 \end{pmatrix}, X = \begin{pmatrix} X_0^0 & 0 \\ 0 & X_1^1 \end{pmatrix} \right\}.$$

The $(\mathfrak{m}_0 + \mathfrak{m}_1 + \mathfrak{m}_2)$-component of the Maurer–Cartan form ω is

$$\omega_{\mathfrak{m}_0+\mathfrak{m}_1+\mathfrak{m}_2} = \begin{pmatrix} \eta & {}^t\Sigma & 0 \\ \theta & 0 & \Sigma L \\ 0 & L^t\theta & -L^t\eta L \end{pmatrix},$$

where

$$\eta = \begin{pmatrix} \omega_0^0 & \theta^3 \\ \theta^2 & \omega_1^1 \end{pmatrix}, \quad \theta = \begin{pmatrix} 0 & \theta^2 \\ \theta^3 & 0 \end{pmatrix}, \quad {}^t\Sigma = \begin{pmatrix} \omega_2^0 & 0 \\ 0 & \omega_3^1 \end{pmatrix}.$$

Letting $k = k(c, c, Z, b) : U \to G_2$ and denoting the $(\mathfrak{m}_0 + \mathfrak{m}_1 + \mathfrak{m}_2)$-component of $\mathrm{Ad}(k^{-1})\omega_{\mathfrak{m}_0+\mathfrak{m}_1+\mathfrak{m}_2}$ with the same letters covered by a tilde, we get

$$\tilde{\theta}^2 = \epsilon\delta\theta^2, \quad \tilde{\theta}^3 = \epsilon\delta\theta^3,$$
$$\tilde{\omega}_0^0 = \omega_0^0 - \delta s\theta^3, \quad \tilde{\omega}_1^1 = \omega_1^1 - \epsilon r\theta^2,$$
$$\tilde{\omega}_2^0 = \omega_2^0 + \epsilon r\theta^3 + \epsilon b_5^1\theta^2, \quad \tilde{\omega}_3^1 = \omega_3^1 + \delta s\theta^2 + \delta b_4^0\theta^3.$$

Hence, the coefficients in (15.42) relative to the new frame satisfy

$$\tilde{D}_2 = \epsilon\delta D_2 + \delta b_5^1, \quad \tilde{D}_3 = \epsilon\delta D_3 + \delta r, \quad \tilde{E}_2 = \epsilon\delta E_2 + \epsilon s, \tag{15.43}$$

from which we see that we can choose $k(c, c, Z, b) : U \to G_2$ to make $\tilde{D}_2 = \tilde{D}_3 = \tilde{E}_2 = 0$ at every point of U.

Definition 15.46. A *third order Lie frame field* $T : U \to G$ along λ is a second order Lie frame field for which

$$\omega_2^0 = 0 = \omega_3^1 \tag{15.44}$$

on U.

It is evident from (15.43) that a third order Lie frame field exists on some neighborhood of any point of M. These same equations also show that if $\tilde{T} = Tk$ are both third order Lie frame fields on U, then $k = k(c, c, 0, 0)$ takes values in the discrete subgroup

$$G_3 = \{k(c, c, 0, 0) \in G_2 : c = \begin{pmatrix} \epsilon & 0 \\ 0 & \delta \end{pmatrix}, \ \epsilon, \delta \in \{\pm 1\}\}.$$

Summary: A third order Lie frame field $T : U \to G$ along a Type A Legendre immersion $\lambda : M \to \Lambda$ is characterized by $[\mathbf{T}_0] = S_0$ and $[\mathbf{T}_1] = S_1$ are the curvature spheres of λ and

1. $\omega_0^2 = 0 = \omega_1^3$, $\theta^2 = \omega_1^2$, $\theta^3 = \omega_0^3$, and $\theta^2 \wedge \theta^3 \neq 0$.
2. $\omega_0^1 = \theta^2$, $\omega_1^0 = \theta^3$, $\omega_3^2 = 0$.
3. $\omega_2^0 = 0 = \omega_3^1$.

Let $T : U \to G$ be a third order Lie frame field. It follows from (15.44) and (15.42) that now

$$\omega_0^0 = q\theta^2 - 2p\theta^3, \quad \omega_1^1 = 2q\theta^2 - p\theta^3, \tag{15.45}$$

for smooth functions $p, q : U \to \mathbf{R}$. These functions are globally defined on M up to sign. From the structure equations of G,

$$d\theta^2 = d\omega_1^2 = p\theta^2 \wedge \theta^3, \quad d\theta^3 = d\omega_0^3 = q\theta^2 \wedge \theta^3,$$

so the functions p, q are determined by the coframe field θ^2, θ^3. Taking the exterior differential of the forms in (15.44) and using the structure equations of G, we get

$$0 = d\omega_2^0 = \omega_2^1 \wedge \theta^3 + \theta^2 \wedge \omega_4^0, \quad 0 = d\omega_3^1 = -\theta^2 \wedge \omega_3^0 + \omega_4^0 \wedge \theta^3,$$

from which we conclude that

$$\omega_2^1 = r\theta^2 + s\theta^3, \quad \omega_3^0 = t\theta^2 + u\theta^3, \quad \omega_4^0 = u\theta^2 - r\theta^3, \tag{15.46}$$

for smooth functions $r, s, t, u : U \to \mathbf{R}$. All other Maurer–Cartan forms can now be determined from the identities in (15.21). Any other third order Lie frame field on U is given by $\tilde{T} = Tk$, where $k = k(c, c, 0, 0) : U \to G_3$ is a smooth map. Then $\tilde{\omega} = \mathrm{Ad}(k^{-1})\omega$ implies that

$$\tilde{\theta}^2 = \epsilon\delta\theta^2, \quad \tilde{\theta}^3 = \epsilon\delta\theta^3, \quad \tilde{\omega}_0^0 = \omega_0^0, \quad \tilde{\omega}_1^1 = \omega_1^1,$$
$$\tilde{\omega}_3^0 = \epsilon\delta\omega_3^0, \quad \tilde{\omega}_2^1 = \epsilon\delta\omega_2^1, \quad \tilde{\omega}_4^0 = \epsilon\delta\omega_4^0$$

from which it follows that the 1-forms ω_0^0 and ω_1^1, relative to a third order Lie frame field, are globally defined on M, the third order invariants

$$\tilde{r} = r, \quad \tilde{s} = s, \quad \tilde{t} = t, \quad \tilde{u} = u$$

are globally defined smooth functions on M, the invariants

$$\tilde{p} = \epsilon\delta p, \quad \tilde{q} = \epsilon\delta q$$

are defined only up to sign, while $\tilde{p}\tilde{q} = pq$ is a globally defined function on M. In addition, the area element

$$\tilde{\theta}^2 \wedge \tilde{\theta}^3 = \theta^2 \wedge \theta^3$$

is globally defined on M. In the literature, see for example Blaschke [8] or Ferapontov [68], the fundamental invariants of a Legendre immersion of Type A are taken to be the symmetric quadratic form $\mathscr{F}$ and cubic form $\mathscr{P}$ defined by the symmetric products

$$\mathscr{F} = -\theta^2\theta^3, \quad \mathscr{P} = -(\theta^2)^3 + (\theta^3)^3.$$

The form $\mathscr{F}$ is globally defined on M, but the form $\mathscr{P}$ is defined only up to sign.

Taking the exterior derivative of (15.45) and (15.46), using the structure equations of G and the identities in (15.21), we obtain the structure equations relative to the third order Lie frame field T,

$$\begin{gathered} d\theta^2 = \omega_1^1 \wedge \theta^2, \quad d\theta^3 = \omega_0^0 \wedge \theta^3, \\ d\omega_0^0 = (\theta^2 - \omega_3^0) \wedge \theta^3, \quad d\omega_1^1 = (\theta^3 - \omega_2^1) \wedge \theta^2, \\ d\omega_2^1 = \omega_2^1 \wedge \omega_1^1, \quad d\omega_3^0 = \omega_3^0 \wedge \omega_0^0, \quad d\omega_4^0 = \omega_4^0 \wedge (\omega_0^0 + \omega_1^1). \end{gathered}$$

These equations imply the invariants p, q, r, s, t, and u satisfy

$$\begin{aligned} dq \wedge \theta^2 - 2dp \wedge \theta^3 &= (1 + pq - t)\theta^2 \wedge \theta^3, \\ 2dq \wedge \theta^2 - dp \wedge \theta^3 &= (s - 1 - pq)\theta^2 \wedge \theta^3, \\ dt \wedge \theta^2 + du \wedge \theta^3 &= -(3pt + 2qu)\theta^2 \wedge \theta^3, \\ dr \wedge \theta^2 + ds \wedge \theta^3 &= -(3qs + 2pr)\theta^2 \wedge \theta^3, \\ du \wedge \theta^2 - dr \wedge \theta^3 &= 4(rq - up)\theta^2 \wedge \theta^3. \end{aligned}$$

15.9 Frame reductions for Dupin immersions

Suppose now that $\lambda : M \to \Lambda$ is a Legendre immersion of Type D. Then for any first order frame field $T : U \to G$, the functions $A_2 = 0 = B_3$ in (15.38). By (15.40), we may choose T such that $A_3 = 0 = B_2$ on U, so $\omega_3^2 = 0$ as well, by (15.39).

Definition 15.47. A *second order Lie frame field* $T : U \to G$ along λ is a first order Lie frame field for which

$$\omega_0^1 = 0, \quad \omega_1^0 = 0, \quad \omega_3^2 = 0 \tag{15.47}$$

on U. Thus, $\omega_{\mathfrak{m}_1} = 0$ on U characterizes a second order Lie frame field.

Let $T : U \to G$ be a second order Lie frame field along λ. Taking the exterior differential of the forms in (15.47) and using the structure equations of G, we get

$$\omega_3^1 \wedge \theta^3 = 0, \quad \omega_2^0 \wedge \theta^2 = 0, \quad \theta^2 \wedge \omega_3^1 + \omega_2^0 \wedge \theta^3 = 0,$$

where $\theta^2 = \omega_1^2$, $\theta^3 = \omega_0^3$ is the coframe field of T. It follows that

$$\omega_2^0 = D\theta^2, \quad \omega_3^1 = -D\theta^3, \tag{15.48}$$

for some smooth function $D : U \to \mathbf{R}$. If T and $\tilde{T}$ are both second order Lie frame fields on U, then (15.40) implies

$$Z_2^0 = 0 = Z_3^1$$

on U. Thus, any other second order Lie frame field on U is given by

$$\tilde{T} = Tk(c,B,Z,b) : U \to G,$$

where $k(c,B,Z,b) : U \to G_2$ is a smooth map into the closed subgroup

$$G_2 = \{k(c,B,Z,b) \in G_1 : {}^tZ = \begin{pmatrix} 0 & Z^0_3 \\ Z^1_2 & 0 \end{pmatrix},\ Z^0_3, Z^1_2 \in \mathbf{R}\}.$$

We note that if $k(c,B,Z,b) \in G_2$, then

$$b = \begin{pmatrix} b^0_4 & \frac{1}{2r}(Z^0_3)^2 \\ \frac{1}{2s}(Z^1_2)^2 & -\frac{s}{r}b^0_4 \end{pmatrix},$$

where $b^0_4 \in \mathbf{R}$ is arbitrary. The Lie algebra of G_2 is

$$\mathfrak{g}_2 = \left\{ \begin{pmatrix} X & {}^tY & X^0_4 I_{1,1} \\ 0 & 0 & YL \\ 0 & 0 & -J{}^tXJ \end{pmatrix} : X = \begin{pmatrix} X^0_0 & 0 \\ 0 & X^1_1 \end{pmatrix}, {}^tY = \begin{pmatrix} 0 & Y^0_3 \\ Y^1_2 & 0 \end{pmatrix} \right\}.$$

Equations (15.48) suggest that for the vector space complement of $\mathfrak{g}_2$ in $\mathfrak{g}_1$ we should choose

$$\mathfrak{m}_2 = \left\{ \begin{pmatrix} 0 & {}^tY & 0 \\ 0 & 0 & YL \\ 0 & 0 & 0 \end{pmatrix} : {}^tY = \begin{pmatrix} Y^0_2 & 0 \\ 0 & Y^1_3 \end{pmatrix} \in \mathbf{R}^{2\times 2} \right\}.$$

Relative to the second order Lie frame field $T : U \to G$, we have

$$\omega_{\mathfrak{m}_0+\mathfrak{m}_1+\mathfrak{m}_2} = \begin{pmatrix} 0 & {}^t\Sigma & 0 \\ \theta & 0 & \Sigma L \\ 0 & L{}^t\theta & 0 \end{pmatrix},$$

where

$$\theta = \begin{pmatrix} 0 & \omega^2_1 \\ \omega^3_0 & 0 \end{pmatrix}, \quad {}^t\Sigma = \begin{pmatrix} \omega^0_2 & 0 \\ 0 & \omega^1_3 \end{pmatrix}.$$

As usual, $\theta^2 = \omega^2_1$, $\theta^3 = \omega^3_0$ is the coframe field on U determined by T. If $\tilde{T} = Tk(c,B,Z,b)$ is any other second order Lie frame field on U, then

$$\tilde{\theta}^2 = \epsilon s\theta^2, \quad \tilde{\theta}^3 = \delta r\theta^3,$$
$$\tilde{\omega}^0_2 = \frac{\epsilon}{r}(\omega^0_2 - sb^0_4\theta^2), \quad \tilde{\omega}^1_3 = \frac{\delta}{s}(\omega^1_3 + sb^0_4\theta^3).$$

Using (15.48) for both frame fields, we thus have

$$\tilde{D} = \frac{1}{rs}(D - sb_4^0). \tag{15.49}$$

Definition 15.48. A *third order Lie frame field* $T : U \to G$ along λ is a second order Lie frame field for which

$$\omega_2^0 = 0 = \omega_3^1. \tag{15.50}$$

It follows from (15.49) that smooth third order Lie frame fields exist on some neighborhood of any point of M. These same equations also show that if $\tilde{T} = Tk(c, B, Z, b)$ are both third order, then $b_4^0 = 0$.

Let $T : U \to G$ be a third order Lie frame field along λ. Taking the exterior differential of the forms in (15.50), we get

$$\omega_4^0 = 0 \tag{15.51}$$

on U. Any other third order Lie frame field on U is given by

$$\tilde{T} = Tk : U \to G,$$

where $k : U \to G_3$ is a smooth map into the closed subgroup of G_2,

$$G_3 = \left\{ \begin{pmatrix} r & 0 & 0 & t & 0 & t^2/2r \\ 0 & s & u & 0 & u^2/2s & 0 \\ 0 & 0 & \epsilon & 0 & \epsilon u/s & 0 \\ 0 & 0 & 0 & \delta & 0 & \delta t/r \\ 0 & 0 & 0 & 0 & 1/s & 0 \\ 0 & 0 & 0 & 0 & 0 & 1/r \end{pmatrix} : r, s, t, u \in \mathbf{R},\ rs \neq 0,\ \epsilon, \delta = \pm 1 \right\}.$$

The connected component of the identity of G_3 is the subgroup G_3^0 of all elements for which $r > 0$, $s > 0$, and $\epsilon = 1 = \delta$. The Lie algebra of G_3 is

$$\mathfrak{g}_3 = \left\{ \begin{pmatrix} x & 0 & 0 & z & 0 & 0 \\ 0 & y & w & 0 & 0 & 0 \\ 0 & 0 & 0 & 0 & w & 0 \\ 0 & 0 & 0 & 0 & 0 & z \\ 0 & 0 & 0 & 0 & -y & 0 \\ 0 & 0 & 0 & 0 & 0 & -x \end{pmatrix} : x, y, z, w \in \mathbf{R} \right\}.$$

Equation (15.51) suggests that for a complementary vector subspace to $\mathfrak{g}_3$ in $\mathfrak{g}_2$ we take

$$\mathfrak{m}_3 = \left\{ \begin{pmatrix} 0 & 0 & X_4^0 I_{1,1} \\ 0 & 0 & 0 \\ 0 & 0 & 0 \end{pmatrix} : X_4^0 \in \mathbf{R} \right\}.$$

But then (15.51) implies that

$$\omega_{\mathfrak{m}_0+\mathfrak{m}_1+\mathfrak{m}_2+\mathfrak{m}_3} = \omega_{\mathfrak{m}_0+\mathfrak{m}_1+\mathfrak{m}_2} = \tilde{\omega}_{\mathfrak{m}_0+\mathfrak{m}_1+\mathfrak{m}_2} = \tilde{\omega}_{\mathfrak{m}_0+\mathfrak{m}_1+\mathfrak{m}_2+\mathfrak{m}_3}$$

on U, which means that the frame reduction ends at the third order. Observe that taking the exterior derivative of the form in (15.51) imposes no further conditions.

Summary: A third order Lie frame field $T : U \to G$ along a Dupin Legendre immersion $\lambda : M \to \Lambda$ is characterized by $[\mathbf{T}_0] = S_0$ and $[\mathbf{T}_1] = S_1$ are the curvature spheres of λ and

$$\begin{aligned} &\text{Order 1: } \omega_0^2 = 0 = \omega_1^3,\ \theta^2 = \omega_1^2,\ \theta^3 = \omega_0^3,\ \theta^2 \wedge \theta^3 \neq 0, \\ &\text{Order 2: } 0 = \omega_0^1 = \omega_1^0 = \omega_3^2 = -\omega_2^3, \\ &\text{Order 3: } 0 = \omega_2^0 = \omega_3^1 = \omega_4^0. \end{aligned} \tag{15.52}$$

By the structure equations of G,

$$d\theta^2 = p\theta^2 \wedge \theta^3, \quad d\theta^3 = q\theta^2 \wedge \theta^3,$$

for smooth functions $p, q : U \to \mathbf{R}$. The remaining entries of ω are given by

$$\omega_0^0 = q\theta^2 + t\theta^3, \quad \omega_1^1 = u\theta^2 - p\theta^3, \quad \omega_3^0 = c_i\theta^i, \quad \omega_2^1 = d_i\theta^i, \tag{15.53}$$

for smooth functions $t, u, c_i, d_i : U \to \mathbf{R}$, where $i = 2, 3$. Taking the exterior differential of these forms, we get

$$\begin{aligned} dq \wedge \theta^2 + dt \wedge \theta^3 &= -(c_2 + q(p+t))\theta^2 \wedge \theta^3, \\ du \wedge \theta^2 - dp \wedge \theta^3 &= (d_3 + p(q-u))\theta^2 \wedge \theta^3. \end{aligned}$$

Lemma 15.49. *The 6-dimensional distribution $\mathscr{D}$ defined on G by the coframe of left-invariant 1-forms*

$$\mathscr{D}^\perp = \{\omega_0^2, \omega_1^3, \omega_0^1, \omega_1^0, \omega_3^2, \omega_2^0, \omega_3^1, \omega_4^0, \omega_0^4\}$$

satisfies the Frobenius condition and

$$\mathfrak{h} = \{\mathscr{T} \in \mathfrak{g} : \alpha(\mathscr{T}) = 0, \textit{ for all } \alpha \in \mathscr{D}^\perp.\}. \tag{15.54}$$

is a Lie subalgebra of $\mathfrak{g}$. Its maximal integrable manifolds are the right cosets of the connected 6-dimensional Lie subgroup H of G whose Lie algebra is $\mathfrak{h}$.

Proof. It is a simple calculation to show that $d\alpha \equiv 0 \mod \mathscr{D}^\perp$ for every $\alpha \in \mathscr{D}^\perp$. Since the nine left invariant 1-forms in $\mathscr{D}^\perp$ are linearly independent over $\mathbf{R}$, the subspace $\mathfrak{h} \subset \mathfrak{g}$ defined in (15.54) is a Lie subalgebra of dimension six. □

Example 15.50 (Legendre lift of a great circle). We obtain an explicit description of H by finding the third order frame fields along the Legendre lift of the great circle

$$\mathbf{f}:\mathbf{R}\to\mathbf{S}^3,\quad \mathbf{f}(u)=\cos u\epsilon_0+\sin u\epsilon_3.$$

By Example 15.35, a Legendre lift of $\mathbf{f}$ is

$$\lambda:\mathbf{R}^2\to\Lambda,\quad \lambda(u,v)=[\mathbf{S}_0,\mathbf{S}_1],$$

where

$$\mathbf{S}_0=\cos u\epsilon_0+\sin u\epsilon_3+\epsilon_4,\quad \mathbf{S}_1=\cos v\epsilon_1+\sin v\epsilon_2+\epsilon_5. \tag{15.55}$$

Then λ is Dupin and a third order frame field along it is $\mathbf{S}_0$, $\mathbf{S}_1$, and

$$\mathbf{S}_2=-\sin v\epsilon_1+\cos v\epsilon_2,\quad \mathbf{S}_3=-\sin u\epsilon_0+\cos u\epsilon_3,$$
$$\mathbf{S}_4=-\frac{1}{2}\cos v\epsilon_1-\frac{1}{2}\sin v\epsilon_2+\frac{1}{2}\epsilon_5,\quad \mathbf{S}_5=-\frac{1}{2}\cos u\epsilon_0-\frac{1}{2}\sin u\epsilon_3+\frac{1}{2}\epsilon_4.$$

The matrix of $(\mathbf{S}_0,\dots,\mathbf{S}_5)(u,v)$ in the standard Lie frame is $S:\mathbf{R}^2\to G$, where

$$S(u,v)=\begin{pmatrix} 1+\cos u & 0 & 0 & -\sin u & 0 & \frac{1-\cos u}{2}\\ 0 & 1+\cos v & -\sin v & 0 & \frac{1-\cos v}{2} & 0\\ 0 & \sin v & \cos v & 0 & -\frac{\sin v}{2} & 0\\ \sin u & 0 & 0 & \cos u & 0 & -\frac{\sin u}{2}\\ 0 & \frac{1-\cos v}{2} & \frac{\sin v}{2} & 0 & \frac{1+\cos v}{4} & 0\\ \frac{1-\cos u}{2} & 0 & 0 & \frac{\sin u}{2} & 0 & \frac{1+\cos u}{4} \end{pmatrix}.$$

From $\omega^0_a=-\langle d\mathbf{S}_a,\mathbf{S}_5\rangle$, $\omega^1_a=-\langle d\mathbf{S}_a,\mathbf{S}_4\rangle$, $\omega^2_a=\langle d\mathbf{S}_a,\mathbf{S}_2\rangle$, $\omega^3_a=\langle d\mathbf{S}_a,\mathbf{S}_3\rangle$, $\omega^4_a=-\langle d\mathbf{S}_a,\mathbf{S}_1\rangle$, and $\omega^5_a=-\langle d\mathbf{S}_5,\mathbf{S}_0\rangle$, for $a=0,\dots,5$,

$$\omega=S^{-1}dS=\begin{pmatrix} 0 & 0 & 0 & -du/2 & 0 & 0\\ 0 & 0 & -dv/2 & 0 & 0 & 0\\ 0 & dv & 0 & 0 & -dv/2 & 0\\ du & 0 & 0 & 0 & 0 & -du/2\\ 0 & 0 & dv & 0 & 0 & 0\\ 0 & 0 & 0 & du & 0 & 0 \end{pmatrix},$$

which is $\mathfrak{h}$-valued. The set $\{S(u,v)G_3:(u,v)\in\mathbf{R}^2\}$, of all third order frames along λ, is thus an integral submanifold of the distribution $\mathscr{D}$ defined in Lemma 15.49. It passes through the origin $I_6=S(0,0)k$, where the diagonal matrix $k=\mathrm{diag}(1/2,1/2,1,1,2,2)$ is an element of G^0_3, the connected component of the identity of G_3. The subgroup H in Lemma 15.49 is $H=S(\mathbf{R}^2)G^0_3$, and the spherical projection of Definition 15.14 of $\pi(H)=H[\boldsymbol{\lambda}_0,\boldsymbol{\lambda}_1]=[\mathbf{S}_0,\mathbf{S}_1]$ is the great circle $f_+\circ\mathbf{f}(\mathbf{R})$ in $\mathscr{M}$.

Theorem 15.51 (Dupin immersions). *If $\lambda : M \to \Lambda$ is a Dupin Legendre immersion of a connected surface M, then up to a Lie sphere transformation $\lambda(M)$ is an open submanifold of*

$$\pi(H) = \{[\mathbf{T}_0, \mathbf{T}_1] : T \in H\} \cong H/G_3,$$

where $\pi : G \to \Lambda$ is the projection (15.23) *and H is the subgroup of Lemma 15.49.*

Proof. The proof is a special case of Proposition 3.11. Let $\lambda : M \to \Lambda$ be a Dupin Legendre immersion of a connected surface. By (15.52), any third order Lie frame field $T : U \to G$ along λ is an integral surface of $\mathscr{D}$. Thus, the immersion

$$F : U \times G_3 \to G, \quad F(m,K) = T(m)K$$

is an integral submanifold of $\mathscr{D}$ of maximal dimension. Assuming the open subset $U \subset M$ connected, we then know that there is a unique right coset SH, where $S \in G$, such that $F(U \times G_3) \subset SH$. Thus, $S^{-1}T : U \to G$ is a third order Lie frame field along $S^{-1} \circ \lambda$ such that $S^{-1}T(U) \subset H$. The element $S \in G$ can be computed from the value of T at any point of U. In fact, if $m_0 \in U$, then $T(m_0)^{-1}T : U \to G$ is a third order frame field along the Legendre immersion $T(m_0)^{-1}\lambda : M \to \Lambda$ whose value at m_0 is the identity element of G. Hence, the right coset into which $T(m_0)^{-1}T : U \to G$ takes values must be H itself, and thus $T(U) \subset T(m_0)H$ shows that we may take $S = T(m_0)$.

If $\tilde{T} : \tilde{U} \to G$ is another third order Lie frame field along λ on a connected open subset $\tilde{U} \subset M$, and if $U \cap \tilde{U} \neq \emptyset$, then on $U \cap \tilde{U}$ we have $\tilde{T} = TK$, where

$$K : U \cap \tilde{U} \to G_3 \subset H,$$

which shows that for any point $m \in U \cap \tilde{U}$,

$$\tilde{T}(m) = T(m)K \in SH,$$

and thus $\tilde{T}(\tilde{U}) \subset SH$, since this connected integral manifold must go into a unique right coset of H. It follows that $\lambda(M) \subset \pi(SH) = S\pi(H)$. □

As a consequence of this Theorem, any connected Dupin immersion is Lie congruent to (an open subset of) $\pi(H)$, which is a Legendre lift of a great circle in $\mathbf{S}^3$, by Example 15.50. Moreover, any two connected Dupin immersions into $\mathbf{R}^3$ are Lie equivalent, in the sense that their Legendre lifts are Lie congruent to open submanifolds of $\pi(H)$. The third order frames give an explicit construction of the Lie transformation that provides the equivalence. Theorem 12.51 classified the connected Dupin immersions into Möbius space, up to Möbius transformation, as the nonumbilic isoparametric immersions in the classical geometries. In particular, it follows that the Legendre lifts of these isoparametric immersions are Lie congruent.

Lie sphere geometry and the method of moving frames have been used successfully to study Dupin hypersurfaces in spheres of higher dimensions. See [38, 39, 41, 135].

The following examples illustrate the construction of the third order Lie frame field along the Legendre lift of a circular torus in $\mathbf{S}^3$ and of a circular cylinder in $\mathbf{R}^3$. A congruence sending the cylinder into an open dense subset of the torus is then constructed from these frames.

Example 15.52 (Circular tori). Consider the circular torus of Example 5.12

$$\mathbf{x}_{(r,s)} : \mathbf{R}^2 \to \mathbf{S}^3, \quad \mathbf{x}_{(r,s)}(x,y) = r\cos\frac{x}{r}\boldsymbol{\epsilon}_0 + r\sin\frac{x}{r}\boldsymbol{\epsilon}_1 + s\cos\frac{y}{s}\boldsymbol{\epsilon}_2 + s\sin\frac{y}{s}\boldsymbol{\epsilon}_3,$$

where $r = \cos\alpha$, $s = \sin\alpha$, and $0 < \alpha < \pi/2$ is fixed. Use the unit normal vector field

$$\mathbf{e}_3 = s\cos\frac{x}{r}\boldsymbol{\epsilon}_0 + s\sin\frac{x}{r}\boldsymbol{\epsilon}_1 - r\cos\frac{y}{s}\boldsymbol{\epsilon}_2 - r\sin\frac{y}{s}\boldsymbol{\epsilon}_3$$

to define the Legendre lift

$$\lambda_{(r,s)} : \mathbf{R}^2 \to \Lambda, \quad \lambda_{(r,s)} = [\mathbf{x} + \boldsymbol{\epsilon}_4, \mathbf{e}_3 + \boldsymbol{\epsilon}_5].$$

The principal curvatures of $\mathbf{x} = \mathbf{x}_{(r,s)}$ are $c = \cot\alpha = r/s$ and $a = \cot(\alpha + \pi/2) = -\tan\alpha = -s/r$. The curvature sphere with principal curvature c is the oriented tangent sphere with center $r\mathbf{x} + s\mathbf{e}_3$ and signed radius α, while the curvature sphere with principal curvature a has center $-s\mathbf{x} + r\mathbf{e}_3$ and signed radius $\alpha + \pi/2$. As smooth maps into the space of oriented spheres Q, these are

$$[r\mathbf{x} + s\mathbf{e}_3 + r\boldsymbol{\epsilon}_4 + s\boldsymbol{\epsilon}_5],\ [-s\mathbf{x} + r\mathbf{e}_3 - s\boldsymbol{\epsilon}_4 + r\boldsymbol{\epsilon}_5] : \mathbf{R}^2 \to Q.$$

If we define smooth null-vector fields into $\mathbf{R}^{4,2}$ by

$$\mathbf{T}_0 = \cos\frac{x}{r}\boldsymbol{\epsilon}_0 + \sin\frac{x}{r}\boldsymbol{\epsilon}_1 + r\boldsymbol{\epsilon}_4 + s\boldsymbol{\epsilon}_5,$$
$$\mathbf{T}_1 = \cos\frac{y}{s}\boldsymbol{\epsilon}_2 + \sin\frac{y}{s}\boldsymbol{\epsilon}_3 + s\boldsymbol{\epsilon}_4 - r\boldsymbol{\epsilon}_5,$$

then $[\mathbf{T}_0]$ and $[\mathbf{T}_1]$ are the curvature spheres. The derivatives

$$d\mathbf{T}_0 = (-\sin\frac{x}{r}\boldsymbol{\epsilon}_0 + \cos\frac{x}{r}\boldsymbol{\epsilon}_1)\frac{1}{r}dx,$$
$$d\mathbf{T}_1 = (-\sin\frac{y}{s}\boldsymbol{\epsilon}_2 + \cos\frac{y}{s}\boldsymbol{\epsilon}_3)\frac{1}{s}dy$$

show that both curvature sphere maps have rank one at every point of $\mathbf{R}^2$, and thus λ is a Dupin immersion. These derivatives prompt us to let

$$\mathbf{T}_2 = -\sin\frac{y}{s}\boldsymbol{\epsilon}_2 + \cos\frac{y}{s}\boldsymbol{\epsilon}_3, \quad \mathbf{T}_3 = -\sin\frac{x}{r}\boldsymbol{\epsilon}_0 + \cos\frac{x}{r}\boldsymbol{\epsilon}_1,$$

which is a pair of space-like orthonormal vectors in $\mathbf{R}^{4,2}$ both orthogonal to $\mathbf{T}_0$ and $\mathbf{T}_1$. With this choice it follows that

$$\omega_0^3 = \frac{1}{r}dx, \quad \omega_0^0 = \omega_0^1 = \omega_0^2 = \omega_0^4 = 0$$

$$\omega_1^2 = \frac{1}{s}dy, \quad \omega_1^0 = \omega_1^1 = \omega_1^3 = \omega_1^5 = 0.$$

Thus, no matter how these four vector fields are completed to a Lie frame field $T : \mathbf{R}^2 \to G$ by adding the null vector fields $\mathbf{T}_4$ and $\mathbf{T}_5$, it will be at least of second order. Note, column i of the matrix T is the vector $\mathbf{T}_i$ expressed in the Lie frame $\boldsymbol{\lambda}_0, \dots, \boldsymbol{\lambda}_5$ defined in (15.20). Reasonable candidates for these last two null vector fields are

$$\mathbf{T}_4 = \frac{1}{2}(-\cos\frac{y}{s}\boldsymbol{\epsilon}_2 - \sin\frac{y}{s}\boldsymbol{\epsilon}_3 + s\boldsymbol{\epsilon}_4 - r\boldsymbol{\epsilon}_5),$$

$$\mathbf{T}_5 = \frac{1}{2}(-\cos\frac{x}{r}\boldsymbol{\epsilon}_0 - \sin\frac{x}{r}\boldsymbol{\epsilon}_1 + r\boldsymbol{\epsilon}_4 + s\boldsymbol{\epsilon}_5).$$

It is easily verified that $T = (\mathbf{T}_0, \dots, \mathbf{T}_5)$ is a smooth map from $\mathbf{R}^2$ into the Lie sphere group G and that all conditions in (15.52) are satisfied, so T is a third order Lie frame field along λ. Relative to it the non-zero forms in (15.53) are

$$\omega_3^0 = -\frac{1}{2}\theta^3, \quad \omega_2^1 = -\frac{1}{2}\theta^2,$$

as can be verified by computing $\langle d\mathbf{T}_3, \mathbf{T}_5 \rangle$ and $\langle d\mathbf{T}_2, \mathbf{T}_4 \rangle$, respectively. To find the columns of T, we must express each vector $\mathbf{T}_i$ in terms of the Lie frame $\boldsymbol{\lambda}_0, \dots, \boldsymbol{\lambda}_5$, so $\boldsymbol{\epsilon}_0 = \boldsymbol{\lambda}_0 - \frac{\boldsymbol{\lambda}_5}{2}$, $\boldsymbol{\epsilon}_1 = \boldsymbol{\lambda}_1 - \frac{\boldsymbol{\lambda}_4}{2}$, $\boldsymbol{\epsilon}_2 = \boldsymbol{\lambda}_2$, $\boldsymbol{\epsilon}_3 = \boldsymbol{\lambda}_3$, $\boldsymbol{\epsilon}_4 = \boldsymbol{\lambda}_0 + \frac{\boldsymbol{\lambda}_4}{2}$, $\boldsymbol{\epsilon}_5 = \boldsymbol{\lambda}_1 + \frac{\boldsymbol{\lambda}_4}{2}$. For use in Example 15.54 below, we record here the value of T at the origin $(0,0) \in \mathbf{R}^2$,

$$T(0,0) = \begin{pmatrix} 1+r & s & 0 & 0 & s/2 & (r-1)/2 \\ s & -r & 0 & 1 & -r/2 & s/2 \\ 0 & 1 & 0 & 0 & -1/2 & 0 \\ 0 & 0 & 1 & 0 & 0 & 0 \\ s/2 & -r/2 & 0 & -1/2 & -r/4 & s/4 \\ (r-1)/2 & s/2 & 0 & 0 & s/4 & (r+1)/4 \end{pmatrix} \in G.$$

By Theorem 15.51, the third order frame field $T : \mathbf{R}^2 \to G$ must take values in the right coset $T(0,0)H$. In addition, T is periodic, so descends to the appropriate torus, which means that its image in G is compact, thus closed. It follows that

$$\lambda_{(r,s)}(\mathbf{R}^2) = \pi(T(\mathbf{R}^2)) = \pi(T(0,0)H) = T(0,0)\pi(H). \tag{15.56}$$

Example 15.53 (Circular cylinders). A circular cylinder of radius $R > 0$ in $\mathbf{R}^3$ is parametrized by

$$\mathbf{x} : \mathbf{R}^2 \to \mathbf{R}^3, \quad \mathbf{x}(x,y) = (R\cos x, R\sin x, y),$$

with second order frame field $(\mathbf{x}, (\mathbf{e}_1, \mathbf{e}_2, \mathbf{e}_3)) : \mathbf{R}^2 \to \mathbf{E}(3)$, where $\mathbf{e}_1 = \frac{1}{R}\mathbf{x}_x = (-\sin x, \cos x, 0)$, $\mathbf{e}_2 = \mathbf{x}_y = \boldsymbol{\epsilon}_3$, and $\mathbf{e}_3 = \mathbf{e}_1 \times \mathbf{e}_2 = (\cos x, \sin x, 0)$. Then $f = f_0 \circ \mathbf{x} : \mathbf{R}^2 \to \mathscr{M}$ is

$$f = [\boldsymbol{\delta}_0 + R\cos x\, \boldsymbol{\delta}_1 + R\sin x\, \boldsymbol{\delta}_2 + y\boldsymbol{\delta}_3 + \frac{1}{2}(R^2 + y^2)\boldsymbol{\delta}_4],$$

which is a conformal Dupin immersion with Möbius frame field

$$Y = F_0 \circ (\mathbf{x}, e) = \begin{pmatrix} 1 & 0 & 0 & 0 & 0 \\ \mathbf{x} & \mathbf{e}_1 & \mathbf{e}_2 & \mathbf{e}_3 & 0 \\ \frac{1}{2}(R^2 + y^2) & 0 & y & R & 1 \end{pmatrix},$$

as explained in Subsection 12.5.2. The Legendre lift of f is

$$\lambda_{\text{cyl}} = [\mathbf{Y}_0, \mathbf{Y}_3 + \boldsymbol{\epsilon}_5] : \mathbf{R}^2 \to \Lambda,$$

where $\mathbf{Y}_i$ is column i of Y. Then

$$\mathbf{Y}_0 = \boldsymbol{\delta}_0 + \mathbf{x} + \frac{1}{2}(R^2 + y^2)\boldsymbol{\delta}_4, \quad \mathbf{Y}_3 = \mathbf{e}_3 + y\boldsymbol{\delta}_4,$$

in terms of the standard Möbius frame (12.15) of $\mathbf{R}^{4,1}$,

$$\boldsymbol{\delta}_0 = \frac{1}{2}(\boldsymbol{\epsilon}_4 + \boldsymbol{\epsilon}_0), \quad \boldsymbol{\delta}_i = \boldsymbol{\epsilon}_i,\ i = 1,2,3, \quad \boldsymbol{\delta}_4 = \boldsymbol{\epsilon}_4 - \boldsymbol{\epsilon}_0.$$

These are related to the standard Lie frame of $\mathbf{R}^{4,2}$ by

$$\boldsymbol{\lambda}_0 = \boldsymbol{\delta}_0, \quad \boldsymbol{\lambda}_1 = \frac{\boldsymbol{\epsilon}_5 + \boldsymbol{\delta}_1}{2}, \quad \boldsymbol{\lambda}_2 = \boldsymbol{\delta}_2, \quad \boldsymbol{\lambda}_3 = \boldsymbol{\delta}_3, \quad \boldsymbol{\lambda}_4 = \boldsymbol{\epsilon}_5 - \boldsymbol{\delta}_1, \quad \boldsymbol{\lambda}_5 = \boldsymbol{\delta}_4,$$

so

$$\mathbf{Y}_0 = \boldsymbol{\lambda}_0 + R\cos x\, (\boldsymbol{\lambda}_1 - \frac{\boldsymbol{\lambda}_4}{2}) + R\sin x\, \boldsymbol{\lambda}_2 + y\boldsymbol{\lambda}_3 + \frac{R^2 + y^2}{2}\boldsymbol{\lambda}_5$$

$$\mathbf{Y}_3 + \boldsymbol{\epsilon}_5 = (1 + \cos x)\boldsymbol{\lambda}_1 + \sin x\, \boldsymbol{\lambda}_2 + \frac{1 - \cos x}{2}\boldsymbol{\lambda}_4 + R\boldsymbol{\lambda}_5.$$

Differentiating $\mathbf{Y}_0$ and $\mathbf{Y}_3$, we see by inspection that the curvature spheres along λ_{cyl} are

$$\mathbf{S}_0 = \mathbf{Y}_0 - R(\mathbf{Y}_3 + \boldsymbol{\epsilon}_5) = \boldsymbol{\lambda}_0 - R\boldsymbol{\lambda}_1 + y\boldsymbol{\lambda}_3 - \frac{R}{2}\boldsymbol{\lambda}_4 + \frac{y^2 - R^2}{2}\boldsymbol{\lambda}_5,$$

$$\mathbf{S}_1 = \mathbf{Y}_3 + \boldsymbol{\epsilon}_5 = (1 + \cos x)\boldsymbol{\lambda}_1 + \sin x\,\boldsymbol{\lambda}_2 + \frac{1 - \cos x}{2}\boldsymbol{\lambda}_4 + R\boldsymbol{\lambda}_5.$$

Taking the derivatives of $\mathbf{S}_0$ and $\mathbf{S}_1$, we are lead to let

$$\mathbf{S}_2 = -\sin x\,\boldsymbol{\lambda}_1 + \cos x\,\boldsymbol{\lambda}_2 + \frac{\sin x}{2}\boldsymbol{\lambda}_4, \quad \mathbf{S}_3 = \boldsymbol{\lambda}_3 + y\boldsymbol{\lambda}_5,$$

an orthonormal pair of vectors, orthogonal to $\mathbf{S}_0$ and $\mathbf{S}_1$, so that

$$\omega_0^2 = \langle d\mathbf{S}_0, \mathbf{S}_2\rangle = 0 = \langle d\mathbf{S}_1, \mathbf{S}_3\rangle = \omega_1^3.$$

Differentiating $\mathbf{S}_2$ and $\mathbf{S}_3$, we can complete $\mathbf{S}_0, \dots, \mathbf{S}_3$ to a third order frame field along λ_{cyl} by choosing

$$\mathbf{S}_4 = \frac{1 - \cos x}{2}\boldsymbol{\lambda}_1 - \frac{\sin x}{2}\boldsymbol{\lambda}_2 + \frac{1 + \cos x}{4}\boldsymbol{\lambda}_4 + \frac{R}{2}\boldsymbol{\lambda}_5,$$

$$\mathbf{S}_5 = \boldsymbol{\lambda}_5.$$

It is an elementary calculation to verify that $S = (\mathbf{S}_0, \dots, \mathbf{S}_5)$ is a third order Lie frame field along λ_{cyl}. For use in Example 15.54 below, we record here

$$S(0,0) = \begin{pmatrix} 1 & 0\,0\,0 & 0 & 0 \\ -R & 2\,0\,0 & 0 & 0 \\ 0 & 0\,1\,0 & 0 & 0 \\ 0 & 0\,0\,1 & 0 & 0 \\ -R/2 & 0\,0\,0 & 1/2 & 0 \\ -R^2/2 & R\,0\,0 & R/2 & 1 \end{pmatrix} \in G.$$

By Theorem 15.51, the third order frame field $S : \mathbf{R}^2 \to G$ must take values in the right coset $S(0,0)H$. It follows that

$$\lambda_{\mathrm{cyl}}(\mathbf{R}^2) = \pi(S(\mathbf{R}^2)) \subset \pi(S(0,0)H) = S(0,0)\pi(H). \tag{15.57}$$

Example 15.54 (Lie congruence of circular cylinders and tori). Continue the notation of the preceding two examples. Let $U = T(0,0)S(0,0)^{-1} \in G$. By equations (15.56) and (15.57), we have

$$U\lambda_{\mathrm{cyl}}(\mathbf{R}^2) = \pi(US(\mathbf{R}^2)) \subset \pi(US(0,0)H) = \pi(T(0,0)H) = \lambda_{(r,s)}(\mathbf{R}^2).$$

That is, the Legendre lift of the circular cylinders are Lie congruent to an open subset of the Legendre lift of any of the circular tori.

Example 15.55. Continue the notation of the preceding three examples. Verify that

$$U\lambda_{\text{cyl}}(x,y) = \begin{bmatrix} 1+r+(r-1)y^2/4 & s \\ s+y+sy^2/4) & -r \\ 0 & \cos x \\ 0 & \sin x \\ s/2-y/2+sy^2/8 & -r/2 \\ (r-1)/2+(r+1)y^2/8 & s/2 \end{bmatrix},$$

whose spherical projection of Definition 15.14 is $\pi[U\lambda_{\text{cyl}}] =$

$$[\frac{r}{4}(4-y^2)\epsilon_0 + ry\epsilon_1 + \frac{s}{4}(4+y^2)\cos x\,\epsilon_2 + \frac{s}{4}(4+y^2)\sin x\,\epsilon_3 + \frac{4+y^2}{4}\epsilon_4] \in \mathcal{M},$$

and $f_+^{-1} \circ \pi[U\lambda_{\text{cyl}}] =$

$$r\frac{4-y^2}{4+y^2}\epsilon_0 + r\frac{4y}{4+y^2}\epsilon_1 + s\cos x\,\epsilon_2 + s\sin x\,\epsilon_3.$$

Verify this is contained in the circular (r,s)-torus, $\mathbf{x}_{(r,s)}\mathbf{R}^2 \subset \mathbf{S}^3$.

Problems

15.56. Prove that if r is not an integer multiple of π, then the vector field $\mathbf{n}$ defined in (15.1) is normal and smooth.

15.57. It was observed in Section 12.6 that for any $\mathbf{m} \in \mathbf{S}^3$, the spheres $S_{\tilde{r}}(\mathbf{m}) = S_r(\mathbf{m})$ if and only if $\tilde{r} = 2\pi a \pm r$, for some integer a. Moreover, $S_r(\mathbf{m}) = S_{\pi-r}(-\mathbf{m})$, for any $r \in \mathbf{R}$.

1. Prove that the orientation induced on $S_r(\mathbf{m})$ by $\tilde{r} = 2\pi a + r$ is the same as that induced by r.
2. Prove that the orientation induced on $S_r(\mathbf{m})$ by $\tilde{r} = 2\pi a - r$ is the opposite to that induced by r.
3. Prove that the orientation induced by r on $S_r(\mathbf{m})$ is the opposite to that induced by $\pi - r$ on $S_{\pi-r}(-\mathbf{m})$. See Figure 15.6.

15.58. Prove that a point $[\mathbf{u}] \in Q$ for which $u^5 = 0$ is a point sphere in $\mathbf{S}^3$, and for which $u^4 = 0$ is an oriented great sphere in $\mathbf{S}^3$.

15.59. Prove that if $[\mathbf{u}] \in Q$, then there exists a line through $[\mathbf{u}]$ contained in Q. Hint: You may assume that $\mathbf{u} = \mathbf{m} + a\epsilon_4 + b\epsilon_5$, where $a^2 + b^2 = 1$ and $\mathbf{m} \in \mathbf{S}^3$. Let $\mathbf{v} = \mathbf{n} - b\epsilon_4 + a\epsilon_5$, where $\mathbf{n} \in T_{\mathbf{m}}\mathbf{S}^3 \subset \mathbf{R}^4$ and $|\mathbf{n}| = 1$.

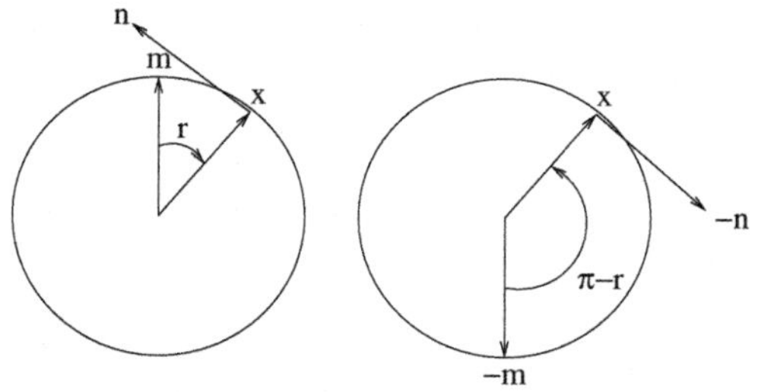

Fig. 15.6 $S_r(\mathbf{m})$ oriented by r and $S_{\pi-r}(-\mathbf{m})$ oriented by $\pi - r$.

15.60. Using the notation of the proof of Proposition 15.9, verify that $\mathbf{x} \in \mathbf{S}^3$ and $\mathbf{n}$ is a unit vector in $T_{\mathbf{x}}\mathbf{S}^3$.

15.61. Prove that Q contains no linear subspace of dimension greater than one. Namely, prove that if $[\mathbf{u}]$, $[\mathbf{v}]$, and $[\mathbf{w}]$ are three non-collinear points in Q, then for some $a, b, c \in \mathbf{R}$, not all zero, the point $[a\mathbf{u} + b\mathbf{v} + c\mathbf{w}]$ is not in $\mathcal{Q}$.

15.62. The *unit tangent bundle of* $\mathbf{S}^3$ is

$$U\mathbf{S}^3 = \{(\mathbf{x}, \mathbf{n}) \in \mathbf{S}^3 \times \mathbf{S}^3 : \mathbf{x} \cdot \mathbf{n} = 0\}.$$

Prove that the map

$$F : U\mathbf{S}^3 \to \Lambda, \quad F(\mathbf{x}, \mathbf{n}) = [\mathbf{x} + \epsilon_4, \mathbf{n} + \epsilon_5]$$

is a diffeomorphism.

15.63. Prove that the spherical projection $\pi : \Lambda \to \mathcal{M}$ is a smooth map onto $\mathcal{M}$. Prove that this is a fiber bundle with standard fiber $\mathbf{S}^2$.

15.64. Prove that the matrix T_t^+ in (15.17) sends the point $[q] \in Q$ in (15.6), which corresponds to the oriented sphere $S_r(\mathbf{m})$ in $\mathbf{S}^3$ of center $\mathbf{m}$ and signed radius $r \in \mathbf{R}$, to the point $[T_t^+ q] \in Q$ that corresponds to $S_{r+t}(\mathbf{m})$ in $\mathbf{S}^3$.

15.65. Prove that the matrix T_t^0 in (15.18) sends the point $[q] \in Q$ in (15.7), which corresponds to the oriented sphere $S_r(\mathbf{p})$ in $\mathbf{R}^3$ with center $\mathbf{p}$ and signed radius $r \in \mathbf{R}$, to the point $[T_t^0 q] \in Q$ that corresponds to $S_{r+t}(\mathbf{p})$ in $\mathbf{R}^3$.

15.66. Prove that the matrix T_t^- in (15.19) sends the point $[q] \in Q$ in (15.9), which corresponds to the oriented sphere $S_r(\mathbf{m})$ in $\mathbf{H}^3$ with center $\mathbf{m}$ and signed radius $r \in \mathbf{R}$, to the point $[T_t^- q] \in Q$ that corresponds to $S_{r+t}(\mathbf{m})$ in $\mathbf{H}^3$.

15.67. Prove that there is no $\mathrm{Ad}(G_0)$-invariant complementary subspace $\mathfrak{m}$ to $\mathfrak{g}_0$ in $\mathfrak{g}$.

15.68. Prove that any $T \in G$ acts on Λ as a contact transformation.

15.69. Prove that if $\mathbf{x} : M^2 \to \mathbf{S}^3$ is an immersion with unit normal vector field $\mathbf{n}$ along $\mathbf{x}$, then

$$\lambda : M \to U\mathbf{S}^3, \quad \lambda(m) = (\mathbf{x}(m), \mathbf{n}(m))$$

is a Legendre immersion for the contact structure defined by β in Example 15.26.

15.70. Prove that through any point of $U\mathbf{S}^3$ there exists a Legendre immersion of β of dimension 2, but none of higher dimension.

15.71 (Legendre lift of surfaces in $\mathbf{S}^3$). Let $\mathbf{x} : M \to \mathbf{S}^3$ be an immersion of an oriented surface, with unit normal vector field $\mathbf{n} : M \to \mathbf{S}^3$. Then $S = \mathbf{n} : M \to \mathbf{S}^{3,1}$ is a smooth tangent sphere map along the conformal immersion $f = f_+ \circ \mathbf{x} : M \to \mathscr{M}$. Prove that the Legendre lift of $\mathbf{x}$ determined by $\mathbf{n}$ is the same as the Legendre lift of f determined by S for the constructions of Corollary 15.33.

15.72 (Legendre lift of immersions into $\mathbf{R}^3$). If $\mathbf{x} : M^2 \to \mathbf{R}^3$ is an immersion with unit normal vector field $\mathbf{n} : M \to \mathbf{S}^2 \subset \mathbf{R}^3$, then $f = f_0 \circ \mathbf{x} : M \to \mathscr{M}$ is a conformal immersion with smooth tangent sphere map $S = -(\mathbf{x}\cdot\mathbf{n})\boldsymbol{\epsilon}_0 + \mathbf{n} + (\mathbf{x}\cdot\mathbf{n})\boldsymbol{\epsilon}_4 : M \to \mathbf{S}^{3,1}$. Prove that the Legendre lift of $\mathbf{x}$ determined by $\mathbf{n}$ is the same as the Legendre lift of f determined by S, by the constructions of Corollary 15.33.

15.73 (Legendre lift of surfaces in $\mathbf{H}^3$). Let $\mathbf{x} : M \to \mathbf{H}^3 \subset \mathbf{R}^{3,1}$ be an immersed surface in hyperbolic space, with unit normal vector field $\mathbf{n} : M \to \mathbf{R}^{3,1}$. Then $S = \mathbf{n} : M \to \mathbf{S}^{3,1}$ is a smooth tangent sphere map along the conformal immersion $f_- \circ \mathbf{x} : M \to \mathscr{M}$. Prove that the Legendre lift of the conformal immersion $f_- \circ \mathbf{x} : M \to \mathscr{M}$, with smooth tangent sphere map $S = \mathbf{n} : M \to \mathbf{S}^{3,1}$, is the same as the Legendre lift of $\mathbf{x}$ determined by $\mathbf{n}$ by the constructions of Corollary 15.33.

15.74. Let $\mathbf{x} : M \to \mathbf{S}^3$ be an immersion with unit normal vector field $\mathbf{n}$ and Legendre lift $\lambda = [\mathbf{x} + \boldsymbol{\epsilon}_4, \mathbf{n} + \boldsymbol{\epsilon}_5] : M \to \Lambda$ given in Corollary 15.33. For $r \in \mathbf{R}$, let $T_r^+ \in \mathbf{SO}(4,2)$ be the Lie parallel transformation (15.17). Prove that the spherical projection of $T_r^+ \circ \lambda : M \to \Lambda$ is $f_+ \circ \tilde{\mathbf{x}}$, where $\tilde{\mathbf{x}} = \cos r\, \mathbf{x} - \sin r\, \mathbf{n}$ is the parallel transformation of $\mathbf{x}$ by $-r$.

15.75. Let $\mathbf{x} : M \to \mathbf{H}^3$ be an immersion with unit normal vector field $\mathbf{n}$ and Legendre lift $\lambda : M \to \Lambda$ given by Corollary 15.33. For $r \in \mathbf{R}$, let $T_r^- \in \mathbf{SO}(4,2)$ be the Lie parallel transformation (15.19). Prove that the spherical projection of $T_r^- \circ \lambda : M \to \Lambda$ is $f_- \circ \tilde{\mathbf{x}}$, where $\tilde{\mathbf{x}} = \cosh r\, \mathbf{x} - \sinh r\, \mathbf{n}$ is the parallel transformation of $\mathbf{x}$ by $-r$.

15.76. Prove that the action induced on $\mathbf{RP}^1$ by the standard linear action of $\mathbf{GL}(2,\mathbf{R})$ on $\mathbf{R}^2$,

$$\mathbf{GL}(2,\mathbf{R}) \times \mathbf{RP}^1 \to \mathbf{RP}^1, \quad (c, \begin{bmatrix} r \\ s \end{bmatrix}) \mapsto c\begin{bmatrix} r \\ s \end{bmatrix} = [c\begin{pmatrix} r \\ s \end{pmatrix}],$$

is *three point transitive*, which means that any three points of $\mathbf{RP}^1$ can be sent to any three points by some $c \in \mathbf{GL}(2,\mathbf{R})$, which is unique up to scalar multiple. In addition, prove that if $\mathfrak{p} : U \to \mathbf{RP}^1$ is a smooth map on an open set $U \subset M$, then there exists a smooth map $c : U \to \mathbf{GL}(2,\mathbf{R})$ such that $c(m)\mathfrak{p}(m) = {}^t[1,0]$, for every $m \in U$.

15.77. Prove that if $\lambda : M \to \Lambda$ is the Legendre lift of an immersion $f : M \to \mathscr{M}$ with tangent sphere map $S : M \to \mathbf{S}^{3,1}$, then $m \in M$ is an umbilic point of λ if and only if it is an umbilic point of f.

15.78. Prove that if $S_0, S_1 : M \to Q$ are the distinct curvature spheres of a Legendre immersion $\lambda : M \to \Lambda$, then λ is of Type

A) if and only if S_0 and S_1 are both immersions;
B) if and only if S_0 is an immersion and S_1 has rank one at every point of M;
C) if and only if S_1 is an immersion and S_0 has rank one at every point of M;
D) if and only if S_0 and S_1 both have rank one at every point of M.

15.79. Prove that if $\lambda : M \to \Lambda$ is the Legendre lift of an immersion $f : M \to \mathscr{M}$ with tangent sphere map $S : M \to \mathbf{S}^{3,1}$, then λ is canal if and only if f is canal, and λ is Dupin if and only if f is Dupin.

15.80. Prove that any isoparametric immersion (6.44) into $\mathbf{H}^3$ is Lie sphere equivalent to an open subset of any circular torus in $\mathbf{S}^3$.

15.81. Prove that any two circular tori $\mathbf{x}_{(r,s)} : \mathbf{R}^2 \to \mathbf{S}^3$ are Lie sphere equivalent; that is, their Legendre lifts are Lie sphere congruent.

15.82. Let $\lambda = T_t^-[\mathbf{S}_0, \mathbf{S}_1] : \mathbf{R}^2 \to \Lambda$, where $\mathbf{S}_0, \mathbf{S}_1 : \mathbf{R}^2 \to \mathbf{R}^{4,2}$ in (15.55) define the Legendre lift of a great circle in $\mathbf{S}^3$ and $T_t^- \in \mathbf{SO}(4,2)$ is defined in (15.19), for any positive $t \in \mathbf{R}$. Prove that stereographic projection of the spherical projection,

$$\mathscr{S} \circ f_+^{-1} \circ \pi \circ T_t^-[\mathbf{S}_0, \mathbf{S}_1],$$

is the surface obtained by rotating the circle in the $y = 0$ plane, $(x - \frac{1}{\sinh t})^2 + z^2 = \coth^2 t$, about the ϵ_3-axis. See Figure 15.5.

15.83. Classify the canal immersions up to Lie sphere congruence.

Solutions to Select Problems

Problems of Chapter 2

2.29 The orbits of K are circles centered at the origin of $\mathbf{D}$. The orbits of A are ultra circles passing through ± 1 as $t \to \pm\infty$. The orbits of N are the horocircles passing through 1 as $t \to \pm\infty$.

2.30 $\mathbf{H}^2$ with the metric $I = \frac{1}{y^2}(dx^2 + dy^2)$ is the hyperbolic plane. The group $\mathbf{SL}(2,\mathbf{R})$ acts on it by isometries. The orbits of K are the hyperbolic circles whose hyperbolic center is at the point i. The orbits of A are the Euclidean rays issuing from the origin $\mathbf{0}$. The orbits of N are the Euclidean horizontal lines.

Problems of Chapter 3

3.14 If $p < 0$, then there are no radial points. If $p > 0$, then the radial points occur where $t^2 = 1/p$. $\kappa = -2p/(1-pt^2)^3$.

3.15 Use the centro-affine parameter s. Then,

1. $\mathbf{x}(s) = \epsilon_1 + s\epsilon_2$, when $\kappa = 0$,
2. $\mathbf{x}(s) = \epsilon_1 \cosh(\sqrt{\kappa}\, s) + \epsilon_2 \frac{\sinh(\sqrt{\kappa}\, s)}{\sqrt{\kappa}}$, when $\kappa > 0$,
3. $\mathbf{x}(s) = \epsilon_1 \cos(\sqrt{-\kappa}\, s) + \epsilon_2 \frac{\sin(\sqrt{-\kappa}\, s)}{\sqrt{-\kappa}}$, when $\kappa < 0$.

G.R. Jensen et al., *Surfaces in Classical Geometries*, Universitext,
DOI 10.1007/978-3-319-27076-0

Problems of Chapter 4

4.62 In the first eight cases

$$\begin{aligned}\tilde{\omega}^1 &= \delta\omega^1, \quad \tilde{\omega}^2 = \rho\delta\omega^2, \quad \tilde{\omega}^1\wedge\tilde{\omega}^2 = \rho\omega^1\wedge\omega^2,\\ \tilde{a} &= \epsilon a, \quad \tilde{c} = \epsilon c, \quad \tilde{p} = \rho\delta p, \quad \tilde{q} = \delta q.\end{aligned} \tag{15.58}$$

In the second eight cases

$$\begin{aligned}\tilde{\omega}^1 &= \delta\omega^2, \quad \tilde{\omega}^2 = \rho\delta\omega^1, \quad \tilde{\omega}^1\wedge\tilde{\omega}^2 = -\rho\omega^1\wedge\omega^2,\\ \tilde{a} &= \epsilon c, \quad \tilde{c} = \epsilon a, \quad \tilde{p} - -\rho\delta q, \quad \tilde{q} = -\delta p.\end{aligned} \tag{15.59}$$

4.65 The frame field defined along $\mathbf{x}$ by

$$\mathbf{e}_1 = T, \quad \mathbf{e}_2 = -\boldsymbol{\epsilon}_3, \quad \mathbf{e}_3 = \mathbf{N} = \mathbf{e}_1\times\mathbf{e}_2$$

is second order, with dual coframe field $\omega^1 = ds$, $\omega^2 = dt$, and principal curvatures $a = \kappa$ and $c = 0$.

4.68 A second order frame field $(\mathbf{x}, e)$ has

$$\begin{aligned}\mathbf{e}_1 &= \frac{1}{\sqrt{\cosh^2 s}}{}^t(\sinh s\cos t, \sinh s\sin t, \cosh s),\\ \mathbf{e}_2 &= {}^t(-\sin t, \cos t, 0),\\ \mathbf{e}_3 &= \mathbf{e}_1\times\mathbf{e}_2 = \frac{1}{\sqrt{\cosh 2s}}{}^t(-\cosh s\cos t, -\cosh s\sin t, \sinh s).\end{aligned}$$

The principal curvatures are

$$a = -\frac{1}{(\cosh 2s)^{3/2}}, \quad c = \frac{1}{\sqrt{\cosh 2s}}.$$

4.73 Prove that $a = c$ for some u if and only if the polynomial $p(u) = (1+u^2)^4 + 10L^2u^2 - 2L^2$ is zero at u. Show that $p(u)$ is increasing on $u \geq 0$, has its minimum at $u = 0$ and goes to $+\infty$ as $u \to +\infty$.

4.85 Let $(\mathbf{x}, e)$ be a local first order frame field along $\mathbf{x}$. Prove that:

$$\begin{aligned}d\tilde{\mathbf{x}} &= \frac{d\mathbf{x}}{|\mathbf{x}|^2} - \frac{2\mathbf{x}\cdot d\mathbf{x}}{|\mathbf{x}|^4}\mathbf{x},\\ \tilde{I} &= d\tilde{\mathbf{x}}\cdot d\tilde{\mathbf{x}} = \frac{1}{|\mathbf{x}|^4}I,\end{aligned}$$

that $(\tilde{\mathbf{x}}, \tilde{e})$ is a first order frame field along $\tilde{\mathbf{x}}$, where

$$\tilde{\mathbf{e}}_1 = \mathbf{e}_i - \frac{2\mathbf{x}\cdot\mathbf{e}_i}{|\mathbf{x}|^2}\mathbf{x}, \quad \tilde{\mathbf{e}}_3 = \mathbf{e}_3 - \frac{2\mathbf{x}\cdot\mathbf{e}_3}{|\mathbf{x}|^2}\mathbf{x},$$

for $i = 1,2$, and

$$\widetilde{II} = \frac{1}{|\mathbf{x}|^2} II + \frac{2\mathbf{x}\cdot\mathbf{e}_3}{|\mathbf{x}|^4} I.$$

Problems of Chapter 5

5.32 Consider the parametrization of a dense open subset of $\mathbf{S}^3$

$$\mathbf{x} : (0,2\pi)\times(0,\pi)\times(0,\pi)\to\mathbf{S}^3, \quad \mathbf{x}(\theta,\varphi,\psi) = \begin{pmatrix}\cos\theta\sin\varphi\sin\psi\\ \sin\theta\sin\varphi\sin\psi\\ \cos\varphi\sin\psi\\ \cos\psi\end{pmatrix}.$$

Calculate $d\mathbf{x}\cdot d\mathbf{x}$ to prove $\mathbf{x}^* dV = \sin\varphi\sin^2\psi\; d\theta\wedge d\varphi\wedge d\psi$.

5.33 Solution 1). There exists a constant vector $\mathbf{v}\in\mathbf{R}^4$ and a neighborhood $V\subset U$ of m such that the vectors $\mathbf{x}, \mathbf{x}_x, \mathbf{x}_y, \mathbf{v}$ are linearly independent. Apply Gram-Schmidt to these vectors. This construction produces a smooth unit normal vector field $\mathbf{e}_3$ along $\mathbf{x}$ on V.

Solution 2). At each point of U, the orthogonal complement $\mathbf{x}^\perp$ is isomorphic to $\mathbf{R}^3$ with the orientation given by the interior product $\iota_{\mathbf{x}}(dx^1\wedge dx^2\wedge dx^3\wedge dx^4)$, where $x^1,\dots,x^4$ are the standard coordinate functions on $\mathbf{R}^4$. Thus, there is a cross product defined on $\mathbf{x}^\perp$ and $\mathbf{x}_x\times\mathbf{x}_y$ is in $\mathbf{x}^\perp$ and normal to $\mathbf{x}_x$ and $\mathbf{x}_y$, so it is a nonzero normal vector to $\mathbf{x}$ at each point of U. Use Gram-Schmidt on $\mathbf{x}_x$ and $\mathbf{x}_y$ to construct a smooth first order frame field on all of U.

5.37 See Figure 5.3 and prove that

$$\frac{\mathbf{m}-\cos r\,\mathbf{x}}{\sin r} = -\left(\frac{-\mathbf{m}-\cos(\pi-r)\,\mathbf{x}}{\sin(\pi-r)}\right).$$

5.38 The inclusion $\subset$ follows from the definition of tangent space, and then equality follows since both sides are 2-dimensional. Prove the opposite inclusion directly by showing that if $\mathbf{z}\cdot\mathbf{x}_0 = 0 = \mathbf{z}\cdot\mathbf{m}$, then the curve

$$\mathbf{x}(t) = \mathbf{m}\cos r + (\mathbf{x}_0 - \mathbf{m}\cos r)\cos(\frac{|\mathbf{z}|t}{\sin r}) + \frac{\mathbf{z}}{|\mathbf{z}|}\sin(\frac{|\mathbf{z}|t}{\sin r})\sin r$$

lies in $S_r(\mathbf{m})$ and satisfies $\mathbf{x}(0) = \mathbf{x}_0$ and $\dot{\mathbf{x}}(0) = \mathbf{z}$.

5.40 If $\tilde{\mathbf{n}} = -\mathbf{n}$, then principal curvatures change sign, so if $\tilde{a} = -a$, then $\tilde{r} = \cot^{-1}(-a) = \pi - r$ and thus the center is

$$\tilde{\mathbf{m}} = \cos\tilde{r}\, \mathbf{x}(m) + \sin\tilde{r}\, \tilde{\mathbf{n}} = -(\cos r\, \mathbf{x}(m) + \sin r\, \mathbf{n}) = -\mathbf{m}$$

and $S_{\tilde{r}}(\tilde{\mathbf{m}}) = S_{\pi-r}(-\mathbf{m}) = S_r(\mathbf{m})$.

5.46 Find a second order frame field along $\mathbf{x}$ to show that the principal curvatures are

$$a = \frac{\kappa(s)\cos t}{\cos r - \kappa(s)\sin r \cos t}, \qquad c = -\frac{1}{\sin r}$$

5.48 The principal curvatures $\tilde{a}$ and $\tilde{c}$ of $\tilde{\mathbf{x}}$ relative to its unit normal vector field $\tilde{\mathbf{e}}_3$ are

$$\tilde{a} = \frac{a\cos r + \sin r}{\cos r - a\sin r}, \qquad \tilde{c} = \frac{c\cos r + \sin r}{\cos r - c\sin r}.$$

5.49 Use Problem 4.69.

5.50 Use $A\mathbf{x}\cdot\mathbf{p} = \mathbf{x}\cdot {}^t\!A\mathbf{p}$.

5.54

$$\mathscr{S}\circ\mathbf{y} : J\times\mathbf{R} \to \mathbf{R}^3, \quad \mathscr{S}\circ\mathbf{y}(s,t) = e^{-t}\sigma(s),$$

is the cone in $\mathbf{R}^3$ over the same curve in $\mathbf{S}^2$ used to define the cylinder in $\mathbf{S}^3$.

5.52 Start with the default stereographic projection.

5.53 $\mathrm{Ad}(AB) = \mathrm{Ad}(A)\circ\mathrm{Ad}(B)$, for any $A, B \in \mathbf{SU}(2)$ and $\pi_h(\mathbf{z}) = \mathrm{Ad}(\mathbf{z}, \mathbf{z}^*)\mathscr{I}$.

Problems of Chapter 6

6.35 Do the case $\mathbf{x} = \boldsymbol{\epsilon}_4$ first, then use the transitivity of the action of $\mathbf{O}_+(3,1)$.

6.38 Using the transitivity of $\mathbf{SO}_+(3,1)$ on $\mathbf{H}^3$, you may assume $\mathbf{p} = \boldsymbol{\epsilon}_4$. Then $\mathbf{q} = q^4\boldsymbol{\epsilon}_4 + \mathbf{q}$ and if you assume that $\mathbf{q} \neq 0$, then $\mathbf{v} = \frac{\mathbf{q}}{|\mathbf{q}|}$ is a unit vector in $T_{\boldsymbol{\epsilon}_4}\mathbf{H}^3$. If $r = \cosh^{-1} q^4 > 0$, then $\mathbf{q} = \cosh r\, \boldsymbol{\epsilon}_4 + \sinh r\, \mathbf{v}$. Consider the geodesic $\gamma(s) = \cosh s\, \boldsymbol{\epsilon}_4 + \sinh s\, \mathbf{v}$.

6.45 If $|a| \neq 1$, complete the square to show that

$$Q(\mathbf{p}) = \{\mathbf{y} \in \mathbf{R}^{3,1} : (a^2-1)(y^4 + \frac{a^2}{1-a^2})^2 + (y^1)^2 + (y^2)^2 = \frac{1}{a^2-1}\},$$

which is an ellipsoid if $|a| > 1$, and a hyperboloid if $|a| < 1$. If $|a| = 1$, then

$$Q(\mathbf{p}) = \{\mathbf{y} \in \mathbf{R}^{3,1} : y^4 = 1 + \frac{1}{2}((y^1)^2 + (y^2)^2)\},$$

which is a paraboloid.

6.47 First do the planar case of $\mathbf{x}'$ in the Poincaré disk, $\mathbf{B}^2$, and $\mathbf{v}'$ in its boundary circle $\partial\mathbf{B}^2 = \mathbf{S}^1$. Then explain why this case is sufficient.

6.48 If $\tilde{\mathbf{v}} = -\mathbf{v}$, then the principal curvatures change sign, so apply Remark 6.10.

6.49 This is a special case of Proposition 3.11. Equation (6.44) is the solution $e = (\mathbf{e}_1, \mathbf{e}_2, \mathbf{e}_3, \mathbf{x}) : \mathbf{R}^2 \to \mathbf{SO}_+(3,1)$ of

$$e^{-1}de = \begin{pmatrix} 0 & 0 & -a\,ds & ds \\ 0 & 0 & -\frac{1}{a}dt & dt \\ a\,ds & \frac{1}{a}dt & 0 & 0 \\ ds & dt & 0 & 0 \end{pmatrix}$$

satisfying the initial condition

$$e(0,0) = \begin{pmatrix} 0 & 0 & -1/b & a/b \\ 0 & 1 & 0 & 0 \\ 1 & 0 & 0 & 0 \\ 0 & 0 & -a/b & 1/b \end{pmatrix}.$$

Problems of Chapter 7

7.48 $e^u = f$ and $h = ((\ddot{g}\dot{f} - \dot{g}\ddot{f})/w^2 - \dot{g}/f)/(2w)$, where $w = \sqrt{\dot{f}^2 + \dot{g}^2}$.

Problems of Chapter 8

8.64 If complex numbers a and b satisfy $a^2 + b^2 = 1$, then there exists a unique $\zeta \in \mathbf{C}$ such that $a = \cos\zeta$ and $b = \sin\zeta$. Also recall that $\cos iy = \cosh y$ and $\sin iy = i\sinh y$, for any number y.

8.69 Its curvature is the constant $k = 0$.

8.70 The curvature is constant $k = -1/2$. Its element of pseudoarc is $\pm i dw$, where the sign depends on the choice of square root of f.

8.71 Its curvature is $k = \frac{3z^2}{z^4 - 1}$. Its element of pseudoarc is $\sqrt{\frac{2}{z^4-1}}dz$.

8.72 What happens when you scale the minimal curve of the catenoid by a nonzero complex constant?

Problems of Chapter 9

9.35 If $\sigma \to \mathbf{S}^2$ is the profile curve with arclength parameter x and curvature $\kappa(x)$, if $M = J \times \mathbf{R}$, and if $\mathbf{x} : M \to \mathbf{R}^3$ is $\mathbf{x}(x,y) = e^{-y}\sigma(x)$ is our cone, then $z = x + iy$ is a principal complex coordinate in M relative to which the conformal factor is e^{-y}, and the Hopf invariant h and mean curvature H of $\mathbf{x}$ satisfy

$$h = \frac{\kappa(x)e^y}{2} = H.$$

9.36 The existence of a principal complex chart implies the existence of a second order frame field along $\mathbf{x}$. Then refer to Example 4.11.

9.37 If $(\mathbf{x}, e) : U \to \mathbf{E}(3)$ is a second order frame field, then differentiating $\omega_1^3 - a\omega^1 = 0$ and $\omega_2^3 - c\omega^2 = 0$ and using Definition 4.36 of Dupin: $da = a_2\omega^2$ and $dc = c_1\omega^1$, gives

$$\alpha = q\omega^1 - p\omega^2 = -\frac{d(a-c)}{a-c}.$$

9.38 For Enneper's surface, $\hat{\mathbf{x}}$ parametrizes a sphere not centered at the origin. Compare these examples with Corollary 9.26.

9.39 If the profile curve is embedded, so also is the curve $(1/f(s)^2, 0, g(s))$. Apply Green's Theorem to the region this curve encloses to prove that (9.16) does not hold. See [86, p 201].

Problems of Chapter 10

10.51 Use the fact that a holomorphic function with constant modulus on a connected open set must be constant.

10.54 Note that $\{A_n(x)\mathbf{v}_n - \mathbf{v}_n : x > 0\}$ is the circle centered at $-\mathbf{v}_n$ of radius $|\mathbf{v}_n| = 1/(1+n^2)$. Check that $\boldsymbol{\gamma}_n$ satisfies the initial conditions

$$\boldsymbol{\gamma}_n(1) = \begin{pmatrix}0\\0\end{pmatrix}, \quad \dot{\boldsymbol{\gamma}}_n(1) = \begin{pmatrix}1\\0\end{pmatrix}, \quad \text{so} \quad \mathbf{N}_n(1) = \begin{pmatrix}0\\1\end{pmatrix}.$$

$\mathbf{T}$ must satisfy the Euler equation

$$x^2\ddot{\mathbf{T}} + x\dot{\mathbf{T}} + n^2\mathbf{T} = 0.$$

Problems of Chapter 11

11.30

1. The entries of $\sigma(A)$ are quadratic functions of the real and imaginary parts of the entries of A.
2. If $a = e^{i\theta}$, then

$$\sigma(A) = \begin{pmatrix} \cos 2\theta & -\sin 2\theta & 0 & 0 \\ \sin 2\theta & \cos 2\theta & 0 & 0 \\ 0 & 0 & 1 & 0 \\ 0 & 0 & 0 & 1 \end{pmatrix}.$$

3.

$$\sigma(A) = \begin{pmatrix} -\cos 2\theta & \sin 2\theta & 0 & 0 \\ \sin 2\theta & \cos 2\theta & 0 & 0 \\ 0 & 0 & -1 & 0 \\ 0 & 0 & 0 & 1 \end{pmatrix}.$$

11.34 Fix a point $m \in \tilde{M}$ and prove that $dA_m : T_m\tilde{M} \to T_m\tilde{M}$ is the identity map. Then $dA_m : T_m\mathbf{H}(3) \to T_m\mathbf{H}(3)$ must be the identity map, since A preserves orientation of $\mathbf{H}(3)$. Now use the exponential map of the Riemannian manifold $\mathbf{H}(3)$ at m to prove that if an isometry of $\mathbf{H}(3)$ fixes a point m and if its derivative at m is the identity map, then the isometry is the identity map.

11.42 Let H be the positive definite square root of the positive definite, hermitian matrix BB^*, and then show $K = H^{-1}B$ is unitary.

11.45

$$G = \frac{1}{2}\begin{pmatrix} -e^{-w/2} & -(4+w)e^{-w/2} \\ e^{w/2} & we^{w/2} \end{pmatrix}.$$

Then

$$\tilde{G} \circ g_n(w) = \tilde{G}(w)D^n,$$

where the monodromy matrix

$$D = -\begin{pmatrix} 1 & 2\pi i \\ 0 & 1 \end{pmatrix}.$$

Problems of Chapter 12

12.54 $T_{(1-r^2)\mathbf{m}} \circ h_{r^2} \circ \mathscr{I}_{\mathbf{m}}$.

12.55 $\mathscr{I}_{\mathbf{m}} = T_{\mathbf{m}} \circ \mathscr{I} \circ T_{-\mathbf{m}}$.

12.59 Prove that the vector $\mathbf{v} = 2u^4\boldsymbol{\delta}_0 - \sum_1^3 u^i\boldsymbol{\delta}_i + \frac{1}{2}u^0\boldsymbol{\delta}_4$ is in $\mathscr{L}^+$ and that $\langle \mathbf{v}, \mathbf{Y}_0\rangle = -r$, for some $r > 0$. Let $\mathbf{Y}_4 = \frac{1}{r}\mathbf{v}$ and then show that the orthogonal complement of the span of $\mathbf{Y}_0, \mathbf{Y}_4$ is a 3-dimensional subspace on which the inner product is positive definite, so there exists an orthonormal basis of it.

12.62 Prove there exists a frame field $Y : U \subset M \to \mathbf{M\ddot{o}b}_+$ along f, then prove the existence of a smooth map $K(r,A,\mathbf{y}) : U \to G_0$, with $r > 0$, such that $\tilde{Y} = YK$ is first order. The action on the Grassmannian

$$\mathbf{R}^+ \times \mathbf{SO}(3) \times G_2(\mathbf{R}^3) \to G_2(\mathbf{R}^3) = \mathbf{R}^{3\times 2*}/\mathbf{GL}(2,\mathbf{R}), \quad (r,A)[P] = [rAP]$$

is transitive.

12.63 In fact, $\tilde{w} = \tilde{Y}^{-1}d\tilde{Y} = K^{-1}\omega K + K^{-1}dK$ gives

$$\tilde{\omega}_{\mathfrak{m}_0+\mathfrak{m}_1} = (K(r,A,\mathbf{y})^{-1}\omega_{\mathfrak{m}_0+\mathfrak{m}_1}K(r,A,\mathbf{y}))_{\mathfrak{m}_0+\mathfrak{m}_1}, \tag{15.60}$$

which calculated gives

$$\begin{aligned} \begin{pmatrix} \tilde{\omega}_0^1 \\ \tilde{\omega}_0^2 \end{pmatrix} &= \frac{1}{r}\begin{pmatrix} \cos t & \sin t \\ -\sin t & \cos t \end{pmatrix}\begin{pmatrix} \omega_0^1 \\ \omega_0^2 \end{pmatrix}, \\ (\tilde{\omega}_1^3, \tilde{\omega}_2^3) &= (\omega_1^3 - y^3\omega_0^1, \omega_2^3 - y^3\omega_0^2)\begin{pmatrix} \cos t & -\sin t \\ \sin t & \cos t \end{pmatrix}, \end{aligned} \tag{15.61}$$

which imply

$$\begin{aligned} \tilde{\omega}_0^1 + i\tilde{\omega}_0^2 &= \frac{1}{r}e^{-it}(\omega_0^1 + i\omega_0^2), \\ \tilde{\omega}_1^3 - i\tilde{\omega}_2^3 &= e^{it}(\omega_1^3 - i\omega_2^3) - y^3e^{it}(\omega_0^1 - i\omega_0^2) \\ &= re^{2it}h(\tilde{\omega}_0^1 + i\tilde{\omega}_0^2) + r(H - y^3)(\tilde{\omega}_0^1 - i\tilde{\omega}_0^2), \end{aligned}$$

from which (12.42) follows.

12.64 If $\tilde{Y} = YK$ is another first order frame field, then $\tilde{h} = re^{2it}h$ by (12.42) and $\tilde{\omega}_0^1 + i\tilde{\omega}_0^2 = \frac{1}{r}e^{-it}(\omega_0^1 + i\omega_0^2)$ by (15.61). It follows that $\Omega_{\tilde{Y}} = \Omega_Y$, since $\omega_0^1\wedge\omega_0^2 = \frac{i}{2}(\omega_0^1 + i\omega_0^2)\wedge(\omega_0^1 - i\omega_0^2)$.

12.71 Use the standard orthonormal basis of $\mathbf{R}^{4,1}$.

12.74 If $e : \mathbf{R}^2 \to \mathbf{SO}(4)$ is the second order frame field along $\mathbf{x}$ constructed in Example 5.12, then $Y = F_+ \circ e : \mathbf{R}^2 \to \mathbf{M\ddot{o}b}$ is first order along f, by Exercise 52, and $\tilde{Y} = YK(rs, e^{i\pi/2}, {}^t(0,0,H))$ is central, where $H = \frac{r^2-s^2}{rs}$.

12.75 $Y = F_0 \circ (\mathbf{x}, e)$ is first order along f, by Exercise 53, and

$$\tilde{Y} = YK(2, e^{i\pi/2}, {}^t(0,0,-1/2))$$

is central.

12.76 The frame field

$$e = (\mathbf{e}_1, \mathbf{e}_2, \mathbf{e}_3, \mathbf{x}) : \mathbf{R}^2 \to \mathbf{SO}_+(3,1),$$

where $\mathbf{e}_1 = \mathbf{x}_x$ and $\mathbf{e}_2 = \mathbf{x}_y$, is second order along $\mathbf{x}$ and

$$Y = F_- \circ e : \mathbf{R}^2 \to \mathbf{M\ddot{o}b}_+$$

is first order by Exercise 55, and $\tilde{Y} = YK(-1/h, e^{i\pi/2}, {}^t(0,0,H))$ is central, where $h = a - a^{-1} < 0$ and $H = a + a^{-1}$.

12.77 Find a central frame field along $f_0 \circ \mathbf{x}_n$.

Problems of Chapter 13

13.42 Prove that $\tilde{Y} = YK(e^u, I_3, \mathbf{0})$ is a time oriented first order frame relative to z and that $\tilde{\omega}_1^3 - i\tilde{\omega}_2^3 = \omega_1^3 - i\omega_2^3$.

13.43 Prove that the frame

$$\tilde{Y} = YK(1, A, \mathbf{0}),$$

where

$$A = \begin{pmatrix} \cos v & -\sin v & 0 \\ \sin v & \cos v & 0 \\ 0 & 0 & 1 \end{pmatrix} : U \to \mathbf{SO}(3),$$

is a time oriented frame field satisfying the conditions of Problem 13.42 and that $\tilde{\omega}_1^3 - i\tilde{\omega}_2^3 = e^{iv}(\omega_1^3 - i\omega_2^3)$.

13.45 $k = \frac{1}{2R^2}$ and $b = \frac{1}{4R^4}$ relative to z are constant.

13.46 $k = \tan\alpha$ by (13.30) and $b = \frac{2+\kappa^2}{4}$, where κ is the curvature of $\boldsymbol{\sigma}$.

13.47 $\hat{k} = 1$, and $\hat{b} = -b/k^2 = r^2 - s^2 = \cos 2\alpha$.

13.49 Use Exercise 56.

13.53 Relative to the complex coordinate $z = x + iy$ for $\mathbf{x}_n : \mathbf{R}^+ \times \mathbf{R} \to \mathbf{R}^3$, we have Hopf invariant $h = \frac{n}{2x}$, mean curvature $H = \frac{n}{2x}$, and conformal factor $e^u = 1$. Then the conformal invariants of $f_0 \circ \mathbf{x}$ are $k = \frac{n}{2x}$ and $b = k^2 = \frac{n^2}{4x^2}$, by Theorem 13.22. Now calculate the Möbius invariants using Proposition 13.37.

Problems of Chapter 14

14.38 Write $\Delta(\log(k)) - \frac{s}{k^2} + k^2 = -C/k^2$, where $C = s + (k_x)^2 + (k_y)^2 - k(k^3 + \Delta k)$, and let $F = \Delta(\frac{k_{xy}}{k}) + 2(k^2)_{xy}$. Prove that $k^3 F = k_x C_y + k_y C_x - k C_{xy}$.

14.39

$$k = \frac{1}{2}(\kappa + i), \quad b = \frac{\kappa}{4}(\kappa + i).$$

$$Q = \frac{1}{16}(\kappa\ddot{\kappa} + \frac{1}{4}(\kappa^4 - 1) - \dot{\kappa}^2 + i(\ddot{\kappa} + \frac{\kappa}{2}(\kappa^2 + 1))),$$

and

$$W = \frac{1}{8}(\ddot{\kappa} + \frac{\kappa}{2}(\kappa^2 + 1)).$$

$$p_1 - p_3 = -\frac{4(\ddot{\kappa} + \frac{1}{2}\kappa(\kappa^2 + 1))}{(\kappa^2 + 1)^{3/2}},$$

$$p_1 + p_3 - 2ip_2 = -\frac{4(\kappa^2 - 1 - i2\kappa)(\kappa\ddot{\kappa} + \frac{1}{4}(\kappa^4 - 1) - \dot{\kappa}^2 + i(\ddot{\kappa} + \frac{\kappa}{2}(\kappa^2 + 1)))}{(\kappa^2 + 1)^3}$$

$$q_1 - iq_2 = \frac{-\dot{\kappa}(3\kappa + \sqrt{\kappa^2 + 1} + i(1 - \kappa(2\sqrt{\kappa^2 + 1} + 1)))}{\sqrt{2}(\sqrt{\kappa^2 + 1} + \kappa)^{1/2}(\kappa^2 + 1)^{7/4}}.$$

Problems of Chapter 15

15.70 Suppose $\lambda : M^3 \to U\mathbf{S}^3$ is a Legendre immersion of dimension 3. For any point $p \in M$, there exists a coframe field $\theta^1, \dots \theta^5$ on an open subset of $U\mathbf{S}^3$ containing $\lambda(p)$ such that $\theta^5 = \beta$ and

$$\lambda^*(\theta^1 \wedge \theta^2 \wedge \theta^3)_p \neq 0.$$

Calculate $d\beta$ in terms of this coframe field and then use the fact that $\lambda^* d\beta = 0$ to get enough information on the coefficients of $d\beta$ to contradict the hypothesis $(\beta \wedge d\beta \wedge d\beta)_{\lambda(p)} = 0$.

15.73 $f_- \circ \mathbf{x} = [\boldsymbol{\epsilon}_0 + \mathbf{x}]$ in (12.62).

15.76 Given distinct points

$$\mathfrak{r} = \begin{bmatrix} r^1 \\ r^2 \end{bmatrix}, \quad \mathfrak{s} = \begin{bmatrix} s^1 \\ s^2 \end{bmatrix}, \quad \mathfrak{t} = \begin{bmatrix} t^1 \\ t^2 \end{bmatrix} \in \mathbf{RP}^1,$$

it suffices to show there exists $c \in \mathbf{GL}(2,\mathbf{R})$, unique up to scalar multiple, such that $c[\boldsymbol{\epsilon}_1] = \mathfrak{r}$, $c[\boldsymbol{\epsilon}_2] = \mathfrak{s}$, and $c[\boldsymbol{\epsilon}_1 + \boldsymbol{\epsilon}_2] = \mathfrak{t}$, where $\boldsymbol{\epsilon}_1, \boldsymbol{\epsilon}_2$ is the standard basis of $\mathbf{R}^2$. The first two equations imply that

$$c = \begin{pmatrix} xr^1 & ys^1 \\ xr^2 & ys^2 \end{pmatrix},$$

for some non-zero real numbers x and y. Then the third equation implies that

$$\begin{pmatrix} r^1 & s^1 \\ r^2 & s^2 \end{pmatrix} \begin{pmatrix} x \\ y \end{pmatrix} = z \begin{pmatrix} t^1 \\ t^2 \end{pmatrix},$$

for some non-zero real number z. Up to a non-zero factor the solution is

$$\begin{pmatrix} x \\ y \end{pmatrix} = \begin{pmatrix} s^2 t^1 - s^1 t^2 \\ r^1 t^2 - r^2 t^1 \end{pmatrix},$$

which determines c up to a non-zero multiple.

15.78 If $T : U \to G$ is a first order Lie frame field along λ, then $S_0 = [\mathbf{T}_0]$ and $S_1 = [\mathbf{T}_1]$ on U. Show that $d\mathbf{T}_0 = \sum_0^5 \omega_0^i \mathbf{T}_i$ has rank two modulo $\mathbf{T}_0$ if and only if $\omega_0^1 \wedge \theta^3 \neq 0$.

References

1. Ahlfors, L.V.: Complex analysis. An Introduction to the Theory of Analytic Functions of One Complex Variable. McGraw-Hill, New York/Toronto/London (1953)
2. Barbosa, J.L.M., Colares, A.G.: Minimal Surfaces in $\mathbf{R}^3$. Lecture Notes in Mathematics, vol. 1195. Springer, Berlin (1986). Translated from the Portuguese
3. Bernstein, H.: Non-special, non-canal isothermic tori with spherical lines of curvature. Trans. Am. Math. Soc. **353**(6), 2245–2274 (electronic) (2001). doi:10.1090/S0002-9947-00-02691-X. http://www.dx.doi.org/10.1090/S0002-9947-00-02691-X
4. Bernstein, S.: Sur un théorèm de géométrie et ses applications aux équations aux dérivées partielles du type elliptique. Commun. de la Soc. Math. de Kharkov 2ème Sér. **15**, 38–45 (1915–1917)
5. Bers, L.: Riemann Surfaces. Mimeographed Lecture Notes, New York University (1957–1958). Notes by Richard Pollack and James Radlow
6. Bianchi, L.: Complementi alle ricerche sulle superficie isoterme. Ann. Mat. Pura Appl. **12**, 19–54 (1905)
7. Bianchi, L.: Ricerche sulle superficie isoterme e sulla deformazione delle quadriche. Ann. Mat. Pura Appl. **11**, 93–157 (1905)
8. Blaschke, W.: Vorlesungen über Differentialgeometrie und geometrische Grundlagen von Einsteins Relativitätstheorie. III: Differentialgeometrie der Kreise und Kugeln. In: Grundlehren der mathematischen Wissenschaften, vol. 29. Springer, Berlin (1929)
9. Bobenko, A., Eitner, U.: Bonnet surfaces and Painlevé equations. J. Reine Angew. Math. **499**, 47–79 (1998). doi:10.1515/crll.1998.061. http://www.dx.doi.org/10.1515/crll.1998.061
10. Bobenko, A.I., Pavlyukevich, T.V., Springborn, B.A.: Hyperbolic constant mean curvature one surfaces: spinor representation and trinoids in hypergeometric functions. Math. Z. **245**(1), 63–91 (2003). doi:10.1007/s00209-003-0511-5. http://www.dx.doi.org/10.1007/s00209-003-0511-5
11. Bohle, C.: Constant mean curvature tori as stationary solutions to the Davey-Stewartson equation. Math. Z. **271**(1–2), 489–498 (2012). doi:10.1007/s00209-011-0873-z. http://www.dx.doi.org/10.1007/s00209-011-0873-z
12. Bohle, C., Peters, G.P.: Bryant surfaces with smooth ends. Commun. Anal. Geom. **17**(4), 587–619 (2009)
13. Bombieri, E., De Giorgi, E., Giusti, E.: Minimal cones and the Bernstein problem. Invent. Math. **7**, 243–268 (1969)
14. Bonnet, P.O.: Mémoire sur la théorie des surfaces applicables sur une surface donnée, première partie. J. l'Ecole Polytech. **41**, 209–230 (1866)

G.R. Jensen et al., *Surfaces in Classical Geometries*, Universitext,
DOI 10.1007/978-3-319-27076-0

15. Bonnet, P.O.: Mémoire sur la théorie des surfaces applicables sur une surface donnée, deuxième partie. J. l'Ecole Polytech. **42**, 1–151 (1867)
16. Boothby, W.M.: An Introduction to Differentiable Manifolds and Riemannian Geometry, 2nd edn. Academic, New York (1986)
17. Bowman, F.: Introduction to Elliptic Functions with Applications. Dover, New York (1961)
18. Boy, W.: Über die Curvatura integra und die Topologie geschlossener Flächen. Math. Ann. **57**(2), 151–184 (1903). doi:10.1007/BF01444342. http://www.dx.doi.org/10.1007/BF01444342
19. Brück, M., Du, X., Park, J., Terng, C.L.: The submanifold geometries associated to Grassmannian systems. Mem. Am. Math. Soc. **155**(735), viii+95 (2002). doi:10.1090/memo/0735. http://www.dx.doi.org.libproxy.wustl.edu/10.1090/memo/0735
20. Bryant, R.L.: A duality theorem for Willmore surfaces. J. Differ. Geom. **20**, 23–54 (1984)
21. Bryant, R.L.: Surfaces of mean curvature one in hyperbolic space. Astérisque **154–155**, 12, 321–347, 353 (1988) (1987). Théorie des variétés minimales et applications (Palaiseau, 1983–1984)
22. Bryant, R.L., Griffiths, P.: Reduction for constrained variational problems and $\int \frac{1}{2}k^2\,ds$. Am. J. Math. **108**(3), 525–570 (1986). doi:10.2307/2374654. http://www.dx.doi.org/10.2307/2374654
23. Bryant, R.L., Kusner, R.: Parametrization of Boy's surface that makes it a Willmore immersion. (2004). https://en.wikipedia.org/wiki/Robert_Bryant_(mathematician)
24. Burstall, F.E.: Isothermic surfaces: conformal geometry, Clifford algebras and integrable systems. In: Integrable Systems, Geometry, and Topology. AMS/IP Studies in Advanced Mathematics, vol. 36, pp. 1–82. American Mathematical Society, Providence (2006)
25. Burstall, F.E., Hertrich-Jeromin, U., Pedit, F., Pinkall, U.: Curved flats and isothermic surfaces. Math. Z. **225**(2), 199–209 (1997). doi:10.1007/PL00004308. http://www.dx.doi.org.libproxy.wustl.edu/10.1007/PL00004308
26. Burstall, F.E., Pedit, F., Pinkall, U.: Schwarzian derivatives and flows of surfaces. In: Differential Geometry and Integrable Systems (Tokyo, 2000). Contemporary Mathematics, vol. 308, pp. 39–61. American Mathematical Society, Providence (2002)
27. Calapso, P.: Sulle superficie a linee di curvatura isoterme. Rend. Circ. Mat. Palermo **17**, 273–286 (1903)
28. Calapso, P.: Sulle trnsformazioni delle superficie isoterme. Ann. Mat. Pura Appl. **24**, 11–48 (1915)
29. Calini, A., Ivey, T.: Bäcklund transformations and knots of constant torsion. J. Knot Theory Ramif. **7**(6), 719–746 (1998). doi:10.1142/S0218216598000383. http://www.dx.doi.org/10.1142/S0218216598000383
30. do Carmo, M.P.: O método do referencial móvel. Instituto de Matemática Pura e Aplicada, Rio de Janeiro (1976)
31. do Carmo, M.P.: Riemannian Geometry. Mathematics: Theory and Applications. Birkhäuser, Boston (1992). Translated from the second Portuguese edition by Francis Flaherty
32. Cartan, E.: La Théorie des groupes finis et continus et la géométrie différentielle traitées par la méthode du repère mobile. Gauthier-Villars, Paris (1937)
33. Cartan, E.: Sur les couples de surfaces applicables avec conservation des courbures principales. Bull. Sci. Math. **66**, 55–85 (1942). Oeuvres Complète, Partie III, vol. 2, pp. 1591–1620
34. Catalan, E.C.: Sur les surfaces réglés dont l'aire est un minimum. J. Math. Pure Appl. **7**, 203–211 (1842)
35. Cayley, A.: On the surfaces divisible into squares by their curves of curvature. Proc. Lond. Math. Soc. **IV**, 8–9 (1871)
36. Cecil, T.E.: Lie Sphere Geometry: With Applications to Submanifolds. Universitext, 2nd edn. Springer, New York (2008).
37. Cecil, T.E., Chern, S.S.: Tautness and Lie sphere geometry. Math. Ann. **278**(1–4), 381–399 (1987). doi:10.1007/BF01458076. http://www.dx.doi.org/10.1007/BF01458076
38. Cecil, T.E., Jensen, G.R.: Dupin hypersurfaces with three principal curvatures. Invent. Math. **132**(1), 121–178 (1998)

39. Cecil, T.E., Jensen, G.R.: Dupin hypersurfaces with four principal curvatures. Geom. Dedicata **79**(1), 1–49 (2000)
40. Cecil, T.E., Ryan, P.J.: Tight and Taut Immersions of Manifolds. Research Notes in Mathematics, vol. 107. Pitman (Advanced Publishing Program), Boston (1985)
41. Cecil, T.E., Chi, Q.S., Jensen, G.R.: Dupin hypersurfaces with four principal curvatures. II. Geom. Dedicata **128**, 55–95 (2007). doi:10.1007/s10711-007-9183-3. http://dx.doi.org/10.1007/s10711-007-9183-3
42. Chern, S.S.: An elementary proof of the existence of isothermal parameters on a surface. Proc. Am. Math. Soc. **6**, 771–782 (1955)
43. Chern, S.S.: On the minimal immersions of the two-sphere in a space of constant curvature. In: Problems in Analysis (Lectures at the Symposium in Honor of Salomon Bochner. Princeton University, Princeton, NJ (1969)), pp. 27–40. Princeton University Press, Princeton (1970)
44. Chern, S.S.: Deformation of surfaces preserving principal curvatures. In: Chavel, I., Farkas, H. (eds.) Differential Geometry and Complex Analysis: A Volume Dedicated to the Memory of Harry Ernest Rauch, pp. 155–163. Springer, Berlin (1985)
45. Chern, S.S.: Moving frames. Astérisque Numero Hors Serie, 67–77 (1985). The Mathematical Heritage of Élie Cartan (Lyon, 1984)
46. Chern, S.S.: Lecture Notes on Differential Geometry. Tech. Rep. UH/MD-72, University of Houston, Houston (1990)
47. Chern, S.S.: Surface theory with Darboux and Bianchi. In: Miscellanea Mathematica, pp. 59–69. Springer, Berlin (1991)
48. Chern, S.S.: Complex Manifolds Without Potential Theory (With an Appendix on the Geometry of Characteristic Classes). Universitext, 2nd edn. Springer, New York (1995).
49. Chern, S.S., Tenenblat, K.: Pseudospherical surfaces and evolution equations. Stud. Appl. Math. **74**(1), 55–83 (1986)
50. Christoffel, E.B.: Über einige allgemeine Eigenschaften der Minimumsflächen. J. Reine Angew. Math. **67**, 218–228 (1867)
51. Cieśliński, J.: The Darboux-Bianchi transformation for isothermic surfaces. Classical results versus the soliton approach. Differ. Geom. Appl. **7**(1), 1–28 (1997). doi:10.1016/S0926-2245(97)00002-8. http://www.dx.doi.org.libproxy.wustl.edu/10.1016/S0926-2245(97)00002-8
52. Cieśliński, J., Goldstein, P., Sym, A.: Isothermic surfaces in $\mathbf{E}^3$ as soliton surfaces. Phys. Lett. A **205**(1), 37–43 (1995). doi:10.1016/0375-9601(95)00504-V. http://www.dx.doi.org.libproxy.wustl.edu/10.1016/0375-9601(95)00504-V
53. Conlon, L.: Differentiable Manifolds. Birkhäuser Advanced Texts: Basler Lehrbücher (Birkhäuser Advanced Texts: Basel Textbooks), 2nd edn. Birkhäuser, Boston (2001). doi:10.1007/978-0-8176-4767-4. http://www.dx.doi.org/10.1007/978-0-8176-4767-4
54. Costa, C.J.: Example of a complete minimal immersion in $\mathbf{R}^3$ of genus one and three embedded ends. Bol. Soc. Brasil. Mat. **15**(1–2), 47–54 (1984). doi:10.1007/BF02584707. http://www.dx.doi.org.libproxy.wustl.edu/10.1007/BF02584707
55. Darboux, G.: Sur les surfaces isothermiques. C. R. Acad. Sci. Paris **128**, 1299–1305 (1899)
56. Darboux, G.: Sur les surfaces isothermiques. Ann. Sci. Ècole Norm. Sup. **3**, 491–508 (1899)
57. Darboux, G.: Sur une classe des surfaces isothermiques liées à la deformations des surfaces du second degré. C. R. Acad. Sci. Paris **128**, 1483–1487 (1899)
58. Darboux, G.: Lecons sur la Théorie Générale des Surfaces et les Applications Géométriques du Calcul Infinitésimal, vol. I–IV, 3rd edn. Chelsea, Bronx (1972)
59. Delaunay, C.: Sur la surface de révolution dont la courbure moyenne est constante. J. Math. Pures Appl. Sér. 1 **6**, 309–320 (1841)
60. Deutsch, M.B.: Equivariant deformations of horospherical surfaces. Ph.D. thesis, Washington University in St. Louis (2010)
61. Dierkes, U., Hildebrandt, S., Küster, A., Wohlrab, O.: Minimal Surfaces, I: Boundary Value Problems. Grundlehren der Mathematischen Wissenschaften [Fundamental Principles of Mathematical Sciences], vol. 295. Springer, Berlin (1992)

62. Dubrovin, B.A., Fomenko, A.T., Novikov, S.P.: Modern Geometry—Methods and Applications, Part I. Graduate Texts in Mathematics, vol. 93. Springer, New York (1984). The Geometry of Surfaces, Transformation Groups, and Fields. Translated from the Russian by Robert G. Burns
63. Eells, J.: The surfaces of Delaunay. Math. Intell. **9**(1), 53–57 (1987). doi:10.1007/BF03023575. http://www.dx.doi.org/10.1007/BF03023575
64. Elghanmi, R.: Spacelike surfaces in Lorentzian manifolds. Differ. Geom. Appl. **6**(3), 199–218 (1996). doi:10.1016/0926-2245(96)82418-1. http://www.dx.doi.org/10.1016/0926-2245(96)82418-1
65. Enneper, A.: Analytisch-geometrische Untersuchungen. Z. Math. Phys. **IX**, 107 (1864)
66. Farkas, H.M., Kra, I.: Riemann Surfaces, 2nd edn. Graduate Texts in Mathematics, vol. 71. Springer, New York (1992)
67. Fels, M., Olver, P.J.: Moving frames and coframes. In: Algebraic Methods in Physics (Montréal, QC, 1997). CRM Series in Mathematical Physics, pp. 47–64. Springer, New York (2001)
68. Ferapontov, E.V.: Lie sphere geometry and integrable systems. Tohoku Math. J. (2) **52**(2), 199–233 (2000)
69. Ferus, D., Pedit, F.: Curved flats in symmetric spaces. Manuscripta Math. **91**(4), 445–454 (1996). doi:10.1007/BF02567965. http://www.dx.doi.org.libproxy.wustl.edu/10.1007/BF02567965
70. Fubini, G.: Sulle metriche definite da una forma hermitiana. Atti Istit. Veneto **6**, 501–513 (1903)
71. Fujimoto, H.: On the number of exceptional values of the Gauss maps of minimal surfaces. J. Math. Soc. Jpn. **40**(2), 235–247 (1988). doi:10.2969/jmsj/04020235. http://www.dx.doi.org.libproxy.wustl.edu/10.2969/jmsj/04020235
72. Gauss, C.F.: General Investigations of Curved Surfaces. Raven Press, Hewlett (1965)
73. Germain, S.: Recherches sur la Théorie des Surfaces Élastiques. Courcier, Paris (1821)
74. Goursat, E.: Sur un mode de transformation des surfaces minima. Acta Math. **11**(1–4), 135–186 (1887). doi:10.1007/BF02418047. http://www.dx.doi.org/10.1007/BF02418047
75. Goursat, E.: Sur un mode de transformation des surfaces minima. Acta Math. **11**(1–4), 257–264 (1887). doi:10.1007/BF02418050. http://www.dx.doi.org/10.1007/BF02418050. Second Mémoire
76. Graustein, W.C.: Applicability with preservation of both curvatures. Bull. Am. Math. Soc. **30**(1–2), 19–23 (1924). doi:10.1090/S0002-9904-1924-03839-7. http://www.dx.doi.org/10.1090/S0002-9904-1924-03839-7
77. Green, M.L.: The moving frame, differential invariants and rigidity theorems for curves in homogeneous spaces. Duke Math. J. **45**(4), 735–779 (1978). http://projecteuclid.org/euclid.dmj/1077313097
78. Greenberg, M.J.: Euclidean and Non-Euclidean Geometries: Development and History, 3rd edn. W.H. Freeman, New York (1993).
79. Griffiths, P.A.: On Cartan's method of Lie groups and moving frames as applied to uniqueness and existence questions in differential geometry. Duke Math. J. **41**, 775–814 (1974)
80. Griffiths, P.A.: Exterior Differential Systems and the Calculus of Variations. Progress in Mathematics, vol. 25. Birkhäuser, Boston (1983)
81. Griffiths, P., Harris, J.: Principles of Algebraic Geometry. Pure and Applied Mathematics. Wiley-Interscience [Wiley], New York (1978)
82. Gunning, R.C.: Lectures on Riemann Surfaces. Princeton Mathematical Notes. Princeton University Press, Princeton (1966)
83. Hatcher, A.: Algebraic Topology. Cambridge University Press, Cambridge (2002)
84. Helgason, S.: Differential Geometry and Symmetric Spaces. Pure and Applied Mathematics, vol. XII. Academic, New York (1962)
85. Hertrich-Jeromin, U.: The surfaces capable of division into infinitesimal squares by their curves of curvature. Math. Intell. **22**, 54–61 (2000)

86. Hertrich-Jeromin, U.: Introduction to Möbius Differential Geometry. London Mathematical Society Lecture Note Series, vol. 300. Cambridge University Press, Cambridge (2003)
87. Hertrich-Jeromin, U., Pinkall, U.: Ein Beweis der Willmoreschen Vermutung für Kanaltori. J. Reine Angew. Math. **430**, 21–34 (1992). doi:10.1515/crll.1992.430.21. http://www.dx.doi.org.libproxy.wustl.edu/10.1515/crll.1992.430.21
88. Hertrich-Jeromin, U., Musso, E., Nicolodi, L.: Möbius geometry of surfaces of constant mean curvature 1 in hyperbolic space. Ann. Glob. Anal. Geom. **19**(2), 185–205 (2001)
89. Hilbert, D., Cohn-Vossen, S.: Geometry and the Imagination. Chelsea, New York (1952). Translated by P. Neményi
90. Hoffman, D., Karcher, H.: Complete embedded minimal surfaces of finite total curvature. In: Geometry, V. Encyclopedia of Mathematical Science, vol. 90, pp. 5–93, 267–272. Springer, Berlin (1997)
91. Hoffman, D.A., Meeks III, W.: A complete embedded minimal surface in $\mathbf{R}^3$ with genus one and three ends. J. Differ. Geom. **21**(1), 109–127 (1985). http://projecteuclid.org/getRecord?id=euclid.jdg/1214439467
92. Hopf, H.: Differential Geometry in the Large. Lecture Notes in Mathematics, vol. 1000. Springer, New York (1983). Seminar lectures New York University 1946 and Stanford University 1956
93. Jensen, G.R.: Higher Order Contact of Submanifolds of Homogeneous Spaces. Lecture Notes in Mathematics, vol. 610. Springer, Berlin (1977)
94. Jensen, G.R.: Deformation of submanifolds of homogeneous spaces. J. Differ. Geom. **16**, 213–246 (1981)
95. Jensen, G.R., Musso, E.: Rigidity of hypersurfaces in complex projective space. Ann. Sci. École Norm. Sup. (4) **27**(2), 227–248 (1994)
96. Jensen, G.R., Rigoli, M.: A class of harmonic maps from surfaces into real Grassmannians. Rend. Sem. Mat. Univ. Politec. Torino **Special Issue**, 99–116 (1984). Conference on differential geometry on homogeneous spaces (Turin, 1983)
97. Kamberov, G., Pedit, F., Pinkall, U.: Bonnet pairs and isothermic surfaces. Duke Math. J. **92**(3), 637–644 (1998)
98. Kamberov, G., Norman, P., Pedit, F., Pinkall, U.: Quaternions, Spinors, and Surfaces. Contemporary Mathematics, vol. 299. American Mathematical Society, Providence (2002)
99. Karcher, H.: Hyperbolic surfaces of constant mean curvature one with compact fundamental domains. In: Global Theory of Minimal Surfaces. Clay Mathematics Proceedings, vol. 2, pp. 311–323. American Mathematical Society, Providence (2005)
100. Kobayashi, S., Nomizu, K.: Foundations of Differential Geometry, vol I. Interscience/A Division of John Wiley & Sons, New York/London (1963)
101. Kobayashi, S., Nomizu, K.: Foundations of Differential Geometry, Vol. II. Interscience Tracts in Pure and Applied Mathematics, vol. 15. Interscience/Wiley, New York/London/Sydney (1969)
102. Korn, A.: Zwei Anwendungen der Methode der sukzessiven Annäherungen. In: H.A. Schwarz (ed.) Gesammelte Mathematische Abhandlungen, I, pp. 215–229. Springer, Berlin (1890)
103. Krantz, S.G.: Complex Analysis: The Geometric Viewpoint. Carus Mathematical Monographs, vol. 23. Mathematical Association of America, Washington, DC (1990)
104. Lagrange, J.L.: Œuvres de Lagrange. Gauthier-Villars, Paris (1867–92). Vol. 1–14, publiées par les soins de m. J.-A. Serret, sous les auspices de Son Excellence le ministre de l'instruction publique
105. Langer, J., Singer, D.A.: The total squared curvature of closed curves. J. Differ. Geom. **20**(1), 1–22 (1984)
106. Langer, J., Singer, D.A.: Lagrangian aspects of the Kirchhoff elastic rod. SIAM Rev. **38**(4), 605–618 (1996)
107. Lawson, H.B.: Lectures on Minimal Submanifolds, vol. I. Publish or Perish, Berkeley (1980)
108. Lawson Jr., H.B.: Complete minimal surfaces in S^3. Ann. Math. (2) **92**, 335–374 (1970)
109. Lawson Jr., H.B., Tribuzy, R.A.: On the mean curvature function for compact surfaces. J. Differ. Geom. **16**(2), 179–183 (1981)

110. Lee, J.M.: Introduction to Smooth Manifolds. Graduate Texts in Mathematics, vol. 218. Springer, New York (2003)
111. Levien, R.: The elastica: a mathematical history. Http://levien.com/phd/elastica_hist.pdf (2008)
112. Liao, R.: Cyclic properties of the harmonic sequence of surfaces in $\mathbf{CP}^n$. Math. Ann. **296**(2), 363–384 (1993). doi:10.1007/BF01445110. http://www.dx.doi.org/10.1007/BF01445110
113. Lichtenstein, L.: Zur Theorie der konformen Abbildung: Konforme Abbildung nichtanalytischer, singularitätenfreier Flächenstücke auf ebene Gebiete. Bull. Int. de L'Acad. Sci. Cracovie, ser. A pp. 192–217 (1916)
114. Lie, S.: Ueber Complexe, insbesondere Linien- und Kugel-Complexe, mit Anwendung auf die Theorie partieller Differential-Gleichungen. Math. Ann. **5**(1), 145–208 (1872). doi:10.1007/BF01446331. http://www.dx.doi.org.libproxy.wustl.edu/10.1007/BF01446331. Ges. Abh. **2**, 1–121
115. Lima, E.L.: Orientability of smooth hypersurfaces and the Jordan-Brouwer separation theorem. Expo. Math. **5**(3), 283–286 (1987)
116. Malliavin, P.: Géométrie Différentielle Intrinsèque. Hermann, Paris (1972)
117. Marques, F.C., Neves, A.: Min-max theory and the Willmore conjecture. Ann. Math. (2) **179**(2), 683–782 (2014). doi:10.4007/annals.2014.179.2.6. http://www.dx.doi.org/10.4007/annals.2014.179.2.6
118. Marsden, J.E.: Basic Complex Analysis. W.H. Freeman, San Francisco (1973)
119. Meusnier, J.B.M.C.: Mémoire sur la courbure des surfaces. Mémoires présentés par div. Etrangers. Acad. Sci. Paris **10**, 477–510 (1785)
120. Mo, X., Osserman, R.: On the Gauss map and total curvature of complete minimal surfaces and an extension of Fujimoto's theorem. J. Differ. Geom. **31**(2), 343–355 (1990). http://projecteuclid.org/getRecord?id=euclid.jdg/1214444316
121. Monge, G.: Applications de l'Analyse à la Géométrie, 5ème edn. Bachelier, Paris (1849). Rev., corr. et annotée par M. Liouville
122. Munkres, J.R.: Topology, 2nd edn. Prentice Hall, Upper Saddle River (2000)
123. Musso, E.: Willmore surfaces in the four-sphere. Ann. Glob. Anal. Geom. **8**(1), 21–41 (1990). doi:10.1007/BF00055016. http://www.dx.doi.org/10.1007/BF00055016
124. Musso, E.: Isothermic surfaces in Euclidean space. In: Cordero, L.A., García-Río, E. (eds.) Workshop on Recent Topics in Differential Geometry. Public. Depto. Geometría y Topología, vol. 89, pp. 219–235. University Santiago de Compostela, Santiago de Compostela (1998)
125. Musso, E.: Variational problems for plane curves in centro-affine geometry. J. Phys. A **43**(30), 305,206, 24 pp. (2010). doi:10.1088/1751-8113/43/30/305206. http://www.dx.doi.org/10.1088/1751-8113/43/30/305206
126. Musso, E., Nicolodi, L.: Willmore canal surfaces in Euclidean space. Rend. Istit. Mat. Univ. Trieste **31**(1–2), 177–202 (1999)
127. Musso, E., Nicolodi, L.: Darboux transforms of Dupin surfaces. In: PDEs, Submanifolds and Affine Differential Geometry (Warsaw, 2000), vol. 57, pp. 135–154. Banach Center/Polish Academy of Sciences, Warsaw (2002)
128. Newlander, A., Nirenberg, L.: Complex analytic coordinates in almost complex manifolds. Ann. Math. (2) **65**, 391–404 (1957)
129. Osserman, R.: Proof of a conjecture of Nirenberg. Commun. Pure Appl. Math. **12**, 229–232 (1959)
130. Osserman, R.: Minimal surfaces in the large. Comment. Math. Helv. **35**, 65–76 (1961)
131. Osserman, R.: Global properties of minimal surfaces in E^3 and E^n. Ann. Math. (2) **80**, 340–364 (1964)
132. Osserman, R.: Global properties of classical minimal surfaces. Duke Math. J. **32**, 565–573 (1965)
133. Osserman, R.: A Survey of Minimal Surfaces, 2nd edn. Dover, New York (1986)
134. Palais, R.S.: The principle of symmetric criticality. Commun. Math. Phys. **69**(1), 19–30 (1979). http://projecteuclid.org.libproxy.wustl.edu/getRecord?id=euclid.cmp/1103905401
135. Pinkall, U.: Dupin hypersurfaces. Math. Ann. **270**, 427–440 (1985)

136. Pinkall, U.: Hopf tori in S^3. Invent. Math. **81**(2), 379–386 (1985). doi:10.1007/BF01389060. http://www.dx.doi.org.libproxy.wustl.edu/10.1007/BF01389060
137. Pinkall, U.: Hamiltonian flows on the space of star-shaped curves. Results Math. **27**(3–4), 328–332 (1995). doi:10.1007/BF03322836. http://www.dx.doi.org/10.1007/BF03322836
138. Ricci-Curbastro, G.: Sulla teoria intrinseca delle superficie ed in ispecie di quelle di secondo grado. Atti R. Ist. Ven. di Lett. ed Arti **6**, 445–488 (1895)
139. Rigoli, M.: The conformal Gauss map of submanifolds of the Möbius space. Ann. Glob. Anal. Geom. **5**(2), 97–116 (1987). doi:10.1007/BF00127853. http://www.dx.doi.org/10.1007/BF00127853
140. Ros, A.: The Gauss map of minimal surfaces. In: Differential Geometry, Valencia, 2001, pp. 235–252. World Science, River Edge (2002)
141. Rossman, W.: Mean curvature one surfaces in hyperbolic space, and their relationship to minimal surfaces in Euclidean space. J. Geom. Anal. **11**(4), 669–692 (2001). doi:10.1007/BF02930762. http://dx.doi.org/10.1007/BF02930762
142. Rothe, R.: Untersuchungen über die Theorie der isothermen Flächen. Berlin thesis, Mayer und Müller, Berlin (1897)
143. Roussos, I.M.: Principal-curvature-preserving isometries of surfaces in ordinary space. Bol. Soc. Brasil. Mat. **18**(2), 95–105 (1987). doi:10.1007/BF02590026. http://www.dx.doi.org/10.1007/BF02590026
144. Roussos, I.M.: The helicoidal surfaces as Bonnet surfaces. Tohoku Math. J. (2) **40**(3), 485–490 (1988). doi:10.2748/tmj/1178227989. http://www.dx.doi.org/10.2748/tmj/1178227989
145. Roussos, I.M.: Global results on Bonnet surfaces. J. Geom. **65**(1–2), 151–168 (1999). doi:10.1007/BF01228686. http://www.dx.doi.org/10.1007/BF01228686
146. Ruh, E., Vilms, J.: The tension field of the Gauss map. Trans. Am. Math. Soc. **149**, 569–573 (1970)
147. Scherk, H.F.: De proprietatibus superficiei quae hac continetur aequatione $(1+q^2)r-2pqs+(1+p^2)t=0$ disquisitiones analyticae. Acta Societatis JablonovianœTome **IV**, 204–280 (1831)
148. Scherk, H.F.: Bemerkungen über die kleinste Fläche innerhalb gegebener Grenzen. J. Reine Angew. Math. **13**, 185–208 (1835). doi:10.1515/crll.1835.13.185. http://dx.doi.org/10.1515/crll.1835.13.185
149. Schwarz, H.A.: Gesammelte Mathematische Abhandlungen, I. Springer, Berlin (1890)
150. Sharpe, R.W.: Differential Geometry. Graduate Texts in Mathematics, vol. 166. Springer, New York (1997). Cartan's Generalization of Klein's Erlangen Program, with a Foreword by S. S. Chern
151. Shepherd, M.D.: Line congruences as surfaces in the space of lines. Differ. Geom. Appl. **10**(1), 1–26 (1999). doi:10.1016/S0926-2245(98)00025-4. http://www.dx.doi.org/10.1016/S0926-2245(98)00025-4
152. Shiohama, K., Takagi, R.: A characterization of a standard torus in E^3. J. Differ. Geom. **4**, 477–485 (1970)
153. Simons, J.: Minimal varieties in Riemannian manifolds. Ann. Math. **88**, 62–105 (1968)
154. Spivak, M.: A Comprehensive Introduction to Differential Geometry, vols. I–V. Publish or Perish, Boston (1970)
155. Struik, D.J.: Lectures on Classical Differential Geometry, 2nd edn. Dover, New York (1988). Unabridged and unaltered republication of the second edition (1961) of the work first published in 1950 by Addison-Wesley, Reading
156. Study, E.: Kürzeste Wege im komplexen Gebiet. Math. Ann. **60**(3), 321–378 (1905). doi:10.1007/BF01457616. http://www.dx.doi.org.libproxy.wustl.edu/10.1007/BF01457616
157. Sulanke, R.: On É. Cartan's method of moving frames. In: Differential Geometry (Budapest, 1979). Colloquium Mathematical Society János Bolyai, vol. 31, pp. 681–704. North-Holland, Amsterdam (1982)
158. Sulanke, R.: Möbius geometry. V. Homogeneous surfaces in the Möbius space S^3. In: Topics in Differential Geometry, vol. I, II (Debrecen, 1984). Mathematical Society János Bolyai, vol. 46, pp. 1141–1154. North-Holland, Amsterdam (1988)

159. Terng, C.L., Uhlenbeck, K.: Bäcklund transformations and loop group actions. Commun. Pure Appl. Math. **53**(1), 1–75 (2000). doi:10.1002/(SICI)1097-0312(200001)53:1<1::AID-CPA1>3.3.CO;2-L. http://www.dx.doi.org.libproxy.wustl.edu/10.1002/(SICI)1097-0312(200001)53:1<1::AID-CPA1>3.3.CO;2-L
160. Thomsen, G.: Über konforme Geometrie I: Grundlagen der konformen Flächentheorie. Abh. Math. Sem. Hamburg **3**, 31–56 (1924)
161. Truesdell, C.A.: The influence of elasticity on analysis: the classic heritage. Bull. Am. Math. Soc. (N.S.) **9**(3), 293–310 (1983). doi:10.1090/S0273-0979-1983-15187-X. http://www.dx.doi.org/10.1090/S0273-0979-1983-15187-X
162. Umehara, M., Yamada, K.: Complete surfaces of constant mean curvature 1 in the hyperbolic 3-space. Ann. Math. (2) **137**(3), 611–638 (1993). doi:10.2307/2946533. http://www.dx.doi.org/10.2307/2946533
163. Umehara, M., Yamada, K.: Surfaces of constant mean curvature c in $H^3(-c^2)$ with prescribed hyperbolic Gauss map. Math. Ann. **304**(2), 203–224 (1996). doi:10.1007/BF01446291. http://www.dx.doi.org/10.1007/BF01446291
164. Voss, K.: Uber vollständige Minimalflächen. L'Enseignement Math. **10**, 316–317 (1964)
165. Voss, K.: Verallgemeinerte Willmore-Flächen. Tech. Rep., Mathematisches Forschungsinstitut Oberwolfach Workshop Report 1985, vol. 42 (1985)
166. Warner, F.W.: Foundations of Differentiable Manifolds and Lie Groups. Scott, Foresman and Company, Glenview (1971)
167. Weber, M.: Classical minimal surfaces in Euclidean space by examples: geometric and computational aspects of the Weierstrass representation. In: Global Theory of Minimal Surfaces. Clay Mathematics Proceedings, vol. 2, pp. 19–63. American Mathematical Society, Providence (2005)
168. Weierstrass, K.: Über eine besondere Gattung von Minimalflächen. Mathematische Werke, vol. 3, pp. 241–247. Mayer & Müller, Berlin (1903)
169. Weierstrass, K.: Untersuchungen über die Flächen, deren mittlere Krümmung überall gleich Null ist. Mathematische Werke, vol. 3, pp. 39–52. Mayer & Müller, Berlin (1903)
170. White, J.H.: A global invariant of conformal mappings in space. Proc. Am. Math. Soc. **38**, 162–164 (1973)
171. Willmore, T.J.: Note on embedded surfaces. An. Şti. Univ. "Al. I. Cuza" Iaşi Secţ. I a Mat. (N.S.) **11B**, 493–496 (1965)
172. Xavier, F.: The Gauss map of a complete non-flat minimal surface cannot omit 7 points on the sphere. Ann. Math. **113**, 211–214 (1981). Erratum, Ann. Math. **115**, 667 (1982)
173. Yang, K.: Frenet formulae for holomorphic curves in the two-quadric. Bull. Aust. Math. Soc. **33**(2), 195–206 (1986). doi:10.1017/S0004972700003063. http://www.dx.doi.org/10.1017/S0004972700003063
174. Yang, K.: Local Hermitian geometry of complex surfaces in $\mathbf{P}^3$: totally parabolic surfaces. Yokohama Math. J. **39**(1), 61–88 (1991)
175. Zheng, Y.: Quantization of curvature of harmonic two-spheres in Grassmann manifolds. Trans. Am. Math. Soc. **316**(1), 193–214 (1989). doi:10.2307/2001280. http:/www.dx.doi.org/10.2307/2001280

Index

G.R. Jensen et al., *Surfaces in Classical Geometries*, Universitext,
DOI 10.1007/978-3-319-27076-0

K

L

M

Printed by Printforce, the Netherlands